DYNAMICS OF STRUCTURES

Dedicated to Yash
and to
Rachna and Rajesh
Abhinav and Priya
Atul and Deepali
and their children
who are a source of immense joy to us

Dynamics of Structures
Second Edition

JAGMOHAN L. HUMAR
Carleton University, Ottawa, Canada

A.A. BALKEMA PUBLISHERS/LISSE/ABINGDON/EXTON (PA)/TOKYO

Transferred to Digital Printing 2005

Library of Congress Cataloging-in-Publication Data

Humar, J. L.
   Dynamics of structures / Jagmohan L. Humar.– 2nd ed.
   p. cm.
   ISBN 9058092453 – ISBN 9058092451 (pbk.)
   1. Structural dynamics. I. Title

   TA654 .H79 2001
   624.1'7–dc21   2001052675

Cover design: Studio Jan de Boer, Amsterdam, The Netherlands.
Typesetting: Macmillan India Ltd., Bangalore, India.

ISBN 90 5809 245 3 (hardback)
ISBN 90 5809 246 1 (paperback)

Contents

Preface

As in the case of its predecessor, the motivation for this second edition of the book is to help engineers and scientists acquire an understanding of the dynamic response of structures and of the analytical tools required for determining such response. The book should be equally helpful to persons working in the field of civil, mechanical or aerospace engineering.

For the proper understanding of an analytical concept, it is useful to develop an appreciation of the mathematical basis. Such appreciation need not depend on a rigorous treatment of the subject matter; a physical understanding of the concepts is in most cases adequate, and perhaps more meaningful to an engineer. The book attempts to explain the mathematical basis for the concepts presented, mostly in physically motivated terms or through heuristic argument. No special mathematical background is required of the reader, except for a basic knowledge of college algebra and calculus and engineering mechanics.

The essential steps in the dynamic analysis of a system are: (a) mathematical modeling (b) formulation of the equations of motion, and (c) solution of the equations. Modeling techniques can be divided into two broad categories. In one technique, the system is modeled as an assembly of rigid body masses and massless deformable elements. Systems modeled in this manner are referred to as discrete parameter systems. In the other technique of modeling, both mass and deformabilty are assumed to be distributed throughout the extent of the system which is treated as continuous. Systems modeled in this manner are called continuous or distributed parameter systems.

In general, a continuous model will better represent the behavior of a dynamical system. However, in most practical situations, the equations of motion of a continuous system are too difficult or impossible to solve. Therefore, in a majority of cases, dynamic analysis of engineering structures must rely on a representation of the structure by a discrete parameter model. The contents of the book reflect this emphasis on the use of discrete models. The first three parts of the book are devoted to the analysis of response of discrete systems. Part 1, consisting of Chapters 2 through 4, deals with the formulation of equations of motion of discrete parameter systems. However, the methods of analytical mechanics presented in Chapter 4 are, equally applicable to continuous systems. Examples of such applications are presented later in the book.

Part 2 of the book, covering Chapters 5 through 9, deals with the solution of equation of motion for a single-degree-of-freedom system. Part 3, consisting of Chapters 10 through 13, discusses the solution of equations of motion for multi degree-of-freedom systems.

Part 4 of the book, covering Chapters 14 through 17, is devoted to the analysis of continuous system. Again, the subject matter is organized so that the formulation of equations of motion is presented first followed by a discussion of the solution techniques.

The book is organized so as to follow the logical succession of steps involved in the analysis. Many readers may prefer to complete a study of the single-degree-of-freedom systems, from formulation of equation to their solution, before embarking on a study of multi-degree-of-freedom systems. This can be easily achieved by selective reading. The book chapters have been planned so that Chapters 3 and 4 relating to the formulation of equations of motion of a general system need not be studied prior to studying the material in Chapters 5 through 9 on the solution of equations of motion for a single-degree-of-freedom system.

A development that has had a profound effect in the recent times on procedures for the analysis of engineering systems is the advent of digital computers. The ability of computers to manage vast amounts of information and the incredible speed with which they can process numerical data has shifted the emphasis from closed from solutions and approximate methods suitable for hand computations to solution of discrete models and numerical techniques of analysis. At the same time, computers have allowed the routine solution of problems vastly greater in size and complexity than was possible only a decade or two ago. The emphasis on discrete methods and numerical solutions is reflected in the contents of the present book. Chapter 8 on single-degree-of-freedom systems and Chapter 13 on multi-degree-of-freedom systems, are devoted exclusively to numerical techniques of solution. A fairly detailed treatment of the frequency domain analysis is included in Chapters 9 and 13, in recognition of the efficiency of this technique in the numeric computation of response. Also, a detailed treatment of the solution of discrete eigenproblems which plays a central role in the numerical analysis of response is included in Chapter 11.

It is recognized that the field of computer hardware as well as software is undergoing revolutionary development. The continuing evolution of personal computers with vastly improved processing speeds and memory capacity and the ongoing development of new programming languages and software tools means that algorithms and programming styles must continue to change to take advantage of the progress made. Program listings or detailed algorithms have not therefore been included in the book. The author believes that in a book like this, it is more useful to provide the necessary background material for an appreciation of the physical behavior and the analytical concepts involved as well as to present the development of methods that are suitable to numeric computations. Hopefully, this will give enough information for the reader to be able to develop his/her own algorithms or to make an informed and intelligent use of existing software.

The material included in this book has been drawn from the vast wealth of available information. Some of it has now become a part of the historical development of structural dynamics, other is more recent. It is difficult to acknowledge the sources for all of the information provided. The author offers his apologies to all researchers who have not been adequately recognized. References have been omitted from the text to avoid distracting the reader. However, where appropriate, a brief list of suitable material for further reading is provided at the end of each chapter.

The style of presentation and the emphasis are the author's own. The contents of the book have been influenced by the author's experience in teaching and research and by the research studies carried out by him and his students. A large number of examples have been included in the text; since they provide the most effective means of developing an understanding of the concepts involved. Exercise problems have also been included at the end of each chapter. They will provide the reader useful practice in the application of techniques presented.

In preparing this second edition, the errors that had inadvertently crept into the first edition have been corrected. The author is indebted to all those readers who brought such errors to his attention. Several sections of the book have been revised and some new concepts and analytical techniques have been included to make the book as comprehensive as possible, within the boundary of its scope. Also included are additional end-of-chapter exercises for the benefit of the reader.

The author wishes to acknowledge the contribution made by his many students and colleagues in the preparation of this book.

List of symbols

The principal symbols used in the text are listed below. All symbols, including those listed here, are defined at appropriate places within the text, usually at the time of their first occurrence. Occasionally, the same symbol may be used to represent more than one parameter, but the meaning should be quite unambiguous when read in context.

Throughout the text, matrices are represented by bold face upper case letters while vectors are represented by bold face lower case letters. An overdot signifies differential with respect to time and a prime stands for differentiation with respect to the argument of the function.

a	acceleration; constant; linear dimension; decay parameter in exponential window method
a_n	coefficient of Fourier series cosine term
a_{ij}	flexibility influence coefficient
$\mathbf{a}_m$	real part of mth eigenvector
A	constant; cross-sectional area
A_a	amplitude of dynamic load factor for acceleration
A_d	amplitude of dynamic load factor for displacement
A_v	amplitude of dynamic load factor for velocity
$\mathbf{A}$	amplification matrix; flexibility matrix; square matrix
$\tilde{\mathbf{A}}$	transformed square matrix
b	constant; linear dimension; width of beam cross section
b_n	coefficient of Fourier series sine term
$\mathbf{b}_m$	imaginary part of mth eigenvector
B	constant; differential operator
$\mathbf{B}$	square matrix
c	damping constant; velocity of wave propagation
c_{cr}	critical damping constant
c_g	velocity of wave group
c_n	coefficient of Fourier series term, constant
c_s	internal damping constant
c_{ij}	damping influence coefficient
$\bar{c}$	damping constant per unit length
c^*	generalized damping constant

$\mathbf{c}$	vector of weighting factors in expansion theorem
C	constant
$\mathbf{C}$	damping matrix; transformation matrix
C_n	modal damping constant for the nth mode
$\mathbf{C}^*, \tilde{\mathbf{C}}$	transformed damping matrix
d	diameter
d_n	constant
D	dynamic load factor
$\mathbf{D}$	diagonal matrix; dynamic matrix
e	eccentricity of unbalanced mass
E	modulus of elasticity
$\mathbf{E}$	dynamic matrix$=\mathbf{D}^{-1}$
EA	axial rigidity
EI	flexural rigidity
E_m	remainder term in numerical integration formula
f	undamped natural frequency in cycles per sec
$f(x)$	eigenfunction of a continuous system
f_d	damped natural frequency
f_D	damping force
f_G	force due to geometric instability
f_I	inertia force
f_S	spring force
f_S^t	total of spring force and damping force for hysteretic damping
f_0	frequency of applied load in cycles per sec
$\mathbf{f}$	vector representing spatial variation of exciting force
$\mathbf{f}_D$	vector of damping forces
$\mathbf{f}_G$	vector of geometric instability forces
$\mathbf{f}_I$	vector of inertia forces
$\mathbf{f}_S$	vector of spring forces
F	force
$\mathbf{F}$	force vector
$\mathbf{F}_a$	vector of applied forces
$\mathbf{F}_c$	vector of constraint forces
F_x, F_y, F_z	components of force vector along Cartesian coordinates
g	acceleration due to gravity
$g(t)$	forcing function
$\hat{g}$	scaled forcing function $e^{-at}g(t)$
G	constant; modulus of rigidity
GJ	torsional rigidity
G_1, G_2	constants
$G(\Omega)$	Fourier transform of $g(t)$
$\hat{G}(\Omega)$	Fourier transform of $\hat{g}(t)$
h	height; time interval

$h(t)$	unit impulse response
$\bar{h}(t)$	periodic unit impulse response
$\hat{h}(t)$	scaled unit impulse function $h(t)e^{-at}$
$H(\omega_0)$, $H(\Omega)$	complex frequency response, Fourier transform of $h(t)$
$\bar{H}(\Omega)$	periodic complex frequency response, Fourier transform of $\bar{h}(t)$
$\hat{H}(t)$	Fourier transform of $\hat{h}(t)$
i	imaginary number; integer
$\mathbf{i}$	unit vector along x axis
I	impulse; moment of inertia
$\mathbf{I}$	identity matrix
I_A	mass moment of inertia for rotation above point A
I_0	functional; mass moment of inertia for rotation about the mass center
j	integer
$\mathbf{j}$	unit vector along y axis
J	polar moment of inertia
k	spring constant; stiffness, integer; wave number
$\mathbf{k}$	unit vector along z axis
k_G	geometric stiffness
k_T	tangent stiffness
k_{ij}	stiffness influence coefficient
k'	shape constant for shear deformation
$\bar{k}$	spring constant per unit length
k^*	generalized stiffness
K	differential operator
$\mathbf{K}$	stiffness matrix
$\mathbf{K}_G$	geometric stiffness matrix
K_n	modal stiffness for the nth mode
$\mathbf{K}^*, \tilde{\mathbf{K}}$	transformed stiffness matrix
l	length
L	Lagrangian; length
$\mathbf{L}_K$	lower triangular factor of stiffness matrix
$\mathbf{L}_M$	lower triangular factor of mass matrix
m	integer; mass; mass per unit length
m_0	mass; unbalanced mass
m_{ij}	mass influence coefficient
$\bar{m}$	mass per unit length; mass per unit area
m^*	generalized mass
M	concentrated mass, differential operator, moment
$\mathbf{M}$	mass matrix
M_I	inertial moment
M_n	modal mass for the nth mode
M_s	moment due to internal damping forces

M_0	concentrated mass
$\mathbf{M}^*, \tilde{\mathbf{M}}$	transformed mass matrix
n	integer
N	normal force; number of degrees of freedom
$\mathbf{N}$	transformation matrix
p	integer; force
$\mathbf{p}$	left eigenvector; force vector
p_n	modal force in the nth mode
$\bar{p}$	force per unit length
p^*	generalized force
$p(\lambda)$	characteristic polynomial
P	axial force; concentrated applied load
$\mathbf{P}$	matrix of left eigenvectors
P_I	inertial force
P_0, p_0	amplitude of applied force
q	integer
$\mathbf{q}$	right eigenvector
q_i	generalized coordinate
$\tilde{\mathbf{q}}$	transformed eigenvector
Q	applied force
$\mathbf{Q}$	matrix of eigenvectors, orthogonal transformation matrix
Q_i	generalized force
r	common ratio; constant; integer; radius of gyration; rank of a matrix; radius vector
$r(t)$	response due to unit initial displacement
$\bar{r}(t)$	response due to periodic unit displacement changes
R	Rayleigh dissipation function; reaction; remainder term
R_a	inertance
R_d	receptance
R_v	mobility
R_i	magnitude of ith corrective force impulse
$\mathbf{R}_a$	inertance matrix
$\mathbf{R}_d$	receptance matrix
$\mathbf{R}_v$	mobility matrix
s	complex eigenvalue
$s(t)$	response due to initial unit velocity
$\bar{s}(t)$	response due to a periodic unit velocity changes
S	axial force
$\mathbf{S}$	matrix of complex eigenvalues
$\mathbf{S}_n$	matrix for sweeping the first n eigenvectors
t	time
t_p	time at peak response
T	torque

T	kinetic energy; tensile force; undamped natural period
T_d	damped natural period
$\mathbf{T}$	transformation matrix; tridiagonal matrix
TR	transmission ratio
T_0	period of applied load
$u(t), u$	displacement
u_g	ground displacement
u_i	constrained coordinate; displacement along degree-of-freedom i
u_x	displacement along x direction
u_y	displacement along y direction
u_0	initial displacement
u^t	absolute displacement
u^s	static displacement
$\bar{u}(t)$	periodic displacement response
U	complex frequency response; strain energy
$\mathbf{U}$	upper triangular matrix, complex frequency response matrix
$U(\Omega)$	Fourier transform of $u(t)$
v	velocity
$\mathbf{v}$	complex eigenvector
$v(x)$	comparison function
v_0	initial velocity
V	potential energy; shear force
$\mathbf{V}$	matrix of complex eigenvectors
V_0	base shear
$w(x)$	comparison function
W_D	energy loss per cycle in viscous damping
W_e	work done by external forces
W_i	energy loss per cycle; work done by internal forces
W_s	work done by elastic force
x	Cartesian coordinate
$\bar{x}$	coordinate of the mass center
$\mathbf{X}$	Lanczos transformation matrix
y	Cartesian coordinate
$\mathbf{y}$	vector of normal coordinates
$\mathbf{y}_0$	vector of initial values of the normal coordinates
y_{0n}	initial value of the nth normal coordinate
z	generalized coordinate
α	angular shear deformation; coefficient; constant; parameter
β	constant; frequency ratio; parameter
γ	angle; inverse eigenvalue; parameter
δ	deflection; eigenvalue; eigenvalue measured from a shifted origin; logarithmic decrement

$\delta(x)$	delta function
Δ_{st}	static deflection
δ_{ij}	Kronecker delta
$\delta\mathbf{r}$	virtual displacement vector
δu	virtual displacement
δW_e	virtual work done by external forces
δW_i	virtual work done by internal forces
δW_{ei}	virtual work done by forces acting on internal elements
δW_S	virtual work done by axial force
δz	virtual displacement
$\delta\theta, \delta\phi$	virtual rotation
Δ	displacement
Δt	increment of time
$\Delta\Omega$	increment of frequency
ε	strain; quantity of a small value
η	hysteretic damping constant, angle
$\eta(t)$	corrective response
η_k	imaginary part of eigenvector
θ	angular displacement; flexural rotation; polar coordinate; parameter
κ	curvature
λ	eigenvalue; Lagrangian multiplier; wave length
Λ	matrix of eigenvalues
μ	coefficient of friction; eigenvalue; eigenvalue shift
$\mu(t)$	unit step function
μ_m	real part of mth eigenvalue
ν_m	imaginary part of mth eigenvalue
ξ	damping ratio; spatial coordinate
ξ_h	equivalent hysteretic damping ratio
ξ_k	real part of eigenvector
ρ	root of difference equation
ρ	amplitude of motion; Rayleigh quotient
$\rho(\mathbf{A})$	spectral radius of $\mathbf{A}$
ρ_h	amplitude of motion for hysteretic damping
σ	stress
σ_D	damping stress
τ	time
$\phi, \boldsymbol{\phi}$	angle; normalized eigenvector or mode shape; phase angle; potential function; spherical coordinate
$\phi(x)$	normalized eigenfunction
ϕ_h	phase angle for hysteretic damping
$\boldsymbol{\Phi}$	modal matrix
χ	response amplitude

$\boldsymbol{\psi}$	shape vector
$\psi(x)$	shape function
ω	undamped natural frequency in rad/s
ω_d	damped natural frequency in rad/s
ω_0	frequency of applied load in rad/s
Ω	frequency of the exciting force
∇	gradient vector

CHAPTER 1

Introduction

1.1 OBJECTIVES OF THE STUDY OF STRUCTURAL DYNAMICS

The response of physical objects to dynamic or time-varying loads is an important area of study in physics and engineering. The physical object whose response is sought may either be treated as rigid-body or considered to be deformable. The subject of rigid-body dynamics treats the physical objects as rigid bodies that undergo motion without deformation when subjected to dynamic loading. The study of rigid-body motion has many applications, including, for example, the movement of machinery, the flight of an aircraft or a space vehicle, and the motion of earth and the planets. In many instances, however, dynamic response involving deformations, rather than simple rigid-body motion, is of primary concern. This is particularly so in the design of structures and structural frames that support manufactured objects. Structural frames form a part of a wide variety of physical objects created by human beings: for example, automobiles, ships, aircraft, space vehicles, offshore platforms, buildings, and bridges. All of these objects, and hence the structure supporting them, are subjected to dynamic disturbances during their service life.

Dynamic response involving deformations is usually oscillatory in nature, in which the structure vibrates about a configuration of stable equilibrium. Such equilibrium configuration may be static, that is, time invariant, or it may be dynamic involving rigid-body motion. Consider, for example, the vibrations of a building under the action of wind. In the absence of wind, the building structure is in a state of static equilibrium under the loads acting on it, such as those due to gravity, earth pressure, and so on. When subjected to wind, the structure oscillates about the position of static equilibrium as shown in Figure 1.1.

An airplane in flight provides an example of oscillatory motion about an equilibrium configuration that involves rigid-body motion. The aircraft can be idealized as consisting of rigid-body masses of fuselage and the engines connected by flexible wing structure (Fig. 1.2). When in flight, the whole system moves as a rigid body and may, in addition, be subjected to oscillatory motion transverse to the flight plane.

Motions involving deformation are caused by dynamic forces or dynamic disturbances. Dynamic forces may, for example, be induced by rotating

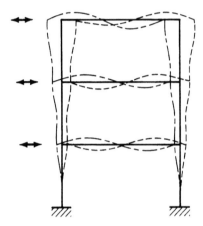

Figure 1.1. Oscillatory motion of a building frame under wind load.

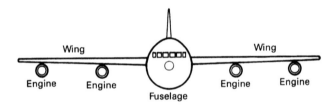

Figure 1.2. Aeroplane in flight.

machinery, wind, water waves, or a blast. A dynamic disturbance may result from an earthquake during which the motion of the ground is transmitted to the supported structure. Later in this chapter, we discuss briefly the nature of some of the dynamic forces and disturbances.

Whatever be the cause of excitation, the resulting oscillatory motion of the structure induces displacements and stresses in the latter. An analysis of these displacements and stresses is the primary objective of a study of the dynamics of structures.

1.2 IMPORTANCE OF VIBRATION ANALYSIS

The analysis of vibration response is of considerable importance in the design of structures that may be subjected to dynamic disturbances. Under certain situations, vibrations may cause large displacements and severe stresses in the structure. As we shall see later, this may happen when the frequency of the exciting force coincides with a natural frequency of the structure. Also, fluctuating stresses, even of moderate intensity, may cause material failure through fatigue if the number of repetitions is large enough. Oscillatory motion may at times cause wearing and malfunction of machinery. Also, the transmission of vibrations to connected structures may lead to undesirable results. Vibrations induced by rotating or reciprocating machinery may, for example, be transmitted through the supporting structure to delicate instruments mounted elsewhere on it, causing such instruments to malfunction. Finally, when the structure is designed for human use, vibratory motion may result in severe discomfort to the occupants.

With progress in engineering design, increasing use is being made of light-weight, high strength materials. As a result, modern structures are more susceptible to critical vibrations. This is as true of mechanical structures as of buildings and bridges. Today's buildings and bridges structures are, for example, lighter, more flexible, and are made of materials that provide much lower energy dissipation, all of which may contribute to more intense vibration response. Dynamic analysis of structures is, therefore, even more important for modern structures, and this trend is likely to continue.

It is apparent from the foregoing discussion that vibrations are undesirable for engineering structures. This is in general true except for certain mechanical machinery which relies on controlled vibration for its functioning. Such machinery includes, for example, vibratory compactors, sieves, vibratory conveyors, certain types of drills, and pneumatic hammers. In any case, whether or not the vibrations arise from natural causes or are induced on purpose, the structure subjected to such vibrations must be designed for the resultant displacements and stresses.

1.3 NATURE OF EXCITING FORCES

As stated earlier, the dynamic forces acting on a structure may result from one or more of a number of different causes, and it may be useful to categorize these forces according to the source of their origin, such as, for example, rotating machinery, wind, blast, or earthquake. The exciting forces may also be classified according to the nature of their variation with time as *periodic, nonperiodic,* or *random.* It is also useful to classify dynamic forces as *deterministic,* being specified as a definite function of time, or *nondeterministic,* being known only in a statistical sense. In the following, we discuss briefly each of these classifications.

1.3.1 *Dynamic forces caused by rotating machinery*

Rotating machinery that is not fully balanced about the center of rotation will give rise to exciting forces that vary with time. Consider, for example, a rotating motor that has an eccentric mass m_0 attached to it at a distance e from the center

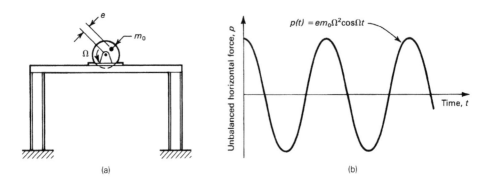

Figure 1.3. (a) Rotating machinery with unbalanced mass; (b) horizontal component of centrifugal force.

of rotation, as shown in Figure 1.3a. If the motor is rotating with a constant angular velocity Ω rad/s, the centrifugal force acting on the unbalanced mass is $em_0\Omega^2$ directed away from center and along the radius connecting the eccentric mass to the center. If time is measured from the instant the radius vector from the center of rotation to the mass is horizontal, the horizontal component of the centrifugal force at time t is given by

$$p(t) = em_0\Omega^2 \cos \Omega t \tag{1.1}$$

Force $p(t)$ is shown as a function of time in Figure 1.3b. If the supporting table is free to translate in a horizontal direction, force $p(t)$ will cause the table to vibrate in that direction.

Dynamic forces arising from unbalanced rotating machinery are quite common in mechanical systems, and the supporting structure must in such cases be designed to withstand the resulting deformations and stresses.

1.3.2   *Wind loads*

Structures subjected to wind experience aerodynamic forces which may be classified as drag forces, which are parallel to the direction of wind, and lift forces which are perpendicular to the wind. Both forces depend on the wind velocity, the wind profile along the height of the structure, and the characteristics of the structure. Winds close to the surface of the earth are affected by turbulence and hence vary with time. The response of the structure to the wind is thus a dynamic phenomenon and a precise estimate of the displacements and stresses induced by the wind can be obtained only through a dynamic analysis. For the purpose of design, wind forces are often converted into equivalent static forces. This approach while reasonable for low rise, comparatively stiff structures, may not be appropriate for structures that are tall, light, flexible, and possess low damping resistance.

Estimates of design wind speeds are obtained by measurements of wind in an open exposure, often at an airport, at a standard height, usually 10 m or 30 ft. Records are kept of maximum daily time-averaged mean wind speeds. Obviously, the mean wind will depend on the time used for the purpose of averaging. Design codes generally specify the use of a maximum mean wind with a given recurrence period. A typical value of recurrence period for strength design of buildings subjected to wind loads is 30 years. The corresponding design wind is usually obtained by a statistical analysis of the recorded data on hourly mean winds.

The variation of wind along the height, called wind profile, is determined on the basis of analytical studies and experimental observation. A similar approach is used to model wind turbulence or the variation of wind with time. The design mean wind speed, the wind profile and the wind turbulence together constitute the input data for a dynamic analysis for wind. It is evident that the effect of wind cannot be represented by a set of forces that are definite functions of time, since the wind loads are known only in a statistical sense.

1.3.3 *Blast loads*

A dynamic load of considerable interest in the design of certain structures is that due to a blast of air striking the structure. The blast or shock wave is usually caused by the detonation of a conventional explosive such as TNT or a bomb. In either case, the explosion results in the rapid release of a large amount of energy. A substantial portion of the energy released is expended in driving a shock wave whose front consists of highly compressed air. The peak overpressure (pressure above atmospheric pressure) in the shock front decreases quite rapidly as the shock wave propagates outward from the center of explosion. The overpressure in a shock wave arriving at a structure will thus depend on both the distance from the center of explosion and the strength of the explosive. The latter is measured in terms of the weight of a standard explosive, usually TNT, required to release the same amount of energy. Thus a 1-kiloton bomb will release the same amount of energy as the detonation of 1000 tons of TNT.

Empirical equations derived on the basis of observations are available for estimating the peak overpressure in a blast caused by the detonation of an explosive of given strength and striking a structure located at a given distance from the center of explosion. The overpressure rapidly decreases behind the front, and at some time after the arrival of the shock wave, the pressure may, in fact, become negative. The duration of the positive phase and the variation of the blast pressure during that phase can also be

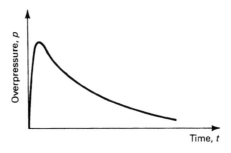

Figure 1.4. Pressure-time curve for a blast.

obtained from available empirical equations. In summary, a blast load can be represented by a pressure wave in which the pressure rises very rapidly or almost instantaneously and then drops off fairly rapidly according to a specified pressure–time relationship. A typical blast load history is shown in Figure 1.4.

1.3.4   *Dynamic forces caused by earthquakes*

Ground motions resulting from earthquakes of sufficiently large magnitude are one of the most severe and disastrous dynamic disturbances that affect human-made structures. Earthquakes are believed to result from a fracture in the earth's crust. The forces that cause such fractures are called tectonic forces. In fact, they are the very forces that have caused the formation of mountains and valleys and the oceans. They arise because of a slow convective motion of the earth's mantle that underlies the crust. This movement sets up elastic strains in the crustal rock. When the ability of the rock to sustain the elastic strain imposed on it is exceeded, a fracture is initiated at a zone of weakness in the rock. Fracturing relieves the elastic strains, causing the opposite sides of the fault to rebound and slip with respect to each other. The consequent release of the elastic strain energy stored in the rock gives rise to elastic waves which propagate outwards from the source fault. Before arriving at a specific location on the earth's surface, these waves may undergo a series of reflections and refractions.

The earthquake wave motion is very complex. The effect of such a motion on the supported structure can best be assessed by obtaining measurements of the time histories of ground displacements or accelerations by means of special measuring instruments called *seismographs*, and then analyzing the structure for the recorded motion. It is generally more convenient and common to obtain measurements of the ground acceleration. Then if required, the velocity and displacement histories are derived from the recorded acceleration history by a process of successive numerical integration. Figure 1.5 shows the

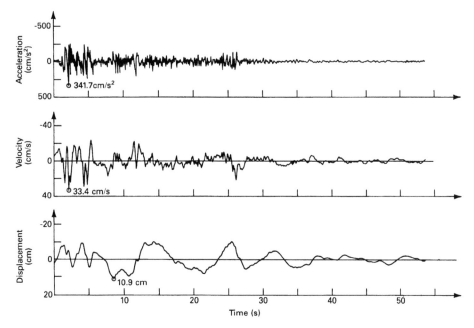

Figure 1.5. Imperial Valley earthquake, El Centro site, May 18, 1940, component N-S.

acceleration history recorded at El Centro, California, during an earthquake that took place in May 1940. Velocity and displacement histories obtained by successive integration are also shown.

Ground motions induced by an earthquake cause dynamic excitation of a supported structure. As we shall see later, the time-varying support motion can be translated into a set of equivalent dynamic forces that act on the structure and cause it to deform relative to its support. If a ground motion history is specified, it is possible to analyze the structure to obtain estimates of the deformations and stresses induced in it. Such analytical studies play an important role in the design of structures expected to undergo seismic vibrations.

1.3.5 *Periodic and nonperiodic loads*

Dynamic loads vary in their magnitude, direction, or position with time. It is, in fact, possible for more than one type of variation to coexist. As an example, earthquake induced forces vary both in magnitude and direction. However, by resolving the earthquake motion into translational components in three orthogonal directions and the corresponding rotational components, the earthquake effect can be defined in terms of six component forces and moments, each of which varies only in the magnitude with time. The constant-magnitude centrifugal force caused by imbalance in a rotating machinery can be viewed as a force of constant magnitude that is continually varying in direction with time. Alternatively, we can interpret the force as consisting of two orthogonal components, each of which varies in magnitude with time. A wheel load rolling along the deck of a bridge provides an instance of a force that varies in its location with time.

A special type of dynamic load that varies in magnitude with time is a load that repeats itself at regular intervals. Such a load is called a periodic load. The era of load duration that is repeated is called a *cycle* of motion and the time taken to complete a cycle is called the *period* of the applied load. The inverse of the period, that is the number of cycles per second, is known as the *frequency* of the load. A general type of periodic load is shown in Figure 1.6 which also identifies the period of the load. The harmonic load caused by an unbalanced rotating machine, shown in Figure 1.3b, is a more regular type of periodic load.

Loads that do not show any periodicity are called nonperiodic loads. A nonperiodic load may be of a comparatively long duration. A rectangular pulse load imposed on a simply supported beam by the sudden application of a weight that remains in contact with the beam from the instance of its initial application is an example of a long-duration nonperiodic load. Such a load is shown in Figure 1.7. Nonperiodic loads may also be of short duration or *transient*,

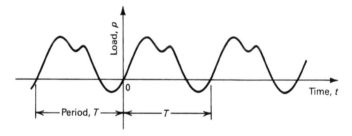

Figure 1.6. General periodic load.

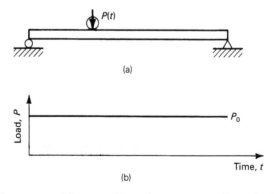

Figure 1.7. Simply supported beam subjected to a rectangular pulse load.

such as, for example, an air blast striking a building. When the duration of the transient load is very short, the load is often referred to as an impulsive load. A load or a disturbance that varies in a highly irregular fashion with time is sometimes referred to as a random load or a random disturbance. Ground acceleration resulting from an earthquake provides one example of a random disturbance.

1.3.6 *Deterministic and nondeterministic loads*

From the discussion in the previous paragraphs, we observe that certain types of loads can be specified as definite functions of time. The time variation may be represented by a regular mathematical function, for example, a harmonic wave, or it may be possible to specify the load only in the form of numerical values at certain regularly spaced intervals of time. Loads that can be specified as definite functions of time, irrespective of whether the time variation is regular or irregular, are called deterministic loads, and the analysis of a structure for the effect of such loads is called deterministic analysis. The harmonic load imposed by unbalanced rotating machinery is an example of a deterministic load that can be specified as a mathematical function of time. A blast load is also a deterministic load, and it may be possible to represent it by a mathematical curve that will closely match the variation. A measured earthquake accelerogram is a deterministic load that can only be specified in the form of numerical values at selected intervals of time.

Certain types of loads cannot be specified as definite functions of time because of the inherent uncertainty in their magnitude and the form of their variation with time. Such loads are known in a statistical sense only and are described through certain statistical parameters such as mean value and spectral density. Loads that cannot be described as definite functions of time are known as nondeterministic loads. The analysis of a structure for nondeterministic loads yields response values that are themselves defined only in terms of certain statistical parameters, and is therefore known as nondeterministic analysis. Earthquake loads are, in reality, nondeterministic because the magnitude and frequency distribution of an acceleration record for a possible future earthquake cannot be predicted with certainty but can be estimated only in a probabilistic sense. Wind loads are quite obviously nondeterministic in nature.

Throughout this book, we assume that the loads are deterministic or can be specified as definite functions of time. The methods of analysis and the resultant response will also therefore be deterministic.

1.4 MATHEMATICAL MODELING OF DYNAMIC SYSTEMS

In an analysis for the response of a system to loads that are applied statically, we need to concern ourselves with only the applied loads and the internal elastic forces that oppose the former. Dynamic response is much more complicated, because in addition to the elastic forces, we must contend with *inertia forces* and the forces of *damping resistance* that oppose the motion. Static response can, in fact, be viewed as a special case of dynamic response in which the accelerations and velocities are so small that the inertia and damping forces are negligible.

Before carrying out any analysis, the physical system considered must be represented by a mathematical model that is most appropriate for obtaining the desired response parameters. In either a static or a dynamic analysis of a structure, the response parameters of interest are displacements, and internal forces or stresses. Critical values of these parameters are required in the design. For a static case, the response is a function of one or more spatial coordinates; for a dynamic case the response depends on both the space variables and the time variable. Apparently, the dynamic response must be governed by partial differential equations involving the space and the time coordinates. This is indeed so. However, in many cases it is possible to model the system as an assembly of rigid bodies that have mass but are not deformable or have no compliance, and massless spring-like elements that deform under load and provide the internal elastic forces that oppose such deformation. The response of such a model is completely defined by specifying the displacements along certain coordinates that determine the position of the rigid-body masses in space. A system modeled as above is referred to as a *discrete system* or a *discrete parameter system*. The number of coordinate directions along which values of the response parameters must be specified in order to determine the behavior of the mathematical model completely is called the number of degrees of freedom. The response of a discrete parameter system is governed by a set of ordinary differential equations whose number is equal to the number of *degrees of freedom*.

As an example of the modeling process, consider a bar clamped at its left-hand end and free at the other, as shown in Figure 1.8a. Let the cross-sectional dimensions of the bar be small as compared to its length, and let the bar be constrained so that its fibers can move in only an axial direction. A cross section such as *AA* will move in the positive and negative directions of the x axis, and the position of the cross section at a time t will be a function of both the spatial location of the cross section, that is, the value of x, and the value of time t. The axial vibrations of the bar are thus governed by a partial differential equation involving the independent variables x and t. It may, however, be reasonable to model this bar by the assembly of a series of rigid masses and interconnecting springs as shown in Figure 1.8b. The selection of the number of mass elements will depend upon the accuracy that is desired. Supposing that this number is N, we must determine the horizontal displacement of each of the N masses to describe completely the response of the system at a given time t. The system is therefore said to have N degrees of freedom, and its response is governed by N ordinary differential equations.

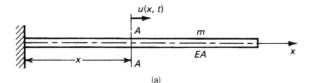

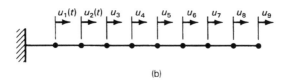

Figure 1.8. (a) Bar undergoing axial vibrations; (b) lumped mass model of the bar in (a).

For the bar model just described, the N degrees of freedom or coordinates correspond to displacements along a Cartesian direction. Alternative choices for coordinates are possible. In fact, in many situations, coordinates known as generalized coordinates may prove to be more effective. We discuss the meaning and application of such coordinates in the subsequent chapters of the book.

For a large majority of physical systems, discrete models consisting of an assembly of rigid mass elements and flexible massless elements are quite adequate for the purpose of obtaining the dynamic response. It should, however, be recognized that, in general, discrete modeling is an idealization because all mass elements will possess certain compliance and all flexible elements will possess some mass. In fact, in certain situations, a model in which both the mass and the flexibility are distributed may be better able to represent the physical system under consideration. Such a model is referred to as a *continuous system* or a *distributed parameter system*. Its response is governed by one or more partial differential equations.

In general, the analysis of a discrete system is much simpler than that of a continuous system. Furthermore, it is usually possible to improve the accuracy of the results obtained from the analysis of a discrete model by increasing the number of degrees of freedom in the model. Discrete modeling is therefore usually the preferred approach in the dynamic analysis of structures. Consequently, a major portion of this book is devoted to the modeling and analysis of discrete systems. Sufficient information is, however, provided on the analysis of simple continuous systems as well.

In the dynamic analysis of engineering structures, it is generally assumed that the characteristics of the system, that is, its *mass, stiffness,* and *damping* properties do not vary with time. It is further assumed that deformations of the structure are small and that the deforming material follows a linear stress–strain relationship. When the foregoing assumptions are made, the structure being analyzed is said to be *linear* and the *principle of superposition* is valid. This principle implies that if u_1 is the response of the structure to an applied load p_1 and u_2 is the response to another load p_2, the total response of the structure under simultaneous application of p_1 and p_2 is obtained by summing the individual responses u_1 and u_2 so that

$$u = u_1 + u_2 \qquad (1.2)$$

The principle of superposition usually results in considerable simplification of the analysis. Fortunately, for a majority of engineering structures in service, it is quite reasonable to assume that the principle of superposition is valid and can be applied to their analysis. In a few situations, however, structural deformations may be quite large. Also, the stress–strain relationships may become nonlinear as the material deforms. Both of these conditions will introduce nonlinearity in the system. The nonlinearity associated with large deformations is called *geometric nonlinearity*, and that associated with nonlinear stress–strain relationship is called *material nonlinearity*. These nonlinearities, particularly the material nonlinearity, may have to be taken into account in analyzing structures that are strained into the postelastic range, which may, for example, be the case under a severe earthquake excitation.

A majority of analysis procedures described in this book assume that the structure is linear. However, a brief discussion of nonlinear analysis procedures has been included under certain topics.

1.5   SYSTEMS OF UNITS

Two different systems of units are in common use in engineering practice. One of these is the International System of metric units, commonly referred to as the SI units. The other system is the system of Imperial units.

In the International System, the basic unit of length is the meter (m), the basic unit of mass is the kilogram (kg), and the basic unit of time is the second (s). The unit of force is a derived unit and is known as a newton. A newton is defined as a force that will produce an acceleration of $1 \, \text{m/s}^2$ on a mass of 1 kg. Since, according to the Newton's law of motion, force is equal to the product of mass and acceleration, we have

$$N = \frac{\text{kg} \cdot \text{m}}{\text{s}^2}$$

Decimal multiples and submultiples of the units used in the International System usually involve a factor of 10^3 and are formed by the addition of prefixes given in Table 1.1. The unit of pressure N/m^2 is also referred to as a pascal (Pa), and a mass of 1000 kg is sometimes called a tonne (or a metric ton).

Table 1.1. Prefixes for multiples and submultiples of basic units in SI.

Factor	Prefix	Symbol
10^9	giga	G
10^6	mega	M
10^3	kilo	k
10^{-3}	milli	m
10^{-6}	micro	μ
10^{-9}	nano	n

In the Imperial system, the basic unit of length is the foot (ft), the basic unit of force is the pound (lb), and the basic unit of time is the second (s). Mass is a derived unit. A unit mass, also known as a slug, is defined as the mass that when subjected to a force of 1 lb will be accelerated at the rate of $1 \, \text{ft/s}^2$. Instead of specifying the mass of a body, it is usual to specify its weight, which is equal to the force of gravitation exerted on the mass. This requires the definition of a standard gravity constant, g, the acceleration due to gravity, which is taken as $32.174 \, \text{ft/s}^2$. The value of g in SI units is $9.8066 \, \text{m/s}^2$. It is readily seen that

$$\text{slug} = \frac{\text{lb} \cdot \text{s}^2}{\text{ft}}$$

Multiples or fractions of the basic Imperial units most commonly used in engineering practice are inches for length and kilopounds or kips (1000 lb) for force. Examples and exercises in this book use both SI and Imperial units. Table 1.2 will assist in conversion from one system of units to the other.

1.6 ORGANIZATION OF THE TEXT

This book is devoted to a study of the analysis of engineering structures excited by time-varying disturbances. The material is divided into four parts and seventeen chapters. Part 1 of the book deals with the formulation of equations of motion, primarily for discrete single- and multi-degree-of-freedom systems. Part 2 is devoted to the solution of the equation of motion for a single-degree-of-freedom system. Part 3 deals with the solution of differential equations of motion governing the response of discrete multi-degree-of-freedom systems. Formulation of the equations of motion for continuous or distributed parameter systems is discussed in Part 4, which also describes the methods used in the solution of such equations. The contents of the individual chapters are described briefly in the following paragraphs:

1. In Chapter 1, we describe the objectives of the study of dynamic response of engineering structures and the importance of such a study in their design.

Table 1.2. Conversion between Imperial and SI units.

Item	Imperial to SI	SI to Imperial
Acceleration	$1\,\text{ft/s}^2 = 0.3048\,\text{m/s}^2$	$1\,\text{m/s}^2 = 3.2808\,\text{ft/s}^2$
Area	$1\,\text{ft}^2 = 0.0929\,\text{m}^2$	$1\,\text{m}^2 = 10.764\,\text{ft}^2$
	$1\,\text{in}^2 = 645.16\,\text{mm}^2$	$1\,\text{mm}^2 = 1.55 \times 10^{-3}\,\text{in}^2$
Force	$1\,\text{lb} = 4.448\,\text{N}$	$1\,\text{N} = 0.2248\,\text{lb}$
	$1\,\text{kip} = 4.448\,\text{kN}$	$1\,\text{kN} = 0.2248\,\text{kip}$
Length	$1\,\text{ft} = 0.3048\,\text{m}$	$1\,\text{m} = 3.2808\,\text{ft}$
	$1\,\text{in.} = 25.4\,\text{mm}$	$1\,\text{mm} = 0.03937\,\text{in.}$
	$1\,\text{mile} = 1.609\,\text{km}$	$1\,\text{km} = 0.6215\,\text{mile}$
Mass	$1\,\text{lb} = 0.4536\,\text{kg}$	$1\,\text{kg} = 2.2046\,\text{lb}$
Mass per unit length	$1\,\text{lb/ft} = 1.488\,\text{kg/m}$	$1\,\text{kg/m} = 0.672\,\text{lb/ft}$
Mass per unit area	$1\,\text{lb/ft}^2 = 4.882\,\text{kg/m}^2$	$1\,\text{kg/m}^2 = 0.2048\,\text{lb/ft}^2$
	$1\,\text{lb/in}^2 = 703.1\,\text{kg/m}^2$	$1\,\text{kg/m}^2 = 1.422 \times 10^{-3}\,\text{lb/in}^2$
Mass density	$1\,\text{lb/ft}^3 = 16.02\,\text{kg/m}^3$	$1\,\text{kg/m}^3 = 0.06242\,\text{lb/ft}^3$
	$1\,\text{lb/in}^3 = 27.680\,\text{Mg/m}^3$	$1\,\text{Mg/m}^3 = 0.03613\,\text{lb/in}^3$
Moment of inertia	$1\,\text{in}^4 = 416.230 \times 10^3\,\text{mm}^4$	$1\,\text{mm}^4 = 2.4 \times 10^{-6}\,\text{in}^4$
Section modulus	$1\,\text{in}^3 = 16387\,\text{mm}^3$	$1\,\text{mm}^3 = 0.06102 \times 10^{-3}\,\text{in}^3$
Pressure or stress	$1\,\text{ksi} = 6.895\,\text{MPa}$	$1\,\text{MPa} = 0.1450\,\text{ksi}$
	$1\,\text{psf} = 47.88\,\text{Pa}$	$1\,\text{Pa} = 0.02089\,\text{psf}$
	$1\,\text{psi} = 6.895\,\text{kPa}$	$1\,\text{kPa} = 0.1450\,\text{psi}$
Torque or moment	$1\,\text{ft} \cdot \text{kip} = 1.356\,\text{kN} \cdot \text{m}$	$1\,\text{kN} \cdot \text{m} = 0.7376\,\text{ft} \cdot \text{kip}$
Volume	$1\,\text{in}^3 = 16387\,\text{mm}^3$	$1\,\text{mm}^3 = 0.06102 \times 10^{-3}\,\text{in}^3$
	$1\,\text{ft}^3 = 28.316 \times 10^{-3}\,\text{m}^3$	$1\,\text{m}^3 = 35.32\,\text{ft}^3$

The nature of dynamic forces acting on engineering structures is discussed and considerations relevant to the mathematical modeling of structures are described.

2. Chapter 2 deals with the formulation of the equations of motion for a single-degree-of-freedom system. The system properties governing the response as well as the internal and external forces acting on a dynamic system are described. D'Alembert's principle, which converts a dynamic problem into an equivalent problem of static equilibrium, is introduced. The governing differential equation is then derived by using either Newton's vectorial mechanics or the principle of virtual displacement. Methods of idealizing a continuous system or a discrete multi-degree-of-freedom system by an equivalent single-degree-of-freedom system are presented. Finally, the effects of gravity load, axial forces, and support motion on the governing equation are discussed.

3. In Chapter 3, we describe the formulation of the equations of motion for a multi-degree-of-freedom system primarily through the principles of vectorial mechanics. As in the case of a single-degree-of-freedom system, the internal and external forces acting on the system are identified. Application

of the Ritz method to the modeling of continuous and discrete systems is discussed. An introductory description of the finite element method is presented and it is shown that the method is, in fact, a specialized form of Ritz analysis. A brief description is also provided of coordinate transformation and the static condensation of stiffness matrix.

4. In Chapter 4, we provide an introduction to the principles of analytical mechanics and their application to the formulation of equations of motion. The concepts of generalized coordinates, constraints, and work function are introduced. It is shown that the response of a dynamical system can be described through the scalar functions of work and energy, and the derivation of the Hamilton's and Lagrange's equations is presented.

5. Chapter 5 deals with the analysis of free-vibration response of a single-degree-of-freedom system. Both undamped and damped systems are treated. Various types of damping mechanisms, such as viscous damping, structural damping, and Coulomb damping, are discussed. The application of the phase plane diagram to the analysis of free-vibration response of damped and undamped system is described.

6. Chapter 6 deals with the response of a single-degree-of-freedom system to a harmonic excitation. The phenomenon of resonance is discussed. Analysis of vibration transmission from a structure to its support, and vice versa, is described. Procedures for computing the energy dissipated through damping resistance are presented. Finally, methods of measurement of damping are described.

7. Response to general dynamic loading and transient response of single-degree-of-freedom systems is presented in Chapter 7. In particular, response to an impulsive force and to shock loading is discussed. The concept of response spectrum is introduced and the application of response spectra to the analysis of the response to ground motion is described.

8. In Chapter 8, we present a detailed review of the approximate and numerical methods for the analysis of single-degree-of-freedom systems. The presentation includes the Rayleigh method, numerical evaluation of the Duhamel's integral, and direct numerical integration of the equation of motion. Errors involved in numerical integration method and the performance of various integration schemes are discussed. Finally, a brief description of the analysis of nonlinear response is presented.

9. Chapter 9 deals with the frequency-domain analysis of single-degree-of-freedom systems. Response to a periodic load and the Fourier series representation of a periodic load are discussed. Analysis of response to a general nonperiodic load through Fourier transform method is described. Applications of discrete Fourier transform and fast Fourier transform are presented.

10. Chapter 10 is devoted to the free-vibration response of multi-degree-of-freedom systems. The eigenvalue problem associated with free-vibration

response is discussed and concepts of mode shapes and frequencies are described. Application of the mode superposition method to the solution of undamped and damped free vibrations of multi-degree-of-freedom systems is presented.

11. In Chapter 11, we describe the various methods for the solution of eigen-value problem of structural dynamics. The solution methods include trans-formation methods, iteration methods, and the determinant search method. The comparative merits of the various methods are discussed and considerations governing the selection of a method are presented.

12. Chapter 12 deals with the forced dynamic response of multi-degree-of-freedom systems and application of the mode superposition method in the analysis of response.

13. In Chapter 13, we present approximate and numerical methods that may be applied to the analysis of multi-degree-of-freedom systems. The methods discussed include the Rayleigh–Ritz method, direct numerical integration, and analysis in the frequency domain.

14. In Chapter 14, we describe formulation of the equation of motion for a simple continuous system. The topics covered include flexural vibrations of a beam, axial and torsional vibrations of a rod, and lateral vibrations of a string and a shear beam.

15. Chapter 15 deals with the analysis of free-vibration response of a simple continuous system. The associated eigenvalue problems are derived and solutions are presented for certain simple systems. Finally, the application of the mode superposition method to the analysis of free vibrations is described.

16. Forced-vibration response of simple continuous systems using modal su-perposition analysis is covered in Chapter 16.

17. In Chapter 17, we describe one-dimensional wave propagation analysis. The one-dimensional wave equation is derived and the propagation of waves in a simple system, including wave reflection and refraction, is discussed. Finally, a brief presentation is given of wave propagation in a simple dis-persive medium.

SELECTED READINGS

Baker, W.E. 1973. *Explosion in Air*. Austin: University of Texas Press.

Crandall, S.H. & Mark W.D. 1963. *Random Vibration in Mechanical Systems*. New York: Academic Press.

Doebelin, E.O. 1980. System Modeling and Response. New York: John Wiley.

Gutman, I. 1968. *Industrial Uses of Mechanical Vibrations*. London, U.K.: Business Books.

Houghton, E.L. & Carruthers, N.B. 1976. *Wind Forces on Building and Structures*. London, U.K.: Edward Arnold.

Housner, G.W. & Jennings, P.C. 1982. *Earthquake Design Criteria*. Berkeley: Earthquake Engineering Research Institute.

Irvine, M. 1986. *Structural Dynamics for the Practicing Engineer*. London, U.K.: Allen & Unwin.

Kornhauser, M. 1964. *Structural Effects of Impact*. Baltimore: Spartan Books.

Lawson, T.V. 1980. *Wind Effects on Buildings, Vol. 1*. London: Applied Science Publishers Ltd.

Lin, Y.K. 1967. *Probabilistic Theory of Structural Dynamics*. New York: McGraw-Hill.

Newmark, N.M. & Rosenblueth, E. 1971. *Fundamentals of Earthquake Engineering*. Englewood Cliffs: Prentice Hall.

Sachs, P. 1978. *Wind Forces in Engineering*. Oxford, U.K.: Pergamon Press. 2nd Edition.

Simiu, E. & Scanlan, R.H. 1996. *Wind Effects on Structures*. New York: John Wiley. 3rd Edition.

Weaver, W. Jr., Timoshenko, S.P. & Young, D.H. 1990. *Vibration Problems in Engineering*. New York: Wiley. 5th Edition.

Yang, C.Y. 1986. *Random Vibrations of Structures*. New York: John Wiley.

PART 1

CHAPTER 2

Formulation of the equations of motion: Single-degree-of-freedom systems

2.1 INTRODUCTION

The displaced configuration of many mechanical systems and structures subject to dynamic loads can be completely described by specifying the time-varying displacement along only one coordinate direction. Such systems are designated as *single-degree-of-freedom systems*. Often, the modeling of a system as a single-degree-of-freedom system is an idealization. How truly the response of the idealized model fits the true behavior depends on several factors, including the characteristics of the system, the initial conditions, the exciting force, and the response quantity of interest. Nevertheless, for a large number of systems, representation as a single-degree-of-freedom model is quite satisfactory from an engineering point of view.

In view of the importance and the simplicity of single-degree-of-freedom systems, this chapter is devoted exclusively to a discussion of the formulation of equations that relate the response of such systems to one or more exciting forces. There is another equally important reason for treating the single-degree-of-freedom systems separately. As we shall see later, the analysis of the response of more complex multi-degree-of-freedom systems is in many cases accomplished by obtaining the response of several related single-degree-of-freedom systems and then superimposing these responses.

In this chapter, the equations of motion are, in general, formulated by using the principles of the vectorial mechanics of Newton, although the principle of virtual displacement is also introduced. The procedures of Newtonian mechanics are quite adequate for simple systems. For more complex single- as well as multi-degree-of-freedom systems, procedures that depend on energy principles or the principles of analytical mechanics are found to be more powerful. These procedures are discussed in detail in Chapter 4.

2.2 INERTIA FORCES

Figure 2.1 shows the simplest of all single-degree-of-freedom systems. It represents a rigid body of mass m constrained to move along the x axis in the plane

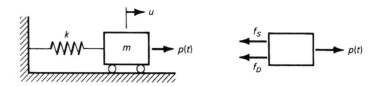

Figure 2.1. Forces on a single-degree-of-freedom system.

of the paper. The mass is attached to a firm support by a spring of stiffness k. At any time, the total time-varying force acting on the mass in a horizontal direction is denoted by $Q(t)$. In general, it is comprised of the externally applied force $p(t)$; the spring force f_S, which depends on the displacement u of the system from a position of equilibrium; and the force of resistance or damping. This last force, denoted as f_D, arises from air resistance and/or internal and external frictions. From Newton's second law of motion, the applied force is equal to the rate of change of momentum

$$Q(t) = \frac{d}{dt}\left(m\frac{du}{dt}\right) \tag{2.1}$$

In Equation 2.1, the momentum is expressed as the product of mass and velocity. In general, the mass of a system may also vary with time. An example of a varying mass system is a rocket in flight, where the mass of the rocket is decreasing continuously as the fuel burns out. For most mechanical systems or structures of interest to us, however, mass does not vary with time and therefore can be taken out of the differentiation. Using overdots to represent differentiation with respect to time, Equation 2.1 can be rewritten as

$$Q(t) = m\ddot{u}$$

or

$$Q(t) - m\ddot{u} = 0 \tag{2.2}$$

Quantity $m\ddot{u}$ has the units of a force. If we define an inertia force as having a magnitude equal to the product of mass and acceleration and a direction opposite to the direction of acceleration, we can view Equation 2.2 as an equation of equilibrium among the forces acting on a body. This principle, known as *d'Alembert's principle*, converts the problem of dynamic response to an equivalent static problem involving equilibrium of forces and permits us to use for its solution all those procedures that we use for solving problems of the latter class.

On a cursory glance, d'Alembert's principle appears simply as a mathematical artifact. Its physical significance can, however, be appreciated by considering the following simple example. Figure 2.2 shows a spring balance bolted to the

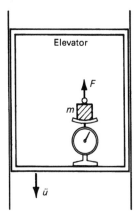

Figure 2.2. Inertia force on a moving mass.

floor of an elevator. A body having mass m is placed on the scale. First, let the elevator be at rest. The reading on the scale will indicate the weight of the body, mg, where g is the acceleration due to gravity. Now, let the body be pulled upward with a force F which is less than the weight mg. The scale will record a new reading equal to $mg - F$. Obviously, the downward force of gravity exerted by the body is being counteracted by an upward force F. Next, let the force F be removed, but let the elevator move downward with an acceleration a. The scale reading will change from mg to $mg - ma$. To an observer inside the elevator, the effect on the scale reading of a downward acceleration is no different from that of the upward force F. The quantity ma thus manifests itself as a virtual force acting in a direction opposite to the direction of acceleration.

2.3   RESULTANTS OF INERTIA FORCES ON A RIGID BODY

Structures or mechanical systems are sometimes modeled as rigid bodies connected to each other and to supports, often through deformable springs. We will find it expedient to replace the inertial forces acting on a rigid body by a set of resultant forces and moments.

 As an example, consider two point masses m_1 and m_2 connected together by a massless rigid rod as shown in Figure 2.3. The motion of this system in a plane can be described by specifying the translations of the center of mass in two mutually perpendicular directions and a rotation about that center. It is readily seen that the total inertial force due to translation in the x direction is equal to $(m_1 + m_2)\ddot{u}_x$ and acts through the center of mass. Similarly, for a y translation the total inertial force is $(m_1 + m_2)\ddot{u}_y$. For rotation about the center

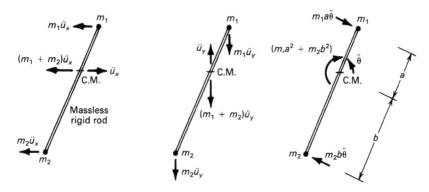

Figure 2.3. Resultant of inertia forces on a rigid body.

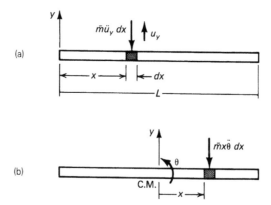

Figure 2.4. Inertia forces on a uniform rigid bar: (a) translation in y direction; (b) rotation about the center of mass.

of mass, the two inertia forces $m_1 a \ddot{\theta}$ and $m_2 b \ddot{\theta}$ are equal and opposite and form a couple whose magnitude is $(m_1 a^2 + m_2 b^2) \ddot{\theta}$.

Next, consider the uniform rigid bar shown in Figure 2.4a. The bar has a mass $\bar{m}$ per unit length. Consider a motion of the bar in the y direction. All particles of the bar have an acceleration $\ddot{u}_y$ and the inertial force on an incremental length dx is equal to $\bar{m}\ddot{u}_y\,dx$. Since all such infinitesimal forces are parallel and act in the negative y direction, they can be replaced by a single resultant force f_I given by

$$f_I = \int_0^L \bar{m}\ddot{u}_y\,dx = \bar{m}L\ddot{u}_y \qquad (2.3)$$

where L is the length of the bar.

The distance of the point of action of this force from the left end of the rod is obtained by taking moments of the inertial forces

$$f_I \bar{x} = \int_0^L \bar{m} \ddot{u}_y x \, dx = \bar{m} \ddot{u}_y \frac{L^2}{2}$$

$$\bar{m} L \ddot{u}_y \bar{x} = \bar{m} \ddot{u}_y \frac{L^2}{2} \qquad (2.4)$$

$$\bar{x} = \frac{L}{2}$$

Thus, the resultant force acts through the center of mass of the bar, which in this case coincides with the midpoint of the bar.

Next, consider a rotation of the bar about its center of mass. As shown in Figure 2.4b, the inertial force acting on an infinitesimal section at a distance x from the center of mass is, in this case, equal to $\bar{m} x \ddot{\theta} \, dx$. The sum of all such forces is given by

$$f_I = \ddot{\theta} \int_{-L/2}^{L/2} \bar{m} x \, dx \qquad (2.5)$$

But since x is measured from the center of mass, the integral in Equation 2.5 is zero and f_I vanishes. The infinitesimal inertial force also contributes a moment about the center of mass whose value is $\bar{m} x^2 \ddot{\theta} \, dx$. The total moment is given by

$$M_I = \ddot{\theta} \int_{-L/2}^{L/2} \bar{m} x^2 \, dx = I_0 \ddot{\theta} \qquad (2.6)$$

where I_0, termed the *mass moment of inertia*, is equal to the value of the integral, which works out to $\bar{m} L^3 / 12 = m L^2 / 12$, m being the total mass of the rigid bar.

A comparison of Equations 2.2 and 2.6 shows that in rotational motion, the mass moment of inertia plays the same role as does mass in translational motion. The force in the latter is replaced by a moment in the former and the translational acceleration $\ddot{u}$ is replaced by the angular acceleration $\ddot{\theta}$. The mass moment of inertia is often expressed as $I_0 = m r^2$, in which r is termed the *radius of gyration*. For a uniform rod rotating about its centroid, the radius of gyration is $L/\sqrt{12}$.

As another example of resultant inertial forces on a rigid body, consider a uniform bar rotating about its left-hand end. This motion is equivalent to a translation $u_y = L\theta/2$ and a rotation θ both measured at the center of mass.

The motions at the center of mass give rise to an inertial force $m L \ddot{\theta}/2$ and a moment $m L^2 \ddot{\theta}/12$ as indicated in Figure 2.5. The force and the moment about the center of mass are equivalent to a force of $m L \ddot{\theta}/2$ and a moment of

$mL^2\ddot{\theta}/12 + L/2 \cdot mL\ddot{\theta}/2 = mL^2\ddot{\theta}/3$ at the left-hand end of the bar. The resultant inertia forces for several different rigid-body shapes are given in Figure 2.6.

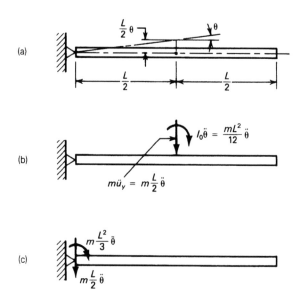

Figure 2.5. Inertia forces on a rigid bar rotating about its end.

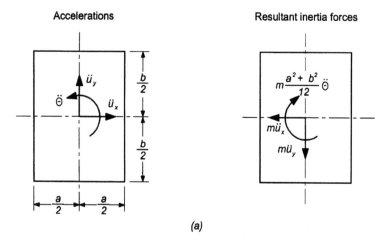

Figure 2.6. Resultant inertia forces in rigid-body shapes: (a) uniform rectangular plate, mass m; (b) uniform bar, mass m; (c) uniform circular plate, mass m; (d) uniform elliptical plate, mass m.

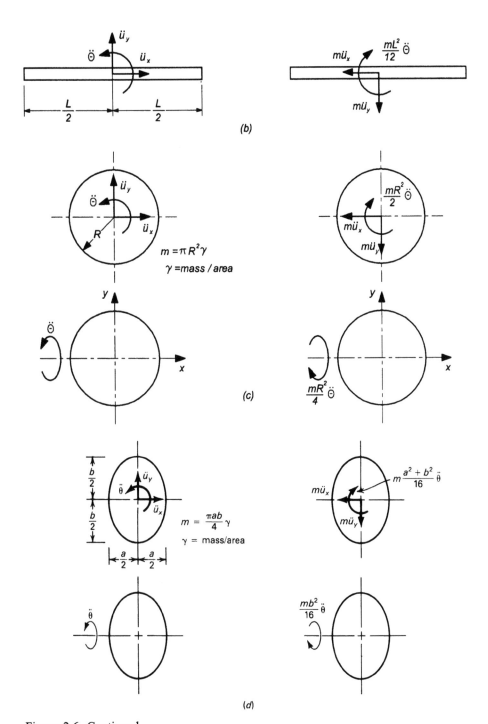

Figure 2.6. Continued.

Example 2.1

Determine the mass moment of inertia of a uniform rectangular rigid plate, having a mass $\bar{m}$ per unit area, for rotation about (i) an axis perpendicular to its plane and passing through its centroid, and (ii) the x axis.

Solution

(i) Referring to Figure E2.1a, the x and y components of the inertial force acting on an infinitesimal area $dx\,dy$ of the plate are $\bar{m}y\ddot{\theta}\,dx\,dy$ and $\bar{m}x\ddot{\theta}\,dx\,dy$, respectively. The resultant force in the x direction is given by

$$F_x = \int_{-b/2}^{b/2} \int_{-a/2}^{a/2} \bar{m}y\ddot{\theta}\,dx\,dy = \bar{m}\ddot{\theta}\int_{-b/2}^{b/2} \int_{-a/2}^{a/2} y\,dx\,dy \tag{a}$$

Because y is measured from the centroid, the integral on the right-hand side of Equation a is zero and F_x vanishes. For a similar reason, F_y is zero. A moment about the centroid however exists and is given by

$$M = I_0\ddot{\theta} = \int_{-b/2}^{+b/2} \int_{-a/2}^{+a/2} \bar{m}\ddot{\theta}(x^2 + y^2)\,dx\,dy$$

$$= \bar{m}ab\frac{a^2 + b^2}{12}\ddot{\theta} \tag{b}$$

$$= m\frac{a^2 + b^2}{12}\ddot{\theta}$$

where m is the total mass of the plate. Equation b gives $I_0 = m(a^2 + b^2)/12$.

(ii) For rotation about the x axis, divide the plate into rigid bars of length b and width dx as shown in Figure E2.1b. The inertial moment contributed by each bar is $\bar{m}bb^2\,dx/12$. The total moment is given by

$$I_{0x}\ddot{\theta} = \ddot{\theta}\int_{-a/2}^{a/2} \bar{m}b\frac{b^2}{12}\,dx$$

$$= \frac{mb^2}{12}\ddot{\theta}$$

giving $I_{0x} = mb^2/12$.

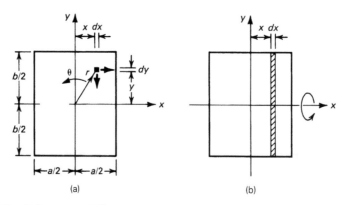

(a)   (b)

Figure E2.1. Inertia forces on a rigid rectangular plate.

2.4   SPRING FORCES

An elastic body undergoing deformation under the action of external forces sets up internal forces that resist the deformation. These *forces of elastic constraints,* alternatively known as *spring forces,* are present irrespective of whether the deformation is a result of static or dynamic forces acting on the body. The simplest example of the force of elastic constraint is provided by a helical spring shown in Figure 2.7. When stretched by a force applied at its end, the spring resists the deformation by an internal force which is related to the magnitude of the deformation. The relationship between the spring force and the displacement of the spring may take the form shown in Figure 2.7. In general, this relationship is nonlinear. However, for many systems, particularly when the deformations are small, the force–displacement relationship can be idealized by a straight line. The slope of this straight line is denoted by k and is called the *spring constant* or the *spring rate.* Thus, for a linear spring, the spring constant can be defined as the force required to cause a unit displacement and the total spring force is given by

$$f_S = ku \tag{2.7}$$

A similar definition of spring force can be applied to the forces of elastic constraints exerted by bodies of other form. For example, consider the cantilever beam shown in Figure 2.8a, and let the coordinate of interest be along a vertical at the end of the beam. A vertical load P applied at the end of the beam will cause a displacement in the direction of the load of $PL^3/3EI$, where L is the length of the beam, E the modulus of elasticity, and I the moment of inertia of the beam cross section. The spring constant, the force required to cause a unit displacement, is therefore $3EI/L^3$. Figure 2.8b shows a portal frame consisting of a rigid beam supported by two similar uniform columns each of length L, both fixed at their base. For a coordinate in the horizontal direction at the level of the beam, the spring constant is easily determined as $24EI/L^3$.

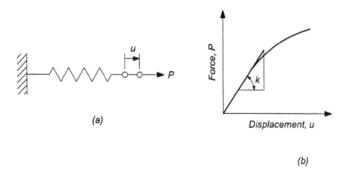

Figure 2.7. Force–displacement relationship for a helical spring.

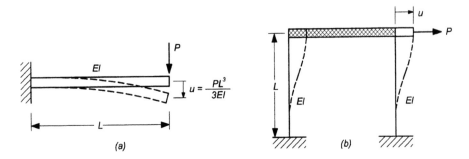

Figure 2.8. Definition of spring forces in elastic structures: (a) cantilever beam; (b) portal frame.

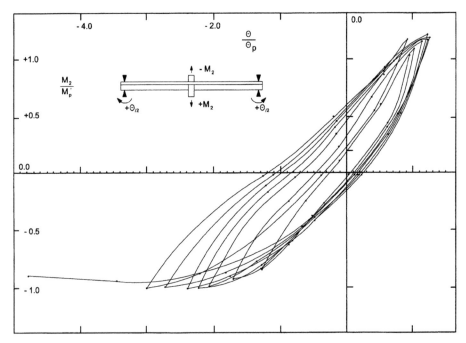

Figure 2.9. Moment-rotation relationship for a steel-concrete composite beam. (From J. L. Humar, "Composite Beams under Cyclic Loading," *Journal of Structural Division*, ASCE, Vol. 105, 1979, pp. 1949–1965).

As stated earlier, for large deformations, the relationship between the spring force and the displacement of the spring becomes nonlinear. For most structural materials, this leads to a softening of the spring, or a reduction in the slope of the force–displacement diagram. In addition, the path followed by the force–displacement diagram during unloading is different from that followed during loading. As an illustration, consider the moment-rotation curves shown in Figure 2.9. These curves were obtained during an experiment on a steel-concrete composite beam subjected to cyclic loading. The moment-rotation relationship is quite complex, the rotation corresponding to a given moment is not unique but depends on the loading history. During each cycle of loading, the moment-rotation relationship forms a closed loop. The area enclosed by this loop represents the energy lost during that cycle of loading. This energy loss is caused by plastic deformation of the material. It is evident that the dynamic analysis of a system in which the structural material is strained into the inelastic range would be quite complex. Nevertheless, analysis in the nonlinear range is often of practical interest. For example, under earthquake loading, structures are usually expected to be strained beyond the elastic limit, and the energy loss in plastic deformation must be considered in design. In Chapters 8 and 13, we briefly discuss the analysis of dynamic response of nonlinear structures.

Example 2.2
Find the equivalent stiffness of a system comprised of two linear springs (Fig. E2.2) having spring constants k_1 and k_2 when the two springs are connected (i) in parallel, and (ii) in series.

Solution
(i) Assume that the system is configured so that the springs undergo equal displacement. For a displacement u, $F_1 = k_1 u$ and $F_2 = k_2 u$. Therefore, $F = (k_1 + k_2)u$ and the spring constant is given by

$$k = \frac{F}{u} = k_1 + k_2$$

(ii) In this case, the force in each spring should be the same and equal to the externally applied force F. The extension of the first spring is $\Delta_1 = F/k_1$. The extension of the second spring is $\Delta_2 = F/k_2$. Thus

$$\Delta = \Delta_1 + \Delta_2 = F\left(\frac{1}{k_1} + \frac{1}{k_2}\right)$$

and

$$k = \frac{F}{\Delta} = \frac{1}{1/k_1 + 1/k_2} = \frac{k_1 k_2}{k_1 + k_2}$$

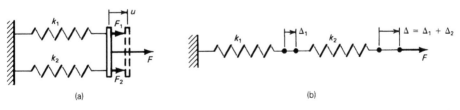

(a)   (b)

Figure E2.2. Definition of equivalent stiffness: (a) springs connected in parallel; (b) springs connected in series.

2.5 DAMPING FORCES

As stated earlier, the motion of a body is in practice resisted by several kinds of damping forces. These forces are always opposed to the direction of motion, but their characteristics are difficult to define or to measure. Usually, the magnitude of a damping force is small in comparison to the force of inertia and the spring force. Despite this, damping force may significantly affect the response. Also, in many mechanical systems, damping devices are incorporated on purpose. These devices help in controlling the vibrations of the system and are usually designed to provide a substantial amount of damping. For example, the shock absorbers in an automobile are devices which by providing large damping forces, cut down the unwanted vibrations.

A damping force may result from the resistance offered by air. When the velocity is small, the resistance offered by a fluid or a gas is proportional to the velocity. A resisting force of this nature is called *viscous damping force* and is given by

$$f_D = c\dot{u} \tag{2.8}$$

where c is a constant of proportionality called the *damping coefficient*. The damping force is, of course, always opposed to the direction of motion. When the velocity is large, the damping force due to air resistance becomes proportional to the square of the velocity. Therefore, a viscous damping force is either a linear or a nonlinear function of the velocity. As we will see later, the velocity of a vibrating system is, in general, proportional to its frequency, and hence a viscous damping force increases with the frequency of vibration.

Forces resisting a motion may also arise from dry friction along a nonlubricated surface. A resisting force of this nature is called the force of *Coulomb friction*. It is usually assumed to be a force of constant magnitude but opposed to the direction of motion. In addition to the forces of air resistance and external friction, damping forces also arise because of imperfect elasticity or internal friction within the body, even when the stresses in the material do not exceed its elastic limit. Observations have shown that the magnitude of such a force is independent of the frequency, but is proportional to the amplitude of vibration or to the displacement. Resisting forces arising from internal friction are called forces of *hysteretic* or *structural damping*.

In real structures, a number of different sources of energy dissipation exist. Energy may be lost due to repeated movements along internal cracks, such as those existing in reinforced concrete and masonry structures. Energy is dissipated through friction when slip takes place in the joints of steel structures. Relative movements at the interfaces between nonstructural elements, such as in-fill walls, and the surrounding structural elements also cause loss of energy. The foregoing discussion indicates that damping forces are complex in nature and difficult to determine. In general, viscous damping forces are easiest to

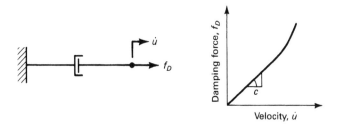

Figure 2.10. Representation of a viscous damping force.

handle mathematically and are known to provide analytical results for the response of a system that conform reasonably well to experimental observations.
It is therefore usual to model the forces of resistance as being caused by viscous damping. The viscous damping mechanism is indicated by a dashpot, as
shown in Figure 2.10. The general nature of damping force is represented by a
nonlinear relationship between the force and the velocity. The relationship can,
however, be approximated by a straight line in which case the damping force
is given by Equation 2.8.

As noted in Section 2.4, energy loss may also occur through repeated cyclic
loading of structural elements in the inelastic range. One may account for such
energy loss by defining a viscous damping model with an appropriately selected
value of the damping coefficient c. However, such a model is unable to correctly
represent the mechanism of energy loss. Energy loss through plastic flow in
repeated cyclic loading is best accounted for by adopting an appropriate model
to represent the nonlinear displacement versus spring force relationship. The
damping mechanism used in the mathematical model then accounts for the loss
of energy through sources other than inelastic deformations.

2.6   PRINCIPLE OF VIRTUAL DISPLACEMENT

As stated earlier, once the inertial forces have been identified and introduced
according to d'Alembert's principle, a dynamic problem can be treated as a
problem of static equilibrium, and the equations of motion can be obtained
by direct formulation of the equations of equilibrium. The latter equations are
obtained by the well-known methods of vectorial mechanics. For complex systems, however, use of these methods may not be straightforward. In such cases,
the application of the principle of *virtual displacement* may greatly simplify the
problem. The principle of virtual displacement, in fact, belongs to the field of
analytical mechanics, which deals with the scalar quantities of work and energy.
It is, however, of sufficient simplicity for us to discuss here, in advance of a
general and more detailed discussion of the methods of analytical mechanics in
Chapter 4.

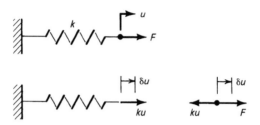

Figure 2.11. Definition of virtual work.

Suppose that a mechanical or structural system is in equilibrium under a set of externally applied forces and the forces of constraints, such as, for example, the reactions at the supports. If the system is subjected to virtual or imaginary displacements that are compatible with the constraints in the system, the applied forces, as well as the internal forces, will do work in riding through the displacements. We denote the work done by the external forces as δW_e and that done by the internal forces as δW_i. The symbol δW is used rather than ΔW to signify that the work done is imaginary or virtual and not real. Let the displacements be applied gradually so that no kinetic energies are developed. Also, assume that no heat is generated in the system and that the process is adiabatic, so that no heat is added or withdrawn. Under these conditions, the law of conservation of energy holds, giving

$$\delta W_e + \delta W_i = 0.0 \tag{2.9}$$

The definition of internal work needs careful consideration. It can be illustrated by a simple example of a particle attached to a linear spring of constant k, shown in Figure 2.11. The system is in equilibrium under an external force F. Now, let the particle undergo a virtual displacement δu as shown in the figure. The virtual work done by the external force is $F\delta u$. Internal work is done by the spring force which results from the deformation, u, of the spring. This force is equal to $-ku$, and acts on the particle. The virtual work done by the spring force is equal to $-ku\,\delta u$ and the virtual work equation takes the form

$$(F - ku)\delta u = 0.0 \tag{2.10}$$

Alternatively, a force equal to ku can be assumed to be acting on the end of the spring. This force is, of course, equal and opposite to the force ku acting on the particle. The virtual work done by this force is $ku\,\delta u$. A work of this nature is sometimes referred to as the work done by forces acting on the internal elements. It is denoted by δW_{ei} and is negative of the internal work. Using this alternative definition, the virtual work equation can be stated as

$$\delta W_e = \delta W_{ei} \tag{2.11}$$

When the system under consideration is rigid, the internal deformations and hence δW_{ei} are zero, and the virtual work equation takes the simpler form

$$\delta W_e = 0.0 \tag{2.12}$$

The principle of virtual displacement can now be stated as follows. If a deformable system in equilibrium under a set of forces is given a virtual displacement that is compatible with the constraints in the system, the sum of the total external virtual work and the internal virtual work is zero. The principle is of sufficient generality and applies equally well to linear or nonlinear elastic bodies. The only restriction is that the displacements be compatible with the constraints in the system. In calculating the internal virtual work in deformable bodies, it simplifies the problem if the internal forces can be assumed to remain constant as they ride through the virtual displacements. In general, this assumption is true only if the virtual displacements are small, so that the geometry of the system is not materially altered. It should be noted, however, that virtual displacements do not really exist; their imposition is simply a mathematical experiment that helps us determine the equations of equilibrium. Since the magnitude of a virtual displacement is arbitrary, it is only logical to assume that it is small. The virtual work equations obtained by imposing appropriate virtual displacements on the system lead to the desired equations of equilibrium. In the following examples, we illustrate the application of the principle of virtual displacement to the formulation of such equations.

Example 2.3
Determine the reaction at A in the simply supported beam shown in Figure E2.3.

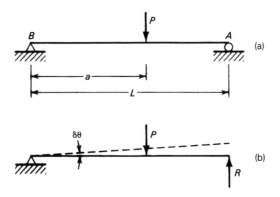

Figure E2.3. (a) Simply supported beam; (b) virtual displacements in the released beam.

Solution
A virtual displacement that is compatible with the constraints in the system will involve no vertical displacement at support A. The reaction R at support will not thus appear in any virtual work equation and cannot therefore be determined by the principle of virtual displacement. To be able to determine the reaction at A, we release the displacement constraint at A and replace it by the unknown force R. There is now only one constraint in the released structure, that is, a zero displacement at support B. We select a virtual displacement pattern caused by a rigid-body rotation of the beam about B. This displacement is compatible with the constraint in the system. Further, it causes no internal deformations so that δW_i is zero and the virtual work equation becomes

$$\delta W_e = 0.0$$

or

$$RL\delta\,\theta - Pa\delta\,\theta = 0.0$$

which gives

$$R = \frac{Pa}{L}$$

Example 2.4
For the frame shown in Figure E2.4a, determine the spring force due to the applied loads W and H.

Solution
To solve the problem, we first replace the spring by a pair of equal and opposite forces X and then give the resulting system a small virtual displacement $\delta\phi$. The virtual displacements of the entire system are shown in Figure E2.4b, and we note that they are compatible with the constraints in the system. The height of the load W above the base AC is given by

$$h = (a+b)\sin\,\phi \tag{a}$$

The change in h as a result of the change in ϕ is given by

$$\delta h = (a+b)\cos\,\phi\,\delta\phi \tag{b}$$

In a similar manner, the horizontal distance of B from A is $(a+b)\cos\,\phi$, and changes by $-(a+b)\sin\,\phi\,\delta\phi$. The horizontal distance to D from a vertical axis at A is $a\cos\,\phi$ and this changes by $-a\sin\,\phi\,\delta\phi$ as ϕ increases by $\delta\phi$. The horizontal distance to E from the vertical axis at A is $(a+2b)\cos\,\phi$. This changes by $-(a+2b)\sin\,\phi\,\delta\phi$.

Since we have replaced the spring by a pair of forces, the resulting system consists of rigid bodies and the simpler form of the virtual work equation (Eq. 2.12) can be used, giving

$$-W(a+b)\cos\,\phi\,\delta\phi + X(-a\,\sin\,\phi\,\delta\phi) + X(a+2b)\sin\,\phi\,\delta\phi$$

$$- H(a+b)\sin\,\phi\,\delta\phi = 0 \tag{c}$$

Equation c gives

$$X = \frac{W(a+b)}{(2b)}\cot\,\phi + H\frac{a+b}{2b} \tag{d}$$

It is instructive to solve the same problem by a direct formulation of the equation of equilibrium. By equating the moment of all the forces about A to zero, the vertical support reaction

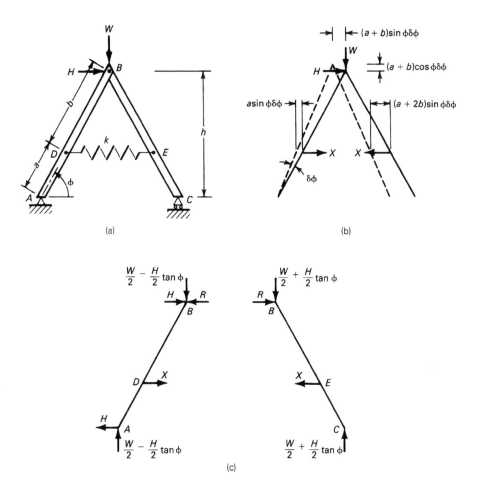

Figure E2.4. Formulation of the equations of equilibrium by the principle of virtual displacement: (a) A-frame in equilibrium; (b) virtual displacements; (c) free body diagrams.

at C is obtained as $W/2 + H \sin \phi/(2 \cos \phi)$. The vertical and the horizontal reactions at A are $W/2 - \{H \tan \phi\}/2$ and H, respectively. Free-body diagrams can now be drawn for rods AB and BC and are shown in Figure E2.4c. The horizontal reaction at the hinge at B is obtained by considering the equilibrium of rod BC and taking moments about E.

$$R = \left(\frac{W}{2} + \frac{H}{2} \tan \phi \right) \frac{a+b}{b} \frac{\cos \phi}{\sin \phi}$$

$$= \frac{W}{2b}(a+b) \cot \phi + H \frac{a+b}{2b} \tag{e}$$

The spring force X is equal to R.

Both the virtual displacement and the direct equilibrium solutions to the problem are based on the assumption that the initial angle ϕ which the leg AB makes with the horizontal does not change appreciably with the application of the two forces.

In this simple example, the advantage of using the method of virtual displacement is not at once evident. We may note, however, that in direct formulation of the equation of equilibrium, we had to determine the support reactions, even though they were of no interest to us. On the other hand, these forces did not appear in our virtual work equations, because the compatible virtual displacements in the directions of the support reactions were zero. The possibility of avoiding the determination of the forces of constraints can simplify the problem significantly when the system being analyzed is complex.

2.7   FORMULATION OF THE EQUATIONS OF MOTION

Having discussed the characteristics of the forces acting in a dynamic system, we are now in a position to formulate the equations of motion. Following d'Alembert's principle, the dynamic problem is first converted to a problem of the equilibrium of forces by introducing appropriate inertia forces. The equations of dynamic equilibrium are then obtained either by the direct methods of vectorial mechanics or by the application of the principle of virtual displacements. In discussing the formulation of equation of motion, it is convenient to classify the system into one of the following four categories:
1. Systems with localized mass and localized stiffness.
2. Systems with localized mass but distributed stiffness.
3. Systems with distributed mass but localized stiffness.
4. Systems with distributed mass and distributed stiffness.
The formulation of the equation of motion for each of the foregoing categories is discussed in the following paragraphs.

2.7.1   *Systems with localized mass and localized stiffness*

Figure 2.12a shows the simplest single-degree-of-freedom system. The mechanism providing the force of elastic constraint is localized in the massless spring. The mass can be assumed to be concentrated at the mass center of the rigid block. The mechanism providing the force of resistance is represented by a dashpot. The spring force, the damping force, the force of inertia, and the externally applied force are all identified on the diagram. The equation of motion is obtained directly on equating the sum of the forces acting along the x axis to zero. Thus,

$$f_I + f_D + f_S = p(t) \tag{2.13}$$

As an alternative, we may reason that the spring forces f_S must be balanced by an equal and opposite force, $p_1 = f_S$, as shown in Figure 2.12b. In a similar manner, a force $p_2 = f_D$ is required to counter the force of damping, and force $p_3 = f_I$ is required to balance the inertia force. The total external force must equal $p_1 + p_2 + p_3$, leading to Equation 2.9. We will find that this line of reasoning makes it simpler to visualize the formulation of the equations of

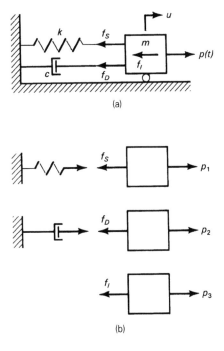

Figure 2.12. Dynamic equilibrium of a simple single-degree-of-freedom system.

motion for multi-degree-of-freedom systems. If the spring is assumed to be linear with spring constants k and the damping is viscous with a damping coefficient c, Equation 2.13 can be written as

$$m\ddot{u} + c\dot{u} + ku = p(t) \tag{2.14}$$

Equation 2.14 is linear because coefficients m, k, and c are constant and because the power of u and its time derivatives is no more than 1. Further, the equation is a second order differential equation. Its solution will give u as a function of time.

2.7.2 *Systems with localized mass but distributed stiffness*

A simple system in which the stiffness is distributed but the mass is localized is shown in Figure 2.13. It consists of a light cantilever beam of flexural rigidity EI, to the end of which is attached a point mass m. The beam is assumed to be axially rigid and is constrained to deform in the plane of the paper. The system shown in Figure 2.13 is a single-degree-of-freedom system, and as in the case of a system with localized mass as well as localized stiffness, we define the single degree of freedom along the coordinate direction in which the mass is free

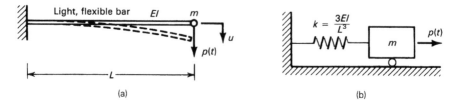

Figure 2.13. Vibration of a point mass attached to a light cantilever beam.

to move. In the present case, this direction is along a vertical through the tip of the beam. The inertia force in the system acts in the coordinated direction we have chosen and is proportional to the vertical acceleration of the mass. Further, we assume that the external force also acts in the same direction. Once the displacement along the selected coordinate direction is known, the displaced shape of the beam can be determined by standard methods of structural analysis. These methods also permit us to obtain the force of elastic restraint acting along the selected coordinate direction. This is the case, for example, for the cantilever beam and the portal frame shown in Figure 2.8.

The foregoing discussion shows that the system of Figure 2.13a is entirely equivalent to the spring–mass system shown in Figure 2.13b. The equation of motion is obtained directly by equating the sum of vertical forces to zero

$$m\ddot{u} + \frac{3EI}{L^3}u = p(t) \qquad (2.15)$$

Again, the force of gravity has been ignored in the formulation. This is admissible provided it is understood that u is measured from the position of static equilibrium.

Example 2.5
The torque pendulum shown in Figure E2.5 consists of a light shaft of uniform circular cross section having diameter d and length L. The material used to build the shaft has a modulus of rigidity G. A heavy flywheel having a mass m and a radius of gyration r is attached to the end of the shaft. Find the equation of free vibrations of the torque pendulum.

Solution
We select the angle of twist at the end of the shaft, θ, as the single degree of freedom of the system. A torque T applied at the end of the shaft will twist it by an angle $\theta = TL/GJ$ in which J, the polar moment of inertia of the shaft, is equal to $\pi d^4/32$. The torque required to cause a twist θ can therefore be expressed as $T = GJ\theta/L$.

The inertial moment induced by an angular acceleration $\ddot{\theta}$ of the flywheel is $mr^2\ddot{\theta}$. When there is no externally applied torque, the equation of equilibrium for angular motion is given by

$$mr^2\ddot{\theta} + \frac{GJ}{L}\theta = 0$$

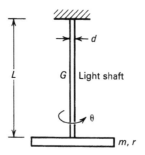

Figure E2.5. Vibrations of flywheel.

2.7.3   *Systems with distributed mass but localized stiffness*

An example of such a system is shown in Figure 2.14. The system consists of a uniform rigid bar of mass m hinged at its left-hand end and suspended by a spring at the right end. The force due to elastic constraint is localized in a spring. However, the mass is uniformly distributed along the rigid bar. As a result, the inertia forces are also distributed along the length. We may be able to replace the distributed inertia forces by their resultants using the methods described in Section 2.3. However, in most cases we are still left with more than one resultant force.

   In spite of the fact that the inertia forces are distributed, the system of Figure 2.14 has only one degree of freedom, because we can completely define the displaced shape of the system in terms of the displacement along a single coordinate. Thus, if $z(t)$ is the unknown displacement along the selected coordinate and $\psi(x)$ is a selected shape function, the displacement of the system is given by

$$u(x,t) = z(t)\psi(x) \tag{2.16}$$

For the bar shown in Figure 2.14, we may select the rotation at the hinge, or the vertical displacement at the tip of the rod as the degree of freedom of the system. Let us choose the latter and denote the displacement along the selected coordinate direction as $z(t)$; then because the given bar is rigid, the displacement shape function is readily seen to be

$$\psi(x) = \frac{x}{L} \tag{2.17}$$

where x is the distance from the hinge. Once we solve for the unknown displacement $z(t)$, the displacement of the bar is fully determined. A coordinate such as $z(t)$ is known as a *generalized coordinate* and the associated system model as a generalized single-degree-of-freedom system. The displaced shape of the system is determined in terms of the generalized coordinate through the

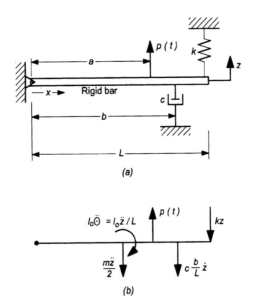

Figure 2.14. Vibration of a rigid bar.

selected shape function. When the system consists of rigid body assemblages, the exact shape function is readily found.

Continuing with the example of rigid bar, it is readily seen that the bar has an angular acceleration of $\ddot{z}/L$ and a vertical translational acceleration of $\ddot{z}/2$ at its mass center. The corresponding forces of inertia are obtained from Section 2.3 and are indicated on the diagram. Taking moments about the hinge, we obtain

$$I_0\ddot{\theta} + m\frac{\ddot{z}}{2}\frac{L}{2} + c\dot{z}\frac{b^2}{L} + kzL = p(t)a \tag{2.18a}$$

or

$$\left(\frac{mL}{12a} + \frac{mL}{4a}\right)\ddot{z} + \left(\frac{cb^2}{La}\right)\dot{z} + \left(\frac{kL}{a}\right)z = p(t)$$

$$m^*\ddot{z} + c^*\dot{z} + k^*z = p^*(t) \tag{2.18b}$$

where $\ddot{\theta} = \ddot{z}/L$ is the angular acceleration, $m^* = mL/3a$ is known as the *generalized mass*, $c^* = cb^2/La$ is the *generalized damping constant*, $k^* = kL/a$ is the *generalized spring constant*, and $p^* = p$ is the *generalized force*. It should be noted that in addition to the forces indicated in Figure 2.14b, the force of gravity equal to mg acts downward through the center of mass of the rigid rod. This force, and the spring force set up to balance it, can both be ignored in the formulation of the equation of motion, as long as it is assumed that the displacement u is measured from the position of equilibrium of the rod when acted upon only by the force of gravity. This point is dealt with in some detail in a later section of this chapter.

Example 2.6
The swinging frame of a balancing machine is shown diagrammatically in Figure E2.6a. Lever *ABC*, which has a nonuniform section, has a mass of 5.5 kg and radius of gyration 75 mm about its mass center, which is located at *B*. Lever *DF* is of uniform section and has a mass of 1 kg. A mass of 0.7 kg is attached to lever *DF* at point *E*. The mass of the connecting link *CF* may be ignored. The springs at *A* and *C* both have a stiffness of 5.6 kN/m. Obtain the equation of motion for small vibrations of the frame about a position of equilibrium.

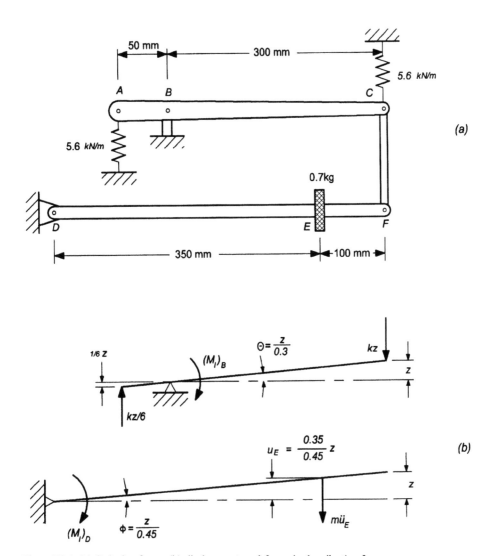

Figure E2.6. (a) Swinging frame; (b) displacements and forces in the vibrating frame.

Solution

The displaced shape of the frame is completely determined by specifying the displacement along one coordinate direction. Therefore, the frame has a single degree of freedom. We select the vertical deflection at F as the single coordinate. The equation of motion will be obtained in terms of the selected coordinate, which we shall denote by z. Also, in our discussion, we shall use the basic units of measurement: meters for length, kilograms for mass, and newtons for force.

The displaced shape of the frame for a vertical deflection z at F is shown in Figure E2.6b, along with all the forces acting on the frame. Corresponding to a vertical upward displacement u meters at C, the spring attached to that point exerts a downward force of $5600u$ newtons. Point A moves a distance $z/6$ meter downward. The spring at A resists this movement by an upward force of $5600 \times u/6 = 933.3u$ N. The inertial moment opposing the rotation of bar ABC acts at B and is given by

$$(M_I)_B = I_B \ddot{\theta}$$

$$= 5.5 \times 0.075^2 \times \frac{\ddot{z}}{0.3}$$

$$= 0.1031\ddot{z}$$

where θ is the rotation of bar ABC about B. The inertial moment at D due to the rotation of bar DEF is obtained from

$$(M_I)_D = I_D \ddot{\phi}$$

$$= \frac{1 \times 0.45^2}{3} \times \frac{\ddot{z}}{0.45}$$

$$= 0.15\ddot{z}$$

where ϕ represents the rotation of bar DEF about D. Finally, the downward force of inertia at E due to the motion of mass attached at that point is given by

$$P_I = m\ddot{u}_E$$

$$= 0.7 \times \frac{0.35}{0.45}\ddot{z}$$

$$= 0.5444\ddot{z}$$

To obtain the equation of motion, we impart a virtual displacement that is compatible with the constraints on the frame and equate the resulting virtual work to zero. A small vertical deflection δz at point F is an appropriate virtual deflection for our purpose. The displacements produced in the frame due to the deflection δz are identical to those shown in Figure E2.5b with z replaced by δz. The virtual work equation becomes

$$0.1031\ddot{z} \times \frac{\delta z}{0.3} + 0.15\ddot{z} \times \frac{\delta z}{0.45} + 0.5444\ddot{z}\,\delta z \times \frac{0.35}{0.45}$$

$$+933.3z \times \frac{\delta z}{6} + 5600z\,\delta z = 0$$

On canceling out δz from the virtual work equation, we get the following equation of motion

$$1.1004\ddot{z} + 5755.6z = 0$$

It is worth noting that we did not have to account for the reaction forces at supports B and D or for the force in link CF. These forces do not, in fact, appear in the virtual work expression and are not of interest to us.

2.7.4 *Systems with distributed stiffness and distributed mass*

Consider the horizontal bar of Figure 2.15. The bar has a mass $\bar{m}$ per unit length, an axial rigidity EA, and is vibrating in the axial direction under the action of a tip force $P(t)$. The inertia forces are now distributed along the length of the bar. Because the flexibility is also distributed along the length, one needs to know the exact distribution of the inertia forces to be able to determine the displaced shape of the bar. One can imagine the bar as being composed of a large number of sections each of a very small length $\Delta x = L/N$, where N is the total number of sections. The inertia force on the ith section is equal to $\bar{m} \, \Delta x \, \ddot{u}_i$. To obtain these inertia forces, one must determine the acceleration $\ddot{u}_i$ for all values of i from 1 to N. The number of unknown displacements or accelerations that must be determined is equal to the number of degrees of freedom. Therefore, the system has as many degrees of freedom as there are sections in it. Theoretically, the bar has an infinite number of degrees of freedom. We can, however, obtain an approximate picture of the behavior of this vibrating bar by dividing it into a finite number of sections and then assuming that displacements of all points

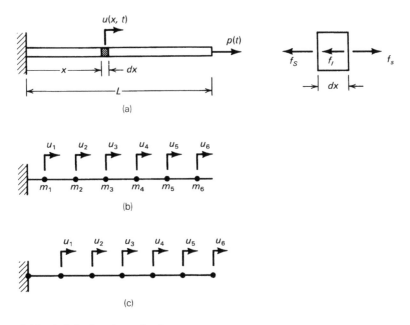

Figure 2.15. Axial vibration of a bar.

within a section are the same and are equal to the displacement of the center of the section. This is equivalent to lumping the mass of a section at its midpoint. We now have a model whose behavior will approximate the true behavior of the bar. This model, called a *lumped mass model,* is shown in Figure 2.15b. Figure 2.15c shows an alternative method of lumping the masses. The number of displacement coordinates that we must know in order to analyze a lumped mass model, and hence the number of degrees of freedom the model has, is equal to the number of mass points. To improve the accuracy of the model, we must increase the number of subdivisions. This will, however, increase the number of degrees of freedom and hence the complexity of the analysis.

As an alternative to the lumped mass model, we can describe the motion of the system by assuming that the displacement of the rod u is given by

$$u(x, t) = z(t)\psi(x) \tag{2.19}$$

where the $z(t)$ is a function of time and $\psi(x)$ a function of the spatial coordinate x. If we make an appropriate choice for the *shape function* $\psi(x)$, the only unknown is $z(t)$, and the system reduces to a single-degree-of-freedom system. As in the case of a rigid-body system with distributed mass, z is referred to as a generalized coordinate. Unlike the case of a rigid-body assemblage, the exact shape function is not determined and judgment must be used in selecting the shape function. It may be noted that z is not necessarily a coordinate which we can physically measure. Nevertheless, once z is determined, Equation 2.19 leads us to u. The concept of the generalized coordinates is explored more fully in Chapter 4.

Continuing with our example of the axial vibration of a rod, let us assume that a shape function $\psi(x)$ has been selected and that the displacement of the rod is therefore given by Equation 2.19. The inertia force acting on a small section of the rod is now given by

$$f_I = \bar{m}\,dx\,\ddot{u}(x, t) = \bar{m}\,dx\,\ddot{z}(t)\psi(x) \tag{2.20}$$

and the spring force by

$$f_S = EA(x)\frac{\partial u}{\partial x} = EA(x)z(t)\frac{d\psi}{dx} \tag{2.21}$$

We now give an admissible virtual displacement to the rod. Note that such an admissible displacement function should lead to small displacements. In addition, it should satisfy the constraint on the system, that is, it should give a zero displacement at the fixed end. Now suppose that $\psi(x)$ was selected to satisfy a similar constraint; then $\delta z\,\psi(x)$ is clearly an admissible virtual displacement. The elongation of the infinitesimal section associated with the virtual displacement $\delta z\,\psi(x)$ is

$$\delta z\frac{d\psi}{dx}dx \tag{2.22}$$

The virtual work done by the elastic spring forces acting on the section is given by

$$d(\delta W_{ei}) = f_S \, \delta z \frac{d\psi}{dx} dx \qquad (2.23)$$

Substituting for f_S from Equation 2.21 and integrating over the length, we obtain the total virtual work done on the internal elements

$$\delta W_{ei} = \delta z \, z(t) \int_0^L EA(x)\{\psi'(x)\}^2 \, dx \qquad (2.24)$$

where a prime denotes differentiation with respect to x.

The external forces are comprised of body forces due to inertia (Eq. 2.20) and the tip force $P(t)$. The virtual work done by these forces is given by

$$\delta W_e = -\delta z \ddot{z}(t) \int_0^L \bar{m}(x)\{\psi(x)\}^2 \, dx + P(t) \, \delta z \, \psi(L) \qquad (2.25)$$

In accordance with the principle of virtual displacement (Eq. 2.11), the total external virtual work must be equal to the virtual work done on the internal elements. On equating Equations 2.24 and 2.25 and canceling out δz, we get

$$\ddot{z}(t) \int_0^L \bar{m}(x)\{\psi(x)\}^2 \, dx + z(t) \int_0^L EA(x)\{\psi'(x)\}^2 \, dx = P(t)\psi(L) \qquad (2.26)$$

or

$$m^* \ddot{z}(t) + k^* z(t) = p^* \qquad (2.27)$$

where

$$m^* = \int_0^L \bar{m}(x)\{\psi(x)\}^2 \, dx$$

$$k^* = \int_0^L EA(x)\{\psi'(x)\}^2 \, dx$$

and

$$p^* = P(t)\psi(L)$$

Mass m^*, which is associated with the generalized coordinate z, is the generalized mass; similarly, k^* is the generalized stiffness and p^* is the generalized force.

Example 2.7
Write the equation of motion for the free axial vibration of the uniform bar shown in Figure E2.7a using a displacement shape function given by (i) $\psi(x) = x/L$, and (ii) $\psi(x) = \sin(\pi x/2L)$.

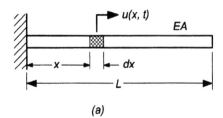

$$EA\frac{\partial u}{\partial x} = \frac{EA}{L}z(t) \leftarrow \qquad \rightarrow EA\left(\frac{\partial u}{\partial x} + \frac{\partial^2 u}{\partial x^2}dx\right) = \frac{EA}{L}z(t)$$

$$f_i = \bar{m}dx\ddot{z}(t)\frac{x}{L}$$

(b)

$$EA\, z(t)\frac{\pi}{2L}\cos\frac{\pi x}{2L} \leftarrow \qquad \rightarrow EA\, z(t)\left(\frac{\pi}{2L}\cos\frac{\pi x}{2L} - \frac{\pi^2}{4L^2}\sin\frac{\pi x}{2L}dx\right)$$

$$f_i = \bar{m}\, dx\, \ddot{z}(t)\sin\frac{\pi x}{2L} = -\frac{\pi^2}{4L^2}EA\, z(t)\sin\frac{\pi x}{2L}dx$$

(c)

Figure E2.7. Axial vibration of a uniform bar: (a) uniform bar; (b) shape function $\psi = x/L$; (c) shape function $\psi = \sin \pi x/2L$.

Solution

(i) The generalized mass is given by

$$m^* = \int_0^L \bar{m}\left(\frac{x}{L}\right)^2 dx = \frac{1}{3}m \tag{a}$$

where m is the total mass of the bar. The generalized stiffness k^* is given by

$$k^* = \int_0^L EA\left(\frac{1}{L}\right)^2 dx = \frac{EA}{L} \tag{b}$$

The equation of motion is

$$\frac{1}{3}m\ddot{z}(t) + \frac{EA}{L}z(t) = 0 \tag{c}$$

The forces acting on a small section of the bar are indicated in Figure E2.7b. It is noted that regardless of the value of z, the forces are not in equilibrium. In general, with the procedure being used, equilibrium will be satisfied only in an average sense and the results obtained are therefore necessarily approximate.

(ii) The generalized mass is in this case given by

$$m^* = \int_0^L \bar{m} \left(\sin \frac{\pi x}{2L} \right)^2 dx = \bar{m} \frac{L}{2} = \frac{m}{2} \tag{d}$$

The generalized stiffness k^* is obtained from

$$k^* = \int_0^L EA \left(\frac{\pi}{2L} \cos \frac{\pi x}{2L} \right)^2 dx = \frac{\pi^2 EA}{8L} \tag{e}$$

The equation of motion is

$$m\ddot{z}(t) + \frac{\pi^2}{4} \frac{EA}{L} z(t) = 0 \tag{f}$$

By substitution, it can be seen that $z(t) = G \sin(\omega t + \phi)$ is a solution of Equation f provided that $\omega = (\pi/2)\sqrt{EA/mL}$. Parameters G and ϕ are as yet undetermined constants. The displaced shape of the bar is given by

$$u(x, t) = G \sin(\omega t + \phi) \sin \frac{\pi x}{2L} \tag{g}$$

The inertia force acting on a section of the bar is now given by

$$f_I = z(\ddot{t})\bar{m} \sin \frac{\pi x}{2L} dx$$

$$= -\omega^2 z(t)\bar{m} \sin \frac{\pi x}{2L} dx \tag{h}$$

$$= -\frac{\pi^2}{4L^2} EAz(t) \sin \frac{\pi x}{2L} dx$$

The forces acting on a section of the bar (Fig. E2.7c) are seen to be in equilibrium. The result, which appears to be a coincidence, has been obtained because the shape function $\psi(x) = \sin(\pi x/2L)$ is one of the exact displacement mode shapes in which the bar can vibrate. The satisfaction of equilibrium does not, however, guarantee that the solution obtained is accurate. In fact, as we shall see later, the true vibration shape depends on the displacements and velocities imparted to the bar at time $t = 0$.

The procedure used to represent a distributed parameter system by a single-degree-of-freedom model, illustrated by reference to the axial vibrations of a bar, is equally applicable to a beam, a two-dimensional structure such as a flat plate, or to a three-dimensional solid. In each case, the displaced shape of the structure is expressed as the product of an appropriately selected shape function of the spatial coordinate(s) multiplied by a generalized coordinate, which is a function of time and is the only unknown in the system. A virtual work equation is then written covering the work done by the inertia forces, the elastic forces, the forces of damping, and the applied forces when moving through a small admissible virtual displacement. An admissible displacement is one that is compatible with the constraints of the system. When the selected shape function of spatial coordinates satisfies these constraints, the virtual displacement is best chosen to be equal to the shape function multiplied by a variation δz of the generalized coordinate z.

As a further example of the procedure, consider the beam shown in Figure 2.16a through d. The beam is supported on an elastic foundation with a spring constant of $\bar{k}(x)$ per unit length as well as by several local springs. It is subjected to a distributed force $\bar{p}(x)$ and one or more concentrated loads.

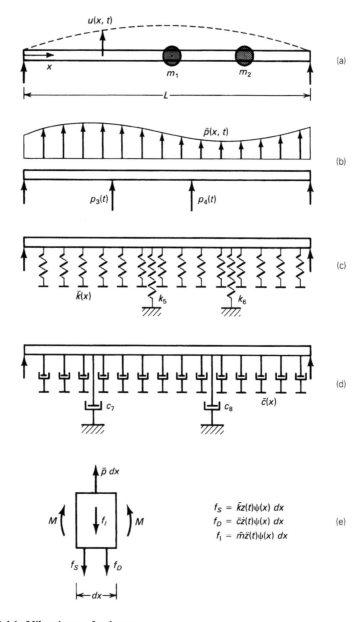

Figure 2.16. Vibrations of a beam.

The mass of the beam is $\bar{m}(x)$ per unit length. In addition, the beam supports several concentrated masses as shown. The damping forces are represented by viscous dampers distributed along the length and having damping constant $\bar{c}(x)$ per unit length. Concentrated dampers are also specified.

Let the displaced shape of the beam be represented by $z(t)\psi(x)$, where $\psi(x)$ is an appropriately chosen shape function and $z(t)$ is the unknown generalized coordinate. The forces acting on a small section of the beam are indicated in Figure 2.16e. They include the reaction provided by the foundation, the distributed damping force, the applied load, the force of inertia, and the elastic moments acting on the ends of the section. The beam is now subjected to a virtual displacement given by $\delta u = \delta z\, \psi(x)$. The relative rotation between the two faces of the section associated with the virtual displacement is $\delta z\psi''(x)dx$. The virtual work done by the elastic moments is given by

$$d(\delta W_{ei}) = M\, \delta z\, \psi''(x)\, dx \qquad (2.28)$$

Now, in accordance with the elementary beam theory, moment M is given by

$$M = EI\frac{\partial^2 u}{\partial x^2} = EIz(t)\psi''(x) \qquad (2.29)$$

Substituting Equation 2.29 in Equation 2.28, the virtual work done by the elastic forces becomes

$$d(\delta W_{ei}) = \delta z\, z(t)EI\{\psi''(x)\}^2\, dx \qquad (2.30)$$

As explained in Section 2.6, the internal virtual work $d(\delta W_i)$ is the work done by the restoring elastic moments in the element and is negative of $d(\delta W_{ei})$ given by Equation 2.30. It should also be noted that in calculating the virtual work of elastic forces, we have ignored the work done by the shear forces on true shear deformations. For normal proportions of the beam section, when the cross-sectional dimensions are small in comparison to the length of the beam, the work done by the shear forces is negligible in comparison to that done by the flexural forces.

On adding the work done by the other forces indicated in Figure 2.16 to the internal virtual work done by the elastic moments and integrating over the length, we obtain the following expression for the virtual work done by the distributed forces

$$\delta W = -\delta z\left[\int_0^L \bar{m}\{\psi(x)\}^2\, dx\, \ddot{z}(t) + \int_0^L \bar{c}\{\psi(x)\}^2\, dx\, \dot{z}(t)\right.$$

$$\left. + \int_0^L \bar{k}\{\psi(x)\}^2\, dx\, z(t) + \int_0^L EI\{\psi''(x)\}^2\, dx\, z(t)\right]$$

$$+ \delta z\int_0^L \bar{p}\psi(x)\, dx \qquad (2.31)$$

The virtual work done by the concentrated masses is given by

$$\delta W = - \left[\delta z \sum_i m_i \{\psi(x_i)\}^2 \ddot{z}(t) + \delta z \sum_i I_{0i} \{\psi'(x_i)\}^2 \ddot{z}(t) \right] \qquad (2.32)$$

where I_{0i} is the mass moment of inertia of the *i*th mass, $\psi(x_i)$ is the value of the shape function at the location of the *i*th mass, and $\psi'(x_i)$ is the value of the corresponding first derivative. If the concentrated masses are assumed to be point masses, the mass moment of inertia term drops out. In a similar manner, the virtual work of the concentrated spring forces is given by

$$\delta W = - \left[\delta z \sum_i k_i \{\psi(x_i)\}^2 z(t) \right] \qquad (2.33)$$

and that done by the concentrated dampers by

$$\delta W = - \left[\delta z \sum_i c_i \{\psi(x_i)\}^2 \dot{z}(t) \right] \qquad (2.34)$$

Finally, the virtual work done by the concentrated external forces is

$$\delta W = \delta z \sum_i p_i \psi(x_i) \qquad (2.35)$$

The total virtual work done by all the forces is obtained by summing Equations 2.31 through 2.35. The equation of motion is obtained by equating the total virtual work to zero (Eq. 2.9)

$$m^* \ddot{z}(t) + c^* \dot{z}(t) + k^* z(t) = p^* \qquad (2.36)$$

where the generalized mass m^*, the generalized damping c^*, the generalized stiffness k^*, and the generalized force p^* are given by

$$m^* = \int_0^L \bar{m} \{\psi(x)\}^2 \, dx + \sum_i m_i \{\psi(x_i)\}^2 + \sum_i I_{0i} \{\psi'(x_i)\}^2 \qquad (2.37a)$$

$$c^* = \int_0^L \bar{c} \{\psi(x)\}^2 \, dx + \sum_i c_i \{\psi(x_i)\}^2 \qquad (2.37b)$$

$$k^* = \int_0^L EI \{\psi''(x)\}^2 \, dx + \int_0^L \bar{k} \{\psi(x)\}^2 \, dx + \sum_i k_i \{\psi(x_i)\}^2 \qquad (2.37c)$$

$$p^* = \int_0^L \bar{p} \psi(x) \, dx + \sum_i p_i \psi(x_i) \qquad (2.37d)$$

Example 2.8
A uniform simply supported beam of mass *m* supports a point mass *M* at its center as shown in Figure E2.8. By selecting the deflected shape assumed by the beam under a central concentrated load as the shape function $\psi(x)$, obtain the equation of motion.

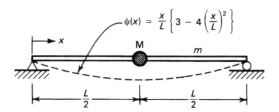

Figure E2.8. Simply supported beam.

Solution

The elastic curve is symmetrical about the center line of the beam. For x less than $L/2$ it is given by

$$\bar{\psi}(x) = C\frac{x}{L}\left\{3 - 4\left(\frac{x}{L}\right)^2\right\} \qquad\qquad x \leq L/2 \tag{a}$$

where C is a constant. The constant C can be included in the generalized coordinate $z(t)$, so that the shape function $\psi(x)$ can be taken as $(x/L)\{3 - 4(x/L)^2\}$. Also, $\psi(L/2) = 1$. The generalized mass is now given by

$$m^* = 2\int_0^{L/2} \bar{m}\frac{x^2}{L^2}\left(3 - 4\frac{x^2}{L^2}\right)^2 dx + M = \frac{17}{35}\bar{m}L + M = \frac{17}{35}m + M \tag{b}$$

The generalized stiffness is given by

$$k^* = 2\int_0^{L/2} EI\left(\frac{24x}{L^3}\right)^2 dx = \frac{48EI}{L^3} \tag{c}$$

The equation of motion is

$$\left(\frac{17}{35}m + M\right)\ddot{z}(t) + \frac{48EI}{L^3}z(t) = 0 \tag{d}$$

It is interesting to note that with the foregoing idealization, the uniform bar of mass m supporting a central mass M is entirely equivalent to a massless bar having the same elastic properties but supporting a central mass of $\left(M + \frac{17}{35}m\right)$.

It is apparent from the foregoing discussion that the success of the procedure described for representing a distributed parameter system by a single-degree-of-freedom model depends on an appropriate selection of the shape function $\psi(x)$. This is often a difficult task because the vibration shape depends not only on the physical characteristics of the system but also on the type of loading. Ideally, the selected shape function should satisfy all boundary conditions of the system. These boundary conditions fall into two categories: *essential* or *geometric boundary conditions*, and *natural boundary conditions*. As a minimum, the shape function should satisfy the essential boundary conditions. As an example, consider a simply supported beam. The two essential boundary conditions specify that the deflection of the beam should be zero at each end. The natural boundary conditions prescribe zero moments at the two ends. The shape

function selected in Example 2.8 satisfied all four conditions. Another shape
function that would satisfy all four boundary conditions is $\psi(x) = \sin(\pi x/L)$.
For the bar built-in at one end and free at the other and vibrating in the axial
direction, the essential condition requires zero axial displacement at the built-in
end, while the natural boundary condition calls for zero normal stress at the free
end. The shape function $\psi(x) = x/L$ satisfies the essential boundary condition
but not the natural one.

In addition to satisfying the essential boundary conditions, the shape function and its derivatives should satisfy certain conditions of continuity. From
the examples presented here, it is apparent that virtual work expressions for
distributed parameter systems involve integration of the shape function and its
derivatives. In the case of axial vibrations of a bar, for example, the highest-
order derivative of the shape function appears in the expression for virtual work
done by the elastic forces. The order of this derivative is 1. Apparently, the
shape function should be once differentiable for the virtual work expression to
be evaluated. This condition is satisfied if the function itself is continuous over
the length of the bar.

For flexural vibrations of a beam, the highest derivative appearing in the
virtual work expression is of order 2. In this case, therefore, the shape function should be twice differentiable, implying that the first derivative of the
shape function should be continuous over the length of the beam. In general,
if the highest-order derivative in the virtual work expression is of order m, the
$(m-1)$th derivative of the shape function should be continuous over the domain. An alternative expression used to specify this condition is to state that
the shape function should satisfy C^{m-1} continuity. The shape function method
of modeling a distributed parameter system is useful only when the general
nature of the vibration shape is known and a function that fits the shape can be
selected. The difficulty in selecting a function increases with the dimensionality
of the problem, and it may become impracticable to select a suitable function for a two-dimensional or a three-dimensional problem. In practical terms,
therefore, the method has limited usefulness.

2.8   MODELING OF MULTI-DEGREE-OF-FREEDOM DISCRETE
PARAMETER SYSTEM

In the preceding section, we used an appropriately selected shape function to
convert a distributed parameter system which has an infinite number of degrees
of freedom to an equivalent single-degree-of-freedom system. A similar procedure can be used to represent a discrete parameter system having more than
one degree of freedom by an equivalent single-degree-of-freedom system. As in
the case of a distributed parameter system, the displaced shape of the discrete
system is expressed as the product of an appropriately selected shape function

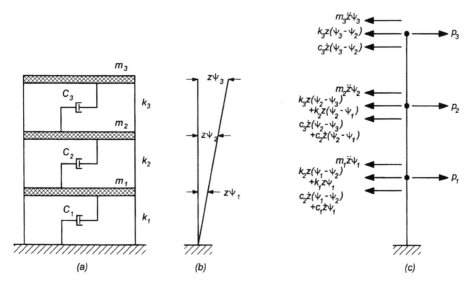

Figure 2.17. (a) Three-story shear frame; (b) displaced shape; (c) forces acting along the degrees of freedom.

and a generalized coordinate. In this case, however, the shape function is a vector rather than a continuous function of the spatial coordinate(s). A virtual work equation is then written covering the work done by the inertia forces, the elastic forces, the forces of damping, and the applied forces in moving through a small virtual displacement. This virtual work equation directly gives the equation of motion.

As an example of the procedure involved, consider the three-story shear frame shown in Figure 2.17a. The mass of the frame is assumed to be lumped at the floor levels. The floor beams are assumed to be rigid and the story stiffnesses are therefore provided by the flexure of the columns. The mass and stiffness properties are indicated in the figure. Damping resistance in the system is represented by interfloor dashpots. The shear frame shown in Figure 2.17a is constrained to vibrate in the plane of the paper and the columns are considered axially rigid. The frame has three degrees of freedom, as indicated. Each degree of freedom represents a possible lateral translation at a floor level. To model the frame by a single-degree-of-freedom system, we assume that the vibration shape is given by $\mathbf{u} = z(t)\boldsymbol{\psi}$, where $\boldsymbol{\psi}$ is a vector with three elements, representing the lateral translations of the floors, and $z(t)$ is the unknown generalized coordinate. The inertia forces, the spring forces, and the damping forces acting at the floor levels are shown in Figure 2.17c. Also shown are the external forces applied at these levels. We now assume that the system is given a virtual displacement that is compatible with the constraints on the system. Such a virtual displacement can be represented by $\delta z \, \psi$. The virtual work equation becomes

$$(m_3 \ddot{z} \psi_3) \, \delta z \, \psi_3 + (m_2 \ddot{z} \psi_2) \, \delta z \, \psi_2 + (m_1 \ddot{z} \psi_1) \, \delta z \, \psi_1$$

$$+ k_3 z(\psi_3 - \psi_2) \, \delta z \, \psi_3 + \{k_3 z(\psi_2 - \psi_3) + k_2 z(\psi_2 - \psi_1)\} \, \delta z \, \psi_2$$

$$+ \{k_2 z(\psi_1 - \psi_2) + k_1 z \psi_1\} \, \delta z \, \psi_1 + \{c_3 \dot{z}(\psi_3 - \psi_2)\} \, \delta z \, \psi_3$$

$$+ \{c_3 \dot{z}(\psi_2 - \psi_3) + c_2 \dot{z}(\psi_2 - \psi_1)\} \, \delta z \, \psi_2 + \{c_2 \dot{z}(\psi_1 - \psi_2) + c_1 \dot{z} \psi_1\} \, \delta z \, \psi_1$$

$$- p_3 \, \delta z \, \psi_3 - p_2 \, \delta z \, \psi_2 - p_1 \, \delta z \, \psi_1 = 0 \qquad (2.38)$$

Equation 2.38 can be written in a more compact form by using matrix notation

$$\boldsymbol{\psi}^T \mathbf{M} \boldsymbol{\psi} \ddot{z} + \boldsymbol{\psi}^T \mathbf{C} \boldsymbol{\psi} \dot{z} + \boldsymbol{\psi}^T \mathbf{K} \boldsymbol{\psi} z = \boldsymbol{\psi}^T \mathbf{p} \qquad (2.39)$$

where

$$\mathbf{M} = \text{mass matrix} = \begin{bmatrix} m_1 & 0 & 0 \\ 0 & m_2 & 0 \\ 0 & 0 & m_3 \end{bmatrix}$$

$$\mathbf{C} = \text{damping matrix} = \begin{bmatrix} c_1 + c_2 & -c_2 & 0 \\ -c_2 & c_2 + c_3 & -c_3 \\ 0 & -c_3 & c_3 \end{bmatrix}$$

$$\mathbf{K} = \text{stiffness matrix} = \begin{bmatrix} k_1 + k_2 & -k_2 & 0 \\ -k_2 & k_2 + k_3 & -k_3 \\ 0 & -k_3 & k_3 \end{bmatrix}$$

$$\mathbf{p} = \text{applied force vector} = \begin{bmatrix} p_1 \\ p_2 \\ p_3 \end{bmatrix}$$

Equation 2.39 can be expressed in the form

$$m^* \ddot{z} + c^* \dot{z} + k^* z = p^* \qquad (2.40)$$

where $m^* = \boldsymbol{\psi}^T \mathbf{M} \boldsymbol{\psi}$ is the generalized mass, $c^* = \boldsymbol{\psi}^T \mathbf{C} \boldsymbol{\psi}$ is the generalized damping, $k^* = \boldsymbol{\psi}^T \mathbf{K} \boldsymbol{\psi}$ is the generalized stiffness, and $p^* = \boldsymbol{\psi}^T \mathbf{p}$ is the generalized force.

As in the case of a distributed parameter system, the success of the method described here depends on how close the selected shape function is to the true vibration shape. Except in simple cases, a single shape function may not be sufficient to represent adequately the response of a multi-degree-of-freedom system.

2.9 EFFECT OF GRAVITY LOAD

In formulating the equations of motion, we have in general ignored the presence of gravity loads. In this section, we examine the effect gravity has on the equations of motion by reference to the simple system shown in Figure 2.18a. The system consists of a mass m suspended from the ceiling by a spring of stiffness k. The damping resistance is represented by a damper of stiffness c inserted between the ceiling and the mass. The system is vibrating under the action of force $p(t)$. When the system mass is at rest under its own load, the spring will be stretched by an amount Δ_{st} given by

$$mg = k\Delta_{st} \tag{2.41}$$

Let the position of vibrating mass be specified by its distance u^t from the end of the spring when the latter is unstretched. The forces acting on the mass are shown by the free-body diagram of Figure 2.18b. The equation of motion is readily obtained by applying the virtual work equation to the system shown in Figure 2.18b.

$$m\ddot{u}^t + c\dot{u}^t + ku^t = p(t) + mg \tag{2.42}$$

By substituting $u^t = u + \Delta_{st}$, where u is the displacement coordinate measured from the position the mass occupies when it is at rest under gravity load alone, Equation 2.42 becomes

$$m\ddot{u} + c\dot{u} + k(u + \Delta_{st}) = p(t) + mg \tag{2.43}$$

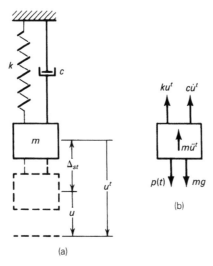

Figure 2.18. Vibrating mass suspended from ceiling.

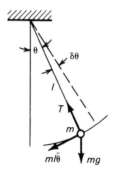

Figure 2.19. Simple pendulum.

Substituting Equation 2.41 into Equation 2.43, we get

$$m\ddot{u} + c\dot{u} + ku = p(t) \tag{2.44}$$

The gravity load does not enter into Equation 2.44. This indicates that when the displacement coordinate is measured from the position of equilibrium under gravity, the forces arising due to gravity may be ignored in formulation of the equation of motion. Solution of such an equation will then give the relative displacement. If the total displacement is of interest, it can be obtained by adding the gravity load displacement to the relative displacement obtained by a solution of the dynamic problem.

In the example just cited, the spring force consists of two components: the component $k\Delta_{st}$ which balances the gravity load mg, and the component ku, which opposes the vibrating motion. Further, the virtual work done by the component $k\Delta_{st}$ exactly balances the virtual work done by the gravity force. The virtual work equation therefore does not contain any term involving the gravity force. This means that the gravity load can be omitted from the formulation provided that displacements are measured from the position of static equilibrium. The virtual work done by the gravity forces will not cancel out in every case, and when it does not, the gravity forces cannot be disregarded. In situations where the gravity forces must be considered, they act as either restoring forces or as destabilizing forces. Examples of each of the two cases are presented in the following paragraphs.

A *simple pendulum* provides an example of the situation where the gravity load acts as a restoring force. As shown in Figure 2.19, the pendulum consists of a point mass m suspended by a light string of length l. When the point mass is displaced by an angle θ from its position of rest, the forces acting on it are as shown in the figure. To obtain the equation of motion, we give a small virtual displacement to the system. A displacement that is compatible with the constraints on the system should be along the tangent to the circular path described by the mass and can be represented by $l\,\delta\theta$. The virtual work

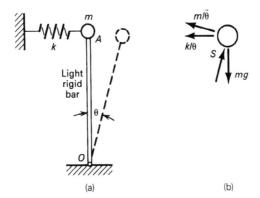

Figure 2.20. Inverted pendulum.

done by the forces of gravity is $-mgl\,\delta\theta\,\sin\theta$. The virtual work performed by the inertia force is $-ml^2\ddot{\theta}\,\delta\theta$. The tension in the string that would balance the gravity force in the position of equilibrium does not perform any work. The equation of virtual work is

$$(ml^2\ddot{\theta} + mgl\,\sin\theta)\,\delta\theta = 0 \tag{2.45}$$

Canceling out $ml\,\delta\theta$, and noting that for vibrations of small amplitude $\sin\theta = \theta$, we get the following equation of motion

$$\ddot{\theta} + \frac{g}{l}\theta = 0 \tag{2.46}$$

A situation where the gravity load acts as a destabilizing force exists in the case of an *inverted pendulum*, shown in Figure 2.20a. In the position of equilibrium, the gravity load mg is balanced by a thrust in the rod OA. The pendulum executes small vibrations about its position of equilibrium. The forces acting on the mass in its displaced position are as shown in Figure 2.20b. The equation of dynamic equilibrium can be obtained by taking moments about O. This gives

$$ml^2\ddot{\theta} + kl^2\theta\cos\theta - mgl\sin\theta = 0 \tag{2.47}$$

For small θ, Equation 2.47 becomes

$$m\ddot{\theta} + \left(k - \frac{mg}{l}\right)\theta = 0 \tag{2.48}$$

The gravity load reduces the effective stiffness of the system. In fact, when $m = kl/g$, the system becomes unstable and buckles under its own weight.

Example 2.9

The heavy flywheel shown in Figure E2.9 is rotating about a vertical axis at a constant angular speed of Ω rad/s. A small mass m is attached by a spring of stiffness k to the axle and is restrained by guides attached to the flywheel, so that the mass can move along the surface of the flywheel only in a radial direction. Obtain the equation of motion for radial vibrations of mass m.

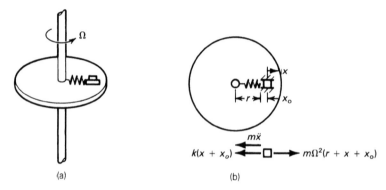

(a)   (b)

Figure E2.9. Mass moving on a rotating flywheel.

Solution

Let r be the unstretched length of spring. As the flywheel starts rotating, mass m moves away from the center of the wheel and eventually comes to rest at a distance x_0 such that the spring force exactly balances the centrifugal force acting on the mass. Thus

$$kx_0 = m\Omega^2(r + x_0) \tag{a}$$

Suppose that the mass executes vibrations about the position of rest, its displacement from the position of rest being denoted by x. The forces acting on the mass at any instant of time are shown in Figure E2.9b and consist of the centrifugal force $m\Omega^2(r + x_0 + x)$, the inertia force $m\ddot{x}$, and the spring force $k(x + x_0)$. The equation of dynamic equilibrium becomes

$$m\ddot{x} - m\Omega^2(r + x_0 + x) + k(x + x_0) = 0 \tag{b}$$

On substituting Equation a in Equation b, we get

$$m\ddot{x} + (k - m\Omega^2)x = 0 \tag{c}$$

In this example, the centrifugal force acts as a destabilizing force. When the speed of rotation is such that $\Omega^2 = k/m$, the system becomes unstable, since the spring is not stiff enough to restrain the mass from flying away off the wheel.

2.10   AXIAL FORCE EFFECT

In Section 2.9, we presented an example where the gravity load acted as a destabilizing force, which effectively reduced the stiffness of the system. In Example 2.9, the destabilizing force was seen to be the centrifugal force. Axial forces acting in a system may also reduce its effective stiffness and may cause

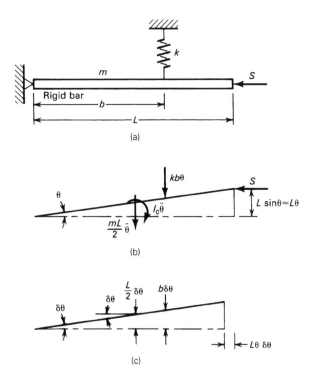

Figure 2.21. Axial force effect on a vibrating bar: (a) rigid bar; (b) free-body diagram; (c) virtual displacement.

it to become unstable. Consider, for example, the simple rigid-body assemblage shown in Figure 2.21a. It consists of a rigid uniform bar of mass m and length L hinged at the left-hand end and supported at a distance b by a spring of stiffness k. The bar is subjected to a constant axial force S acting at its tip and is undergoing small vibrations in the plane of the paper.

The forces acting on the bar when it is displaced by an angle θ about the hinge are indicated in Figure 2.21b. They include the inertia forces, the spring force, and the axial force. The reaction at the hinge is not shown. Also, the gravity forces have been ignored; they do not affect the motion in this case. The equation of dynamic equilibrium is obtained quite simply by taking moments about the hinge.

$$I_0\ddot{\theta} + \frac{mL^2}{4}\ddot{\theta} + kb^2\theta - SL\theta = 0 \tag{2.49}$$

Substituting $I_0 = mL^2/12$ and setting $u \simeq L\theta$, we get

$$\frac{m}{3}\ddot{u} + \left(\frac{kb^2}{L^2} - \frac{S}{L}\right)u = 0 \tag{2.50}$$

or

$$m^* \ddot{u} + (k^* - k_G)u = 0 \qquad (2.51)$$

where, $m^* = m/3$, $k^* = kb^2/L^2$, and $k_G = S/L$. The axial load S has the effect of reducing the stiffness of the system by k_G. The stiffness term k_G is normally referred to as the *geometric stiffness*. When $k_G = k^*$, the system becomes unstable. The load at which instability will take place, called the *buckling load*, is given by

$$S = \frac{kb^2}{L} \qquad (2.52)$$

The equation of motion could also be obtained by writing the virtual work equation. The distance by which the axial load moves toward the hinge is given by

$$y = L(1 - \cos\theta) \qquad (2.53)$$

When the system is subjected to a virtual displacement $\delta\theta$, the axial force undergoes a virtual displacement δy, where

$$\delta y = L \sin\theta\, \delta\theta \qquad (2.54a)$$

which for small θ becomes

$$\delta y = L\theta\, \delta\theta \qquad (2.54b)$$

The other virtual displacements are easily obtained and are shown in Figure 2.21c. The virtual work equation is given by

$$-I_0 \ddot{\theta}\, \delta\theta - \frac{mL^2}{4} \ddot{\theta}\, \delta\theta - kb^2\theta\, \delta\theta + SL\theta\, \delta\theta = 0 \qquad (2.55)$$

On canceling out $\delta\theta$, substituting $I_0 = mL^2/12$, and setting $u \equiv L\theta$, we get the equation of motion, which is identical to Equation 2.50.

The effect of axial force on the motion of distributed parameter systems idealized as a single-degree-of-freedom system can be accounted for in a manner very similar to that described for the rigid-body assemblage described in the foregoing paragraphs. Consider, for example, the vibrations of the beam shown in Figure 2.16. The infinitesimal section of Figure 2.16e is redrawn in Figure 2.22 but with only the axial forces shown on it. The virtual displacement imposed on the beam will cause the section to rotate so that the forces $S(x)$ shown in Figure 2.22 will move closer together by a distance δy which is obtained

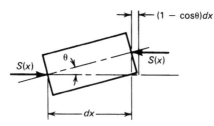

Figure 2.22. Axial load on a section of a beam.

from

$$\delta y = \frac{d}{d\theta}(1 - \cos\theta)\,\delta\theta\,dx$$

$$= \sin\theta\,\delta\theta\,dx \tag{2.56}$$

$$\approx \theta\,\delta\theta\,dx$$

Since the displacement of the beam is $u = z\psi(x)$, the slope of the beam θ is given by

$$\theta = z\psi'(x) \tag{2.57}$$

The infinitesimal virtual work done by the axial force $S(x)$ is now obtained from

$$d(\delta W_S) = \delta z \left[S(x)\{\psi'(x)\}^2\,dx \right] z \tag{2.58}$$

On integrating Equation 2.56 over the length of the beam, we get the following expression for the virtual work done by the axial force

$$\delta W_S = \delta z \left[\int_0^L S(x)\{\psi'(x)\}^2\,dx \right] z \tag{2.59}$$

The geometric stiffness is now given by

$$k_G = \int_0^L S(x)\{\psi'(x)\}^2\,dx \tag{2.60}$$

and the equation of motion is modified to

$$m^*\ddot{z}(t) + c^*\dot{z}(t) + (k^* - k_G)z(t) = p^* \tag{2.61}$$

For a static case, both z and p^* are independent of time; hence $\dot{z}(t)$ and $\ddot{z}(t)$ vanish. The resulting equation leads to an expression for deflection under static load and the condition $k^* = k_G$ provides the value of axial load $S(x)$ at which the system will buckle. In the special case of constant axial force, S can be taken out of the integral sign in Equation 2.58; the geometric stiffness is then given by

$$k_G = S \int_0^L \{\psi'(x)\}^2 \, dx \tag{2.62}$$

Example 2.10
The roof structure shown in Figure E2.10 consists of a uniform light column of flexural rigidity EI supporting a uniform reinforced concrete circular slab of radius R and mass m. Write the equation of motion for small vibrations of the structure in the plane of the paper (i) ignoring the axial force effect, and (ii) including the axial force effect.

Solution
If the column is considered axially rigid, the system has two degrees of freedom: a tip deflection, and a rotation, as shown in the figure. To represent the system by a single-degree-of-freedom system, a suitable displacement shape function must be selected. Let us assume that the displaced shape is given by

$$u = z \left(1 - \cos \frac{\pi x}{2L}\right)$$

where z is the generalized coordinate and $\psi(x) = 1 - \cos(\pi x/2L)$ is the displacement shape function.

(i) The generalized mass m^* is obtained from Equation 2.37a

$$m^* = m\{\psi(L)\}^2 + I_0\{\psi'(L)\}^2$$

where $I_0 = mR^2/4$ is the mass moment of inertia of the roof slab about its diameter. Noting that $\psi(L) = 1$ and $\psi'(L) = \pi/2L$, we get

$$m^* = m + \frac{mR^2}{4}\left(\frac{\pi}{2L}\right)^2$$

$$= m\left\{1 + \frac{\pi^2}{16}\left(\frac{R}{L}\right)^2\right\} \tag{a}$$

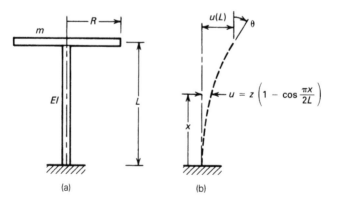

(a)   (b)

Figure E2.10. Vibrations of an umbrella roof.

The generalized stiffness is obtained from Equation 2.37c

$$k^* = \int_0^L EI \left\{ \psi''(x) \right\}^2 dx$$

$$= \int_0^L EI \frac{\pi^4}{16L^4} \left(\cos \frac{\pi x}{2L} \right)^2 dx \tag{b}$$

$$= \frac{\pi^4 EI}{32L^3}$$

The equation of motion becomes

$$m \left\{ 1 + \frac{\pi^2}{16} \left(\frac{R}{L} \right)^2 \right\} \ddot{z} + \frac{\pi^4 EI}{32L^3} z = 0 \tag{c}$$

(ii) The weight of the roof slab imposes an axial force mg on the column. The geometric stiffness is obtained from Equation 2.62

$$k_G = mg \int_0^L \frac{\pi^2}{4L^2} \sin^2 \frac{\pi x}{2L} dx \tag{d}$$

$$= \frac{mg\pi^2}{8L} \tag{d}$$

The modified equation of motion becomes

$$m \left\{ 1 + \frac{\pi^2}{16} \left(\frac{R}{L} \right)^2 \right\} \ddot{z} + \left(\frac{\pi^4 EI}{32L^3} - \frac{mg\pi^2}{8L} \right) z = 0 \tag{e}$$

The column supporting the roof will buckle under the weight of the roof when $k^* = k_G$, that is, when

$$mg = \frac{\pi^2 EI}{4L^2} \tag{f}$$

The buckling load given by Equation f is the true buckling load of a uniform vertical cantilever loaded at its end. This happens to be so because the assumed displacement shape is, in fact, the true displaced shape at buckling.

2.11 EFFECT OF SUPPORT MOTION

Dynamical systems are often excited by the motion of their supports. Earthquake-induced motion provides one example of support excitation. For a single-degree-of-freedom system, the equation governing a support excited motion is easily obtained. A system of this type can be modeled as shown in Figure 2.23a. Let u^t denote the displacement of the mass with reference to a fixed frame in space and u_g denote the displacement of the ground with reference to such a frame. The displacement of the mass relative to the ground is represented by u so that

$$u^t = u_g + u \tag{2.63}$$

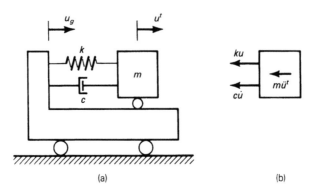

Figure 2.23. Effect of support motion.

The forces acting on the moving mass are shown in the free-body diagram of Figure 2.23b. It should be noted that the inertia force is given by the product of the mass and the absolute acceleration relative to the fixed frame. The spring and damping forces, on the other hand, depend on the displacement and velocity, respectively, relative to the ground. The equation of motion for the system is given by

$$m\ddot{u}^t + c\dot{u} + ku = 0 \tag{2.64}$$

On substituting for u^t from Equation 2.63, Equation 2.64 becomes

$$m\ddot{u} + c\dot{u} + ku = -m\ddot{u}_g \tag{2.65}$$

Equation 2.63 expresses the equation of motion in terms of the relative displacement coordinate and its derivatives, with the term $-m\ddot{u}_g$, in effect, acting as an exciting force.

SELECTED READINGS

Best, C.W. 1973. Material Damping: An Introductory Review of Mathematical Models, Measures and Experimental Techniques. *Journal of Sound and Vibration*, Vol. 29: 129–153.

Clough, R.W. & Penzien, J. 1993. *Dynamics of Structures*. New York: McGraw-Hill. 2nd Edition.

Crandall, S.H. 1970. The Role of Damping in Vibration Theory. *Journal of Sound and Vibration*, Vol. 11: 3–18.

Fowles, G.R. 1986. *Analytical Mechanics*. New York: Holt, Rinehart and Winston. 4th Edition.

Lazan, B.J. 1968. *Damping of Materials and Members in Structural Mechanics*. Oxford, U.K.: Pergamon Press.

Nashif, A.D., Jones, D.I.G. & Henderson, J.P. 1985. *Vibration Damping.* New York: John Wiley.
Pinsker, W. 1949. Structural Damping. *Journal of the Aeronautical Sciences,* Vol. 16: 619.
Scanlan, R.H. & Mendelson, A. 1963. Structural Damping. *American Institute of Aeronautics and Astronautics Journal,* Vol. 1: 938–939.

PROBLEMS

2.1 Design an arrangement of three springs with stiffness 40, 60, and 90 N/mm, respectively, to have an effective stiffness of 76 N/mm.

2.2 Derive expressions for the resultant inertia forces for translation along the x, and y axes, and rotation about the mass center for the rigid bodies shown in Figures 2.6c and d. In each case, assume that the body has a uniform thickness and mass density.

2.3 A rigid block of mass M is suspended from a uniform pulley of mass m and radius R as shown in Figure P2.3. If the cord is inextensible, of negligible weight, and does not slip on the pulley, and the spring stiffness is k, determine the equation of motion for vertical vibrations of the system.

2.4 A wheel-axle assembly weighing 10 lb is mounted at the end of a cantilever strip of length 1 ft, width 1 in., and thickness 1/4 in. (Fig. P2.4). The strip is rigidly attached to a solid circular shaft of length 3 ft and diameter 1 in. The other end of the shaft is rigidly fixed to a heavy vehicular body. The wheel-axle assembly has a radius of gyration of 3.5 in. $E = 30 \times 10^6$ psi, and $G = 12 \times 10^6$ psi.

Obtain the equation of motion for vertical vibrations of the wheel-axle assembly assuming that the wheel is free to rotate about the axle and the friction in the bearings is negligible.

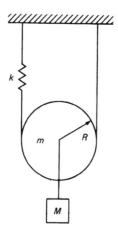

Figure P2.3.

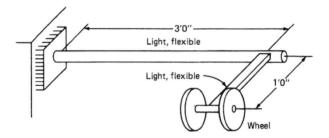

Figure P2.4.

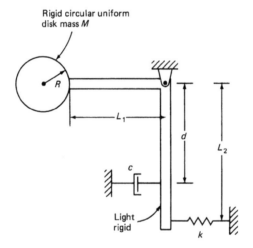

Figure P2.5.

2.5 Obtain the equation of motion for small vibrations of the system shown in Figure P2.5.

2.6 A uniform flywheel of mass M and radius R rotates freely about an axle carried by two identical rigid bars of total mass m as shown in Figure P2.6. The rigid bars are pinned at their left-hand end and suspended by a vertical spring at the right-hand end. A dashpot with damping coefficient c is attached to the bars at a distance a from the pinned end. Assuming that there is no friction in the bearings of the wheel, obtain the equation of motion for small vertical vibrations of the system. What will be the equation of motion, if the flywheel was braked to its axle.

2.7 Shrouds are often used at the tips of turbine blades to improve performance and to control vibration of the blades (Fig. P2.7). For purposes of vibration analysis, the shrouded blade can be considered to behave as a cantilevered beam with the tip pinned.

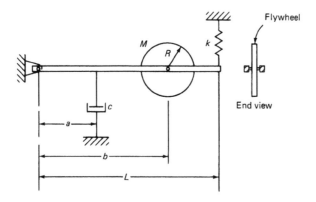

Figure P2.6.

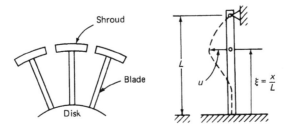

Figure P2.7.

In order to model the blade as a single-degree-of-freedom system, the following function is used to represent its vibration shape:

$$\psi(\xi) = 2\xi^4 - 5\xi^3 + 3\xi^2, \qquad \xi = \frac{x}{L}$$

(a) To what extent does $\psi(\xi)$ meet the boundary conditions?
(b) Determine the generalized mass m^* and the generalized stiffness k^* and hence obtain the equation of free vibration of the blade. Assume that the blade is of uniform section, the mass per unit length is $\bar{m}$, and the flexural rigidity is EI.

2.8 Ultrasonic transducers often contain a "transformer" which has the purpose of magnifying the input vibration amplitude as shown in Figure P2.8. Assuming the shape function for axial vibration as $\psi(x) = 2\xi - \xi^2$, where $\xi = x/L$ and that the cross-sectional area varies linearly so that $A(\xi) = A_1 + (A_2 - A_1)\xi$, obtain expressions for m^* and k^*. If the input is a base motion given by $u_g(t)$, formulate the equation of motion.

2.9 A vertical chimney of length L (Fig. P2.9) has a uniform cross-section, moment of inertia I, and mass per unit length $\bar{m}$. The modulus of elasticity

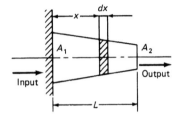

Figure P2.8.

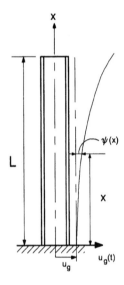

Figure P2.9.

of the material is E. The chimney is subjected to lateral ground motion having acceleration $\ddot{u}_g(t)$. Find the equation of motion: **(a)** neglecting the gravity load effect produced by the self weight of the chimney; **(b)** taking gravity load effect into account. Ignore damping and assume that the vibration shape is given by

$$\psi(x) = \frac{3x^2}{2L^2} - \frac{x^3}{2L^3}$$

2.10 For the simplified analysis of the response of a bridge to moving loads, the bridge deck is idealized as a simply supported beam of span L, mass per unit length $\bar{m}$, and flexural rigidity EI. A single wheel load of magnitude F traverses the bridge at a speed v (Fig. P2.10). Assuming a displacement shape function given by $\psi = \sin(\pi x/L)$, obtain the equation of motion for flexural vibrations of the bridge deck. Ignore the mass of the wheel.

2.11 The following properties are given for the frame shown in Figure 2.17: $m_1 = m_2 = 2$ kip s^2/in., $m_3 = 1$ kip s^2/in., $k_1 = k_2 = 500$ kips/in., and $k_3 = 250$ kips/in. The frame is subjected to lateral ground motion whose acceleration varies as $\ddot{u}_g = 130 \sin(6\pi t)$ in./s^2. Obtain the equations governing the motion of the frame. Ignore damping and assume that the vibration shape is given by

$$\psi^T = [1 \ 2 \ 3]$$

2.12 A mass m is attached to a massless rigid bar of length l. The bar is suspended from a ceiling and is restrained by two springs each of stiffness k, as shown in Figure P2.12. Find the equation of motion for small vibrations of the bar in a vertical plane.
If the whole assembly were in a horizontal plane, what would be the equation of motion.

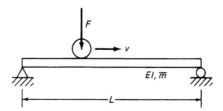

Figure P2.10.

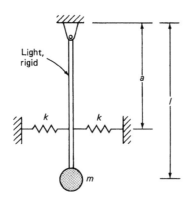

Figure P2.12.

CHAPTER 3

Formulation of the equations of motion: Multi-degree-of-freedom systems

3.1 INTRODUCTION

For an accurate description of its displaced configuration, a structural or mechanical system subjected to dynamic disturbances may require the specification of displacements along more than one coordinate direction. Such a system is known as a multi-degree-of-freedom system. In this chapter, we describe procedures for formulation of the equations of motion of multi-degree-of-freedom systems. These procedures are similar in principle to those used for the single-degree-of-freedom systems, even though, in detail, they are quite a bit more involved.

In Chapter 2, we discussed a method by which the vibrations of a multi-degree-of-freedom system could be adequately represented by a single shape function. Such a representation is, in fact, an idealization that is only occasionally valid. In a general case, accurate description of the displaced configuration of a vibrating system is possible only through the superposition of a number of different shapes. Even when the use of a single shape function is adequate, the selection of the shape function is not, in general, easy, and if an inappropriate choice is made, the results obtained may be completely unreliable. The difficulty is compounded by the fact that in procedures that use shape function idealization, there is no simple way to verify the reliability of the results obtained.

As a simple example of the difficulties involved in using a single-degree-of-freedom idealization of what is truly a multi-degree-of-freedom system, consider the two-mass system shown in Figure 3.1. To describe the displaced configuration of the system completely, we need to specify the displacements along two coordinate directions: u_1 and u_2. We may attempt to idealize the system as a single-degree-of-freedom system by choosing a shape function ψ such that

$$\psi^T = [1 \quad 1] \tag{3.1}$$

With this idealization, the response of the system is given by

$$\mathbf{u} = z(t)\psi \tag{3.2}$$

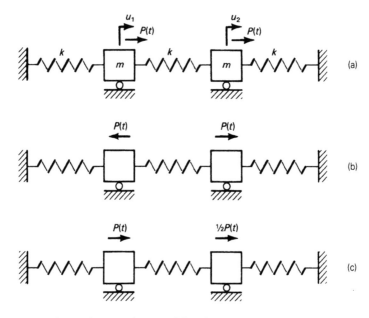

Figure 3.1. Vibrations of a two-degree-of-freedom system.

The only unknown in Equation 3.2 is $z(t)$ and the shape function assumption has allowed us to represent a two-degree-of-freedom system by a single-degree-of-freedom model.

Now, let the system be excited from rest by the application of two equal forces, each equal to $p(t)$, and acting in the same direction, as indicated in Figure 3.1a. By using the method described in Section 2.8, we obtain the following equation for the single-degree-of-freedom idealization of the system

$$m^*\ddot{z} + k^*z = p^* \tag{3.3}$$

where $m^* = 2m$, $k^* = 2k$, and $p^* = 2p$.

In the present case, the solution of Equation 3.3 gives the exact response. However, let us now assume that instead of acting in the same direction, the two equal forces act in opposite directions, as in Figure 3.1b. For such forces, p^* in Equation 3.3 will be equal to zero and that equation will entirely fail to predict the response of the system. The problem is that the shape function $\psi^T = [1 \ 1]$ is no longer appropriate; instead, we must use $\psi^T = [1 \ -1]$. In fact, for the general loading shown in Figure 3.1c, a single shape function will not be adequate to provide the response of the system; a superposition of two shape functions must be used.

Any number of examples can be cited of a multi-degree-of-freedom system, but consider the simple flexible cantilever beam of Figure 3.2 supporting three lumped masses. First, let the masses be point masses and let the beam be

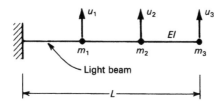

Figure 3.2. Cantilever beam with lumped masses.

constrained to vibrate in the plane of the paper. If the cantilever is axially rigid, the masses can undergo vibrations only in the vertical direction and the system has three degrees of freedom. If the beam is axially flexible, horizontal motion of the masses is also possible, making the system a six-degree-of-freedom system. Next, if the masses are of finite size instead of being concentrated at a point, they have rotational inertia, too, and three more rotational degrees of freedom must be specified, one for each mass. In the most general case, when the cantilever can undergo motion in a three-dimensional space, each mass has six degrees of freedom, three translational, and three rotational, and the cantilever has a total of 18 degrees of freedom.

In conclusion, we may state that it is essential in many cases for the system being analyzed to be represented by a multi-degree-of-freedom system. In this chapter, we present techniques for the formulation of the equations of motion for such systems.

3.2 PRINCIPAL FORCES IN MULTI-DEGREE-OF-FREEDOM DYNAMIC SYSTEM

As in the case of a single-degree-of-freedom system, the principal forces acting on a multi-degree-of-freedom system are 1. the forces of inertia, 2. the elastic forces, and 3. the forces of resistance or damping. Once these forces have been identified and calculated, formulation of the equations of motion reduces to a problem of writing the equations of equilibrium. In this section, we describe methods of obtaining these forces.

3.2.1 *Inertia forces*

Consider the lumped mass idealization of a simply supported beam shown in Figure 3.3. The model consists of N point masses which are free to vibrate in the vertical direction in the plane of the paper. To define the displaced configuration of this system, we need to specify the displacements along N coordinates. For example, if we know the vertical displacement at each point mass, the displaced shape of the beam is uniquely determined. The system

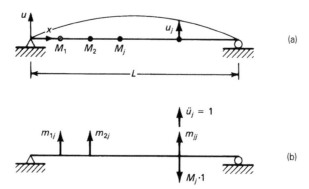

Figure 3.3. Inertia forces in a lumped mass model of a simply supported beam.

is therefore an N-degree-of-freedom system and we may choose the vertical displacements at the point masses as the N coordinates.

Next, let us imagine that the system has a unit acceleration along one of the coordinate directions, say j, while the accelerations at all other coordinates are zero. Since the coordinates are independent of each other, such a state is possible. For our choice of coordinates, this state implies that only the mass M_j is undergoing acceleration, and therefore the force of magnitude $-M_j \cdot 1$ shown is the only inertia force in the system. The negative sign implies that the inertia force is directed away from the direction of the positive unit acceleration. We now determine the external forces that must be applied along the N coordinate directions to equilibrate the inertia force. Denoting by m_{ij} the force that must be applied along coordinate i to equilibrate the inertia forces produced by a unit acceleration at coordinate j, we note that

$$m_{ij} = 0 \quad i \neq j$$
$$m_{jj} = M_j \tag{3.4}$$

By superposition of the forces, it can easily be shown that the external force vector required to equilibrate the inertia forces arising from simultaneous acceleration of all masses, the mass accelerations being $\ddot{u}_j$, $j = 1$ to N, is given by

$$
\mathbf{f}_I =
\begin{bmatrix}
f_{I1} \\
f_{I2} \\
\cdot \\
\cdot \\
\cdot \\
f_{IN}
\end{bmatrix}
=
\begin{bmatrix}
m_{11} & m_{12} & \cdot & \cdot & \cdot & m_{1N} \\
m_{21} & m_{22} & \cdot & \cdot & \cdot & m_{2N} \\
\cdot & \cdot & \cdot & \cdot & \cdot & \cdot \\
\cdot & \cdot & \cdot & \cdot & \cdot & \cdot \\
\cdot & \cdot & \cdot & \cdot & \cdot & \cdot \\
m_{N1} & m_{N2} & \cdot & \cdot & \cdot & m_{NN}
\end{bmatrix}
\begin{bmatrix}
\ddot{u}_1 \\
\ddot{u}_2 \\
\cdot \\
\cdot \\
\cdot \\
\ddot{u}_N
\end{bmatrix}
\tag{3.5a}
$$

$$\mathbf{f}_I = \mathbf{M}\ddot{\mathbf{u}} \tag{3.5b}$$

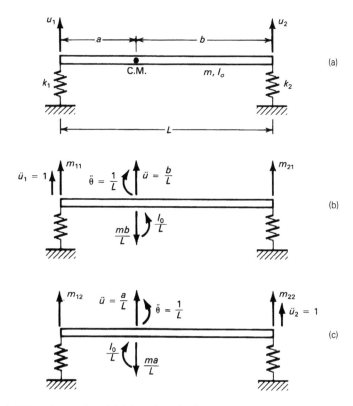

Figure 3.4. Vibrations of a rigid bar, inertia forces.

The matrix on the right-hand side of Equation 3.5a is called the *mass matrix* and is denoted by $\mathbf{M}$. Individual elements m_{ij} of the mass matrix are referred to as the *mass influence coefficients*. In our particular case, because of relationship in Equation 3.4, the mass matrix is diagonal, the terms on the diagonal being M_j, $j = 1$ to N.

It will readily be appreciated that the mass matrix need not always be diagonal. As an example, consider the rigid bar shown in Figure 3.4a. The bar, which is nonuniform, has a total mass m and is supported at its two ends by massless springs of stiffness k_1 and k_2. The mass center of the bar is at a distance a from the left-hand end, and the mass moment of inertia about this center is I_0. The bar is constrained so that it can move only in a vertical direction in the plane of the paper. In order that the position of the bar be completely determined, we need to specify the displacements along two coordinates. There are a number of possible choices for the coordinates; we choose the vertical displacements at the bar ends as our two coordinates.

To obtain the inertia forces acting on the bar, we first apply a unit acceleration along coordinate 1, the acceleration at the other coordinate being zero. This will cause a translational acceleration of b/L at the center of mass along with a clockwise rotational acceleration of $1/L$ about the same center. The resulting inertia forces acting at the mass center are indicated in Figure 3.4b. They consist of a vertical force mb/L directed downward and an anticlockwise moment I_0/L. If the bar is to be kept in dynamic equilibrium, the inertia forces must be balanced by appropriate forces applied at each of the two coordinate directions. Using the notation already introduced, the force at coordinate 1 is denoted by m_{11}, being the external force required at coordinate 1 to equilibrate the inertia forces arising from the application of a unit acceleration also at coordinate 1. In a similar manner, the force at coordinate 2 is denoted by m_{21}, being the external force required at coordinate 2 to equilibrate the inertia forces arising from the application of a unit acceleration at coordinate 1. The magnitudes of m_{11} and m_{21} are easily obtained by considering the equilibrium of the bar. Thus, taking moments about the right-hand end yields

$$m_{11}L = m\frac{b}{L}b + I_0\frac{1}{L}$$

$$m_{11} = \frac{mb^2}{L^2} + \frac{I_0}{L^2}$$

(3.6)

In a similar manner

$$m_{21} = \frac{mab}{L^2} - \frac{I_0}{L^2}$$

(3.7)

Next, we apply a unit acceleration at coordinate 2 with the acceleration at coordinate 1 being zero. The resulting inertia forces at the center of mass are indicated in Figure 3.4c. The external forces required to equilibrate the inertia forces are denoted by m_{12} and m_{22}, respectively, and are given by

$$m_{12} = \frac{mab}{L^2} - \frac{I_0}{L^2}$$

(3.8)

$$m_{22} = \frac{ma^2}{L^2} + \frac{I_0}{L^2}$$

(3.9)

If the actual accelerations in the system are $\ddot{u}_1$ and $\ddot{u}_2$, the external forces required to balance the resulting inertia forces are easily seen to be

$$\begin{bmatrix} f_{I1} \\ f_{I2} \end{bmatrix} = \begin{bmatrix} m_{11} & m_{12} \\ m_{21} & m_{22} \end{bmatrix} \begin{bmatrix} \ddot{u}_1 \\ \ddot{u}_2 \end{bmatrix}$$

$$= \begin{bmatrix} \frac{mb^2}{L^2} + \frac{I_0}{L^2} & \frac{mab}{L^2} - \frac{I_0}{L^2} \\ \frac{mab}{L^2} - \frac{I_0}{L^2} & \frac{ma^2}{L^2} + \frac{I_0}{L^2} \end{bmatrix} \begin{bmatrix} \ddot{u}_1 \\ \ddot{u}_2 \end{bmatrix}$$

(3.10)

or

$$\mathbf{f}_I = \mathbf{M}\ddot{\mathbf{u}}$$

(3.11)

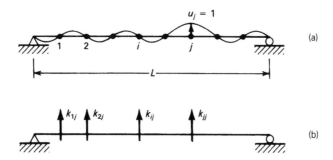

Figure 3.5. Lumped mass model of a beam: (a) imposed displacement pattern; (b) elastic forces required to maintain displacement in (a).

Several interesting facts can be noted from Equation 3.10. First, the mass matrix is no longer diagonal; that is, the-off diagonal terms have nonzero values. Second, the matrix is symmetric, that is, $m_{12} = m_{21}$. The latter is a direct consequence of Maxwell's reciprocal theorem and, in general, $m_{ij} = m_{ji}$.

3.2.2 Forces arising due to elasticity

As in the case of a single-degree-of-freedom system, whenever an elastic body is subjected to deformations, internal forces are set up that counter such deformations. Therefore, if the deformations are to be maintained, external forces must be applied along the coordinate directions of the system. We will refer to these external forces as the spring forces or the forces arising from elasticity; in fact, they are forces that must be applied to counter the internal elastic forces.

When the system is linear, that is, when the stress–strain relationship of the material used in the structure is linear and the displacements are small, the elastic forces can be obtained by the method of superposition. To illustrate the procedure, we consider again the example of lumped mass beam model shown in Figure 3.3. We apply a unit displacement along coordinate direction j, holding all other displacements to zero as shown in Figure 3.5a. Internal elastic forces will oppose these displacements, and to maintain the displacements, we must apply external forces along all coordinate directions as indicated in Figure 3.5b. We denote the external force required at coordinate i as k_{ij}.

If the actual displacements in the system are u_j, $j = 1$ to N, it is easily shown using the principle of superposition that the external forces along the coordinate directions required to balance the internal elastic forces are

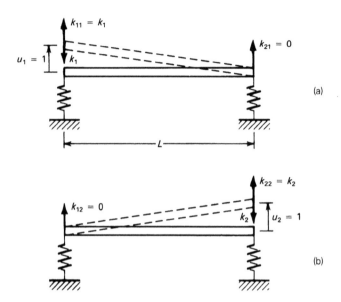

Figure 3.6. Vibrations of a rigid bar, elastic forces.

given by

$$
\mathbf{f}_S = \begin{bmatrix} f_{S1} \\ f_{S2} \\ \cdot \\ \cdot \\ \cdot \\ f_{SN} \end{bmatrix} = \begin{bmatrix} k_{11} & k_{12} & \cdot & \cdot & k_{1N} \\ k_{21} & k_{22} & \cdot & \cdot & k_{2N} \\ \cdot & \cdot & \cdot & \cdot & \cdot \\ \cdot & \cdot & \cdot & \cdot & \cdot \\ \cdot & \cdot & \cdot & \cdot & \cdot \\ k_{N1} & k_{N2} & \cdot & \cdot & k_{NN} \end{bmatrix} \begin{bmatrix} u_1 \\ u_2 \\ \cdot \\ \cdot \\ \cdot \\ u_N \end{bmatrix} \tag{3.12}
$$

or

$$
\mathbf{f}_S = \mathbf{K}\mathbf{u} \tag{3.13}
$$

where the matrix on the right-hand side of Equation 3.12 is referred to as the *stiffness matrix* and is denoted by **K**. Individual elements k_{ij} of the stiffness matrix are referred to as the *stiffness influence coefficients*. As in the case of mass matrix, the stiffness matrix is also symmetrical.

As another example, we derive the spring forces for the rigid bar shown in Figure 3.4. For the coordinate directions selected, the spring forces are shown in Figure 3.6, and are given by

$$
\begin{bmatrix} f_{S1} \\ f_{S2} \end{bmatrix} = \begin{bmatrix} k_1 & 0 \\ 0 & k_2 \end{bmatrix} \begin{bmatrix} u_1 \\ u_2 \end{bmatrix} \tag{3.14}
$$

Thus for the coordinates selected, the stiffness matrix is diagonal.

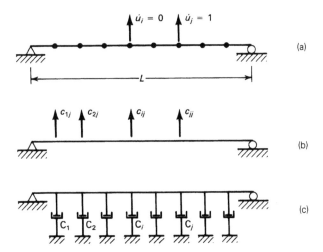

Figure 3.7. Lumped mass model of a beam: (a) velocity pattern; (b) external forces required to balance resistance forces caused by damping; (c) damping model.

3.2.3 *Damping forces*

As pointed out in Section 2.5, the motion of a body is opposed by several types of resisting forces. These forces may, for example, arise from air resistance or from internal and external frictions. As noted earlier, the characteristics of the resisting forces are difficult to define. From a mathematical point of view, viscous damping forces which are proportional to the velocities in the system but are opposed to the direction of motion are easiest to handle and give analytical results that for small amounts of damping conform reasonably well to experimental observations. In this section, we deal only with viscous damping forces. Later, we will have occasion to discuss other types of damping forces.

Consider again the lumped mass model of a simply supported beam. Let us impart a unit velocity at coordinate direction j, the velocities at other coordinates being zero (Fig. 3.7a). The specified motion will be opposed by damping forces in the system, and if the motion is to be maintained, external forces as shown in Figure 3.7b must be applied at the coordinate directions to balance the forces of resistance. As in the case of inertia and spring forces, we denote the external force at coordinate i by c_{ij}. If the real velocities along the coordinates are $\dot{u}_j$, $j = 1$ to N, the damping forces are obtained by superposition. Thus

$$\mathbf{f}_D = \begin{bmatrix} f_{D1} \\ f_{D2} \\ \cdot \\ \cdot \\ \cdot \\ f_{DN} \end{bmatrix} = \begin{bmatrix} c_{11} & c_{12} & \cdot & \cdot & \cdot & c_{1N} \\ c_{21} & c_{22} & \cdot & \cdot & \cdot & c_{2N} \\ \cdot & \cdot & \cdot & \cdot & \cdot & \cdot \\ \cdot & \cdot & \cdot & \cdot & \cdot & \cdot \\ \cdot & \cdot & \cdot & \cdot & \cdot & \cdot \\ c_{N1} & c_{N2} & \cdot & \cdot & \cdot & c_{NN} \end{bmatrix} \begin{bmatrix} \dot{u}_1 \\ \dot{u}_2 \\ \cdot \\ \cdot \\ \cdot \\ \dot{u}_N \end{bmatrix} \qquad (3.15)$$

or

$$\mathbf{f}_D = \mathbf{C}\dot{\mathbf{u}} \tag{3.16}$$

where the matrix on the right-hand side of Equation 3.15 is called the *damping matrix* and is denoted by **C**. Elements c_{ij} of the damping matrix are referred to as the *damping influence coefficients*.

The damping influence coefficients c_{ij} are similar in concept to the mass influence coefficients m_{ij} and the stiffness influence coefficients k_{ij}. We could derive influence coefficient m_{ij}'s from a knowledge of the internal mass distribution in the system. In a similar manner, k_{ij}'s could be derived from the internal stiffness characteristics or the stress–strain relationship of the material. However, internal damping characteristics are difficult or impossible to define, and therefore the coefficients c_{ij}'s can rarely be obtained from the consideration of internal damping characteristics. In later chapters, we shall discuss alternative methods of constructing damping matrices or defining damping resistances. For the present, we assume that c_{ij}'s have been specified or can be computed.

As a specific example, consider the beam model with the damping characteristics shown in Figure 3.7c. For this particular model

$$\begin{aligned} c_{ij} &= 0 \quad i \neq j \\ c_{ii} &= C_i \end{aligned} \tag{3.17}$$

and the damping matrix is diagonal.

3.2.4   *Axial force effects*

As in the case of a single-degree-of-freedom system, the presence of axial forces usually causes a reduction in the resistance that the system offers to elastic deformations. The axial force effect can be represented by a force vector $\mathbf{f}_G$ that is opposite in direction to the spring force vector $\mathbf{f}_S$.

Consider, for example, the lumped mass model of a beam. Suppose that we wish to impose a unit displacement along coordinate direction j while holding to zero the displacements along other coordinate directions. As stated earlier, the resulting deformations will be resisted by internal elastic forces. At the same time, the presence of axial forces will tend to amplify the deformations. The net external force required along coordinate direction i to maintain the displacements will be $k_{ij} - k_{Gij}$, where k_{ij} is the elastic stiffness influence coefficient, discussed earlier, and k_{Gij} is the *geometric stiffness influence coefficient*. If the displacements along the coordinates are u_j, $j = 1$ to N, the forces of instability that tend to amplify the deformations are

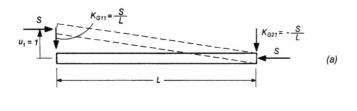

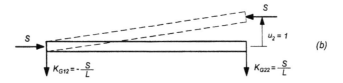

Figure 3.8. Axial force effect on the vibration of a rigid bar.

given by

$$
\mathbf{f}_G = \begin{bmatrix} f_{G1} \\ f_{G2} \\ \vdots \\ f_{GN} \end{bmatrix} = \begin{bmatrix} k_{G11} & k_{G12} & \cdots & k_{G1N} \\ k_{G21} & k_{G22} & \cdots & k_{G2N} \\ \vdots & \vdots & \ddots & \vdots \\ k_{GN1} & k_{GN2} & \cdots & k_{GNN} \end{bmatrix} \begin{bmatrix} u_1 \\ u_2 \\ \vdots \\ u_N \end{bmatrix}
\tag{3.18}
$$

or

$$
\mathbf{f}_G = \mathbf{K}_G \mathbf{u}
\tag{3.19}
$$

in which $\mathbf{K}_G$ is referred to as the *geometric stiffness matrix*.

As an example of the forces of instability, consider the rigid bar of Figure 3.6. Let the bar be subjected to an axial force S. Figure 3.8a shows the displaced shape of the bar when a unit displacement has been imposed along coordinate direction 1, the displacement along coordinate 2 being zero. The resulting internal spring forces and the external forces required to counter them were derived earlier and are not shown. However, the axial forces S give rise to a couple $S \times 1$. To maintain equilibrium, this couple must be balanced by external forces k_{G11} and k_{G21} acting along coordinate directions 1 and 2 as shown. Using equations of static equilibrium, we get

$$
k_{G11} = \frac{S}{L}
$$

$$
\tag{3.20}
$$

$$
k_{G21} = -\frac{S}{L}
$$

In a similar manner, by imposing a unit displacement along coordinate 2, as shown in Figure 3.8b, we get

$$
\begin{aligned}
k_{G12} &= -\frac{S}{L} \\
k_{G22} &= \frac{S}{L}
\end{aligned}
\tag{3.21}
$$

If the real displacements along the coordinate are u_1 and u_2, the total forces of instability are given by

$$
\mathbf{f}_G = \begin{bmatrix} f_{G1} \\ f_{G2} \end{bmatrix} = \begin{bmatrix} \frac{S}{L} & -\frac{S}{L} \\ -\frac{S}{L} & \frac{S}{L} \end{bmatrix} \begin{bmatrix} u_1 \\ u_2 \end{bmatrix}
$$

$$
= \mathbf{K}_G \mathbf{u}
\tag{3.22}
$$

It should be noted that the signs of the geometric stiffness influence coefficients have been determined with the understanding that $\mathbf{f}_G$ will be deducted from $\mathbf{f}_S$.

3.3   FORMULATION OF THE EQUATIONS OF MOTION

Once the external forces required to equilibrate the inertia forces, damping forces, the elastic forces, and the forces of instability have been obtained by the procedures discussed in Section 3.2, the formulation of the equations of motion is straightforward. For equilibrium, the sum of the above-mentioned forces should be equal to the actual external forces acting along the coordinates of the system. If these external forces are denoted by $\mathbf{p}(t)$, the equations of motion become

$$
\mathbf{f}_I + \mathbf{f}_D + \mathbf{f}_S - \mathbf{f}_G = \mathbf{p}(t)
\tag{3.23}
$$

or

$$
\mathbf{M}\ddot{\mathbf{u}} + \mathbf{C}\dot{\mathbf{u}} + \mathbf{K}\mathbf{u} - \mathbf{K}_G\mathbf{u} = \mathbf{p}(t)
\tag{3.24}
$$

When the external forces in the system act at locations other than the coordinates defined for the system, the former must be replaced by equivalent forces acting at the coordinates. As discussed in succeeding paragraphs, this is usually best accomplished by using the principle of virtual displacement.

In dealing with more specific examples of the formulation of the equations of motion for multi-degree-of-freedom systems, we will find it convenient to classify the latter into one of the following categories, just as we did in the case of single-degree-of-freedom systems.
1. Systems with localized mass and localized stiffness
2. Systems with localized mass but distributed stiffness
3. Systems with distributed mass but localized stiffness
4. Systems with distributed mass and distributed stiffness

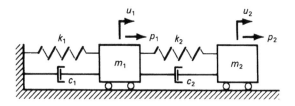

Figure 3.9. System with localized mass and localized stiffness.

3.3.1 *Systems with localized mass and localized stiffness*

Figure 3.9 shows a system consisting of two rigid blocks restrained by springs and vibrating in a horizontal direction in the plane of the paper. The forces of elastic constraints are localized in the massless springs, while the masses can be assumed to be concentrated at the mass centers of the rigid blocks. The system, which has two degrees of freedom, identified in Figure 3.9 by the two displacement coordinates, is vibrating under the action of externally applied forces p_1 and p_2, also shown in Figure 3.9. Using the procedure outlined in Section 3.2, the mass, stiffness, and damping matrices are easily obtained and are given by

$$\mathbf{M} = \begin{bmatrix} m_1 & 0 \\ 0 & m_2 \end{bmatrix} \tag{3.25a}$$

$$\mathbf{C} = \begin{bmatrix} c_1 + c_2 & -c_2 \\ -c_2 & c_2 \end{bmatrix} \tag{3.25b}$$

$$\mathbf{K} = \begin{bmatrix} k_1 + k_2 & -k_2 \\ -k_2 & k_2 \end{bmatrix} \tag{3.25c}$$

The equations of motion, which are of the form of Equation 3.24, can now be written as

$$\begin{bmatrix} m_1 & 0 \\ 0 & m_2 \end{bmatrix} \begin{bmatrix} \ddot{u}_1 \\ \ddot{u}_2 \end{bmatrix} + \begin{bmatrix} c_1 + c_2 & -c_2 \\ -c_2 & c_2 \end{bmatrix} \begin{bmatrix} \dot{u}_1 \\ \dot{u}_2 \end{bmatrix}$$

$$+ \begin{bmatrix} k_1 + k_2 & -k_2 \\ -k_2 & k_2 \end{bmatrix} \begin{bmatrix} u_1 \\ u_2 \end{bmatrix} = \begin{bmatrix} p_1 \\ p_2 \end{bmatrix} \tag{3.26}$$

3.3.2 *Systems with localized mass but distributed stiffness*

Figure 3.10a shows a simply supported massless beam, with point masses concentrated at one-third points along the span. The beam is axially rigid and is constrained to move in the plane of the paper. The system shown in Figure 3.10a can, in fact, be viewed as a lumped mass model of a simply supported

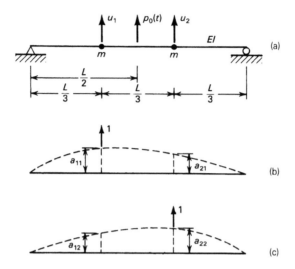

Figure 3.10. (a) Lumped mass model of a simply supported beam; (b), (c) flexibility influence coefficients.

uniform beam whose mass has been lumped at one-third points. The beam is vibrating under a central load $p_0(t)$, and it is required to obtain the equations of motion. The given beam can be classified as a system having localized masses but distributed stiffness. It has two degrees of freedom corresponding to the vertical translation of each mass as shown in the figure. The inertia forces act along these two degrees of freedom. The external force does not act through the selected degrees of freedom, but can be replaced by equivalent forces acting through these degrees of freedom by using the principle of virtual displacement as discussed later in this section. The forces of elastic restraint are distributed throughout the length of the beam, but can be replaced by resultant forces acting along the two degrees of freedom using standard methods of structural analysis. Once the response along the two selected degrees of freedom has been determined, the deformed shape of the beam can be determined by methods of elastic analysis. The mass matrix of the beam model is given by

$$\mathbf{M} = \begin{bmatrix} m & \\ & m \end{bmatrix} \tag{3.27}$$

Instead of calculating the stiffness matrix of the beam directly, it is more convenient first to obtain the *flexibility matrix* of the beam in the coordinates shown and then to take the inverse of this matrix.

The *flexibility influence coefficient* a_{ij} is defined as the displacement at coordinate i due to a unit load acting along coordinate j. Thus to obtain a_{11} and a_{21}, we apply a unit load at coordinate 1 and compute the deflections produced due to this load at coordinates 1 and 2, respectively (Fig. 3.10b). These deflections are easily obtained by standard procedures of structural analysis and lead to the following values for the influence coefficients

$$a_{11} = \frac{4L^3}{243EI}$$

$$a_{21} = \frac{7L^3}{486EI}$$

(3.28)

In a similar manner, to obtain a_{12} and a_{22}, we apply a unit load at coordinate 2 and compute the deflections produced at coordinates 1 and 2, respectively, (Fig. 3.10c). This gives

$$a_{12} = \frac{7L^3}{486EI}$$

$$a_{22} = \frac{4L^3}{243EI}$$

(3.29)

The off-diagonal elements a_{12} and a_{21} are equal, as would be expected on account of the *Maxwell's reciprocal theorem*. Coefficients a_{11} and a_{22} are also equal in this special case because of symmetry. The flexibility matrix is now given by

$$\mathbf{A} = \frac{L^3}{243EI} \begin{bmatrix} 4 & \frac{7}{2} \\ \frac{7}{2} & 4 \end{bmatrix}$$

(3.30)

The stiffness matrix is obtained by taking the inverse of the flexibility matrix

$$\mathbf{K} = \mathbf{A}^{-1} = \frac{324}{5} \frac{EI}{L^3} \begin{bmatrix} 4 & -\frac{7}{2} \\ -\frac{7}{2} & 4 \end{bmatrix}$$

(3.31)

The principle of virtual displacement is used to obtain the equivalent applied force at the two coordinates. Let a virtual displacement δu_1 be applied at coordinate 1 while the displacement at coordinate 2 is zero. Let the resulting displacement at the load point be Δ_{10}. The virtual work equation then gives

$$p_1 \delta u_1 = p_0 \Delta_{10}$$

(3.32)

where p_1 is the equivalent force at coordinate 1. To obtain Δ_{10}, we first recognize that the forces F_1 and F_2 that must act at coordinates 1 and 2 to produce the displacement pattern $\delta u_1, 0$ are given by

$$\begin{bmatrix} F_1 \\ F_2 \end{bmatrix} = \frac{324EI}{5L^3} \begin{bmatrix} 4 & -\frac{7}{2} \\ -\frac{7}{2} & 4 \end{bmatrix} \begin{bmatrix} \delta u_1 \\ 0 \end{bmatrix}$$

$$= \begin{bmatrix} \frac{1296}{5} \frac{EI}{L^3} \delta u_1 \\ -\frac{2268}{10} \frac{EI}{L^3} \delta u_1 \end{bmatrix}$$

(3.33)

The deflection produced at midspan due to the set of forces F_1 and F_2 is obtained by standard methods of structural analysis and is

$$\Delta_{10} = \frac{23}{40} \delta u_1 \tag{3.34}$$

Substitution of Equation 3.34 in Equation 3.32 gives

$$p_1 = \frac{23}{40} p_0 \tag{3.35}$$

In a similar manner, it can be shown that

$$p_2 = \frac{23}{40} p_0 \tag{3.36}$$

The equivalent loads p_1 and p_2 are often referred to as *consistent loads*, and it is interesting to note that in this case, they are not statically equivalent to the applied load. Often, however, static equivalents of the applied load provide a reasonable approximation of the load effect and are used in place of consistent loads. In our example, the static equivalents will be $p_1 = p_2 = p/2$. Using the load vector and ignoring damping, the equations of motion can be written as

$$\begin{bmatrix} m & \\ & m \end{bmatrix} \begin{bmatrix} \ddot{u}_1 \\ \ddot{u}_2 \end{bmatrix} + \frac{324EI}{5L^3} \begin{bmatrix} 4 & -\frac{7}{2} \\ -\frac{7}{2} & 4 \end{bmatrix} \begin{bmatrix} u_1 \\ u_2 \end{bmatrix} = p_0 \begin{bmatrix} \frac{23}{40} \\ \frac{23}{40} \end{bmatrix} \tag{3.37}$$

3.3.3 *Systems with distributed mass but localized stiffness*

The rigid bar of Figure 3.4 is an example of a system that can be categorized as having a distributed mass but localized stiffnesses. It is redrawn in Figure 3.11a, where the external forces consisting of a direct force and a moment are also shown. It is assumed that the damping is negligible. The spring forces on the bar are concentrated at the two locations where the springs are attached to the bar. However, the inertia forces are distributed throughout the length. In spite of this, the system has only two degrees of freedom, because we can completely define the displaced position of the bar in terms of the displacements along two selected degrees of freedom. An infinite number of choices exist when selecting the two degrees of freedom. In the present case, we select vertical displacements at the two ends of the bar, u_1 and u_2, as shown in Figure 3.11. The displaced shape of the bar is related to u_1 and u_2 as follows

$$u(x,t) = u_1(t)\psi_1(x) + u_2(t)\psi_2(x) \tag{3.38}$$

where u is the vertical displacement of the bar at a distance x from the left-hand end and the shape functions $\psi_1(x)$ and $\psi_2(x)$ are given by

$$\psi_1(x) = 1 - \frac{x}{L} \tag{3.39a}$$

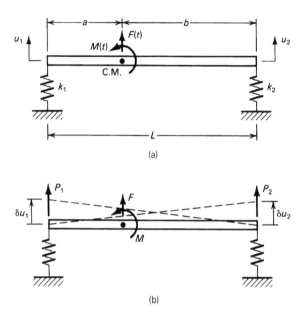

Figure 3.11. Vibrations of a rigid bar: (a) applied loads; (b) equivalent loads.

$$\psi_2(x) = \frac{x}{L} \tag{3.39b}$$

The two shape functions are shown by dashed lines in Figure 3.11b, and are exact because the bar is rigid. As in the case of a single-degree-of-freedom system, coordinates u_1 and u_2 are known as generalized coordinates.

The mass matrix of the system corresponding to the two-degree-of-freedom system shown in Figure 3.11a was obtained in Section 3.2.1 (Eq. 3.10). The stiffness matrix for the same two-degree-of-freedom system is given by Equation 3.14. The externally applied forces do not act along the coordinate directions. Therefore, before formulating the equations of motion, we must convert the applied forces into equivalent forces along the coordinates. These equivalent forces are identified as P_1 and P_2 in Figure 3.11b. The principle of virtual work can be effectively utilized to calculate the magnitude of forces P_1 and P_2. Thus if we apply a virtual displacement δu_1 along coordinate direction 1, the work done by the applied forces is given by

$$\Delta W = F\frac{b}{L}\delta u_1 - M\frac{\delta u_1}{L} \tag{3.40}$$

On the other hand, the virtual work done by the equivalent forces is given by

$$\Delta W_E = P_1\,\delta u_1 + P_2 \times 0 \tag{3.41}$$

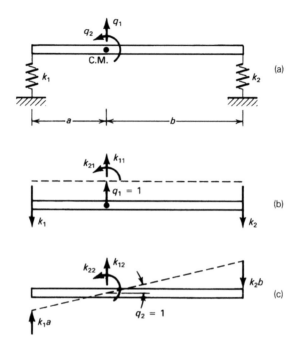

Figure 3.12. Vibrating rigid bar; coordinates defined at the mass center.

Because the work obtained by the two alternative methods should be the same, we equate Equations 3.40 and 3.41 and get

$$P_1 = F\frac{b}{L} - \frac{M}{L} \tag{3.42}$$

In a similar manner, by applying a virtual displacement δu_2, we obtain

$$P_2 = F\frac{a}{L} + \frac{M}{L} \tag{3.43}$$

It will be noted that forces P_1 and P_2 are, in this case, statically equivalent to the applied actions F and M. The equations of motion for the system of Figure 3.11 are now obtained as

$$\begin{bmatrix} \frac{mb^2}{L^2} + \frac{I_0}{L^2} & \frac{mab}{L^2} - \frac{I_0}{L^2} \\ \frac{mab}{L^2} - \frac{I_0}{L^2} & \frac{ma^2}{L^2} + \frac{I_0}{L^2} \end{bmatrix} \begin{bmatrix} \ddot{u}_1 \\ \ddot{u}_2 \end{bmatrix} + \begin{bmatrix} k_1 & 0 \\ 0 & k_2 \end{bmatrix} \begin{bmatrix} u_1 \\ u_2 \end{bmatrix} = \begin{bmatrix} F\frac{b}{L} - \frac{M}{L} \\ F\frac{a}{L} + \frac{M}{L} \end{bmatrix} \tag{3.44}$$

It is of interest to re-solve this problem with a new set of coordinates. As indicated in Figure 3.12a, these coordinates consist of vertical translation q_1 at the mass center and rotation q_2 about the mass center. With this set of coordinates, the mass matrix is easily shown to be given by

$$\mathbf{m} = \begin{bmatrix} m & 0 \\ 0 & I_0 \end{bmatrix} \tag{3.45}$$

To obtain the stiffness matrix, we first apply a unit displacement along coordinate 1, identify the internal elastic forces produced by such a displacement, and then calculate the external forces required to equilibrate the internal elastic forces. The relevant forces have all been indicated in Figure 3.12b. Using the conditions of equilibrium, we get

$$k_{11} = k_1 + k_2$$
$$k_{21} = k_2 b - k_1 a \tag{3.46}$$

Next, we apply a unit displacement along coordinate 2 and calculate the external forces required to maintain the displacement. Referring to Figure 3.12c, these external forces are given by

$$k_{12} = k_2 b - k_1 a$$
$$k_{22} = k_1 a^2 + k_2 b^2 \tag{3.47}$$

The stiffness matrix is now assembled from the influence coefficients defined in Equations 3.46 and 3.47

$$\mathbf{K} = \begin{bmatrix} k_1 + k_2 & k_2 b - k_1 a \\ k_2 b - k_1 a & k_1 a^2 + k_2 b^2 \end{bmatrix} \tag{3.48}$$

Since the applied forces act along the coordinates already defined, the force vector is obtained directly

$$\mathbf{p} = \begin{bmatrix} F \\ M \end{bmatrix} \tag{3.49}$$

The equations of motion can be assembled using Equations 3.45, 3.48 and 3.49.

$$\begin{bmatrix} m & 0 \\ 0 & I_0 \end{bmatrix} \begin{bmatrix} \ddot{q}_1 \\ \ddot{q}_2 \end{bmatrix} + \begin{bmatrix} k_1 + k_2 & k_2 b - k_1 a \\ k_2 b - k_1 a & k_1 a^2 + k_2 b^2 \end{bmatrix} \begin{bmatrix} q_1 \\ q_2 \end{bmatrix} = \begin{bmatrix} F \\ M \end{bmatrix} \tag{3.50}$$

A comparison of Equations 3.44 and 3.50 shows that while the stiffness matrix is diagonal in the first case, it is the mass matrix that is diagonal in the second case. A natural question that might arise is whether there is a set of coordinates in which both the mass matrix and the stiffness matrix are diagonal. There is indeed such a set of coordinates. Coordinates in the set are called *normal coordinates* and play a very important role in the field of structural dynamics. In later chapters, we discuss such coordinate sets in considerable detail.

Example 3.1
The uniform rigid rectangular slab of total mass m shown in Figure E3.1a is supported by three massless columns rigidly attached to the slab and fixed at the base. The columns have a flexural rigidity EI about each of the two principal axes, which are oriented so that they are parallel to the adjacent sides of the rectangle. Evaluate the mass and stiffness matrices for the system in the coordinates u_1, u_2, and u_3 defined at the mass center.

Solution
Because the columns are massless, they can be modeled by linear springs shown in Figure E3.1b. Each spring has a stiffness $k = (12EI)/(L^3)$.

To obtain the first column of stiffness matrix, we impose a unit displacement in the direction of coordinate 1 and identify the internal spring forces that oppose the displacement. The external forces required to balance these spring forces, shown in Figure E3.1c, are now obtained from the conditions of equilibrium.

$$k_{11} = 3k$$
$$k_{21} = 0 \tag{a}$$
$$k_{31} = 2k\frac{a}{2} - k\frac{a}{2} = \frac{ka}{2}$$

The second column of the stiffness matrix is obtained in a similar manner by imposing a unit displacement along coordinate direction 2. The resulting internal forces are shown in Figure E3.1d. The stiffness coefficients are given by

$$k_{12} = 0$$
$$k_{22} = 3k \tag{b}$$
$$k_{32} = -2k\frac{b}{2} + k\frac{b}{2} = -\frac{kb}{2}$$

The third column of stiffness matrix is obtained by imposing a unit rotation along coordinate 3. Referring to Figure E3.1e, we see that

$$k_{13} = \frac{ka}{2}$$
$$k_{23} = -\frac{kb}{2} \tag{c}$$
$$k_{33} = 3k\left(\frac{b}{2}\right)^2 + 3k\left(\frac{a}{2}\right)^2 = \frac{3}{4}k(a^2 + b^2)$$

The stiffness matrix is now assembled from Equations a, b, and c

$$\mathbf{K} = k\begin{bmatrix} 3 & 0 & \frac{a}{2} \\ 0 & 3 & -\frac{b}{2} \\ \frac{a}{2} & -\frac{b}{2} & \frac{3}{4}(a^2 + b^2) \end{bmatrix} \tag{d}$$

where $k = 12EI/L^3$. Since the selected coordinates are located at the mass center, the mass matrix is obtained directly and is given by

$$\mathbf{M} = \begin{bmatrix} m & & \\ & m & \\ & & I_0 \end{bmatrix} \tag{e}$$

where $I_0 = m(a^2 + b^2)/12$.

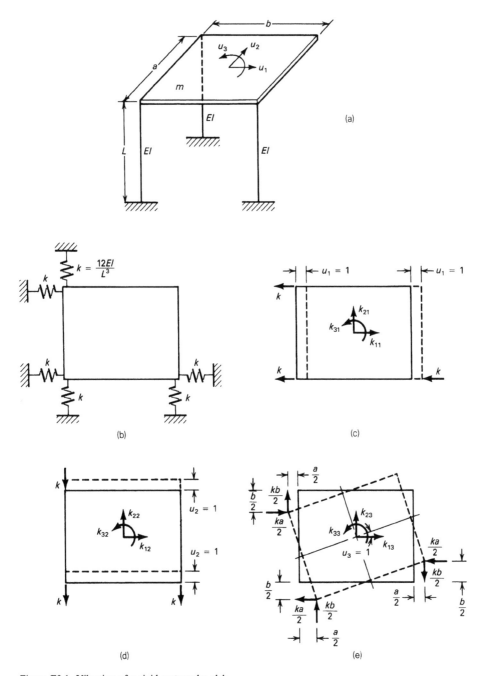

Figure E3.1. Vibration of a rigid rectangular slab.

3.3.4   *Systems with distributed mass and distributed stiffness*

Systems having distributed mass as well as distributed stiffness are also referred to as continuous systems or distributed parameter systems. Theoretically, they have an infinite number of degrees of freedom and their motion is represented by partial differential equations. They can, however, be idealized as multi-degree-of-freedom systems either by mass lumping or by expressing their displaced shape as a superposition of a series of shape functions of the spatial coordinates each multiplied by its own generalized coordinate. Mass lumping was described briefly in Section 2.7.4. Equations of motion for lumped mass systems are obtained by the procedure described in Section 3.3.2. In this section, we describe the procedure used to obtain the equations of motion for a distributed parameter system represented by a superposition of shape functions.

The shape functions chosen to represent the displaced shape should be independent of each other; that is, it should not be possible to derive any one of them by a linear combination of one or more of the remaining. The functions should also, as a minimum, satisfy the geometric or essential boundary conditions of the system. Each shape function is multiplied by a generalized coordinate. The generalized coordinates then serve as the unknowns in the system, whose values must be determined by a solution of the equations of motion. The number of such coordinates is equal to the number of degrees of freedom in the system.

As an illustration of the procedure, consider again the flexural vibrations of a beam supported at its ends. The vibration shape of the beam shown in Figure 3.13a is represented by a superposition of N shape functions $\psi_1, \psi_2, \ldots, \psi_N$, so that

$$u(x,t) = z_1\psi_1(x) + z_2\psi_2(x) + \cdots + z_N\psi_N(x) \tag{3.51}$$

The forces acting on an element of length dx are shown in Figure 3.13b. These forces include the flexural moments M, the inertia force $\bar{m}\ddot{u}\,dx$, and the externally applied force $\bar{p}\,dx$. We now obtain the virtual work equations for all admissible virtual displacements of the system. The admissible virtual displacements should satisfy the constraints in the system and should be independent. Since the shape functions $\psi_i(x)$ are selected to satisfy the geometric or essential boundary conditions and are independent of each other, functions $\delta z_i\psi_i(x)$, $i = 1, 2, \ldots, N$ form a set of N admissible virtual displacement shapes.

A virtual displacement applied to the beam causes the two ends of the infinitesimal element to rotate relative to each other by an angle $\delta z_i\psi_i''(x)\,dx$. Now, in accordance with the elementary beam theory, moment $M = EI(\partial^2 u/\partial x^2)$. The virtual work done by the elastic moments acting on the element is therefore

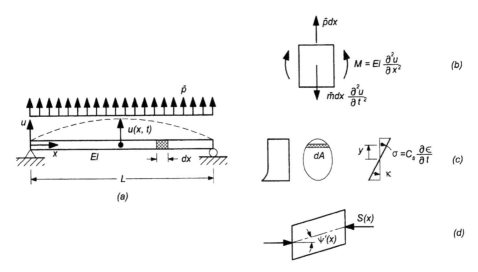

Figure 3.13. (a) Flexural vibrations of a beam; (b) forces acting on an element; (c) internal damping resistance; (d) axial force effect.

given by

$$d(\delta W_{ei}) = \delta z_i \psi_i''(x) EI \frac{\partial^2 u}{\partial x^2}\, dx$$

$$= \delta z_i \sum_{j=1}^{N} z_j EI \psi_i''(x)\psi_j''(x)\, dx \qquad (3.52)$$

The total virtual work done by the elastic moments is obtained by integrating Equation 3.52 over the length

$$\delta W_{ei} = \delta z_i \sum_{j=1}^{N} z_j \int_0^L EI\psi_i''(x)\psi_j''(x)\, dx$$

$$= \delta z_i \sum_{j=1}^{N} z_j k_{ij} \qquad (3.53)$$

where

$$k_{ij} = \int_0^L EI\psi_i''(x)\psi_j''(x)\, dx \qquad (3.54)$$

As explained in Section 2.6, the internal virtual work $d(\delta W_i)$ is, in fact, the work done by the restoring elastic moments in the element. Since the restoring moments are opposite in sign to the moments shown in Figure 3.13b, work $d(\delta W_i)$ is the negative of work $d(\delta W_{ei})$ given by Equation 3.52. The virtual work done by the externally applied force is

$$\delta W_p = \delta z_i \int_0^L \bar{p}\psi_i(x)\, dx = p_i \delta z_i \qquad (3.55)$$

where

$$p_i = \int_0^L \bar{p}\psi_i(x)\, dx \tag{3.56}$$

In a similar manner, the virtual work of the inertia force is

$$\delta W_I = -\delta z_i \int_0^L \psi_i(x)\bar{m}\frac{\partial^2 u}{\partial t^2}\, dx$$

$$= -\delta z_i \sum_{j=1}^N \int_0^L \ddot{z}_j \bar{m}\psi_i(x)\psi_j(x)\, dx$$

$$= -\delta z_i \sum_{j=1}^N \ddot{z}_j m_{ij} \tag{3.57}$$

where

$$m_{ij} = \int_0^L \bar{m}\psi_i(x)\psi_j(x)\, dx \tag{3.58}$$

Equating the sum of virtual works δW_i, δW_p, and δW_I to zero, we get the following virtual work equation

$$-\delta z_i \left(\sum_{j=1}^N \ddot{z}_j m_{ij} + \sum_{j=1}^N z_j k_{ij} \right) + \delta z_i\, p_i = 0 \tag{3.59}$$

Since δz_i is arbitrary, it can be canceled from Equation 3.59.

Corresponding to the N independent virtual displacements $\delta z_i \psi_i(x)$, $i = 1$, $2,\ldots,N$, there are N virtual work equations of the form of Equation 3.59. Together, they can be expressed in matrix notations as

$$\mathbf{M}\ddot{\mathbf{z}} + \mathbf{K}\mathbf{z} = \mathbf{p} \tag{3.60}$$

where $\mathbf{z}$ is a vector of N generalized coordinates, $\mathbf{M}$ is a mass matrix whose elements are given by Equation 3.58, $\mathbf{K}$ is a stiffness matrix whose elements are defined by Equation 3.54, and $\mathbf{p}$ is the vector of generalized forces given by Equation 3.56. By referring to Equation 3.58 for mass influence coefficients, it is seen that $m_{ij} = m_{ji}$. In a similar manner, Equation 3.54 shows that $k_{ij} = k_{ji}$. Thus, both mass matrix $\mathbf{M}$ and stiffness matrix $\mathbf{K}$ are symmetric. In formulating the virtual work done by internal spring forces, we have neglected the work done by the internal shear forces on virtual shear deformations. For normal proportions of beams, that is, when the cross-sectional dimensions are small in relation to the span length, the work done by shear forces is negligible in comparison to that done by the flexural moments.

Equation 3.60 does not include damping forces present in the system. Such forces may either be external or internal. External damping can be provided by distributed viscous damping forces, as indicated in Figure 2.16d. Denoting the distributed viscous damping coefficient by $\bar{c}$, the virtual work done by external damping forces is obtained as

$$\delta W_{DE} = -\delta z_i \int_0^L \bar{c}\psi_i(x)\frac{\partial u}{\partial t}dx$$

$$= -\delta z_i \sum_{j=1}^N \int_0^L \dot{z}_j\bar{c}\psi_i(x)\psi_j(x)\,dx \tag{3.61}$$

Internal damping forces resist deformations within the element and their magnitudes depend on the strain rate. Thus if the strain rate is $\partial\varepsilon/\partial t$, the damping resistance can be represented by a stress σ_D which is proportional to the strain rate, the constant of proportionality being a damping constant c_s. Thus

$$\sigma_D = c_s\frac{\partial\varepsilon}{\partial t} \tag{3.62}$$

Referring to Figure 3.13c and using elementary beam theory, which assumes that plane sections remain plane under bending, we get the following kinematic relationship

$$\varepsilon = \kappa y$$

$$= \frac{\partial^2 u}{\partial x^2}y \tag{3.63}$$

where κ is the curvature of the beam. From Equations 3.62 and 3.63

$$\sigma_D = c_s\frac{\partial^3 u}{\partial x^2\partial t}y \tag{3.64}$$

The resisting moment due to internal damping is given by

$$M_D = \int_A \sigma_D y\,dA$$

$$= \int_A c_s y^2\frac{\partial^3 u}{\partial x^2\partial t}\,dA$$

$$= c_s I\frac{\partial^3 u}{\partial x^2\partial t} \tag{3.65}$$

Finally, recognizing that under a virtual displacement the ends of the element undergo a relative rotation of $\delta z_i\,\psi_i''(x)\,dx$, the virtual work done by the internal damping moment works out to

$$\delta W_{DI} = -\delta z_i \int_0^L c_s I \psi_i''(x) \frac{\partial^3 u}{\partial x^2 \partial t} dx$$

$$= -\delta z_i \sum_{j=1}^N \int_0^L c_s I \dot{z}_j \psi_i''(x) \psi_j''(x) \, dx \tag{3.66}$$

Combining Equations 3.61 and 3.66, the total work done by the damping forces can be expressed as

$$\delta W_D = -\delta z_i \sum_{j=1}^N \dot{z}_j \left\{ \int_0^L \bar{c} \psi_i(x) \psi_j(x) \, dx + \int_0^L c_s I \psi_i''(x) \psi_j''(x) \, dx \right\}$$

$$= -\delta z_i \sum_{j=1}^N \dot{z}_j c_{ij} \tag{3.67}$$

where

$$c_{ij} = \int_0^L \bar{c} \psi_i(x) \psi_j(x) \, dx + \int_0^L c_s I \psi_i''(x) \psi_j''(x) \, dx \tag{3.68}$$

When damping is included, the equations of motion (Eq. 3.60) are revised to

$$\mathbf{M\ddot{z} + C\dot{z} + Kz = p} \tag{3.69}$$

where $\mathbf{C}$ is a damping matrix whose elements are defined by Equation 3.68.

In practice, it is difficult to define damping constants $\bar{c}$ and c_s on the basis of physical characteristics of the system. Alternative methods are therefore used to include damping resistance in the model. The suggested methods lead to a response characteristic that correlates well with experimental or observed behavior. The methods of defining the damping forces or the damping matrix $\mathbf{C}$ are considered in the subsequent chapters. When concentrated applied forces, masses, springs, or dampers are present in the system, they are handled in a manner similar to that described in Section 2.7.4. If axial forces are present, the equations of motion (Eq. 3.69) will need modification. As discussed earlier, the effect of axial forces can be allowed for by deriving a geometric stiffness matrix which must be deducted from the elastic stiffness matrix of the beam.

The derivation of the geometric stiffness matrix follows a procedure very similar to that used in the single-degree-of-freedom representation of the beam. Thus referring to Figure 3.13d, the virtual displacement $\delta z_i \, \psi_i(x)$ will cause the two axial forces $S(x)$ to move closer to each other by a distance $\delta z_i (\partial u/\partial x) \psi_i'(x) \, dx$, in which $\partial u/\partial x$ is obtained from Equation 3.51. The virtual work done by the axial forces is therefore given by

$$d(\delta W_S) = \delta z_i \sum_{j=1}^N z_j S(x) \psi_j'(x) \psi_i'(x) \, dx \tag{3.70}$$

The total virtual work is obtained by integrating Equation 3.70 over the length

$$\delta W_S = \delta z_i \sum_{j=1}^{N} \int_0^L z_j S(x) \psi_i'(x) \psi_j'(x) \, dx$$

$$= \delta z_i \sum_{j=1}^{N} z_j k_{Gij} \tag{3.71}$$

where

$$k_{Gij} = \int_0^L S(x) \psi_i'(x) \psi_j'(x) \, dx \tag{3.72}$$

It should be noted that δW_S is positive while δW_i was negative. The equation of motion, including the axial force effect, becomes

$$\mathbf{M\ddot{z}} + \mathbf{C\dot{z}} + (\mathbf{K} - \mathbf{K}_G)\mathbf{z} = \mathbf{p} \tag{3.73}$$

where $\mathbf{K}_G$ is the geometric stiffness matrix whose elements are defined by Equation 3.72.

The procedure described in the foregoing paragraphs in which the displaced shape is expressed as a superposition of a series of appropriately selected shape functions is also known as the *Ritz method*. The shape functions selected to represent the displaced configuration of the system are called the *Ritz shapes*. While conceptually elegant, the Ritz method poses several difficulties in its practical application. These difficulties are outlined below.

1. It is apparent that the success of the Ritz method depends on the selection of shape functions. This is, in general, a difficult task. As a minimum, the shape functions should satisfy the essential boundary conditions of the problem. In addition, if the highest-order differential appearing in the virtual work equation is of order m, the shape functions should be m times differentiable. In other words, shape functions should satisfy C^{m-1} continuity; that is, their $(m-1)$th differential should be continuous. The choice of shape functions should thus be guided by the nature of the problem and the boundary conditions, and it is not always apparent what shape functions would be appropriate in a particular case.

2. The shape functions should span the entire domain of the system, yet the displacement being represented may vary in widely different manner in different regions of the domain. As an example, even in the simple case of the flexural vibrations of a beam, if the moments of inertia in different sections of the length are significantly different, shape functions that are appropriate for one region may not be appropriate for another. A flat plate with edge beams is another example where displacements may vary in widely differing manners.

3. It is not always clear how the Ritz method should be refined, or what additional Ritz shapes should be included to improve the accuracy of the solution.

4. The virtual work expressions used in setting up the equations of motion involve integration of the shape function derivatives and their products. Unless the shape functions are simple mathematical functions, such integration may not be straightforward.
5. The property matrices obtained in a Ritz formulation are fully populated. As a result, when the number of Ritz vectors is large, the solution becomes computationally inefficient.
6. The generalized coordinates used as the unknown weights on the shape functions do not always have a physical meaning. It is therefore difficult to interpret the behavior of the system being analyzed from the generalized coordinate values obtained in the analysis.

The difficulties involved in the Ritz function approach are largely overcome in a finite element formulation, which is discussed in a subsequent section.

Example 3.2
The vibration shape of the simply supported uniform beam of length L, flexural rigidity EI, and mass $\bar{m}$ per unit length shown in Figure E3.2 is approximated by

$$u(x,t) = z_1(t)\psi_1(x) + z_2(t)\psi_2(x) + z_3(t)\psi_3(x) \tag{a}$$

where

$$\psi_1(x) = \sin\frac{\pi x}{L}$$

$$\psi_2(x) = \sin\frac{2\pi x}{L} \tag{b}$$

$$\psi_3(x) = \sin\frac{3\pi x}{L}$$

Obtain the equations of motion when the beam is vibrating under the action of a uniformly distributed load $\bar{p}(t)$.

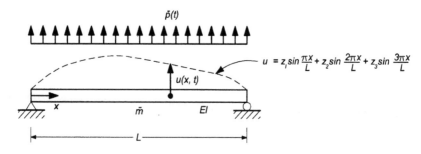

Figure E3.2. Flexural vibrations of a uniform beam.

Solution

We have

$$\psi_m(x) = \sin\frac{m\pi x}{L}$$

$$\psi_m''(x) = -\frac{m^2\pi^2}{L^2}\sin\frac{m\pi x}{L}$$

(c)

Also

$$\int_0^L \sin\frac{m\pi x}{L}\sin\frac{n\pi x}{L}dx = \begin{cases} 0 & m \neq n \\ \frac{L}{2} & m = n \end{cases}$$

(d)

The elements of stiffness matrix are obtained from Equation 3.54. Substituting from Equations c and d in Equation 3.54, we get

$$k_{mn} = 0 \quad m \neq n$$

$$k_{mm} = \frac{m^4\pi^4 EI}{2L^3}$$

(e)

The complete stiffness matrix is given by

$$\mathbf{K} = \frac{\pi^4 EI}{2L^3}\begin{bmatrix} 1 & 0 & 0 \\ 0 & 16 & 0 \\ 0 & 0 & 81 \end{bmatrix}$$

(f)

The mass matrix is obtained by using Equation 3.58 and can be shown to be

$$\mathbf{M} = \frac{\bar{m}L}{2}\begin{bmatrix} 1 & 0 & 0 \\ 0 & 1 & 0 \\ 0 & 0 & 1 \end{bmatrix}$$

(g)

The elements of the load vector obtained from Equation 3.56 are

$$p_m = \bar{p}\int_0^L \sin\frac{m\pi x}{L}dx$$

(h)

The load vector reduces to

$$\mathbf{p} = \frac{2\bar{p}L}{\pi}\begin{bmatrix} 1 \\ 0 \\ \frac{1}{3} \end{bmatrix}$$

(i)

The equation of motion is

$$\mathbf{M\ddot{z}} + \mathbf{Kz} = \mathbf{p}$$

(j)

where $\mathbf{M}$, $\mathbf{K}$, and $\mathbf{p}$ are defined by Equations g, f, and i, respectively.

It is a coincidence that in this case, both the mass matrix and the stiffness matrix are diagonal and Equation j is therefore uncoupled; that is, it is equivalent to three independent equations, each of which can be solved separately. In fact, uncoupling of the equations of motion has been made possible by the selection of shape functions that are orthogonal to each other. Coordinates that uncouple the equations of motion are called normal coordinates. Such coordinates exist for every dynamic system, whether discrete or continuous. It is obvious that a formulation of the equations of motion in terms of normal coordinates will make the solution of such equations much simpler. However, it is not always easy to recognize the normal coordinates of a system. In later chapters, we describe in considerable detail procedures for obtaining the normal coordinates.

Example 3.3

The uniform beam in Figure E3.3a has four independent displacement degrees of freedom, as indicated in the figure. Two of these are the rotations at the ends of the beam, while the other two are vertical displacements at the same locations. By applying appropriate external forces along the four degrees of freedom, it is possible to obtain a displaced configuration of the beam in which there is a unit displacement along one of the four degrees of freedom while the displacements along the remaining three degrees of freedom are zero. Corresponding to the four degrees of freedom, there are four such independent displaced shapes. By using elementary beam theory, obtain these four shapes. Then, by expressing the displacement of the beam as a superposition of the four basic shapes, obtain the stiffness matrix of the beam.

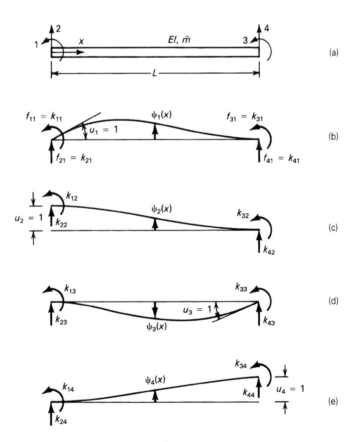

Figure E3.3. Flexural displacements of a prismatic beam.

Solution

According to elementary beam theory, the deflections of a uniform beam subjected to loads only at its ends are governed by the following differential equation

$$EI\frac{d^4\psi}{dx^4} = 0 \tag{a}$$

Equation a has a solution of the form

$$\psi = Ax^3 + Bx^2 + Cx + D \tag{b}$$

where A, B, C, and D are constants to be determined from the boundary conditions. For the configuration shown in Figure E3.3b, these boundary conditions are

$$x = 0 \quad \psi = 0 \quad \text{and} \quad \frac{\partial\psi}{\partial x} = 1$$

$$x = L \quad \psi = 0 \quad \text{and} \quad \frac{\partial\psi}{\partial x} = 0 \tag{c}$$

When constants A, B, C, and D are obtained by using Equation c, displacement ψ is given by

$$\psi_1(x) = x\left(1 - \frac{x}{L}\right)^2 \tag{d}$$

Displaced shapes shown in Figure E3.3c, d, and e are obtained in a similar manner and are given by

$$\psi_2(x) = 1 - 3\left(\frac{x}{L}\right)^2 + 2\left(\frac{x}{L}\right)^3$$

$$\psi_3(x) = \frac{x^2}{L}\left(\frac{x}{L} - 1\right) \tag{e}$$

$$\psi_4(x) = 3\left(\frac{x}{L}\right)^2 - 2\left(\frac{x}{L}\right)^3$$

The displaced shape of the beam can be expressed as a superposition of the shape functions given by Equations d and e, so that

$$u = z_1\psi_1(x) + z_2\psi_2(x) + z_3\psi_3(x) + z_4\psi_4(x) \tag{f}$$

The elements of the stiffness matrix are now obtained by using Equation 3.54. The resulting stiffness matrix is given by

$$\mathbf{K} = \begin{bmatrix} \frac{4EI}{L} & \frac{6EI}{L^2} & \frac{2EI}{L} & -\frac{6EI}{L^2} \\[2mm] \frac{6EI}{L^2} & \frac{12EI}{L^3} & \frac{6EI}{L^2} & -\frac{12EI}{L^3} \\[2mm] \frac{2EI}{L} & \frac{6EI}{L^2} & \frac{4EI}{L} & -\frac{6EI}{L^2} \\[2mm] -\frac{6EI}{L^2} & -\frac{12EI}{L^3} & -\frac{6EI}{L^2} & \frac{12EI}{L^3} \end{bmatrix} \tag{g}$$

We now derive the external forces that must be applied along the four degrees of freedom defined earlier to maintain a specified displaced configuration, say the one in Figure E3.3b. These forces can be obtained either by direct application of the beam theory or by the virtual work method. Thus, if the beam in Figure E3.3b is given a virtual displacement of $\delta z_2\,\psi_2(x)$, the virtual work done by the moments acting on internal elements is

$$\delta W_{ei} = \delta z_2 \int_0^L EI\psi_1''(x)\psi_2''(x)\,dx \tag{h}$$

At the same time, of the four external forces shown in Figure E3.3b, only the force f_{21} does any work, the virtual displacements along the other three forces being zero. This can be verified from Figure E3.3c, which also represents the virtual displacement shape. The virtual work done by the external forces becomes

$$\delta W_e = \delta z_2 f_{21} \tag{i}$$

Using the virtual work equation yields

$$\delta z_2 f_{21} = \delta W_{ei} \tag{j}$$

or

$$f_{21} = \int_0^L EI\psi_1''(x)\psi_2''(x)\,dx = k_{21} \tag{k}$$

Other forces shown in Figure E3.3b can be derived similarly and are seen to be equal to the elements on the first column of the stiffness matrix. Elements on other columns of the stiffness matrix are interpreted in a similar manner.

In conclusion, it can be stated that when functions given by Equations d and e are used as the shape functions, the stiffness matrix derived by the procedure of Equation 3.54 is, in fact, the stiffness matrix corresponding to the four degrees of freedom defined in Figure E3.3a, and the weighting factors z_1 through z_4 in Equation f represent displacements along these degrees of freedom.

Example 3.4
The uniform bar shown in Figure E3.4 has an area A and mass per unit length $\bar{m}$. The bar is vibrating in the axial directions under the action of a distributed axial force $\bar{p}(x)$. Express the vibration shape of the bar as a superposition of appropriate shape functions, selecting for the generalized coordinates the longitudinal displacements at the two ends. Then, obtain the stiffness and mass matrices and the force vector corresponding to the two generalized coordinates defined above.

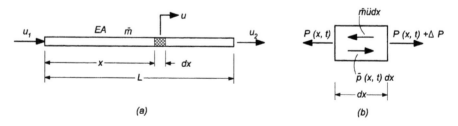

(a)　　　　　　　　　　　　　　　　(b)

Figure E3.4. Axial vibrations of a uniform bar: (a) elevation of the bar showing displacement coordinates; (b) forces acting on the bar element.

Solution
The vibration shape of the bar is represented by a superposition of two shape functions, ψ_1 and ψ_2, weighted by the generalized coordinates u_1 and u_2 shown in Figure E3.4a.

$$u(x,t) = u_1(t)\psi_1(x) + u_2(t)\psi_2(x) \tag{a}$$

We select the following shape functions:

$$\psi_1 = 1 - \frac{x}{L}$$
$$\psi_2 = \frac{x}{L} \tag{b}$$

The shape functions have been selected so that at $x = 0$, $\psi_1 = 1$ and $\psi_2 = 0$ while at $x = L$, $\psi_1 = 0$ and $\psi_2 = 1$. With the selected shape functions, Equation a gives a displacement $u = u_1$ at $x = 0$ and a displacement $u = u_2$ at $x = L$.

We now apply appropriate virtual displacements to the bar and obtain an equation of virtual work for each applied virtual displacement. A set of admissible virtual displacements of the system consists of $\delta u_i \psi_i(x)$, $i = 1, 2$. A virtual work equation is obtained for each displacement in the set.

The forces acting on an element of length dx are shown in Figure E3.4b. They consist of the elastic axial forces P and the inertia forces $m\ddot{u}\,dx$. A virtual displacement applied to the bar causes the two ends of the infinitesimal element to move relative to each other in an axial direction by a distance $\delta u_i \psi_i'\,dx$. Since the axial force P is given by $P = EA(\partial u/\partial x)$, the virtual work done by the elastic forces acting on the element is given by

$$d(\delta W_{ei}) = \delta u_i \psi_i'(x) EA \frac{\partial u}{\partial x} dx$$

$$= \delta u_i \sum_{j=1}^{2} u_j EA \psi_i'(x) \psi_j'(x)\, dx \tag{c}$$

The total virtual work done by the elastic forces is obtained by integrating Equation c over the length

$$\delta W_{ei} = \delta u_i \sum_{j=1}^{2} u_j \int_0^L EA \psi_i'(x) \psi_j'(x)\, dx$$

$$= \delta u_i \sum_{j=1}^{2} u_j k_{ij} \tag{d}$$

where the stiffness influence coefficients k_{ij} are given by

$$k_{ij} = \int_0^L EA \psi_i'(x) \psi_j'(x)\, dx \tag{e}$$

Substitution for ψ_i and ψ_j in Equation e gives the following values for the stiffness influence coefficients

$$k_{11} = \int_0^L EA \left(-\frac{1}{L}\right)^2 dx = \frac{EA}{L}$$

$$k_{12} = k_{21} = \int_0^L EA \left(-\frac{1}{L}\right)\frac{1}{L} dx = -\frac{EA}{L} \tag{f}$$

$$k_{22} = \int_0^L EA \left(\frac{1}{L}\right)^2 dx = \frac{EA}{L}$$

Assembly of the stiffness influence coefficients gives the following stiffness matrix

$$\mathbf{K} = \frac{EA}{L}\begin{bmatrix} 1 & -1 \\ -1 & 1 \end{bmatrix} \qquad (g)$$

The virtual work of the inertia forces is obtained from

$$\delta W_I = -\delta u_i \int_0^L \bar{m}\psi_i(x)\frac{\partial^2 u}{\partial t^2}\,dx$$

$$= -\delta u_i \sum_{j=1}^{2}\int_0^L \bar{m}\ddot{u}_j\psi_i(x)\psi_j(x)\,dx$$

$$= -\delta u_i \sum_{j=1}^{2}\ddot{u}_j m_{ij} \qquad (h)$$

where the mass influence coefficients m_{ij} are given by

$$m_{ij} = \int_0^L \bar{m}\psi_i(x)\psi_j(x)\,dx \qquad (i)$$

Substitution for ψ_i and ψ_j in Equation i leads to the following value for the mass influence coefficients

$$m_{11} = \int_0^L \bar{m}\left(1 - \frac{x}{L}\right)^2 dx = \bar{m}\frac{L}{3}$$

$$m_{12} = m_{21} = \int_0^L \bar{m}\left(1 - \frac{x}{L}\right)\frac{x}{L}dx = \bar{m}\frac{L}{6} \qquad (j)$$

$$m_{22} = \int_0^L \bar{m}\left(\frac{x}{L}\right)^2 dx = \bar{m}\frac{L}{3}$$

Assembly of the mass influence coefficients gives the following mass matrix

$$\mathbf{M} = \bar{m}L\begin{bmatrix} \frac{1}{3} & \frac{1}{6} \\ \frac{1}{6} & \frac{1}{3} \end{bmatrix} \qquad (k)$$

The virtual work done by the externally applied force is

$$\delta W_p = \delta u_i \int_0^L \bar{p}(x,t)\psi_i(x)\,dx$$

$$= p_i\,\delta u_i \qquad (l)$$

where

$$p_i = \int_0^L \bar{p}(x,t)\psi_i(x)\,dx \qquad (m)$$

For a uniformly distributed load $\bar{p}(t)$, Equation m yields

$$
\begin{aligned}
p_1 &= \bar{p}(t) \int_0^L \left(1 - \frac{x}{L}\right) dx \\
&= \bar{p}(t)\frac{L}{2} \\
p_2 &= \bar{p}(t) \int_0^L \frac{x}{L} dx \\
&= \bar{p}(t)\frac{L}{2}
\end{aligned}
\tag{n}
$$

The generalized force vector becomes

$$
p = \bar{p}\frac{L}{2}\begin{bmatrix} 1 \\ 1 \end{bmatrix}
\tag{o}
$$

3.4 TRANSFORMATION OF COORDINATES

Methods of formulating the equations of motion in a selected set of coordinates, either physical or generalized, were discussed in the preceding sections. Often, it may be necessary or convenient for the analysis of response to express the equations in a set of coordinates that are different from the ones in which the equations were initially formulated. Such a transformation can easily be achieved by using the principle of virtual work. Suppose, for instance, that the equations have originally been formulated in a set of N coordinates denoted by $\mathbf{u}$, and let the corresponding mass matrix, stiffness matrix, and applied force vector be denoted by $\mathbf{M}$, $\mathbf{K}$, and $\mathbf{p}$ respectively. Now let it be required to transform the equations of motion to a set of N independent coordinates $\mathbf{q}$ which are related to the set $\mathbf{u}$ by the equation

$$
\mathbf{u} = \mathbf{T}\mathbf{q}
\tag{3.74}
$$

where $\mathbf{T}$ is a transformation matrix, which is square. Let the transformed mass and stiffness matrices and the applied force vector be denoted by $\tilde{\mathbf{M}}$, $\tilde{\mathbf{K}}$, and $\tilde{\mathbf{p}}$, respectively. The spring forces in $\mathbf{u}$ set of coordinates are

$$
\mathbf{f}_S = \mathbf{K}\mathbf{u}
\tag{3.75}
$$

while those in $\mathbf{q}$ set of coordinates are

$$
\tilde{\mathbf{f}}_S = \tilde{\mathbf{K}}\mathbf{q}
\tag{3.76}
$$

Now, let the system be subjected to a set of virtual displacement $\delta\mathbf{q}$. The corresponding displacements in the $\mathbf{u}$ set are $\delta\mathbf{u}$, where

$$
\delta\mathbf{u} = \mathbf{T}\,\delta\mathbf{q}
\tag{3.77}
$$

If the forces $\tilde{\mathbf{f}}_S$ and $\mathbf{f}_S$ are equivalent, the virtual work done by each set should be the same. Thus

$$\delta\mathbf{q}^T \tilde{\mathbf{K}}\mathbf{q} = \delta\mathbf{u}^T \mathbf{K}\mathbf{u} \tag{3.78}$$

Substituting for $\mathbf{u}$ and $\delta\mathbf{u}$ from Equations 3.74 and 3.77, respectively, we get

$$\delta\mathbf{q}^T \tilde{\mathbf{K}}\mathbf{q} = \delta\mathbf{q}^T \mathbf{T}^T\mathbf{K}\mathbf{T}\mathbf{q} \tag{3.79}$$

Since Equation 3.79 should hold for arbitrary $\mathbf{q}$ and $\delta\mathbf{q}$, we get

$$\tilde{\mathbf{K}} = \mathbf{T}^T\mathbf{K}\mathbf{T} \tag{3.80}$$

In a similar manner, considering the virtual work done by inertia forces, we can show that

$$\tilde{\mathbf{M}} = \mathbf{T}^T\mathbf{M}\mathbf{T} \tag{3.81}$$

Consideration of virtual work done by the applied forces gives

$$\tilde{\mathbf{p}} = \mathbf{T}^T\mathbf{p} \tag{3.82}$$

If damping is present in the system, the damping matrix can be transformed in a similar manner, giving

$$\tilde{\mathbf{C}} = \mathbf{T}^T\mathbf{C}\mathbf{T} \tag{3.83}$$

Example 3.5
In Section 3.3.3, equations of motion were obtained for a spring-supported rigid bar in two different sets of coordinates, shown in Figures 3.4a and 3.12a, respectively. Starting from the formulation in the $\mathbf{u}$ set of coordinates (Fig. 3.4a) and a transformation relationship, obtain the equations of motion in the $\mathbf{q}$ set of coordinates shown in Figure 3.12a.

Solution
The transformation from the $\mathbf{q}$ set to the $\mathbf{u}$ set can be expressed as

$$\mathbf{u} = \mathbf{T}\mathbf{q} \tag{a}$$

where $\mathbf{T}$ is a transformation matrix given by

$$\mathbf{T} = \begin{bmatrix} 1 & -a \\ 1 & b \end{bmatrix} \tag{b}$$

The transformed stiffness matrix $\tilde{\mathbf{K}}$ is obtained from Equation 3.80

$$\tilde{\mathbf{K}} = \mathbf{T}^T\mathbf{K}\mathbf{T}$$

$$= \begin{bmatrix} 1 & 1 \\ -a & b \end{bmatrix} \begin{bmatrix} k_1 & 0 \\ 0 & k_2 \end{bmatrix} \begin{bmatrix} 1 & -a \\ 1 & b \end{bmatrix}$$

$$= \begin{bmatrix} k_1 + k_2 & bk_2 - ak_1 \\ bk_2 - ak_1 & a^2k_1 + b^2k_2 \end{bmatrix} \tag{c}$$

which is the same as that obtained directly in Equation 3.48. The transformed mass matrix is obtained from Equation 3.81

$$\tilde{\mathbf{M}} = \mathbf{T}^T \mathbf{M} \mathbf{T}$$

$$= \begin{bmatrix} 1 & 1 \\ -a & b \end{bmatrix} \begin{bmatrix} \frac{mb^2}{l^2} + \frac{I_0}{l^2} & \frac{mab}{l^2} - \frac{I_0}{l^2} \\ \frac{mab}{l^2} - \frac{I_0}{l^2} & \frac{ma^2}{l^2} + \frac{I_0}{l^2} \end{bmatrix} \begin{bmatrix} 1 & -a \\ 1 & b \end{bmatrix}$$

$$= \begin{bmatrix} m & \\ & I_0 \end{bmatrix} \tag{d}$$

and is the same as that given by Equation 3.45. The transformed applied force vector is given by

$$\tilde{\mathbf{p}} = \mathbf{T}^T \mathbf{p}$$

$$= \begin{bmatrix} 1 & 1 \\ -a & b \end{bmatrix} \begin{bmatrix} \frac{Fb}{l} - \frac{M}{l} \\ \frac{Fa}{l} + \frac{M}{l} \end{bmatrix}$$

$$= \begin{bmatrix} F \\ M \end{bmatrix} \tag{e}$$

which is equal to that obtained directly in Equation 3.49.

Example 3.6
The uniform bar shown in Figure E3.6 undergoes axial vibrations. Obtain the property matrices and the force vector of the bar in terms of generalized coordinates q_1 through q_4 shown in the figure. The longitudinal axis of the bar makes an angle α with respect to the coordinate direction q_1.

Solution
The stiffness and mass matrices as well as the force vector for the bar were obtained in Example 3.4 corresponding to the coordinates u_1 and u_3 shown in Figure E3.6. We can obtain the structural property matrices and the force vector in the generalized coordinate set $\mathbf{q}$ by means of a simple

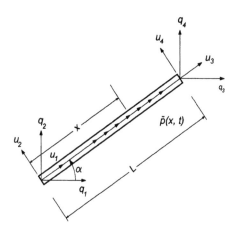

Figure E3.6. Transformation of coordinates for a bar in axial vibrations.

transformation. Before carrying out the transformation, we need to expand the property matrices and the force vector obtained earlier to an order 4 to correspond to the four coordinates u_1 through u_4. This is easily achieved by adding appropriate number of rows and columns of zero. The revised stiffness matrix is given by

$$
\mathbf{K}_u = \begin{bmatrix} \frac{EA}{L} & 0 & -\frac{EA}{L} & 0 \\ 0 & 0 & 0 & 0 \\ -\frac{EA}{L} & 0 & \frac{EA}{L} & 0 \\ 0 & 0 & 0 & 0 \end{bmatrix}
\tag{a}
$$

where we have added a row and a column of zeros in each of positions 2 and 4. Physically, this represents the fact that movements of the bar along coordinate directions 2 and 4 represent rigid-body motions which induce no stresses.

The mass matrix is expanded in the same manner as the stiffness matrix

$$
\mathbf{M}_u = \bar{m}L \begin{bmatrix} \frac{1}{3} & 0 & \frac{1}{6} & 0 \\ 0 & 0 & 0 & 0 \\ \frac{1}{6} & 0 & \frac{1}{3} & 0 \\ 0 & 0 & 0 & 0 \end{bmatrix}
\tag{b}
$$

The force vector becomes

$$
\frac{\bar{p}(t)L}{2} \begin{bmatrix} 1 \\ 0 \\ 1 \\ 0 \end{bmatrix}
\tag{c}
$$

The transformation between the $\mathbf{q}$ and $\mathbf{u}$ set of coordinates is easily shown to be

$$
\mathbf{u} = \begin{bmatrix} \cos\alpha & \sin\alpha & 0 & 0 \\ -\sin\alpha & \cos\alpha & 0 & 0 \\ 0 & 0 & \cos\alpha & \sin\alpha \\ 0 & 0 & -\sin\alpha & \cos\alpha \end{bmatrix} \mathbf{q}
\tag{d}
$$

or

$$
\mathbf{u} = \mathbf{Tq}
\tag{e}
$$

where $\mathbf{T}$ represents the transformation matrix. The transformed stiffness matrix is given by

$$
\mathbf{K}_q = \mathbf{T}^T \mathbf{K}_u \mathbf{T}
\tag{f}
$$

$$
= \frac{EA}{L} \begin{bmatrix} c^2 & cs & -c^2 & -cs \\ cs & s^2 & -cs & -s^2 \\ -c^2 & -cs & c^2 & cs \\ -cs & -s^2 & cs & s^2 \end{bmatrix}
\tag{g}
$$

where $c \equiv \cos \alpha$ and $s \equiv \sin \alpha$. The transformed mass matrix is obtained from

$$\mathbf{M}_q = \mathbf{T}^T \mathbf{M}_u \mathbf{T} \tag{h}$$

$$= \bar{m}L \begin{bmatrix} \frac{c^2}{3} & \frac{cs}{3} & \frac{c^2}{6} & \frac{cs}{6} \\ \frac{cs}{3} & \frac{s^2}{3} & \frac{cs}{6} & \frac{s^2}{6} \\ \frac{c^2}{6} & \frac{cs}{6} & \frac{c^2}{3} & \frac{cs}{3} \\ \frac{cs}{6} & \frac{s^2}{6} & \frac{cs}{3} & \frac{s^2}{3} \end{bmatrix} \tag{i}$$

and the transformed force vector is given by

$$\mathbf{p}_q = \mathbf{T}^T \mathbf{p}_u \tag{j}$$

$$= \frac{\bar{p}L}{2} \begin{bmatrix} c \\ s \\ c \\ s \end{bmatrix} \tag{k}$$

3.5 FINITE ELEMENT METHOD

The development of finite element method represents a major milestone in the field of applied mechanics. A large volume of literature now exists on the general formulation of the finite element method, as on its application to dynamic problems. In the following sections, we provide a brief discussion of the method. The method is equally applicable to assemblages of one-dimensional elements, planar systems, and general three-dimensional systems. However, for the purpose of illustration, we restrict our discussion to assemblages of one-dimensional elements.

In the finite element approach, the structural or the mechanical system being analyzed is divided into a number of subregions interconnected at a finite number of nodes. A selected number of nodal parameters are now defined at each of these nodes. These parameters consist of the displacements and their derivatives and serve as the generalized coordinates in the problem. Within each subregion, the displacements are expressed as the superposition of a set of shape functions of spatial coordinates, each function being weighted by a generalized coordinate. For example, consider the vibrations of the beam shown in Figure 3.14a. The beam has been divided into a number of elements of finite dimensions. There is a single node at each end of an element, and the elements are assumed to be interconnected at these nodes. At each node, we define two parameters: the lateral displacement and the slope. An isolated element of the beam is shown in Figure 3.14b in which the four nodal parameters, two displacements and two slopes, have been represented by the generalized coordinates u_1, u_2, u_3, and u_4. The lateral displacement within the element is

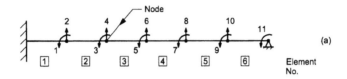

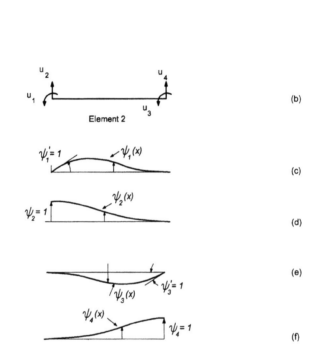

Figure 3.14. Finite element representation of a beam under lateral vibrations.

now given by

$$u(x,t) = u_1(t)\psi_1(x) + u_2(t)\psi_2(x) + u_3(t)\psi_3(t) + u_4(t)\psi(x) \qquad (3.84)$$

The shape functions used in Equation 3.84 are shown in Figure 3.14c through 3.14f. It will be noted that the shape function $\psi_i(x)$ or its first derivative has a magnitude of 1 along coordinate u_i, but is zero along all other coordinates. This selection for shape functions ensures that the nodal displacements and displacement derivatives obtained from Equation 3.84 are, in fact, equal to the corresponding nodal parameters chosen earlier. The representation of displacements within the element is entirely analogous to the Ritz shape formulation discussed in Section 3.3.4, except that the Ritz shapes now span only a sub-region of the structure. The element mass, stiffness, and damping matrices and the element load vector can therefore be developed by the procedures already discussed with reference to the Ritz vector representation.

At each interconnecting node, the elements joining it share their nodal parameters. Compatibility of displacements and displacement derivatives included in the nodal parameters list is therefore automatically ensured at the nodes. Thus, for the beam in Figure 3.14a, at any node, the two elements meeting at the node will have identical values for the lateral displacement and rotation. These displacement and rotation values serve as unknowns in the global equilibrium equations. In fact, an equilibrium equation is obtained corresponding to each unknown nodal parameter using a method described in the subsequent paragraphs.

Since the nodal parameters consisting of displacements and their spatial derivatives for elements interconnected at a node are required to be matched with each other, they must all be measured in a common system of coordinates. In a similar manner, in writing the equations of dynamic equilibrium, the nodal forces must first be resolved along the global coordinate system. On the other hand, for individual elements it may be most expedient to formulate the stiffness, damping, and mass matrices as well as the load vector in a local system of coordinates attached to the element. If the local system is different from the global system of coordinates, a transformation must be performed before the equations are formed.

The division of a system into several subregions or elements, the selection of shape functions and generalized coordinates, and the derivation of element property matrices and load vectors are illustrated in the following simple example.

Example 3.7
The simple pin-connected truss shown in Figure E3.7a has a pin support at A and a roller support at B. Each bar element of the truss has an axial rigidity EA and length L and a mass per unit

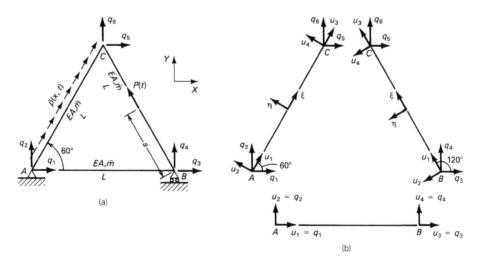

Figure E3.7. Vibrations of a pin-connected truss: (a) truss geometry, applied forces, and global coordinates; (b) local and global coordinates for individual members.

length $\bar{m}$. Member AC of the truss is acted upon by a uniformity distributed axial load $\bar{p}(x,t)$ per unit length. Member BC has an axial concentrated load $P(t)$ acting at a distance a from end B. The applied forces are shown in Figure E3.7a. Obtain the stiffness and mass matrices and the force vectors for individual members of the truss.

Solution
The division of the truss into component elements is obtained quite naturally in this case. The truss is composed of three bar elements, AC, BC, and AB, interconnected at nodes A, B, and C. The parameters associated with each node consist of two displacements along the X and Y coordinate directions. Thus, there are six nodal displacement parameters, corresponding to the three nodes of the truss as identified in Figure E3.7a.

Displacements within a member are expressed in terms of the nodal parameters pertaining to the end of the member. Take, for example, member AC. The nodal parameters at the end of the member are, in this case, q_1, q_2, q_5, and q_6. Displacements ξ and η as shown in Figure E3.7b therefore depend on q_1, q_2, q_5, and q_6. As stated earlier, it is usually more convenient to work with a local set of coordinate directions attached to the member. In the present case, we find parameters u_1, u_2, u_3, and u_4 to be the most convenient. Displacements ξ and η are therefore expressed in terms of u_1 through u_4 as

$$\xi = \psi_1 u_1 + \psi_2 u_3$$
$$\eta = \psi_1 u_2 + \psi_2 u_4$$

(a)

where ψ_1 and ψ_2 are functions of the spatial coordinate x given by

$$\psi_1 = 1 - \frac{x}{L}$$
$$\psi_2 = \frac{x}{L}$$

(b)

Note that the shape functions have been selected so that at A, $\psi_1 = 1$ and $\psi_2 = 0$, while at C, $\psi_1 = 0$ and $\psi_2 = 1$. As a result, Equation a signifies that the displacements at end A are u_1 and u_2, while those at the end C are u_3 and u_4. Correspondingly, along the global directions the displacements at A are q_1 and q_2, while those at the end C are q_5 and q_6. In a similar manner, displacements at the end C of member BC are q_5 and q_6, while those at end B of the same member are q_3 and q_4. By expressing the displacements in the members in terms of the nodal displacements as above, we ensure that at each node the displacements of all the members connected to the node are equal. Compatability of displacements is thus maintained throughout the truss.

Corresponding to the local coordinates u_1, u_2, u_3, and u_4, the stiffness matrix of member AC is equal to that given by Equation a of Example 3.6. Transformation of the matrix to coordinates **q** can be obtained by using Equations f and g of Example 3.6. Recognizing that $\alpha = 60°$, we get

$$\mathbf{K}^{AC} = \frac{EA}{L} \begin{bmatrix} \frac{1}{4} & \frac{\sqrt{3}}{4} & -\frac{1}{4} & -\frac{\sqrt{3}}{4} \\ \frac{\sqrt{3}}{4} & \frac{3}{4} & -\frac{\sqrt{3}}{4} & -\frac{3}{4} \\ -\frac{1}{4} & -\frac{\sqrt{3}}{4} & \frac{1}{4} & \frac{\sqrt{3}}{4} \\ -\frac{\sqrt{3}}{4} & -\frac{3}{4} & \frac{\sqrt{3}}{4} & \frac{3}{4} \end{bmatrix}$$

(c)

The transformed mass matrix is obtained from Equation i of Example 3.6.

$$\mathbf{M}^{AC} = \bar{m}L \begin{bmatrix} \frac{1}{12} & \frac{\sqrt{3}}{12} & \frac{1}{24} & \frac{\sqrt{3}}{24} \\ \frac{\sqrt{3}}{12} & \frac{3}{12} & \frac{\sqrt{3}}{24} & \frac{3}{24} \\ \frac{1}{24} & \frac{\sqrt{3}}{24} & \frac{1}{12} & \frac{\sqrt{3}}{12} \\ \frac{\sqrt{3}}{24} & \frac{3}{24} & \frac{\sqrt{3}}{12} & \frac{3}{12} \end{bmatrix} \tag{d}$$

The load vector, calculated from Equation k of Example 3.6, becomes

$$\mathbf{p}^{AC} = \frac{\bar{p}L}{2} \begin{bmatrix} \frac{1}{2} \\ \frac{\sqrt{3}}{2} \\ \frac{1}{2} \\ \frac{\sqrt{3}}{2} \end{bmatrix} \tag{e}$$

For member AB, the local coordinate directions coincide with global coordinate direction, hence $\alpha = 0$, and the stiffness and mass matrices are given, respectively, by Equations a and b of Example 3.6.

For member BC, $\alpha = 120°$ and the stiffness matrix becomes

$$\mathbf{K}^{BC} = \frac{EA}{L} \begin{bmatrix} \frac{1}{4} & -\frac{\sqrt{3}}{4} & -\frac{1}{4} & \frac{\sqrt{3}}{4} \\ -\frac{\sqrt{3}}{4} & \frac{3}{4} & \frac{\sqrt{3}}{4} & -\frac{3}{4} \\ -\frac{1}{4} & \frac{\sqrt{3}}{4} & \frac{1}{4} & -\frac{\sqrt{3}}{4} \\ \frac{\sqrt{3}}{4} & -\frac{3}{4} & -\frac{\sqrt{3}}{4} & \frac{3}{4} \end{bmatrix} \tag{f}$$

The mass matrix for member BC is given by

$$\mathbf{M}^{BC} = \bar{m}L \begin{bmatrix} \frac{1}{12} & -\frac{\sqrt{3}}{12} & \frac{1}{24} & -\frac{\sqrt{3}}{24} \\ -\frac{\sqrt{3}}{12} & \frac{3}{12} & -\frac{\sqrt{3}}{24} & \frac{3}{24} \\ \frac{1}{24} & -\frac{\sqrt{3}}{24} & \frac{1}{12} & -\frac{\sqrt{3}}{12} \\ -\frac{\sqrt{3}}{24} & \frac{3}{24} & -\frac{\sqrt{3}}{12} & \frac{3}{12} \end{bmatrix} \tag{g}$$

The load vector for member BC along its local coordinates is obtained by using the following virtual work equation

$$p_i = P(t)\psi_i(a) \tag{h}$$

Substitution for ψ_i from Equation b gives

$$\begin{aligned} p_1 &= P(t)\left(1 - \frac{a}{L}\right) \\ p_3 &= P(t)\frac{a}{L} \end{aligned} \tag{i}$$

Transformation to the global coordinate direction using Equation j of Example 3.6 yields

$$
\mathbf{p} = P(t) \begin{bmatrix} (1 - \frac{a}{L}) \cos \alpha \\ (1 - \frac{a}{L}) \sin \alpha \\ \frac{a}{L} \cos \alpha \\ \frac{a}{L} \sin \alpha \end{bmatrix}
\tag{j}
$$

On substituting $\alpha = 120°$ in Equation j, we get

$$
\mathbf{p} = P(t) \begin{bmatrix} -\frac{1}{2}(1 - \frac{a}{L}) \\ \frac{\sqrt{3}}{2}(1 - \frac{a}{L}) \\ -\frac{1}{2}\frac{a}{L} \\ \frac{\sqrt{3}}{2}\frac{a}{L} \end{bmatrix}
\tag{k}
$$

3.5.1   *Formulation of the equations of motion*

The formulation of the equations of motion in a finite element analysis can be viewed as a process of assembly of the element matrices into the corresponding global matrices. Consider, for example, the ith element in the finite element system. The nodal spring forces for the element are given by

$$
\mathbf{f}_S^i = \mathbf{K}^i \mathbf{q}^i
\tag{3.85}
$$

where all quantities are written in the transformed coordinate system. In Equation 3.85, $\mathbf{q}^i$ is the vector of element nodal parameters in the global coordinate system, $\mathbf{K}^i$ is the element stiffness matrix in the same system, and $\mathbf{f}_S^i$ is the element spring force vector, also in the global system.

Now, define a vector $\mathbf{q}$ which lists all unknown nodal parameters in the system. Included in the vector are the nodal parameters of element i, $\mathbf{q}^i$. Equation 3.85 can then be written in the alternative form

$$
\hat{\mathbf{f}}_S^i = \hat{\mathbf{K}}^i \mathbf{q}
\tag{3.86}
$$

Matrix $\hat{\mathbf{K}}^i$ is an expanded form of $\mathbf{K}^i$. The size of $\hat{\mathbf{K}}^i$ corresponds to the number of elements in $\mathbf{q}$; but at most, only those elements in it are nonzero that correspond to the parameters $\mathbf{q}^i$ included in $\mathbf{q}$. In a similar manner, $\hat{\mathbf{f}}_S^i$ is an expanded form of $\mathbf{f}_S^i$. Again, only those elements of $\hat{\mathbf{f}}_S^i$ are nonzero that correspond to $\mathbf{q}^i$.

Similar to Equation 3.86, an expression can be derived for the nodal inertia forces of element i

$$
\hat{\mathbf{f}}_I^i = \hat{\mathbf{M}}^i \ddot{\mathbf{q}}
\tag{3.87}
$$

The element nodal force vector is expanded in a similar manner and represented by $\hat{\mathbf{p}}^i$. Ignoring the damping forces for the time being, the global equations of equilibrium now become

$$\mathbf{f}_I + \mathbf{f}_S = \mathbf{p} \tag{3.88}$$

where

$$\mathbf{f}_I = \sum_i \hat{\mathbf{f}}_I^i$$

$$\mathbf{f}_S = \sum_i \hat{\mathbf{f}}_S^i \tag{3.89}$$

$$\mathbf{p} = \sum_i \hat{\mathbf{p}}^i$$

and $\hat{\mathbf{p}}^i$ is the expanded form of element nodal force vector. It may be noted that in actual computer implementation, it is not necessary to obtain the expanded form of element matrices. Instead, individual terms in the element matrices can be directly added to a global matrix by noting the correspondence between the global degrees of freedom of the structure and the transformed local degrees of freedom of an element.

On using Equations 3.89, 3.86 and 3.87, we can write Equation 3.88 in the alternative form

$$\mathbf{M}\ddot{\mathbf{q}} + \mathbf{K}\mathbf{q} = \mathbf{p} \tag{3.90}$$

where

$$\mathbf{M} = \sum_i \hat{\mathbf{M}}^i$$

$$\mathbf{K} = \sum_i \hat{\mathbf{K}}^i \tag{3.91}$$

Damping can easily be incorporated in the equations provided that damping matrices have been formulated at the element level. In a similar manner, axial force effects can be included once geometric stiffness matrices have been formulated at the element level.

Before Equation 3.90 is solved, appropriate boundary conditions must be applied and any external forces acting directly at the nodes must be included in the equilibrium equations. Boundary conditions are incorporated by specifying that the nodal parameters restrained by such conditions be set as zero. This is done by omitting the corresponding rows and columns in the global equilibrium equations. The procedure for the formulation of equations of motion and the application of boundary conditions is illustrated in the following example.

Example 3.8
The truss of Example 3.7 is subjected to dynamic forces $F(t)$ and $W(t)$ at node C as shown in Figure E3.8, in addition to the distributed load $\bar{p}(x,t)$ and concentrated load $P(t)$ shown in Figure E3.7a. Obtain the equations of motion for the truss assuming that damping is negligible.

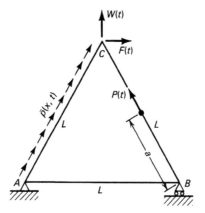

Figure E3.8. Vibrations of a pin-connected truss; applied nodal and member forces.

Solution
Corresponding to the six nodal displacements shown in Figure E3.8, six equations of dynamic equilibrium will be obtained. These equation can be expressed as

$$\mathbf{f}_I + \mathbf{f}_S = \mathbf{p} + \mathbf{p}_e \tag{a}$$

where $\mathbf{f}_I$, $\mathbf{f}_S$, and $\mathbf{p}$ are obtained by superposing the contributions from individual elements according to Equation 3.89 and $\mathbf{p}_e$ is the vector of loads applied directly at the nodes. To obtain the components of $\mathbf{f}_I$ and $\mathbf{f}_S$, we use Equations 3.87 and 3.86, respectively. The first step is to define a vector of all nodal parameters. This is given by

$$\mathbf{q}^T = [\, q_1 \quad q_2 \quad q_3 \quad q_4 \quad q_5 \quad q_6 \,] \tag{b}$$

Next, we consider in turn each element of the truss. Take, for example, member AC. For this member, only the displacements q_1, q_2, q_5, and q_6 are effective. The expanded stiffness matrix for the member is therefore given by

$$\hat{\mathbf{K}}^{AC} = \frac{EA}{L} \begin{bmatrix} \frac{1}{4} & \frac{\sqrt{3}}{4} & 0 & 0 & -\frac{1}{4} & -\frac{\sqrt{3}}{4} \\ \frac{\sqrt{3}}{4} & \frac{3}{4} & 0 & 0 & -\frac{\sqrt{3}}{4} & -\frac{3}{4} \\ 0 & 0 & 0 & 0 & 0 & 0 \\ 0 & 0 & 0 & 0 & 0 & 0 \\ -\frac{1}{4} & -\frac{\sqrt{3}}{4} & 0 & 0 & \frac{1}{4} & \frac{\sqrt{3}}{4} \\ -\frac{\sqrt{3}}{4} & -\frac{3}{4} & 0 & 0 & \frac{\sqrt{3}}{4} & \frac{3}{4} \end{bmatrix} \tag{c}$$

Rows and columns 3 and 4 of the expanded matrix are zero because displacements along those directions do not affect the member. In a similar manner, the expanded mass matrix of the element becomes

$$\hat{\mathbf{M}}^{AC} = \bar{m}L \begin{bmatrix} \frac{1}{12} & \frac{\sqrt{3}}{12} & 0 & 0 & \frac{1}{24} & \frac{\sqrt{3}}{24} \\ \frac{\sqrt{3}}{12} & \frac{1}{4} & 0 & 0 & \frac{\sqrt{3}}{24} & \frac{1}{8} \\ 0 & 0 & 0 & 0 & 0 & 0 \\ 0 & 0 & 0 & 0 & 0 & 0 \\ \frac{1}{24} & \frac{\sqrt{3}}{24} & 0 & 0 & \frac{1}{12} & \frac{\sqrt{3}}{12} \\ \frac{\sqrt{3}}{24} & \frac{1}{8} & 0 & 0 & \frac{\sqrt{3}}{12} & \frac{1}{4} \end{bmatrix} \qquad (d)$$

The load vector is given by

$$\hat{\mathbf{p}}^{AC} = \frac{\bar{p}L}{2} \begin{bmatrix} \frac{1}{2} \\ \frac{\sqrt{3}}{2} \\ 0 \\ 0 \\ \frac{1}{2} \\ \frac{\sqrt{3}}{2} \end{bmatrix} \qquad (e)$$

The expanded stiffness matrix for member *BC* is obtained by noting that nodal displacements q_1 and q_2 have no effect on the member, so that rows and columns 1 and 2 of the matrix will be zero.

$$\hat{\mathbf{K}}^{BC} = \frac{EA}{L} \begin{bmatrix} 0 & 0 & 0 & 0 & 0 & 0 \\ 0 & 0 & 0 & 0 & 0 & 0 \\ 0 & 0 & \frac{1}{4} & -\frac{\sqrt{3}}{4} & -\frac{1}{4} & \frac{\sqrt{3}}{4} \\ 0 & 0 & -\frac{\sqrt{3}}{4} & \frac{3}{4} & \frac{\sqrt{3}}{4} & -\frac{3}{4} \\ 0 & 0 & -\frac{1}{4} & \frac{\sqrt{3}}{4} & \frac{1}{4} & -\frac{\sqrt{3}}{4} \\ 0 & 0 & \frac{\sqrt{3}}{4} & -\frac{3}{4} & -\frac{\sqrt{3}}{4} & \frac{3}{4} \end{bmatrix} \qquad (f)$$

In a similar manner, the mass matrix is obtained as

$$\hat{\mathbf{M}}^{BC} = \bar{m}L \begin{bmatrix} 0 & 0 & 0 & 0 & 0 & 0 \\ 0 & 0 & 0 & 0 & 0 & 0 \\ 0 & 0 & \frac{1}{12} & -\frac{\sqrt{3}}{12} & \frac{1}{24} & -\frac{\sqrt{3}}{24} \\ 0 & 0 & -\frac{\sqrt{3}}{12} & \frac{1}{4} & -\frac{\sqrt{3}}{24} & \frac{1}{8} \\ 0 & 0 & \frac{1}{24} & -\frac{\sqrt{3}}{24} & \frac{1}{12} & -\frac{\sqrt{3}}{12} \\ 0 & 0 & -\frac{\sqrt{3}}{24} & \frac{1}{8} & -\frac{\sqrt{3}}{12} & \frac{1}{4} \end{bmatrix} \qquad (g)$$

while the load vector is given by

$$
\hat{\mathbf{p}}^{BC} = P(t) \begin{bmatrix} 0 \\ 0 \\ -\frac{1}{2}(1 - \frac{a}{L}) \\ \frac{\sqrt{3}}{2}(1 - \frac{a}{L}) \\ -\frac{1}{2}\frac{a}{L} \\ \frac{\sqrt{3}}{2}\frac{a}{L} \end{bmatrix} \tag{h}
$$

For member AB, displacements q_5 and q_6 have no effect; hence the expanded stiffness and mass matrices are given by

$$
\hat{\mathbf{K}}^{AB} = \frac{EA}{L} \begin{bmatrix} 1 & 0 & -1 & 0 & 0 & 0 \\ 0 & 0 & 0 & 0 & 0 & 0 \\ -1 & 0 & 1 & 0 & 0 & 0 \\ 0 & 0 & 0 & 0 & 0 & 0 \\ 0 & 0 & 0 & 0 & 0 & 0 \\ 0 & 0 & 0 & 0 & 0 & 0 \end{bmatrix} \tag{i}
$$

$$
\hat{\mathbf{M}}^{AB} = \bar{m}L \begin{bmatrix} \frac{1}{3} & 0 & \frac{1}{6} & 0 & 0 & 0 \\ 0 & 0 & 0 & 0 & 0 & 0 \\ \frac{1}{6} & 0 & \frac{1}{3} & 0 & 0 & 0 \\ 0 & 0 & 0 & 0 & 0 & 0 \\ 0 & 0 & 0 & 0 & 0 & 0 \\ 0 & 0 & 0 & 0 & 0 & 0 \end{bmatrix} \tag{j}
$$

The member stiffness and mass matrices are now assembled to form the global matrices, giving

$$
\mathbf{K} = \frac{EA}{L} \begin{bmatrix} \frac{5}{4} & \frac{\sqrt{3}}{4} & -1 & 0 & -\frac{1}{4} & -\frac{\sqrt{3}}{4} \\ \frac{\sqrt{3}}{4} & \frac{3}{4} & 0 & 0 & -\frac{\sqrt{3}}{4} & -\frac{3}{4} \\ -1 & 0 & \frac{5}{4} & -\frac{\sqrt{3}}{4} & -\frac{1}{4} & \frac{\sqrt{3}}{4} \\ 0 & 0 & -\frac{\sqrt{3}}{4} & \frac{3}{4} & \frac{\sqrt{3}}{4} & -\frac{3}{4} \\ -\frac{1}{4} & -\frac{\sqrt{3}}{4} & -\frac{1}{4} & \frac{\sqrt{3}}{4} & \frac{1}{2} & 0 \\ -\frac{\sqrt{3}}{4} & -\frac{3}{4} & \frac{\sqrt{3}}{4} & -\frac{3}{4} & 0 & \frac{3}{2} \end{bmatrix} \tag{k}
$$

$$\mathbf{M} = \bar{m}L \begin{bmatrix} \frac{5}{12} & \frac{\sqrt{3}}{12} & \frac{1}{6} & 0 & \frac{1}{24} & \frac{\sqrt{3}}{24} \\ \frac{\sqrt{3}}{12} & \frac{1}{4} & 0 & 0 & \frac{\sqrt{3}}{24} & \frac{1}{8} \\ \frac{1}{6} & 0 & \frac{5}{12} & -\frac{\sqrt{3}}{12} & \frac{1}{24} & -\frac{\sqrt{3}}{24} \\ 0 & 0 & -\frac{\sqrt{3}}{12} & \frac{3}{12} & -\frac{\sqrt{3}}{24} & \frac{1}{8} \\ \frac{1}{24} & \frac{\sqrt{3}}{24} & \frac{1}{24} & -\frac{\sqrt{3}}{24} & \frac{1}{6} & 0 \\ \frac{\sqrt{3}}{24} & \frac{1}{8} & -\frac{\sqrt{3}}{24} & \frac{1}{8} & 0 & \frac{1}{2} \end{bmatrix} \tag{1}$$

Assembly of the load vectors gives

$$\mathbf{p} = \begin{bmatrix} \frac{\bar{p}L}{4} \\ \frac{\sqrt{3}\bar{p}L}{4} \\ -\frac{P}{2}\left(1 - \frac{a}{L}\right) \\ \frac{P\sqrt{3}}{2}\left(1 - \frac{a}{L}\right) \\ \frac{\bar{p}L}{4} - \frac{Pa}{2L} \\ \frac{\bar{p}L\sqrt{3}}{4} + \frac{P\sqrt{3}}{2}\frac{a}{L} \end{bmatrix} \tag{m}$$

The vector of loads applied to the nodes is given by

$$\mathbf{p}_e = \begin{bmatrix} 0 \\ 0 \\ 0 \\ 0 \\ F \\ W \end{bmatrix} \tag{n}$$

The equations of motion can now be written as

$$\mathbf{M}\ddot{\mathbf{q}} + \mathbf{K}\mathbf{q} = \mathbf{p} + \mathbf{p}_e \tag{o}$$

Boundary conditions must be applied before these equations can be solved. Supports at A and B require that displacements q_1, q_2, and q_4 be zero. These conditions can be directly incorporated in Equation o by omitting rows and columns 1, 2, and 4 from the matrices $\mathbf{M}$ and $\mathbf{K}$ and rows 1, 2, and 4 from vectors $\mathbf{q}$, $\ddot{\mathbf{q}}$, $\mathbf{p}$, and $\mathbf{p}_e$. This will leave the following three equations in the unknown displacements q_3, q_5, and q_6

$$\bar{m}L \begin{bmatrix} \frac{5}{12} & \frac{1}{24} & -\frac{\sqrt{3}}{24} \\ \frac{1}{24} & \frac{1}{6} & 0 \\ -\frac{\sqrt{3}}{24} & 0 & \frac{1}{2} \end{bmatrix} \begin{bmatrix} \ddot{q}_3 \\ \ddot{q}_5 \\ \ddot{q}_6 \end{bmatrix} + \frac{EA}{L} \begin{bmatrix} \frac{5}{4} & -\frac{1}{4} & \frac{\sqrt{3}}{4} \\ -\frac{1}{4} & \frac{1}{2} & 0 \\ \frac{\sqrt{3}}{4} & 0 & \frac{3}{2} \end{bmatrix} \begin{bmatrix} q_3 \\ q_5 \\ q_6 \end{bmatrix}$$

$$= \begin{bmatrix} -\frac{P}{2}\left(1 - \frac{a}{L}\right) \\ \frac{\bar{p}L}{4} - \frac{Pa}{2L} \\ \frac{\bar{p}L\sqrt{3}}{4} + \frac{P\sqrt{3}}{2}\frac{a}{L} \end{bmatrix} + \begin{bmatrix} 0 \\ F \\ W \end{bmatrix} \tag{p}$$

3.5.2   *Selection of shape functions*

From the discussion presented in foregoing paragraphs, it is apparent that as in the case of the Ritz method, a finite element formulation relies on the representation of displaced shape by a superposition of selected shape functions. Unlike in the Ritz method, however, where the shape functions span the entire domain, the Ritz shapes of the finite element method span only the subregions. For a complete equivalence between the two methods, displacements as well as their derivatives up to order $m - 1$ should be continuous over the entire domain when the highest-order derivative included in the virtual work expression is of order m. In addition, the shape functions must satisfy the boundary condition. Because in a finite element formulation, the nodal parameters of element that are interconnected to each other are assumed to have the same values, such nodal parameters are automatically continuous at the nodes. At a minimum, therefore, the nodal parameters should include displacement derivatives of order up to $m - 1$. However, elements may be interconnected not just at the common nodes but also along adjoining boundary. A C^{m-1} continuity should also be satisfied along such interelement boundaries and ideally, the shape functions should be selected in such a manner as to fulfill this condition. In practice, it has been found that convergence can be achieved even when some of the continuity requirements are violated, and *nonconforming elements* in which the shape functions do not satisfy all of the continuity requirements have often been used with success. The shape functions must satisfy an additional requirement for convergence; they must permit a constant strain state, including zero strain, to be achieved. When shape functions consist of polynomials, which is invariably the case, this requirement is easily satisfied.

As an example of the continuity requirements, consider the truss of Examples 3.7 and 3.8. The virtual work equation for the axial vibrations of an individual element of the truss contains the first-order derivative of the displacement. Therefore, at a minimum, displacement continuity should be maintained throughout the structure. The selected shape functions ensure that displacements are continuous within an element. Further, by choosing the displacements in two orthogonal directions as the nodal parameters, displacement continuity is also maintained across the elements meeting at a node. In Equation a of Example 3.4, because the shape functions do not include powers of x higher than 1, the derivative of the displacement u, and hence the strain, is always constant throughout the element. For the special case when $u_1 = u_2 = c$, where c is a constant representing a rigid body displacement of the element, $u = c$ and the strain is zero throughout the element. In addition, the boundary conditions for the truss are explicitly satisfied by restraining the appropriate displacements. The procedure outlined in the examples therefore satisfies all the conditions that must be satisfied for convergence. In fact, for uniform bar elements, the linear shape functions selected also satisfy the differential equation of the element; hence the solution obtained is exact. The effectiveness of the solution procedure, however, lies in the fact that even when the same shape functions are used for nonuniform bars, the results obtained are still reasonably accurate.

3.5.3 *Advantages of the finite element method*

The advantages of the finite element formulation can now be summarized as follows:
1. The continuity requirements for ensuring convergence are easily satisfied by including appropriate nodal parameters in the formulation. The boundary conditions are satisfied simply by setting the restrained nodal parameters to zero.
2. Variation of material or geometric properties across the structure poses no special problem. Since the structure is subdivided into elements, individual elements can be selected so as to span only that region of the structure across which the geometric and material properties do not vary substantially.
3. The accuracy of the solution can be improved by increasing the number of elements and hence the number of generalized coordinates.
4. Since the shape functions consist of simple polynomials, integration of mathematical expressions involving the shape functions and their derivatives is straightforward.
5. The global property matrices obtained in a finite element analysis are always sparse, and with appropriate numbering of the degrees of freedom, it can be ensured that they are also banded. As a result, the method is computationally efficient. In fact, the property matrices for elements that are similar need be obtained only once, so when many elements are similar, the computational efficiency is further improved.
6. The generalized coordinates in the solution represent displacements and their derivatives at the nodes of the elements, and therefore, have a physical meaning.

3.6 FINITE ELEMENT FORMULATION OF THE FLEXURAL VIBRATIONS OF A BEAM

The essential steps involved in the application of finite element method to a dynamic problem can be further illustrated by considering the lateral vibrations of a beam. If the beam is of nonuniform section and is subjected to complex loading conditions, a straightforward Ritz approach may not be successful. In such a case, as shown in Figure 3.15a, we divide the beam into a number of subelements which are themselves flexural members, although of a much smaller length. The individual elements are connected to each other only at nodes that are common to them and because continuity among nodal

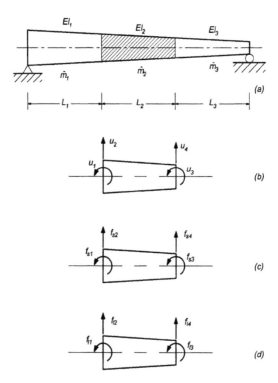

Figure 3.15. (a) Finite element representation of a beam under lateral vibrations; (b) element coordinates; (c) spring forces; (d) inertia forces.

parameters is automatically satisfied, the displacement and its derivatives included in the nodal parameters become continuous throughout the length of the beam. Since the virtual work equation for the flexural vibrations of a beam contains a derivative of order 2, we must include in our nodal parameters the displacement as well as its first derivative, the slope.

Based on the foregoing, we assign four nodal degrees of freedom to each element: two end rotations and two end translations as shown in Figure 3.15b. Within each element, the lateral displacement u is expressed as the superposition of four shape functions each weighted by a nodal parameter. We can use as our shape functions, the polynomials derived in Example 3.3 and restated below for ease of reference.

$$\psi_1(x) = x\left(1 - \frac{x}{L}\right)^2$$

$$\psi_2(x) = 1 - 3\left(\frac{x}{L}\right)^2 + 2\left(\frac{x}{L}\right)^3$$

$$\psi_3(x) = \frac{x^2}{L}\left(\frac{x}{L} - 1\right)$$

$$\psi_4(x) = 3\left(\frac{x}{L}\right)^2 - 2\left(\frac{x}{L}\right)^3 \tag{3.92}$$

The cubic polynomials of Equation 3.92 are called *Hermitian polynomials*. They satisfy the differential equation for the displacement of a uniform beam. We may, however, use Equation 3.92 to represent the shape functions even for a nonuniform beam. The error involved in doing this will be small provided that a large number of elements is used. The accuracy of the solution can, in fact, always be increased by using a finer division.

3.6.1   *Stiffness matrix of a beam element*

The stiffness matrix of a beam element corresponding to the four degrees of freedom shown in Figure 3.15b can be derived by the virtual work method, as explained in Section 3.3.4 and Example 3.3. The resulting matrix for the *i*th element is given by

$$\mathbf{K}^i = \begin{bmatrix} \frac{4EI}{L} & \frac{6EI}{L^2} & \frac{2EI}{L} & -\frac{6EI}{L^2} \\ \frac{6EI}{L^2} & \frac{12EI}{L^3} & \frac{6EI}{L^2} & -\frac{12EI}{L^3} \\ \frac{2EI}{L} & \frac{6EI}{L^2} & \frac{4EI}{L} & -\frac{6EI}{L^2} \\ -\frac{6EI}{L^2} & -\frac{12EI}{L^3} & -\frac{6EI}{L^2} & \frac{12EI}{L^3} \end{bmatrix} \tag{3.93}$$

where EI_i is the average flexural rigidity of the *i*th element and L_i is its length. For the sake of clarity, subscript *i* on the flexural rigidity and length have been omitted from Equation 3.93. The stiffness matrix given by Equation 3.93 is exact for a beam of uniform cross section.

If the unknown nodal displacements are represented by

$$(\mathbf{u}^i)^T = [u_1 \quad u_2 \quad u_3 \quad u_4] \tag{3.94}$$

the nodal forces are obtained from

$$\mathbf{f}_S^i = \mathbf{K}^i \mathbf{u}^i \tag{3.95}$$

These forces are shown in Figure 3.15c.

3.6.2   *Mass matrix of a beam element*

Consistent mass matrix
The mass matrix of a beam element can be obtained by using Equation 3.58. When the shape functions used in Equation 3.58 are the same as those used

in deriving the stiffness matrix, the resulting matrix is called *consistent mass matrix*. It can easily be shown that for a uniform beam, the shape functions given by Equation 3.92 will lead to the following mass matrix for the *i*th element

$$
\mathbf{M}^i = \frac{\bar{m}L}{420}
\begin{bmatrix}
4L^2 & 22L & -3L^2 & 13L \\
22L & 156 & -13L & 54 \\
-3L^2 & -13L & 4L^2 & -22L \\
13L & 54 & -22L & 156
\end{bmatrix}
\tag{3.96}
$$

where $\bar{m}_i$ is the average mass per unit length of the *i*th element and L_i is its length. Again, for the sake of clarity, subscripts i have been omitted from Equation 3.96. The nodal inertia forces are obtained from

$$
\mathbf{f}_I^i = \mathbf{M}^i \ddot{\mathbf{u}}^i
\tag{3.97}
$$

and are shown in Figure 3.15d.

Lumped mass matrix
Comparison of Equations 3.52 and 3.57 shows that while the expression for virtual work done by spring forces involves the second derivative of the displacement shape function, the expression for the virtual work of inertia forces uses only the shape function itself. This suggests that for finding the virtual work of inertia forces in an element, it may be possible to use different shape functions that are of lower order than those used for finding the virtual work of spring forces. This has been found to be a practical alternative in the finite element formulation; and although the formulation no longer remains equivalent to a Ritz analysis, a reasonable approximations to the true solution is still obtained. For the flexural vibration of a beam, for example, we may use the following shape functions

$$
\begin{aligned}
\psi_1 &= 0 \\
\psi_2 &= 1 - \frac{x}{L} \\
\psi_3 &= 0 \\
\psi_4 &= \frac{x}{L}
\end{aligned}
\tag{3.98}
$$

The resulting mass matrix becomes

$$\mathbf{M}^i = \bar{m}L \begin{bmatrix} 0 & 0 & 0 & 0 \\ 0 & \frac{1}{3} & 0 & \frac{1}{6} \\ 0 & 0 & 0 & 0 \\ 0 & \frac{1}{6} & 0 & \frac{1}{3} \end{bmatrix} \tag{3.99}$$

The use of shape functions defined by Equation 3.98 has eliminated the rotational degrees of freedom and, in fact, implies that the kinetic energy component associated with the rotational degrees of freedom is negligible. This is a reasonable assumption for lateral vibration of beams. As we shall see later, when rotational degrees of freedom are absent from the mass matrix, they can be eliminated from the entire formulation by a process of static condensation of the matrices. The resulting problem is of a smaller size than the original, and is thus easier to solve.

The mass matrix of an element can, however, be further simplified by assuming that the distributed mass of the element can be lumped as point masses along the translational degrees of freedom at the ends. For example, if the mass of the beam element, considered uniform, is $\bar{m}$ per unit length a point mass of magnitude $(\bar{m}L)/2$ will be assigned to each end. The resulting inertia forces will be $(\bar{m}L)/2 \times \ddot{u}_2$ and $(\bar{m}L)/2 \times \ddot{u}_4$, where u_2 and u_4 are the two translational degrees of freedom. It is readily seen that this is equivalent to assuming that the left half of the beam is vibrating with an acceleration $\ddot{u}_2$, while the right half is vibrating with an acceleration $\ddot{u}_4$. The same results will thus be obtained by using the following shape functions

$$\psi_1 = 0$$

$$\psi_2 = \begin{cases} 1 & x \le \frac{L}{2} \\ 0 & x > \frac{L}{2} \end{cases}$$

$$\psi_3 = 0$$

$$\psi_4 = \begin{cases} 0 & x < \frac{L}{2} \\ 1 & x \ge \frac{L}{2} \end{cases} \tag{3.100}$$

The resulting mass matrix is given by

$$\mathbf{M}^i = \bar{m}L \begin{bmatrix} 0 & 0 & 0 & 0 \\ 0 & \frac{1}{2} & 0 & 0 \\ 0 & 0 & 0 & 0 \\ 0 & 0 & 0 & \frac{1}{2} \end{bmatrix} \tag{3.101}$$

Now, not only have the rotational degrees of freedom been eliminated, but the mass matrix is diagonal. The matrix of Equation 3.101 is called a *lumped mass matrix*. Use of a lumped mass matrix in place of a consistent mass matrix results in considerable savings in computation costs. This is because 1. the rotational degrees of freedom have been eliminated, and 2. the mass matrix is diagonal.

3.6.3   *Nodal applied force vector for a beam element*

Consistent load vector
The vector of equivalent nodal applied forces can be obtained by using Equation 3.56. Again, if the shape functions used in Equation 3.56 are the same as those used in obtaining the virtual work due to spring forces, the resulting vector of nodal forces is called a *consistent load vector*. For a beam subjected to a uniformly distributed load $\bar{p}$ per unit length, the nodal applied force vector obtained by using shape functions of Equation 3.92 can be shown to be

$$(\mathbf{p}^i)^T = \left[\frac{\bar{p}L^2}{12} \quad \frac{\bar{p}L}{2} \quad -\frac{\bar{p}L^2}{12} \quad +\frac{\bar{p}L}{2} \right] \tag{3.102}$$

Statically equivalent load vector
As in the case of virtual work of inertia forces, the virtual work expression for applied forces involves only the displacement shape functions, and not their derivatives. Lower-order functions may therefore be employed in forming these expressions. For a beam element, linear functions of Equation 3.98 are commonly used. If the beam element is loaded with a uniformly distributed load of $\bar{p}$ per unit length, use of shape functions given by Equation 3.98, will lead to the following nodal applied force vector.

$$(\mathbf{p}^i)^T = \left[0 \quad \frac{\bar{p}L}{2} \quad 0 \quad \frac{\bar{p}L}{2} \right] \tag{3.103}$$

It will be noted that the nodal loads given by Equation 3.103 are statically equivalent to a uniform load acting on simply supported beam. Statically equivalent nodal loads may also be used for other types of loading, and although such loads may be derived by forming the equations of static equilibrium, use of the virtual work method along with the shape functions of Equation 3.98 is often more convenient.

3.6.4   *Geometric stiffness matrix for a beam element*

Consistent geometric stiffness matrix
The geometric stiffness matrix is derived by using Equation 3.72. When the shape functions used in that equation are the same as those used in deriving the elastic stiffness matrix, the resulting geometric stiffness matrix is said to

be consistent. For a constant axial force S, the geometric stiffness matrix for element i will work out to

$$\mathbf{K}_G^i = \frac{S}{L}\begin{bmatrix} \frac{2}{15}L^2 & \frac{1}{10}L & -\frac{1}{30}L^2 & -\frac{1}{10}L \\[2mm] \frac{1}{10}L & \frac{6}{5} & \frac{1}{10}L & -\frac{6}{5} \\[2mm] -\frac{1}{30}L^2 & \frac{1}{10}L & \frac{2}{15}L^2 & -\frac{1}{10}L \\[2mm] -\frac{1}{10}L & -\frac{6}{5} & -\frac{1}{10}L & \frac{6}{5} \end{bmatrix} \tag{3.104}$$

in which S_i is the axial force in the element. For the sake of clarity, subscripts i on S and L have been omitted in Equation 3.104.

Linear geometric stiffness matrix

As in the case of a mass matrix, we may neglect the virtual work done by the axial forces on displacements induced by rotational degrees of freedom and use the linear shape functions given by Equation 3.98 in deriving the geometric stiffness matrix. For a constant axial force, the geometric stiffness matrix can be easily shown to be

$$\mathbf{K}_G = \frac{S}{L}\begin{bmatrix} 0 & 0 & 0 & 0 \\ 0 & 1 & 0 & -1 \\ 0 & 0 & 0 & 0 \\ 0 & -1 & 0 & 1 \end{bmatrix} \tag{3.105}$$

Example 3.9

The simply supported beam shown in Figure E3.9a and b is vibrating in the lateral direction. Section AB of the beam has a uniform depth, but its width varies linearly from a minimum of b at B to $2b$ at A. Section BC has a constant cross section. The moment of inertia of section BC and the minimum moment of inertia of section AB are both equal to I. Use two finite elements AB and BC, to represent the beam and obtain the equations of motion for its lateral vibrations under the loads indicated in the figure.

Solution

The two finite elements as well as their nodal coordinates are shown in Figure E3.9d. Section AB is of nonuniform section. However, we may still use the shape functions derived for a uniform section (Eq. 3.92) to represent the displacement. The corresponding stiffness matrix is obtained by using Equation 3.54, in which the moment of inertia, I_x, is variable and is given by

$$I_x = \frac{1}{12}bd^3\left(2 - \frac{x}{L}\right)$$

$$= I\left(2 - \frac{x}{L}\right)$$

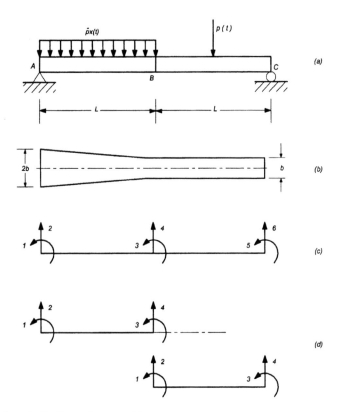

Figure E3.9. Lateral vibrations of a non-uniform beam: (a) longitudinal elevation; (b) plan; (c) global coordinates; (d) local coordinates.

On carrying out the necessary integrations, the stiffness matrix works out to

$$\mathbf{K}^1 = \frac{EI}{L} \begin{bmatrix} 7 & \frac{10}{L} & 3 & -\frac{10}{L} \\ \frac{10}{L} & \frac{18}{L^2} & \frac{8}{L} & -\frac{18}{L^2} \\ 3 & \frac{8}{L} & 5 & -\frac{8}{L} \\ -\frac{10}{L} & -\frac{18}{L^2} & -\frac{8}{L} & \frac{18}{L^2} \end{bmatrix} \qquad (a)$$

The global coordinates are shown in Figure E3.9c and the global displacement vector is given by

$$\mathbf{q}^T = [\, q_1 \quad q_2 \quad q_3 \quad q_4 \quad q_5 \quad q_6 \,] \qquad (b)$$

We note that the global displacement vector contains the local vector for element *AB*, so that

$$
\begin{aligned}
u_1^1 &= q_1 \\
u_2^1 &= q_2 \\
u_3^1 &= q_3 \\
u_4^1 &= q_4
\end{aligned}
\qquad (c)
$$

In this case, since both the local and the global vectors are measured in the same set of Cartesian coordinates, no transformation of coordinates is required. We now expand $\mathbf{K}^l$ to correspond to the dimension of $\mathbf{q}$. Taking account of relationships (Eq. c) the expanded matrix is seen to be

$$\hat{\mathbf{K}}^l = \frac{EI}{L}\begin{bmatrix} 7 & \frac{10}{L} & 3 & -\frac{10}{L} & 0 & 0 \\ \frac{10}{L} & \frac{18}{L^2} & \frac{8}{L} & -\frac{18}{L^2} & 0 & 0 \\ 3 & \frac{8}{L} & 5 & -\frac{8}{L} & 0 & 0 \\ -\frac{10}{L} & -\frac{18}{L^2} & -\frac{8}{L} & \frac{18}{L^2} & 0 & 0 \\ 0 & 0 & 0 & 0 & 0 & 0 \\ 0 & 0 & 0 & 0 & 0 & 0 \end{bmatrix} \tag{d}$$

The consistent mass matrix for element AB is obtained from Equation 3.58, where $\bar{m}(x)$ is now a variable given by

$$\bar{m}(x) = m_0\left(2 - \frac{x}{L}\right)$$

m_0 is the mass per unit length at section B, and ψ_i are the shape functions of Equation 3.92. Mass matrix $\mathbf{M}$ is obtained by carrying out the necessary integrations. When the resulting matrix is expanded to the size of $\mathbf{q}$, we get

$$\hat{\mathbf{M}}^l = \frac{m_0 L}{840}\begin{bmatrix} 13L^2 & 284L & -9L^2 & 38L & 0 & 0 \\ 284L & 552 & -40L & 162 & 0 & 0 \\ -9L^2 & -40L & 11L^2 & -58L & 0 & 0 \\ 38L & 162 & -58L & 384 & 0 & 0 \\ 0 & 0 & 0 & 0 & 0 & 0 \\ 0 & 0 & 0 & 0 & 0 & 0 \end{bmatrix} \tag{e}$$

If a lumped mass matrix is to be used instead, it can be derived by lumping the mass of the left half of the element at A and that of the right half at B. The lumped mass at A is given by

$$m_A = \int_0^{L/2} m_0(2 - xL)dx$$

$$= \frac{7m_0 L}{8}$$

In a similar manner, the mass to be lumped at B is given by

$$m_B = \int_{L/2}^{L} m_0\left(2 - \frac{x}{L}\right)dx$$

$$= \frac{5M_0 L}{8}$$

The lumped mass matrix thus works out to

$$\mathbf{M}^1 = \frac{m_0 L}{8} \begin{bmatrix} 0 & 0 & 0 & 0 \\ 0 & 7 & 0 & 0 \\ 0 & 0 & 0 & 0 \\ 0 & 0 & 0 & 5 \end{bmatrix} \tag{f}$$

The consistent load vector for section AB is obtained from Equation 3.56. When expanded to the size of $\mathbf{q}$, it becomes

$$[\hat{\mathbf{p}}^{(1)}]^T = \begin{bmatrix} \frac{\bar{p}L^2}{12} & \frac{\bar{p}L}{2} & -\frac{\bar{p}L^2}{12} & \frac{\bar{p}L}{2} & 0 & 0 \end{bmatrix} \tag{g}$$

If an equivalent static load vector is to be used instead of a consistent load vector, then

$$[\hat{\mathbf{p}}^{(1)}]^T = \begin{bmatrix} 0 & \frac{\bar{p}L}{2} & 0 & \frac{\bar{p}L}{2} & 0 & 0 \end{bmatrix} \tag{h}$$

The shape functions of Equation 3.92 are used again in deriving the property matrices of element BC, which is of a uniform section. The stiffness matrix for the element is the same as in Equation 3.93. Noting that

$$u_1^{(2)} = q_3$$

$$u_2^{(2)} = q_4$$

$$u_3^{(2)} = q_5$$

$$u_4^{(2)} = q_6$$

we may expand matrix $\mathbf{K}^{(2)}$ to the size of $\mathbf{q}$

$$\hat{\mathbf{K}}^{(2)} = \frac{EI}{L} \begin{bmatrix} 0 & 0 & 0 & 0 & 0 & 0 \\ 0 & 0 & 0 & 0 & 0 & 0 \\ 0 & 0 & 4 & \frac{6}{L} & 2 & -\frac{6}{L} \\ 0 & 0 & \frac{6}{L} & \frac{12}{L^2} & \frac{6}{L} & -\frac{12}{L^2} \\ 0 & 0 & 2 & \frac{6}{L} & 4 & -\frac{6}{L} \\ 0 & 0 & -\frac{6}{L} & -\frac{12}{L^2} & -\frac{6}{L} & \frac{12}{L^2} \end{bmatrix} \tag{i}$$

The consistent mass matrix for BC is given by Equation 3.96; when expanded to the size of $\mathbf{q}$, it becomes

$$\hat{\mathbf{M}}^{(2)} = \frac{m_0 L}{420} \begin{bmatrix} 0 & 0 & 0 & 0 & 0 & 0 \\ 0 & 0 & 0 & 0 & 0 & 0 \\ 0 & 0 & 4L^2 & 22L & -3L^2 & 13L \\ 0 & 0 & 22L & 156 & -13L & 54 \\ 0 & 0 & -3L^2 & -13L & 4L^2 & -22L \\ 0 & 0 & 13L & 54 & -22L & 156 \end{bmatrix} \tag{j}$$

The lumped mass matrix is the same as Equation 3.101.

The consistent load vector for a central concentrated load is given by

$$p_i = \psi_i \left(\frac{L}{2} \right) P \tag{k}$$

where ψ_i is the ith shape function as in Equation 3.92. Evaluation of Equation k gives

$$[\mathbf{p}^{(2)}]^T = \begin{bmatrix} \frac{PL}{8} & \frac{P}{2} & -\frac{PL}{8} & \frac{P}{2} \end{bmatrix} \tag{l}$$

On expanding the load vector to the size of $\mathbf{q}$, we get

$$[\hat{\mathbf{p}}^{(2)}]^T = \begin{bmatrix} 0 & 0 & \frac{PL}{8} & \frac{P}{2} & -\frac{PL}{8} & \frac{P}{2} \end{bmatrix} \tag{m}$$

On the other hand, the equivalent static load vector is given by

$$[\hat{\mathbf{p}}^{(2)}]^T = \begin{bmatrix} 0 & 0 & 0 & \frac{P}{2} & 0 & \frac{P}{2} \end{bmatrix} \tag{n}$$

The global stiffness matrix is now obtained by assembling the element stiffness matrices

$$
\mathbf{K} = \sum_{i=1}^{2} \hat{\mathbf{K}}^i
$$

$$
= \frac{EI}{L}
\begin{bmatrix}
7 & \frac{10}{L} & 3 & -\frac{10}{L} & 0 & 0 \\
\frac{10}{L} & \frac{18}{L^2} & \frac{8}{L} & -\frac{18}{L^2} & 0 & 0 \\
3 & \frac{8}{L} & 9 & -\frac{2}{L} & 2 & -\frac{6}{L} \\
-\frac{10}{L} & -\frac{18}{L} & -\frac{2}{L} & \frac{30}{L^2} & \frac{6}{L} & -\frac{12}{L^2} \\
0 & 0 & 2 & \frac{6}{L} & 4 & -\frac{6}{L} \\
0 & 0 & -\frac{6}{L} & -\frac{12}{L^2} & -\frac{6}{L} & \frac{12}{L^2}
\end{bmatrix}
\tag{o}
$$

The global mass matrix can be obtained by summing the consistent mass matrices of the elements

$$
\mathbf{M} = \sum_{i=1}^{2} \hat{\mathbf{M}}^i
$$

$$
= \frac{m_0 L}{840}
\begin{bmatrix}
13L^2 & 284L & -9L^2 & 38L & 0 & 0 \\
284L & 552 & -40L & 162 & 0 & 0 \\
-9L^2 & -40L & 19L^2 & -14L & -6L^2 & 26L \\
38L & 162 & -14L & 696 & -26L & 108 \\
0 & 0 & -6L^2 & -26L & 8L^2 & -44L \\
0 & 0 & 26L & 108 & -44L & 312
\end{bmatrix}
\tag{p}
$$

As an alternative, we may use the lumped mass matrices of the elements, in which case

$$\mathbf{M} = \begin{bmatrix} 0 & 0 & 0 & 0 & 0 & 0 \\ 0 & \frac{7m_0L}{8} & 0 & 0 & 0 & 0 \\ 0 & 0 & 0 & 0 & 0 & 0 \\ 0 & 0 & 0 & \frac{9m_0L}{8} & 0 & 0 \\ 0 & 0 & 0 & 0 & 0 & 0 \\ 0 & 0 & 0 & 0 & 0 & \frac{m_0L}{2} \end{bmatrix} \tag{q}$$

The consistent applied force vector is given by

$$\mathbf{P} = \sum_{i=1}^{2} \hat{\mathbf{p}}^i$$

where $\hat{\mathbf{p}}^{(1)}$ is as in Equation g and $\hat{\mathbf{p}}^{(2)}$ is as in Equation m, so that

$$\mathbf{p}^T = \left[\frac{\bar{p}L^2}{12} \quad \frac{\bar{p}L}{2} \quad \left(\frac{PL}{8} - \frac{\bar{p}L^2}{12} \right) \quad \left(\frac{P}{2} + \frac{\bar{p}L}{2} \right) \quad -\frac{PL}{8} \quad \frac{P}{2} \right] \tag{r}$$

The equivalent static load vector is obtained by superposing Equations h and n so that

$$\mathbf{p}^T = \left[0 \quad \frac{\bar{p}L}{2} \quad 0 \quad \left(\frac{P}{2} + \frac{\bar{p}L}{2} \right) \quad 0 \quad \frac{P}{2} \right] \tag{s}$$

Before writing the equations of motion, we must apply the boundary conditions and include in the formulation, loads that act directly at the nodes. Simple supports at A and C require that the displacements q_2 and q_6 be zero. The only external loads applied directly to the joints are the unknown reactions at A and C. Denoting these reactions by A_2 and A_6 and using a consistent formulation, the equations of motion become

$$\mathbf{M}\ddot{\mathbf{q}} + \mathbf{K}\mathbf{q} = \mathbf{p} \tag{t}$$

where $\mathbf{M}$ and $\mathbf{K}$ are given by Equations p and o, respectively, and

$$\mathbf{q}^T = [\, q_1 \quad 0 \quad q_3 \quad q_4 \quad q_5 \quad 0 \,]$$

$$\mathbf{p}^T = \left[\frac{\bar{p}L^2}{12} \quad \left(A_2 + \frac{\bar{p}L}{2} \right) \quad \left(\frac{PL}{8} - \frac{\bar{p}L^2}{12} \right) \quad \left(\frac{P}{2} + \frac{\bar{p}L}{2} \right) \quad -\frac{PL}{8} \quad \left(\frac{P}{2} + A_6 \right) \right]$$

By deleting the second and the sixth rows and columns from matrices $\mathbf{M}$ and $\mathbf{K}$ as well as the second and sixth row from vectors $\mathbf{q}$, $\ddot{\mathbf{q}}$, and $\mathbf{p}$, we obtain four simultaneous differential equations in the unknowns q_1, q_3, q_4, and q_5.

Example 3.10
The roof structure shown in Figure E3.10 consists of a circular reinforced concrete slab 6 m in diameter and 100 mm thick supported by a steel pipe column 3.5 m in height. The mass of the concrete slab is 6500 kg, and the column has a moment of inertia of 25×10^6 mm^4. Neglecting the mass of the column, obtain the equations of motion (a) ignoring the axial load effect, and (b) taking axial load into account.

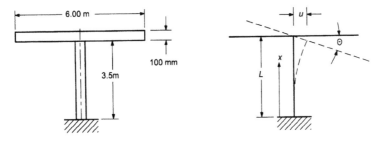

Figure E3.10. Vibrations of a cantilever roof structure.

Solution
(a) The system has two degrees of freedom. As shown in the figure, the two can be selected as the tip deflection u and the rotation at the tip, θ. For a uniform column, the deflection u_x can be represented in terms of u and θ as

$$u_x = \theta\psi_3 + u\psi_4$$

where ψ_3 and ψ_4 are identical to those given by Equation 3.92. To determine the mass matrix, we note that

$$m_{11} = I_0\{\psi_3'(L)\}^2$$

where I_0 is the mass moment of inertia of the circular slab about its diameter. Since $I_0 = mR^2/4$

$$m_{11} = \frac{mR^2}{4}$$
$$= 14.625 \text{ tm}^2$$

Also

$$m_{22} = m\{\psi_4(L)\}^2$$
$$= m = 6.5 \text{ t}$$

and

$$m_{12} = m_{21} = 0$$

The mass matrix is therefore given by

$$\begin{bmatrix} 14.625 & 0 \\ 0 & 6.500 \end{bmatrix} \tag{a}$$

The stiffness matrix is derived from Equation 3.54 and can be obtained from Equation 3.93 by deleting the first and the second rows and columns.

$$\mathbf{K} = \begin{bmatrix} \frac{4EI}{L} & -\frac{6EI}{L^2} \\ -\frac{6EI}{L^2} & \frac{12EI}{L^3} \end{bmatrix} \tag{b}$$

On substituting the values of E, I, and L and adjusting the units to kN and meters, we get

$$\mathbf{K} = \begin{bmatrix} 5714.3 & -2449.0 \\ -2449.0 & 1399.4 \end{bmatrix} \tag{c}$$

When the axial force effect is neglected the equations of motion become

$$\begin{bmatrix} 14.625 & 0 \\ 0 & 6.500 \end{bmatrix} \begin{bmatrix} \ddot{\theta} \\ \ddot{u} \end{bmatrix} + \begin{bmatrix} 5714.3 & -2449.0 \\ -2449.0 & 1399.4 \end{bmatrix} \begin{bmatrix} \theta \\ u \end{bmatrix} = \begin{bmatrix} 0 \\ 0 \end{bmatrix} \tag{d}$$

(b) To take the axial force effect into account, we need to calculate the geometric stiffness matrix. A consistent matrix is obtained from Equation 3.104 after deleting rows and columns 1 and 2. The resulting matrix is given by

$$\mathbf{K}_G = \frac{S}{L} \begin{bmatrix} \frac{2}{15}L^2 & -\frac{1}{10}L \\ -\frac{1}{10}L & \frac{6}{5} \end{bmatrix} \tag{e}$$

The axial load is $S = 6.5 \times 9.81 = 63.8$ kN Substituting for S and L in Equation e, we get

$$\mathbf{K}_G = \begin{bmatrix} 29.75 & -6.38 \\ -6.38 & 21.88 \end{bmatrix} \tag{f}$$

The equations of motion are now given by

$$\mathbf{M}\ddot{\mathbf{q}} + (\mathbf{K} - \mathbf{K}_G)\mathbf{q} = 0$$

$$\begin{bmatrix} 14.625 & 0 \\ 0 & 6.500 \end{bmatrix} \begin{bmatrix} \ddot{\theta} \\ \ddot{u} \end{bmatrix} + \begin{bmatrix} 5684.00 & -2442.62 \\ -2442.62 & 1377.52 \end{bmatrix} \begin{bmatrix} \theta \\ u \end{bmatrix} = \begin{bmatrix} 0 \\ 0 \end{bmatrix} \tag{g}$$

3.7   STATIC CONDENSATION OF STIFFNESS MATRIX

As discussed in the finite element formulation of the equations of motion, the rotational degrees of freedom can be eliminated from the mass matrix because the kinetic energy component corresponding to these degrees of freedom is generally negligible in comparison to that corresponding to the translational degrees of freedom. Thus, if there are no other loads acting in the direction of the rotational degrees of freedom, the spring forces along these degrees of freedom should also be zero. This condition allows us to eliminate the rotational coordinates from the stiffness matrix as well, permitting a significant reduction in the size of the problem. The reduced problem usually is much easier and less expensive to solve.

For example, if the displacement vector is partitioned so that $\mathbf{u}_t$ represents the translational degrees of freedom and $\mathbf{u}_\theta$ represents the rotational degrees of freedom, and the stiffness and mass matrices are also partitioned accordingly, the equations of motion can be written as

$$\begin{bmatrix} \mathbf{M}_{tt} & 0 \\ 0 & 0 \end{bmatrix} \begin{bmatrix} \ddot{\mathbf{u}}_t \\ \ddot{\mathbf{u}}_\theta \end{bmatrix} + \begin{bmatrix} \mathbf{K}_{tt} & \mathbf{K}_{t0} \\ \mathbf{K}_{0t} & \mathbf{K}_{00} \end{bmatrix} \begin{bmatrix} \mathbf{u}_t \\ \mathbf{u}_\theta \end{bmatrix} = \begin{bmatrix} \mathbf{p}_t \\ 0 \end{bmatrix} \tag{3.106}$$

where it is assumed that the rotational degrees of freedom have not been included in the mass formulation and that no applied loads act along such degrees of freedom. Expansion of Equation 3.106 gives

$$\mathbf{M}_{tt}\ddot{\mathbf{u}}_t + \mathbf{K}_{tt}\mathbf{u}_t + \mathbf{K}_{t\theta}\mathbf{u}_\theta = \mathbf{p}_t \tag{3.107}$$

$$\mathbf{K}_{\theta t}\mathbf{u}_t + \mathbf{K}_{\theta\theta}\mathbf{u}_\theta = 0 \tag{3.108}$$

Equation 3.108 leads to

$$\mathbf{u}_\theta = -\mathbf{K}_{\theta\theta}^{-1}\mathbf{K}_{\theta t}\mathbf{u}_t \tag{3.109}$$

Substitution of Equation 3.109 in Equation 3.107 gives

$$\mathbf{M}_{tt}\ddot{\mathbf{u}}_t + \mathbf{K}^*\mathbf{u}_t = \mathbf{p}_t \tag{3.110a}$$

where

$$\mathbf{K}^* = \mathbf{K}_{tt} - \mathbf{K}_{t\theta}\mathbf{K}_{\theta\theta}^{-1}\mathbf{K}_{\theta t} \tag{3.110b}$$

Matrix K^* is called the *condensed stiffness matrix*. The reduced problem of Equation 3.110a is generally significantly smaller in size than the original problem represented by Equation 3.106.

Example 3.11
The stiffness matrix for a simply supported beam of Figure 3.10a is initially formulated in the four degrees of freedom shown in Figure E3.11a. These include two rotational degrees of freedom indicated in the figure. The mass matrix (Eq. 3.27) and the load vector (Eq. 3.35 and 3.36), however, include only the translational degrees of freedom. Obtain first the stiffness matrix corresponding to the four degrees of freedom shown in Figure E3.11a, and then eliminate the two rotational degrees of freedom to obtain a reduced stiffness matrix.

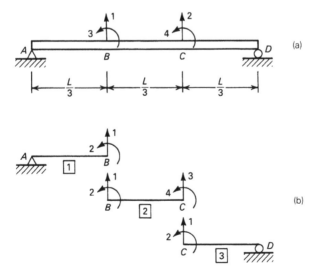

Figure E3.11. Vibrations of a simply supported beam: matrix condensation.

Solution

Stiffness matrices of the three sections of the beam: *AB*, *BC*, and *CD* corresponding to the local coordinates shown in Figure E3.11b are obtained by standard methods of structural analysis and are given by

$$
\mathbf{K}^{AB} = EI \begin{array}{c} (1) \\ (3) \end{array}
\begin{pmatrix}
\overset{(1)}{\frac{81}{L^3}} & \overset{(3)}{-\frac{27}{L^2}} \\[2mm]
-\frac{27}{L^2} & \frac{9}{L}
\end{pmatrix}
\tag{a}
$$

$$
\mathbf{K}^{BC} = EI \begin{array}{c} (1) \\ (3) \\ (2) \\ (4) \end{array}
\begin{pmatrix}
\overset{(1)}{\frac{324}{L^3}} & \overset{(3)}{\frac{54}{L^2}} & \overset{(2)}{-\frac{324}{L^3}} & \overset{(4)}{\frac{54}{L^2}} \\[2mm]
\frac{54}{L^2} & \frac{12}{L} & -\frac{54}{L^2} & \frac{6}{L} \\[2mm]
-\frac{324}{L^3} & -\frac{54}{L^2} & \frac{324}{L^3} & -\frac{54}{L^2} \\[2mm]
\frac{54}{L^2} & \frac{6}{L} & -\frac{54}{L^2} & \frac{12}{L}
\end{pmatrix}
\tag{b}
$$

$$
\mathbf{K}^{CD} = EI \begin{array}{c} (2) \\ (4) \end{array}
\begin{pmatrix}
\overset{(2)}{\frac{81}{L^3}} & \overset{(4)}{\frac{27}{L^2}} \\[2mm]
\frac{27}{L^2} & \frac{9}{L}
\end{pmatrix}
\tag{c}
$$

In Equations a, b, and c, the positions that individual elements of the given matrices will occupy after assembly in the global stiffness matrix have been indicated by specifying the global matrix row and column numbers that correspond to the rows and columns of the component matrices. This information is easily obtained by noting the correspondence between the element and global degrees of freedom. For example, in Equation b, element $k_{23} = -54/L^2$ will be placed in row 3 and column 2 of the global matrix. The 4×4 global stiffness matrix is now obtained by direct stiffness assembly, giving

$$
\mathbf{K} = EI \begin{bmatrix}
\frac{405}{L^3} & -\frac{324}{L^3} & \frac{27}{L^2} & \frac{54}{L^2} \\[2mm]
-\frac{324}{L^3} & \frac{405}{L^3} & -\frac{54}{L^2} & -\frac{27}{L^2} \\[2mm]
\frac{27}{L^2} & -\frac{54}{L^2} & \frac{21}{L} & \frac{6}{L} \\[2mm]
\frac{54}{L^2} & -\frac{27}{L^2} & \frac{6}{L} & \frac{21}{L}
\end{bmatrix}
\tag{d}
$$

Partitioning the stiffness matrix Equation d along translational and rotational degrees of freedom, we get

$$
\mathbf{K}_{tt} = EI \begin{bmatrix}
\frac{405}{L^3} & -\frac{324}{L^3} \\[2mm]
-\frac{324}{L^3} & \frac{405}{L^3}
\end{bmatrix}
$$

$$
\mathbf{K}_{t\theta} = \mathbf{K}_{\theta t}^T = EI \begin{bmatrix}
\frac{27}{L^2} & \frac{54}{L^2} \\[2mm]
-\frac{54}{L^2} & -\frac{27}{L^2}
\end{bmatrix}
\tag{e}
$$

$$
\mathbf{K}_{\theta\theta} = EI \begin{bmatrix}
\frac{21}{L} & \frac{6}{L} \\[2mm]
\frac{6}{L} & \frac{21}{L}
\end{bmatrix}
$$

Substitution into Equation 3.110b gives the condensed stiffness matrix

$$\mathbf{K}^* = \frac{324EI}{5L^3} \begin{bmatrix} 4 & -\frac{7}{2} \\ -\frac{7}{2} & 4 \end{bmatrix} \tag{f}$$

The stiffness matrix in Equation f is the same as that obtained in Equation 3.31 by first forming the flexibility matrix corresponding to the two translational degrees of freedom and then taking its inverse. In practice, the latter approach in which the stiffness matrix is obtained as an inverse of the flexibility matrix is usually simpler and computationally more efficient than direct static condensation using Equation 3.110b, particularly when the size of the reduced matrix is significantly smaller than that of the original matrix.

3.8 APPLICATION OF THE RITZ METHOD TO DISCRETE SYSTEMS

In Section 3.3.4, we described the application of the Ritz method to the representation of a distributed parameter system having an infinite number of degrees of freedom by a system having a finite number of degrees of freedom. The Ritz procedure can also be used to represent the displaced shape of a discrete multi-degree-of-freedom system by the superposition of a few appropriately selected shape functions. Analogous to Equation 3.51, we have

$$\mathbf{u} = \sum_{i=1}^{M} z_i \boldsymbol{\psi}_i \tag{3.111}$$

where $\mathbf{u}$ is the displacement shape vector of the original N-degree-of-freedom system, $\boldsymbol{\psi}_i$ are shape function vectors, and z_i are generalized coordinates. We note that instead of being continuous function of spatial coordinates, $\boldsymbol{\psi}_i$'s are now vectors of size N. In matrix notation, Equation 3.111 can be expressed as

$$\mathbf{u} = \boldsymbol{\Psi} \mathbf{z} \tag{3.112}$$

where $\boldsymbol{\Psi}$ is an N by M matrix of shape function vectors and $\mathbf{z}$ is a vector of size M.

When the number of shape vectors, M, is equal to N, Equation 3.112 is entirely equivalent to the standard transformation equation (Eq. 3.74). Vectors $\boldsymbol{\Psi}$ are in this case, said to span the entire N-dimensional space and the solution of transformed equations will lead to the same response as the original equations. As stated earlier, transformation of this type may be useful if the transformed equations are easier to solve. In general, however, the usefulness of the Ritz procedure is most evident when only a few shape function vectors are adequate to represent the response. In such a case, M is much smaller than N. Transformation equations 3.80 through 3.83 are still applicable, except that the transformation matrix $\boldsymbol{\Psi}$ is rectangular of size N by M and the transformed property matrices are of size M by M. Thus

$$\tilde{\mathbf{M}} = \mathbf{\Psi}^T \mathbf{M} \mathbf{\Psi} \tag{3.113}$$

$$\tilde{\mathbf{K}} = \mathbf{\Psi}^T \mathbf{K} \mathbf{\Psi} \tag{3.114}$$

$$\tilde{\mathbf{C}} = \mathbf{\Psi}^T \mathbf{C} \mathbf{\Psi} \tag{3.115}$$

$$\tilde{\mathbf{p}} = \mathbf{\Psi}^T \mathbf{p} \tag{3.116}$$

and the transformed equations of motion are

$$\tilde{\mathbf{M}} \ddot{z} + \tilde{\mathbf{C}} \dot{z} + \tilde{\mathbf{K}} z = \tilde{\mathbf{p}} \tag{3.117}$$

Equations 3.117 are of order M and because $M \ll N$, the reduced problem is of a significantly smaller size than the original.

Example 3.12
The four-story shear-type building frame shown in Figure E3.12 has a concentrated mass m at each story·level. The total column stiffness for each story is k. The frame is subjected to a lateral dynamic force of $p_0 \sin \Omega t$ applied at story level 4. Using the following two shape function vectors, obtain the equations of motion of a reduced system of order 2.

$$\psi_1 = \begin{bmatrix} 0.250 \\ 0.500 \\ 0.750 \\ 1.000 \end{bmatrix} \qquad \psi_2 = \begin{bmatrix} -0.643 \\ -0.929 \\ -0.500 \\ 1.000 \end{bmatrix}$$

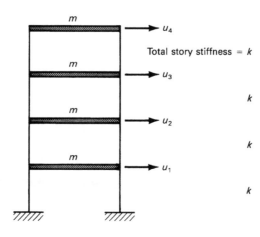

Figure E3.12. Lateral vibrations of a shear frame.

Solution

Corresponding to the four degrees of freedom shown in Figure E3.12, the mass and stiffness matrices are given by the following equations

$$\mathbf{M} = m \begin{bmatrix} 1 & 0 & 0 & 0 \\ 0 & 1 & 0 & 0 \\ 0 & 0 & 1 & 0 \\ 0 & 0 & 0 & 1 \end{bmatrix} \tag{a}$$

$$\mathbf{K} = k \begin{bmatrix} 2 & -1 & 0 & 0 \\ -1 & 2 & -1 & 0 \\ 0 & -1 & 2 & -1 \\ 0 & 0 & -1 & 1 \end{bmatrix} \tag{b}$$

The applied force vector is given by

$$\mathbf{p}^T = p_0 \sin \Omega t [0 \quad 0 \quad 0 \quad 1] \tag{c}$$

The transformation matrix $\boldsymbol{\Psi}$ is obtained from the given values of $\boldsymbol{\psi}_1$ and $\boldsymbol{\psi}_2$, so that

$$\boldsymbol{\Psi} = \begin{bmatrix} 0.250 & -0.643 \\ 0.500 & -0.929 \\ 0.750 & -0.500 \\ 1.000 & 1.000 \end{bmatrix} \tag{d}$$

The transformed mass and stiffness matrices are now obtained from Equations 3.113 and 3.114, respectively

$$\tilde{\mathbf{M}} = \begin{bmatrix} 1.875 & 0 \\ 0 & 2.526 \end{bmatrix} \tag{e}$$

$$\tilde{\mathbf{K}} = \begin{bmatrix} 0.250 & 0.250 \\ 0.250 & 2.929 \end{bmatrix} \tag{f}$$

The transformed applied force vector obtained from Equation 3.116 becomes

$$\tilde{\mathbf{p}} = \begin{bmatrix} 1 \\ 1 \end{bmatrix} p_0 \sin \Omega t \tag{g}$$

It may be of interest to note that the Ritz vectors used in this case leave the mass matrix diagonal but do not diagonalize the stiffness matrix. This is because the procedure that was used to obtain the Ritz vectors shown here orthogonalizes the vectors produced with respect to the mass matrix. We discuss this procedure in a later chapter.

SELECTED READINGS

Clough, R.W. 1971. Analysis of Structural Vibrations and Dynamic Response. In R.H. Gallagher, Y. Yamada & J.T. Oden (eds) *Recent Advances in Matrix Methods of*

Structural Analysis and Design: 441–486. Huntsville, Ala.: University of Alabama Press.

Clough, R.W. & Bathe, K.J. 1972. Finite Element Analysis of Dynamic Response. *2nd US-Japan Seminar on Matrix Methods of Structural Analysis and Design, Advances in Computational Methods in Structural Mechanics and Design*: Huntsville, Ala.: University of Alabama Press.

Clough, R.W. & Penzien, J. 1993. *Dynamics of Structures*. New York: McGraw-Hill. 2nd Edition.

Meirovitch, L. 1967. *Analytical Methods in Vibrations*. London: Macmillan.

Warburton, G.B. 1976. *The Dynamical Behavior of Structures*. Oxford, U.K.: Pergamon Press. 2nd Edition.

Warburton, G.B. 1979. Response Using the Rayleigh Ritz Method. *Earthquake Engineering and Structural Dynamics*, Vol. 7: 327–334.

Weaver, W. Jr & Johnston, P.R. 1987. *Structural Dynamics by Finite Elements*. Englewood Cliffs: Prentice Hall.

Zienkiewicz, O.C. 1988. *The Finite Element Method*. London: McGraw-Hill. 4th Edition.

PROBLEMS

3.1 An automobile and its suspension system are idealized as shown in Figure P3.1. The vehicle body along with its payload is considered to be a rigid body with total weight W, and radius of gyration r for pitching motion about an axis perpendicular to the paper and passing through the centroid. The suspension system supporting each axle is represented by a spring of stiffness k_S and a viscous damper with damping coefficient c_S. Weights w

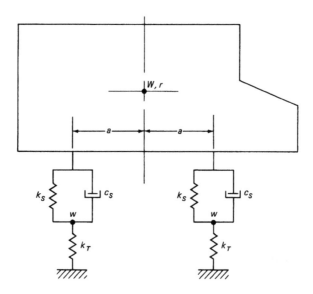

Figure P3.1.

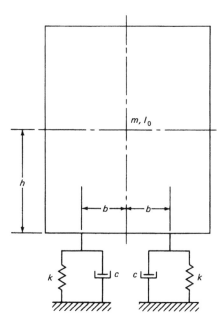

Figure P3.2.

applied to the top of front and rear tires represent the weight of chassis and axles. The stiffness of a pair of tires is k_T. The vehicle undergoes vibrations involving bounce and pitch. Identify the degrees of freedom of the system and obtain the equations of motion.

3.2 The rocking behavior of certain types of structures under the action of earthquake motion can be studied by analyzing the model shown in Figure P3.2. The structure is represented by a block of mass m and mass moment of inertia about its centroid I_0. The foundation soil is modeled by a pair of springs each of stiffness k and a pair of viscous dampers of damping constants c. Assuming that the structure does not slide laterally with respect to the foundation or lift off its foundation, obtain the equations of motion for a ground acceleration $\ddot{v}_g$ in the vertical direction.

3.3 A platform consisting of precast concrete planks is supported by three steel frames as shown in Figure P3.3. Frames A and B, which are identical, run parallel to the X direction. Frame C is parallel to the Y direction. The lateral stiffness of each of frames A and B is k_1 while that of frame C is k_2. Obtain the equations of motion for free vibrations of the deck in terms of the three coordinates defined at the center of mass as shown in Figure P3.3.

3.4 Figure P3.4 shows the plan views of a small two story building. The structure consists of three frames A, B and C oriented as shown. The stiffness matrices of the frames, defined along the coordinate directions

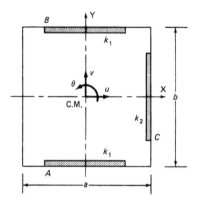

Figure P3.3.

indicated, are given below. The floor weight is uniformly distributed and
has a magnitude of $100 \, \text{lb}/\text{ft}^2$. The building is subjected to a time varying
uniform pressure $p_0 \sin \Omega t \, \text{lb}/\text{ft}^2$ on its face XY. Obtain the equations of
motion governing the response of the building in terms of coordinates θ_1,
θ_2, θ_3 and θ_4.

3.5 An industrial crane of span 80 ft runs along two crane girders which are
supported on a span of 20 ft (Fig. 3.5). The mass of the crane girders and
rails may be neglected. The mass of the crane itself may be assumed to be
lumped at the center of its span and at the two ends as shown. Obtain the
equations of free vibrations of the crane and the hoist trolley, assuming
that the trolley is at the center of the span and does not carry any load
and that the crane is in the middle of the crane girder. The following data
is supplied.

Moment of inertia of crane girder, $I_1 = 4090 \, \text{in.}^4$
Moment of inertia of crane structure, $I_2 = 36000 \, \text{in.}^4$
Total weight of the crane $= 55000 \, \text{lb}$
Weight of hoist trolley $= 7500 \, \text{lb}$
$E = 30000 \, \text{ksi}$

3.6 Obtain the equations of motion for the free vibrations of the system shown
in Figure P3.6.

3.7 The double pendulum shown in Figure P3.7 consists of two simple pen-
dulums connected by a spring of stiffness k attached at a distance a from
the point of suspension. The length of each pendulum is l and the pendulum masses are m_1 and m_2, respectively. Obtain the equations of motion
for small vibrations of the double pendulum.

3.8 Obtain the equations of motion for the frame in Figure P3.8, which is
vibrating under the action of a vertical load P(t) acting as shown.

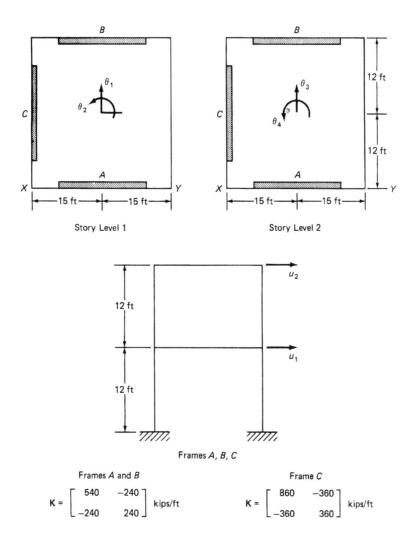

Figure P3.4.

$$K = \begin{bmatrix} 540 & -240 \\ -240 & 240 \end{bmatrix} \text{ kips/ft}$$

Frames A and B

$$K = \begin{bmatrix} 860 & -360 \\ -360 & 360 \end{bmatrix} \text{ kips/ft}$$

Frame C

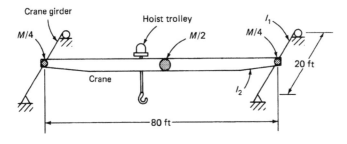

Figure P3.5.

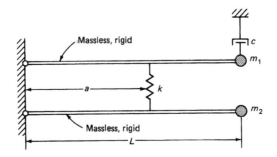

Figure P3.6.

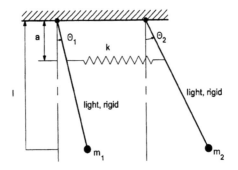

Figure P3.7.

3.9 The chimney referred to in Problem 2.9 is vibrating under the action of a lateral ground motion having an acceleration $\ddot{u}_g(t)$. The vibrations of the chimney can be represented by superposition of the following two mode shapes:

$$\psi_1 = 1 - \cos \frac{\pi x}{2L}$$

$$\psi_2 = 1 - \cos \frac{3\pi x}{2L}$$

Obtain the equations of motion: (a) ignoring the gravity load effect; (b) taking gravity load effect into account. Assume that damping is negligible.

3.10 Obtain the equations of motion for free vibrations of the haunched beam shown in Figure P3.10 by dividing its length into three segments and applying the finite element method. Assume that in the haunched portion, the flexural rigidity varies linearly from $3EI$ to EI while the mass per unit length varies linearly from $2m$ to m.

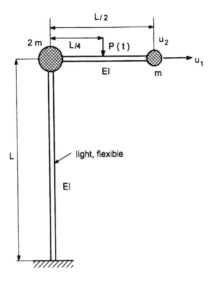

Figure P3.8.

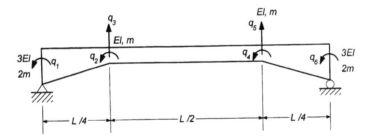

Figure P3.10.

3.11 A uniform slab is supported by three identical frames as shown in Figure P3.11. Derive the stiffness matrix of an individual frame with reference to the three degrees of freedom indicated in the figure. Assume that the beam as well as the columns are axially rigid. Obtain the lateral stiffens of a frame by static condensation of the stiffness matrix to eliminate degrees of freedom u_1 and u_2. Then, obtain the equations of motion for free vibrations of the slab-frame system in terms of the three degrees-of-freedom defined about the mass center.

The weight of the slab including the superimposed dead load is $100\,\text{lb/ft}^2$.

3.12 A vehicle traveling across a bridge deck is represented by an unsprung mass m_t connected to a sprung mass m_v through a spring of stiffness k and a damper of coefficient c (Figure P3.12). The bridge deck has a span L, a uniform mass $\bar{m}$ per unit length and flexural rigidity EI. The vehicle

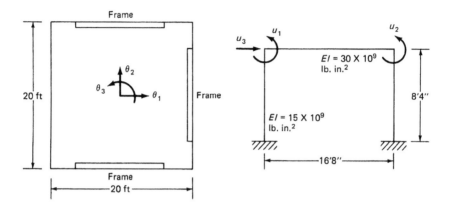

Figure P3.11.

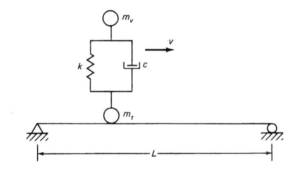

Figure P3.12.

is traveling across the deck at a constant speed of v. If the vibration shape of the deck is $\psi(x) = \sin(\pi x/L)$, obtain the equations of motion for the system assuming that the vehicle damping c and unsprung mass m_t are negligible.

3.13 Repeat Problem 3.12 taking vehicle damping c and unsprung mass m_t into account.

3.14 Obtain the equations of motion for free vibrations of the system of Problem 2.4 assuming that the wheel is braked to its axle.

3.15 Derive the equations of free vibration for the concrete plane frame ABC shown in Figure P3.15. The moment of inertia of leg AB may be assumed to vary linearly from 10×10^6 mm^4 at A to twice that value at B. The mass of AB also varies linearly from 69 kg/m at A to 138 kg/m at B. Beam BC has a uniform cross-section with a moment of inertia 20×10^6 mm^4 and a mass of 138 kg/m. In addition, BC supports a superimposed dead load of 6 kN/m. Assume that both members AB and BC are axially rigid. $E = 20,000$ MPa.

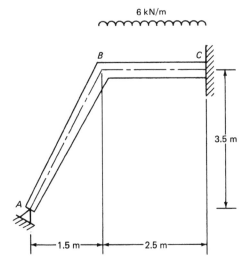

Figure P3.15.

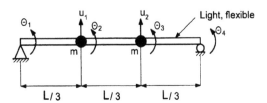

Figure P3.16.

3.16 A simply supported beam of uniform cross-section is represented by the lumped mass model shown in Figure P3.16. Obtain the stiffness matrix of the beam for the six degrees-of-freedom shown in the figure. Eliminate the rotational degrees-of-freedom by a static condensation of the stiffness matrix to obtain a two by two matrix, then derive the equations of motion for the free vibrations of the beam with reference to the two translational degrees-of-freedom. The flexural rigidity of the beam is EI.

CHAPTER 4

Principles of analytical mechanics

4.1 INTRODUCTION

The development of the science of mechanics has relied on two different sets of fundamental principles. The first of these is based on Newton's laws of motion. These laws deal with the motion of a particle under the action of forces acting on it, and the quantities of primary interest in the analysis are: the forces, which may include applied forces, forces of interaction between particles, and the forces of constraints; and the momentum. Since both the forces and the momentum are vector quantities, the branch of mechanics based on the principles enunciated by Newton is called *vectorial mechanics*. When the methods of vectorial mechanics are used for the solution of a problem, the forces of constraints and the forces of interaction must be explicitly accounted for and evaluated even when there is no interest in obtaining their values.

The second branch of mechanics, called *analytical mechanics*, is based on the works of Bernoulli, Euler, d'Alembert, Lagrange, Poisson, Hamilton, Jacobi, and Gauss. As contrasted to force and momentum, the parameters of interest in analytical mechanics are scalar functions, and therefore the fundamental equations used in it do not depend on the choice of coordinates. Also, it is not necessary that the forces of constraints be explicitly accounted for or evaluated. The methods of analytical mechanics are therefore very effective in the solution of complex systems with multiple constraints.

This chapter deals with the basic principles of analytical mechanics and their application to the problems of structural dynamics. The concepts of generalized coordinates and generalized forces are introduced at the beginning. The important principle of virtual work is described next. This is followed by a presentation of Hamilton's principle and Lagrange's equation of motion. The notations and the methods of variational calculus are used throughout the chapter.

4.2 GENERALIZED COORDINATES

In vectorial mechanics, the position of a particle mass in the geometric space is specified by the three *rectangular coordinates* of Descarte. It is not however, necessary to use the rectangular coordinates; other appropriately selected

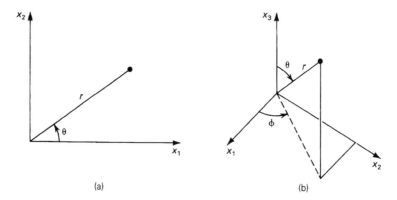

Figure 4.1. (a) Polar coordinates; (b) spherical coordinates.

parameters may be used as effectively to locate the position of the particle. For example, if the particle is constrained to move in a plane, we may use the two *polar coordinates*, r and θ, shown in Figure 4.1a. The transformation from the polar to the rectangular coordinates x_1 and x_2 is given by the following equations

$$x_1 = r \cos \theta$$
$$x_2 = r \sin \theta \tag{4.1}$$

In a three-dimensional space, one might use the *spherical coordinates* r, θ, and ϕ, shown in Figure 4.1b. The transformation is in this case given by

$$x_1 = r \sin \theta \cos \phi$$
$$x_2 = r \sin \theta \sin \phi \tag{4.2}$$
$$x_3 = r \cos \theta$$

The polar or spherical coordinates are by no means the only types of co-ordinates that may be used to specify the position of a particle in space. To illustrate the point, consider a particle P moving in a plane (Fig. 4.2). Also consider the line joining a point A having coordinates $(2,4)$ and a point B having coordinates $(4,2)$. The particle P can then be located by specifying the angles q_1 and q_2 that lines AP and BP, respectively, make with line AB. It is easily proved that coordinates sets (q_1, q_2) and (x, y) are related by the following transformation equations

$$x = 4 + 2 \frac{\sin q_1 (\sin q_2 + \cos q_2)}{\sin (q_2 - q_1)}$$
$$y = 2 + 2 \frac{\sin q_1}{\sin (q_2 - q_1)} (\sin q_2 - \cos q_2) \tag{4.3}$$

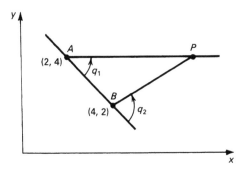

Figure 4.2. Generalized coordinates.

Note that when P lies on the line AB, q_1 and q_2 are both zero and functions given by Equation 4.3 are no longer single valued. Coordinates q_1 and q_2 cannot, therefore, be used for describing the motion of P near line AB.

The arbitrary coordinates used above to specify the location of a particle are examples of what are called generalized coordinates. Their number is the minimum required to specify the position of a particle or particles in space. In general, a mechanical system may consist of a number of particles. Let this number be N. If there are no constraints, $3N$ coordinates, rectangular or generalized, will be required to specify the position of the system in space. The $3N$ rectangular coordinates, $x_1, y_1, z_1, x_2, y_2, z_2, \ldots, z_N$, are related to $3N$ generalized coordinates $q_1, q_2, \ldots, q_{3N}$ by equations of the form

$$x_1 = f_1(q_1, q_2, q_3, \ldots, q_{3N})$$

$$y_1 = f_2(q_1, q_2, q_3, \ldots, q_{3N})$$

$$z_1 = f_3(q_1, q_2, q_3, \ldots, q_{3N}) \qquad (4.4)$$

$$\ldots\ldots\ldots\ldots\ldots\ldots\ldots$$

$$z_N = f_{3N}(q_1, q_2, q_3, \ldots, q_{3N})$$

It is necessary that functions given by Equation 4.4 be finite, single valued, continuous and differentiable.

In the examples cited so far, the number of generalized coordinates was equal to the number of rectangular coordinates and there was no particular advantage in using the former. However, when constraints exist in a system, the number of generalized coordinates may be significantly less than the number of rectangular coordinates, and the use of the former may considerably reduce the complexity of the problem. As an example of constraints, consider the double pendulum shown in Figure 4.3. The position of the two point masses can be defined by specifying the rectangular coordinates x_1, y_1 and x_2, y_2. However, the four

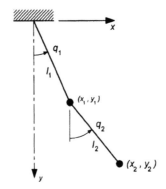

Figure 4.3. Double pendulum.

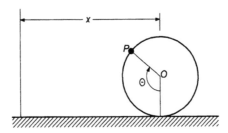

Figure 4.4. Cylinder rolling on a plane.

coordinates are related by the following two *equations of constraint*

$$x_1^2 + y_1^2 = l_1^2$$
$$(x_2 - x_1)^2 + (y_2 - y_1)^2 = l_2^2$$

$$(4.5)$$

These two constraint equations imply that the four coordinates x_1, y_1 and x_2, y_2 cannot all be independently specified. In fact, if two of them are given, the other two are determined automatically. It is obvious that the minimum number of coordinates required to specify the position of the system is 2. Although we may choose x_1 and x_2 as the two generalized coordinates, it is probably easier to use the two angles q_1 and q_2 shown in Figure 4.3.

A particle moving on a plane is constrained so that the displacement normal to the plane is always zero. This condition gives one equation of constraint so that the minimum number of independent coordinates needed to specify the position of the particle is two rather than three.

The position of a cylinder rolling on a horizontal plane can be specified by the two coordinates shown in Figure 4.4: the x distance of the center of cylinder from a prescribed vertical reference axis and the angle that a given radius OP makes with the vertical. However, when the cylinder rolls without slipping, these two coordinates are related by one equation of constraint. This equation obtained from the condition that the instantaneous velocity of the point of contact should be zero is

$$r\dot{\theta} = \dot{x} \tag{4.6}$$

Integration of Equation 4.6 yields

$$r\theta = x \tag{4.7}$$

Coordinates θ and x are called *constrained coordinates*. The minimum number of coordinates required to specify the position of the cylinder is one; and either x or θ may be chosen as the generalized coordinate. If θ is chosen as the independent coordinate, the transformation equations can be written as

$$
\begin{aligned}
x &= r\theta \\
\theta &= \theta
\end{aligned} \tag{4.8}
$$

Based on the foregoing discussion, we can now provide a formal definition for the term generalized coordinates. The term can be defined as the minimum number of independent coordinates required to specify the position of a mechanical system in space. For systems of interest to us, the number of such coordinates is equal to the number of degrees of freedom of the system. Thus, a particle moving on a plane has two degrees of freedom while a cylinder rolling on a rough plane has only one degree of freedom. It is obvious that generalized coordinates are not unique; rather, there are an infinite number of different coordinate sets. The choice of a particular set will depend on the ease with which a problem can be formulated in that set.

In the examples cited so far, the generalized coordinates were identified with physical quantities that could be measured. This is not always necessary, however, and we may in fact choose generalized coordinates that have no physical meaning. As an example, consider three particle masses attached to a massless flexible cantilever shown in Figure 4.5. If the bar is axially rigid and constrained to move in the plane of the paper, the positions of the particle masses can be determined by specifying the three coordinates x_1, x_2, and x_3. Figure 4.5 also shows three possible displaced shapes of the cantilever. The choice of displacement shapes is arbitrary except that they should be independent, so that none of them can be derived by a linear combination of the remaining two. Let these shapes be specified by three vectors ϕ_1, ϕ_2 and ϕ_3 given by

$$
\begin{aligned}
\phi_1^T &= [1 \quad 2 \quad 3] \\
\phi_2^T &= [1 \quad 0 \quad -1] \\
\phi_3^T &= [1 \quad -1 \quad 1]
\end{aligned} \tag{4.9}
$$

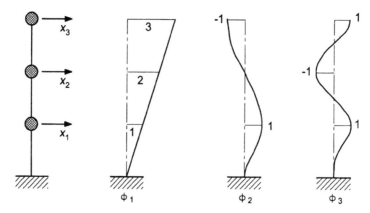

Figure 4.5. Representation of displaced shape through superposition of modes.

It is possible to describe any displaced shape of the cantilever bar by a super-position of the shape functions ϕ_1, ϕ_2 and ϕ_3 as follows

$$\mathbf{x} = \begin{bmatrix} x_1 \\ x_2 \\ x_3 \end{bmatrix} = \alpha_1\phi_1 + \alpha_2\phi_2 + \alpha_3\phi_3 \qquad (4.10)$$

or in the equivalent form

$$\mathbf{x} = \begin{bmatrix} 1 & 1 & 1 \\ 2 & 0 & -1 \\ 3 & -1 & 1 \end{bmatrix} \begin{bmatrix} \alpha_1 \\ \alpha_2 \\ \alpha_3 \end{bmatrix} = \mathbf{C}\boldsymbol{\alpha} \qquad (4.11)$$

in which $\mathbf{x}^T = (x_1, x_2, x_3)$ and α_1, α_2, and α_3 are the coefficients to be determined. For instance, a displacement shape in which $x_1 = 1$, $x_2 = 2.5$ and $x_3 = 2.0$ can equally well be described by specifying $\alpha_1 = 1$, $\alpha_2 = \frac{1}{2}$, and $\alpha_3 = -\frac{1}{2}$. Since vectors ϕ_1, ϕ_2 and ϕ_3 were chosen to be independent, the transformation matrix $\mathbf{C}$ is nonsingular and a unique value of vector $\boldsymbol{\alpha}$ can be determined for every value of $\mathbf{x}$. The coefficients α_1, α_2, and α_3 can be interpreted as generalized coordinates. It is not, however, possible to assign any physical meaning to them. Later, in Chapters 10 and 12, it will be shown that the normal coordinate transformation which plays a central role in problems of structural dynamics is indeed a special case of the transformation given by Equation 4.11.

4.3 CONSTRAINTS

The idea of constraints was briefly introduced in Section 4.2. However, because constraints play an important role in analytical mechanics and the particular

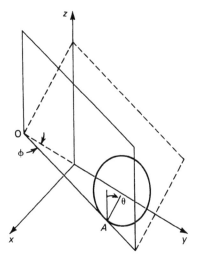

Figure 4.6. Thin disk rolling on a plane.

method of solution used may depend on the type of constraints in the system, it is of interest to take a closer look at the nature of constraints and the resulting constraint equations. In the following discussion, the symbol u is employed to designate constrained coordinates, while q is used for generalized coordinates. Constraint equations 4.5 and 4.7 are special cases of the general form

$$f(u_1, u_2, \ldots, u_N, t) = 0 \tag{4.12}$$

Equation 4.12 prescribes a relationship that involves the constrained coordinates and the time. The equations of constraint may involve differentials of coordinates rather than the coordinates themselves. An example is provided by Equation 4.6, rewritten in the form

$$r\, d\theta - dx = 0 \tag{4.13}$$

Equation 4.13, which involves differentials of the coordinates, can be converted to the form of Equation 4.12 by simple integration. Such reduction may not, however, always be possible. A constraint of the form of Equation 4.12 or reducible to that form is called a *holonomic* constraint. A constraint that cannot be reduced to the form of Equation 4.12 is called a *nonholonomic* constraint.

As an example of nonholonomic constraints, consider the motion of a thin disk rolling on a rough plane. As shown in Figure 4.6, the position of the disk can be described by specifying the coordinates of the point of contact A, the angle between a vertical plane and the plane containing the disk, and the angle that the tangent to the circumference of the disk at the point of the contact makes with the y axis. The four coordinates described in the foregoing

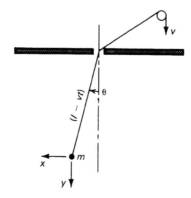

Figure 4.7. Pendulum of variable length.

paragraph are, in fact, independent. This is easily verified by noting that even when any three of them have been prescribed, the fourth one can take any arbitrary value from within its admissible range of values. Thus, when x, y, and ϕ are given, θ can take any value within the range $-\pi/2$ to $\pi/2$. However, even though the coordinates are independent, a constraint equation must exist. This equation is derived by noting that the instantaneous motion should be along the tangent line *OA*, so that

$$\dot{x} = \dot{y} \tan \phi \qquad\qquad\qquad (4.14\text{a})$$

or

$$dx = dy \tan \phi \qquad\qquad\qquad (4.14\text{b})$$

Since $\tan \phi$ is also a variable, it is not possible to integrate Equation 4.14 to reduce it to the form of Equation 4.12; the constraint described by Equation 4.14 is therefore nonholonomic. This constraint reduces the number of degrees of freedom of the system to three, which is one less than the number of independent coordinates. Thus, for systems having nonholonomic constraints, the number of degrees of freedom is less than, and not equal to, the number of generalized coordinates.

In a majority of engineering problems, the constraints are holonomic. Only systems with holonomic constraints will therefore be treated in this book. Holonomic constraints can be further subdivided into two categories: *rhenomic* and *scleronomic*. When a constraint equation contains time t explicitly, the constraint is said to be rhenomic, when time t is not present, the constraint is scleronomic. The holonomic constraints described by Equations 4.5 and 4.7 are both scleronomic. As an example of a rhenomic constraint, consider the simple pendulum shown in Figure 4.7 in which the string from which the mass is suspended is being pulled upward with a constant velocity v. Let the length of string at time $t = 0$ be l. Also, let x and y be the rectangular coordinates describing the position of point mass m, and let θ be the angle that the string makes with the vertical. The rectangular coordinates are not independent but satisfy a constraint equation of the form

$$x^2 + y^2 = (l - vt)^2 \tag{4.15}$$

Equation 4.15 contains time t explicitly and describes a rhenomic holonomic constraint. Note that angle θ can be used as the generalized coordinate describing the system. The transformation between the rectangular and the generalized coordinate θ is given by

$$
\begin{aligned}
x &= (l - vt)\sin\theta \\
y &= (l - vt)\cos\theta
\end{aligned}
\tag{4.16}
$$

4.4 VIRTUAL WORK

If a particle is at rest under the action of a set of forces, Newton's second law requires that the vector sum of all the forces acting on the particle be zero. Mathematically, this can be stated as

$$\mathbf{F} = \mathbf{0} \tag{4.17}$$

where $\mathbf{F}$ represents the vector sum of forces. In general, the forces acting on the particle can be categorized as applied forces and the forces of constraints. As an example, if the particle is constrained to move on a plane, the reaction normal to the plane is classified as a *force of constraint*. If the vector sum of applied forces is denoted by $\mathbf{F}_a$ and that of the forces of constraints by $\mathbf{F}_c$, Equation 4.17 can be rewritten in the form

$$\mathbf{F}_a + \mathbf{F}_c = \mathbf{0} \tag{4.18}$$

Now, suppose that the particle in equilibrium is given a virtual displacement $\delta\mathbf{r}$. The term virtual displacement signifies that the displacement is imaginary and not due to any real cause. We use the symbol $\delta\mathbf{r}$ to denote such an imaginary displacement. The work done by the real force in riding through the imaginary displacement is termed *virtual work* and is given by

$$\overline{\delta W} = \mathbf{F} \cdot \delta\mathbf{r} \tag{4.19}$$

where a dot represents the scalar product of two vectors. Again, we use the symbol δW rather than ΔW or dW to signify that the work done is not performed by real forces undergoing real displacements. Since $\mathbf{F}$ is zero, $\overline{\delta W}$ should also be zero, giving

$$\overline{\delta W} = 0 \tag{4.20}$$

Equation 4.20 can be treated as an expression of the principle of virtual work. In this form, it is simply a restatement of the equations of static equilibrium. But now, let the arbitrary virtual displacement $\delta\mathbf{r}$ be chosen such that the forces of constraint do not work in moving through $\delta\mathbf{r}$. Such a virtual displacement is said to be compatible with the constraints of the system. For example, in the case of a particle moving on a plane, a virtual displacement parallel to the plane is compatible with the constraint, and the normal reaction will do no work in moving through such a displacement. The virtual work equation can then be written as

$$\overline{\delta W} = \mathbf{F}_a \cdot \delta\mathbf{r} + \mathbf{F}_c \cdot \delta\mathbf{r} = 0 \tag{4.21}$$

and since $\mathbf{F}_c \cdot \delta\mathbf{r}$ is zero, this gives

$$\overline{\delta W} = \mathbf{F}_a \cdot \delta\mathbf{r} = 0 \tag{4.22}$$

Equation 4.22 does not contain the forces of constraint. Also, $\mathbf{F}_a$ is no longer zero; only the scalar product of $\mathbf{F}_a$ and $\delta\mathbf{r}$ is zero. In this form, Equation 4.22 is quite different from Equation 4.17.

To illustrate the points mentioned above, consider again the case of a particle moving in the xy plane. The particle is in equilibrium under an applied force and the reaction exerted by the plane. Let the components of applied force be F_x, F_y, and F_z, and let the reaction normal to the plane be denoted by N. For an arbitrary displacement with component δx, δy and δz, the virtual work equation becomes

$$F_x \delta x + F_y \delta y + (F_z + N)\delta z = 0 \tag{4.23}$$

Now, because δx, δy, and δz are independent, Equation 4.23 is entirely equivalent to the three equations $F_x = 0$, $F_y = 0$, and $(F_z + N) = 0$, which are readily seen to be the equations of equilibrium. If the virtual displacement is consistent with the constraint, component δz should be zero and Equation 4.23 reduces to

$$F_x \delta x + F_y \delta y = 0 \tag{4.24}$$

Equation 4.22 is easily extended to a system consisting of a number of particles. Consider, for example, the case of two particles moving on the xy plane. The virtual work equation is

$$\mathbf{F_1} \cdot \delta\mathbf{r_1} + \mathbf{F_2} \cdot \delta\mathbf{r_2} = 0 \tag{4.25}$$

When the virtual displacements are consistent with the constraints, $\delta\mathbf{r_1}$ and $\delta\mathbf{r_2}$ will have no components normal to the plane of motion. Equation 4.25 can therefore be written as

$$F_{x_1}\delta x_1 + F_{y_1}\delta y_1 + F_{x_2}\delta x_2 + F_{y_2}\delta y_2 = 0 \tag{4.26}$$

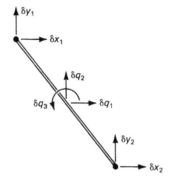

Figure 4.8. Virtual displacements of two interconnected particles constrained to lie on a plane.

The virtual displacement components in Equation 4.26 are all independent; Equation 4.26 is therefore equivalent to four separate equations which can easily be identified as the equations of equilibrium, two per particle.

When additional constraints are introduced, the number of independent virtual displacement components will be reduced correspondingly. For example, if the two particles moving in the xy plane are tied together by a rigid bar, the number of independent virtual displacements is reduced to three. As shown in Figure 4.8, the virtual displacements can be expressed in terms of δq_1, and δq_2, the x and y displacements of the center of the rod, respectively, and δq_3, the rotation of the rod about its center. Since a rigid body is simply a collection of an infinite number of particles constrained to move together, the principle of virtual work is easily extended to such a body. A rigid body moving in a plane has three independent displacement components, two translations and one rotation. A rigid body moving in space has three independent translation components and three independent rotational components.

A system consisting of an assemblage of particles and rigid bodies is referred to as a mechanical system. The *principle of virtual work* for such a system can be stated as follows. If a mechanical system in equilibrium under a set of forces is given an arbitrary virtual displacement that is compatible with the constraints on the system, the sum of work done by the applied forces in undergoing the virtual displacement is zero. The principle of virtual work can also be extended to deformable bodies, but now we must take account of the work done by the internal forces in riding through the virtual deformation. To illustrate the point, consider the simple example of two particles attached to each other by a spring of stiffness k and constrained to move along the x axis as shown in Figure 4.9a. Let the system be in equilibrium under applied forces F_1 and F_2 shown in the figure, and let the spring be stretched by Δx under the action of these forces. If the system is given virtual displacements δx_1 and δx_2, the virtual work done

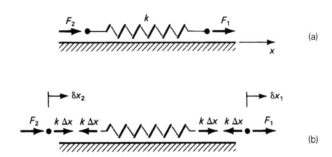

Figure 4.9. Two particles connected by a spring lying on the x axis and in equilibrium under forces F_1 and F_2.

by the applied forces F_1 and F_2 will be

$$\overline{\delta W_e} = F_1 \delta x_1 + F_2 \delta x_2 \tag{4.27}$$

Now, although δx_1 and δx_2 are independent, the virtual work obtained from Equation 4.27 is no longer zero. However, if we add to $\overline{\delta W_e}$ the work done by the spring force, the total work so obtained will be zero. The forces exerted by the spring on the two particles are indicated in Figure 4.9b. The work done by the spring forces, called the internal work, is denoted by $\overline{\delta U}$ and is given by

$$\overline{\delta U} = -k\,\Delta x\,\delta x_1 + k\,\Delta x\,\delta x_2 \tag{4.28}$$

where Δx is the stretch of the spring. The modified virtual work equations becomes

$$\overline{\delta W_e} + \overline{\delta U} = 0 \tag{4.29}$$

or

$$(F_1 - k\,\Delta x)\,\delta x_1 + (F_2 + k\,\Delta x)\,\delta x_2 = 0 \tag{4.30}$$

Because δx_1 and δx_2 are independent, Equation 4.30 is equivalent to the two equations $F_1 = k\,\Delta x$ and $F_2 = -k\,\Delta x$. These relationships imply that for equilibrium $F_1 = -F_2$ and that at the position of rest, the spring will be stretched by $\Delta x = F_1/k$.

As an alternative to defining the internal work as in Equation 4.28, we may identify the internal forces acting on the spring rather than on the two particles and calculate the work done on the internal elements. We shall denote it by

$\overline{\delta W_{ie}}$. For our example

$$\overline{\delta W_{ie}} = k \, \Delta x(\delta x_1 - \delta x_2) \tag{4.31}$$

It is obvious that

$$\overline{\delta W_{ie}} = -\overline{\delta U} \tag{4.32}$$

On substituting Equation 4.32 into Equation 4.29, the virtual work equation can be written in the alternative form

$$\overline{\delta W_e} = \overline{\delta W_{ie}} \tag{4.33}$$

In the example that we just considered, the deformable body was a simple spring and the calculation of the internal work was quite straightforward. In general, the deformable bodies will be of more complex shape. However, internal virtual work expressions for elastic bodies of various shapes, including bars, beams, and plates, are derived in standard textbooks on applied mechanics.

The principle of virtual work for deformable bodies can now be stated as follows. If a deformable system in equilibrium under a set of forces is given virtual displacements that are consistent with the constraints on the system, the sum of the internal virtual work and the virtual work done by the applied forces is zero. The principle of virtual work occupies the same position in analytical mechanics as Newton's second law does in vectorial mechanics. They are both fundamental postulates on which all further developments in the respective branches of mechanics are based. One advantage of using the methods of analytical mechanics is at once obvious. Vectorial mechanics requires that the forces acting on each individual particle of the system, including those due to constraints, be identified and evaluated. In analytical mechanics, the forces of constraints need not be accounted for explicitly. This advantage will be evident from the example that follows.

Example 4.1
A uniform bar of mass m and length l is constrained to move in a vertical plane by guides as shown in Figure E4.1a. A mass M is attached to one end of the bar and the other end is restrained by a spring of stiffness k. The spring is unstretched when the bar is horizontal. Find the position of equilibrium of the bar.

Solution
The free-body diagram shown in Figure E4.1b indicates all the forces acting on the bar. These forces include the two reactions R_1 and R_2. When the virtual work method is used to solve the problem, it is not necessary to evaluate these reactions.

The system has a single degree of freedom and its position is completely described by one generalized coordinate. We may choose any one of the coordinates θ, x_1, y_1, and y_2 shown in Figure E4.1a. Let angle θ be chosen as the generalized coordinate. A small change $\delta\theta$ in this

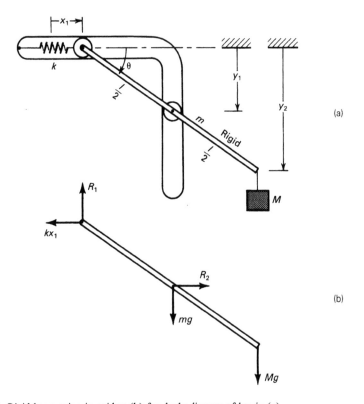

Figure E4.1. Rigid bar moving in guides; (b) free-body diagram of bar in (a).

angle will serve as the virtual displacement. The corresponding variations in x_1, y_1, and y_2 are evaluated as follows

$$x_1 = \frac{l}{2}(1 - \cos \theta)$$

$$y_1 = \frac{l}{2}(\sin \theta) \tag{a}$$

$$y_2 = l \sin \theta$$

$$\delta x_1 = \frac{l}{2} \sin \theta \, \delta \theta$$

$$\delta y_1 = \frac{l}{2} \cos \theta \, \delta \theta \tag{b}$$

$$\delta y_2 = l \cos \theta \, \delta \theta$$

The equation of virtual work becomes

$$\delta W = -k\frac{l}{2}(1 - \cos \theta)\frac{l}{2} \sin \theta \, \delta \theta + mg\frac{l}{2} \cos \theta \, \delta \theta + Mgl \cos \theta \, \delta \theta = 0 \tag{c}$$

Equation c leads to the transcendental equation

$$\frac{kl}{4}(1 - \cos \theta) \tan \theta = \frac{mg}{2} + Mg \tag{d}$$

the solution of which gives the value of angle θ.

4.5 GENERALIZED FORCES

Just as the position of a system can be described in terms of generalized coordinates, the forces acting on the system can be represented by generalized forces. These forces are entirely equivalent to the real forces acting on the system in that the work done by the real forces on a set of virtual displacements is equal to the work done by the generalized forces on the corresponding virtual displacements along the generalized coordinates. This last statement permits us to express the generalized forces in terms of the real forces acting on the system. Suppose that the system subjected to real forces $\mathbf{F}_i$ $(i = 1, 2, \ldots, M)$ is given virtual displacements $\delta \mathbf{r}_i$ $(i = 1, 2, 3, \ldots, M)$ The virtual work is then

$$\overline{\delta W} = \sum_{i=1}^{M} \mathbf{F}_i \cdot \delta \mathbf{r}_i \tag{4.34}$$

The overbar on W means that it is simply infinitesimal work, not the variation of a scalar function W.

Now the radius vectors $\mathbf{r}$ are related to the N generalized coordinates $q_1, q_2, \ldots, q_N$ by equations of the form

$$\mathbf{r}_1 = f_1(q_1, q_2, \ldots, q_N)$$
$$\mathbf{r}_2 = f_2(q_1, q_2, \ldots, q_N)$$
$$\ldots\ldots\ldots\ldots\ldots\ldots \tag{4.35}$$
$$\mathbf{r}_M = f_M(q_1, q_2, \ldots, q_N)$$

It follows that

$$\delta \mathbf{r}_i = \sum_{j=1}^{N} \frac{\partial \mathbf{r}_i}{\partial q_j} \delta q_j \qquad (i = 1, 2, \ldots, M) \tag{4.36}$$

The equations of virtual work can therefore be written as

$$\overline{\delta W} = \sum_{i=1}^{M} \left(\mathbf{F}_i \cdot \sum_{j=1}^{N} \frac{\partial \mathbf{r}_i}{\partial q_j} \delta q_j \right)$$

$$= \sum_{j=1}^{N} \left(\sum_{i=1}^{M} \mathbf{F}_i \cdot \frac{\partial \mathbf{r}_i}{\partial q_j} \right) \delta q_j \tag{4.37}$$

By the definition of generalized forces, the virtual work should also be given by

$$\overline{\delta W} = \sum_{j=1}^{N} Q_j \, \delta q_j \qquad (4.38)$$

where Q_j are the generalized forces. Equations 4.37 and 4.38 should be entirely equivalent. Also, since δq_j's are all independent, they can be assigned arbitrary values. Comparing Equations 4.37 and 4.38, and setting all δq_j's except the mth to zero

$$Q_m = \sum_{i=1}^{M} \mathbf{F}_i \cdot \frac{\partial \mathbf{r}_i}{\partial q_m} \qquad (4.39)$$

giving an expression for the mth generalized force. If the system is in equilibrium, $\overline{\delta W}$ should be zero, and since δq_j's are independent, it follows from Equation 4.38 that each Q_j must be zero. Unlike real forces, it is not always possible to assign a physical meaning to the generalized forces. Also, the latter need not have the units of a force. However, the products of a generalized force and the corresponding generalized displacement will always have the unit of work.

Example 4.2

A counterbalanced door of weight P rotates freely about a frictionless hinge as shown in Figure E4.2. The balancing weight Q is attached to the tip of the door by a string as shown. Determine the position of the door when it is in equilibrium.

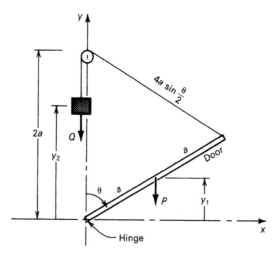

Figure E4.2. Counterbalanced door.

Solution

The system has a single degree of freedom and needs one coordinate to describe its position. The angle that the door makes with the vertical is selected as the generalized coordinate. The constrained coordinates y_1 and y_2 are related to the generalized coordinates by the following equations

$$\mathbf{y}_1 = a \cos \theta \, \mathbf{j} \tag{a}$$

$$\mathbf{y}_2 = \left(C + 4a \sin \frac{\theta}{2} \right) \mathbf{j} \tag{b}$$

where C is a constant and $\mathbf{j}$ is a unit vector along the y axis.

The virtual work done by the real forces $-Q\mathbf{j}$ and $-P\mathbf{j}$ is given by

$$\overline{\delta W} = -Q\mathbf{j} \cdot \delta \mathbf{y}_2 - P\mathbf{j} \cdot \delta \mathbf{y}_1$$

$$= -Q\mathbf{j} \cdot \frac{\partial \mathbf{y}_2}{\partial \theta} \delta\theta - P\mathbf{j} \cdot \frac{\partial \mathbf{y}_1}{\partial \theta} \delta\theta$$

$$= \left[-Q \left(2a \cos \frac{\theta}{2} \right) + Pa \sin \theta \right] \delta\theta \tag{c}$$

In terms of generalized forces, the virtual work is

$$\overline{\delta W} = Q_\theta \, \delta\theta \tag{d}$$

Comparison of Equations c and d gives the generalized force Q

$$Q_\theta = Pa \sin \theta - 2aQ \cos \frac{\theta}{2} \tag{e}$$

The condition that Q_θ should be zero for equilibrium gives the following transcendental equation in θ

$$P \sin \theta = 2Q \cos \frac{\theta}{2}$$

or

$$\sin \frac{\theta}{2} = \frac{Q}{P}$$

Note that the generalized force Q_θ has the units of a moment.

Example 4.3

A rigid uniform bar of mass m supported at its ends by springs of stiffnesses k_1 and k_2 is subjected to a force P as shown in Figure E4.3a. If the unstretched length of each spring is d, find the position of equilibrium.

Solution

Figure E4.3b shows a free-body diagram of the rod. A displacement coordinate is defined along each of the four forces acting on the bar. These coordinates designated y_1, y_2, y_3, and y_4 are

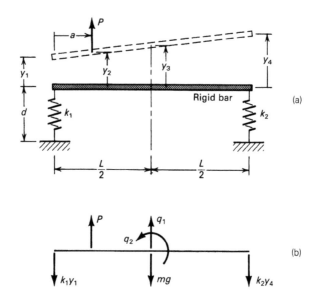

Figure E4.3. Rigid bar supported on springs.

obviously not independent. The system has two degrees of freedom and therefore, needs only two generalized coordinates to specify its position in space. The vertical translation q_1 of the center of gravity and the rotation q_2 about it are selected as the two generalized coordinates. The constrained coordinates can be expressed in terms of q_1 and q_2 as follows

$$\mathbf{y}_1 = \left(q_1 - \frac{L}{2} q_2 \right) \mathbf{j}$$

$$\mathbf{y}_2 = \left\{ q_1 - \left(\frac{L}{2} - a \right) q_2 \right\} \mathbf{j}$$

$$\mathbf{y}_3 = q_1 \mathbf{j}$$ (a)

$$\mathbf{y}_4 = \left(q_1 + \frac{L}{2} q_2 \right) \mathbf{j}$$

The virtual work equation is given by

$$\overline{\delta W} = \sum_{i=1}^{4} \mathbf{F}_i \cdot \left(\frac{\partial \mathbf{y}_i}{\partial q_1} \delta q_1 + \frac{\partial \mathbf{y}_i}{\partial q_2} \delta q_2 \right)$$

$$= -k_1 \left(q_1 - \frac{L}{2} q_2 \right) \left(\delta q_1 - \frac{L}{2} \delta q_2 \right) + P \left\{ \delta q_1 - \left(\frac{L}{2} - a \right) \delta q_2 \right\}$$

$$- mg \, \delta q_1 - k_2 \left(q_1 + \frac{L}{2} q_2 \right) \left(\delta q_1 + \frac{L}{2} \delta q_2 \right)$$

On collecting terms, we obtain

$$\overline{\delta W} = \left(-k_1 q_1 + \frac{L}{2} k_1 q_2 + P - mg - k_2 q_1 - k_2 \frac{L}{2} q_2 \right) \delta q_1$$

$$+ \left(k_1 \frac{L}{2} q_1 - k_1 \frac{L^2}{4} q_2 - \frac{PL}{2} + Pa - k_2 \frac{L}{2} q_1 - k_2 \frac{L^2}{4} q_2 \right) \delta q_2 \qquad \text{(b)}$$

The virtual work equation in terms of generalized forces is

$$\overline{\delta W} = Q_1 \, \delta q_1 + Q_2 \, \delta q_2 \qquad \text{(c)}$$

On comparing Equations b and c, we have

$$Q_1 = -(k_1 + k_2)q_1 + (k_1 - k_2)\frac{L}{2} q_2 + (P - mg)$$

$$Q_2 = \frac{L}{2}(k_1 - k_2)q_1 - \frac{L^2}{4}(k_1 + k_2)q_2 + P\left(a - \frac{L}{2} \right) \qquad \text{(d)}$$

For equilibrium, each of Q_1 and Q_2 should be zero. This condition leads to the following equations in q_1 and q_2

$$\begin{bmatrix} k_1 + k_2 & -\frac{L}{2}(k_1 - k_2) \\ -\frac{L}{2}(k_1 - k_2) & \frac{L^2}{4}(k_1 + k_2) \end{bmatrix} \begin{bmatrix} q_1 \\ q_2 \end{bmatrix} = \begin{bmatrix} P - mg \\ P(a - \frac{L}{2}) \end{bmatrix} \qquad \text{(e)}$$

The solution of Equation e will give q_1 and q_2. Note that Q_1 has the units of force, while Q_2 has the units of a moment.

Example 4.4
The vertical cantilever of Figure E4.4 is subjected to the forces as shown. If the coordinates described by Equation 4.11 are used as the generalized coordinates, find the corresponding generalized forces.

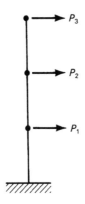

Figure E4.4. Vertical cantilever.

Solution

The virtual work equation is given by

$$\overline{\delta W} = P_1\,\delta x_1 + P_2\,\delta x_2 + P_3\,\delta x_3$$

$$= [P_1 \quad P_2 \quad P_3] \begin{bmatrix} 1 & 1 & 1 \\ 2 & 0 & -1 \\ 3 & -1 & 1 \end{bmatrix} \begin{bmatrix} \delta\alpha_1 \\ \delta\alpha_2 \\ \delta\alpha_3 \end{bmatrix}$$

$$= [(P_1 + 2P_2 + 3P_3) \quad (P_1 - P_3) \quad (P_1 - P_2 + P_3)] \begin{bmatrix} \delta\alpha_1 \\ \delta\alpha_2 \\ \delta\alpha_3 \end{bmatrix} \tag{a}$$

from which the three generalized forces are readily identified as

$$Q_1 = P_1 + 2P_2 + 3P_3$$

$$Q_2 = P_1 - P_3 \tag{b}$$

$$Q_3 = P_1 - P_2 + P_3$$

Note that in this case, it is not possible to give a physical meaning to the generalized forces, although they all have the units of force.

4.6   CONSERVATIVE FORCES AND POTENTIAL ENERGY

Consider a particle moving in space under the action of a force **F** as shown in Figure 4.10. The work done by the force **F** in moving from point A to B is given by

$$W = \int_A^B \mathbf{F} \cdot d\mathbf{r} \tag{4.40}$$

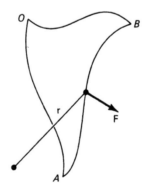

Figure 4.10. Work done by a force in moving along a given path.

where **r** denotes the radius vector. In general, the work done will depend on the path the particle follows from A to B. However, as a special case, consider a force that can be derived from a scalar function ϕ as follows

$$\mathbf{F} = \frac{\partial \phi}{\partial x}\mathbf{i} + \frac{\partial \phi}{\partial y}\mathbf{j} + \frac{\partial \phi}{\partial z}\mathbf{k}$$

$$= \nabla \phi \tag{4.41}$$

in which the operator ∇, called the *gradient vector*, is given by

$$\nabla = \frac{\partial}{\partial x}\mathbf{i} + \frac{\partial}{\partial y}\mathbf{j} + \frac{\partial}{\partial z}\mathbf{k} \tag{4.42}$$

On substituting Equation 4.41 into Equation 4.40 and noting that $d\mathbf{r} = dx\mathbf{i} + dy\mathbf{j} + dz\mathbf{k}$

$$W = \int_A^B \left(\frac{\partial \phi}{\partial x}dx + \frac{\partial \phi}{\partial y}dy + \frac{\partial \phi}{\partial z}dz \right)$$

$$= \int_A^B d\phi \tag{4.43}$$

$$= \phi_B - \phi_A$$

Equation 4.43 shows that the work done in moving the particle from point A to B depends only on the values of the function ϕ at those points and is independent of the path along which the particle moves. When the work done by a force during the motion of its point of application is independent of the path followed but depends only on the position of the terminal points of the path, that force is said to be conservative. A force that can be derived from a scalar function according to Equation 4.41 is thus a *conservative force*.

The work done by a conservative force in moving from a point A to datum O is designated as the *potential energy* of the force at A. Thus

$$V_A = \int_A^O \mathbf{F} \cdot d\mathbf{r} \tag{4.44}$$

where V denotes the potential energy. Now, let the particle move from A to B along a path that includes the datum O (Fig. 4.10). Then, since the work done is independent of the path, Equation 4.40 can be written as

$$W = \int_A^O \mathbf{F} \cdot d\mathbf{r} + \int_O^B \mathbf{F} \cdot d\mathbf{r}$$

$$= \int_A^O \mathbf{F} \cdot d\mathbf{r} - \int_B^O \mathbf{F} \cdot d\mathbf{r} \tag{4.45}$$

$$= -(V_B - V_A)$$

Figure 4.11. Gravitational force of the earth.

Figure 4.12. Work done against the force of a spring.

Equation 4.45 implies that the work done is the negative of the change of potential energy. In differential form, this can be stated as

$$dW = d\phi = -dV \tag{4.46}$$

implying that dW is a complete differential. From Equations 4.46 and 4.41, it follows that

$$\mathbf{F} = -\nabla V \tag{4.47}$$

As an example of a conservative force, consider the gravitational force near the surface of the earth. As shown in Figure 4.11, the force acting on a particle of mass m is given by

$$\mathbf{F} = -mg\mathbf{k} \tag{4.48}$$

Now, define a scalar function $\phi = -mgz$. It is seen that

$$\nabla\phi = -mg\mathbf{k} = \mathbf{F} \tag{4.49}$$

The gravitational force can thus be expressed as the gradient of a scalar function and is therefore conservative. If the surface of the earth is taken as the datum, the potential energy of the mass at a height z above the surface is given by

$$V_A = mgz \tag{4.50}$$

The force exerted by a linear spring of stiffness k is another example of a conservative force (Fig. 4.12). The work done by the spring force when the end of the spring is moved a distance x from the unstretched position, is given by

$$W = \int_0^x -kx\,dx = -\frac{1}{2}kx^2 \tag{4.51}$$

The negative of this work is the potential energy stored in the spring. Thus

$$V = \frac{1}{2}kx^2 \tag{4.52}$$

and

$$F = -\frac{dV}{dx} = -kx \tag{4.53}$$

The energy stored in an elastic body when it is deformed is called *strain energy*. Expressions for strain energies stored due to flexure, axial, shear, and torsional deformations can be found in any standard textbook on applied mechanics.

There are many forces that are *nonconservative* in nature and cannot be derived from a scalar function. The force of friction is an example of a nonconservative force, since the work done by it is not independent of the path through which its point of application moves.

Example 4.5
  (i) Find the conditions under which a force $F = A\mathbf{i} + B\mathbf{j} + C\mathbf{k}$ is conservative.
 (ii) A particle moving in space is subjected to a force of constant magnitude, always directed toward the origin. Using the result obtained in part (a), show that the force is conservative.

Solution
(i) The infinitesimal work is given by

$$\overline{dW} = \mathbf{F} \cdot \mathbf{dr} = A\,dx + B\,dy + C\,dz \tag{a}$$

If the force $\mathbf{F}$ is conservative, then from Equation 4.46, $dW = d\phi$. In other words, dW must be a complete differential. From elementary calculus, we know that for dW to be a complete differential, the following conditions must be satisfied:

$$\frac{\partial A}{\partial y} - \frac{\partial B}{\partial x} = 0$$

$$\frac{\partial A}{\partial z} - \frac{\partial C}{\partial x} = 0 \tag{b}$$

$$\frac{\partial B}{\partial z} - \frac{\partial C}{\partial y} = 0$$

(ii) A force of magnitude G directed toward the origin is given by

$$-G\left(\frac{x}{r}\mathbf{i} + \frac{y}{r}\mathbf{j} + \frac{z}{r}\mathbf{k}\right) \tag{c}$$

where

$$r = (x^2 + y^2 + z^2)^{1/2}$$

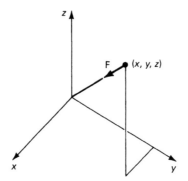

Figure E4.5. Central force.

Setting $A = -Gx/r$, $B = -Gy/r$, and $C = -Gz/r$ gives us

$$\frac{\partial A}{\partial y} - \frac{\partial B}{\partial x} = \frac{xyG}{(x^2 + y^2 + z^2)^{3/2}} - \frac{yxG}{(x^2 + y^2 + z^2)^{3/2}}$$

$$= 0 \tag{d}$$

In a similar manner, it can be proved that the other two conditions specified in Equation b are also satisfied. The force F is therefore conservative. A force that is always directed toward a fixed point in space, as shown in Figure E4.5, is called a *central force*. It can be shown that if the magnitude of the central force is proportional to r^n, the force is conservative. Gravitational forces between two bodies and the force between two electric charges, both proportional to $1/r^2$, are examples of conservative central forces.

4.7   WORK FUNCTION

Equation 4.38 is an expression of the work done when a system is subjected to infinitesimal virtual displacements δq_j's. Now, among the set of admissible virtual displacements are also the differentials dq_j's. On replacing displacements δq_j's by dq_j's, the expression for the infinitesimal work becomes

$$\overline{dW} = Q_1\,dq_1 + Q_2\,dq_2 + \cdots + Q_N\,dq_N \tag{4.54}$$

Now, if a scalar function U can be found such that

$$\frac{\partial U}{\partial q_i} = Q_i \qquad i = 1, 2, \ldots, N \tag{4.55}$$

then the right-hand side of Equation 4.54 is equal to dU and $\overline{dW}$ becomes a complete differential

$$\overline{dW} = dU \tag{4.56}$$

Following the reasoning advanced in Section 4.6, it can be shown that be-cause the infinitesimal work dW is a complete differential, the generalized forces, Q_i, are conservative. The scalar function U is called the *work function* and its negative is equal to the potential energy V, so that

$$U(q_1,q_2,q_3,\dots,q_N) = -V(q_1,q_2,q_3,\dots,q_N)$$ (4.57)

and

$$-\frac{\partial V}{\partial q_i} = Q_i$$ (4.58)

Further, if the system is in equilibrium, then as proved in Section 4.5, all generalized forces must be equal to zero and the following relations hold

$$\frac{\partial V}{\partial q_i} = 0 \qquad i = 1,2,\dots,N$$ (4.59)

Conditions expressed by Equation 4.59 are precisely the conditions for a stationary value of the function V. The stationary value may represent a minimum, a maximum, or a saddle point. It can be shown that a minimum value represents a state of stable equilibrium, a maximum value represents a state of unstable equilibrium, and a saddle point corresponds to a position of neutral equilibrium.

Example 4.6
For the counterbalanced door of Example 4.2, determine the position of equilibrium by finding a stationary value of the potential energy. Is the equilibrium stable?

Solution
Ignoring a constant additive, the potential energy is given by

$$V = 4aQ \sin \frac{\theta}{2} + Pa\cos\theta$$ (a)

This potential energy function can also be obtained from the expression for infinitesimal virtual work derived in Example 4.2 using the reasoning outlined in Section 4.7. Thus

$$\overline{dW} = \left\{-Q\left(2a\cos\frac{\theta}{2}\right) + Pa\sin\theta\right\} d\theta$$ (b)

Hence

$$Q_\theta = -Q\left(2a\cos\frac{\theta}{2}\right) + Pa\sin\theta$$ (c)

It is readily seen that the work function U which satisfies the condition $\partial U/\partial\theta = Q_\theta$, is given by

$$U = -Q\left(4a\sin\frac{\theta}{2}\right) - Pa\cos\theta$$ (d)

The forces acting on the door are thus conservative and the potential energy function which is negative of the work function is given by Equation a.

For a stationary value of the potential energy,

$$\frac{\partial V}{\partial \theta} = 0$$

which gives

$$2aQ \cos \frac{\theta}{2} - Pa \sin \theta = 0 \tag{e}$$

or

$$\sin \frac{\theta}{2} = \frac{Q}{P} \tag{f}$$

Equation f shows that equilibrium is possible if Q, the counterbalancing weight is less than P, the weight of the door.

To determine whether equilibrium is stable, find $\partial^2 V/\partial \theta^2$ at the position of equilibrium.

$$\frac{\partial^2 V}{\partial \theta^2} = -aQ \sin \frac{\theta}{2} - Pa \cos \theta \tag{g}$$

On substituting the value of $\sin(\theta/2)$ from Equation f, Equation g gives

$$\frac{\partial^2 V}{\partial \theta^2} = -aQ \frac{Q}{P} - Pa \left(1 - 2\frac{Q^2}{P^2} \right)$$

$$= \frac{a}{P}(Q^2 - P^2) \tag{h}$$

For Q less than P, Equation h will always give a negative value for $\partial^2 V/\partial \theta^2$. This means that the position of equilibrium corresponds to the maximum value of the potential energy and the equilibrium is unstable. This can be verified from the graph of potential energy versus angle θ plotted in Figure E4.6 for $Q/P = \frac{1}{2}$.

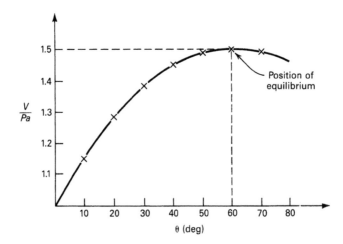

Figure E4.6. Variation of potential energy.

Example 4.7

A rigid bar of mass m and length l is connected at one end to another rigid bar, also of mass m and length l. The free end of the first bar is attached to the ceiling by a hinge. If a horizontal force P acts on the free end of the second bar, find the position of equilibrium of the system.

Solution

The system has two degrees of freedom and can be described by the generalized coordinates q_1 and q_2 as shown in Figure E4.7. If the potential energy is assumed to be zero when the parts are in a vertical position, the potential energy for any other position can be expressed in terms of q_1 and q_2 as follows

$$V = \frac{l}{2}mg(1 - \cos q_1) + mg\left\{ l(1 - \cos q_1) + \frac{l}{2}(1 - \cos q_2) \right\}$$

$$- Pl(\sin q_1 + \sin q_2) \tag{a}$$

The conditions for the stationary value of V will give the position of equilibrium as follows

$$\frac{\partial V}{\partial q_1} = 0$$

or

$$\frac{l}{2}mg \sin q_1 + lmg \sin q_1 - Pl \cos q_1 = 0$$

$$\tan q_1 = \frac{2}{3}\frac{P}{mg} \tag{b}$$

and

$$\frac{\delta V}{\delta q_2} = 0$$

$$\frac{l}{2}mg \sin q_2 - Pl \cos q_2 = 0 \tag{c}$$

$$\tan q_2 = \frac{2P}{mg}$$

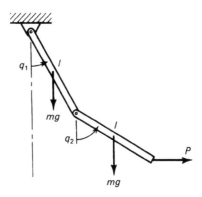

Figure E4.7. Multiple links under equilibrium.

4.8   LAGRANGIAN MULTIPLIERS

In the solution of a problem, it may at times be simpler to work with constrained coordinates. The differential work Equation 4.54 when written in the M constrained coordinates becomes

$$\overline{dW} = F_1 \, du_1 + F_2 \, du_2 + \cdots + F_M \, du_M \tag{4.60}$$

If forces $F_1, F_2, \ldots, F_M$ are conservative, Equation 4.60 reduces to

$$dW = -\frac{\partial V}{\partial u_1} du_1 - \frac{\partial V}{\partial u_2} du_2 - \cdots - \frac{\partial V}{\partial u_m} du_m \tag{4.61}$$

in which the overbar on dW has been removed because dW is now a complete differential. For equilibrium, dW must be zero. However, since the coordinates $u_1, u_2, \ldots, u_M$ are not independent, Equation 4.61 does not imply that the coefficients of the differential $du_1, du_2, \ldots, du_M$ are zero. If the number of generalized coordinates is N, where $N < M$, there must exist $(M - N) = L$ constraint equations of the form

$$\phi_1(u_1, u_2, \ldots, u_M) = 0$$
$$\phi_2(u_1, u_2, \ldots, u_M) = 0$$
$$\cdots \cdots \cdots \cdots \cdots \tag{4.62}$$
$$\phi_L(u_1, u_2, \ldots, u_M) = 0$$

Equation 4.62 can be written in the alternative form

$$\frac{\partial \phi_1}{\partial u_1} du_1 + \frac{\partial \phi_1}{\partial u_2} du_2 + \cdots + \frac{\partial \phi_1}{\partial u_M} du_M = 0$$

$$\frac{\partial \phi_2}{\partial u_1} du_1 + \frac{\partial \phi_2}{\partial u_2} du_2 + \cdots + \frac{\partial \phi_2}{\partial u_M} du_M = 0 \tag{4.63}$$

$$\cdots \cdots \cdots \cdots \cdots \cdots \cdots \cdots \cdots \cdots$$

$$\frac{\partial \phi_L}{\partial u_1} du_1 + \frac{\partial \phi_L}{\partial u_2} du_2 + \cdots + \frac{\partial \phi_L}{\partial u_M} du_M = 0$$

If each of Equations 4.63 is multiplied by an arbitrary function, the multiplier for the ith equation being denoted by λ_i, and the resulting products are added to Equation 4.61, then at equilibrium, the following equation is obtained:

$$\left(-\frac{\partial V}{\partial u_1} + \lambda_1 \frac{\partial \phi_1}{\partial u_1} + \cdots + \lambda_L \frac{\partial \phi_L}{\partial u_1} \right) du_1$$

$$+ \left(-\frac{\partial V}{\partial u_2} + \lambda_1 \frac{\partial \phi_1}{\partial u_2} + \cdots + \lambda_L \frac{\partial \phi_L}{\partial u_2} \right) du_2$$

$$+ \cdots + \left(-\frac{\partial V}{\partial u_M} + \lambda_1 \frac{\partial \phi_1}{\partial u_M} + \cdots + \lambda_L \frac{\partial \phi_L}{\partial u_M} \right) du_M = 0 \qquad (4.64)$$

Now, since $\lambda_1, \lambda_2, \ldots, \lambda_L$ are arbitrary, they can be so chosen that the expressions in the first L set of parentheses of Equation 4.64 are zero. This will leave $M - L = N$ expressions in parentheses in that equation, but since the system has N degrees of freedom, the coefficients of these expressions du_i can be assigned arbitrarily and it follows that the expressions in these remaining sets of parentheses should also each be zero. The foregoing reasoning leads to M equations, which together with the L constraint conditions will permit a solution for the M displacement coordinates $u_1, u_2, \ldots, u_M$ and the L multipliers, $\lambda_1, \lambda_2, \ldots, \lambda_L$. The arbitrary multipliers used in the method above are called *Lagrangian multipliers*.

Example 4.8

A massless rigid bar is suspended from the ceiling by three wires as shown in Figure E4.8. The elastic properties of the wires are

$$T_1 = b_1 u_1^{1/2}$$

$$T_2 = b_2 u_2$$

$$T_3 = b_3 u_3$$

where T represents tension in a wire, u the displacement of the end of the wire, and b a material property constant. If the bar supports a central vertical load F, find the tension in each wire.

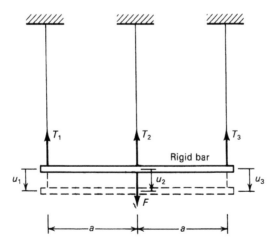

Figure E4.8. Rigid bar suspended by three wires.

Solution

The system has two degrees of freedom, but the three coordinates u_1, u_2, and u_3 will be used in the solution of the problem. These coordinates satisfy a constraint equation of the form

$$\phi = u_1 - 2u_2 + u_3 = 0 \tag{a}$$

The potential energy of the system comprises the strain energy stored in the wires and the potential energy of the load. It can easily be shown to be given by

$$V = \frac{2}{3}b_1 u_1^{3/2} + \frac{1}{2}b_2 u_2^2 + \frac{1}{2}b_3 u_3^2 - Fu_2 \tag{b}$$

The virtual work equation can be written as

$$dW = -\frac{\partial V}{\partial u_1}du_1 - \frac{\partial V}{\partial u_2}du_2 - \frac{\partial V}{\partial u_3}du_3 = 0$$

or

$$b_1 u_1^{1/2} + (b_2 u_2 - F)du_2 + b_3 u_3\,du_3 = 0 \tag{c}$$

The virtual work equation can also be obtained directly as

$$dW = -T_1\,du_1 - T_2\,du_2 - T_3\,du_3 + F_3\,du_2 = 0$$

which will lead to Equation c. The differentials du_1, du_2, and du_3 in Equation c are not independent since u_1, u_2, and u_3 are related by constraint equation a. That equation can be expressed as

$$d\phi = \frac{\partial \phi}{\partial u_1}du_1 + \frac{\partial \phi}{\partial u_2}du_2 + \frac{\partial \phi}{\partial u_3} = 0$$

or

$$du_1 - 2du_2 + du_3 = 0 \tag{d}$$

On multiplying Equation d by an arbitrary multiplier λ and adding to Equation c, we obtain

$$(b_1 u_1^{1/2} + \lambda)du_1 + (b_2 u_2 - F - 2\lambda)du_2 + (b_3 u_3 + \lambda)du_3 = 0 \tag{e}$$

If λ is chosen so that the expression in the first set of parentheses in Equation e is zero, that equation is left with only two differential coordinates: du_2 and du_3. Since the system has two degrees of freedom, these differentials can be assigned arbitrary values. It follows that the expressions in the remaining two sets of parentheses in Equation e should also be zero. We thus obtain the following equations

$$b_1 u_1^{1/2} + \lambda = 0$$
$$b_2 u_2 - F - 2\lambda = 0 \tag{f}$$
$$b_3 u_3 + \lambda = 0$$

Solution of Equation f along with Equation a gives

$$\lambda = \frac{b_1^2}{2}\left[\frac{b_2 + 4b_3}{b_2 b_3} - \sqrt{\left(\frac{b_2 + 4b_3}{b_2 b_3}\right)^2 + 8\frac{F}{b_1^2 b_2}}\right]$$

$$T_1 = -\lambda$$
$$T_2 = F + 2\lambda \tag{g}$$
$$T_3 = -\lambda$$

4.9 VIRTUAL WORK EQUATION FOR DYNAMICAL SYSTEMS

The principal of d'Alembert, described in Chapter 2, converts a problem of dynamics into an equivalent problem of static equilibrium. This means that the methods of analytical mechanics used for static problems can be applied to dynamical problems as well. For a system composed of M particles, the principle of d'Alembert can be stated as

$$\mathbf{F}_i - m_i\ddot{\mathbf{r}}_i = 0 \qquad i = 1, M \tag{4.65}$$

where $\mathbf{F}_i$ is the vector sum of the impressed forces and the forces of constraint acting on particle i, and $m_i\ddot{\mathbf{r}}_i$ is the force of inertia. At a particular instant of time, Equation 4.65 represents an equation of equilibrium. Therefore, if the system is given virtual displacements $\delta\mathbf{r}_i$ which are compatible with the constraints at that instant of time, the virtual work equation can be written as

$$\sum_{i=1}^{M}(\mathbf{F}_{ai} + \mathbf{F}_{ci} - m_i\ddot{\mathbf{r}}_i) \cdot \delta\mathbf{r}_i = 0 \tag{4.66}$$

where $\mathbf{F}_{ai}$ is the vector sum of the impressed forces and $\mathbf{F}_{ci}$ is the vector sum of the forces of constraints. Since virtual displacements do no work on the forces of constraints, Equation 4.66 becomes

$$\sum_{i=1}^{M}(\mathbf{F}_{ai} - m_i\ddot{\mathbf{r}}_i) \cdot \delta\mathbf{r}_i = 0 \tag{4.67}$$

It should be noted that virtual displacements do not occupy any time. In other words, the mathematical experimentation in which the system is given the virtual displacement takes place while the system is frozen in its state described by Equation 4.65. The real displacements, on the other hand, take place over real time and may no longer be admissible as virtual displacements. If that is so, $\delta\mathbf{r}_i$'s in Equation 4.67 cannot be replaced by $d\mathbf{r}_i$'s. As an example of a situation in which the real displacements are not admissible as virtual displacements, consider the simple pendulum shown in Figure 4.13 in which the string is being pulled up at a constant velocity so that the distance from the point of suspension to the point mass reduces with time.

The position of the pendulum at a certain instance of time is shown in Figure 4.13a. The virtual displacement $\delta\mathbf{r}$ given to the system at that time is along the tangent to an arc with center O and radius OA. Since this displacement is perpendicular to the force of constraint exerted by the string, the string tension does no work on it. The real displacement that takes place in a short time is shown as AB in Figure 4.13b. This displacement is not perpendicular to OA and is therefore not admissible as a virtual displacement. The force of constraint will, in fact, do work on the real displacement.

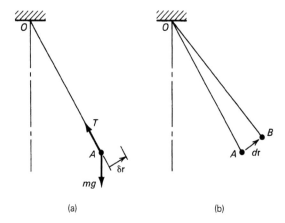

Figure 4.13. Pendulum of variable length: (a) virtual displacement; (b) real displacement.

The constraint condition for the pendulum of Figure 4.13 as given by Equation 4.15 is rhenomic because it contains time t explicitly. Whenever constraints are rhenomic, infinitesimal real displacements are not admissible as virtual displacements. Consider now a situation in which the constraints are scleronomic, so that dr_i's become admissible virtual displacements. Equation 4.67 can then be written as

$$\sum_{i=1}^{M} (\mathbf{F}_{ai} - m_i\ddot{\mathbf{r}}_i) \cdot d\mathbf{r}_i = 0 \qquad (4.68)$$

The second term in the expression above can be reduced as follows

$$\sum_{i=1}^{M} m_i\ddot{\mathbf{r}}_i \cdot d\mathbf{r}_i = \frac{d}{dt}\sum_{i=1}^{M} \left(\frac{1}{2}m_i\dot{\mathbf{r}}_i \cdot \dot{\mathbf{r}}_i\right) dt$$

$$= \frac{d}{dt}\left(\sum_{i=1}^{M} \frac{1}{2}m_i\dot{r}_i^2\right) dt \qquad (4.69)$$

$$= dT$$

where T is the kinetic energy of the system.

If the impressed forces can be derived from a potential energy function V of the type $V = V(r_1, r_2, r_3, \ldots, r_M)$

$$\sum_{i=1}^{M} \mathbf{F}_{ai} \cdot d\mathbf{r}_i = \sum_{i=1}^{M} -\frac{\partial V}{\partial \mathbf{r}_i} \cdot d\mathbf{r}_i$$

$$= -dV \qquad (4.70)$$

Note that if V contains t explicitly, Equation 4.70 no longer holds good because then

$$dV = \sum \frac{\partial V}{\partial \mathbf{r}_i} \cdot d\mathbf{r}_i + \frac{\partial V}{\partial t} dt \qquad (4.71)$$

As an example, the pendulum of Figure 4.13 has a potential energy function given by

$$V = l - (l - vt) \cos \theta \qquad (4.72)$$

which contains time explicitly. Equation 4.70 is not applicable to this pendulum. Substitution of Equations 4.69 and 4.70 into Equation 4.68 gives

$$-(dV + dT) = 0 \qquad (4.73a)$$

or

$$V + T = \text{constant} \qquad (4.73b)$$

Equation 4.73 states that in a system which is scleronomic in both the work function and the constraints, the sum of the potential and the kinetic energies is a constant. This principle of conservation of energy, although not sufficient for obtaining a solution to multi-degree-of-freedom systems, does give a complete solution for a single-degree-of-freedom system.

Example 4.9
A uniform rigid bar of mass M and length l is pinned at one end to a support and at the other end to a uniform cylinder of radius r and mass m as shown in Figure E4.9. The cylinder rolls without slipping inside a cylindrical surface of radius $l + r$. Find the equation of motion.

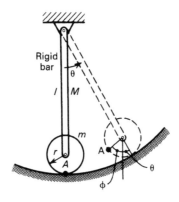

Figure E4.9. Cylinder rolling inside another cylinder.

Solution

If ϕ is the absolute rotation of the cylinder (see Fig. E4.9), then since the cylinder rolls without slip,

$$r(\theta + \phi) = (l + r)\theta$$
$$r\phi = l\theta \tag{a}$$

The potential energy of the system measured from the vertical position is

$$V = \frac{l}{2}(1 - \cos\theta)Mg + lmg(1 - \cos\theta) \tag{b}$$

The kinetic energy is given by

$$T = \frac{1}{2}\frac{Ml^2}{3}(\dot\theta)^2 + \frac{1}{2}\frac{mr^2}{2}(\dot\phi)^2 + \frac{1}{2}ml^2(\dot\theta)^2$$

$$= \frac{1}{6}Ml^2\dot\theta^2 + \frac{3}{4}ml^2\dot\theta^2 \tag{c}$$

Since the system is scleronomic, the energy is conserved so that

$$\frac{d}{dt}(V + T) = 0$$

or

$$\left(\frac{Mgl}{2} + mgl\right)\sin\theta\,\dot\theta + \left(\frac{Ml^2}{3} + \frac{3}{2}ml^2\right)\dot\theta\frac{d\dot\theta}{dt} = 0 \tag{d}$$

Since this would be true for all values of t, $\dot\theta$ can be canceled out, giving the following equation of motion

$$\left(\frac{Ml^2}{3} + \frac{3}{2}ml^2\right)\ddot\theta + \left(\frac{Mgl}{2} + mgl\right)\sin\theta = 0 \tag{e}$$

When the system executes small oscillations, $\sin\theta$ in Equation e can be replaced by θ. A possible solution to the equation of motion is then

$$\theta = A\sin\omega t \tag{f}$$

where ω, called the frequency of oscillation, obtained by substituting for θ and $\ddot\theta$ from Equation f into Equation e, is

$$\omega = \sqrt{\frac{g(m + M/2)}{l(M/3 + 3m/2)}} \tag{g}$$

Example 4.10

The system shown in Figure E4.10 is mounted in a vertical plane. It is set vibrating about its position of equilibrium. Find the equations of motion. The spring is unstretched when the bar is in a horizontal position.

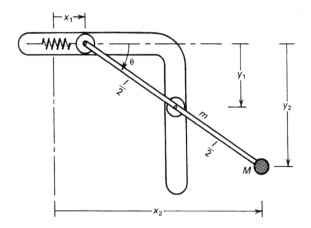

Figure E4.10. Vibrations of rigid bar moving in guides.

Solution

The potential energy of the system is given by

$$V = \frac{1}{2}kx_1^2 - mgy_1 - Mgy_2 \tag{a}$$

and the kinetic energy is

$$T = \frac{1}{2}m\dot{y}_1^2 + \frac{1}{2}\frac{ml^2}{12}\dot{\theta}^2 + \frac{1}{2}M\dot{y}_2^2 + \frac{1}{2}M\dot{x}_2^2 \tag{b}$$

Referring to Figure E4.10, we see that

$$x_1 = \frac{l}{2}(1 - \cos\theta)$$

$$x_2 = \frac{l}{2}(1 + \cos\theta)$$

$$y_1 = \frac{l}{2}\sin\theta \tag{c}$$

$$y_2 = l\sin\theta$$

From Equation c

$$\dot{x}_1 = +\frac{l}{2}\sin\theta\dot{\theta}$$

$$\ddot{x}_1 = +\frac{l}{2}\cos\theta\dot{\theta}^2 + \frac{l}{2}\sin\theta\ddot{\theta}$$

$$\dot{x}_2 = -\frac{l}{2}\sin\theta\dot{\theta}$$

$$\ddot{x}_2 = -\frac{l}{2}\cos\theta\dot{\theta}^2 - \frac{l}{2}\sin\theta\ddot{\theta}$$

$$\dot{y}_1 = \frac{l}{2}\cos\theta\dot{\theta}$$

(d)

$$\ddot{y}_1 = -\frac{l}{2}\sin\theta\dot{\theta}^2 + \frac{l}{2}\cos\theta\ddot{\theta}$$

$$\dot{y}_2 = l\cos\theta\dot{\theta}$$

$$\ddot{y}_2 = -l\sin\theta\dot{\theta}^2 + l\cos\theta\ddot{\theta}$$

Since the system is scleronomic, energy is conserved and

$$\frac{d}{dt}(V + T) = 0$$

or

$$kx_1\dot{x}_1 - mg\dot{y}_1 - Mg\dot{y}_2 + m\dot{y}_1\ddot{y}_1 + \frac{ml^2}{12}\dot{\theta}\ddot{\theta} + M\dot{x}_2\ddot{x}_2 + M\dot{y}_2\ddot{y}_2 = 0$$

(e)

On substituting from Equation d and canceling out $\dot{\theta}$, the following equation of motion is obtained

$$\left\{\left(\frac{m}{12} + \frac{M}{4}\right) + \left(\frac{m}{4} + \frac{3M}{4}\right)\cos^2\theta\right\}\ddot{\theta} - \left(\frac{m}{4} + \frac{3M}{4}\right)\sin\theta\cos\theta\dot{\theta}^2$$

$$+ \frac{k}{4}\sin\theta(1 - \cos\theta) - \frac{g}{l}\left(\frac{m}{2} + M\right)\cos\theta = 0$$

(f)

If the system is not in motion, $\dot{\theta}$ and $\ddot{\theta}$ are zero and Equation f will give the condition for static equilibrium (compare with Example 4.1).

4.10   HAMILTON'S EQUATION

In Section 4.9, it was pointed out that d'Alembert's principle transforms the dynamical problem into a problem of static equilibrium, so that the method of virtual work can be applied to its solution. The resulting virtual work equation is given by Equation 4.67. In that equation, the impressed forces $\mathbf{F}_{ai}$ may be derivable from a scalar work function, but the inertia forces $m_i\mathbf{r}_i$ cannot be similarly derived. Hamilton showed that by integrating the virtual work equation over time, the virtual work term $\sum m_i\ddot{\mathbf{r}}_i \cdot \delta\mathbf{r}_i$ can be replaced by the variation of another scalar function: the kinetic energy of the system. *Hamilton's principle* occupies an important position in analytical mechanics because it reduces the formulation of a dynamic problem to the variation of two scalar quantities: the work function and the kinetic energy, and because it is invariant under coordinate transformation. Multiplication of Equation 4.67 by dt and integration between limits t_1 and t_2 gives

$$\int_{t_1}^{t_2} \sum (\mathbf{F}_i - m_i\ddot{\mathbf{r}}_i) \cdot \delta\mathbf{r}_i \, dt = 0 \tag{4.74}$$

or

$$\int_{t_1}^{t_2} \sum \mathbf{F}_i \cdot \delta\mathbf{r}_i \, dt - \int_{t_1}^{t_2} \sum m_i\ddot{\mathbf{r}}_i \cdot \delta\mathbf{r}_i \, dt = 0 \tag{4.75}$$

The first integral on the left-hand side of Equation 4.75 can be written as $\int_{t_1}^{t_2} \overline{\delta W} \, dt$, where $\overline{\delta W}$ represents infinitesimal work. The second integral is re-duced as follows

$$\int_{t_1}^{t_2} \sum m_i\ddot{\mathbf{r}}_i \cdot \delta\mathbf{r}_i \, dt = \sum m_i\dot{\mathbf{r}}_i \cdot \delta\mathbf{r}_i \Big|_{t_1}^{t_2} - \int_{t_1}^{t_2} \sum m_i\dot{\mathbf{r}}_i \cdot \frac{d}{dt}(\delta\mathbf{r}_i) \, dt \tag{4.76}$$

If it is assumed that the path of all particles is prescribed at time t_1 and t_2, the variations δr_i's are zero at these times and the first term on the right-hand side of Equation 4.76 vanishes giving

$$\int_{t_1}^{t_2} \sum m_i\ddot{\mathbf{r}}_i \cdot \delta\mathbf{r}_i \, dt = - \int_{t_1}^{t_2} \sum m_i\dot{\mathbf{r}}_i \cdot \delta\left(\frac{d\mathbf{r}_i}{dt}\right) dt$$

$$= - \int_{t_1}^{t_2} \delta \sum \frac{1}{2} m_i\dot{\mathbf{r}}_i \cdot \dot{\mathbf{r}}_i \, dt$$

$$= - \int_{t_1}^{t_2} \delta \sum \frac{1}{2} m_i\dot{r}_i^2 \, dt$$

$$= - \int_{t_1}^{t_2} \delta T \, dt \tag{4.77}$$

Substitution of Equation 4.77 into Equation 4.75 gives

$$\int_{t_1}^{t_2} \overline{\delta W} + \delta T \, dt = 0 \tag{4.78}$$

Equation 4.78 is generally known as the *extended Hamilton's principle*. In the special case when the impressed forces can be derived from a scalar work function, $\overline{\delta W}$ becomes a complete variation and Equation 4.78 becomes

$$\int_{t_1}^{t_2} \delta(U + T) \, dt = 0$$

or

$$\delta \int_{t_1}^{t_2} (U + T) \, dt = 0 \tag{4.79}$$

Recalling that the potential energy function V is the negative of work function, Equation 4.79 can be written as

$$\delta \int_{t_1}^{t_2} (T - V) dt = 0 \qquad (4.80a)$$

or

$$\delta \int_{t_1}^{t_2} L \, dt = 0 \qquad (4.80b)$$

where $L = T - V$ is called the *Lagrangian*. In a majority of engineering systems, the impressed forces are derivable from a potential energy function. The special form of Hamilton's principle given by Equation 4.80 therefore applies to all such systems. Hamilton's principle is a variational principle which states that of all possible paths, a mechanical system under motion will take that path which makes the integral in Equation 4.80 a minimum.

4.11   LAGRANGE'S EQUATION

Hamilton's principle provides a complete formulation of a dynamical problem. However, for obtaining a solution to the problem, the Hamilton's integral formulation should be converted into one or more differential equations of motion. These equations are of second order in the generalized coordinates and their number is equal to the number of degrees of freedom of the system.

For holonomic constraints, the physical coordinates of the system can be expressed in terms of the generalized coordinates and the time variable as follows

$$\mathbf{r}_i = f_i(q_1, q_2, \ldots, q_N, t) \qquad (4.81)$$

Differentiation of Equation 4.81 gives

$$\dot{\mathbf{r}}_i = \sum_{j=1}^{N} \frac{\partial f_i}{\partial q_j} \dot{q}_j + \frac{\partial f_i}{\partial t} \qquad (4.82)$$

The kinetic energy T is given by

$$T = \sum_i \frac{1}{2} m_i \dot{\mathbf{r}}_i \cdot \dot{\mathbf{r}}_i \qquad (4.83)$$

Substitution of Equation 4.82 into Equation 4.83 indicates that the kinetic energy is a function of the following form

$$T = T(q_1, q_2, q_3, \ldots, q_N, \dot{q}_1, \dot{q}_2, \dot{q}_3, \dot{q}_N, t) \qquad (4.84)$$

Also, Equations 4.38 and 4.39 show that the infinitesimal virtual work is given by

$$\overline{\delta W} = \sum_{j=1}^{N} Q_j\, \delta q_j \qquad\qquad (4.38)$$

where $Q_j = \sum_{i=1}^{M} \mathbf{F}_i \cdot \partial \mathbf{r}_i/\partial q_j$.

Using Equations 4.84 and 4.38, the Hamilton's integral Equation 4.78 can be reduced to the following form by the methods of variational calculus

$$\int_{t_1}^{t_2} \sum_{j=1}^{N} \left\{ -\frac{d}{dt}\left(\frac{\partial T}{\partial \dot{q}_j}\right) + \frac{\partial T}{\partial q_j} + Q_j \right\} \delta q_j\, dt = 0 \qquad\qquad (4.85)$$

Since the virtual displacements δq_j's are independent and arbitrary, it is possible to select a set of values in which δq_j is zero for all values of j except $j = m$. Equation 4.85 will then reduce to

$$\int_{t_1}^{t_2} \left\{ -\frac{d}{dt}\left(\frac{\partial T}{\partial \dot{q}_m}\right) + \frac{\partial T}{\partial q_m} + Q_m \right\} \delta q_m\, dt = 0 \qquad\qquad (4.86)$$

Again, since δq_m is arbitrary, the integral in Equation 4.86 can be zero only when the integrand is zero. The reasoning above leads to the following N *Euler–Lagrange equations*

$$\frac{d}{dt}\left(\frac{\partial T}{\partial \dot{q}_j}\right) - \frac{\partial T}{\partial q_j} - Q_j = 0 \qquad \text{for } j = 1, 2, \ldots, N \qquad\qquad (4.87)$$

Next, consider the special case when the generalized forces can be derived from a potential energy function V, so that

$$\frac{\partial V}{\partial q_j} = -Q_j$$

$$= -\sum_{j=1}^{M} \mathbf{F}_i \cdot \frac{\partial \mathbf{r}_i}{\partial q_j} \qquad\qquad (4.88)$$

It is apparent from Equations 4.81 and 4.88 that V will be of the form

$$V = V(q_1, q_2, \ldots, q_N, t)$$

The *Lagranges equations* are now obtained from the special form of Hamilton's equation (Eq. 4.80)

$$\frac{d}{dt}\left(\frac{\partial L}{\partial \dot{q}_j}\right) - \frac{\partial L}{\partial q_j} = 0 \qquad j = 1, 2, \ldots, N \qquad\qquad (4.89)$$

On setting $L = T - V$ and noting that V does not depend on the velocities, $\dot{q}_j$, Equation 4.89 can be written in the alternative form

$$\frac{d}{dt}\left(\frac{\partial T}{\partial \dot{q}_j}\right) - \frac{\partial T}{\partial q_j} + \frac{\partial V}{\partial q_j} = 0 \qquad j = 1, 2, \ldots, N \tag{4.90}$$

In general, a system will be subjected to some forces that can be derived from a scalar function and some others that cannot be so derived. In the latter class are forces of friction, air resistance, and some types of impressed forces. The Lagrange equation for such systems can be written as

$$\frac{d}{dt}\left(\frac{\partial T}{\partial \dot{q}_j}\right) - \frac{\partial T}{\partial q_j} + \frac{\partial V}{\partial q_j} - Q_j = 0 \qquad j = 1, 2, \ldots, N \tag{4.91}$$

where the Q_j's now denote those forces that cannot be derived from a scalar function.

Example 4.11
Obtain the equation of motion for the single-degree-of-freedom system shown in Figure E4.11.

Solution
The kinetic energy, potential energy, and the infinitesimal work are given by the following expressions

$$T = \frac{1}{2}m\dot{q}^2$$

$$V = \frac{1}{2}kq^2 \tag{a}$$

$$\overline{\delta W} = (-c\dot{q} + p\sin\Omega t)\,\delta q = Q\,\delta q$$

The equation of motion is obtained by substituting Equation a into the Lagrange's equation

$$\frac{d}{dt}\left(\frac{\partial T}{\partial \dot{q}}\right) - \frac{\partial T}{\partial q} + \frac{\partial V}{\partial q} - Q = 0 \tag{b}$$

$$m\ddot{q} + c\dot{q} + kq - p\sin\Omega t = 0$$

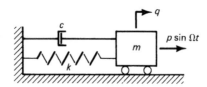

Figure E4.11. Motion of a simple single-degree-of-freedom system.

Example 4.12

Obtain the equation of motion for the simple pendulum shown in Figure E4.12, in which the inextensible string supporting the mass m is being pulled up at a constant velocity v.

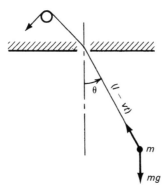

Figure E4.12. Simple pendulum of varying length.

Solution

Angle θ between the string and the vertical is chosen as the generalized coordinate. The kinetic energy is given by

$$T = \frac{1}{2}m\{(l - vt)\dot{\theta}\}^2 \qquad (a)$$

The potential energy is a function of θ and t, so that the total energy is not conserved.

$$V = \{l - (l - vt)\cos\theta\}mg \qquad (b)$$

Also

$$\frac{\partial T}{\partial \dot{\theta}} = m(l - vt)^2\dot{\theta}$$

$$\frac{d}{dt}\left(\frac{\partial T}{\partial \dot{\theta}}\right) = m(l - vt)^2\ddot{\theta} + 2m(l - vt)(-v)\dot{\theta} \qquad (c)$$

$$\frac{\partial V}{\partial \theta} = mg(l - vt)\sin\theta$$

Substitution of Equations c into the Lagrange's equation gives

$$m(l - vt)^2\ddot{\theta} - 2mv(l - vt)\dot{\theta} + mg(l - vt)\sin\theta = 0$$

or

$$(l - vt)\ddot{\theta} - 2v\dot{\theta} + g\sin\theta = 0 \qquad (d)$$

Example 4.13

Three interconnected rigid links, each of length l, are suspended from the ceiling by a hinge and subjected to a horizontal force $P\sin\Omega t$ at the free end of the lowest bar. Find the equations of motion.

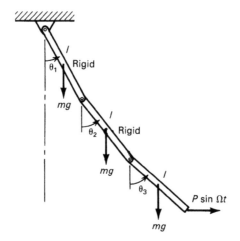

Figure E4.13. Motion of three interconnected links.

Solution

The system has three degrees of freedom. The three angles θ_1, θ_2 and θ_3 shown in Figure E4.13 are used as the generalized coordinates. The kinetic energy of each bar is composed of the energies due to the vertical and horizontal translations of the bar centroid, and the rotation of the bar about its centroid. The total kinetic energy of the system is given by

$$T = \frac{7}{6}ml^2\dot{\theta}_1^2 + \frac{4}{6}ml^2\dot{\theta}_2^2 + \frac{1}{6}ml^2\dot{\theta}_3^2 + \frac{3}{2}ml^2\dot{\theta}_2\dot{\theta}_1\cos(\theta_2 - \theta_1)$$

$$+ \frac{1}{2}ml^2\dot{\theta}_3\dot{\theta}_1\cos(\theta_3 - \theta_1) + \frac{1}{2}ml^2\dot{\theta}_3\dot{\theta}_2\cos(\theta_3 - \theta_2) \qquad \text{(a)}$$

The potential energy measured from a horizontal datum through the point of suspension is

$$V = -\frac{5}{2}mgl\cos\theta_1 - \frac{3}{2}mgl\cos\theta_2 - \frac{1}{2}mgl\cos\theta_3 \qquad \text{(b)}$$

The virtual work is given by

$$\delta W = \delta[P\sin\Omega t\{l\sin\theta_1 + l\sin\theta_2 + l\sin\theta_3\}]$$

$$= Pl\sin\Omega t[\cos\theta_1\,\delta\theta_1 + \cos\theta_2\,\delta\theta_2 + \cos\theta_3\,\delta\theta_3] \qquad \text{(c)}$$

so that the generalized forces are

$$Q_1 = Pl\sin\Omega t\cos\theta_1$$
$$Q_2 = Pl\sin\Omega t\cos\theta_2 \qquad \text{(d)}$$
$$Q_3 = Pl\sin\Omega t\cos\theta_3$$

Substitution of Equations a, b, and d in Lagrange's equations (Eq. 4.91) gives the following equations of motion

$$\frac{7}{3}ml^2\ddot{\theta}_1 + \frac{3}{2}ml^2\ddot{\theta}_2\cos(\theta_2 - \theta_1) + \frac{1}{2}ml^2\ddot{\theta}_3\cos(\theta_3 - \theta_1)$$

$$-\frac{3}{2}ml^2\dot{\theta}_2^2\sin(\theta_2 - \theta_1) - \frac{1}{2}ml^2\dot{\theta}_3^2\sin(\theta_3 - \theta_1) \tag{e}$$

$$+\frac{5}{2}mgl\sin\theta_1 - Pl\sin\Omega t\cos\theta_1 = 0$$

$$\frac{4}{6}ml^2\ddot{\theta}_2 + \frac{3}{2}ml^2\ddot{\theta}_1\cos(\theta_2 - \theta_1) + \frac{1}{2}ml^2\ddot{\theta}_3\cos(\theta_3 - \theta_2)$$

$$\frac{3}{2}ml^2\dot{\theta}_1^2\sin(\theta_2 - \theta_1) - \frac{1}{2}ml^2\dot{\theta}_3^2\sin(\theta_3 - \theta_1) \tag{f}$$

$$+\frac{3}{2}mgl\sin\theta_2 - Pl\sin\Omega t\cos\theta_2 = 0$$

$$\frac{1}{3}ml^2\ddot{\theta}_3 + \frac{1}{2}ml^2\ddot{\theta}_1\cos(\theta_3 - \theta_1) + \frac{1}{2}ml^2\ddot{\theta}_2\cos(\theta_3 - \theta_2)$$

$$\frac{1}{2}ml^2\dot{\theta}_1^2\sin(\theta_3 - \theta_1) + \frac{1}{2}ml^2\dot{\theta}_2^2\sin(\theta_3 - \theta_2) \tag{g}$$

$$+\frac{1}{2}mgl\sin\theta_3 - Pl\sin\Omega t\cos\theta_3 = 0$$

Equations e, f, and g are highly nonlinear. If the system undergoes small vibrations, terms of second and higher order in θ can be neglected, and $\sin\theta \approx \theta$, $\cos\theta \approx 1$. The resulting equations of motion are now linear and can be written in matrix form as follows

$$ml^2\begin{bmatrix} 7/3 & 3/2 & 1/2 \\ 3/2 & 2/3 & 1/2 \\ 1/2 & 1/2 & 1/3 \end{bmatrix}\begin{bmatrix} \ddot{\theta}_1 \\ \ddot{\theta}_2 \\ \ddot{\theta}_3 \end{bmatrix} + mgl\begin{bmatrix} 5/2 & & \\ & 3/2 & \\ & & 1/2 \end{bmatrix}\begin{bmatrix} \theta_1 \\ \theta_2 \\ \theta_3 \end{bmatrix}$$

$$= Pl\sin\Omega t\begin{bmatrix} 1 \\ 1 \\ 1 \end{bmatrix} \tag{h}$$

Example 4.14

A uniform cylinder of mass m rolls without slip on a cart of mass M to which it is connected by a spring of stiffness k_2 and a dashpot of coefficient c_2. The cart itself rolls on a horizontal surface and is connected to a wall by a spring of stiffness k_1 and a dashpot of constant c_1. Find the equations of free vibrations of the system.

Solution

As shown in Figure E4.14, the system has two degrees of freedom, the absolute translations of the cart and the cylinder. The angular velocity of the cylinder $\dot{\theta}$ is related to the two generalized

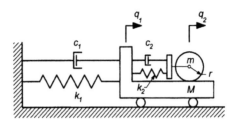

Figure E4.14. Cylinder rolling on a moving cart.

coordinates as follows

$$r\dot{\theta} = \dot{q}_2 - \dot{q}_1 \tag{a}$$

The kinetic and the potential energies of the system are given by

$$T = \frac{1}{2}M\dot{q}_1^2 + \frac{1}{2}m\dot{q}_2^2 + \frac{1}{2}\frac{mr^2}{2}\left(\frac{\dot{q}_2 - \dot{q}_1}{r}\right)^2 \tag{b}$$

$$V = \frac{1}{2}k_1 q_1^2 + \frac{1}{2}k_2(q_2 - q_1)^2 \tag{c}$$

The virtual work equation is

$$\begin{aligned}
\overline{\delta W} &= -c_2(\dot{q}_2 - \dot{q}_1)\,\delta q_2 + c_2(\dot{q}_2 - \dot{q}_1)\,\delta q_1 - c_1\dot{q}_1\,\delta q_1 \\
&= (c_2\dot{q}_2 - c_2\dot{q}_1 - c_1\dot{q}_1)\,\delta q_1 + (-c_2\dot{q}_2 + c_2\dot{q}_1)\,\delta q_2
\end{aligned} \tag{d}$$

The generalized forces are thus

$$\begin{aligned}
Q_1 &= c_2(\dot{q}_2 - \dot{q}_1) - c_1\dot{q}_1 \\
Q_2 &= -c_2(\dot{q}_2 - \dot{q}_1)
\end{aligned} \tag{e}$$

Substitution of Equations b, c, and e into Lagrange's equations gives the following equations of motion in the matrix form

$$\begin{aligned}
&\begin{bmatrix} M + \frac{m}{2} & -\frac{m}{2} \\ -\frac{m}{2} & \frac{3m}{2} \end{bmatrix} \begin{bmatrix} \ddot{q}_1 \\ \ddot{q}_2 \end{bmatrix} + \begin{bmatrix} c_1 + c_2 & -c_2 \\ -c_2 & c_2 \end{bmatrix} \begin{bmatrix} \dot{q}_1 \\ \dot{q}_2 \end{bmatrix} \\
&+ \begin{bmatrix} k_1 + k_2 & -k_2 \\ -k_2 & k_2 \end{bmatrix} \begin{bmatrix} q_1 \\ q_2 \end{bmatrix} = \begin{bmatrix} 0 \\ 0 \end{bmatrix}
\end{aligned} \tag{f}$$

4.12   CONSTRAINT CONDITIONS AND LAGRANGIAN MULTIPLIERS

At times, it is more convenient to work with constrained coordinates rather than with generalized coordinates. When coordinates q are related through one or more constraint conditions, the variations δq_j in Equation 4.85 are not independent and that equation does not lead to Lagrange's equations (Eq. 4.87). In such situations, the method of Lagrange's multipliers can be used effectively in the solution of the problem. The method is illustrated by the example that follows.

Example 4.15
A circular cylinder of mass m and radius R_1 rolls without slip on a fixed cylinder of radius R_2 as shown in Figure E4.15. Find the equations of motion and the condition when the rolling cylinder will leave the surface of the fixed cylinder.

Solution
Let the motion be described by the two constrained coordinates, given by, θ, the angle that the line joining the centers of the cylinder makes with the vertical, and ϕ, the absolute rotation of the rolling cylinder. These coordinates are related by the following constraint equation, which applies as long as contact persists between the two cylinders

$$R_2\theta - R_1(\phi - \theta) = 0 \tag{a}$$

The kinetic energy of the rolling cylinder is

$$T = \frac{1}{2}m(R_1 + R_2)^2(\dot{\theta})^2 + \frac{1}{2}\frac{mR_1^2}{2}(\dot{\phi})^2 \tag{b}$$

and the potential energy is

$$V = A + mg(R_1 + R_2)\cos\theta \tag{c}$$

where A is a constant.

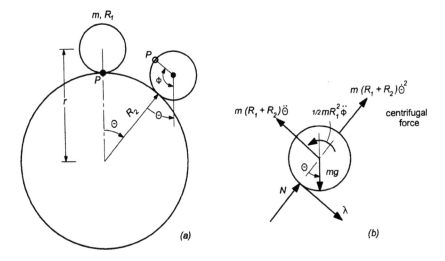

Figure E4.15. Cylinder rolling on a cylindrical surface.

The equations of motion are obtained by minimizing the functional

$$\int_{t_1}^{t_2} \frac{1}{2} m(R_1 + R_2)^2(\dot{\theta})^2 + \frac{mR_1^2}{4}(\dot{\phi})^2 - A - mg(R_1 + R_2)\cos\theta$$

$$+\lambda[(R_1 + R_2)\theta - R_1\phi]\,dt \tag{d}$$

The corresponding Lagrange's equations are

$$m(R_1 + R_2)\ddot{\theta} - mg\sin\theta - \lambda = 0 \tag{e}$$

$$\frac{1}{2}mR_1^2\ddot{\phi} + \lambda R_1 = 0 \tag{f}$$

Substituting for ϕ in terms of θ from Equation a and eliminating λ between Equations e and f, we get

$$\frac{3}{2}(R_1 + R_2)\ddot{\theta} - g\sin\theta = 0 \tag{g}$$

which is the required equation of motion.

Referring to Figure E4.15b, which shows the forces acting on the rolling cylinder, including those due to inertia, Equations e and f are easily recognized as the equations of equilibrium and λ can be interpreted as the force of friction at the contact between the two cylinders. Also, the normal reaction N that the stationary cylinder exerts on the rolling cylinder is given by

$$N = mg\cos\theta - m(R_1 + R_2)(\dot{\theta})^2 \tag{h}$$

This force is positive as long as $g\cos\theta > (R_1 + R_2)\dot{\theta}^2$, after which contact ceases.

4.13 LAGRANGE'S EQUATIONS FOR DISCRETE MULTI-DEGREE-OF-FREEDOM SYSTEMS

The equations of motion for a linear discrete multi-degree-of-freedom system can, of course, also be obtained by the applications of Lagrange's equations. Consider, for example, an N degree-of-freedom system with generalized coordinates $q_1, q_2, \ldots, q_N$. The strain energy stored in the system is given by

$$V = \frac{1}{2}\mathbf{q}^T\mathbf{f}_S$$

$$= \frac{1}{2}\mathbf{q}^T\mathbf{K}\mathbf{q} \tag{4.92}$$

where $\mathbf{f}_S$ is the vector of forces that cause the displacements and $\mathbf{K}$ is the stiffness matrix.

Equation 4.92 is called a *quadratic form*. Since the strain energy is always positive, the quadratic form of Equation 4.92 is always positive and is called a positive definite quadratic form. The kinetic energy is given by an expression similar to that for the strain energy but involving the generalized velocities instead of the displacements. Thus

$$T = \frac{1}{2}\dot{\mathbf{q}}^T\mathbf{M}\dot{\mathbf{q}} \tag{4.93}$$

Equation 4.93 is also a positive definite quadratic form. Differentiation of Equation 4.92 with respect to q_j gives

$$\frac{\partial V}{\partial q_j} = \frac{1}{2}[0 \quad 0 \quad \cdots \quad 1 \quad \cdots \quad 0]\mathbf{Kq} + \frac{1}{2}\mathbf{q}^T\mathbf{K}\begin{bmatrix} 0 \\ 0 \\ \vdots \\ 1 \\ \vdots \\ 0 \end{bmatrix} \tag{4.94}$$

where the row vector in the first term on the right-hand side has a 1 in column j with zeros elsewhere, while the column vector in the second term has a 1 in row j with zeros elsewhere.

Because $\mathbf{K}$ is symmetric, the two terms on the right-hand side of Equation 4.94 are the transpose of each other, and since both are scalars, they must be equal. Equation 4.94 therefore becomes

$$\frac{\partial V}{\partial q_j} = [k_{j1} \quad k_{j2} \quad \cdots \quad k_{jN}]\mathbf{q} \tag{4.95}$$

In a similar manner

$$\frac{\partial T}{\partial \dot{q}_j} = [m_{j1} \quad m_{j2} \quad \cdots \quad m_{jN}]\dot{\mathbf{q}} \tag{4.96}$$

and

$$\frac{d}{dt}\frac{\partial T}{\partial \dot{q}_j} = [m_{j1} \quad m_{j2} \quad \cdots \quad m_{jN}]\ddot{\mathbf{q}} \tag{4.97}$$

Since $\partial T/\partial q_j = 0$, Lagrange's equations become

$$[m_{j1} \quad m_{j2} \quad \cdots \quad m_{jN}]\ddot{\mathbf{q}} + [k_{j1} \quad k_{j2} \quad \cdots \quad k_{jN}]\mathbf{q} = Q_j$$
$$\text{for} \quad j = 1, 2, \ldots, N \tag{4.98a}$$

or

$$\mathbf{M}\ddot{\mathbf{q}} + \mathbf{Kq} = \mathbf{Q} \tag{4.98b}$$

where $\mathbf{Q}$ is the vector of applied forces as well as the other nonconservative forces. Equation 4.98b represents the equations of motion of a linear discrete multi-degree-of-freedom system.

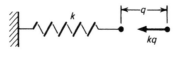

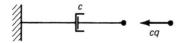

Figure 4.14. Analogy between spring force and damping force.

4.14   RAYLEIGH'S DISSIPATION FUNCTION

When a spring of stiffness k is stretched, the virtual work done by the spring force in moving through a distance δq is given by

$$\delta W = -kq\,\delta q = \frac{\partial}{\partial q}\left(-\frac{1}{2}kq^2\right)\delta q = \frac{\partial U}{\partial q}\,\delta q \qquad (4.99)$$

where U, the work function, is equal to $-\frac{1}{2}kq^2$. As mentioned earlier, the negative of the work done is designated as the potential energy function or the strain energy function of the spring, and the spring force is obtained from

$$F_s = -\frac{\partial V}{\partial q} = -kq \qquad (4.100)$$

The dissipative forces arising in mechanical systems due to such sources as air resistance, internal friction, and acoustic vibrations are usually assumed to be proportional to the velocities along the physical coordinates and opposed to the motion. For a single coordinate, for example, the dissipative force or the damping force can be represented as arising from the motion of a dashpot in a viscous medium, as shown in Figure 4.14. If the dashpot constant is denoted by c, the damping force, F_d, is given by

$$F_d = -c\dot{q} \qquad (4.101)$$

The virtual work done by the damping force is then

$$\overline{\delta W} = -c\dot{q}\,\delta q \qquad (4.102)$$

Considering the analogy between the spring forces and the forces of damping, we can obtain a scalar function R of the generalized velocity $\dot{q}$ from which the damping force F_d is obtained by an expression similar to Equation 4.100

$$F_d = -\frac{\partial R}{\partial \dot{q}} \qquad (4.103)$$

Function R is called the *Rayleigh dissipation function*; for the single dashpot shown, it is easily seen to be equal to $\frac{1}{2}c\dot{q}^2$. In a general case, the damping forces will be functions of more than one generalized velocity, and the dissipation function R should be such that

$$\overline{\delta W} = \sum_{i=1}^{N} -\frac{\partial R}{\partial \dot{q}_i}\delta q_i \qquad (4.104)$$

Assuming that the generalized forces consist only of forces derived from a potential energy function and damping forces derived from a dissipation function as in Equation 4.104, Equation 4.87 becomes

$$\frac{d}{dt}\left(\frac{\partial T}{\partial \dot{q}_j}\right) - \frac{\partial T}{\partial q_j} + \frac{\partial V}{\partial q_j} + \frac{\partial R}{\partial \dot{q}_j} = 0 \qquad \text{for } j = 1, 2, \dots, N$$

or

$$\frac{d}{dt}\left(\frac{\partial L}{\partial \dot{q}_j}\right) - \frac{\partial L}{\partial q_j} + \frac{\partial R}{\partial \dot{q}_j} = 0 \qquad (4.105)$$

Example 4.16
The mechanical system shown in the plan view in Figure E4.16 consists of three masses constrained to move between guides. Find the equations of motion.

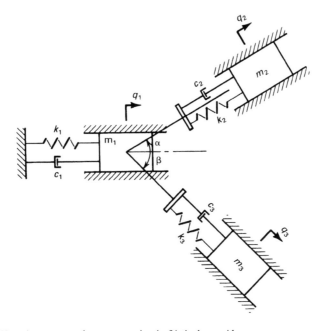

Figure E4.16. Three interconnected masses moving in frictionless guides.

Solution

The motion of the system is described by the three generalized coordinates q_1, q_2, and q_3 shown in Figure E4.16. The strain energy of a spring depends on the difference of the displacements parallel to the spring of its terminal points. The potential energy function of the system is thus given by

$$V = \frac{1}{2}k_1 q_1^2 + \frac{1}{2}k_2(q_2 - q_1 \cos \alpha)^2 + \frac{1}{2}k_3(q_3 - q_1 \cos \beta)^2 \tag{a}$$

In an entirely analogous manner, the dissipation function is obtained as

$$R = \frac{1}{2}c_1 \dot{q}_1^2 + \frac{1}{2}c_2(\dot{q}_2 - \dot{q}_1 \cos \alpha)^2 + \frac{1}{2}(\dot{q}_3 - \dot{q}_1 \cos \beta)^2 \tag{b}$$

The kinetic energy of the system is

$$T = \frac{1}{2}m_1 \dot{q}_1^2 + \frac{1}{2}m_2 \dot{q}_2^2 + \frac{1}{2}m_3 \dot{q}_3^2 \tag{c}$$

Lagrange's equations are therefore

$$
\begin{aligned}
&m_1 \ddot{q}_1 + k_1 q_1 - k_2(q_2 - q_1 \cos \alpha)\cos \alpha - k_3(q_3 - q_1 \cos \beta)\cos \beta \\
&\quad + c_1 \dot{q}_1 - c_2(\dot{q}_2 - \dot{q}_1 \cos \alpha)\cos \alpha - c_3(\dot{q}_3 - \dot{q}_1 \cos \beta)\cos \beta = 0 \\
&m_2 \ddot{q}_2 + k_2(q_2 - q_1 \cos \alpha) - c_2(\dot{q}_2 - \dot{q}_1 \cos \beta) = 0 \\
&\quad + m_3 \ddot{q}_3 + k_3(q_3 - q_1 \cos \beta) + c_3(\dot{q}_3 - \dot{q}_1 \cos \beta) = 0
\end{aligned} \tag{d}
$$

The equations above can be written in matrix form as follows

$$\mathbf{M\ddot{q}} + \mathbf{C\dot{q}} + \mathbf{Kq} = 0$$

$$\mathbf{M} = \begin{bmatrix} m_1 & & \\ & m_2 & \\ & & m_3 \end{bmatrix}$$

$$\mathbf{K} = \begin{bmatrix} k_1 + k_2 \cos^2 \alpha + k_3 \cos^2 \beta & -k_2 \cos \alpha & -k_3 \cos \beta \\ -k_2 \cos \alpha & k_2 & 0 \\ -k_3 \cos \beta & 0 & k_3 \end{bmatrix} \tag{e}$$

$$\mathbf{C} = \begin{bmatrix} c_1 + c_2 \cos^2 \alpha + c_3 \cos^2 \beta & -c_2 \cos \alpha & -c_3 \cos \beta \\ -c_2 \cos \alpha & c_2 & 0 \\ -c_3 \cos \beta & 0 & c_3 \end{bmatrix}$$

$$\mathbf{q}^T = [q_1 \quad q_2 \quad q_3]$$

SELECTED READINGS

Fowles, G.R. 1986. *Analytical Mechanics*. Philadelphia, PA: Saubders College Publishing. 4th Edition.

Goldstein, H. 1980. *Classical Mechanics*. Don Mills, Ontario: Addison-Wesley. 2nd Edition.

Greenwood, D.T. 1977. *Classical Dynamics.* Englewood Cliffs: Prentice Hall.

Lanczos, C. 1970. *The Variational Principles of Mechanics.* University of Toronto Press. 4th Edition.

Leipholtz, H.H.E. 1978. *Six Lectures on Variational Principles in Structural Engineering.* University of Waterloo: Solid Mechanics Division.

Reddy, J.N. 1984. *Energy and Variational Methods in Applied Mechanics.* New York: John Wiley.

Rosenberg, R.M. 1977. *Analytical Dynamics of Discrete Systems.* New York: Plenum Press.

PROBLEMS

4.1   A hollow cylinder of mass M, internal radius r_1 and external radius r_2 rolls without slipping inside a cylindrical surface of radius R (Fig. P4.1). By applying Lagrange's equation, obtain the equation of motion for small vibrations of the system about a position of equilibrium.

4.2   Obtain expressions for the potential and kinetic energies of the system shown in Figure P4.2. Using the energy expressions derived by you, formulate the equations of motion. Linearize these equations assuming small vibrations.

4.3   In an attempt to reduce the vibrations of a floor deck, a vibration absorber is suspended from midspan of the deck. The floor deck can be idealized as a simply supported uniform beam of span L, mass $\bar{m}$ per unit length and flexural rigidity EI. The vibration absorber consists of a spring of stiffness k and a mass M attached to the deck as shown in Figure P4.3. Taking u_1 and u_2 as the generalized coordinates and assuming that the vibration shape of the floor deck is $u = u_1 \sin(\pi x/L)$, obtain the Lagrange's equations of motion.

4.4   Solve Problem 3.1 by using energy expressions and Lagrange's equations.

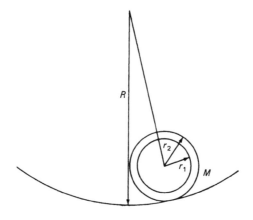

Figure P4.1.

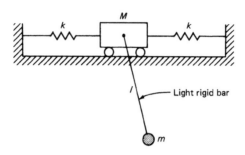

Figure P4.2.

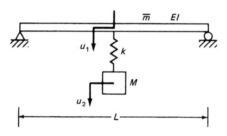

Figure P4.3.

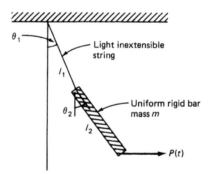

Figure P4.5.

4.5   Obtain the equations of motion for the system shown in Figure P4.5 by using Hamilton's principle.

4.6   A slider of mass 3 kg slides in vertical frictionless guides and is restrained by a spring of stiffness 10 kN/m as shown in Figure P4.6. The slider is attached by a rigid link of mass 1 kg and length 1 m to a circular disc of mass 5 kg which rolls without slipping on a horizontal plane. In the position of equilibrium, the rigid link makes an angle of 45° with the vertical. Obtain the equations of motion for small vibrations of the system about its equilibrium position.

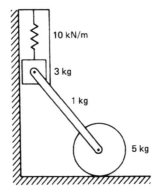

Figure P4.6.

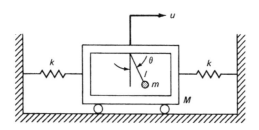

Figure P4.7.

4.7 Obtain the equations of motion for the system shown in Figure P4.7 in terms of generalized coordinates u and θ. Linearize the equations assuming small vibrations.

4.8 The axial mass moment of inertia of a wheel-axle assembly of weight W is determined by performing an experiment in which the assembly is allowed to roll without slipping on a pair of curved rails of radius R as shown in Figure P4.8. The motion of the system can be described in terms of the constrained coordinates θ and ϕ shown in the figure. Write the energy expressions in terms of the coordinates shown. Then using the concept of Lagrangian multiplier, obtain the equations of motion. Can a physical meaning be assigned in this case to the Lagrangian multiplier?

4.9 The point of attachment A of the simple pendulum shown in Figure P4.9 slides in a vertical slot; its displacement is given by $y = G \sin \Omega t$. Obtain the equation of motion for the pendulum mass assuming that the vibration amplitude is small.

4.10 A cylinder of mass M and radius R rolls without slipping on a horizontal surface (Fig. P4.10). It is restrained by a spring of stiffness k and a damper with coefficient c. A simple pendulum consisting of mass m

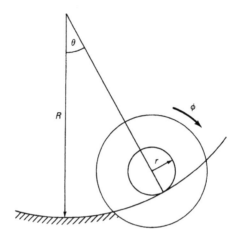

Figure P4.8.

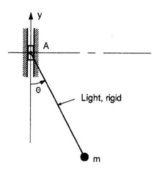

Figure P4.9.

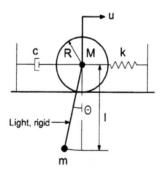

Figure P4.10.

and a light rigid bar is attached by a pin to the center of the cylinder. Obtain the equations of motion for the cylinder and the pendulum considering small vibrations and assuming that: (**a**) the pin connecting the cylinder and the pendulum is locked so that the swing of the pendulum is equal to the roll of the cylinder; (**b**) the pendulum is free to rotate about the pin.

PART 2

CHAPTER 5

Free-vibration response: Single-degree-of-freedom system

5.1 INTRODUCTION

The procedures used to formulate the equation of motion for a single-degree-of-freedom system were described in Chapter 2. When damping is viscous, the equation of motion invariably takes the form

$$m\ddot{u} + c\dot{u} + ku = p \tag{5.1}$$

Provided that m, c, and k do not vary with time, Equation 5.1 represents a linear second-order differential equation. Its solution depends not only on the nature of applied force, but also on initial conditions from which motion is started. In fact, even when p is zero, the system will undergo vibrations if it is given an initial displacement or an initial velocity or both.

The response of a system that is not subjected to any external force but is excited by initial disturbances alone is called *free-vibration response*. It is important to study such a response, not only because practical situations may arise when vibrations are excited by initial disturbances, but also because, when response to an applied force is desired, the complete solution must include a free-vibration component.

When p is zero, Equation 5.1 becomes a homogeneous second-order differential equation of motion. In this chapter, we study the solution of such an equation. We begin by analyzing the free-vibration response of a system in which the damping resistance is absent or negligible. Although systems without damping resistance do not exist in practice, the analysis of their response provides an insight into the nature of damped vibrations. Following the presentation of undamped free-vibration response, we discuss the free-vibration response of a damped system. As described in Chapter 2, damping forces may be of several different kinds. The easiest to handle mathematically is viscous damping. Other types of damping are hysteretic damping and Coulomb damping. We discuss the analysis of free damped response under each of the three types of damping resistances mentioned above.

5.2   UNDAMPED FREE VIBRATION

When damping is absent, the equation of motion becomes

$$m\ddot{u} + ku = 0 \tag{5.2}$$

A possible solution of Equation 5.2 is of the form

$$u = Ge^{\lambda t} \tag{5.3}$$

where G and λ are arbitrary constants to be determined. Substitution of Equation 5.3 in Equation 5.2 gives

$$G\lambda^2 m e^{\lambda t} + Gke^{\lambda t} = 0 \tag{5.4}$$

Equation 5.4 will be satisfied provided that $G = 0$, but as evident from Equation 5.3, this leads to $u = 0$ and no motion takes place. For motion to take place, G must be nonzero and can be canceled from Equation 5.4. On canceling $e^{\lambda t}$ as well, we obtain the *characteristic equation*

$$\lambda^2 m + k = 0 \tag{5.5a}$$

or

$$\lambda = \pm i\omega \tag{5.5b}$$

where $\omega = \sqrt{k/m}$.

The general solution of Equation 5.2 now becomes

$$u = G_1 e^{i\omega t} + G_2 e^{-i\omega t} \tag{5.6}$$

By using de Moivre's theorem, Equation 5.6 can be written in the alternative form

$$u = A \cos \omega t + B \sin \omega t \tag{5.7}$$

where A and B are arbitrary constants. To determine the arbitrary constants, we use the specified initial conditions of displacement and velocity at time $t = 0$.

$$\left. \begin{array}{c} u = u_0 \\ \dot{u} = v_0 \end{array} \right\} \quad \text{at} \quad t = 0 \tag{5.8}$$

Substitution of the initial conditions gives $A = u_0$ and $B = v_0/\omega$, so that Equation 5.7 becomes

$$\begin{aligned} u &= u_0 \cos \omega t + \frac{v_0}{\omega} \sin \omega t \\ &= \rho \sin(\omega t + \phi) \end{aligned} \tag{5.9a}$$

where

$$\rho = \sqrt{u_0^2 + \left(\frac{v_0}{\omega}\right)^2}$$

$$\tan \phi = \frac{u_0 \omega}{v_0}$$

(5.9b)

Equation 5.9 has been plotted in Figure 5.1c. It shows an oscillatory motion which repeats itself. Specifically, on examining the response at two instances of time $2\pi/\omega$ apart, namely t_1 and $(2\pi/\omega) + t_1$, we get

$$u(t_1) = \rho \sin(\omega t_1 + \phi)$$

(5.10a)

$$u\left(\frac{2\pi}{\omega} + t_1\right) = \rho \sin\left\{\left(\frac{2\pi}{\omega} + t_1\right)\omega + \phi\right\}$$
$$= \rho \sin(\omega t_1 + \phi)$$
$$= u(t_1)$$

(5.10b)

It is easily proved that the velocity $\dot{u}$ is also the same at the two instances of time. This implies that the motion repeats itself after $T = 2\pi/\omega$ seconds. This interval of time is called the period of motion. It takes T seconds from the time the system passes through a certain configuration moving in a certain direction until it next passes through the same configuration moving in an identical direction. This phase of motion is referred to as one *cycle of motion*. For the simple block–spring system of Figure 5.1a, the motion from the time the block is, say, in its zero-displacement position and moving to the right until the next instant of time when it is again in that position and is moving to the right can be described as one cycle of motion. The system executes f cycles of motion in 1 s, where f, called the *natural frequency*, is given by

$$f = \frac{1}{T} = \frac{\omega}{2\pi} \text{ hertz (Hz)}$$

(5.11)

The maximum displacement of the system is equal to ρ and is called the *amplitude* of vibration. It remains constant with time. Angle ϕ in Equation 5.9a is called the *phase angle*. The motion represented by Equation 5.9a and shown in Figure 5.1 is a harmonic motion of the simplest type and is referred to as a *simple harmonic motion*.

5.2.1 *Phase plane diagram*

For conceptual visualization of a simple harmonic motion, it is useful to imagine a vector of length ρ starting off at an angle ϕ from the horizontal time axis and rotating anticlockwise with a constant angular velocity ω rad/s as shown in

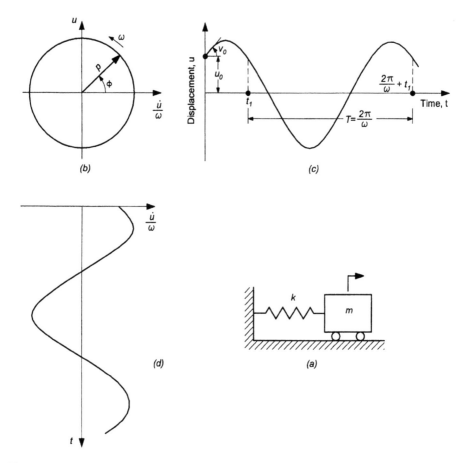

Figure 5.1. Free undamped vibrations of a single-degree-of-freedom system.

Figure 5.1b. The projection of the rotating vector on the vertical axis at time t is equal to $\rho \sin(\omega t + \phi)$ and gives the displacement of the system at that time. It is readily seen that as the rotating vector goes through one complete round, the system completes one cycle of motion. The angular velocity of the rotating vector, ω, is called the *circular frequency* of the system and is related to T and to f as follows

$$\omega = 2\pi f$$

$$= \frac{2\pi}{T} \tag{5.12}$$

The rotating vector representation provides a convenient graphical method of obtaining the free-vibration response. It can also be extended to obtain the response under the action of an applied force.

As already pointed out, the displacement u of a freely-vibrating system is given by Equation 5.9a, while the velocity $\dot{u}$ is obtained by differentiating Equation 5.9a. It is useful to define a velocity function defined by $\dot{u}/\omega$ and having the units of displacement. An expression for the velocity function can be obtained by differentiating Equation 5.9a and dividing by ω. Thus

$$\frac{\dot{u}}{\omega} = \rho \cos(\omega t + \phi) \tag{5.13}$$

Equations 5.9a and 5.13 are parametric equations of a circle of radius ρ. They represent the circle described by the rotating vector. The projection of this vector on horizontal axis is equal to the velocity function $\dot{u}/\omega$ and can be used to construct the velocity–time diagram shown in Figure 5.1d. As already noted, the vertical component of the rotating vector is equal to the displacement u and gives the displacement–time diagram shown in Figure 5.1c.

The curve described by the rotating vector is also referred to as the *phase plane diagram*, in which $\dot{u}/\omega$ is measured along the abscissa and u is measured along the ordinate. It can be used to obtain the response for different kinds of initial conditions, for impulses applied during a response era, for sudden support motion, and even for an arbitrary applied load. In the following paragraphs, we illustrate several different applications of the phase plane diagram.

As a simple example of the use of phase plane diagram, consider the free vibration response after an initial displacement of u_0. In this case, the length of the rotating vector obtained from Equation 5.9b is $\rho = u_0$ and the phase angle $\phi = \pi/2$. The vector rotates at an angular speed of $\omega = \sqrt{k/m}$ rad/s. The phase plane diagram is shown in Figure 5.2b. The response can be obtained directly from the phase plane diagram as

$$u = u_0 \sin\left(\omega t + \frac{\pi}{2}\right) = u_0 \cos \omega t$$

$$\frac{\dot{u}}{\omega} = u_0 \cos\left(\omega t + \frac{\pi}{2}\right) = -u_0 \sin \omega t \tag{5.14}$$

Alternatively, the displacement versus time and velocity function versus time relationships can be constructed graphically as shown in Figure 5.2c and d, respectively.

Next, consider the case when the motion is caused by an impulse I applied at time $t = 0$. From Newton's law of motion, we know that the application of an impulse I results in an increase of velocity by I/m. In the case under consideration, therefore, the motion starts with an initial velocity $v_0 = I/m$. The length of the rotating vector is v_0/ω and $\phi = 0$. The resulting phase plane diagram is as shown in Figure 5.3b. The response is obtained either by graphical construction or by the following mathematical expressions derived from the

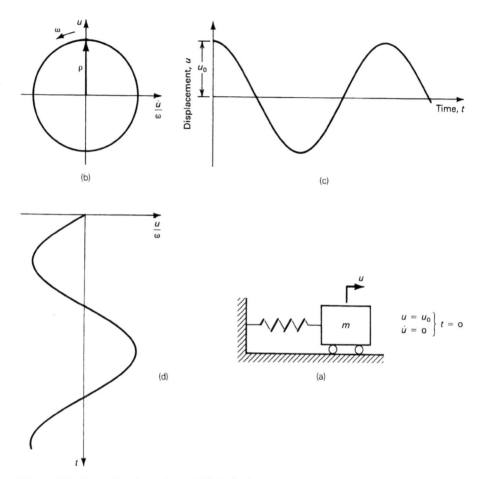

Figure 5.2. Free vibrations from initial displacement.

phase plane diagram

$$u = \frac{v_0}{\omega} \sin \omega t$$

$$\frac{\dot{u}}{\omega} = \frac{v_0}{\omega} \cos \omega t$$

(5.15)

The displacement and velocity responses are shown in Figure 5.3c and d, respectively.

As another example, consider a system undergoing free vibrations from initial conditions u_0 and v_0. After t_1 seconds, an impulse I_1 is applied to the system. As a result, its velocity changes by $v_1 = I_1/m$. It is required to obtain the response of the system. The phase plane diagram for the system is shown in

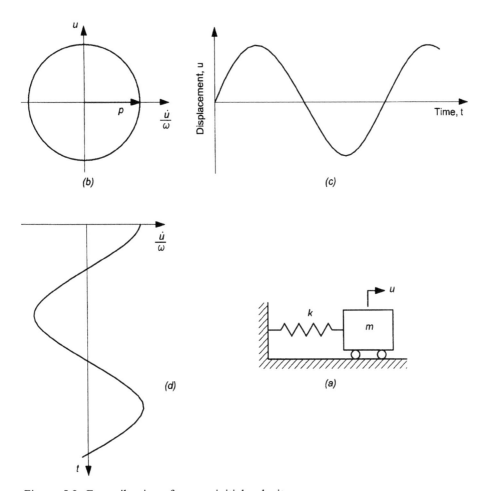

Figure 5.3. Free vibrations from an initial velocity.

Figure 5.4b. The starting length of the rotating vector $\rho = \sqrt{u_0^2 + (v_0/\omega)^2}$ and its initial position is established by scaling v_0/ω on the horizontal axis and u_0 on the vertical axis. In t_1 seconds, the vector has rotated through an angle ωt_1 radians. At this time, impulse I_1 causes the velocity to change by I_1/m. On the phase plane diagram, a horizontal line of length $I_1/(m\omega)$ is drawn to the right from the tip of the current position of the rotating vector. The new position of the rotating vector as well as its new length is now obtained by joining the origin to the end of the horizontal line just drawn. The second era of response begins from this new position of the vector. Since the frequency of the system is unchanged, the vector continues to rotate at the initial speed of ω radians per second. The response can be obtained from the phase plane diagram either

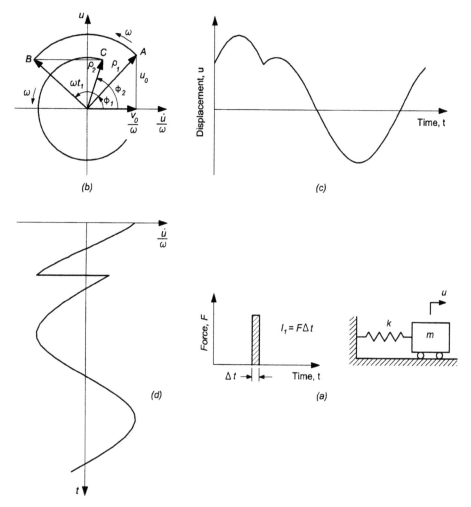

Figure 5.4. Vibrations of a single-degree-of-freedom system subjected to an impulse.

by graphical means or in the form of mathematical expressions

$$u = \rho_1 \sin(\omega t + \phi_1) \qquad 0 \le t \le t_1$$
$$u = \rho_2 \sin\{\omega(t - t_1) + \phi_2\} \quad t_1 < t$$

(5.16)

where ρ_1, ρ_2, ϕ_1, and ϕ_2 are as indicated in Figure 5.4b. The displacement and velocity responses are shown in Figure 5.4c and d, respectively.

As a final example, suppose that the simple single-degree-of-freedom system shown in Figure 5.5a starts vibrating due to a sudden displacement u_g of the support to the right. It is evident that the resulting motion of the block is the same as if the block were vibrating about a new origin shifted a distance u_g to

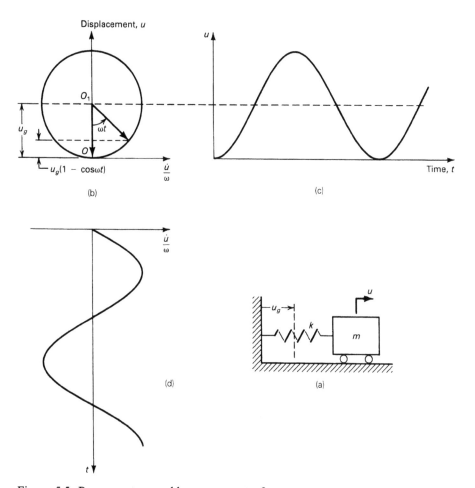

Figure 5.5. Response to a sudden movement of support.

the right after it was given an initial displacement of $-u_g$ toward that origin. The resulting phase plane diagram is shown in Figure 5.5b, from which the response is seen to be

$$u = u_g(1 - \cos \omega t)$$

$$\frac{\dot{u}}{\omega} = u_g \sin \omega t$$

(5.17)

The displacement and velocity responses are shown in Figure 5.5c and d, respectively. Later, we shall see that these responses are exactly equal to those caused by the sudden application of a constant force of magnitude ku_g.

The potential energy of a system undergoing free vibration is equal to the strain energy stored in the spring, so that

$$V = \frac{1}{2}ku^2$$

$$= \frac{1}{2}k\rho^2 \sin^2(\omega t + \phi) \tag{5.18}$$

At the same time, the kinetic energy of the mass is given by

$$T = \frac{1}{2}m\dot{u}^2$$

$$= \frac{1}{2}m\omega^2\rho^2 \cos^2(\omega t + \phi)$$

$$= \frac{1}{2}k\rho^2 \cos^2(\omega t + \phi) \tag{5.19}$$

Equations 5.18 and 5.19 give

$$V + T = \frac{1}{2}k\rho^2 \tag{5.20}$$

Since no energy is dissipated in a system undergoing free vibrations, the sum of the potential energy and the kinetic energy should be constant. This implies that as long as no external energy is input, the length of the rotating vector, ρ, should be a constant. Circles in a phase plane diagram thus represent constant-energy states. Application of external excitation such as an impulse or a support motion represent input of energy which causes a change in the length of the rotating vector.

Example 5.1
With the aid of a phase plane diagram, obtain the displacement and velocity response of the simple single-degree-of-freedom system of Figure E5.1a to the support motion shown.

Solution
The natural frequency of the system is given by

$$\omega = \sqrt{\frac{k}{m}} = \sqrt{\frac{50}{0.5}} = 10 \text{ rad/s} \tag{a}$$

In the first era of response, the system vibrates about an origin located at 1.5 in. to the right. Since the rotating vector has a speed of 10 rad/s, in 0.1 s, it will rotate through 1 rad or $57.3°$ as shown in Figure E5.1b.

The phase plane diagram for the first era of motion is shown by arc AB in Figure E5.1b. The displacement response during this era is given by

$$u = 1.5(1 - \cos 10t) \tag{b}$$

and the velocity response is

$$\frac{\dot{u}}{\omega} = 1.5 \sin 10t \tag{c}$$

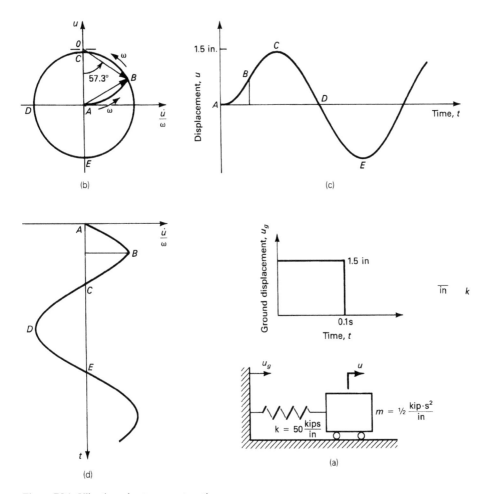

Figure E5.1. Vibrations due to support motion.

At $t = 0.1$ s, the displacement and velocity functions are

$$u(1.0) = 0.69 \text{ in.}$$

$$\frac{\dot{u}}{\omega}(1.0) = 1.26 \text{ in.}$$

(d)

The response in the second era of motion is represented in the phase plane diagram by the circle BCD with center at A. The length of the rotating vector in this second era is

$$\rho = \sqrt{(0.69)^2 + (1.26)^2} = 1.438 \text{ in.}$$

(e)

while the new phase angle is

$$\phi_1 = \tan^{-1} \frac{0.69}{1.26} = 0.501 \text{ rad} = 28.7°$$

(f)

The displacement and velocity responses in the second era are given by

$$u = 1.438 \sin\{10(t - 0.1) + 0.501\}$$

$$\frac{\dot{u}}{\omega} = 1.438 \cos\{10(t - 0.1) + 0.501\}$$

(g)

The displacement–time and velocity–time relationships are shown in Figure E5.1c and d, respectively.

5.3   FREE VIBRATIONS WITH VISCOUS DAMPING

The equation governing the free vibration of a system with viscous damping is obtained from Equation 5.1 by setting $p = 0$.

$$m\ddot{u} + c\dot{u} + ku = 0$$

(5.21)

Equation 5.21 also has a solution of the form of Equation 5.3. Substitution of this solution in Equation 5.21 leads to the following characteristic equation in λ

$$m\lambda^2 + c\lambda + k = 0$$

(5.22)

Solution of Equation 5.22 gives the following two values for λ

$$\lambda_1 = -\frac{c}{2m} + \sqrt{\frac{c^2}{4m^2} - \frac{k}{m}}$$

$$\lambda_2 = -\frac{c}{2m} - \sqrt{\frac{c^2}{4m^2} - \frac{k}{m}}$$

(5.23)

The characteristics of the damped free-vibration response will depend on the value of damping coefficient c. Three different cases arise:
1. Critically damped system.
2. Overdamped system.
3. Underdamped system.
Each of the three cases is discussed in the following paragraphs.

5.3.1   *Critically damped system*

A system in which the magnitude of damping is such that the discriminant in Equation 5.23 is zero is called a *critically damped system*. The damping constant c is, in this case, denoted by c_{cr} and its value is given by

$$c = c_{cr} = 2\sqrt{km}$$

$$= 2m\omega$$

(5.24)

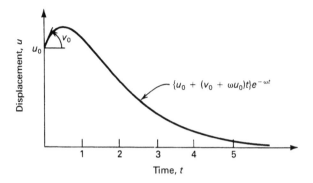

Figure 5.6. Free-vibration response of a critically damped system.

where $\omega = \sqrt{k/m}$ is the natural circular frequency of the associated undamped system. The two roots of Equation 5.22 are now equal, so that

$$\lambda_1 = \lambda_2 = -\frac{c_{cr}}{2m}$$

$$= -\omega \tag{5.25}$$

For repeated roots as in Equation 5.25, the general solution of Equation 5.21 is given by

$$u = (G_1 + G_2 t)e^{-\omega t} \tag{5.26}$$

where G_1 and G_2 are arbitrary constants to be determined from initial conditions. Substitution of initial displacement and initial velocity from Equation 5.8 leads to the following values for G_1 and G_2

$$G_1 = u_0$$

$$G_2 = v_0 + \omega u_0 \tag{5.27}$$

The general solution thus becomes

$$u = \{u_0 + (v_0 + \omega u_0)t\}e^{-\omega t} \tag{5.28}$$

Equation 5.28 has been plotted in Figure 5.6. It is noted that motion is not os-cillatory. The system displacement decays exponentially with time and becomes very nearly zero after a while, although theoretically it takes an infinite time for the displacement to become zero. As we shall see later, critical damping is the least amount of damping for which the motion is nonoscillatory. Thus, whenever damping is less than critical, motion becomes oscillatory.

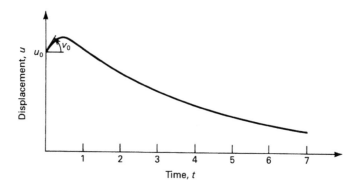

Figure 5.7. Free-vibration response of an over damped system.

5.3.2   *Overdamped system*

When damping is greater than c_{cr}, the system is said to be *overdamped*. For convenience, we define a *damping ratio* ξ given by

$$\xi = \frac{c}{c_{cr}} \tag{5.29}$$

For overdamped systems, $\xi > 1$. Also, the damping constant c can be expressed as

$$c = c_{cr}\xi$$

$$= 2m\omega\xi \tag{5.30}$$

Substitution of Equation 5.30 in Equation 5.23 gives

$$\lambda_1 = -\omega\xi + \omega\sqrt{\xi^2 - 1}$$

$$\lambda_2 = -\omega\xi - \omega\sqrt{\xi^2 - 1} \tag{5.31}$$

If we denote $\omega\sqrt{\xi^2 - 1}$ by $\bar{\omega}$, the general solution of Equation 5.21 becomes

$$u = e^{-\omega\xi t}(G_1 e^{\bar{\omega}t} + G_2 e^{-\bar{\omega}t}) \tag{5.32}$$

where the arbitrary constants G_1 and G_2 are again determined by initial conditions.

Equation 5.32 can be expressed in the alternative form

$$u = e^{-\omega\xi t}(A \cosh \bar{\omega}t + B \sinh \bar{\omega}t) \tag{5.33}$$

where A and B are also arbitrary constants. Equation 5.33 has been plotted in Figure 5.7. In this case, too, the motion is nonoscillatory. The displacement decays exponentially with time, although in comparison to a critically damped system, it takes longer for the system to return to a zero-displacement position.

5.3.3 Underdamped system

In all structural systems and in a majority of mechanical systems, the damping is less than critical. Such systems are said to be *underdamped*. Mechanical systems for which it is required that the system return to a zero-displacement position in the least amount of time are designed to have critical damping. Examples are a recoiling gun and a weighing scale. Certain other recoil mechanisms, for example an automatic door closer, are designed to have overdamping. Leaving aside the few examples of the type cited above, most real systems are underdamped, and a study of underdamped vibrations is therefore of considerable importance. The damping ratio ξ is again defined as c/c_{cr}, and the two solutions λ_1 and λ_2 take the form of Equation 5.31. The damping ratio ξ is however less than 1 in this case, the discriminant in Equation 5.31 is negative, and the two roots becomes imaginary. Thus

$$\begin{aligned} \lambda_1 &= -\omega\xi + i\omega\sqrt{1-\xi^2} \\ \lambda_2 &= -\omega\xi - i\omega\sqrt{1-\xi^2} \end{aligned} \tag{5.34}$$

Denoting $\omega\sqrt{1-\xi^2}$ by ω_d, the general solution to Equation 5.21 can be written as

$$u = e^{-\omega\xi t}(G_1 e^{i\omega_d t} + G_2 e^{-i\omega_d t}) \tag{5.35}$$

By using de Moivre's theorem, Equation 5.35 can be expressed in the alternative form

$$u = e^{-\omega\xi t}(A\cos\omega_d t + B\sin\omega_d t) \tag{5.36}$$

where A and B are arbitrary constants to be determined by the initial conditions. When A and B are determined from the conditions given in Equation 5.8, Equation 5.36 becomes

$$u = e^{-\omega\xi t}\left(u_0 \cos\omega_d t + \frac{v_0 + \omega\xi u_0}{\omega_d}\sin\omega_d t\right) \tag{5.37}$$

Equation 5.37 can be expressed as

$$u = \rho e^{-\omega\xi t}\sin(\omega_d t + \phi) \tag{5.38}$$

where the amplitude ρ and the phase angle ϕ are given by

$$\rho = \left\{(u_0)^2 + \left(\frac{v_0 + \omega\xi u_0}{\omega_d}\right)^2\right\}^{1/2} \tag{5.39}$$

$$\tan\phi = \frac{u_0\omega_d}{v_0 + u_0\omega\xi}$$

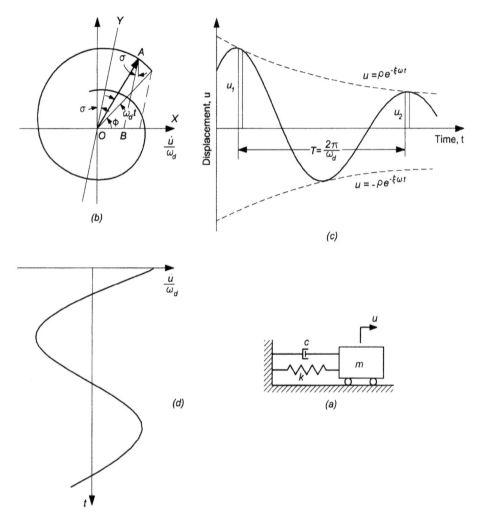

Figure 5.8. Free vibrations of a damped system.

Equation 5.38 has been plotted in Figure 5.8c. The motion is oscillatory or cyclic and repeats itself after a period $T_d = 2\pi/\omega_d$, referred to as the *damped period*. Referring to the simple system of Figure 5.8a, a cycle of motion can be described as the motion taking place from the instant the block is at its extreme right position to the next instant of time, when it is again in that position. However, in this case, the amplitude of displacement at the second instant of time is lower than that at the first. This amplitude decay is caused by the exponential term in Equation 5.38. The natural frequency of the system in cycles per second, denoted by f_d, is given by $f_d = 1/T_d$.

Figure 5.8c also shows curves representing two other functions of time: $\rho e^{-\omega \xi t}$ and $-\rho e^{-\omega \xi t}$. These curves envelope the displacement–time relationship and touch the latter at those points where $\sin(\omega_d t + \phi)$ is equal to $+1$ and -1, respectively. These points, however, do not represent maximas on the displacement–time relationship; the actual maximas lie just a bit to the left of these points. The time at which a maximum occurs is obtained by equating the derivative of Equation 5.38 to zero. The relationship giving the time at maximum displacement is found to be

$$\tan(\omega_d t + \phi) = \frac{\sqrt{1 - \xi^2}}{\xi} \tag{5.40}$$

or

$$\sin(\omega_d t + \phi) = \pm \sqrt{1 - \xi^2} \tag{5.41}$$

From Equation 5.41, it is evident that for small values of ξ, $\sin(\omega_d t + \phi)$ is very nearly equal to ± 1 at peaks in the displacement curve.

5.3.4 *Phase plane diagram*

As in the case of undamped free-vibration response, a rotating vector representation can be used for damped free vibration, too. Imagine a vector of length $\rho e^{-\omega \xi t}$ starting off at an angle ϕ from the time axis and rotating anticlockwise at a constant angular velocity of ω_d radians per second. At any instant of time, the projection of such a vector on the vertical axis gives the displacement of the system at that instant. Unlike in the case of an undamped system, the size of the vector is not constant, but decays exponentially with time. The speed of angular rotation, ω_d, is called the damped circular frequency and it takes the vector one period $= 2\pi/\omega_d$ seconds to complete one full rotation. The velocity expression for the system is obtained by differentiating Equation 5.38. Thus

$$\dot{u} = -\omega \xi \rho e^{-\omega \xi t} \sin(\omega_d t + \phi) + \omega_d \rho e^{-\omega \xi t} \cos(\omega_d t + \phi) \tag{5.42}$$

Equation 5.42 can be written in the alternative form

$$\frac{\dot{u}}{\omega_d} = \rho e^{-\omega \xi t} \cos(\omega_d t + \phi) - \frac{\omega \xi}{\omega_d} \rho e^{-\omega \xi t} \sin(\omega_d t + \phi) \tag{5.43}$$

Now suppose that we define an oblique pair of axes OX and OY, with axis OY inclined from the vertical by an angle σ as shown in Figure 5.8b. The component of the rotating vector of length $\rho e^{-\xi \omega t}$ on the vertical axis is $\rho e^{-\xi \omega t} \sin(\omega_d t + \phi)$. The component OB along the OX axis obtained by drawing

AB parallel to the axis *OY* is

$$OB = \rho e^{-\xi \omega t} \cos(\omega_d t + \phi) - \rho e^{-\xi \omega t} \tan \sigma \sin(\omega_d t + \phi) \qquad (5.44)$$

On comparing Equations 5.43 and 5.44, we note that *OB* will be equal to $\dot{u}/\omega_d$, provided that we select σ such that

$$\tan \sigma = \frac{\omega \xi}{\omega_d} = \frac{\xi}{\sqrt{1 - \xi^2}} \qquad (5.45)$$

Figure 5.8b represents the phase plane diagram for a damped single-degree-of-freedom system. In this case, since the length of rotating vector decreases as $e^{-\xi \omega t}$, its trace is a spiral rather than a circle. The speed of the rotating vector is ω_d instead of ω. As noted earlier, the projection of the rotating vector on the vertical axis is equal to the displacement and can be used to draw the displacement–time diagram of Figure 5.8c. The velocity function $\dot{u}/\omega_d$ is obtained by taking an oblique projection of *OA* on the horizontal axis. The oblique projection is obtained by drawing *AB* parallel to a line inclined at an angle σ from the vertical, where σ is given by Equation 5.45. The projected length *OB* can be used to construct the velocity function versus time relationship shown in Figure 5.8d.

As in the case of an undamped system, the speed of angular rotation ω_d is called circular frequency. It takes the vector one period, that is, $T_d = 2\pi/\omega_d$ seconds, to complete one full rotation. The damped frequency $\omega_d = \omega\sqrt{1 - \xi^2}$ is always less than the undamped frequency ω. However, for small values of ξ, the difference between the two is quite small, and in such cases, ω_d can be taken to be equal to ω without much error. As an example, for $\xi = 0.1$, ω_d is 99.0% of ω.

The construction of a phase plane diagram for a damped system involves somewhat more work than it does for an undamped system because the curve traced by the rotating vector is a spiral rather than a circle. The equation of the spiral $r = \rho e^{-\omega \xi t}$ shows that the shape of the spiral depends on the damping ratio ξ. For a given value of ξ, a spiral has to be drawn only once. Using this drawing as a template, the spiral can be transferred to the phase plane diagram by selecting the required value of ρ on the spiral. Thus, referring to Figure 5.9, if the starting length of vector is $\rho = \rho_1$, the spiral will begin from line *OA*, which is of length ρ_1, and the angle $\theta = \omega_d t$ will be measured from that line. At the end of time *t*, the vector will be in position *OB* and its length will be given by

$$\begin{aligned} OB &= \rho_1 e^{-\xi \omega t} \\ &= \rho_1 e^{-\xi \theta \omega / \omega_d} \\ &= \rho_1 e^{-\theta \xi / \sqrt{1 - \xi^2}} \end{aligned} \qquad (5.46)$$

For small values of ξ, $OB = \rho_1 e^{-\theta \xi}$.

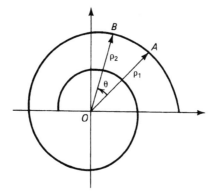

Figure 5.9. Spiral $r = \rho e^{-\theta \xi / \sqrt{1 - \xi^2}}$.

5.3.5 *Logarithmic decrement*

In the free vibration of an underdamped system, displacement amplitude decays exponentially with time. The rate of decrease depends on the damping ratio ξ. If we denote the displacement at time t_1 by $u_1 \equiv u(t_1)$, then

$$u(t_1) = \rho e^{-\xi \omega t_1} \sin(\omega_d t_1 + \phi) \tag{5.47}$$

The displacement at time $t_1 + 2\pi/\omega_d$ is given by

$$u\left(t_1 + \frac{2\pi}{\omega_d}\right) = \rho e^{-\xi \omega (t_1 + 2\pi/\omega_d)} \sin(\omega_d t_1 + \phi) \tag{5.48}$$

The ratio of $u(t_1)$ to $u(t_1 + 2\pi/\omega_d)$ provides a measure of the decrease in displacement over one cycle of motion. This ratio is constant and does not vary with time; its natural log is called *logarithmic decrement* and is denoted by δ. The value of δ is given by

$$\delta = \ln\left\{\frac{e^{-\xi \omega t_1}}{e^{-\xi \omega (t_1 + 2\pi/\omega_d)}}\right\}$$

$$= 2\pi\xi \frac{\omega}{\omega_d} = 2\pi \frac{\xi}{\sqrt{1 - \xi^2}} \tag{5.49}$$

For small values of ξ, $\delta \approx 2\pi\xi$. If δ is obtained from measurements and ξ is to be evaluated, we can use

$$\xi = \frac{\delta}{2\pi} \tag{5.50}$$

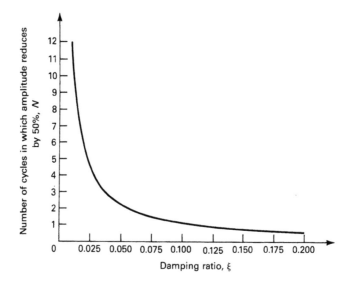

Figure 5.10. Amplitude decay versus damping.

or more accurately

$$\xi = \frac{\delta}{\sqrt{4\pi^2 + \delta^2}} \tag{5.51}$$

Over n cycles of motion, the displacement will decrease from u_1 to u_n, the ratio of the two values being

$$\frac{u_1}{u_n} = e^{2\pi\xi n\omega/\omega_d} \tag{5.52}$$

Equation 5.52 gives

$$\ln \frac{u_1}{u_n} = n\delta \tag{5.53}$$

The number of cycles N over which the displacement amplitude will decay to half its value at the beginning can be obtained from Equation 5.53 and is given by

$$N = \frac{\ln 2}{\delta} \tag{5.54}$$

The value of N obtained from Equation 5.54 is plotted in Figure 5.10 as a function of ξ.

Example 5.2
The single-degree-of-freedom system with viscous damping shown in Figure E5.2a is displaced from its position of rest by a distance u_0 and released. Obtain a plot of u/u_0 versus ωt when the damping in the system is given by (i) $\xi = 2$, (ii) $\xi = 1$, and (iii) $\xi = 0.1$.

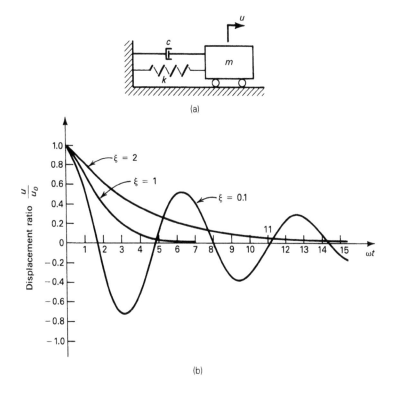

(a)

(b)

Figure E5.2. Free-vibration response of a damped system with different levels of damping.

Solution

(i) For $\xi = 2$, the system is overdamped and the response is given by Equation 5.33

$$u(t) = e^{-\omega \xi t}(A \cosh \bar{\omega} t + B \sinh \bar{\omega} t) \qquad (a)$$

where

$$\bar{\omega} = \omega \sqrt{\xi^2 - 1} = 1.732\, \omega$$

The arbitrary constants A and B are evaluated by using the following initial conditions

$$\text{At } t = 0, \quad \begin{cases} u = u_0 \\ \dot{u} = 0 \end{cases} \qquad (b)$$

The resulting values of A and B are

$$A = u_0$$

$$B = \frac{u_0 \xi \omega}{\bar{\omega}} = 1.155 u_0$$

Equation a now reduces to

$$\frac{u}{u_0} = e^{-2\omega t}(\cosh 1.732\omega t + 1.155 \sinh 1.732\omega t) \qquad (c)$$

Equation c is plotted in Figure E5.2b for ωt from 0 to 15. The response is nonoscillatory and decays exponentially with time.

(ii) For $\xi = 1$, the system is critically damped and the response is given by Equation 5.28 with $v_0 = 0$

$$\frac{u}{u_0} = (1 + \omega t)e^{-\omega t} \tag{d}$$

Equation d has also been plotted in Figure E5.2b. Again, the response is nonoscillatory and decays exponentially with time. As compared to the overdamped system, the rate of decay is faster.

(iii) For $\xi = 0.1$, the system is underdamped and will undergo cyclic motion. The response is given by Equation 5.37. Setting $v_0 = 0$ in that equation, we get

$$u = e^{-\omega \xi t}\left(u_0 \cos \omega_d t + \frac{\omega \xi u_0}{\omega_d} \sin \omega_d t\right) \tag{e}$$

where

$$\omega_d = \omega \sqrt{1 - \xi^2}$$

$$= 0.995\omega$$

Substituting for ξ and ω_d in Equation e, we get

$$\frac{u}{u_0} = e^{-0.1\omega t}(\cos 0.995\omega t + 0.1005 \sin 0.995\omega t) \tag{f}$$

Equation f is also plotted in Figure E5.2b for ωt from 0 to 15. As expected, the response is oscillatory and decays with time.

Example 5.3

For the system shown in Figure E5.2a, mass $m = 91,000$ kg and $u_0 = 30$ mm. If the maximum displacement on the return swing is 20 mm at 0.5 s, determine (i) the spring stiffness k, (ii) the damping ratio ξ, and (iii) the damping constant c.

Solution

Using the definition of logarithmic decrement

$$\delta = \ln \frac{u_1}{u_2} = \frac{2\pi\xi}{\sqrt{1 - \xi^2}}$$

or

$$\ln \frac{30}{20} = \frac{2\pi\xi}{\sqrt{1 - \xi^2}} \tag{a}$$

On solving Equation a for ξ, we get

$$\xi = 0.0644 \tag{b}$$

Also

$$T_d = \frac{2\pi}{\omega_d} = 0.5 \text{ s}$$

Hence

$$\omega_d = \omega \sqrt{1 - \xi^2} = \frac{2\pi}{0.5} \tag{c}$$

and

$$\omega = \frac{2\pi}{0.5\sqrt{1-\xi^2}} = 12.593 \text{ rad/s}$$

$$k = m\omega^2 = 91,000 \times (12.593)^2$$

$$= 1.443 \times 10^7 \text{ N/m} \tag{d}$$

The critical damping constant is obtained from

$$c_{cr} = 2m\omega$$
$$= 2 \times 91,000 \times 12.593$$
$$= 2.292 \times 10^6 \text{ Ns/m} \tag{e}$$

The damping constant c is given by

$$c = \xi c_{cr} = 147,600 \text{ Ns/m} \tag{f}$$

5.4 DAMPED FREE VIBRATION WITH HYSTERETIC DAMPING

As noted earlier, viscous damping force is given by $c\dot{u}$, where c is a constant. Equation 5.42 shows that for small amounts of damping, velocity is proportional to ω. The damping force is thus seen to be proportional to the frequency of vibration and increases with the latter. As a result, the energy loss per cycle is also proportional to the frequency. Measurements of response, on the other hand, show that for most structural and mechanical systems, this is not true and the energy loss is either independent of frequency or in some cases, decreases with increasing frequency. This is because in such systems, a major part of damping occurs from internal friction, localized plastic deformation, or plastic flow, and strictly speaking, a viscous damping mechanism is not applicable.

Damping resistance occurring from internal friction is referred to as *hysteretic damping*, structural damping, or solid damping. It results from the thermal effect of repeated elastic strain imposed on the material. Heat flowing across boundaries of grains causes dissipation of energy and hence damping of the motion. The loss of energy during repeated straining can be measured from a load–deformation curve of the type shown in Figure 5.11. The loop formed by the load–deformation curve is called a hysteresis loop and the area enclosed by it represents the loss of energy per cycle. The name "hysteretic damping" is thus derived from its relationship to the hysteresis loop.

For a perfectly elastic material strained within its elastic limit, the hysteresis loop degenerates into a straight line; consequently, the area enclosed by the loop and hence the energy loss becomes zero. In practice, a perfectly elastic material does not exist and hysteresis energy loss is always present. It may be noted that when a material is strained beyond its elastic limit, the stress–strain relationship becomes significantly nonlinear, and as explained in Section 2.4, the

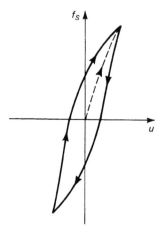

Figure 5.11. Hysteresis curve.

area enclosed by the hysteresis loop becomes quite substantial. The energy loss caused through global plastic deformation under stresses beyond the elastic limit is usually not modeled by a damping term in the equation of motion. Instead, a nonlinear force-displacement relation is used to determine the spring force f_S. The resulting equation becomes nonlinear and a numerical method must be used for its solution. Methods that may be used in the solution of a nonlinear equation are briefly described in Chapter 8. For the present discussion, we will assume that the material is not strained beyond its elastic limit and hysteresis energy loss can be accounted for by including a damping term in the equation of motion.

If hysteretic damping is the only type of damping present in the system, the equation of motion takes the form

$$m\ddot{u} + f_S^t(u, \dot{u}) = p \tag{5.55}$$

where p is zero for free vibrations. In a general case, the solution of Equation 5.55 is quite complex. For harmonic motion, such as, for example, in the case of free vibrations with small amounts of damping, or, as we shall see later, in steady-state forced vibration under a harmonic excitation, hysteretic damping can be accounted for by assuming that in the equation of damped vibration, f_S^t consists of two components: $f_S = ku$ in which k is an average stiffness, and a damping force given by

$$f_D = \frac{\eta k}{\Omega}\dot{u} \tag{5.56}$$

where Ω is the frequency of vibration and η is a constant. Since for harmonic motion, $\dot{u}$ is proportional to the frequency of vibration, the latter cancels out from Equation 5.56, making the damping force independent of frequency. For free vibration, $\Omega \approx \omega$, and if we define $c_h = \eta k / \omega$, and $\xi_h = c_h/(2m\omega) = \eta/2$, then because $f_D = c_h \dot{u}$ from Equation 5.56, the free vibration equations (Eqs. 5.37 through 5.39) still apply, with c replaced by c_h and ξ replaced by ξ_h. The free vibration response under hysteretic damping is thus given by

$$u = \rho e^{-\omega \xi_h t} \sin(\omega_d t + \phi) \tag{5.57}$$

where

$$\omega_d = \omega \sqrt{1 - \xi_h^2}$$

$$\rho = \left\{ u_0^2 + \left(\frac{u_0 + \omega \xi_h u_0}{\omega_d} \right)^2 \right\}^{1/2} \tag{5.58}$$

$$\tan \phi = \frac{u_0 \omega_d}{v_0 + u_0 \omega \xi_h}$$

Also, the logarithmic decrement is obtained from

$$\delta \approx 2\pi \xi_h = \pi \eta \tag{5.59}$$

5.5 DAMPED FREE VIBRATION WITH COULOMB DAMPING

Damping resistance may at times be provided by friction against sliding along a dry surface. The force of sliding friction, called *Coulomb damping* force, is proportional to the normal force acting on the contact surface, but is opposed to the direction of motion. Consider the simple block and spring system shown in Figure 5.12a. When the block is moving to the left (Fig. 5.12c), the friction is directed to the right and its magnitude is μN, where μ is the *coefficient of friction*. The equation of motion is given by

$$m\ddot{u} + ku = \mu N \tag{5.60}$$

When the block is moving toward the right (Fig. 5.12b), the friction force is directed toward the left and the equation of motion is

$$m\ddot{u} + ku = -\mu N \tag{5.61}$$

Equation 5.60 has a solution of the form

$$u = A \cos \omega t + B \sin \omega t + \frac{\mu N}{k} \tag{5.62}$$

where A and B are arbitrary constants to be determined from initial conditions. In a similar manner, Equation 5.61 has the solution

$$u = C \cos \omega t + D \sin \omega t - \frac{\mu N}{k} \tag{5.63}$$

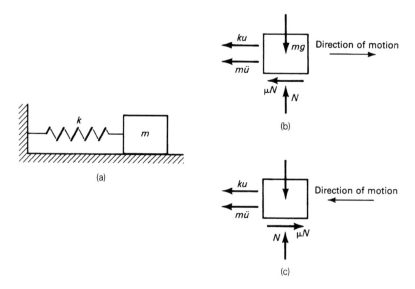

Figure 5.12. Free vibration under Coulomb damping.

where C and D are again arbitrary constants whose values depend on the initial conditions.

To illustrate the essentials of free-vibration response under dry Coulomb friction, consider the case when the block has been displaced to the right a distance u_0, and released. During the first half cycle of motion that follows, the block is moving to the left and its motion is governed by Equation 5.62. To determine A and B, we substitute the initial conditions.

$$\text{At } t=0, \qquad \begin{cases} u=u_0 \\ \dot{u}=0 \end{cases} \tag{5.64}$$

and obtain

$$\begin{aligned} A &= u_0 - \frac{\mu N}{k} \\ B &= 0 \end{aligned} \tag{5.65}$$

Equation 5.62 now reduces to

$$u = \left(u_0 - \frac{\mu N}{k} \right) \cos \omega t + \frac{\mu N}{k} \tag{5.66}$$

Equation 5.66 is valid until the next instant of time, when the velocity becomes zero, that is, up to $t = \pi/\omega$. At that instant, the block is at its extreme left position and its displacement is $-(u_0 - 2\mu N/k)$. The block now starts moving

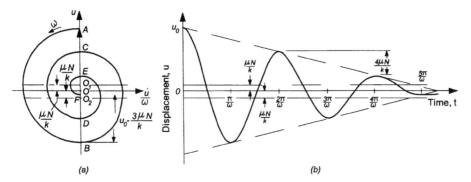

(a)   (b)

Figure 5.13. Free vibration under Coulomb damping: (a) phase plane diagram; (b) displacement-time curve.

to the right and its motion is governed by Equation 5.63 with the following conditions

At $t = \dfrac{\pi}{\omega}$, $\begin{cases} u = -(u_0 - \frac{2\mu N}{k}) \\ \dot{u} = 0 \end{cases}$ (5.67)

When C and D are determined by using the conditions in Equation 5.67, Equation 5.63 reduces to

$$u = \left(u_0 - \frac{3\mu N}{k}\right)\cos \omega t - \frac{\mu N}{k}$$ (5.68)

Equation 5.68 is valid until the block again reaches its extreme right position, that is, when $t = 2\pi/\omega$. At this instant, the block has completed one cycle of motion and its displacement is $u_0 - (4\mu N/k)$. The complete motion of the block is described by the displacement–time curve of Figure 5.13b. The period of motion, that is, the time taken to complete one cycle of motion, is $2\pi/\omega$, which means that Coulomb friction does not change the frequency or the period of vibration. The amplitude reduces by $4\mu N/k$ during each cycle of motion. At any instant when the velocity is zero, that is, when the block is either at its extreme left or extreme right position, if the displacement u is equal to or less than $\mu N/k$, the spring force ku will be equal to or smaller than the friction μN and the block will cease to move. Thus, the block will come to rest at the end of a half-cycle in a position which is displaced from its original position of rest.

In the foregoing discussion, we have not made any distinction between μ_s, the coefficient of static friction, and μ_d, the coefficient of dynamic friction. In general, μ_s is greater than μ_d. The dynamic coefficient of friction must be used in the equation of motion. However, the static friction is applicable in determining the position in which the block will come to rest. Thus, at the end of any half-cycle when the spring force ku is less than the force of static friction $\mu_s N$, the block will cease to move.

The amplitude decay per cycle due to Coulomb friction is constant and is equal to $4\mu N/k$. Thus, the envelopes to the displacement–time curve are straight lines. It is of interest to note that with either viscous or hysteretic damping, amplitude decay is exponential, and theoretically, the system never comes to rest.

5.5.1 Phase plane representation of vibrations under Coulomb damping

A phase plane diagram can be used quite effectively to represent the vibrations of a system with Coulomb damping. Let us develop such a representation for the simple block shown in Figure 5.12. Let the block undergo vibrations after it has been displaced to the right a distance u_0 and released. During the first half-cycle of its motion when the block is moving to the left, the motion is governed by Equation 5.66, which can be expressed as

$$u - \frac{\mu N}{k} = \rho_1 \sin(\omega t + \phi_1) \qquad (5.69)$$

where

$$\rho_1 = u_0 - \frac{\mu N}{k}$$

$$\phi_1 = \frac{\pi}{2} \qquad (5.70)$$

The velocity function is obtained by differentiating Equation 5.69, so that

$$\frac{\dot{u}}{\omega} = \rho_1 \cos(\omega t + \phi_1) \qquad (5.71)$$

In the phase plane, Equations 5.69 and 5.71 represent the parametric equations of a circle as shown in Figure 5.13a. The circle has a radius $\rho_1 = O_1 A$ and its center is located at $(0, \mu N/k)$. The motion during the first half-cycle is represented by arc AB. The arc is drawn with O_1 as the center and a radius of ρ_1.

The second half-cycle of motion is governed by Equation 5.68, which can be expressed as

$$u = \rho_2 \sin(\omega t + \phi_2) - \frac{\mu N}{k} \qquad (5.72)$$

where

$$\rho_2 = u_0 - \frac{3\mu N}{k}$$

$$\phi_2 = \frac{\pi}{2} \qquad (5.73)$$

The velocity function is obtained by differentiating Equation 5.72 so that

$$\frac{\dot{u}}{\omega} = \rho_2 \cos(\omega t + \phi_2) \qquad (5.74)$$

Equations 5.72 and 5.74 represent a circle with radius ρ_2 and the center located at $(0, -\mu N/k)$. In the phase plane diagram, the second half-cycle of motion is therefore represented by arc BC which is drawn with the center at O_2 and a radius of $O_2 B$.

It can easily be shown that the third half-cycle of motion is represented by the arc CD drawn from center O_1 with a radius $O_1 C$. In summary, on a phase plane diagram, the motion of a system with Coulomb damping is represented by arcs of circles with origins located at $\pm(\mu N/k)$ on the displacement axis. When the block is moving to the left, the center is on the positive direction of u axis, when it is moving to the right, the center is on the negative direction of u axis. The block comes to rest when at the end of a half-cycle of motion, the absolute value of the displacement is less than $\mu N/k$. In the example of Figure 5.13, this happens at the instant represented by point F which lies within the region bounded by $O_1 O_2$. Projection of the rotating vector on the vertical axis gives the displacement–time diagram (Fig. 5.13b). The velocity function versus time diagram is obtained from the projection of the vector on the horizontal axis.

Example 5.4
The mass of 15 kg shown in Figure E5.4a is restrained by a spring of stiffness $k = 1800$ N/m and slides on a rough surface with a coefficient of friction $\mu = 0.1$. The mass is given an initial displacement of 40 mm and an initial velocity of 250 mm/s and allowed to vibrate freely. At $t = 0.25$ s, the mass receives an impulse acting to the left of $2N$ s. At $t = 0.5$ s, the support suddenly moves to the right a distance of 30 mm. With the aid of a phase plane diagram, obtain the displacement–time relationship for the entire duration of motion of the block.

Solution
The phase plane diagram is shown in Figure E5.4b. In free vibrations, the rotating vector will swing about origins located at $\pm \mu N/k = \pm 8.175$ mm. The origins are indicated in Figure E5.4b by O_2 and O_1. In the first era of vibration, the block is moving to the right; hence the rotating vector swings about O_1 with an angular velocity given by

$$\omega = \sqrt{k/m} = 10.95 \text{ rad/s}$$

The initial position of the vector is obtained by locating the tip of the vector at coordinates $u = 40$ mm and $\dot{u}/\omega = v_0/\omega = 22.83$ mm. The vector is shown by $O_1 A$ and its length is given by

$$\rho_1 = \sqrt{(40 + 8.175)^2 + 22.83^2}$$

$$= 53.31 \text{ mm}$$

The inclination of the rotating vector from the vertical, that is, angle AO_1B can either be calculated or measured from the diagram and is found to be 25.36°.

In the second era of vibration, the block is moving to the left and the rotating vector swings about O_2. The length of the vector is $\rho_2 = O_2 B = 36.96$ mm. In 0.25 s, the vector should rotate through an angle of 0.25×10.95 rad $= 156.85°$. Angle BO_2C should therefore be $156.85 - 25.36 = 131.49°$. At this instant, the block is still moving to the left. The impulse

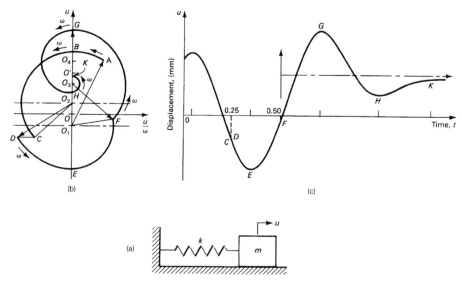

Figure E5.4. Vibrations under Coulomb friction.

imparted to the block gives it an additional velocity of $2 \times 1000/15 = 133.33$ mm/s to the left. On the phase plane diagram, this is represented by the horizontal line CD of length $133.33/10.95 = 12.18$ mm.

The length of the vector now changes to O_2D, but it continues to swing about O_2 until it reaches the extreme left position represented by point E. Angle O_2DE can be measured or calculated to be $58.60°$. The total angle traversed by the vector up to this instant is $156.85 + 58.60 = 215.45°$.

Having reached the extreme left position represented by point E, the block will now start moving to the right, and the next era of motion is therefore represented by vector O_1E swinging about O_1. In 0.5 s, the vector should swing a total of $0.5 \times 10.95 = 313.69°$. The angle of swing O_1EF is therefore $313.69 - 215.45 = 98.24°$.

At 0.5 s, the support moves by a distance of 30 mm to the right. The origin on the phase plane diagram shifts an equal amount along the displacement axis from O to O'. Simultaneously, O_1 moves to O_3 and O_2 moves to O_4. The next phase of motion is represented by arc FG drawn with center at O_3. Following this phase, the rotating vector swings about O_4, and arc GH represents another half-cycle of motion. At H, the block is in its extreme left position. The final half-cycle of motion is represented by arc HK drawn with center at O_3. Since K lies within the region O_3O_4, the force of friction is greater than the spring force and the block now comes to rest. The displacement–time relationship for the entire duration of motion is shown in Figure E5.4c.

SELECTED READINGS

Crandall, S.H. 1970. The role of Damping in Vibration Theory. *Journal of Sound and Vibration*, Vol. 11: 3–18.

Rao, S.S. 1995. *Mechanical Vibrations*. Reading, Massachusetts: Addison-Wesley. 3rd Edition.

Thomson, W.T. 1998. *Theory of Vibration with Applications*. Upper Saddle River, New Jersey: Prentice Hall. 5th Edition.

PROBLEMS

5.1 A gun weighing 1500 lb is restrained by a spring and a damper (Fig. P5.1). The spring stiffness is selected so as to restrict the recoil of the gun to 4 ft. The damper is designed to engage only during the return to the firing position and the damping coefficient is adjusted so that the time taken for return to the firing position is a minimum. If the initial recoil velocity is 64 ft/s., determine the recoil spring stiffness and the damping required.

5.2 A machine weighs 200 lb and is supported on four springs and dampers such that the static deflection is 3/4 in. For vertical vibrations of the system, the dampers are adjusted to reduce oscillations to 1/4 of initial amplitude after two complete cycles. Find the damping coefficient and compare the frequency of damped and undamped oscillations.

5.3 A vehicle for landing on the moon has a mass of 4500 kg. The damped spring-undercarriage system of the vehicle has a stiffness of 450 kN/m and a damping ratio of 0.20. The rocket lift engines are cut off when the vehicle is hovering at an altitude of 10 m. Calculate the maximum deflection of the undercarriage system when the vehicle hits the surface. On the moon, the gravitational acceleration is 1.6 m/s^2.

5.4 The packaging for a delicate instrument can be modeled as shown in Figure P5.4 in which the instrument of mass m is restrained by springs of total stiffness k inside a container of mass M. If the container is dropped on the ground through a height h and does not bounce on contact, obtain the maximum acceleration experienced by the instrument. Assume that during free fall, the relative motion between masses M and m is negligible.

5.5 A wooden block of mass 3 kg is restrained by a spring of stiffness 2 N/mm (Fig. P5.5). A bullet of mass 0.2 kg is fired at a speed of 20 m/s into the block and embeds itself into the latter. Obtain the maximum displacement of the block, (a) neglecting damping; (b) assuming that damping is 10% of critical.

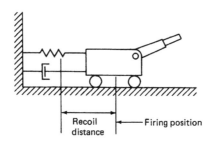

Figure P5.1.

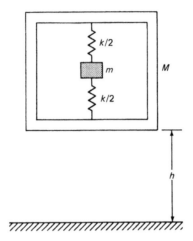

Figure P5.4.

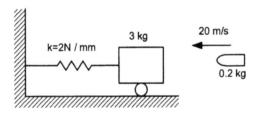

Figure P5.5.

5.6 The single story frame shown in Figure P5.6 has a total story stiffness of 100 kips/in. The floor mass is 1 kip.s^2/in. The frame is subjected to a ground displacement as shown in the figure. Using a phase plane diagram, obtain the displacement history for the first 0.3s. Neglect damping.

5.7 An automobile is modeled as a single-degree-of-freedom system free to vibrate in the vertical direction. In a free-vibration test, the amplitude of oscillation is observed to decrease from 20 mm to 2 mm in one cycle which takes 1/3 s. When carrying passengers with total mass 300 kg, the free-vibration amplitude is observed to decay from 20 mm to 2.75 mm in one cycle. Find the mass of the vehicle and the stiffness and damping of the suspension system.

5.8 A water meter consists of a float attached to a rigid light bar of length *L* (Fig. P5.8). The bar is attached to a wall by means of a hinge and is restrained by a damper of coefficient *c* attached at a distance *a* from the pin. The mass of the float is *m*, its cross-sectional area is *A*, and the mass density of water is ρ. If the damper is designed to provide critical

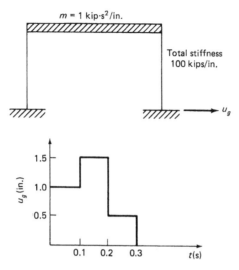

$m = 1$ kip·s²/in.

Total stiffness
100 kips/in.

u_g

u_g (in.)

1.5

1.0

0.5

0.1 0.2 0.3 t(s)

Figure P5.6.

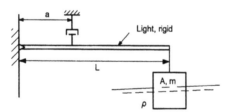

a

Light, rigid

L

A, m

ρ

Figure P5.8.

damping, find its damping constant c. Neglect added mass of the water vibrating along with the float.

5.9 A mass M is supported on a firm base by a spring of stiffness k and a damper with damping constant c (Fig. P5.9). A mass m falls on the larger mass M from a height h and sticks to the latter without rebounding on impact. Find the equation governing the motion of the two masses after impact and determine their maximum displacement with reference to the position at impact. The following data are given: $M = 20$ kg, $m = 5$ kg, $k = 1920$ N/m, $c = 48$ Ns/m, and $h = 1$ m.

5.10 A 1 m, long diving board with a diver of mass 75 kg standing at its tip oscillates with a frequency of 3 Hz. When the diver is standing still, the amplitude of oscillation is observed to decrease from 150 mm to 80 mm in 10 cycles. What is the modulus of rigidity EI for the board and what is the damping factor? Neglect the mass of the diving board.

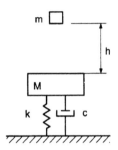

Figure P5.9.

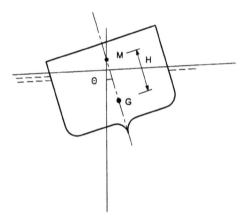

Figure P5.11.

5.11   The metacenter of a ship is defined as the point at which the resultant buoyant force intersects the centerline of the ship (Fig. P5.11). The distance from the center of gravity of the ship to the metacenter is referred to as the metacentric height. While in water, the ship executes a rolling motion about the metacenter. The frequency of roll is observed to be ω. The mass of the ship is M and the mass moment of inertia for rotation about the center of gravity is J_0. Obtain an expression for the metacentric height assuming small vibrations.

5.12   An undamped system is observed to have a natural frequency of 20rad/s. The system is equipped with a damper designed to provide critical damping. If it is now given an initial displacement, how much time will it take for the displacement to reduce to one-tenth its initial value.

5.13   A block of mass 3 kg is restrained by a spring of stiffness 2 N/mm and slides along a rough surface with a coefficient of friction of 0.2. An impulsive force applied to the block gives it an initial velocity of 1 m/s to the right. How far to the right will the block move? How much will be the maximum displacement of the block on its return swing towards the left?

5.14 How much time from the beginning of motion will it take for the block in Problem 5.13 to come to rest, and how much will be the offset between the starting point and the point at which the block comes to rest?

5.15 A single degree-of-freedom system of mass m and stiffness k is allowed to vibrate freely. Measurements show that over four cycles of motion, the amplitude of vibration reduces by a factor of 3.52. If the mass of the system is increase to $1.5m$, by how much would the amplitude of vibration decrease over 4 cycles of motion when the damping in the system is (a) viscous; (b) hysteretic?

CHAPTER 6

Forced harmonic vibrations: Single-degree-of-freedom system

6.1 INTRODUCTION

The vibrations in a dynamic system are usually caused by the presence of a time-varying force. The response that results from the application of such a force is called forced vibration response. When damping is absent or negligible, the vibrations are commonly referred to as *undamped forced vibrations*; when damping is taken into account, the term *damped forced vibration* is used.

As discussed in Chapter 1, a system may be subjected to forces of several different types. When the applied force or forces are of a short duration, the resulting response is also of a comparatively short duration. After the applied force has ceased acting, damping in the system causes the vibrations to decay and the system returns to rest after a while. Response of this type is called transient response. Even though transient response is of a short duration, it may be quite significant from an engineering point of view. It is possible that certain components of the structural or mechanical system may be stressed beyond their capacity during the course of transient vibrations, resulting in a failure. In certain cases, failure may also result from a few repeated applications of a high stress level, a phenomenon called *low cycle fatigue*.

Dynamic forces that act for a long duration cause the system to undergo sustained vibrations. Such vibrations have two components: a transient component and a steady-state component. The transient component is present at the beginning of the vibrations and its magnitude is governed by the initial conditions. Damping in the system causes it to decay rapidly. The steady-state component lasts as long as the exciting force. After the transient component has died, only the steady-state component remains. In certain situations, the *steady-state response* may build up to very large amplitudes, causing severe overstressing and possible failure. When such a condition occurs, the system is said to have achieved *resonance*. Even when resonance is not present, steady-state response may cause failure through fatigue. This is particularly significant for mechanical systems, which under steady vibrations could be subjected to a large number of stress cycles during their service life. There may also be cases where the total response in the initial phases consisting of the transient and steady-state components may be the most critical.

An important class of long-term loads are those that can be expressed as harmonic functions of time. Such loads may, for example, arise due to imbalance in a rotating machine. Study of response under harmonic excitation, while useful by itself, also provides an insight into the nature of forced vibrations of a more general type. Therefore, after a brief introduction to the procedures used for solving the equations of motion under different types of loading situations, we discuss in this chapter the response to a *harmonic excitation*. In fact, any load that is periodic in nature can be treated by resolving it into its harmonic components. The response to each such component is obtained by methods presented in this chapter. These individual responses are then superimposed to obtain the total response. We defer the discussion of response to a periodic load to Chapter 9, where the general topic of analysis in frequency domain is discussed in detail. The analysis of transient response caused by short-term or impulsive loads as well as the response to a general dynamic loading are presented in Chapter 7.

6.2 PROCEDURES FOR THE SOLUTION OF THE FORCED VIBRATION EQUATION

When damping is viscous in nature, the equation of motion for a single-degree-of-freedom system reduces to the form

$$m\ddot{u} + c\dot{u} + ku = p(t) \tag{6.1}$$

Provided that m, c, and k do not vary with time, Equation 6.1 represents a second-order linear differential equation. Linearity is a useful property because it simplifies the solution of differential equations. The principle of superposition holds for linear equations. Thus, as an example, suppose that the applied force $p(t)$ can be expressed as a sum of its components, say $p_1(t)$ and $p_2(t)$

$$p(t) = p_1(t) + p_2(t) \tag{6.2}$$

Also suppose that the solution of Equation 6.1 with $p_1(t)$ as the exciting force is $u_1(t)$ and that with $p_2(t)$ as the exciting force is $u_2(t)$. Then, if the principle of superposition holds, the response $u(t)$ to an exciting force $p(t)$ is given by

$$u(t) = u_1(t) + u_2(t) \tag{6.3}$$

When Equation 6.1 is linear, its solution can be expressed as the sum of two parts: a *complementary function* and a *particular integral*. The complementary function is obtained by solving Equation 6.1 with $p(t) = 0$. The resulting equation is called a homogeneous equation; its solution was discussed in Chapter 5.

As shown there, the solution of a homogeneous equation involves two arbitrary constants whose values are chosen to satisfy the initial conditions. The particular solution of Equation 6.1 depends on the exciting force, but does not involve any arbitrary constants. By itself, the particular solution will not satisfy the initial conditions of displacement and velocity at time $t = 0$.

Complementary solutions are transient in nature because they decay with time as a result of the damping. On the other hand, a particular solution persists as long as the exciting force does and therefore represents the steady-state part of the solution. When the exciting force is a simple mathematical function, the particular solution can be obtained by a process of trial. In this chapter, we use this procedure to obtain solutions for several different types of loads.

Linear equations can also be solved by using the transform methods of mathematics. The two transform methods that can be most effectively used are 1. the Laplace transform, and 2. the Fourier transform. By taking a Laplace transform, the differential equation of motion can be converted to an algebraic equation in the transfer function. This equation can easily be solved to obtain the transfer function. Inverse transform of the solution then gives the desired response. Success of the Laplace transform method thus depends on the availability of the values of direct and inverse transforms, and the usefulness of the method is therefore limited to cases where the direct and inverse transforms are known or are easily determined.

In the Fourier transform method, an arbitrary applied load is expressed as the sum of a series of harmonic components. Except in the case of a periodic load, the number of such components is theoretically infinite. The response to the component loads is then synthesized to obtain the resultant response in the time domain. The resolution and synthesis are accomplished by using direct and inverse Fourier transforms. The advantage of Fourier transform method lies in the fact that efficient numerical techniques are available for obtaining the required transforms. The Fourier transform method has therefore proved to be a very powerful tool in the response analysis of linear systems subjected to arbitrary loads. Details of the method are discussed in Chapter 9.

Response to general dynamic loads can also be obtained by using the convolution theorem discussed in Chapter 7. When the convolution theorem is used, the solution process reduces to the evaluation of an integral, part of whose integrand is the forcing function. When the latter function is of a complicated nature, or when it is specified only in the form of numerical values at certain discrete intervals of time, numerical methods are used for the evaluation of the convolution integral. Some of these methods are discussed in Chapter 8. In addition to the procedures outlined, it is also possible to obtain the response of a system to any arbitrary loading by a direct numerical integration technique. Such techniques are also discussed in detail in Chapter 8.

Finally, it is useful to discuss the effect of the nature of damping on response as well as on the solution procedure. In a majority of structural and mechanical system, the level of damping is quite small; as a result, the response is oscillatory. The presence of damping ensures that oscillations will eventually die out. However, when damping is small, its short-term effect on response is negligible. In the case of transient response, it is the maximum level of response rather than its duration that is of importance, at least from an engineering point of view. Because such a maximum value will be attained during the initial stages, there would not be enough time for the damping forces to absorb any significant amount of energy from the system and the level of maximum response will not be affected appreciably by the presence of damping. Damping forces may therefore usually be neglected in the evaluation of response to impulsive forces.

For steady-state vibration under long-term loads, the presence of damping does affect response. This effect is generally quite small except in the case of resonance when damping reduces the response very significantly. When damping is either hysteretic or of Coulomb type, the equation of motion becomes nonlinear and its solution is quite complex. For steady-state vibration under harmonic excitation, approximate solutions can be obtained for both hysteretic and Coulomb damping by defining an equivalent viscous damping in each case. In this chapter, we discuss procedures for determining equivalent damping of this type.

6.3   UNDAMPED HARMONIC VIBRATION

Neglecting damping and assuming that the excitation is given by $p(t) = p_0 \sin \Omega t$, the equation of motion becomes

$$m\ddot{u} + ku = p_0 \sin \Omega t \tag{6.4}$$

The complementary solution to Equation 6.4 is the same as that given by Equation 5.7. For a particular solution, we try

$$u = G \sin \Omega t \tag{6.5}$$

where G is a constant chosen so as to satisfy Equation 6.4. Substitution into Equation 6.4 gives

$$G(k - m\Omega^2) \sin \Omega t = p_0 \sin \Omega t \tag{6.6}$$

Equation 6.6 will be true for all values of t, provided that $G = p_0/(k - m\Omega^2)$. The complete solution is now obtained by adding the complementary and particular solutions. Thus

$$u = A \cos \omega t + B \sin \omega t + \frac{p_0}{k - m\Omega^2} \sin \Omega t \tag{6.7}$$

where, A and B are arbitrary constants to be determined such that the total solution given by Equation 6.7 satisfies the initial conditions. The complementary function represents the transient part of the solution, and for the slightest amount of damping resistance, it dies out with time. This will become more evident when we consider damped harmonic response.

If the initial displacement is u_0 and the initial velocity is v_0, Equation 6.7 reduces to

$$u = u_0 \cos \omega t + \frac{v_0}{\omega} \sin \omega t + \frac{p_0}{k} \frac{1}{1 - \beta^2} (\sin \Omega t - \beta \sin \omega t) \tag{6.8}$$

where β is the *frequency ratio* given by

$$\beta = \frac{\Omega}{\omega} \tag{6.9}$$

The transient part of the response is given by

$$u_1 = u_0 \cos \omega t + \frac{v_0}{\omega} \sin \omega t - \frac{p_0}{k} \frac{\beta}{1 - \beta^2} \sin \omega t \tag{6.10}$$

while the steady-state component of the displacement is

$$u_2 = \frac{p_0}{k} \frac{1}{1 - \beta^2} \sin \Omega t \tag{6.11}$$

The static deflection $\delta_{\max}$ under a load p_0 equal to the amplitude of harmonic excitation is obtained from

$$\delta_{\max} = \frac{p_0}{k} \tag{6.12}$$

The transient and steady-state displacements for a system with $\beta = 0.5$ and zero initial conditions are plotted in Figure 6.1 against t/T, where T is the natural period. Also shown is the total displacement, which is equal to the sum of the steady-state and transient displacements. In each case, the displacement has been normalized by the static displacement $\delta_{\max}$. It will be noted that the transient response is not zero even though the initial conditions are zero. In fact, the value of transient response is such that the total response obtained by a superposition of the transient response and the steady-state response satisfies the given initial conditions, which in this case are zero.

We now define the ratio of dynamic deflection to the static deflection $\delta_{\max}$ as the *dynamic load factor D*. For steady-state response of an undamped system subjected to harmonic excitation, the dynamic load factor becomes

$$D(t) = \frac{1}{1 - \beta^2} \sin \Omega t \tag{6.13}$$

The amplitude of dynamic load factor, $A_d = 1/(1 - \beta^2)$, has been plotted in Figure 6.2 as a function of β, the frequency ratio. For small values of β, $A_d \simeq 1$ and the maximum displacement is very nearly equal to that produced

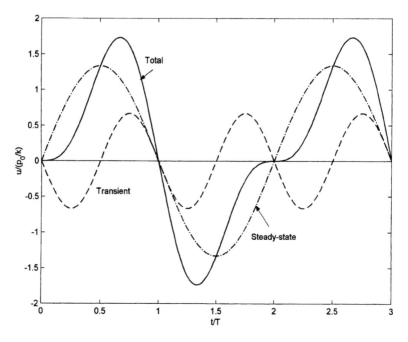

Figure 6.1. Transient, steady-state, and total response of an undamped system to harmonic load; $\beta = 0.5$, $u_0 = 0$, $\dot{u}_0 = 0$.

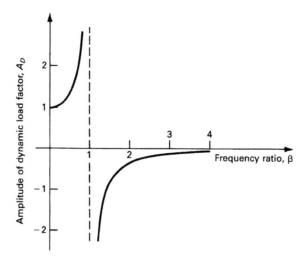

Figure 6.2. Variation of response amplitude with frequency ratio.

under a static load of the same amplitude. The applied load in this case varies so gradually that the response is very nearly equal to that produced in a static case. When the frequency of the applied load is very high, that is, when β is very large, the displacement approaches zero. In this case, the load is varying so rapidly that the system has no time to respond and remains at rest.

For $\beta = 1$, the response is infinitely large. This condition is commonly referred to as resonance. It is clear that at or near resonance, that is, when the frequency of the applied force is equal to or near the natural frequency of the system, the resulting response may be catastrophic. In practical design, such a situation should of course be avoided. Further, we note that for $\Omega < \omega$, that is, for $\beta < 1$, A_d is positive and the system is said to be vibrating in phase with the applied force. In other words, the applied force and the displacement are always in the same direction. For $\beta > 1$, A_d is negative and the system vibration is out of phase with the excitation, so that the displacement is always in an opposite direction to the applied force.

To illustrate the nature of response, the static deflections are compared in Figure 6.3 with the steady-state dynamic deflections produced by a harmonic load for two different values of β. The static deflection is obtained when ω is very large compared to Ω, so that $\beta \approx 0$, and Equation 6.11 gives

$$\delta = \frac{p_0}{k} \sin \Omega t \tag{6.14}$$

Equation 6.14 is plotted in Figure 6.3a. The displacement–time relationship of Figure 6.3a can also be obtained from a rotating vector representation in which we imagine a vector of length p_0/k starting off from along the horizontal axis and rotating anticlockwise at an angular velocity of Ω rad/s. The projection of this vector on the vertical axis gives the static displacement at any instant of time.

In Figure 6.3b, we show the steady-state displacement produced by a dynamic load with frequency ratio $\beta = 0.5$. The amplitude of displacement is, in this case, 1.33 times that in the static case. The dynamic displacement follows the applied load exactly so that both are in identical direction and both reach their peaks simultaneously. If a rotating vector representation is used, the response is seen to be equal to the projection on a vertical axis of a vector of length $1.33 p_0/k$, starting off from along the horizontal axis and rotating anticlockwise with an angular velocity of Ω rad/s. On comparing with Figure 6.3a, we note that the force vector or static displacement vector and the steady-state dynamic displacement vector move along with each other and are therefore said to be in phase.

Figure 6.3c shows the steady-state displacement under a dynamic load with $\beta = 2$. The amplitude of displacement is, in this case, one-third that of the static displacement and the direction of displacement is opposite to the direction of applied load. The rotating vector has a length of $p_0/3k$, and its starting position makes an angle of $180°$ from the horizontal axis. The displacement vector is thus $180°$ out of phase with the force vector.

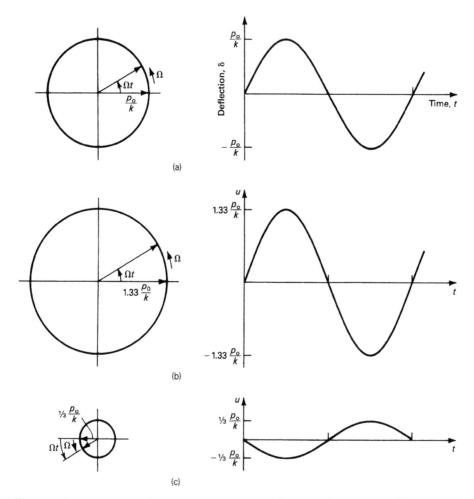

Figure 6.3. Steady-state displacement response: (a) static displacement; (b) response for $\beta = 0.5$; (c) response for $\beta = 2$.

By following a procedure identical to that outlined in the foregoing paragraph, it can be shown that when the excitation is $p(t) = p_0 \cos \Omega t$, the steady-state response is given by

$$u = \frac{p_0}{k} \frac{1}{1 - \beta^2} \cos \Omega t \qquad\qquad (6.15)$$

The rotating vector representation is still applicable except that the displacement is now equal to the projection of the rotating vector on the horizontal axis.

6.4 RESONANT RESPONSE OF AN UNDAMPED SYSTEM

It is of interest to examine, at resonance, the nature of the complete response given by Equation 6.8. We do this for the simple case when $u_0 = v_0 = 0$. With these initial conditions, Equation 6.8 reduces to

$$u = \frac{p_0}{k} \frac{1}{1 - \beta^2} (\sin \beta \omega t - \beta \sin \omega t) \tag{6.16}$$

Resonance occurs when $\beta = 1$. For this value of β, the numerator and the denominator in Equation 6.16 are both zero and the displacement becomes indeterminate. Its limiting values can, however, be determined by using L'Hospital's rule. According to this rule, both the numerator and the denominator are separately differentiated with respect to β. The limiting value of the resulting expression as β approaches 1 gives the required displacement. Thus

$$\lim_{\beta \to 1} u = \lim_{\beta \to 1} \frac{p_0}{k} \frac{\omega t \cos \beta \omega t - \sin \omega t}{-2\beta}$$

$$= \frac{1}{2} \frac{p_0}{k} (\sin \omega t - \omega t \cos \omega t) \tag{6.17}$$

Equation 6.17 is plotted in Figure 6.4. The response is periodic with a period of $2\pi/\omega$. The amplitude of response continues to grow indefinitely. A measure of the rate of growth can be obtained by taking the difference of amplitudes at two successive peaks.

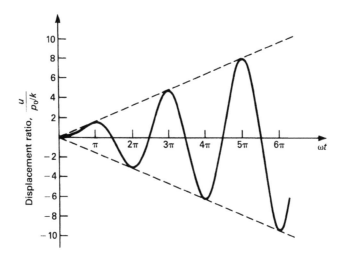

Figure 6.4. Response of an undamped system at resonance ($\beta = 1$).

The time at a peak is obtained by equating the differential of Equation 6.17 to zero, giving

$$\dot{u} = \frac{p_0}{2k}\omega^2 t \sin \omega t = 0 \tag{6.18a}$$

or

$$\omega t = n\pi \quad n = 0, 1, 2, \ldots \tag{6.18b}$$

The difference between two successive peaks on the same side of the axis is then given by

$$u\left(\frac{2\pi}{\omega} + \frac{n\pi}{\omega}\right) - u\left(\frac{n\pi}{\omega}\right) = -\frac{p_0}{2k} 2\pi \cos n\pi$$

$$= \pm \frac{p_0}{k} \pi \tag{6.19}$$

6.5 DAMPED HARMONIC VIBRATION

When viscous damping is included, the equation of motion for a system subjected to excitation $p(t) = p_0 \sin \Omega t$ becomes

$$m\ddot{u} + c\dot{u} + ku = p_0 \sin \Omega t \tag{6.20}$$

The complementary part of the response is obtained by solving Equation 6.20 with the right-hand side equal to zero. For a damping value less than the critical, this solution is given by Equation 5.36. For the particular integral, we try a solution of the form

$$u = G_1 \cos \Omega t + G_2 \sin \Omega t \tag{6.21}$$

where G_1 and G_2 are as yet unknown constants. Substitution of u and its derivatives from Equation 6.21 in Equation 6.20 gives

$$(-G_1\Omega^2 m + G_2\Omega c + G_1 k)\cos \Omega t$$

$$+ (-G_2\Omega^2 m - G_1\Omega c + G_2 k)\sin \Omega t = p_0 \sin \Omega t \tag{6.22}$$

If Equation 6.21 is to be a valid solution, Equation 6.22 should hold for all values of t. In particular, it should hold when $\sin \Omega t$ is zero, so that $\cos \Omega t = \pm 1$. This implies that the coefficient of $\cos \Omega t$ in Equation 6.22 should be zero. By similar reasoning, we conclude that the coefficient of $\sin \Omega t$ in that equation should be equal to p_0. We thus obtain the following two equations in G_1 and G_2

$$-G_1\Omega^2 m + G_2\Omega c + G_1 k = 0$$

$$-G_2\Omega^2 m - G_1\Omega c + G_2 k = p_0 \tag{6.23}$$

On substituting $c = 2\xi\omega m$ and $k = m\omega^2$ in Equations 6.23, dividing each side of the two equations by $m\omega^2$, and noting that $\Omega/\omega = \beta$, we get

$$(1 - \beta^2)G_1 + 2\xi\beta G_2 = 0$$

$$-2\xi\beta G_1 + (1 - \beta^2)G_2 = \frac{p_0}{k} \quad (6.24)$$

On solving Equation 6.24 for G_1 and G_2, we obtain

$$G_1 = \frac{p_0}{k} \frac{-2\xi\beta}{(1 - \beta^2)^2 + (2\xi\beta)^2}$$

$$G_2 = \frac{p_0}{k} \frac{(1 - \beta^2)}{(1 - \beta^2)^2 + (2\xi\beta)^2} \quad (6.25)$$

The particular solution is now obtained from Equations 6.21 and 6.25

$$u = \frac{p_0}{k} \frac{1}{(1 - \beta^2)^2 + (2\xi\beta)^2}\{(1 - \beta^2)\sin\Omega t - 2\xi\beta\cos\Omega t\} \quad (6.26)$$

Equation 6.26 can be written in the alternative form

$$u = \rho\sin(\Omega t - \phi) \quad (6.27)$$

where

$$\rho = \frac{p_0}{k} \frac{1}{\sqrt{(1 - \beta^2)^2 + (2\xi\beta)^2}} \quad (6.28a)$$

$$\tan\phi = \frac{2\xi\beta}{1 - \beta^2} \quad (6.28b)$$

It is evident from the definition of angle ϕ that $\sin\phi = 2\xi\beta/\sqrt{(1 - \beta^2)^2 + (2\xi\beta)^2}$ will always be positive. On the other hand, $\cos\phi = (1 - \beta^2)/\sqrt{(1 - \beta^2)^2 + (2\xi\beta)^2}$ will be positive when $\beta < 1$ and negative when $\beta > 1$. Angle ϕ should therefore lie between 0 and 180°.

The complete solution is obtained by adding the complementary solution (Eq. 5.36) and the particular solution from Equation 6.26, so that

$$u = e^{-\xi\omega t}\{A\cos\omega_d t + B\sin\omega_d t\}$$

$$+ \frac{p_0}{k} \frac{1}{(1 - \beta^2)^2 + (2\xi\beta)^2}\{(1 - \beta^2)\sin\Omega t - 2\xi\beta\cos\Omega t\} \quad (6.29)$$

where A and B are arbitrary constant determined such that the total solution given by Equation 6.29 satisfies the initial conditions. The complementary part of the solution represents the transient response which decays rapidly with time because of the exponential term $e^{-\xi\omega t}$.

If the initial displacement is u_0 and the initial velocity is v_0, the arbitrary constants A and B can be shown to be

$$A = \frac{p_0}{k} \frac{2\xi\beta}{(1-\beta^2)^2 + (2\xi\beta)^2} + u_0 \tag{6.30a}$$

$$B = \frac{p_0}{k} \frac{\omega}{\omega_d} \left\{ \frac{2\beta\xi^2 - \beta(1-\beta^2)}{(1-\beta^2)^2 + (2\xi\beta)^2} \right\} + \frac{v_0 + u_0\omega\xi}{\omega_d} \tag{6.30b}$$

The transient part of the response thus becomes

$$u_t = e^{-\xi\omega t}\left(u_0 \cos \omega_d t + \frac{v_0 + u_0\omega\xi}{\omega_d} \sin \omega_d t \right) + \frac{p_0}{k} \frac{e^{-\omega\xi t}}{(1-\beta^2)^2 + (2\xi\beta)^2}$$

$$\left[2\xi\beta \cos \omega_d t + \frac{\omega}{\omega_d}\{2\beta\xi^2 - \beta(1-\beta^2)\} \sin \omega_d t \right] \tag{6.31}$$

while the steady-state part is given by Equation 6.26 or alternatively, by Equation 6.27. The transient and steady-state displacement for a system with $\beta = 0.5$, $\xi = 0.1$ and zero initial conditions are plotted in Figure 6.5. Also shown is the total displacement, which is equal to the sum of the steady-state and transient displacements. In each case, the displacement has been normalized by the static displacement δ_{max}. The transient response diminishes with time because of damping in the system. The total response obtained by a superposition of the transient response and the steady-state response satisfies the given initial conditions, which in this case are zero.

As in the case of an undamped system, the dynamic load factor $D(t)$ is defined as the ratio of dynamic deflection to the static deflection under load p_0. For steady-state response of a damped system, $D(t)$ is given by

$$D(t) = A_d \sin(\Omega t - \phi) \tag{6.32}$$

where

$$A_d = \frac{1}{\sqrt{(1-\beta^2)^2 + (2\xi\beta)^2}} \tag{6.33}$$

and ϕ is obtained from Equation 6.28b.

Amplitude A_d is plotted in Figure 6.6 as a function of β for different values of ξ. A plot of the type of Figure 6.6 showing the variation of the amplitude of steady-state response with the exciting frequency is referred to as a *frequency response function*. As the exciting frequency approaches zero, the amplitude of steady-state displacement approaches the static displacement.

$$\rho = p_0/k \tag{6.34}$$

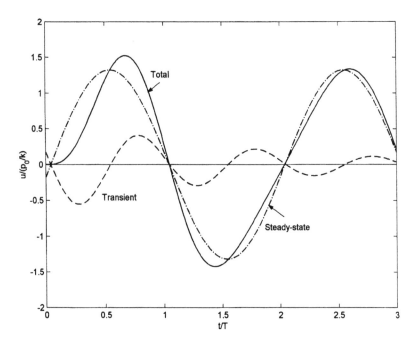

Figure 6.5. Transient, steady-state, and total response of a damped system to harmonic load; $\beta = 0.5$, $\xi = 0.1$, $u_0 = 0$, $\dot{u}_0 = 0$.

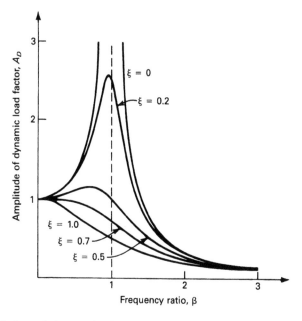

Figure 6.6. Variation of the amplitude of dynamic load factor with frequency ratio and viscous damping ratio.

Thus, for small values of exciting frequency, the response is controlled by the stiffness. For large values of the exciting frequency, that is large β, the limiting value of the displacement amplitude is given by

$$\rho = \frac{p_0}{k\beta^2} = \frac{p_0}{m\Omega^2} \tag{6.35}$$

so that the response is controlled by the mass of the system. Of course, as Ω approaches infinity, the displacement becomes zero. The value of β at which A_d is a maximum can be obtained by differentiating Equation 6.33 with respect to β and equating the differential to zero. This gives

$$\frac{dA_d}{d\beta} = -\frac{1}{2} \frac{-4\beta(1-\beta^2) + 4\beta(2\xi^2)}{\{(1-\beta^2)^2 + (2\xi\beta)^2\}^{3/2}} = 0 \tag{6.36a}$$

or

$$\beta = \sqrt{1 - 2\xi^2} \tag{6.36b}$$

Substitution of Equation 6.36b in Equation 6.33 gives the maximum value of A_d

$$(A_D)_{\max} = \frac{1}{2\xi\sqrt{1-\xi^2}} \tag{6.37}$$

Equation 6.37 shows that in the vicinity of a maximum, the response is controlled by the damping.

As in the case of an undamped system, resonance is said to occur when the exciting frequency is equal to the natural frequency, that is, when $\beta = 1$, and the exciting frequency $\Omega = \omega$ is called the *resonant frequency*. Unlike the undamped case, however, maximum amplitude does not occur at resonance but at a value of β given by Equation 6.36b. This value of β is slightly less than 1. The corresponding value of the amplitude is $(p_0/k)(1/2\xi\sqrt{1-\xi^2})$. When ξ is small, β for maximum amplitude is approximately equal to 1 and the maximum amplitude is $(p_0/k)(1/2\xi)$. It may be noted that in the literature, some authors define resonant frequency as that frequency at which a response quantity, for example the displacement, is a maximum. As seen from Equation 6.36b, the frequency at which the displacement is a maximum is slightly less that the frequency corresponding to $\beta = 1$, although, for the amount of damping normally present in structures, the two are not very different. In the present book, we will continue to define a resonant frequency as the frequency corresponding to $\beta = 1$.

Equation 6.27 shows that the displacement response of a damped system subjected to a harmonic load is itself harmonic, with a frequency equal to that of the exciting force but lags the latter by an angle ϕ called the phase angle. The phase angle, given by Equation 6.28b, is a function of both ξ and β. It is plotted in Figure 6.7 as a function of β for different values of ξ. At resonance,

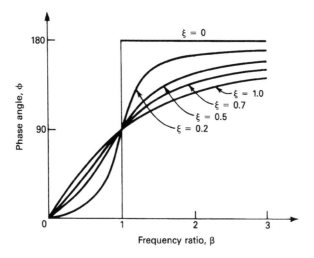

Figure 6.7. Variation of phase angle with frequency ratio and damping.

that is, when $\beta = 1$, ϕ is equal to 90° for all values of the damping ratio. This can also be verified from Equation 6.28b.

To illustrate the nature of steady-state response, the static displacements are compared in Figure 6.8 with the steady-state displacements for three different values of β: 0.5, 1.0 and 2.0 and $\xi = 0.1$. For $\beta = 0.5$, the amplitude of displacement is $1.322 p_0/k$. The phase angle is 0.133 rad or 7.6°. On the time axis, this translates to a lag of $(t/T) = 0.133/\Omega T = 0.133/(\beta 2\pi) = 0.042$. For $\beta = 1.0$, the displacement amplitude is $5 p_0/k$. The phase angle is $\pi/2$ rad or 90°, which translates to a lag of $t/T = 0.25$ on the time axis. For $\beta = 2$, the displacement amplitude is $0.330 p_0/k$. The phase angle is 3.01 rad or 172.4°, which is equivalent to a lag of $t/T = 0.24$ on the time axis.

The response of an undamped system can be obtained directly from that of a damped system by setting $\xi = 0$ in Equations 6.28a and 6.28b giving

$$u = \frac{p_0}{k} \frac{1}{|1 - \beta^2|} \sin(\Omega t - \phi) \tag{6.38}$$

where

$$\phi = \begin{cases} 0 & \text{for } \beta < 1 \\ \pi & \text{for } \beta > 1 \end{cases}$$

Equation 6.38 is the same as Equation 6.11. It also shows that the response is in phase with the applied force when $\beta < 1$ and lags the applied force by 180° when $\beta > 1$. As in the case of an undamped system, a rotating vector can be used to represent the displacement response for a damped system, too. In Figure 6.9, the static displacement vector of length p_0/k is shown

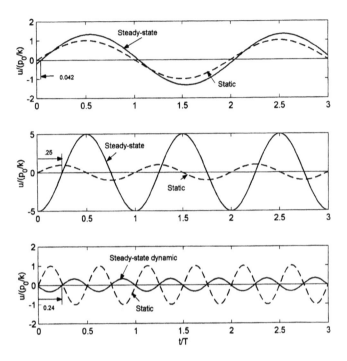

Figure 6.8. Steady-state response of a damped system with $\xi = 0.1$ to harmonic load:
(a) $\beta = 0.5$; (b) $\beta = 1.0$; (c) $\beta = 2.0$.

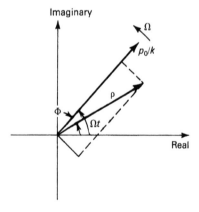

Figure 6.9. Rotating vector representation of steady-state displacements.

rotating counterclockwise at Ω rad/s. At any instant of time t, its inclination
from the horizontal axis is Ωt, and its projection on the vertical axis represents the static displacement at that time. The dynamic displacement vector of
length ρ lags the static displacement vector by an angle ϕ. Its projection on

the vertical axis is $\rho \sin(\Omega t - \phi)$ and represents the dynamic displacement at time t.

As shown in Figure 6.9, the dynamic displacement vector can be resolved into two components. One of these component vectors is in phase with the static displacement vector and hence with the applied force; its magnitude is $\rho \cos \phi$. The other component vector has a magnitude of $\rho \sin \phi$ and lags the applied force vector by 90°. Figure 6.9 also applies to the case when the exciting force is $p_0 \cos \Omega t$, except that the displacement is now represented by the projection of the rotating vector on the horizontal axis. A rotating vector representation can also be used effectively to show the balance of forces acting on a system undergoing steady-state harmonic response. The force balance is expressed by the equation of motion, rewritten as

$$p - f_I - f_D - f_S = 0 \tag{6.39a}$$

where

$$
\begin{aligned}
p &= p_0 \sin \Omega t \\
f_I &= -m\Omega^2 \rho \sin(\Omega t - \phi) \\
f_D &= c\Omega \rho \cos(\Omega t - \phi) \\
f_S &= k\rho \sin(\Omega t - \phi)
\end{aligned}
\tag{6.39b}
$$

The force vectors are indicated in Figure 6.10. They all rotate in an anticlockwise direction at Ω rad/s.

At time t, the applied force vector makes an angle of Ωt from the horizontal axis. The spring force vector is of length $k\rho$ and is opposed to the displacement vector, which lags the applied force vector by angle ϕ. The projection of spring

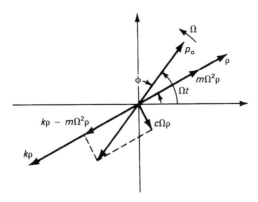

Figure 6.10. Force balance diagram of steady-state response.

force vector on the vertical axis gives the magnitude of spring force at time t. The damping force vector is of length $c\Omega\rho = 2\xi k\beta\rho$; it is opposed to the velocity vector and leads the spring force vector by $90°$. Its projection on the vertical axis gives the damping force. The inertia force vector is of length $m\Omega^2\rho = k\beta^2\rho$ and, as seen from Equation 6.39b, it is opposed to the spring force vector. Its projection on the vertical axis gives the inertia force. The sum of the vectors of the inertia force, damping force, and spring force is a vector that is equal and opposite to the applied force vector. The force balance representation applies as well in the case when $p = p_0 \cos \Omega t$, except that all projections must be taken on the horizontal axis. When required, the steady-state velocity and acceleration response of a system subjected to harmonic load can be obtained by successive differentiation of Equation 6.32. The velocity response is obtained from

$$\frac{\dot{u}}{p_0\omega/k} = A_d\beta \cos(\Omega t - \phi) \tag{6.40}$$

The amplitude of the velocity response is thus given by

$$A_v = \beta A_d \tag{6.41}$$

In a similar manner, the acceleration response is given by

$$\frac{\ddot{u}}{p_0\omega^2/k} = -A_d\beta^2 \sin(\Omega t - \phi) \tag{6.42}$$

and the amplitude of acceleration is

$$A_a = \beta^2 A_d \tag{6.43}$$

It is easily shown that A_v becomes a maximum at $\beta = 1$, while A_a attains its maximum value at $\beta = 1/\sqrt{1 - 2\xi^2}$. The maximum values of the velocity and acceleration amplitudes are

$$(A_v)_{max} = \frac{1}{2\xi} \qquad (A_a)_{max} = \frac{1}{2\xi\sqrt{1 - \xi^2}}$$

The displacement, velocity, and acceleration amplitudes are plotted as functions of β in Figure 6.11 for several different values of ξ. Because of the relationship among the three response quantities given by Equations 6.41 and 6.43, all of them can be plotted on a single four-way logarithmic graph as shown in Figure 6.12. In such a plot, $\log A_v$ is plotted on the ordinate against $\log \beta$ on the abscissae. Equation 6.41 gives

$$\log A_v - \log \beta = \log A_d \tag{6.44}$$

For a constant A_d, Equation 6.44 represents a straight line on the four-way log graph sloping to the right with a slope of 1. Lines sloping to the right having a

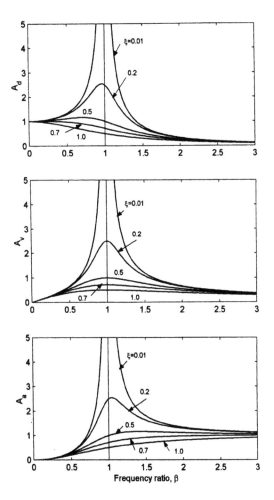

Figure 6.11. Amplitudes of displacement, velocity, and acceleration response to a harmonic load for different damping ratios.

slope of 1 thus represent constant A_d lines. A line perpendicular to these lines provides the log scale for displacement. In a similar manner, Equation 6.43 gives

$$\log A_v + \log \beta = \log A_a \qquad (6.45)$$

For a constant A_a, Equation 6.45 represents a straight line on the four-way log graph sloping to the left with a slope of -1. Lines sloping to the left with a slope of -1 thus represent constant A_a lines and a line perpendicular to these lines provides the log scale for acceleration. The four-way logarithmic graph

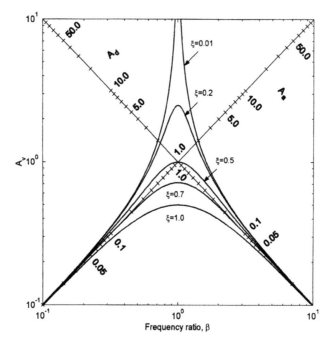

Figure 6.12. Four-way logarithmic graph showing response amplitudes for a system subjected to harmonic load.

in Figure 6.12 shows the variation of displacement, velocity, and acceleration amplitudes with the frequency ratio β for several different values of ξ.

Example 6.1

The seating in a football stadium is mounted on precast prestressed concrete T-beams simply supported on a span of 11.7 m. The cross-sectional properties of a T-beam are indicated in Figure E6.1. The superimposed load due to fixed seats and spectators can be taken as $2.4\,\text{kN/m}^2$. Clapping and stamping by spectators during a sporting event impress a harmonic dynamic load on the beams. Previous field observations have shown that the frequency of the harmonic load due to stamping is 3 Hz, the load amplitude is $0.4\,\text{kN/m}^2$, and the damping is 3% of critical. By assuming that the vibration shape function is

$$\psi(x) = \left\{ \frac{x}{L} - 2\left(\frac{x}{L}\right)^3 + \left(\frac{x}{L}\right)^4 \right\}$$

convert a T-beam and the seating supported by it into an equivalent single-degree-of-freedom system. Then, determine the natural frequency of the system, the maximum dynamic deflection at midspan, and the magnitude of the maximum acceleration experienced by the spectators.

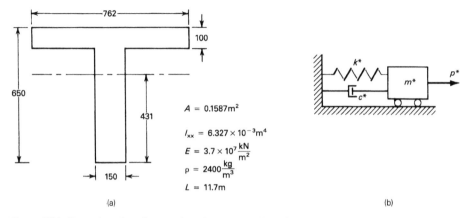

Figure E6.1. Dynamics of stadium seating: (a) cross section of precast concrete T-beam; (b) equivalent single-degree-of-freedom system.

Solution

The generalized mass m^*, generalized stiffness k^*, and the generalized load p^* are given by

$$m^* = \int_0^L \bar{m}\{\psi(x)\}^2 \, dx = \frac{31}{630}\bar{m}L$$

$$k^* = \int_0^L EI\{\psi''(x)\}^2 \, dx = \frac{24}{5}\frac{EI}{L^3} \qquad \text{(a)}$$

$$p^* = \int \bar{p}\psi(x) \, dx = \frac{\bar{p}L}{5}$$

Mass per unit length, $\bar{m}$

Due to mass of the beam: $\quad 0.1587 \times 2400 = 380.9 \text{ kg/m}$

Due to seats and spectators: $\quad \dfrac{2400 \times 0.762}{9.81} = 186.4 \text{ kg/m}$

$\bar{m} = 567.3 \text{ kg/m}$

$$m^* = \frac{31 \times 567.3 \times 11.73}{630} = 326.6 \text{ kg}$$

$$k^* = \frac{24 \times 3.7 \times 10^7 \times 6.327 \times 10^{-3}}{5 \times 11.7^3}$$

$$= 701.6 \text{ kN/m}$$

The uniformly distributed harmonic load $\bar{p}$ caused by stamping and clapping is obtained from

$$\bar{p} = 0.40 \times 0.762 \sin \Omega t = 0.3048 \sin \Omega t \text{ kN/m} \qquad \text{(b)}$$

where $\Omega = 2\pi \times 3$ rad/s. The effective force p^* is thus given by

$$p^* = p_0 \sin \Omega t = \frac{0.3048 \times 11.7}{5} \sin \Omega t = 0.7137 \sin \Omega t \text{ kN}$$

The equivalent single-degree-of-freedom system is shown in Figure E6.1b. Its motion is governed by the following equations

$$u = z(t)\psi(x) \tag{c}$$

$$m^* \ddot{z}(t) + c^* \dot{z}(t) + k^* z(t) = p^* = p_0 \sin \Omega t \tag{d}$$

The solution of Equation d is obtained from Equation 6.27. Thus, the displacement of the beam is given by

$$u(x, t) = z(t)\psi(x)$$

$$= \rho \sin(\Omega t - \phi)\psi(x) \tag{e}$$

where

$$\rho = \frac{p_0}{k^*}[(1 - \beta^2)^2 + (2\beta\xi)^2]^{-1/2} \tag{f}$$

The frequency ratio β is obtained from the exciting frequency and the natural frequency.

$$\text{Natural frequency:} \quad \omega = \sqrt{\frac{k^*}{m^*}} = \sqrt{\frac{701.6 \times 1000}{326.6}}$$

$$= 46.35 \, \text{rad/s} = 7.38 \, \text{Hz}$$

$$\text{Frequency ratio:} \quad \beta = \frac{\Omega}{\omega} = \frac{3}{7.38} = 0.4065$$

Substitution for β and ξ in Equation f gives

$$\rho = \frac{p_0}{k^*} 1.197 = \frac{0.7137}{701.6} \times 1.197 = 1.218 \times 10^{-3} \, \text{m}$$

The midspan deflection is obtained from Equation e with $x = L/2$, so that

$$u\left(\frac{L}{2}, t\right) = \rho \sin(\Omega t - \phi)\left\{\frac{x}{L} - 2\left(\frac{x}{L}\right)^3 + \left(\frac{x}{L}\right)^4\right\}_{x=L/2}$$

$$= 1.218 \times 10^{-3} \times 0.3125 \sin(\Omega t - \phi) \tag{g}$$

$$= 3.805 \times 10^{-4} \sin(\Omega t - \phi) \, \text{m}$$

The maximum deflection at midspan is 3.805×10^{-4} m. The maximum acceleration at midspan is equal to $\Omega^2 u_{max} = (6\pi)^2 \times 3.805 \times 10^{-4} = 0.135 \, \text{m/s}^2 = 1.38\%$ g, where $g = 9.81 \, \text{m/s}^2$ is the acceleration due to gravity. Experience has shown that during a sports event, the spectators will not have a feeling of discomfort provided that the maximum acceleration experienced by them is below 5% g. The design is satisfactory from this point of view. If w kN/m^2 is the equivalent static load that will produce a deflection of 3.805×10^{-4},

$$\frac{5}{384} \frac{w \times 0.762 L^4}{EI} = 3.805 \times 10^{-4}$$

from which we obtain

$$w = 1.197 \times 0.4 = 0.4788 \, \text{kN/m}^2$$

The precast beams should therefore be designed to carry a total superimposed load of $2.4 + 0.479 = 2.88 \, \text{kN/m}^2$ in addition to their self-weight.

6.6 COMPLEX FREQUENCY RESPONSE

The notation of complex algebra can be conveniently used to express the response to a harmonic force. If the forcing function is of the form

$$p = p_0 e^{i\Omega t}$$
$$= p_0(\cos \Omega t + i \sin \Omega t) \tag{6.46}$$

the real part of the solution will represent the response to a cosine function, while the imaginary part will represent the response to a sine function. The steady-state response to the exciting force given by Equation 6.46 has the same frequency as the exciting force and can be expressed as

$$u = U e^{i\Omega t} \tag{6.47}$$

where U is the complex amplitude, also referred to as the *complex frequency response*. To obtain the value of U, we substitute for displacement and its derivatives from Equation 6.47 into the equation of motion (Eq. 6.1). This gives

$$(-m\Omega^2 + ic\Omega + k)U e^{i\Omega t} = p_0 e^{i\Omega t} \tag{6.48}$$

Equation 6.48 leads to the following value of U

$$U = \frac{p_0}{-m\Omega^2 + ic\Omega + k} \tag{6.49}$$

By using the relationships: $c = 2\xi\omega m$, $\omega^2 = k/m$, and $\beta = \Omega/\omega$, we can express Equation 6.49 in the alternative form

$$U = \frac{p_0}{k(1 - \beta^2 + 2i\xi\beta)} \tag{6.50}$$

The complex frequency response given by Equations 6.49 or 6.50 contains information about both the amplitude and the phase of the response. This can be seen by expressing U in the following alternative form

$$U = \frac{p_0}{k} \frac{1}{(1 - \beta^2)^2 + (2\xi\beta)^2} [(1 - \beta^2) - i2\xi\beta]$$
$$= \rho e^{-i\phi} \tag{6.51}$$

where ρ is the real amplitude and ϕ is the phase angle, given by

$$\rho = \frac{p_0}{k} \frac{1}{\sqrt{(1 - \beta^2)^2 + (2\xi\beta)^2}} \tag{6.52a}$$

$$\tan \phi = \frac{2\xi\beta}{1 - \beta^2} \qquad 0 \le \phi \le \pi \tag{6.52b}$$

Substitution of Equation 6.51 in 6.47 gives the following expression for the displacement response

$$u = \rho e^{i(\Omega t - \phi)}$$
$$= \rho \{\cos(\Omega t - \phi) + i \sin(\Omega t - \phi)\} \tag{6.53}$$

On comparing Equations 6.46 and 6.53, we conclude that the response to $p_0 \sin \Omega t$ is $\rho \sin(\Omega t - \phi)$, while the response to $p_0 \cos \Omega t$ is $\rho \cos(\Omega t - \phi)$.

The complex frequency displacement response per unit force, R_d, is generally referred to as the *receptance* and is given by

$$R_d = \frac{u}{p_0 e^{i\Omega t}} = \frac{\rho}{p_0} e^{-i\phi} = \frac{U}{p_0} \tag{6.54}$$

Two other related output parameters are also used to define the response to a harmonic force. The velocity response per unit force, R_v, is called *mobility* and is given by

$$R_v = \frac{\dot{u}}{p_0 e^{i\Omega t}} = i\Omega R_d = \Omega R_d e^{i\pi/2} \tag{6.55}$$

The amplitude of R_v is thus Ω times the amplitude of R_d while its phase angle is $\phi - \frac{\pi}{2}$. The acceleration response per unit force is called *inertance* and is given by

$$R_a = \frac{\ddot{u}}{p_0 e^{i\Omega t}} = -\Omega^2 R_d = \Omega^2 R_d e^{-i\pi} \tag{6.56}$$

The amplitude of R_a is thus Ω^2 times that of R_d, while the phase angle is $\phi + \pi$. It will be noted that parameters U, V, and A are all complex and possess an amplitude and a phase angle, both of which are functions of the exciting frequency.

Graphical representations of the complex frequency response functions must include three different parameters: the exciting frequency, the amplitude, and the phase angle. One form of graphical presentation consists of a plot of amplitude versus frequency, as shown in Figures 6.11 and 6.12, and a plot of phase angle versus frequency as shown in Figure 6.7. In another form, the imaginary part of a function is plotted against the real part of the same function. Plots of this type are called Nyquist plots. They do not explicitly show the value of the exciting frequency. However, each point on the Nyquist graph corresponds to a specific value of the exciting frequency, which can therefore be identified on the plot. Figure 6.13 shows a typical Nyquist plot of receptance in which the

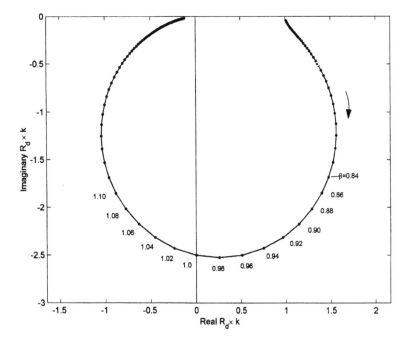

Figure 6.13. Nyquist plot of receptance; $\xi = 0.2$.

imaginary and real parts are both obtained from Equation 6.51.

$$
\begin{aligned}
\text{Real}(R_d) &= \frac{1}{k} \frac{1 - \beta^2}{(1 - \beta^2)^2 + (2\xi\beta)^2} \\
\text{Imag}(R_d) &= -\frac{1}{k} \frac{2\xi\beta}{(1 - \beta^2)^2 + (2\xi\beta)^2}
\end{aligned}
\tag{6.57}
$$

The Nyquist plot shown in Figure 6.13 is close to a circle. Data points on the graph have been evaluated at regular increments of β and have been identified on the plot. Points near resonance are well separated and clearly distinguishable. Away from the resonant frequency, the data points are very close together. This property of a Nyquist plot by which regions close to resonance are enlarged is very useful in the experimental determination of vibration characteristics, such as the natural frequency and damping.

Figure 6.10 can be interpreted as a force balance diagram for the steady-state response of a system subjected to a complex harmonic force. The projections of the rotating vectors on the vertical axis, labeled imaginary axis, represent the imaginary parts of the forces, while those on the horizontal axis represent the real parts. In this form, the force balance diagram of Figure 6.10 is referred to as the Argand diagram.

Example 6.2

The single-degree-of-freedom system shown in Figure E6.2a has a mass $m = 4,500$ kg, stiffness $k = 450$ kN/m, and damping constant $c = 18.0$ kN.m/s. It is subjected to a harmonic force $p_0 \sin \Omega t$. Plot the frequency response function on a log-log graph for exciting frequencies ranging from 0.1 to five-times the natural frequency. Also, obtain the Nyquist plots for receptance, mobility and inertance.

Solution

Natural frequency: $\omega = \sqrt{k/m} = \sqrt{450 \times 10^3/4500} = 10$ rad/s

Damping ratio: $\xi = c/(2\sqrt{km}) = 18/(2\sqrt{450 \times 4.5}) = 0.2$

Frequency ratio: $\beta = \Omega/10$

The amplitude of the frequency response function U/p_0 obtained from Equation 6.52a is plotted on a log-log graph in Figure E6.2b for Ω ranging from 1.0 to 50 rad/s. For very small

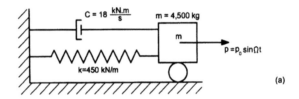

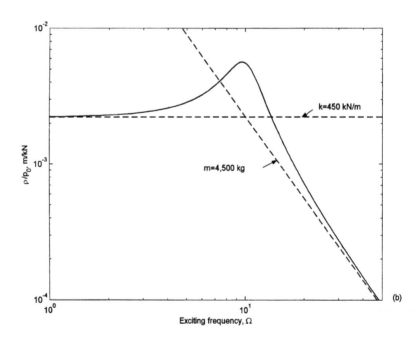

Figure E6.2. (a) Single-degree-of-freedom system subjected to a harmonic force; (b) amplitude of frequency response function versus exciting frequency; (c) Nyquist plot of receptance; (d) Nyquist plot of mobility; (e) Nyquist plot of inertance.

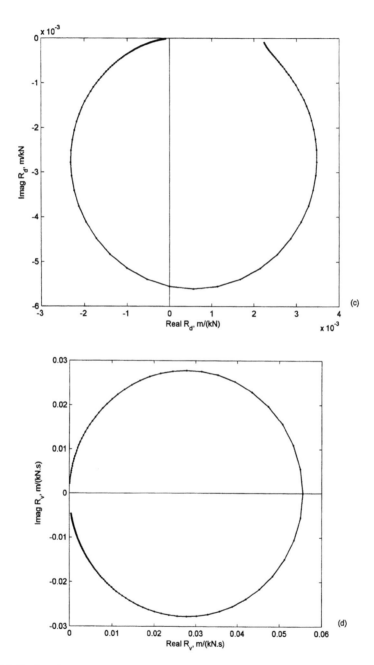

Figure E6.2. Continued.

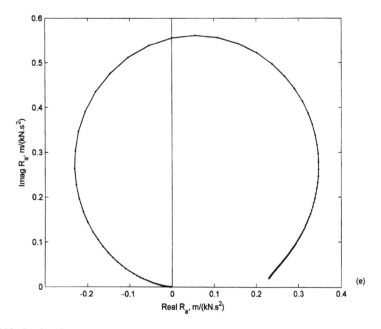

Figure E6.2. Continued.

values of Ω, the frequency response should approach $1/k$. On the log-log graph, $1/k$ plots as a horizontal straight line. This line is also shown on the graph. For large values of Ω, the response should approach $1/(m\Omega^2)$, which can also be represented by a straight line as shown on the log-log graph.

The Nyquist plots for receptance, mobility, and inertance are shown in Figures E6.2c, d, and e, respectively. The mobility graph is a perfect circle, the others are close to a circle. The real and imaginary parts of the receptance are obtained from Equation 6.57. Mobility and inertance are obtained from receptance using the relationships given in Equations 6.55 and 6.56.

6.7   RESONANT RESPONSE OF A DAMPED SYSTEM

The complete response of a damped system is equal to the sum of a steady-state part given by Equation 6.26 and a transient part given by Equation 6.31. At resonance, that is, when $\beta = 1$ and hence $\Omega = \omega$, the steady-state solution, u_s, becomes

$$u_s = -\frac{p_0}{k}\frac{1}{2\zeta}\cos \omega t \qquad (6.58)$$

while the transient part of the response, u_t, is obtained by setting $\beta = 1$ in Equation 6.31. Assuming that the initial displacement as well as the initial velocity is zero, Equation 6.31 gives

$$u_t = \frac{p_0}{k} \frac{e^{-\omega \xi t}}{2\xi} \left(\cos \omega_d t + \frac{\omega \xi}{\omega_d} \sin \omega_d t \right) \tag{6.59}$$

For small values of damping, the contribution of the second term inside the parentheses in Equation 6.59 is negligible. As well, $\omega_d \approx \omega$, and Equation 6.59 therefore reduces to

$$u_t = \frac{p_0}{k} \frac{e^{-\omega \xi t}}{2\xi} \cos \omega t \tag{6.60}$$

The steady-state component of the response given by Equation 6.58 is plotted in Figure 6.14b, for a value of $\xi = 0.1$. The transient response for the same damping is obtained from Equation 6.59 and is plotted in Figure 6.14a. Superposition of the two responses gives the total response shown in Figure 6.14c. As would be expected, the transient term cancels the initial displacement and initial velocity values obtained from the steady-state part so that the net values of these quantities is zero.

It is evident from Figure 6.14c that the response rapidly builds up to its maximum value of $p_0/(2\xi k)$. For a low damping ratio, the total response is equal to the sum of Equations 6.58 and 6.60. The amplitude of response is, in this case, given by

$$\rho = \frac{p_0}{k} \frac{1}{2\xi} (e^{-\omega \xi t} - 1) \tag{6.61}$$

Equation 6.61 approximates the envelope curves shown by dashed lines in Figure 6.14c. Such envelope curves are plotted in Figure 6.15 for different values of ξ and show the rate of buildup of response.

6.8 ROTATING UNBALANCED FORCE

Imbalance in rotating machinery is a common source of harmonic excitation. As an example, consider the rotation of a motor shown in Figure 6.16. The motor is rotating at a constant angular speed of Ω rad/s. The imbalance in the motor is represented by a small mass m_0 attached to it at a radial distance e from the center. The coordinates x and y of the unbalanced mass are related to angle θ shown in the figure as follows

$$x = e \cos \theta \tag{6.62a}$$

$$y = e \sin \theta \tag{6.62b}$$

Also

$$\theta = \Omega t \tag{6.63}$$

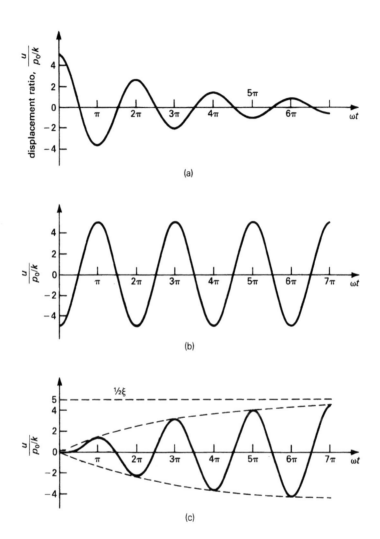

Figure 6.14. Resonant response of a damped system: (a) transient component; (b) steady-state component; (c) total response.

The acceleration of the mass in the x direction is obtained by twice differentiating Equation 6.62a with respect to time

$$\ddot{x} = - e\Omega^2 \cos \Omega t \tag{6.64}$$

In a similar manner, the acceleration in the y direction is given by

$$\ddot{y} = - e\Omega^2 \sin \Omega t \tag{6.65}$$

These accelerations produce inertia forces $m_0 e\Omega^2 \cos \Omega t$ and $m_0 e\Omega^2 \sin \Omega t$ as shown in the figure. The inertia forces must, in turn, be resisted by the support

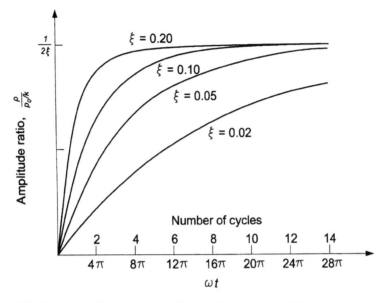

Figure 6.15. Variation of response amplitude with duration of loading at resonance.

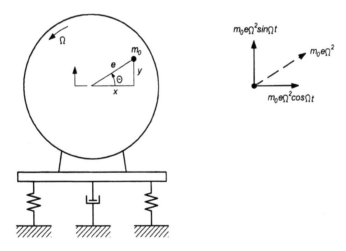

Figure 6.16. Unbalanced rotating mass.

system of the rotating motor. The unbalanced mass thus exerts harmonic forces in both a horizontal and a vertical direction. The resultant of the two forces is seen to be equal to $m_0 e \Omega^2$ acting along the radius vector. It is the well-known centrifugal force that acts on a rotating body. If the system shown in Figure 6.16 is free to vibrate in the vertical direction, its steady-state motion will be governed by the following equation

$$m\ddot{u} + c\dot{u} + ku = m_0 e \Omega^2 \sin \Omega t \tag{6.66}$$

where m is the total mass of the motor and its base, including the unbalanced mass m_0, k is the total spring stiffness, and c represents the damping in the system. The steady-state response is given by Equation 6.26, so that

$$u = \rho \sin(\Omega t - \phi) \tag{6.67}$$

where

$$\rho = \frac{m_0 e \Omega^2}{k \sqrt{(1 - \beta^2)^2 + (2\xi\beta)^2}}$$

$$= \frac{e\beta^2 m_0/m}{\sqrt{(1 - \beta^2)^2 + (2\xi\beta)^2}} \tag{6.68a}$$

$$\tan \phi = \frac{2\xi\beta}{1 - \beta^2} \tag{6.68b}$$

In Equations 6.68, frequency ratio $\beta = \Omega/\omega$, and ω is the natural frequency of the motor and its support system given by $\omega = \sqrt{k/m}$. By equating the derivative of Equation 6.68a to zero, it can be shown that amplitude ρ is a maximum when $\beta = 1/\sqrt{1 - 2\xi^2}$, and the maximum value is given by

$$\rho_{max} = \frac{e(m_0/m)}{2\xi\sqrt{1 - \xi^2}} \tag{6.69}$$

For small values of ξ, the amplitude will be a maximum when $\beta = 1$ and $\rho_{max} = (em_0/m)(1/2\xi)$.

The dimensionless ratio $\rho/(em_0/m)$, has been plotted in Figure 6.17 as a function of β for different values of the damping ratio ξ. It is observed that as would be expected, peaks in the response curve are slightly to the right of $\beta = 1$. Also, when β is large, amplitude ρ approaches a value em_0/m. It is evident that to keep the amplitude down, resonance should be avoided and the ratio em_0/m should be kept small. For practical design situations, this will be achieved by ensuring that the frequency ratio is significantly larger than 1 and the mass ratio m_0/m is kept as small as possible. In practice, low mass ratio is often obtained by mounting the machine on either a heavy frame or a massive concrete base. The expression for the phase angle is the same as Equation 6.28b, which applies

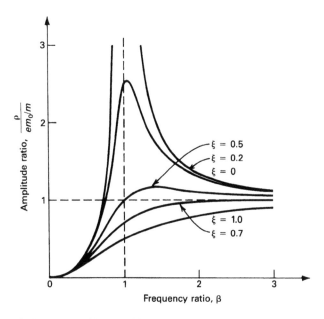

Figure 6.17. Variation of amplitude with frequency ratio and damping in an unbalanced rotating system.

to the case of a forcing function given by $p=p_0 \sin \Omega t$. The variation of the phase angle with β is therefore similar to that shown in Figure 6.7.

Example 6.3
A steel frame consisting of four legs, each a steel wide-flange section W200 × 27, supports a rigid steel table as shown in Figure E6.3a. A rotating motor with an unbalanced mass of 200 kg at an eccentricity of 50 mm is mounted on the table. If the total mass of the table and the motor is 2500 kg, find the range of speed of the motor over which the maximum flexural stress in the legs will exceed 100 MPa. Assume that the legs have a negligible mass, are fixed at the foundation as well as at the table, and neglect damping.

Solution
The lateral stiffness of the four legs is given by

$$k = \frac{4 \times 12EI}{L^3} = \frac{4 \times 12 \times 200,000 \times 25.8 \times 10^6}{(3000)^3}$$

$$= 9173 \times 10^3 \text{ N/m}$$

Natural frequency of the system is obtained from

$$\omega = \sqrt{\frac{k}{m}} = \sqrt{\frac{9173 \times 10^3}{2500}} = 60.57 \text{ rad/s}$$

$$= 9.64 \text{ Hz}$$

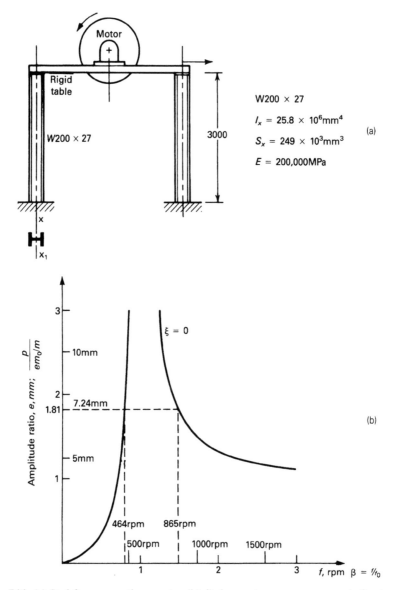

Figure E6.3. (a) Steel frame supporting a motor; (b) displacement response versus speed of motor.

If the lateral deflection of the table is Δ, the maximum moment M at the base as well as at the top of each leg is given by

$$M = \frac{6EI\Delta}{L^2}$$

The flexural stress σ is obtained from

$$\sigma = \frac{M}{S}$$

where S is the section modulus. Thus

$$\sigma = \frac{6EI\Delta}{L^2 S}$$

$$100 = \frac{6 \times 200,000 \times 25.8 \times 10^6 \times \Delta}{(3000)^2 \times 249 \times 10^3}$$

or

$$\Delta = 7.24 \text{ mm}$$

For the maximum stress in the leg to exceed 100 MPa, the maximum lateral deflection must exceed 7.24 mm; that is, ρ in Equation 6.68a must exceed 7.24.
From Equation 6.68a, we obtain

$$\rho = \frac{em_0}{m} \frac{\beta^2}{\sqrt{(1 - \beta^2)^2}}$$

For $\beta < 1$

$$\rho \equiv \frac{em_0}{m} \frac{\beta^2}{1 - \beta^2} > 7.24$$

$$\frac{50 \times 200}{2500} \frac{\beta^2}{1 - \beta^2} > 7.24$$

or

$$\beta > 0.802$$

For $\beta > 1$

$$\rho \equiv \frac{em_0}{m} \frac{\beta^2}{\beta^2 - 1} > 7.24$$

or

$$\beta < 1.495$$

The lower limit of motor speed is $\beta\omega = 0.802 \times 9.64\,\text{Hz} = 464\,\text{rpm}$. The upper limit is $\beta\omega = 1.495 \times 9.64\,\text{Hz} = 865$ rpm. In the range 464 to 865 rpm, the deflection will be larger than 7.24 mm and the flexural stress in the leg will be larger than 100 MPa. The amplitude of response

$$\rho = \frac{em_0}{m} \frac{\beta^2}{\sqrt{(1 - \beta^2)^2}}$$

is plotted in Figure E6.3b as a function of the motor speed. The figure indicates the range of values of β over which the flexural stress will be greater than 100 MPa.

6.9 TRANSMITTED MOTION DUE TO SUPPORT MOVEMENT

A system mounted on a moving support will have some of the support motion transmitted to it. Often, the design of such a system requires that the transmitted motion be minimized. In other situations, the requirement may be to obtain

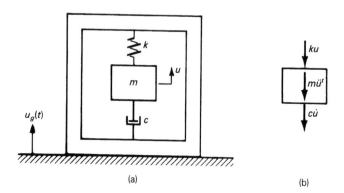

Figure 6.18. Harmonic support excitation: (a) frame supporting a mass; (b) free-body diagram of supported mass.

estimates of the support motion by taking measurements of the transmitted motion. In either case, a relationship must be established between the motion of the support and that of the supported system. For harmonic excitation, this is easily accomplished.

Consider the simple single-degree-of-freedom system shown in Figure 6.18a. The system consists of a mass m mounted on a frame through a dashpot and a spring. Let the frame move in a vertical direction according to the equation

$$u_g = G \sin \Omega t \tag{6.70}$$

Let u represent the vertical displacement of mass m relative to the frame, while u^t represents the same displacement with respect to a fixed frame of reference. Then, the following relationship holds between u^t, u, and u_g

$$u^t = u + u_g \tag{6.71}$$

Referring to the free-body diagram shown in Figure 6.18b, the equation of motion for mass m is given by

$$m\ddot{u}^t + c\dot{u} + ku = 0 \tag{6.72a}$$

or

$$m\ddot{u} + c\dot{u} + ku = -m\ddot{u}_g$$

$$= mG\Omega^2 \sin \Omega t \tag{6.72b}$$

On comparing Equations 6.20 and 6.72b, the steady-state response of the mass in Figure 6.18 is seen to be given by Equations 6.26 or 6.27 with p_0 replaced by $mG\Omega^2$, so that

$$u = \rho \sin(\Omega t - \phi) \tag{6.73}$$

where

$$\rho = \frac{mG\Omega^2}{k} \frac{1}{\sqrt{(1-\beta^2)^2+(2\xi\beta)^2}}$$

$$= G\frac{\beta^2}{\sqrt{(1-\beta^2)^2+(2\xi\beta)^2}} \tag{6.74a}$$

$$\tan\phi = \frac{2\xi\beta}{1-\beta^2} \tag{6.74b}$$

If we are interested in determining the total displacement of mass m, we can use Equation 6.71 and get

$$u^t = G\sin\Omega t + \frac{G\beta^2}{(1-\beta^2)^2+(2\xi\beta)^2}\{(1-\beta^2)\sin\Omega t - 2\xi\beta\cos\Omega t\}$$

$$= \frac{G}{(1-\beta^2)^2+(2\xi\beta)^2}\{(1-\beta^2+4\xi^2\beta^2)\sin\Omega t - 2\xi\beta^3\cos\Omega t\}$$

$$= \chi\sin(\Omega t - \gamma) \tag{6.75}$$

where

$$\chi = G\sqrt{\frac{1+(2\xi\beta)^2}{(1-\beta^2)^2+(2\xi\beta)^2}} \tag{6.76a}$$

$$\tan\gamma = \frac{2\xi\beta^3}{(1-\beta^2)+4\xi^2\beta^2} \tag{6.76b}$$

In Figure 6.19, amplitude ratio χ/G has been plotted as a function of β for several different values of ξ.

The rotating vector representation can also be used quite effectively to obtain the total displacement u^t. Referring to Figure 6.20, the support displacement is represented by a vector of length G rotating at Ω rad/s, so that at time t it makes an angle Ωt with the horizontal axis. The displacement of the supported system relative to the support has an amplitude ρ and lags vector G by an angle ϕ. The total displacement is obtained by taking the vector sum of G and ρ and is given by the trigonometric identity

$$\chi^2 = G^2 + \rho^2 + 2G\rho\cos\phi \tag{6.77}$$

On substituting for ρ from Equation 6.74a and noting that $\cos\phi = (1-\beta^2)/\sqrt{(1-\beta^2)^2+(2\xi\beta)^2}$, Equation 6.77 reduces to the form of Equation 6.76a. Phase angle γ between G and χ is obtained by applying the cosine identity to

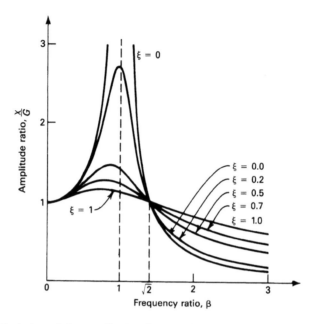

Figure 6.19. Variation of the amplitude of transmitted motion with frequency ratio and damping.

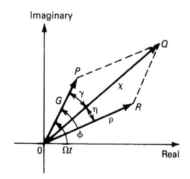

Figure 6.20. Rotating vector representation of the displacement due to support excitation.

triangle OPQ

$$\rho^2 = G^2 + \chi^2 - 2G\chi \cos \gamma \tag{6.78}$$

It is easily shown that Equation 6.78 will lead to the same value of γ as that given by Equation 6.76b. Alternatively, angle η between vectors χ and ρ can be obtained from the relationship

$$G^2 = \chi^2 + \rho^2 - 2\rho\chi \cos \eta \tag{6.79}$$

Substituting for χ and ρ in Equation 6.79, we get

$$\cos \eta = \frac{1}{\sqrt{1 + (2\xi\beta)^2}} \qquad (6.80\text{a})$$

and

$$\tan \eta = 2\xi\beta \qquad (6.80\text{b})$$

Angle γ is now obtained from

$$\gamma = \phi - \eta \qquad (6.81)$$

Example 6.4
If it is assumed that the stiffness of the tires is much greater than that of the springs, an auto-mobile suspension can be idealized by a single-degree-of-freedom system shown in Figure E6.4. As the tire rolls along a washboard road, a vertical sinusoidal motion is imparted to the contact point. As a result, the car body also undergoes vertical vibrations. Assuming that $h = 2$ in., $L = 20$ ft, the suspension heave frequency is 1.2 Hz and the damping ratio is 0.85, determine the amplitude of motion of the car body when the car speed is 20 mph. Neglect the weight of the wheel and the axle. What would the amplitude be if the damper became ineffective?

Solution
The period T of the harmonic vertical motion of the tire is equal to the time it takes to roll a distance L. The speed of the tire, V, is

$$V = \frac{20 \times 5280}{3600} = 29.33 \text{ ft/s}$$

Hence

$$T = \frac{L}{V} = \frac{20}{29.33} = 0.682 \text{ s}$$

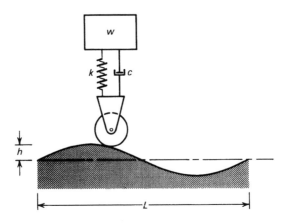

Figure E6.4. Automobile suspension on a washboard road.

and the frequency f is given by

$$f = \frac{1}{T} = 1.467 \, \text{Hz}$$

The frequency ratio $\beta = f/f_0 = 1.467/1.2 = 1.22$.

By substituting for β, ξ, and $G = h = 2$ in. in Equation 6.76a, we obtain the amplitude of motion of the car body

$$\chi = h\sqrt{\frac{1 + (2 \times 0.85 \times 1.22)^2}{\{1 - (1.22)^2\}^2 + (2 \times 0.85 \times 1.22)^2}}$$

$$= 2.16 \, \text{in.}$$

With $\xi = 0$, the amplitude χ will work out to 4.10 in. The shock absorber is thus effective in cutting down the amplitude of motion considerably.

6.10   TRANSMISSIBILITY AND VIBRATION ISOLATION

In practical design, it is frequently required that the dynamic forces transmitted by a machine to its surroundings be minimized. Vibration isolation of this type is referred to as *force isolation* and is usually achieved by inserting suitable springs and damping devices between the machine and its foundation. An inverse problem of isolation exists when it is required that vibrations from the surroundings are not transmitted to a supported system. Such isolation may be required when a delicate instrument is mounted inside a moving body or close to a source of vibrations, for example, in a space shuttle, an airplane, a ship, or near a floor supporting reciprocating machinery. In all these situations, the isolation problem is that of motion rather than of force. The two problems are, however, identical in most respects and the isolating devices used in each case have similar characteristics.

Let us first consider the problem of force isolation. Figure 6.21 shows a machine of mass m mounted on a foundation through a set of springs and dampers, the total spring stiffness being k, and the damping coefficient c. Let the dynamic force acting on the machine in a vertical direction be $p_0 \sin \Omega t$. The steady-state response of the machine is $u = \rho \sin(\Omega t - \phi)$, where ρ and ϕ are given by Equations 6.28. Force is transmitted to the foundation through the spring and the damper. The total transmitted force F is given by

$$F = f_S + f_D \tag{6.82}$$

On substituting for f_S and f_D from Equations 6.39b, we get

$$F = k\rho \sin(\Omega t - \phi) + c\Omega\rho \cos(\Omega t - \phi)$$

$$= \chi \sin(\Omega t - \phi + \eta) \tag{6.83}$$

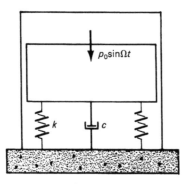

Figure 6.21. Transmitted force under harmonic excitation.

where

$$\chi = \sqrt{(k\rho)^2 + (c\Omega\rho)^2} \qquad\qquad (6.84a)$$

$$\tan \eta = \frac{c\Omega}{k} \qquad\qquad (6.84b)$$

On substituting for ρ from Equation 6.28 and noting that $c/k = 2\xi/\omega$, Equation 6.84 reduces to

$$\chi = p_0 \sqrt{\frac{1 + (2\xi\beta)^2}{(1 - \beta^2)^2 + (2\xi\beta)^2}} \qquad\qquad (6.85a)$$

$$\tan \eta = 2\xi\beta \qquad\qquad (6.85b)$$

The ratio χ/p_0, called the *transmission ratio* or *transmissibility*, and denoted by TR, provides a measure of the force transmitted to the foundation. A comparison of Equations 6.76a and 6.85a shows that Figure 6.19 should also represent the variation of transmission ratio with β for different values of the damping ratio ξ. It is evident from Figure 6.19 that if the force is to be smaller than the applied force p_0, the natural frequency ω should be selected so that the ratio $\beta = \Omega/\omega$ is greater than $\sqrt{2}$. Also, for $\beta > \sqrt{2}$, the transmission ratio decreases with damping, so that, theoretically, zero damping will give the smallest transmitted force. In practice, however, some damping should always be provided to ensure that during startup as the machine passes through the resonant frequency, the response is kept within reasonable limits.

In the design of isolation systems, when the frequency ratio β is greater than $\sqrt{2}$, the damping would usually be kept quite small. When damping forces are negligible, the expression for transmission ratio is greatly simplified, giving

$$\mathrm{TR} \equiv \frac{\chi}{p_0} = \frac{1}{\beta^2 - 1} \qquad\qquad (6.86a)$$

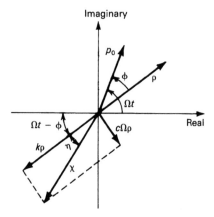

Figure 6.22. Vector diagram for transmitted force under steady-state harmonic excitation.

The force balance diagram can also be utilized effectively to obtain the transmitted force. Such a diagram is shown in Figure 6.22. The transmitted force χ is obtained by taking the vector sum of the spring force vector $k\rho$ and the damping force vector $c\Omega\rho$. Since the last two vectors are perpendicular to each other, their sum is given by Equation 6.84a. Also, the angle η between the spring force vector and the transmitted force vector is readily seen to be given by Equation 6.84b.

We next direct our attention to the isolation of vibratory motion. In this case, we are interested in determining the absolute movement of the supported system caused by a motion of the support. When the support undergoes harmonic motion, the amplitude of transmitted movement is given by Equation 6.76a. The similarity of this equation to Equation 6.85a implies that the design for motion isolation is governed by the same considerations that apply to force isolation. A variety of materials and devices are used in practice for vibration isolation. Isolators may consist of coiled springs of steel or pads of natural rubber, cork, or felt. In general, these materials serve both as springs and dampers. Coiled springs have a very small amount of damping. On the other hand, organic materials such as rubber, cork, or felt provide significant damping resistances. Metal springs are thus quite satisfactory when the supports can be designed so that the ratio of the operating frequency to the natural frequency of the system, β, is greater than $\sqrt{2}$. When this is not possible, metal springs are not satisfactory.

Example 6.5
A machine of mass 100 kg is supported on steel springs that deflect 1.2 mm under the weight of the machine. At the operating speed of the motor of 3000 rpm, imbalance causes a maximum disturbing force of 360 N. What is the maximum force transmitted to the foundation if damping in the steel springs is negligible? If the steel springs were replaced by neoprene pads having the same stiffness but a damping ratio of 0.2, what would be the maximum transmitted force?

Solution
Denoting by Δ the static deflection under the weight of the motor, we have

$$\Delta = \frac{mg}{k} \tag{a}$$

The natural frequency is now given by

$$\omega = \sqrt{\frac{k}{m}} = \sqrt{\frac{g}{\Delta}}$$

$$= \sqrt{\frac{9.81}{1.2 \times 10^{-3}}}$$

$$= 90.4 \text{ rad/s}$$

Hence the frequency in hertz, $f_0 = 14.4$. The operating speed of the motor is 50 Hz, and the frequency ratio is therefore given by $\beta = 50/14.4 = 3.47$. The transmission ratio for zero damping is obtained from Equation 6.86

$$\text{TR} = \frac{1}{\beta^2 - 1} = \frac{1}{(3.47)^2 - 1} = 0.09$$

$$\text{maximum transmitted force} = \text{TR} \times 360$$

$$= 32.4 \text{ N}$$

When damping is significant, the transmitted ratio is obtained from Equation 6.85a. Substituting $\xi = 0.20$ and $\beta = 3.47$ gives

$$\text{TR} = \sqrt{\frac{1 + (2\xi\beta)^2}{(1 - \beta^2)^2 + (2\xi\beta)^2}}$$

$$= 0.154$$

$$\text{transmitted force} = 0.154 \times 360$$

$$= 55.3 \text{ N}$$

Example 6.6
A sensitive instrument that requires to be insulated from vibration is to be installed in a laboratory, where a reciprocating machine is in use. The vibrations of the floor of the laboratory may be assumed to be a simple harmonic motion having a frequency in the range 1000 to 3000 cycles per minute. The instrument is to be mounted on a small platform and supported on four springs arranged to carry equal loads. The combined mass of the instrument and supporting table is 5 kg. Calculate a suitable value for the stiffness of each spring if the amplitude of transmitted vibrations is to be less than 15% of the floor vibrations over the given frequency range. Assume that the damping is negligible.

Solution
The frequency ratio should be larger than $\sqrt{2}$ for TR to be less than 1. The transmission ratio is then given by

$$\text{TR} = \frac{1}{\beta^2 - 1}$$

The condition that TR be less than 0.15 leads to

$$\beta^2 > \frac{1}{TR} + 1 = \frac{1}{0.15} + 1 = 7.67$$

Hence the natural frequency $f_0 < f/\sqrt{7.67} = f/2.77$. The governing value of f will be the lower limit of the range of exciting frequency, that is, equal to 1000 cycles per minute. Hence

$$f_0 = \frac{1000}{60} \times \frac{1}{2.77}$$

$$= 6.02 \text{ Hz}$$

The total stiffness is now obtained from

$$\frac{1}{2\pi} \sqrt{\frac{k}{m}} = 6.02$$

or

$$k = (6.02 \times 2\pi)^2 \times 5$$

$$= 7154 \text{ N/m}$$

The stiffness of each spring is $1/4 \times 7154 = 1788$ N/m.

6.11   VIBRATION MEASURING INSTRUMENTS

Measurement of vibrations plays an important role in engineering practice. The quantity to be measured may be a displacement, an acceleration, or a stress. The applications are quite diverse: from measurements of machine vibrations to the time history record of ground motion during an earthquake. In all cases, however, the underlying principle in the design of the instrument is the same. The instrument would usually consist of a spring–mass–damper system mounted on a rigid frame which is attached to the surface whose motion is to be measured. The quantity measured is the relative displacement between the frame and the instrument mass, and depending on the details of design, this displacement can be related to the displacement or acceleration of the support.

The details of the design of a vibration measuring instrument, also known as a *seismic instrument*, are quite intricate. Usually, the instrument displacement must be magnified in some way before it can be measured. Such magnification may be achieved through mechanical levers or by optical means. A suitable damping device must be provided, and they range from an oil dashpot to an electric coil moving in a magnetic field. In some cases, a permanent record must be obtained of the vibration history. This is usually accomplished by exposing a photographic film to a light beam or by using a stylus to scribe on a waxed paper. Details of design also depend on the application. Instruments for seismological observations have different requirements from those used for

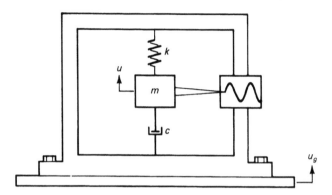

Figure 6.23. Schematic diagram of a seismic instrument.

earthquake engineering measurements. Measurement of mechanical vibrations has its own special needs.

Our primary interest here is to discuss the underlying principle in the design of seismic instruments. Figure 6.23 presents a schematic diagram of such an instrument. The instrument mass m is supported in the rigid frame through a spring of stiffness k and a damper with damping coefficient c. The frame of the instrument is rigidly attached to the surface whose motion is to be measured and moves along with the latter. The motion of the mass m relative to its support frame is measured by a suitable device and is correlated to a support motion parameter as discussed in the following paragraphs.

6.11.1 *Measurement of support acceleration*

If the support is undergoing a harmonic motion, with acceleration $\ddot{u}_g = A \sin \Omega t$, then, as discussed in Section 6.9, the equation governing the displacement of mass m relative to the frame is

$$m\ddot{u} + c\dot{u} + ku = -mA \sin \Omega t \tag{6.87}$$

The solution to Equation 6.87 is given by Equations 6.27 and 6.28, with p_0 replaced by $-mA$. Ignoring the negative sign on the right hand-side of Equation 6.87, we get

$$u = \rho \sin(\Omega t - \phi) \tag{6.88}$$

where

$$\rho = \frac{mA}{k} \frac{1}{\sqrt{(1 - \beta^2)^2 + (2\xi\beta)^2}} \tag{6.89a}$$

$$\tan \phi = \frac{2\xi\beta}{1 - \beta^2} \tag{6.89b}$$

and β is the ratio of the frequency of support motion to the natural frequency, ω, of the instrument.

If the instrument is to be designed to measure an input acceleration which may, in fact, have several harmonic components of different frequencies, the measured displacement u should be proportional to the input for all values of the input frequency. From Equations 6.88 and 6.89a we note that the constant of proportionality is given by

$$\frac{\rho}{A} = \frac{1}{\omega^2} \frac{1}{\sqrt{(1 - \beta^2)^2 + (2\xi\beta)^2}} \tag{6.90}$$

For a satisfactory instrument design $\rho\omega^2/A$ should not vary with β. Figure 6.6 shows the variation of $\rho\omega^2/A$ with β for different values of ξ. It is observed that for $\xi = 0.7$, $\rho\omega^2/A$ stays approximately constant at a value of 1, provided that β lies between about 0 and 0.6. This is borne out by the figures in Table 6.1. If $\xi = 0$, then an instrument calibrated to be correct at a very small input frequency will be in error by 56% at an input frequency that is 60% of the instrument frequency. However, if $\xi = 0.7$, the error in measurement at $\beta = 0.6$ is only about 5.3%.

It is evident from the discussion above that when the motion to be measured has a maximum frequency component with frequency equal to Ω, the instrument should be designed to have a natural frequency $\omega > \Omega/0.6 = 1.67\Omega$ if the error in measurement is not to exceed about 5%. Since, as seen from Equation 6.90, the relative displacement u is inversely proportional to the square of ω, a high instrument frequency will result in very small instrument displacements, and the latter must therefore be substantially magnified for proper measurement.

Equation 6.88 also shows that the measured output of the instrument will lag the input motion by phase angle ϕ, implying that with respect to the excitation, the output will be shifted on the time axis by an amount $t_s = \phi/\Omega$. In terms of β and ξ, this shift is given by

$$t_s = \frac{1}{\Omega} \tan^{-1} \frac{2\xi\beta}{1 - \beta^2}$$

$$= \frac{1}{\beta\omega} \tan^{-1} \frac{2\xi\beta}{1 - \beta^2} \tag{6.91}$$

If the motion being measured comprises a single harmonic, this shift on the time axis will be of no particular significance. However, when the input motion has several harmonic components, the time shift being a function of β is, in general, different for each component output. The resulting shape of the total output, which is a superposition of its components is thus completely distorted and the

Table 6.1. Characteristics of the Output of a Seismic Instrument Designed to Measure Acceleration.

β	$\rho\omega^2/A$		ϕ/β
	$\xi = 0$	$\xi = 0.7$	$\xi = 0.7$
0.0	1.00	1.000	1.400
0.1	1.01	1.000	1.405
0.2	1.04	1.000	1.419
0.3	1.10	1.998	1.442
0.4	1.19	1.991	1.470
0.5	1.33	1.975	1.502
0.6	1.56	1.947	1.533

measured displacement is unable to predict the input motion. It is therefore, a requirement that $t_s = \phi/\Omega$ not vary with β. Fortunately, for $\xi = 0.7$, ϕ/Ω is practically constant, or ϕ is a linear function of Ω and hence of β. This is evident from both Figure 6.7 and Table 6.1.

6.11.2 *Measurement of support displacement*

In the preceding section, we discussed the design of an instrument for measuring support acceleration. If the quantity to be measured is support displacement rather than support acceleration, the design should be suitably modified. The harmonic input motion is now given by Equation 6.70 and the output by Equations 6.73 and 6.74. The amplitude of the input motion is G and that of the output is ρ. The variation of amplitude ratio ρ/G with β and ξ is indicated by Figure 6.17 provided that the ordinate in that figure is interpreted as ρ/G. For a well-designed instrument, this ratio should not vary with β, but should remain constant. From Figure 6.17 it is seen that for large values of β the amplitude ratio remains approximately constant at 1 irrespective of the value of ξ. The requirement that β should be large implies that the natural frequency of the instrument ω should be kept small. This can be achieved by providing a flexible spring and/or a heavy mass, both of which make the instrument unwieldy. For zero damping, Equation 6.74b gives $\phi = 180°$ and Equation 6.73 reduces to $u = -\rho \sin \Omega t$. This implies that although the measured motion is negative of the input motion, there is no shift along the time axis and the output motion comprised of any number of harmonic components will be reproduced correctly.

Example 6.7
An accelerometer is designed to have a natural frequency of 30 rad/s and a damping ratio of 0.7. It is calibrated to read correctly the input acceleration at very small values of the exciting frequency.

(i) What would be the percentage error in the instrument reading if the harmonic motion being measured has a frequency of 20 rad/s?

(ii) The displacement input imparted to the frame of the instrument is given by $u_g = 500 \sin 10t + 400 \sin 20t$. Plot the variation of the input acceleration with time. Obtain the curve of instrument reading versus time and compare it with the input.

Solution

(i) From Equation 6.89a,

$$\frac{\rho \omega^2}{A} = \frac{1}{\sqrt{(1 - \beta^2)^2 + (2\xi\beta)^2}}$$

where A is the amplitude of input acceleration. For $\beta = 0$, $\rho\omega^2/A = 1$. The instrument is calibrated so that its reading is equal to ω^2 times ρ. This ensures that at low frequencies, the instrument will read the amplitude of input acceleration correctly. For $\beta = 20/30 = 0.67$ and $\xi = 0.7$

$$\frac{\rho \omega^2}{A} = 0.921$$

This implies that for an input frequency of 20 Hz, the instrument reading $\rho\omega^2$ is 92.1% of A or 7.9% too low.

(ii) The input acceleration is given by

$$\ddot{u}_g = 500 \sin 10t + 400 \sin 20t \tag{a}$$

The two harmonic components of the input acceleration are plotted in Figure E6.7a and b, respectively. The instrument reading for the input motion with a frequency of 10 rad/s is given by

$$a_1 = 500 \times \frac{1}{\sqrt{(1 - \beta_1)^2 + (2\xi\beta_1)^2}} \sin(10t - \phi_1)$$

where $\beta_1 = 10/30 = 1/3$, $\xi = 0.7$, and $\tan \phi_1 = 2\xi\beta_1/(1 - \beta_1^2)$. Substitution of the given values yields

$$a_1 = 0.996 \times 500 \sin(10t - 0.483) \tag{b}$$

In a similar manner, the instrument reading for the input motion with a frequency of 20 rad/s is given by

$$a_2 = 0.921 \times 400 \sin(20t - 1.034) \tag{c}$$

Relationships given by Equations b and c are shown on Figure E6.7a and b by dashed lines. It is observed that the curve for a_1 is shifted by $0.483/10 = 0.0483$ s with respect to the input acceleration, while the curve for a_2 is shifted by $1.034/20 = 0.0517$ s. Finally, the total input acceleration is compared with the resultant instrument reading in Figure E6.7c. Because the shift for the two components is slightly different, the output signal is a bit distorted, but the distortion is fairly small.

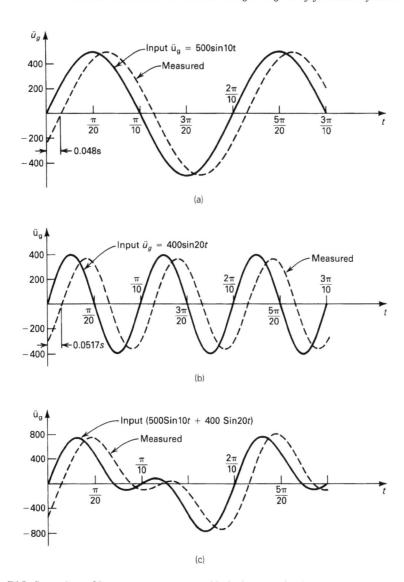

Figure E6.7. Comparison of instrument measurement with the input acceleration.

6.12 ENERGY DISSIPATED IN VISCOUS DAMPING

The energy balance in a system with viscous damping undergoing steady-state harmonic motion is of considerable interest. Energy is input into the system by the applied force $p = p_0 \sin \Omega t$. Denoting the infinitesimal work done by p as the system moves through a small distance du by dW_e, we can write

$$dW_e = p_0 \sin \Omega t \, du \qquad (6.92)$$

Since $u = \rho \sin(\Omega t - \phi)$, $du = \rho\Omega \cos(\Omega t - \phi)\,dt$, and the time taken by the system to undergo one cycle of motion is $2\pi/\Omega$, the energy input per cycle becomes

$$W_e = \int_0^{2\pi/\Omega} p_0 \rho\Omega \sin \Omega t \cos(\Omega t - \phi)\,dt$$

$$= p_0 \rho\pi \sin \phi \tag{6.93}$$

The energy input into the system is dissipated in viscous damping. We can prove this directly by considering the work done by the damping force per cycle of motion. Under steady-state vibration, the damping force is by

$$f_D = c\Omega\rho \cos(\Omega t - \phi) \tag{6.94}$$

The work done by f_D as the system undergoes a small displacement $du = \rho\Omega \cos(\Omega t - \phi)\,dt$ is given by

$$dW_i = c\Omega^2 \rho^2 \cos^2(\Omega t - \phi)\,dt \tag{6.95}$$

The energy dissipated per cycle is obtained by integrating Equation 6.95 from 0 to $2\pi/\Omega$ seconds.

$$W_i = c\Omega^2 \rho^2 \int_0^{2\pi/\Omega} \cos^2(\Omega t - \phi)\,dt$$

$$= c\pi\Omega\rho^2 \tag{6.96}$$

From Equations 6.26 and 6.28 we note that

$$\sin \phi = \frac{2\xi\beta}{\sqrt{(1 - \beta^2)^2 + (2\xi\beta)^2}}$$

$$= \frac{2\xi\beta k}{p_0} \rho \tag{6.97}$$

Substitution of Equation 6.97 into Equation 6.93 gives

$$W_e = 2\xi\beta k\rho^2 \pi \tag{6.98}$$

Since $\xi = c/c_{cr}$, $c_{cr} = 2k/\omega$, and $\beta = \Omega/\omega$, Equation 6.98 reduces to

$$W_e = c\pi\Omega\rho^2 = W_i \tag{6.99}$$

The energy balance represented by Equation 6.99 implies that the work done by the spring and inertial forces per cycle is zero. This can also be proved by noting that $f_S = k\rho \sin(\Omega t - \phi)$ and $f_I = -m\Omega^2 \rho \sin(\Omega t - \phi)$ and integrating the infinitesimal work terms $f_S\,du$ and $f_I\,du$ over one cycle. The following

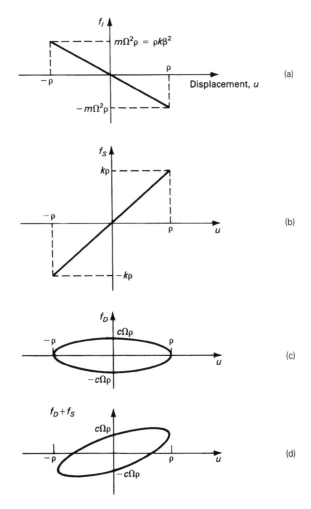

Figure 6.24. Force–displacement relationship under harmonic motion: (a) inertia force versus u; (b) spring force versus u; (c) damping force versus u; (d) spring force plus damping force versus u.

alternative method is probably more illustrative of the concepts involved. The inertia and spring forces are related to the displacement as follows

$$f_I = -m\Omega^2 u \qquad\qquad (6.100a)$$

$$f_S = ku \qquad\qquad (6.100b)$$

The inertia force versus displacement relation is plotted in Figure 6.24a. It consists of a single straight line as shown. Since the area enclosed by the inertia force versus displacement curve over one cycle of motion is zero, the

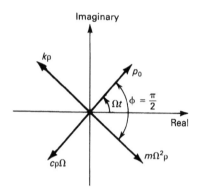

Figure 6.25. Force balance diagram at resonance.

corresponding work done should also be zero. In a similar manner, the spring force versus displacement diagram drawn in Figure 6.24b is a straight line and shows that the work done per cycle by the spring force is also zero.

Using Equation 6.94, the relationship between the damping force and displacement can be expressed as follows

$$\left(\frac{f_D}{c\Omega\rho}\right)^2 + \left(\frac{u}{\rho}\right)^2 = 1 \tag{6.101}$$

Equation 6.101 is the equation of an ellipse which is shown in Figure 6.24c. The area of ellipse is equal to $c\pi\Omega\rho^2$ and represents the energy dissipated per cycle in viscous damping.

The transmitted force $f_D + f_S$ is obtained by superposition of Figure 6.24b and c. The transmitted force versus u diagram shown in Figure 6.24d is an ellipse whose principal axes are inclined with respect to the coordinate axes. The area of the skew ellipse is, of course, equal to the area enclosed by the ellipse shown in Figure 6.24c. In a similar manner, the applied force, $p = f_I + f_D + f_S$, versus the displacement diagram is obtained by the superposition of Figures 6.24a, b, and c. The resulting curve is again a skewed ellipse whose area is equal to the area of ellipse shown in Figure 6.24c.

The energy balance at resonance can be derived from the foregoing discussion. In this case, because $\beta = 1$, the spring force exactly balances the inertia force. This will be evident by comparing Equation 6.100a and b. The damping force should now exactly balance the applied force and the applied force versus displacement diagram should be identical to the damping force versus displacement diagram. The same conclusions can be drawn by referring to the force balance at resonance, Figure 6.25. Since, in this case, the phase angle $\phi = \pi/2$, the spring force balances the inertia force and the damping force balances the applied force. In addition, Equation 6.93 reduces to

$$W_e = p_0 \rho \pi \qquad (6.102)$$

6.13   HYSTERETIC DAMPING

Equation 6.96 shows that in a system with viscous damping, the energy loss per cycle is proportional to the square of the amplitude and directly to the frequency of motion. For a given amplitude, therefore, the energy loss will be proportional to frequency and will increase with the latter. As stated in Section 5.4, practical observations do not corroborate this and the actual energy loss per cycle is, in fact, independent of the frequency of motion. This is because damping forces are not viscous in nature but arise from internal friction, localized plastic deformation, and plastic flow that are present even when the global stress level is within the elastic range. It should be noted that our intent here is not to model the energy dissipated through plastic deformations at a global level introduced by stresses beyond the elastic limit. Such energy dissipation is usually accounted for by using in the analysis a nonlinear stress-strain relationship for the structural material under consideration. Methods of nonlinear analysis are discussed in Chapter 8. Damping caused by internal forces and local plasticity is referred to as *hysteretic damping* or *rate-independent damping*. The equation of motion for forced vibrations under hysteretic damping can be expressed as

$$m\ddot{u} + f_S^t(u, \dot{u}) = p(t) \qquad (6.103)$$

In a general case, the solution of Equation 6.103 is quite complex. However, for steady-state harmonic motion, hysteretic damping can be accounted for by expressing the total spring force $f_S^t(u, \dot{u})$ as the sum of two components: a spring force $f_S = ku$, where k is the average stiffness and a damping force f_D given by

$$f_D = \frac{\eta k}{\Omega} \dot{u} \qquad (5.56)$$

in which η is a constant. The energy loss per cycle due to hysteretic damping is obtained from Equation 6.96 by replacing c with $\eta k/\Omega$

$$W_i = \eta k \pi \rho_h^2 \qquad (6.104)$$

The energy loss is seen to be independent of frequency but proportional to the square of the amplitude. The total spring force f_S^t is given by

$$f_S^t = ku + \frac{\eta k}{\Omega} \dot{u} \qquad (6.105)$$

and the equation of motion becomes

$$m\ddot{u} + \frac{\eta k}{\Omega}\dot{u} + ku = p_0 \sin \Omega t \tag{6.106}$$

We now define $c_h = \eta k/\Omega$ and $\xi_h = c_h/(2\sqrt{km}) = \eta/(2\beta)$, where η is a constant while c_h and ξ_h both vary with β. The steady-state solution to Equation 6.106 is given by Equations 6.28a and b with ξ replaced by ξ_h. Thus noting that $2\xi_h\beta = \eta$

$$u = \rho_h \sin(\Omega t - \phi_h) \tag{6.107}$$

where

$$\rho_h = \frac{p_0}{k} \frac{1}{\sqrt{(1 - \beta^2)^2 + \eta^2}} \tag{6.108a}$$

$$\tan \phi_h = \frac{\eta}{1 - \beta^2} \tag{6.108b}$$

Amplitude ratio $\rho_h/(p_0/k)$ and phase angle ϕ_h have been plotted in Figure 6.26a and b, respectively, as functions of β for several different values of η. The general shape of the curves is similar to that for viscous damping. However, the amplitude ratio is a maximum at $\beta = 1$, unlike the case of viscous damping, where the maximum occurs for a value of β slightly less than 1. Also, for $\beta = 0$, the phase angle ϕ_h is given by $\phi_h = \tan^{-1} \eta$, while in the case of viscous damping $\phi = 0$ for $\beta = 0$. In other words, with hysteretic damping, the response is never in phase with the exciting motion. We can obtain an equivalent viscous damping coefficient by equating the energy loss per cycle in viscous damping to that given by Equation 6.104

$$2\xi_{eq}\beta k\pi\rho^2 = \eta k\pi\rho^2$$
$$\xi_{eq} = \frac{\eta}{2\beta} \tag{6.109}$$

It is apparent from Equation 6.109 that the equivalent viscous damping varies with the exciting frequency, so that we cannot use a single value for it throughout the frequency range of interest. However, as an approximation we may use in our analysis the equivalent viscous damping derived for a frequency equal to the resonant frequency. Thus setting $\beta = 1$ in Equation 6.109 we get

$$\xi_{eq} = (\xi)_{\beta=1} = \frac{\eta}{2} \tag{6.110}$$

The amplitude and phase angle of the frequency response obtained with equivalent viscous damping given by Equation 6.110 are shown in Figure 6.26a and b by dashed lines. The match between the two sets of curves is exact at $\beta = 1$ because the equivalent viscous damping was derived for that value of β.

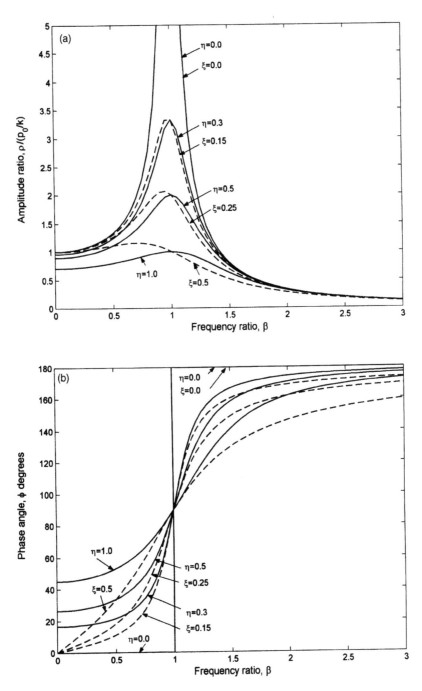

Figure 6.26. (a) Variation of the amplitude ratio with frequency ratio and hysteretic damping ratio; (b) variation of phase angle with frequency ratio and hysteretic damping ratio.

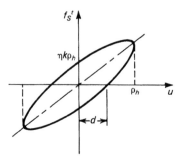

Figure 6.27. Spring force versus displacement in a system with hysteretic damping.

Away from resonance, the two curves depart from each other, although the difference is not too large. It is of interest to plot the relationship between the total spring force f_S^t and the displacement u. From Equation 6.107, $\dot{u} = \rho_h \Omega \cos(\Omega t - \phi_h)$. Substituting this in Equation 6.105, we get

$$\frac{f_S^t - ku}{\eta k \rho_h} = \cos(\Omega t - \phi_h) \tag{6.111}$$

Since $u = \rho_h \sin(\Omega t - \phi_h)$, Equation 6.111 gives

$$\left(\frac{f_S^t - ku}{\eta k \rho_h}\right)^2 + \left(\frac{u}{\rho_h}\right)^2 = 1 \tag{6.112}$$

Equation 6.112 is plotted in Figure 6.27 with force f_S^t on the ordinate and displacement u on the abscissa. The resulting curve is a skewed ellipse as shown. The area enclosed by the curve represents the energy loss per cycle and its value is $\eta k \pi \rho_h^2$. Equation 6.112 does not contain the exciting frequency Ω. This implies that the hysteresis loop represented by Equation 6.112 can be obtained experimentally by carrying out a quasi-static cyclic load test in which the rate of loading is fairly slow, and plotting the resulting load–displacement relationship. The relationship shown in Figure 6.27 can be interpreted as the hysteresis loop of the structure. Intercept d on the displacement axis is given by

$$d = \rho_h \frac{\eta}{\sqrt{1 + \eta^2}} \tag{6.113}$$

from which the damping fraction η is obtained as

$$\eta = \frac{d}{\sqrt{\rho_h^2 - d^2}} \approx \frac{d}{\rho_h} \tag{6.114}$$

For light damping, intercept d is small in comparison to ρ_h and the approximation in Equation 6.114 applies.

6.14 COMPLEX STIFFNESS

For simple harmonic motion, hysteretic damping can conveniently be expressed by using the concept of complex stiffness. Thus if the exciting force is $p_0 e^{i\Omega t}$, the displacement and velocity are given by

$$u = U e^{i\Omega t} \tag{6.115a}$$

and

$$\dot{u} = U i \Omega e^{i\Omega t} \tag{6.115b}$$

The hysteretic damping force can now be expressed as

$$f_D = \frac{\eta k}{\Omega} \dot{u}$$

$$= U i \eta k e^{i\Omega t}$$

$$= i \eta k u \tag{6.116}$$

By substituting Equation 6.116 into Equation 6.106, the equation of motion is expressed as

$$m\ddot{u} + (1 + i\eta)ku = p_0 e^{i\Omega t} \tag{6.117a}$$

or

$$m\ddot{u} + \bar{k}u = p_0 e^{i\Omega t} \tag{6.117b}$$

The quantity $\bar{k} = k(1 + i\eta)$ is called *complex stiffness*. It is useful in taking account of hysteretic damping, particularly when the analysis is carried out in the frequency domain as described in Chapter 9.

6.15 COULOMB DAMPING

When damping resistance in a system arises due to dry friction, the equation of motion for harmonic excitation becomes

$$m\ddot{u} + ku \pm \mu N = p_0 \sin \Omega t \tag{6.118}$$

Referring to Figure 6.28a the positive sign for the friction force applies when the mass m is moving to the right while the negative sign applies when the mass is moving to the left. Equation 6.118 is a nonlinear equation, and in a general case, its solution is quite complex. However, for small amounts of damping, an approximate but simpler solution can be obtained for steady-state harmonic vibrations. From the friction force versus displacement diagram shown in Figure 6.28b, it is apparent that the work done by the friction force per cycle

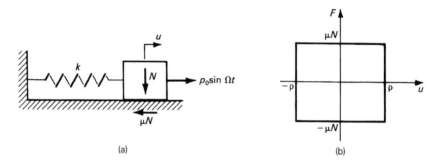

(a)   (b)

Figure 6.28. System with Coulomb friction subjected to harmonic excitation.

of motion is given by

$$W_i = 4\mu N\rho \tag{6.119}$$

We can obtain an equivalent viscous damping coefficient by equating the energy loss per cycle in viscous damping to that given by Equation 6.119

$$\rho^2 \pi \Omega c_{eq} = 4\mu N\rho$$

$$c_{eq} = \frac{4\mu N}{\rho \pi \Omega} \tag{6.120}$$

The equivalent damping ratio becomes

$$\xi_{eq} = \frac{c_{eq}}{2\sqrt{km}} = \frac{2\mu N}{\pi k\rho\beta} \tag{6.121}$$

The displacement response is now given by Equation 6.27, where the amplitude ρ is obtained from Equation 6.28 with ξ_{eq} substituted for ξ

$$\rho = \frac{p_0}{k} \frac{1}{\sqrt{(1-\beta^2)^2 + (4\mu N/\pi k\rho)^2}} \tag{6.122}$$

On solving Equation 6.122 for ρ, we get

$$\rho = \frac{p_0}{k} \sqrt{\frac{1 - (4\mu N/\pi p_0)^2}{(1-\beta^2)^2}} \tag{6.123}$$

Equation 6.123 is valid provided that the ratio of friction force to the amplitude of applied load, $\mu N/p_0$, is less than $\pi/4$. For a larger friction force, the numerator in Equation 6.123 becomes imaginary and the method breaks down. In fact, for reasonable accuracy $\mu N/p_0$ must be less than about $\frac{1}{2}$.

Amplitude ratio $\rho/(p_0/k)$ is plotted in Figure 6.29a as a function of β for several different values of $\mu N/p_0$. It is apparent from the plots and from Equation 6.123 that, at resonance, friction damping does not limit the amplitude,

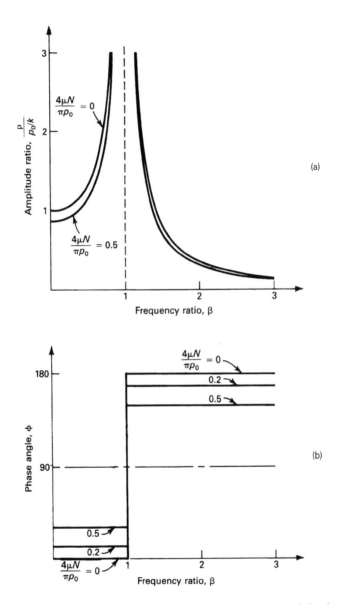

Figure 6.29. (a) Variation of amplitude ratio with frequency ratio and Coulomb friction; (b) variation of phase angle with frequency ratio and Coulomb friction.

which therefore becomes unbounded. Phase angle ϕ is obtained from Equation 6.28b with $\zeta = \zeta_{eq}$ and ρ given by Equation 6.123

$$\tan \phi = \pm \frac{4\mu N/\pi p_0}{\sqrt{1 - (4\mu N/\pi p_0)^2}} \qquad (6.124)$$

where the positive sign applies for $\beta < 1$ and the negative sign for $\beta > 1$. The phase angle is independent of the frequency ratio but changes sign as β passes through 1. The variation of phase angle with the frequency ratio and $\mu N / p_0$ is shown in Figure 6.29b.

6.16   MEASUREMENT OF DAMPING

As stated earlier, the mass and stiffness of a dynamic system can be determined from its physical characteristics. Mass can be calculated from the known geometry and the mass density. Stiffness is determined if the geometry and material properties are known. However, it is difficult or impractical to relate damping resistance to known or measurable physical characteristics. An estimate of damping resistance can, however, be obtained by experimental measurement of the response of the structure to a given excitation. From a practical point of view, harmonic excitation is easiest to impart. It is for this reason that a discussion of the measurement of damping resistance is appropriately included in a section on response to harmonic excitation. A number of different experimental techniques are available. Some of these are described in the following paragraphs. Most techniques are based on the assumption that damping is of viscous type. When damping is not viscous in nature, an equivalent viscous damping is usually obtained.

6.16.1   *Free vibration decay*

In this method, which is related to free vibration described in Chapter 5, the system is excited by any convenient means and then allowed to vibrate freely. Measurements of displacement amplitudes are then made on a few successive peaks. For free vibration with viscous damping, if u_i denotes the displacement at any peak and u_{n+i} is the displacement n cycles later, then, as shown in Section 5.3.5, the following relationship holds

$$
\begin{aligned}
\frac{u_i}{u_{n+i}} &= \frac{e^{-\xi \omega t_i}}{e^{-\xi \omega (2\pi n / \omega_d + t_i)}} \\
&= e^{2\pi n \xi \omega / \omega_d} \\
&= e^{2\pi n \xi / \sqrt{1 - \xi^2}}
\end{aligned} \tag{6.125}
$$

Viscous damping ratio ξ is obtained from Equation 6.125

$$
\xi = \frac{\delta_n}{\sqrt{4\pi^2 n^2 + \delta_n^2}} \tag{6.126}
$$

where $\delta_n = \ln(u_i/u_{n+i})$. For light damping, Equation 6.126 reduces to $\xi = \delta_n/2\pi n$. In the determination of damping resistance through an observation of the free vibration decay, the only measurements required are those of relative displacements, and the instrumentation required is therefore fairly simple. When damping is hysteretic, the damping force is defined as in Equation 5.56. For free vibration, the vibration frequency is approximately equal to ω, provided that damping is light. The equivalent damping constant c_h and damping ratio ξ_h are therefore given by

$$
\begin{aligned}
c_h &= \frac{\eta k}{\omega} \\
\xi_h &= \frac{c_h}{2\sqrt{km}} = \frac{\eta}{2}
\end{aligned}
\tag{6.127}
$$

The hysteretic damping coefficient η is obtained by using Equation 6.126 with ξ replaced by ξ_h, so that

$$
\begin{aligned}
\eta &= \frac{2\delta_n}{\sqrt{4\pi^2 n^2 + \delta_n^2}} \\
&\approx \frac{\delta_n}{\pi n}
\end{aligned}
\tag{6.128}
$$

6.16.2 *Forced vibration response*

Damping resistance can also be obtained by measuring the displacement response of the system to a harmonic excitation. The system is excited by a series of harmonic forces with closely spaced frequencies spanning the resonant frequency. The displacement response to each excitation is measured so that the entire response curve in the vicinity of resonance is obtained. The properties of the response curve can be used in a number of different ways to obtain an estimate of the damping.

Resonant response
At resonance, that is, when $\beta = 1$, the phase angle $\phi = \pi/2$. Resonance can therefore be detected by measuring the phase angle and progressively adjusting the exciting frequency until ϕ is 90°. If the measured displacement amplitude at resonance is denoted by u_r, then

$$
\frac{u_r}{p_0/k} = \frac{1}{2\xi}
\tag{6.129a}
$$

so that

$$
\xi = \frac{1}{2u_r}\frac{p_0}{k}
\tag{6.129b}
$$

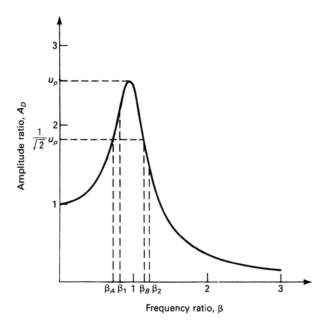

Figure 6.30. Use of response curve in the vicinity of resonance to determine the damping ratio.

The application of Equation 6.129 requires that the static displacement produced by a force of the same amplitude as the dynamic force be ascertained. This can be done in a separate test.

Measurement of the phase angle may be somewhat difficult. Therefore, as an alternative the resonance curve is obtained in the vicinity of resonance and the peak response u_p is measured. For viscous damping

$$\frac{u_p}{p_0/k} = \frac{1}{2\xi\sqrt{1-\xi^2}} \tag{6.130}$$

Equation 6.130 can be used to obtain ξ from the measured value of u_p and the static deflection produced by a force of amplitude p_0. For light damping ξ^2 is negligible in comparison to 1 and $\xi = (p_0/k)(1/2u_p)$. It should be noted that if the damping is hysteretic in nature, u_p and u_r are identical and are given by

$$u_p = u_r = \frac{1}{\eta}\frac{p_0}{k} \tag{6.131}$$

Width of response curve and half-power method
The width of response curve near resonance can be used to obtain an estimate of the damping. The method involves the measurement of frequencies at which the phase angle is $\pm\pi/4$, or $\tan\phi = \pm1$. Referring to Figure 6.30, one such

frequency will be below the resonant frequency and the other above it. The two values of β are obtained from Equation 6.28b

$$\frac{2\xi\beta_1}{1 - \beta_1^2} = 1 \tag{6.132a}$$

$$\frac{2\xi\beta_2}{1 - \beta_2^2} = -1 \tag{6.132b}$$

Equations 6.132a and b give

$$1 - \beta_1^2 - 2\xi\beta_1 = 0 \tag{6.133a}$$

$$1 - \beta_2^2 + 2\xi\beta_2 = 0 \tag{6.133b}$$

Subtracting Equation 6.133a from Equation 6.133b, we obtain

$$\xi = \frac{1}{2}(\beta_2 - \beta_1) = \frac{1}{2}\left(\frac{\Omega_2 - \Omega_1}{\omega}\right) \tag{6.134}$$

Equation 6.134 can be used to obtain ξ provided that Ω_2 and Ω_1 have been measured and natural frequency ω is either known or can be estimated.

The foregoing method relies on the ability to measure the phase angle, which may require sophisticated instrumentation. Another property of the response curve whose use does not rely on the measurement of phase angle is therefore sometimes employed in estimating the damping ratio. If the response curve in the vicinity of resonance has been plotted, the frequencies at which the amplitude is $1/\sqrt{2}$ times that at the peak can be measured. As shown in Figure 6.30, there will be two such frequencies. Let these frequencies be denoted by Ω_A and Ω_B, and the corresponding frequency ratios, by β_A and β_B. For a small amount of damping, the peak response is $(p_0/k)(1/2\xi)$. Hence β_A and β_B are obtained from the equation

$$\frac{1}{\sqrt{2}}\frac{1}{2\xi}\frac{p_0}{k} = \frac{p_0}{k}\frac{1}{\sqrt{(1 - \beta^2)^2 + (2\xi\beta)^2}} \tag{6.135}$$

On solving Equation 6.135 for β, we get

$$\begin{aligned}
\beta_A^2 &= (1 - 2\xi^2) - 2\xi\sqrt{1 + \xi^2} \\
\beta_B^2 &= (1 - 2\xi^2) + 2\xi\sqrt{1 + \xi^2}
\end{aligned} \tag{6.136}$$

For small values of ξ, Equation 6.136 can be simplified as follows

$$\begin{aligned}
\beta_A^2 &\approx 1 - 2\xi - 2\xi^2 \\
\beta_B^2 &\approx 1 + 2\xi - 2\xi^2
\end{aligned} \tag{6.137a}$$

or

$$\beta_A \approx 1 - \xi - \xi^2$$
$$\beta_B \approx 1 + \xi - \xi^2 \tag{6.137b}$$

Equation 6.137b gives

$$\xi = \frac{1}{2}(\beta_B - \beta_A) \tag{6.138}$$

Equation 6.138 is similar to Equation 6.134. In fact, for small amounts of damping, frequencies corresponding to half power points, β_A and β_B are very close to those corresponding to phase angles of $\pm\pi/4$, that is, β_1 and β_2.

Energy loss per cycle
Instead of using the response curve obtained for a range of frequencies, it is possible to determine the damping resistance by running a harmonic excitation test at one specified frequency provided that the complete force–displacement relationship can be measured. When damping is viscous, the applied force versus displacement relationship will be a skewed ellipse whose intercept on the ordinate is equal to $c\Omega\rho$. Since Ω and ρ are both known, c can be determined by measuring the intercept.

For damping that is not viscous in nature, the force–displacement relationship will not in general be an ellipse. An equivalent damping constant can still be determined by measuring W_D, the area enclosed by the force–displacement relationship, and equating it to the theoretical value of energy loss per cycle in viscous damping. Thus from Equation 6.99,

$$W_D = c_{eq} \pi \Omega \rho^2 \tag{6.139a}$$

or

$$c_{eq} = \frac{W_D}{\pi \Omega \rho^2} \tag{6.139b}$$

The damping ratio ξ_{eq} is given by

$$\xi_{eq} = \frac{c_{eq}}{2\sqrt{km}} \tag{6.140a}$$

or

$$\xi_{eq} = \frac{c_{eq}\omega}{2k} \tag{6.140b}$$

To use Equation 6.140a, m must be estimated from the geometry and mass density, and k can either be estimated or measured. Stiffness k can be determined by running a static load test. In such a test, if as shown in Figure 6.31,

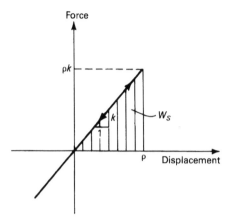

Figure 6.31. Static force–displacement relationship.

the area bounded by the force–displacement relationship, the displacement axis and an ordinate through amplitude ρ is denoted by W_s, we have

$$W_s = \frac{1}{2}\rho^2 k \qquad (6.141)$$

Hence

$$k = \frac{2W_s}{\rho^2} \qquad (6.142)$$

If it is desired that the damping ratio be obtained without having to estimate the mass, the test should be run at resonance. Resonance is achieved by adjusting the input frequency until the phase angle is 90°. The input frequency is now equal to the natural frequency and Equation 6.139b becomes

$$c_{eq} = \frac{W_D}{\pi\omega\rho^2} \qquad (6.143)$$

Substituting Equations 6.140b and 6.142 in Equation 6.143, we get

$$\xi_{eq} = \frac{W_D}{4\pi W_s} \qquad (6.144)$$

It is of interest to note that, at resonance, the viscous damping force exactly balances the applied force, and the force–displacement relationship is the ellipse shown in Figure 6.24c.

Example 6.8
A water tower with a total weight of 50 kips is represented by a single-degree-of-freedom system. The total stiffness of the supporting columns is estimated at 52 kips/in. In a test, the tower is subjected to a harmonic excitation at a frequency of 10 rad/s and the force displacement

relationship is obtained at steady-state. The amplitude of displacement is measured as 2 in. and the energy loss per cycle as 65.2 kip · in.

  (i) If the damping is considered to be viscous in nature, determine c and ξ.
 (ii) If the damping is considered to be hysteretic, determine η.
(iii) If the test is rerun at 20 rad/s, and the amplitude of displacement is still 2 in., what would be the energy loss if the damping is truly viscous in nature? What would the energy loss be, if the damping were hysteretic?

Solution

  (i) For viscous damping, the energy loss per cycle is given by Equation 6.96

$$W_D = c\pi\Omega\rho^2 \tag{a}$$

With $\Omega = 10$ rad/s, $\rho = 2$ in., and $W_D = 65.2$ kip · in., Equation a gives

$$c = 0.519 \frac{\text{kip} \cdot \text{s}}{\text{in.}}$$

The natural frequency of the system is obtained from

$$\omega = \sqrt{\frac{k}{m}} = \sqrt{\frac{52 \times 386.4}{50}}$$
$$= 20.05 \text{ rad/s}$$

The damping ratio is given by

$$\xi = \frac{c}{2m\omega} = \frac{0.519 \times 386.4}{2 \times 50 \times 20.05}$$
$$= 0.1$$

 (ii) In this case, the energy loss is given by Equation 6.104:

$$W = \eta k \pi \rho^2 \tag{b}$$

Substituting for W, k, and ρ in Equation b we get

$$\eta = 0.1$$

(iii) If the damping is viscous, the energy loss per cycle at a test frequency of 20 rad/s will be given by

$$W = c\pi\Omega\rho^2$$
$$= 0.519 \times \pi \times 20 \times 4$$
$$= 130.4 \text{ kip} \cdot \text{in.}$$

If damping were hysteretic, the energy loss per cycle will not change and will still be 65.2 kip · in.

SELECTED READINGS

Clough, R.W. & Penzien, J. 1993. *Dynamics of Structures*. New York: McGraw-Hill. 2nd Edition.

Crede, C.E. 1951. *Vibration and Shock Isolation*. New York: John Wiley.

Crede, C.E. 1965. *Shock and Vibration Concepts in Engineering Design*. Englewood Cliffs: Prentice Hall.

Harris, C.M. (ed.) 1996. *Shock and Vibration Handbook*. New York: McGraw-Hill. 4th Edition.

Eller, E.E. & Whittier, R.M. 1996. Piezoelectric and Piezoresistive Transducers. In C.M. Harris (ed.) *Shock and Vibration Handbook*. New York: McGraw-Hill. 4th Edition.

Hudson, D.E. 1970. Ground Motion Measurements. In R.L. Weigel (ed.) *Earthquake Engineering*. Englewood Cliffs: Prentice Hall.

Hudson, D.E. 1979. *Reading and Interpreting Strong Motion Accelerograms*. Berkeley, California: Earthquake Engineering Research Institute.

Myklestad, N.O. 1952. The Concept of Complex Damping. *Journal of Applied Mechanics*, Vol. 19: 284–286.

Westermo, B. & Udwadia, F. 1983. Periodic Response of a Sliding Oscillator System to Harmonic Excitation. *Earthquake Engineering and Structural Dynamics*, Vol. 11: 135–146.

PROBLEMS

6.1 Two uniform parallel beams *AB* 1.2 m long, weighing 20 kg each, are hinged at *A* and supported at *B* by a spring of stiffness 4.33 N/mm, as shown in Figure P6.1. The beams carry a flywheel of mass 132 kg attached at *D*, 1.0 m away from the hinge. The flywheel is rotating at 200 RPM and its center of gravity is offset 3 mm from the axis of rotation. Find (a) the natural frequency of the system, (b) the maximum vertical movement of end *B* of the beams. Neglect damping.

6.2 A generator is mounted on a bed plate supported on four spring-dampers, the springs each having stiffness 20,000 lb/in. The damping present is estimated to be 20% of critical and the total weight supported is 2240 lb. Measurements show a maximum amplitude of vertical motion of 0.0025 in. at a speed of 2000 RPM of the generator. Calculate the force transmitted to the foundation, assuming the motion to be simple harmonic.

6.3 An electric motor weighing 1160 N is attached to a floor beam which deflects 0.76 mm under the weight. The armature of the motor weighs 40 kg.

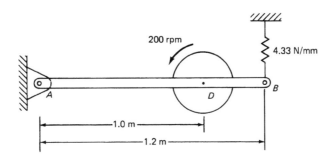

Figure P6.1.

As the motor is run up gradually to operating speed of 1600 RPM, it is observed that the maximum amplitude of oscillation is 2.5 mm, decreasing to 0.5 mm at the operating speed. Calculate the damping coefficient and the eccentricity of the armature from its axis of rotation.

6.4 A vibration test machine consists of a 1200-lb payload mounted on a beam arrangement. The static deflection is noted to be 0.386 in. Damping is small, but the oscillation of the system was observed to decrease from 0.2 in. to 0.05 in. in 18 cycles. If the supports of the beam can be given a harmonic displacement of amplitude 0.1 in. at the natural frequency of the system, estimate the expected amplitude attained by the payload.

6.5 A machine weighing 900 N is supported on springs of total stiffness 700 N/mm. Unbalance results in a disturbing force of 360 N at a speed of 3000 RPM. Damping is estimated at 20% of critical value. Determine
(a) the amplitude of motion
(b) the transmissibility
(c) the transmitted force

6.6 An electronic instrument of mass 45 kg is to be attached to an aircraft bulkhead. Vibration measurements on the bulkhead reveal a motion consisting of three sinusoidals: (i) frequency 20 Hz and amplitude 0.018 mm, (ii) frequency 35 Hz and amplitude 0.030 mm, (iii) frequency 53 Hz and amplitude 0.045 mm. The installation specifications call for a suspension system that will limit the amplitude of vibration to not more than 0.004 mm. What spring rate would you recommend for the suspension?

6.7 A seismic instrument designed to measure vertical acceleration consists of a light beam *AB* pivoted at *A* through a torsional spring of constant 20 Nm (Figure P6.7). A small body of mass 0.6 kg is fixed to the beam at end *B* and a viscous damper is attached at *C*. The pointer *BED*, pivoted to the 0.6 kg mass at *B* and to the instrument casing at *E*, moves over the scale at *D* as shown in the figure. The viscous damper is adjusted to provide a damping of 0.7 times the critical. Obtain the undamped natural frequency

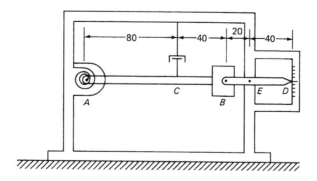

Figure P6.7.

of the instrument and the damping coefficient of the viscous damper. The instrument is calibrated to read the support acceleration correctly at an input frequency of 1 Hz. Obtain the calibration factor. If the support acceleration is 3 m/s² at a frequency of 5 Hz what will be the instrument reading and what is the percentage error?

6.8 For the seismic instrument of Problem 6.7 what are the phase differences between the measured and input signals for each of the two input frequencies: 1 Hz and 5 Hz?

6.9 An accelerometer has a natural frequency of 40 rad/s and a damping ratio of 0.6. It is calibrated to read correctly the input acceleration at very low frequencies. Find the calibration factor. An input signal given by $\ddot{u}_g = 4\sin(4\pi t) + 3\sin(5\pi t)$ m/s² is supplied to the accelerometer. Plot the output signal produced by the instrument. What are the measured amplitudes of the two components of the given signal.

6.10 A displacement meter has a natural frequency of 3 Hz and damping ratio of 0.7. It is calibrated to read correctly the displacements at very high frequencies. What are the errors in readings at input frequencies of 4.5 Hz, 6 Hz, and 9 Hz? What is the phase shift in each case?

6.11 A rotating unbalanced mass m_0 has an eccentricity e and frequency Ω rad/s. It is supported by a system of springs such that the natural frequency of vibration is ω. Show that with hysteretic damping, the amplitude of steady state response is given by

$$\rho = \frac{\frac{em_0}{m}\beta^2}{\{(1-\beta^2)^2 + \eta^2\}^{\frac{1}{2}}}$$

where $\beta = \Omega/\omega$ is the frequency ratio, and η is the hysteretic damping ratio. Hence prove that the maximum amplitude occurs when $\beta = \sqrt{1 + \eta^2}$ and is given by

$$\frac{\rho_{max}}{em_0/m} = \frac{1}{\eta}\sqrt{1 + \eta^2}$$

6.12 To obtain an estimate of the damping in the system shown in Problem 6.1, end B is displaced 32 mm and the system is allowed to vibrate freely. The amplitude decays to 6 mm in 5 cycles. If the damping is assumed to be viscous in nature, what will be the maximum amplitude of B due to the rotation of eccentric flywheel.

6.13 Solve Problem 6.12 assuming the damping to be hysteretic.

6.14 Using an eccentric mass shaker, a dynamic test was run on a structure in which the exciting frequency was varied over a range of values. For each frequency, the response amplitude as well as the phase angle were measured and the following data was obtained. The estimated natural frequency of the structure is 2 Hz. Plot a frequency versus amplitude curve; then using the half power method, obtain an estimate of the damping.

It will be readily seen that the graphical relationship obtained by taking the amplitude ρ and the phase angle ϕ as polar coordinates is a Nyquist plot. Draw such a plot from the data supplied. By estimating the values of Ω_1 and Ω_2 corresponding to $\phi = 45°$ and $135°$ from the Nyquist plot, obtain an estimate of the damping.

Of the two estimates of damping obtained by you, which do you think is more reliable.

Ω (Hz)	ρ (mm)	ϕ (deg)
0.0	5.00	0.0
0.4	5.20	2.4
0.8	5.60	5.4
1.2	7.70	10.6
1.6	12.70	24.0
1.8	19.10	43.5
2.0	25.00	90.0
2.2	16.40	133.2
2.4	10.00	151.4
2.8	5.00	163.7
3.2	3.10	168.4
3.6	2.20	170.9
4.0	1.70	172.4

6.15 Show that the Nyquist plot for the displacement of a damped system with hysteretic damping is a circle with center $(0, \frac{1}{2\eta})$ and radius $\frac{1}{2\eta}$. Also show that half power points occur exactly at phase angles $45°$ and $135°$. Obtain a Nyquist plot from the following data and hence calculate the damping factor η

β	$\frac{\rho}{(p_0/k)}$	ϕ (deg)
0.0	0.98	11.3
0.2	1.02	11.8
0.4	1.16	13.4
0.6	1.49	17.4
0.8	2.43	29.1
0.9	3.63	46.5
1.0	5.00	90.0
1.1	3.45	136.4
1.2	2.07	155.6
1.4	1.02	168.2
1.6	0.64	172.7
1.8	0.44	174.9
2.0	0.33	176.2

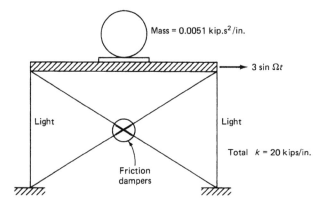

Figure P6.16.

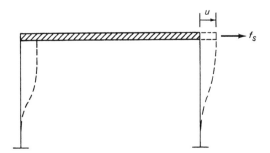

Figure P6.17.

6.16 In order to measure the characteristics of friction dampers used in earth-quake-resistant construction, laboratory tests are conducted on a scaled model of a single-story frame as shown in Figure P6.16. In a free vibration test, the amplitude of motion of the frame reduces from 3/4 in. to 1/64 in. in 6 cycles of motion which take 1.5 s. The lateral stiffness of the frame is estimated at 20 kips/in. An eccentric mass shaker mounted on the frame is then run at a frequency of 6 Hz. If the unbalanced force is 3 kips, find the maximum amplitude of vibration.

6.17 The lateral displacement of the frame shown in Figure P6.17 are measured under a quasi-static cyclic load varying from −12 to 12 kips. The experimental results are shown below. Plot the hysteretic loop of the frame. Obtain η by estimating the area enclosed by the loop. If a harmonic force of amplitude 12 kips is applied to the frame at resonant frequency, what would be the maximum displacement of the frame?

Lateral force f_S^t (kips)	Displacement u, (in.)
3.18	0.2
5.13	0.4
7.04	0.6
8.89	0.8
10.66	1.0
12.00	1.2
9.34	1.0
7.11	0.8
4.96	0.6
2.87	0.4
0.82	0.2
− 1.20	0.0
− 3.18	−0.2
− 5.13	−0.4
− 7.04	−0.6
− 8.89	−0.8
− 10.66	−1.0
−12.0	−1.2
− 9.34	−1.0
− 7.11	−0.8
− 4.96	−0.6
− 2.87	−0.4
− 0.82	−0.2
1.20	0.0

CHAPTER 7

Response to general dynamic loading and transient response

7.1 INTRODUCTION

The analysis of response to a forcing function that is neither harmonic nor periodic is comparatively more complex. However, for linear systems a general method exists which is conceptually quite straightforward. This method is based on the ability to obtain mathematical expressions for the response of the system to an impulsive force. A general forcing function is expressed as the sum of a series of impulses and the response to the applied force is obtained by superposing the responses to the individual impulses. Superposition, in fact, involves the evaluation of an integral, known as a convolution integral or Duhamel's integral, either mathematically or by numerical means of integration.

An important class of nonperiodic loads are those that act for a relatively short duration of time. Such loads are also known as impulsive loads or shock loads, and may have considerable significance in the design of certain systems. Examples are: loads generated due to a blast or explosion, and dynamic loads in automobiles, traveling cranes, and other mobile machinery. The response produced by such loads is transient in nature and decays rapidly after the load has ceased to act. However, from the point of view of engineering design, the duration of transient response is of no particular importance; rather, it is the maximum displacement and stress attained during the response that is of interest. Because of the short duration of response, damping does not have a significant influence and can reasonably be ignored in the analysis.

7.2 RESPONSE TO AN IMPULSIVE FORCE

An *impulsive force* is a large force that acts for a very short duration of time. Although the magnitude of such a force may be infinitely large, its time integral, that is, the area enclosed by the force–time curve and the time axis, is finite. The time integral is referred to as the impulse of the force. Referring to Figure 7.1, the magnitude of impulse, denoted by I, is obtained

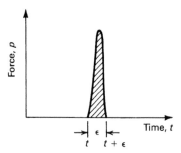

Figure 7.1. Impulse of an impulsive force.

from

$$I = \int_{t}^{t+\varepsilon} p(t)\,dt \tag{7.1}$$

were ε is the very small interval of time during which the impulsive force is acting.

Mathematically, an impulsive force can be expressed in terms of the *delta function*. Function $\delta(t)$ represents a delta function centered at time $t = 0$. The magnitude of the function is zero at all times other than zero. At $t = 0$, the function value is infinitely large. However, the area enclosed by the delta function and the time axis is finite and is equal to unity. Thus an impulsive force applied at time $t = 0$ and having an impulse equal to I can be represented by $I\,\delta(t)$. Analogously, function $I\delta(t - \tau)$ represents a force of impulse I acting at time $t = \tau$. Newton's second law of motion states that the action of an impulsive force on a mass results in a change in the velocity of the mass and hence in its momentum, the change in momentum being equal to the *impulse* of the impulsive force. Thus, representing the change in velocity by Δv

$$m\Delta v = I \tag{7.2}$$

If the mass was initially at rest, it will have a velocity I/m after the action of the impulse.

Suppose that a single-degree-of-freedom system at rest having a mass m, a restraining spring stiffness k, and negligible damping is subjected to an impulsive force with impulse I. The action of the impulsive force will set the system vibrating. The ensuing free-vibration response can be obtained by recognizing that the initial displacement is zero and the initial velocity is given by $v_0 = I/m$. The resulting response is given by Equation 5.9a with $u_0 = 0$

$$u = \frac{I}{m\omega}\sin \omega t \tag{7.3}$$

The response to an impulse of unit magnitude is called *unit impulse response.* It is denoted by $h(t)$ and is given by

$$h(t) = \frac{1}{m\omega} \sin \omega t \qquad (7.4)$$

When damping is present, the unit impulse response is obtained from Equation 5.37 with $u_0 = 0$ and $v_0 = 1/m$

$$h(t) = \frac{1}{m\omega_d} e^{-\omega \xi t} \sin \omega_d t \qquad (7.5)$$

7.3 RESPONSE TO GENERAL DYNAMIC LOADING

The unit impulse response functions given by Equations 7.4 and 7.5 can be used to obtain the response to any general dynamic loading. The system being excited is assumed to be linear, so that the principle of superposition holds. The applied force can be viewed as the sum of a series of impulses such as the one shown shaded in Figure 7.2. The shaded impulse has a magnitude of $p(\tau) d\tau$ and is imparted at time $t = \tau$. At any subsequent time t, when the elapsed time since the impulse is $t - \tau$, the incremental response, du, due to the impulse under consideration is obtained from Equation 7.4 with $t - \tau$ replacing t

$$du = \frac{p(\tau) d\tau}{m\omega} \sin \omega(t - \tau) \qquad (7.6)$$

in which the system has been assumed to be undamped. The total response at time t is obtained by superposing the effects of all impulses from time $\tau = 0$ to $\tau = t$, giving

$$u(t) = \frac{1}{m\omega} \int_0^t p(\tau) \sin \omega(t - \tau) d\tau \qquad (7.7)$$

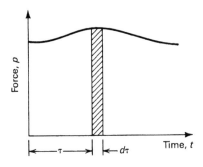

Figure 7.2. Representation of a general dynamic load by a series of impulses.

In a similar manner, the response of a damped system is given by

$$u(t) = \frac{1}{m\omega_d} \int_0^t p(\tau) e^{-\xi\omega(t-\tau)} \sin \omega_d(t-\tau) \, d\tau \qquad (7.8)$$

Superposition integral equations 7.7 and 7.8 can be expressed in the general form

$$u(t) = \int_0^t p(\tau) h(t-\tau) \, d\tau \qquad (7.9)$$

where an appropriate expression is used for the unit impulse function h. Equation 7.9 is known as the *convolution integral* or *Duhamel's integral*. It provides a general method for the analysis of the response of a linear system to an arbitrary loading. It also forms the basis for the development of the Fourier transform method of analysis described in Chapter 9. When $p(\tau)$ is a simple mathematical function, closed-form evaluation of the integral in Equation 7.9 is possible; in other cases, a numerical technique must be used in the evaluation. Some of the numerical techniques are discussed in Chapter 8.

It should be noted that in obtaining Equation 7.9, the system was assumed to be at rest at time $t = 0$. If the system starts with initial condition u_0 and v_0, the free-vibration component of the response given by Equation 5.9a (undamped system) or Equation 5.37 (damped system) must be added to Equation 7.9 to obtain the total response. Thus for a damped system,

$$u(t) = e^{-\omega\xi t} \left(u_0 \cos \omega_d t + \frac{v_0 + \omega\xi u_0}{\omega_d} \sin \omega_d t \right)$$
$$+ \frac{1}{m\omega_d} \int_0^t p(\tau) e^{-\xi\omega(t-\tau)} \sin \omega_d(t-\tau) \, d\tau \qquad (7.10)$$

In the following sections, we present several examples of the use of Duhamel's integral in obtaining the response to a general dynamic load.

7.4   RESPONSE TO A STEP FUNCTION LOAD

A *step function load* is a suddenly applied load that remains constant after application. It is a simple example of a nonperiodic load. Figure 7.3 shows a step function load applied at time $t = 0$. The equation of motion of a single-degree-of-freedom damped system subjected to a step load of magnitude P_0 is

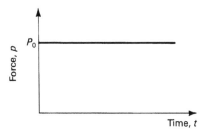

Figure 7.3. Step function load.

given by

$$m\ddot{u} + c\dot{u} + ku = P_0 \tag{7.11}$$

The complementary solution of Equation 7.11 is given by Equation 5.36. For a particular solution, we use the method of trials and assume for our trial solution $u = G$. Substitution in Equation 7.11 shows that $G = P_0/k$. The complete solution then becomes

$$u = e^{-\omega\xi t}(A\cos\omega_d t + B\sin\omega_d t) + \frac{P_0}{k} \tag{7.12}$$

in which A and B are arbitrary constants to be determined from initial conditions. When the system starts from rest, $u_0 = v_0 = 0$ and the arbitrary constants A and B are easily shown to be

$$
\begin{aligned}
A &= -\frac{P_0}{k} \\
B &= -\frac{P_0}{k}\frac{\omega\xi}{\omega_d}
\end{aligned}
\tag{7.13}
$$

The resulting solution becomes

$$u = \frac{P_0}{k}\left[1 - e^{-\omega\xi t}\left(\cos\omega_d t + \frac{\omega\xi}{\omega_d}\sin\omega_d t\right)\right] \tag{7.14}$$

The dynamic load factor $u/(P_0/k)$, which gives the ratio of the dynamic displacement to the static displacement under load P_0, is plotted in Figure 7.4 as a function of t/T, where $T = 2\pi/\omega$ is the natural period of the system, for several different values of the damping ratio ξ. The dynamic load factor attains

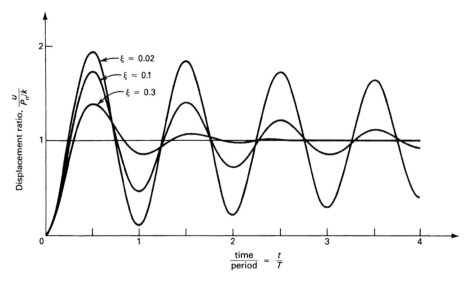

Figure 7.4. Response to a step function load.

its maximum value of 2 when the damping is negligible. For finite damping it is always less than 2.

The time at which the maximum occurs can be obtained by equating the differential of Equation 7.14 to zero. Thus, if t_p represents the time at peak value

$$\frac{P_0}{k} e^{-\omega\xi t_p} \left\{ \frac{(\omega\xi)^2}{\omega_d} + \omega_d \right\} \sin \omega_d t_p = 0 \qquad (7.15)$$

or

$$t_p = \frac{n\pi}{\omega_d} \qquad n = 0, 1, 2, \ldots \qquad (7.16)$$

For $n = 0$, $t_p = 0$, and substitution into Equation 7.14 gives $u = 0$, which represents a minimum. The first peak occurs when $n = 1$ and $t_p = \pi/\omega_d$. Other values of n represent subsequent peaks in the displacement curve, where the magnitude of displacement is less than that at the first peak. The maximum is thus obtained by substituting $t = \pi/\omega_d$ in Equation 7.14.

$$
\begin{aligned}
u_{\max} &= \frac{P_0}{k}(1 + e^{-\omega\xi\pi/\omega_d}) \\[2mm]
&= \frac{P_0}{k}(1 + e^{-\pi\xi/\sqrt{1-\xi^2}})
\end{aligned}
\qquad (7.17)
$$

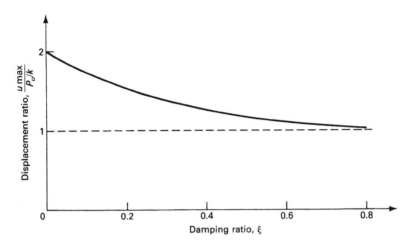

Figure 7.5. Maximum response to a step function load.

The peak response ratio is thus a function of only the damping ratio. Figure 7.5 shows the relationship between $u_{max}/(P_0/k)$ and ξ. As expected, the maximum value of the response ratio is 2 for $\xi = 0$.

The response to step function can also be obtained by using the Duhamel's integral.

$$u(t) = \frac{1}{m\omega_d} \int_0^t P_0 e^{-\omega\xi(t-\tau)} \sin \omega_d(t - \tau)\,d\tau \qquad (7.18)$$

Evaluation of the integral in Equation 7.18 will provide an expression for the response that is identical to Equation 7.14.

7.5 RESPONSE TO A RAMP FUNCTION LOAD

A *ramp function load* is a load that increases linearly with time as shown in Figure 7.6a. Mathematically, it can be expressed as

$$p(t) = \frac{P_0 t}{t_1} \qquad (7.19)$$

The response of a single-degree-of-freedom system to a ramp function load is obtained by substituting for $p(\tau)$ in the Duhamel's integral

$$u(t) = \frac{1}{m\omega_d} \int_0^t \frac{P_0 \tau}{t_1} e^{-\omega\xi(t-\tau)} \sin \omega_d(t - \tau)\,d\tau \qquad (7.20)$$

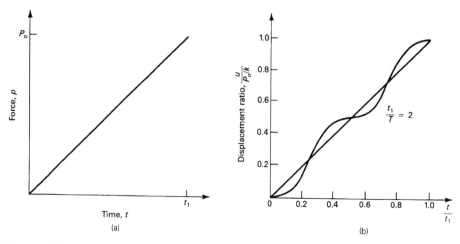

Figure 7.6. Response to a ramp function load: (a) ramp function load; (b) response.

For an undamped system, the Duhamel's integral simplifies to

$$u(t) = \frac{1}{m\omega} \int_0^t \frac{P_0\tau}{t_1} \sin \omega(t - \tau)\, d\tau \tag{7.21a}$$

$$= \frac{P_0}{k}\left(\frac{t}{t_1} - \frac{\sin \omega t}{\omega t_1} \right) \tag{7.21b}$$

Equation 7.21b has been plotted in Figure 7.6b for a specific value of t_1/T. The response is seen to oscillate about the average value $P_0 t/(kt_1)$. The velocity response of the system to the ramp function load is obtained by taking the differential of Equation 7.21.

$$\dot{u}(t) = \frac{P_0}{kt_1}(1 - \cos \omega t) \tag{7.22}$$

Since $\cos \omega t$ ranges between -1 and $+1$, the velocity is either zero or positive. This implies that the system displacement is either stationary or is increasing with time. At some stage the displacement reaches the elastic limit of the restraining spring; the solutions given by Equations 7.20 through 7.22 are then no longer valid.

7.6   RESPONSE TO A STEP FUNCTION LOAD WITH RISE TIME

A load that rises linearly to P_0 in time t_1 and then remains constant at P_0 is shown in Figure 7.7a. Such a load is commonly known as a *step function load*

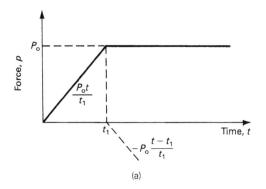

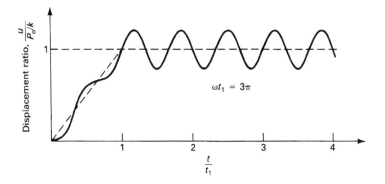

Figure 7.7. (a) Step function load with rise time; (b) response to step function load with rise time.

with a rise time. It can be viewed as the superposition of a ramp function load applied at time $t = 0$, and an equal but negative ramp function load applied at time $t = t_1$. The response of an undamped system to a ramp function load applied at $t = 0$ is given by Equation 7.21b. The response to a ramp function applied at time $t = t_1$ is

$$u(t) = \frac{P_0}{k} \left\{ \frac{t - t_1}{t_1} - \frac{\sin \omega(t - t_1)}{\omega t_1} \right\} \qquad t > t_1 \qquad (7.23)$$

The response to the step function with rise time can be obtained from Equations 7.21 and 7.23. For $t \leq t_1$, Equation 7.21b applies, for $t > t_1$, the response is obtained by subtracting Equation 7.23 from Equation 7.21b.

$$u(t) = \begin{cases} \frac{P_0}{k} \left\{ \frac{t}{t_1} - \frac{\sin \omega t}{\omega t_1} \right\} & t \leq t_1 & (7.24a) \\[2ex] \frac{P_0}{k} \left\{ 1 - \frac{\sin \omega t}{\omega t_1} + \frac{\sin \omega(t - t_1)}{\omega t_1} \right\} & t > t_1 & (7.24b) \end{cases}$$

The history of response is shown in Figure 7.7b for a specific value of ωt_1. In the second era of response, that is, for $t > t_1$, the system executes a simple harmonic motion about the position of equilibrium. For $t < t_1$ the response is similar to that for a ramp function load, so that the displacement continues to grow with time and the velocity at $t = t_1$ is either zero or positive. This implies that the maximum value of response occurs either at $t = t_1$ or in the era $t > t_1$. The time, t_p, at which the maximum occurs is obtained by equating the time differential of Equation 7.24b to zero. Thus

$$\frac{P_0}{kt_1}\{-\cos\omega t + \cos\omega(t - t_1)\} = 0 \tag{7.25}$$

After simplification, Equation 7.25 gives

$$\tan \omega t = \tan \frac{\omega t_1}{2} \tag{7.26a}$$

or

$$t_p = \frac{n\pi}{\omega} + \frac{t_1}{2} \quad n = 0, 1, 2\ldots \tag{7.26b}$$

Obviously, the value of n in Equation 7.26b should be chosen so that $t_p \geq t_1$. Substitution of $t = t_p$ in Equation 7.24b gives the following value for the maximum response

$$u_{\max} = \frac{P_0}{k}\left\{1 + \frac{2\sin(n\pi - \omega t_1/2)}{\omega t_1}\right\} \tag{7.27}$$

The true maximum will be obtained by selecting a value of n such that the second term within the braces in Equation 7.27 is positive, so that

$$u_{\max} = \frac{P_0}{k}\left\{1 + \frac{2|\sin(\omega t_1/2)|}{\omega t_1}\right\} \tag{7.28}$$

It is of interest to note that for the special case when $\omega t_1 = 2n\pi$ the displacement at $t = t_1$ is P_0/k as shown by Equation 7.21b, while the velocity is zero as shown by Equation 7.22. The displacement in the era $t > t_1$, given by Equation 7.24b, remains constant at P_0/k and the system does not oscillate.

The response ratio $u_{\max}/(P_0/k)$ has been plotted as a function of $\omega t_1/2\pi = t_1/T$ in Figure 7.8. Plots of the type of Figure 7.8 which show the relationship between the maximum value of a response parameter and a characteristic of the system are called *response spectra*. The system characteristic used in a response spectrum is either the natural frequency or the natural period. In Figure 7.8 the period T, or rather its inverse, may be considered as the variable system

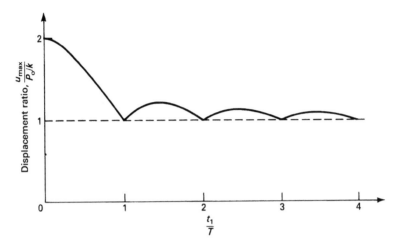

Figure 7.8. Response spectrum for a step function load with rise time.

characteristic, while the response parameter is the maximum displacement. Availability of response spectra often greatly simplifies the design of a system for a given loading. It is clear from both Equation 7.27 and Figure 7.8 that as t_1/T increases, the maximum response tends to a value equal to the static displacement under load P_0. In other words, if the load is applied very gradually, the response is essentially static.

Example 7.1

A suddenly applied load P_0 that decays exponentially after application is shown in Figure E7.1a. Obtain the response of an undamped system to such a load.

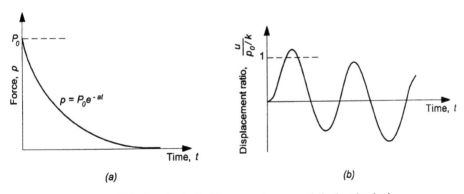

Figure E7.1. (a) Exponentially decaying load; (b) response to exponentially decaying load.

Solution
The equation of motion is

$$m\ddot{u} + ku = P_0 e^{-at} \tag{a}$$

and the response is obtained by using the Duhamel's integral

$$u(t) = \frac{P_0}{m\omega} \int_0^t e^{-a\tau} \sin \omega(t - \tau)\, d\tau \tag{b}$$

in which it has been assumed that the system starts from rest. Evaluation of integral in Equation b gives

$$u = \frac{P_0}{k(1 + a^2/\omega^2)} \left(\frac{a}{\omega} \sin \omega t - \cos \omega t + e^{-at} \right) \tag{c}$$

The displacement response has been plotted in Figure E7.1b. It shows that the system oscillates with decreasing amplitude. As t increases, the term e^{-at} becomes very small and the system attains a steady-state harmonic motion whose amplitude is given by

$$\rho = \frac{P_0}{k} \frac{1}{\sqrt{1 + a^2/\omega^2}}$$

Example 7.2
A blast-induced pressure wave striking a single-story building is represented by $p = 100(e^{-10t} - e^{-100t})$, where p is the pressure in psf. The building face subject to the wind pressure has an area of 144 ft^2. The mass of the building, assumed concentrated at the floor level, is 0.1 kip·s^2/in. and the total stiffness of the columns supporting the floor is 350 kips/in. Find the displacement response of the building and the maximum base shear induced by the pressure wave.

Solution
As seen from Figure E7.2a, the superposition of two exponentially decaying waves can be used effectively to represent a blast-generated wave. The force applied at the floor level of the building due to the pressure wave is given by

$$P = \frac{1}{2} \times 144 \times 100(e^{-10t} - e^{-100t}) \times 10^{-3} \text{ kips} \tag{a}$$

The total response is obtained by summing the response to the two component waves. The response to each component wave is obtained by an expression of the form of Equation c of Example 7.1, so that

$$u = \frac{P_0}{k(1 + a^2/\omega^2)} \left(\frac{a}{\omega} \sin \omega t - \cos \omega t + e^{-at} \right)$$
$$- \frac{P_0}{k(1 + b^2/\omega^2)} \left(\frac{b}{\omega} \sin \omega t - \cos \omega t + e^{-bt} \right) \tag{b}$$

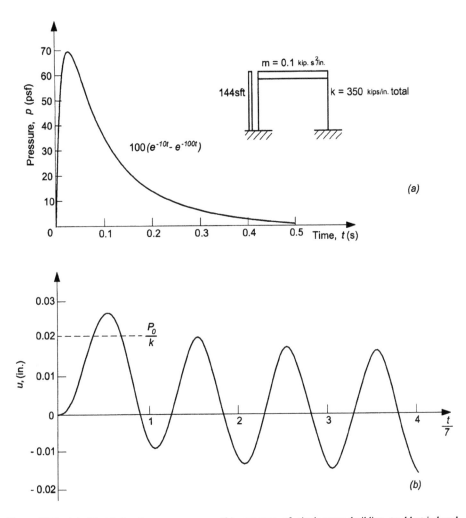

Figure E7.2. (a) Blast-induced pressure wave; (b) response of single-story building to blast-induced pressure wave.

In this case

$$k = 350 \text{ kips/in.}$$

$$\omega = \sqrt{\frac{350}{0.1}} = 59.16 \text{ rad/s}$$

$$P_0 = 7.2 \text{ kips}$$

$$a = 10$$

$$b = 100$$

On substituting the values, we get

$$u = \frac{7.2}{350(1 + 0.028)} \left(\frac{10}{59.16} \sin \omega t - \cos \omega t + e^{-10t} \right)$$

(c)

$$- \frac{7.2}{350(1 + 2.857)} \left(\frac{100}{59.16} \sin \omega t - \cos \omega t + e^{-100t} \right)$$

or

$$u = -0.0056 \sin \omega t - 0.01467 \cos \omega t + 0.02 e^{-10t} - 0.0053 \times e^{-100t}$$

(d)

The response given by Equation d has been plotted in Figure E7.2b. The maximum displacement is 0.02685 in. and hence the maximum base shear $= k u_{max}$ is 9.4 kips.

7.7   RESPONSE TO SHOCK LOADING

In the preceding sections, examples were presented of response to nonperiodic loading. We now discuss several cases of response to *shock loading* or *impulsive loading*. As the name implies, the loading in these cases is of short duration. As a result, the maximum value of the response is not significantly affected by the presence of damping. The effect of damping is therefore ignored in the presentation.

7.7.1   *Rectangular pulse*

A *rectangular pulse load* of duration t_1 seconds is shown in Figure 7.9. For $t \leq t_1$, the response of the system is the same as that to a step function load.

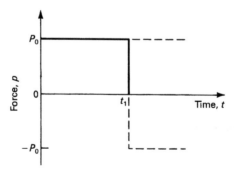

Figure 7.9. Rectangular pulse load.

Thus setting $\xi = 0$ in Equation 7.14, we get

$$u(t) = \frac{P_0}{k}(1 - \cos \omega t) \quad t \le t_1 \tag{7.29}$$

At time $t = t_1$, the displacement and velocity of the system are given by

$$u(t_1) = \frac{P_0}{k}(1 - \cos \omega t_1) \tag{7.30a}$$

$$\dot{u}(t_1) = \frac{P_0 \omega}{k}(\sin \omega t_1) \tag{7.30b}$$

In the second era of response, that is, for $t > t_1$, the system is undergoing free-vibration with initial conditions given by Equations 7.30. The response in this era is therefore obtained from Equation 5.9a.

$$u = u(t_1) \cos \omega \bar{t} + \frac{\dot{u}(t_1)}{\omega} \sin \omega \bar{t} \tag{7.31}$$

where $\bar{t} = t - t_1$ is the time measured from the beginning of the free-vibration era. Substitution of appropriate values in Equation 7.31 gives

$$u = \frac{P_0}{k}\{\cos \omega(t - t_1) - \cos \omega t\} \quad t > t_1 \tag{7.32}$$

As an alternative, the rectangular pulse can be obtained by a superposition of a step function of magnitude P_0 applied at $t = 0$ and another step function of magnitude $-P_0$ applied at time $t = t_1$. The response to the second step function is

$$u = -\frac{P_0}{k}\{1 - \cos \omega(t - t_1)\} \tag{7.33}$$

The response in the second era is then obtained by taking the sum of Equations 7.29 and 7.33, which leads to Equation 7.32.

The histories of displacement response obtained from Equations 7.29 and 7.32 for four different values of the ratio t_1/T are shown in Figure 7.10. The static displacement produced by the rectangular pulse load is also shown in that

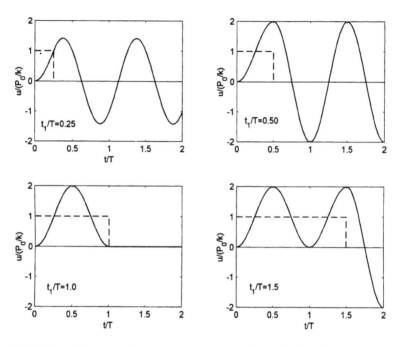

Figure 7.10. Time histories of response to a rectangular pulse load.

figure by dashed lines. For $t_1/T = 0.25$ the maximum response occurs in the free-vibration era, and the maximum value is $1.414P_0/k$. For $t_1/T = 0.5$ the maximum response occurs at $t = t_1$, that is, at the end of the pulse, and the maximum value is $2P_0/k$. For the special case when $\omega t_1 = 2n\pi$ or $t_1/T = n$ the displacement and velocity are both zero at $t = t_1$, and the motion ceases at this point. For both $t_1/T = 1$ and 1.5 the maximum response is $2P_0/k$ and occurs within the duration of the pulse.

As seen from the response histories, the maximum value of response is attained in either the first or second vibration era, depending on the value of t_1. First, assuming that the maximum is achieved for $t < t_1$, the time at maximum, t_p, is obtained by equating the differential of Equation 7.29 to zero.

$$\frac{P_0\omega}{k}\sin\omega t_p = 0$$

$$\omega t_p = n\pi \quad n = 1,3,5,\dots \tag{7.34}$$

In Equation 7.34 n has been taken as being odd, since for even n the displacement is a minimum as will be evident from Equation 7.29. The maximum value of response is reached in the forced-vibration era provided that the smallest value of t_p, obtained by taking $n = 1$, is less than t_1, that is

$$t_p \equiv \frac{\pi}{\omega} < t_1 \tag{7.35a}$$

or

$$\frac{t_1}{T} > \frac{1}{2} \tag{7.35b}$$

The corresponding value of the maximum response obtained from Equation 7.29 with $t = t_p$ is given by

$$u_{max} = \frac{2P_0}{k} \tag{7.36}$$

If the condition given by Equation 7.35 does not hold, the maximum will occur in the free-vibration era, and the time to maximum will be given by equating the differential of Equation 7.32 to zero.

$$\frac{P_0}{k}\omega\{-\sin\omega(t_p - t_1) + \sin\omega t_p\} = 0 \tag{7.37}$$

After some manipulation, Equation 7.37 gives

$$\tan\omega t_p = -\cot\frac{\omega t_1}{2} \tag{7.38a}$$

or

$$\begin{aligned} t_p &= (2n + 1)\frac{\pi}{2\omega} + \frac{t_1}{2} \\ &= (2n + 1)\frac{T}{4} + \frac{t_1}{2} \end{aligned} \qquad n = 0, 1, 2, \ldots \tag{7.38b}$$

The maximum response is obtained by substituting the value of t_p from Equation 7.38b for t in Equation 7.32, giving

$$\begin{aligned} u_{max} &= 2\frac{P_0}{k}\sin\frac{\omega t_1}{2} \\ &= 2\frac{P_0}{k}\sin\frac{\pi t_1}{T} \end{aligned} \tag{7.39}$$

The response ratio $u_{max}/(P_0/k)$ has been plotted as a function of t_1/T in Figure 7.11. Figure 7.11 is another example of a response spectrum. When the loading is impulsive, as in this case, the response spectrum is usually referred to as a *shock spectrum*.

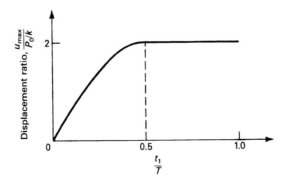

Figure 7.11. Shock spectrum for a rectangular pulse load.

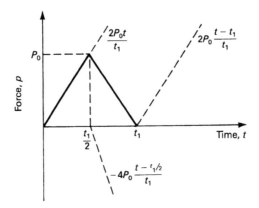

Figure 7.12. Triangular pulse load.

7.7.2   *Triangular pulse*

Figure 7.12 shows a triangular pulse load of total duration t_1. The pulse can be represented as a superposition of three ramp functions indicated in the figure and given by

$$p_1 = \frac{2P_0 t}{t_1} \quad t > 0$$

$$p_2 = -\frac{4P_0(t - t_1/2)}{t_1} \quad t > \frac{t_1}{2} \tag{7.40}$$

$$p_3 = \frac{2P_0(t - t_1)}{t_1} \quad t > t_1$$

The first ramp function applied at time $t = 0$, has a slope of $2P_0/t_1$, the second ramp function applied at time $t = t_1/2$, has a slope of $-4P_0/t_1$, and the third ramp function applied at $t = t_1$, has a slope of $2P_0/t_1$. The response to ramp function p_1 is directly obtained from Equation 7.21b.

$$u_1(t) = \frac{2P_0}{k} \left(\frac{t}{t_1} - \frac{\sin \omega t}{\omega t_1} \right) \tag{7.41}$$

The response to function p_2 is obtained from Equation 7.21b, by substituting $t - \frac{1}{2} t_1$ for t and multiplying the amplitude by -4.

$$u_2(t) = -\frac{4P_0}{k} \left(\frac{t - \frac{1}{2} t_1}{t_1} - \frac{\sin \omega (t - \frac{1}{2} t_1)}{\omega t_1} \right) \tag{7.42}$$

In a similar manner, the response to function p_3 is obtained from Equation 7.21b by substituting $t - t_1$ for t and multiplying the amplitude by 2.

$$u_3(t) = \frac{2P_0}{k} \left(\frac{t - t_1}{t_1} - \frac{\sin \omega (t - t_1)}{\omega t_1} \right) \tag{7.43}$$

The response to the triangular pulse in the era $\frac{1}{2} t_1 < t \leq t_1$ is obtained by superposing $u_1(t)$ and $u_2(t)$, while the response in the era $t > t_1$ is obtained by superposing u_1, u_2, and u_3. Thus, we get

$$u(t) = \begin{cases} \frac{2P_0}{k} \left(\frac{t}{t_1} - \frac{\sin \omega t}{\omega t_1} \right) & t \leq \frac{t_1}{2} \quad (7.44\text{a}) \\[2ex] \frac{2P_0}{k} \left\{ 1 - \frac{t}{t_1} - \frac{\sin \omega t}{\omega t_1} + \frac{2 \sin \omega (t - t_1/2)}{\omega t_1} \right\} & \frac{t_1}{2} < t \leq t_1 \quad (7.44\text{b}) \\[2ex] \frac{2P_0}{k} \left\{ - \frac{\sin \omega t}{\omega t_1} + \frac{2 \sin \omega (t - t_1/2)}{\omega t_1} - \frac{\sin \omega (t - t_1)}{\omega t_1} \right\} & t_1 < t \quad (7.44\text{c}) \end{cases}$$

The histories of displacement response obtained from Equations 7.44a, 7.44b, and 7.44c for six different values of the ratio t_1/T are shown in Figure 7.13. The static displacement produced by the triangular pulse load is also shown in that figure by dashed lines. For $t_1/T = 0.25$, the maximum response occurs in the free-vibration era, and the maximum value is $0.746 P_0/k$. For $t_1/T = 0.5$ the maximum response occurs at $t = t_1$, that is, at the end of the pulse, and the maximum value is $1.273 P_0/k$. For $t_1/T = 1$, 1.5, 2 and 3 the maximum occurs during the forced-vibration era. For the special case when $t_1/T = 2n$ or $\omega t_1 = 4n\pi$ the displacement and velocity are both zero at $t = t_1$; the motion ceases at this point and the maximum response equal to P_0/k occurs within the duration of the pulse at $t = t_1/2$.

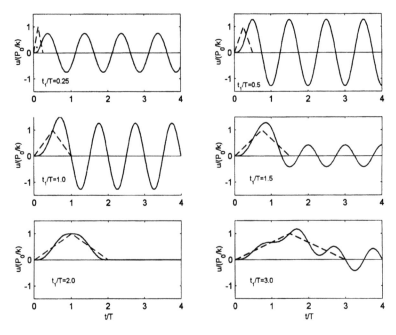

Figure 7.13. Time histories of response to a triangular pulse load.

To obtain the shock spectrum, we first assume that the maximum response occurs in the free-vibration era $t > t_1$. Then the time to maximum, t_p, is obtained by equating the time differential of Equation 7.44c to zero.

$$\frac{2P_0}{k\omega t_1}\left\{-\omega\cos\omega t_p + 2\omega\cos\omega\left(t_p - \frac{t_1}{2}\right) - \omega\cos\omega(t_p - t_1)\right\} = 0 \quad (7.45)$$

After some manipulation, Equation 7.45 gives

$$\cos\left(\omega t_p - \frac{\omega t_1}{2}\right) = 0 \qquad\qquad (7.46a)$$

or

$$t_p = (2n+1)\frac{\pi}{2\omega} + \frac{t_1}{2}$$

$$= (2n+1)\frac{T}{4} + \frac{t_1}{2} \quad n = 0, 1, 2, \ldots \qquad (7.46b)$$

The value of the maximum response is obtained by substituting t_p from Equation 7.46b for t in Equation 7.44c, which gives

$$\frac{u_{\max}}{P_0/k} = \pm\frac{4}{\omega t_1}\left(1 - \cos\frac{\omega t_1}{2}\right) \tag{7.47}$$

where the negative sign in Equation 7.44 applies for odd values of n. The maximum response will occur in the free-vibration era provided that t_p obtained from Equation 7.46b is greater than t_1. Suppose now that t_p obtained from Equation 7.46 with $n = 0$ is less than t_1, but that obtained with $n = 1$ is greater than t_1. The first peak in the free-vibration era is then negative. This implies that the system velocity at the end of the forced-vibration era is negative, and that a peak positive displacement was reached during the forced-vibration era. Obviously, the peak in the forced-vibration era is going to be larger than any peak in the free-vibration era. This reasoning leads to the conclusion that if the maximum displacement is to occur in the free-vibration era, the t_p obtained with $n = 0$ should be larger than t_1, so that the first peak in the free-vibration era is positive. Thus

$$t_p = \frac{T}{4} + \frac{t_1}{2} \geq t_1$$

or

$$\frac{t_1}{T} \leq \frac{1}{2} \tag{7.48}$$

For $t_1/T > 1/2$, the maximum occurs in the forced-vibration era. From our discussion of the response to a ramp function load we know that during the period $t \leq t_1/2$ when the force is rising the displacement continues to grow with time and the velocity at $t = t_1/2$ is either zero or positive. As a consequence, if the maximum occurs in the forced-vibration era it must occur in the period $t_1/2 \leq t \leq t_1$. The time to maximum is obtained by equating the differential of Equation 7.44b to zero. Substitution of this value of time in Equation 7.44b then gives the maximum response. The shock spectrum for the triangular pulse load is shown in Figure 7.14, in which the maximum response ratio $u_{\max}/(P_0/k)$ has been plotted as a function of t_1/T.

7.7.3 Sinusoidal pulse

The sinusoidal pulse shown in Figure 7.15 can be considered to be the superposition of two sine waves, the second of which begins at $t = t_1$ seconds. The period T_f of each sine wave is $2t_1$ seconds, corresponding to a frequency of $\Omega = \pi/t_1$ rad/s. For $t \leq t_1$, the response is obtained from

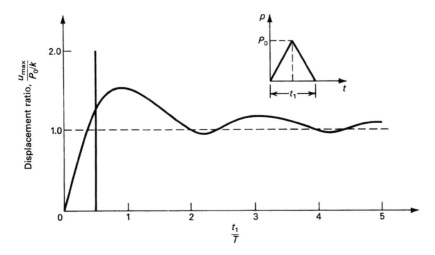

Figure 7.14. Shock spectrum for a triangular pulse load.

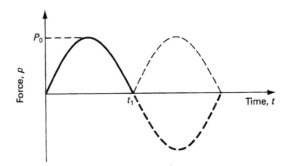

Figure 7.15. Sinusoidal pulse load.

Equation 6.8

$$u(t) = \frac{P_0}{k}\frac{1}{1-\beta^2}(\sin \Omega t - \beta \sin \omega t) \quad t \leq t_1 \tag{7.49}$$

where it has been assumed that the system starts from rest and $\beta = \Omega/\omega = \frac{1}{2}T/t_1$. The response to the second sine wave is similar to that given by Equation 7.49 with t replaced by $t - t_1$. The total response in the era $t > t_1$ is thus obtained from

$$u(t) = \frac{P_0}{k}\frac{1}{1-\beta^2}\ \{(\sin \Omega t - \beta \sin \omega t) + \sin \Omega(t - t_1)$$
$$-\beta \sin \omega(t - t_1)\} \quad t > t_1 \tag{7.50}$$

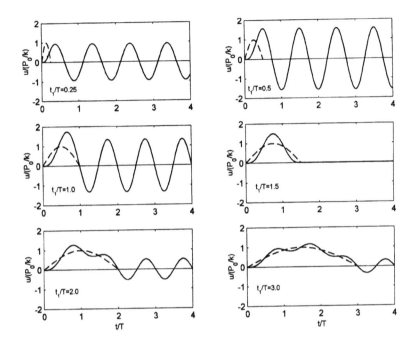

Figure 7.16. Time histories of response to a half-sine pulse load.

The histories of displacement response obtained from Equations 7.49 and 7.50 for six different values of t_1/T are shown in Figure 7.16. The static displacement produced by the sinusoidal pulse is also shown in that figure by dashed lines. It may be noted that for $t_1/T = 0.5$, which corresponds to $\beta = 1$, Equations 7.49 and 7.50 both become indeterminate. The limiting values of these equations can, however, be determined by applying L'Hospital's rule. Thus on differentiating the numerator and denominator of Equation 7.49 with respect to β and setting $\beta = 1$, we get

$$u(t) = \frac{1}{2}\frac{P_0}{k}(\sin \omega t - \omega t \cos \omega t) \quad t \leq t_1 \tag{7.51}$$

which is identical to Equation 6.17. For $t_1/T = 0.5$ the natural frequency $\omega = \pi/t_1$, and Equation 7.51 reduces to

$$u(t) = \frac{1}{2}\frac{P_0}{k}\left\{ \sin \frac{\pi t}{t_1} - \frac{\pi t}{t_1}\cos \frac{\pi t}{t_1} \right\} \quad t \leq t_1 \tag{7.52}$$

In a similar manner Equation 7.50 gives

$$u(t) = \frac{1}{2}\frac{P_0}{k}(\sin \omega t - \omega t \cos \omega t + \sin \omega(t - t_1) \\ - \omega(t - t_1)\cos \omega(t - t_1)) \quad t > t_1 \tag{7.53}$$

On substituting $\omega = \pi/t_1$ Equation 7.53 reduces to

$$u(t) = -\frac{\pi}{2} \cos \frac{\pi t}{t_1} \quad t > t_1 \tag{7.54}$$

Referring to Figure 7.16 when $t_1/T = 0.25$ the maximum response occurs in the free-vibration era and the maximum value is $0.943\,P_0/k$. For $t_1/T = 0.5$ the maximum response occurs at $t = t_1$, that is, at the end of the pulse and the maximum value is $1.414\,P_0/k$. For $t_1/T = 1.0$ and 1.5 there is a single peak in the forced-vibration era; for $t_1/T = 2.0$ there are two peaks, the first one of which is larger; and for $t_1/T = 3.0$ there are three peaks, in which the second is the largest. For the special case of $t_1/T = (2n + 1)/2$ both the displacement and velocity are zero at the end of the pulse, as can be verified from Equation 7.49 and its differential, and the motion ceases at the end of the pulse.

In order to obtain the shock spectrum we need to determine whether the maximum occurs in the forced-vibration era or the free-vibration era. If the maximum displacement occurs in the forced-vibration era $t < t_1$, the time to maximum, t_p, is obtained by equating the differential of Equation 7.49 to zero. This leads to

$$\cos \Omega t_p = \cos \omega t_p \tag{7.55a}$$

and hence to

$$\Omega t_p = 2\pi n \pm \omega t_p \quad n = 0, 1, 2, \dots \tag{7.55b}$$

The positive sign in Equation 7.55b corresponds to local minima, while the negative sign corresponds to local maxima. The time at which maximum occurs is thus given by

$$t_p = \frac{2n\pi}{\Omega + \omega} = \frac{2n}{1/t_1 + 2/T} \quad n = 1, 2, 3, \dots \tag{7.56}$$

For the maximum to occur within the forced-vibration era t_p must be less than or equal to t_1 giving

$$\frac{2n}{1/t_1 + 2/T} \leq t_1 \tag{7.57a}$$

or

$$\frac{t_1}{T} \geq \frac{2n - 1}{2} \tag{7.57b}$$

The smallest value of t_1 for which the maximum will occur in the forced-vibration era is obtained by substituting $n = 1$ in Equation 7.57b

$$\frac{t_1}{T} \geq \frac{1}{2} \tag{7.58}$$

Since $t_1/T = 1/2\beta$, Equation 7.58 is equivalent to the condition $\beta \leq 1$. The maximum response is obtained by substituting for $t = t_p$ in Equation 7.49.

$$\frac{u_{\max}}{P_0/k} = \frac{1}{1 - \beta^2} \left(\sin \frac{2\pi n \beta}{1 + \beta} - \beta \sin \frac{2\pi n}{1 + \beta} \right) \tag{7.59}$$

A peak response is obtained corresponding to each value of n in Equation 7.59. However, the maximum value of n is limited by the need to satisfy Equation 7.57b. Of the several peaks that may be obtained in the forced-vibration era, the one that gives the largest response is used in determining the shock spectrum. For $t_1/T < 1/2$ or $\beta > 1$, the maximum response occurs in the free-vibration era $t > t_1$, for which Equation 7.50 holds. By substituting $t_1 = \pi/\Omega$, Equation 7.50 can be simplified to

$$u = -\frac{P_0}{k} \frac{2\beta}{1 - \beta^2} \cos \frac{\pi}{2\beta} \sin \left(\omega t - \frac{\pi}{2\beta} \right) \tag{7.60}$$

Equation 7.60 represents a harmonic motion with amplitude

$$u = \frac{P_0}{k} \frac{2\beta}{1 - \beta^2} \cos \frac{\pi}{2\beta}$$

Hence the maximum response ratio is given by

$$\frac{u_{\max}}{P_0/k} = \frac{2\beta}{1 - \beta^2} \cos \frac{\pi}{2\beta} \qquad \frac{t_1}{T} < \frac{1}{2} \tag{7.61}$$

The shock spectrum for the sine pulse obtained from Equations 7.59 and 7.61 is plotted in Figure 7.17.

7.7.4 *Effect of viscous damping*

As stated earlier, from the point of view of design, the response parameter of particular interest is the maximum displacement. For a short-duration load, the maximum displacement is not significantly affected by the damping normally present in a structure. As will be evident from the displacement histories shown in Figures 7.10, 7.13, and 7.16, for a short-duration pulse, the maximum displacement is usually reached after about one-fourth of a cycle of motion. For such a short duration of motion, the amount of energy that can be dissipated through damping is quite small.

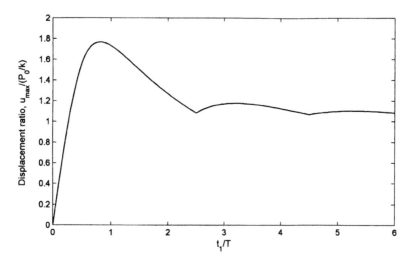

Figure 7.17. Shock spectrum for a sinusoidal pulse load.

As an example of the effect of damping consider the response of a damped system to a rectangular pulse function of duration t_1. Equation 7.16 shows that if $t_1 \geq \pi/\omega_d$, the maximum displacement occurs in the forced-vibration era and its value is given by Equation 7.17. When $t_1 < \pi/\omega_d$, maximum displacement is in the free-vibration era, and to determine its value, we need the displacement and velocity at $t = t_1$. These are obtained from Equation 7.14 and its derivative.

$$u_0(t_1) = \frac{P_0}{k}\left\{1 - e^{-\omega\xi t_1}\left(\cos\omega_d t_1 + \frac{\omega\xi}{\omega_d}\sin\omega_d t_1\right)\right\} \tag{7.62}$$

$$\frac{v_0(t_1)}{\omega_d} = \frac{P_0}{k}\frac{1}{1-\xi^2}e^{-\omega\xi t_1}\sin\omega_d t_1 \tag{7.63}$$

The response in the free-vibration era is obtained from Equation 5.38

$$u(t) = \rho e^{-\omega\xi\bar{t}}\sin(\omega_d\bar{t} + \phi) \tag{7.64}$$

where $\bar{t} = t - t_1$ and ρ and ϕ are given by Equation 5.39. The time at which a peak is reached in the displacement response is obtained from Equation 5.41, so that

$$\sin(\omega_d\bar{t} + \phi) = \sqrt{1 - \xi^2} \tag{7.65}$$

The peak value of the displacement is determined by substituting in Equation 7.64 the value of $\bar{t}$ obtained from Equation 7.65.

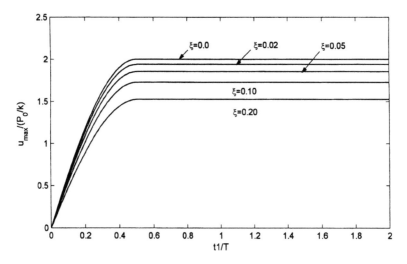

Figure 7.18. Shock spectra for a rectangular pulse load corresponding to different values of the damping ratio.

The shock spectrum obtained by the method described in the previous paragraphs is plotted in Figure 7.18 for five different values of the damping ratio ξ. It will be observed that for $t_1/T = 0.5$ the maximum displacement reduces by only about 13.5% as the damping changes from 0 to 10%.

7.7.5 *Approximate response analysis for short-duration pulses*

For a short-duration pulse, it is possible to obtain a reasonably good estimate of the maximum response by assuming that the force pulse is concentrated at $t = 0$ and has an impulse whose magnitude is equal to the area enclosed by the force pulse under consideration. For the three types of force pulses discussed in the earlier paragraphs, the magnitude of impulse I is given by

   For rectangular pulse: $I = P_0 t_1$
   For triangular pulse: $I = \frac{1}{2} P_0 t_1$
   For sinusoidal pulse: $I = 2 P_0 t_1 / \pi$

where P_0 is the amplitude of each of the three pulses. The response produced by an impulse I imparted at time $t = 0$ is obtained from

$$u = \frac{I}{m\omega} \sin \omega t$$

$$= \frac{I\omega}{k} \sin \omega t$$

(7.66)

The maximum displacement is thus $I\omega/k$.

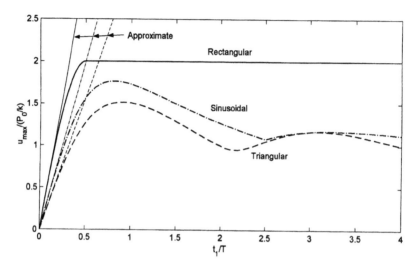

Figure 7.19. Approximate and exact values of the displacement spectra for rectangular, triangular, and sinusoidal pulse loads.

The approximate value of the displacement spectrum calculated from Equation 7.66 is compared in Figure 7.19 with the more exact values presented in earlier sections. For a relatively short duration of the pulse, for example $t_1/T < 0.25$, the approximate analysis provides a reasonably good estimate. For $t_1/T > 0.5$, when the maximum is achieved in the forced-vibration era, Equation 7.66 is meaningless because it is based on the assumption that the peak response occurs in the free-vibration era.

7.8   RESPONSE TO GROUND MOTION

The analysis of dynamic response of a structure to ground motion is of considerable interest because of its application in the design for earthquake forces. Earthquake ground motions are quite random in nature. However, if the time history of ground motion is available, it is possible to determine the response of a single-degree-of-freedom system subjected to such a motion. The equation of motion of an underdamped single-degree-of system subjected to ground motion is given by

$$m\ddot{u}^t + c\dot{u} + ku = 0 \tag{7.67a}$$

or

$$m\ddot{u} + c\dot{u} + ku = -m\ddot{u}_g \tag{7.67b}$$

where u' is the absolute motion measured from a fixed frame of reference, and u is the displacement relative to the ground, so that $u' = u + u_g$. On dividing both sides of Equation 7.67 by m we get

$$\ddot{u} + 2\xi\omega\dot{u} + \omega^2 u = -\ddot{u}_g \qquad (7.68)$$

Equation 7.67b is an equation of forced-motion, where the forcing function $p(t) = -m\ddot{u}_g$. It can be solved for u by a numerical evaluation of the Duhamel's integral or by direct numerical integration using methods described in Chapter 8. It will be observed that for a given ground motion, the response of the single-degree-of-freedom system depends only on the frequency ω and the damping ratio ξ. Therefore, once a ground motion has been selected, we can construct a series of curves that provide the maximum response of a system to the selected ground motion for a range of values of the system natural frequency and damping. Usually, each curve in the series is constructed for one specific value of the damping and plots the maximum displacement versus the frequency, or the period. As stated earlier, such plots are referred to as response spectrum curves. Once the spectra have been determined, the maximum displacement and forces induced in a single-degree-of-freedom system having specified dynamic characteristics by the given earthquake ground motion are easily calculated.

Consider the Duhamel integral solution of a single-degree-of-freedom system subjected to a ground acceleration $\ddot{u}_g$

$$u(\omega, \xi) = \frac{1}{m\omega_d} \int_0^t -m\ddot{u}_g(\tau)e^{-\xi\omega(t-\tau)}\sin\omega_d(t-\tau)\,d\tau \qquad (7.69)$$

The maximum value of u obtained from Equation 7.69 is referred to as the spectral-displacement and denoted by S_d. Thus

$$S_d(\omega, \xi) = u_{\max}(\omega, \xi) \qquad (7.70)$$

We will find it useful to define two related spectral parameters, referred to as the pseudo-spectral-velocity, S_v, and pseudo-spectral-acceleration, S_a. The two spectral parameters are given by

$$S_v = \omega S_d \qquad (7.71a)$$

$$S_a = \omega S_v = \omega^2 S_d \qquad (7.71b)$$

The pseudo-spectral-velocity is not the same as the true spectral velocity. The latter can be obtained by evaluating the differential of Equation 7.69 and taking the maximum value of the differential. Differentiation under the integral sign gives

$$\dot{u}(t) = -\frac{\xi\omega}{\omega_d}u + \int_0^t \ddot{u}_g(\tau)e^{-\xi\omega(t-\tau)}\cos\omega_d(t-\tau)\,d\tau \tag{7.72}$$

The true spectral velocity $V = \dot{u}_{max}$ is determined by taking the maximum value of $\dot{u}$ obtained from Equation 7.72.

The pseudo-spectral-acceleration is approximately equal to the maximum value of the true absolute acceleration. The latter can be obtained from the equation of motion. Thus

$$\ddot{u}^t(t) = -2\xi\omega\dot{u}(t) - \omega^2 u(t) \tag{7.73}$$

For low values of damping $\ddot{u}^t(t) \approx -\omega^2 u(t)$, so that $u(t)$ and $\ddot{u}^t(t)$ attain their maximum values at the same instant of time and $\ddot{u}^t_{max} \approx \omega^2 u_{max} = \omega^2 S_d = S_a$. For an undamped system the pseudo-spectral acceleration and the maximum absolute acceleration are identical; for low amounts of damping the two are nearly equal.

The pseudo-spectral-velocity and pseudo-spectral-acceleration have useful physical properties. The pseudo-spectral velocity provides a measure of the peak value of strain energy $E_{s(max)}$ stored in the system.

$$E_{s(max)} = \frac{1}{2}ku^2_{max} = \frac{1}{2}kS_d^2 = \frac{1}{2}k\left(\frac{S_v}{\omega}\right)^2 = \frac{1}{2}mS_v^2 \tag{7.74}$$

The pseudo-spectral acceleration can be used to determine the peak value of the spring force.

$$f_{S(max)} = ku_{max} = m\omega^2 S_d = mS_a \tag{7.75}$$

Because of the relationship between S_d, S_v, and S_a given by Equation 7.71, all three of them can be plotted on a single four-way logarithmic graph. In such a plot $\log S_v$ is plotted on the ordinate against $\log T$ (or $\log\omega$) on the abscissae. Equation 7.71a gives

$$S_v = \frac{2\pi}{T}S_d$$
$$\log S_v + \log T = \log 2\pi + \log S_d \tag{7.76}$$

For a constant S_d, Equation 7.76 represents a straight line on the four-way log graph sloping to the left with a slope of -1. Lines sloping to the left with a slope of -1 thus represent constant S_d lines. A line perpendicular to these lines provides the log scale for the spectral-displacement. In a similar manner, Equation 7.71b gives

$$S_a = \frac{2\pi}{T} S_v \qquad (7.77)$$

$$\log S_v - \log T = \log S_a - \log 2\pi$$

For a constant S_a, Equation 7.77 represents a straight line on the four-way log graph sloping to the right with a slope of 1. Lines sloping to the right with a slope of 1 thus represent constant S_a lines and a line perpendicular to these lines provides a log scale for the pseudo-spectral-acceleration.

As an example of the application of response spectra, consider the one-story shear frame shown in Figure 7.20a. The columns are considered to be axially rigid and of negligible mass. The floor is infinitely stiff and all of the mass is concentrated at the floor level. The system has only one degree of freedom; the lateral displacement at the floor level. Damping is of viscous type and is represented by the dash pot shown on the diagram. The floor mass is m, the total stiffness of the columns is k and the damping constant is c. Correspondingly, the natural period of vibration is $T = 2\pi\sqrt{m/k}$ and the damping ratio is ξ. These two parameters are varied over a range of values. The frame is subjected to a lateral ground motion. The variation of ground acceleration $\ddot{u}_g$ with time is shown in Figure 7.20b, which also shows the histories of ground velocity and displacement.

The maximum value of the lateral displacement of mass m relative to the ground will be attained either during the time the ground motion lasts or within at most half a cycle of free vibration after the ground ceases to move. For a selected value of the period T and damping ratio ξ, the history of ground displacement is obtained for the duration of ground motion, plus a half-cycle of free vibration by using one of the numerical methods of integration described in Chapter 8. In the present case, the method that relies on a piece-wise linear representation of the excitation is most effective. The spectral-displacement S_d, that is the maximum value of relative displacement, corresponding to the given period and damping is obtained from the time history of motion. The pseudo-spectral-velocity and acceleration are now determined from Equation 7.71. The spectral-velocity has been plotted against the natural period for two values of damping, $\xi = 0.0$ and 0.1, on a four-way logarithmic graph in Figure 7.21. The displacement and acceleration scales are shown in Figure 7.21 along lines sloping to the right and left, respectively.

Several properties of interest can be observed from the response spectrum plots of Figure 7.21. For very short period structures, the stiffness k would be very large, and the mass m would more-or-less move along with the ground. The absolute acceleration of the mass should, therefore, be close to that of the ground. This can be verified from Figure 7.21, where the maximum value of

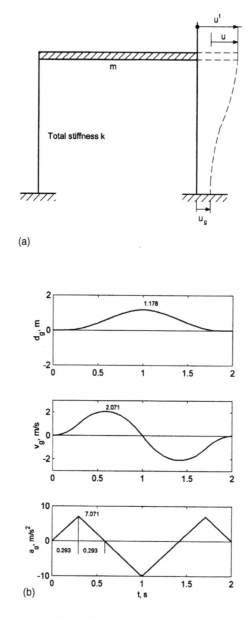

Figure 7.20. (a) Single-story shear frame; (b) variation of ground motion with time.

ground acceleration has been shown by dashed line. For a long period structure k approaches zero, and the structure is so flexible that the mass remains
stationary even as the ground moves. The relative displacement of the mass
should, therefore, be equal to the ground displacement. This is seen to be the

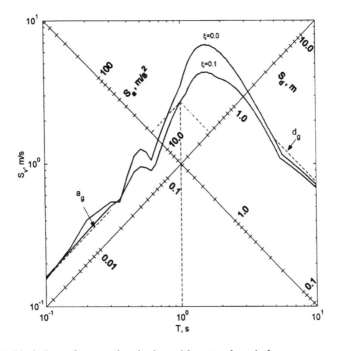

Figure 7.21. Variation of spectral velocity with natural period.

case from Figure 7.21 where the maximum value of ground displacement has also been shown.

It would be of interest to compare the relative values of the true spectral-velocity and the pseudo-spectral-velocity. This has been done in Figure 7.22 for $\xi = 0.1$. In the intermediate period range, the two are quite close. In the long period range, the true spectral-velocity is greater than the pseudo-spectral velocity. This would be expected for the following reason. For a very flexible structure, the relative displacement and relative velocity of the mass m are equal to the ground displacement and ground velocity, respectively. Thus, $S_d = u_{g0}$ and $V = \dot{u}_{g0}$, where V is the true-spectral-velocity and u_{g0}, $\dot{u}_{g0}$ are the maximum ground displacement and velocity, respectively. On the other hand, the pseudo-spectral-velocity approaches zero when T approaches ∞ as would be seen from Equation 7.71a. In the short period range, the true spectral-velocity is smaller than the pseudo-spectral velocity.

The pseudo-spectral-acceleration has been compared with the true spectral-acceleration in Figure 7.23 for a damping ratio $\xi = 0.1$. As would be expected, the two are quite close.

To illustrate the application of response spectrum in design, suppose that it is required to determine the maximum relative displacement as well as the maximum base sear induced in the frame of Figure 7.20a by the design ground

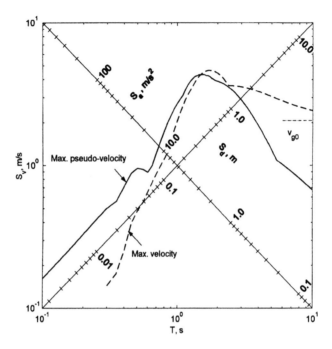

Figure 7.22. Comparison between the true spectral velocity and pseudo-spectral velocity.

motion shown in Figure 7.20b. The natural period of the frame is given as 1.0 s and the damping ratio $\xi = 0.1$. The S_d and S_a values for a period of 1 s and damping ratio 0.1 can be read off from the response spectrum shown in Figure 7.21. From the numerical data used to plot the spectral curves in Figure 7.21 the following values are obtained.

$$S_d = 0.435 \text{ m}, \quad S_a = 17.16 \text{ m/s}^2$$

The maximum relative displacement is thus 0.435 m and the maximum base shear is $V_b = kS_d = m\omega^2 S_d = m(2\pi/T)^2 S_d = 17.16m$. The base shear can also be obtained from the pseudo-spectral acceleration. Thus $V_b = mS_a = 17.16m$.

7.8.1   *Response to a short-duration ground motion pulse*

As stated earlier, damping can reasonably be neglected in determining the response to a short-duration pulse. Neglecting damping, the equation of motion for ground excitation specializes to

$$m\ddot{u}^t + ku = 0 \tag{7.78a}$$

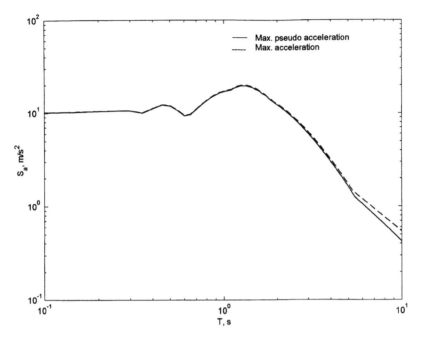

Figure 7.23. Comparison between the true spectral acceleration and pseudo-spectral acceleration.

or

$$m\ddot{u} + ku = -m\ddot{u}_g \qquad (7.78\text{b})$$

Equation 7.78b is an equation of forced-vibration response in which the effective force is given by $p = -m\ddot{u}_g$. When maximum displacement is the quantity of interest, the negative sign in the forcing function is of no consequence, and if $\ddot{u}_{g0}$ represents the maximum amplitude of ground acceleration, the amplitude of effective force is given by $P_0 = m\ddot{u}_{g0}$. The response ratio R is then obtained from

$$R = \frac{u_{max}}{m\ddot{u}_{g0}/k} = \frac{\omega^2 u_{max}}{\ddot{u}_{g0}} \qquad (7.79)$$

Equation 7.79 shows that the shock spectra derived earlier for different types of force impulses also apply to a ground motion pulse when the ordinate is taken to be equal to $\omega^2 u_{max}/\ddot{u}_{g0}$. This statement is, however, true only when the initial conditions represented by $u(0)$ and $\dot{u}(0)$ are both zero. From Equation 7.78a we note that $\ddot{u}^t_{max} = -\omega^2 u_{max}$, hence the shock spectrum also gives the absolute value of $\ddot{u}^t_{max}/\ddot{u}_{g0}$.

Example 7.3

The base of an undamped single-degree-of-freedom system is subjected to a velocity pulse shown in Figure E7.3a. Obtain the response of the system and plot the shock spectrum of relative motion.

Solution

The velocity pulse function is given by

$$\dot{u}_g = v_0 \left(1 - \frac{t}{t_1} \right) \qquad\qquad (a)$$

from which the ground acceleration is obtained as

$$\ddot{u}_g = -\frac{v_0}{t_1} \qquad\qquad (b)$$

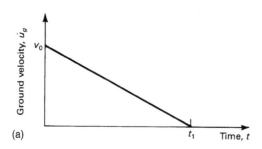

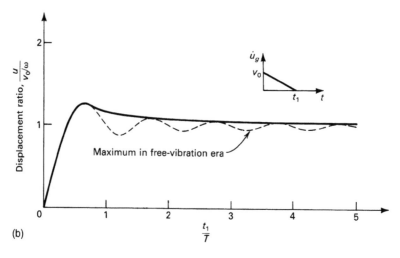

Figure E7.3. Shock spectrum for ground velocity pulse: (a) ground velocity pulse; (b) shock spectrum.

The equation of motion, written in terms of the relative displacement, is

$$m\ddot{u} + ku = \frac{mv_0}{t_1} \tag{c}$$

with the initial condition $u_0 = 0$ and $\dot{u}_0 = \dot{u}^t - \dot{u}_{g0} = -v_0$. Using Duhamel's integral, and the expression for free-vibration response, the total response is obtained as follows

$$u = -\frac{v_0}{\omega} \sin \omega t + \frac{1}{m\omega} \int_0^t \frac{mv_0}{t_1} \sin \omega(t - \tau)\, d\tau \quad t \leq t_1 \tag{d}$$

and

$$u = -\frac{v_0}{\omega} \sin \omega t + \frac{1}{m\omega} \int_0^{t_1} \frac{mv_0}{t_1} \sin \omega(t - \tau)\, d\tau \quad t > t_1 \tag{e}$$

Evaluating the integrals in Equations d and e, we obtain

$$u = \begin{cases} \frac{v_0}{\omega}[-\sin \omega t + \frac{1}{\omega t_1}(1 - \cos \omega t)] & t \leq t_1 \tag{f} \\[2ex] \frac{v_0}{\omega}[-\sin \omega t + \frac{1}{\omega t_1}\{\cos \omega(t - t_1) - \cos \omega t\}] & t > t_1 \tag{g} \end{cases}$$

If the peak response occurs for $t < t_1$, the time to maximum response is obtained by equating the differential of Equation f to zero. This gives

$$\tan \omega t_p = \omega t_1 \tag{h}$$

Substitution of $t = t_p$ in Equation f gives the shock spectrum

$$\frac{u_{max}\omega}{v_0} = \frac{1}{\omega t_1} \pm \frac{\omega t_1}{\sqrt{1 + (\omega t_1)^2}} \pm \frac{1}{\omega t_1}\frac{1}{\sqrt{1 + (\omega t_1)^2}} \tag{i}$$

If the maximum response occurs in the free-vibration era, $t > t_1$, its value should be obtained from Equation g, which can be expressed as

$$\frac{u\omega}{v_0} = \rho \sin(\omega t + \phi) \tag{j}$$

where

$$\rho = \left\{ 1 + \frac{2}{(\omega t_1)^2} - \frac{2}{\omega t_1}\left(\sin \omega t_1 + \frac{\cos \omega t_1}{\omega t_1}\right) \right\}^{1/2}$$

and

$$\tan \phi = \frac{\cos \omega t_1 - 1}{\sin \omega t_1 - \omega t_1}$$

The maximum response is equal to the amplitude ρ. The shock spectrum obtained from Equations i and j is plotted in Figure E7.3b. For $t_1/T < 0.715$, the maximum occurs in the free-vibration era, $t > t_1$; for $t_1/T > 0.715$, the maximum occurs during $t < t_1$. At $t_1/T = 0.715$, the maximum value of response $u\omega/v_0$ is 1.247.

7.9   ANALYSIS OF RESPONSE BY THE PHASE PLANE DIAGRAM

Let an undamped single-degree-of-freedom system be subjected to a base displacement pulse of magnitude u_g and duration t_1, as shown in Figure 7.24a. From Section 5.2.1, we know that the total response of such a system can be obtained from a phase plane diagram. As shown in Figure 7.24b, the phase vector rotates with its center located a distance u_g from the origin. If the system starts from at rest condition, the tip of the vector is at the origin. In the era of motion from 0 to t_1, the vector rotates through an angle ωt_1 to position CA. If the base now returns to its original position, the subsequent era of motion is represented by vector OA rotating about O. The displacement and velocity at A constitute the initial conditions for the second phase of motion. The displacement–time relationship for both eras of motion is shown in Figure 7.24c.

The equation of motion for the system subjected to a base displacement impulse is given by

$$m\ddot{u}^t + ku^t = ku_{g0} \quad t \le t_1$$
$$m\ddot{u}^t + ku^t = 0 \qquad t > t_1 \tag{7.80}$$

The equation of motion of a system subjected to a rectangular pulse of magnitude P_0 and duration t_1 is

$$m\ddot{u} + ku = P_0 \quad t \le t_1$$
$$\ddot{u} + ku = 0 \qquad t > t_1 \tag{7.81}$$

On comparing Equations 7.80 and 7.81, and noting that for no base motion $u \equiv u^t$, we observe that the motion caused by a rectangular pulse is similar

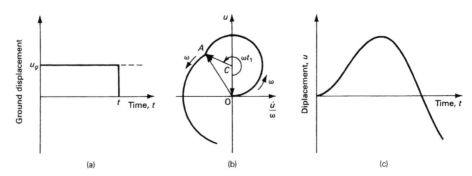

Figure 7.24. Phase plane diagram for a ground displacement pulse.

$\Theta_1 = \omega\Delta t_1$
$\Theta_2 = \omega\Delta t_2$
$\Theta_3 = \omega\Delta t_3$

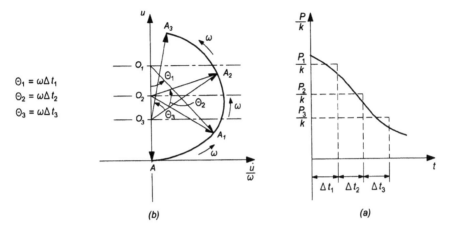

Figure 7.25. Response to an arbitrary load: (a) loading; (b) phase plane diagram for an undamped system.

to that caused by a base displacement, provided that ku_{g0} is replaced by P_0. This suggests that a phase plane diagram may also be effectively utilized to obtain the response to a rectangular pulse force. The center of rotation of the phase vector will, in this case, be located at a distance P_0/k from the origin. In all other respects, the phase plane diagram will be similar to that for base displacement impulse.

The phase plane method can also be extended to the analysis of response of a general dynamic loading. The loading is divided into a series of rectangular pulses of short durations as shown in Figure 7.25a. The center of rotation of the phase vector corresponding to the nth pulse, of magnitude P_n, will be located a distance P_n/k from the origin. The initial conditions for the phase of motion from t_{n-1} to t_n will be those existing at the end of the previous pulse. The resulting phase plane diagram will be as shown in Figure 7.25b. In this diagram, the applied load has been represented by three force pulses of heights P_1, P_2, and P_3 and durations Δt_1, Δt_2, and Δt_3, respectively. The system is assumed to be undamped and initially at rest. In the first phase of response, the center of rotation O_1 is located at a distance P_1/k from the origin, and vector O_1A rotates at ω rad/s through an angle $\omega\Delta t_1$ to O_1A_1. In the second phase of response, the center of rotation is located at O_2, a distance P_2/k from the origin, and the starting position of the vector is O_2A_1. At the end of this phase, the vector has rotated by an angle $\omega\Delta t_2$ to position O_2A_2. In the third phase of response, vector O_3A_2 rotates to position O_3A_3, where $AO_3 = P_3/k$ and angle $A_2O_3A_3 = \omega\Delta t_3$. The displacement u and the velocity function $\dot{u}/\omega$ can now be obtained by taking the projections of the rotating vector on the vertical and horizontal axis, respectively.

SELECTED READINGS

Ayre, R.S. 1996. Transient Response to Step and Pulse Functions. In C.M. Harris (ed.) *Shock and Vibration Handbook*. New York: McGraw-Hill. 4th Edition.

Bishop, R.E.D., Parkinson, A.G. & Pendered, J.W. 1969. Linear Analysis of Transient Vibration, *Journal of Sound and Vibration*, Vol. 9: 313–337.

Crede, C.E. 1951. *Vibration and Shock Isolation*. New York: John Wiley.

Jacobsen, L.S. & Ayre, R.S. 1958. *Engineering Vibrations with applications to structures and machinery*. New York: McGraw-Hill.

Matsuzaki, Y. & Kibe, S. 1983. Shock and Seismic Response Spectra in Design Problems. *Shock and Vibration Digest*, Vol. 15: 3–10.

Thomson, W.T. 1998. *Theory of Vibration with Applications*. Upper Saddle River, New Jersey: Prentice Hall. 5th Edition.

PROBLEMS

7.1 A single degree-of-freedom system with mass m, stiffness k and negligible damping is subjected to alternating step force shown in Figure P7.1. Note that the applied force is periodic with a period equal to the natural period of the system. Plot the response of the system as a function of time. Show that resonance takes place and that the successive peaks of response have a magnitude of $2nP_0/k$ where n is the number of half cycles from the starting point.

7.2 Obtain the response of the system shown in Figure P7.2 to an applied force F which varies as indicated.

7.3 A single degree-of-freedom system is subjected to shock loading shown in Figure P7.3. Determine the condition under which the maximum response will be obtained for $t > 3t_1$ and find the magnitude of such response.

7.4 In Problem 7.3, assume that t_1/T is very small so that the applied force can reasonably be viewed as an impulsive force applied at $t = 0$. Obtain

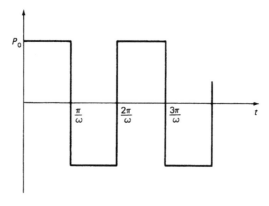

Figure P7.1.

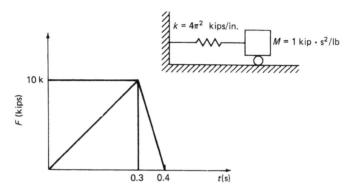

Figure P7.2.

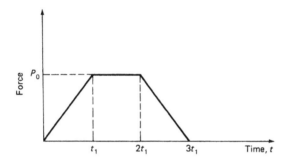

Figure P7.3.

an expression for the resulting response for $t > 3t_1$ and hence find the maximum response. If $t_1/T = 1/32$, how much is the approximate response in error as compared to the exact value of the maximum obtained in Problem 7.3.

7.5 A ground velocity pulse v_0 applied at time $t = 0$ can be interpreted as a sudden change in ground velocity at that time (Figure P7.5). Show that such a change in velocity will be caused by an acceleration input of $v_0 \delta(t)$, where $\delta(t)$ is the delta function. Using Duhamel's integral and the properties of the delta function, prove that the response of the system to the velocity pulse is given by

$$u = -\frac{v_0}{\omega_d} e^{-\xi \omega t} \sin \omega_d t$$

Hence show that the displacement response spectrum is obtained from

$$u_{max} = \frac{v_0}{\omega} \exp\left(-\frac{\xi}{\sqrt{1 - \xi^2}} \sin^{-1} \sqrt{1 - \xi^2} \right)$$

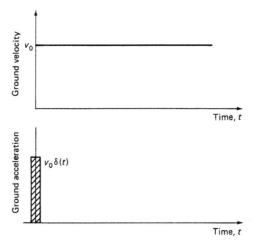

Figure P7.5.

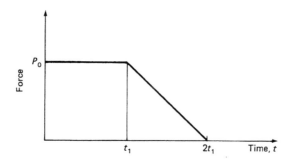

Figure P7.6.

Plot the response spectra (u_{max} as a function of period T) for $\xi = 0$, 0.05 and 0.1.

7.6 A single degree-of-freedom system is subjected to a force which varies as shown in Figure P7.6. Obtain a plot of the maximum value of response as a function of t_1/T, where T is the natural period of the system.

What will be the maximum displacement of the system when $P_0 = 10$kN, and $t_1 = 1/2$ s. The natural period of the system is 1s and the spring stiffness is 250 kN/m.

7.7 Obtain the response of a single degree-of-freedom system subjected to the ground velocity pulse shown in Figure P7.7.

7.8 Obtain expressions for the response to step function load with rise time shown in Figure 7.7 by the Duhamel's integral.

7.9 Using Duhamel's integral, obtain expressions for the response to the triangular pulse force shown in Figure 7.12.

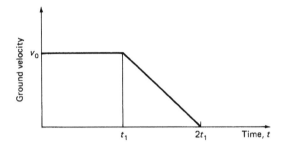

Figure P7.7.

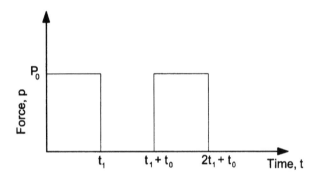

Figure P7.10.

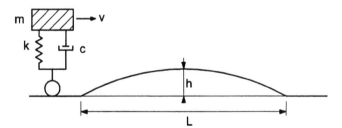

Figure P7.11.

7.10 Obtain the response of an undamped single-degree-of-freedom system to the double rectangular pulse shown in Figure P7.10 for time $t > 2t_1 + t_0$. The duration of each pulse is t_1 and the time lapse between the two pulses is t_0. Determine the value of response when $t_0 = (\pi/\omega) - t_1$.

7.11 An automobile modeled as a single-degree-of-freedom system rides over a hump in a road. The hump can be represented by a half sine wave of length L and height h as shown in Figure P7.11. The speed of the vehicle is v and its natural frequency is ω. Assuming that the shock absorbers of the car are worn, so that the damping is zero, show that the absolute displacement of the vehicle while it is riding over the hump is given by

$$u^t = \frac{h}{1-\beta^2}(\sin \Omega t - \beta \sin \omega t)$$

where $\Omega = \pi v/L$ and $\beta = \Omega/\omega$. Also show that the displacement of the vehicle relative to the ground while the vehicle is riding over the hump is given by

$$u = \frac{h\beta^2}{1-\beta^2}\left(\sin \Omega t - \frac{1}{\beta}\sin \omega t\right)$$

7.12 The following data are supplied for the vehicle and the road referred to in Problem 7.11. $L = 1.5$ m, $h = 50$ mm, $v = 20$ km/hr. The mass of the vehicle is 1400 kg, and the stiffness of its spring is 40,000 N/m. Find the absolute maximum displacement of the vehicle and the maximum spring force. What is the maximum acceleration experienced by the passengers?

7.13 Repeat Problem 7.12 assuming that the damping is 40% of critical.

CHAPTER 8

Analysis of single-degree-of-freedom systems: Approximate and numerical methods

8.1 INTRODUCTION

In the many problems of the vibration of structural or mechanical systems, the quantity of primary interest is the predominant frequency of vibration. When the system has only one degree of freedom, the frequency of vibration is easily determined by the methods described in earlier chapters. Very often, however, the system under consideration is in fact a multi-degree-of-freedom system. Frequency determination of such systems is complex and time consuming. On the other hand, if a reasonable estimate can be made of the vibration shape, the system can be represented by an equivalent single-degree-of-freedom model whose frequency of vibration is more easily determined. Because we cannot expect the estimated vibration shape to be the same as the true vibration shape, the frequency determined by this method is in error. The practical utility of the method, however, lies in the fact that even with a crude estimate of the vibration shape, the error in the calculated frequency is generally small enough so that the frequency estimate is acceptable for all practical purposes.

The frequency of a single-degree-of-freedom system, whether real or idealized, is most conveniently determined by the principle of conservation of energy. The procedure is commonly known as the *Rayleigh method* after Lord Rayleigh, who first proposed it in his book on the theory of sound. Its main application is to a system that can be idealized as a single-degree-of-freedom system by assuming an approximate vibration shape. Rayleigh showed that the frequency obtained by such an assumption is always higher than the true frequency. In this chapter, we present a detailed description of the application of Rayleigh method in estimating the approximate frequency of an idealized single-degree-of-freedom system.

In many instances, the complete history of vibration or at least the maximum displacement amplitude attained during an era of vibration, rather than just the frequency, is of interest. The structural or mechanical system can be set vibrating whenever it is disturbed from its position of rest by being subjected to an initial displacement or an initial velocity or both. The system may also be excited by the application of a dynamic force. Often, the source of excitation will be a combination of initial disturbances and an applied force. In each case, a response history calculation requires the solution of the equation of motion.

Consider first a linear system. When such a system is excited by initial disturbances but is not subjected to any external force, closed-form mathematical solution can be obtained for the equation of motion. These solutions have been presented in the earlier chapters of the book. Thus, for a damped single-degree-of-freedom system subjected to an initial displacement u_0 and an initial velocity v_0, the displacement response is given by

$$u(t) = e^{-\xi \omega t} \left(\frac{v_0 + \xi \omega u_0}{\omega_d} \sin \omega_d t + u_0 \cos \omega_d t \right) \tag{8.1}$$

A mathematical solution can also be obtained for the response of a linear system excited by an applied dynamic force provided that such a force can be represented by a simple mathematical function. The response of a linear system subjected to an exciting force $p(t)$ is given by the Duhamel's integral

$$u(t) = \frac{1}{m \omega_d} \int_0^t e^{-\xi \omega (t - \tau)} p(\tau) \sin \omega_d (t - \tau) \, d\tau \tag{8.2}$$

Exact evaluation of the Duhamel's integral is possible provided that $p(t)$ is a simple function. Response solutions for several such exciting functions have been presented in Chapter 7. However, if $p(t)$ is a complicated function or is defined only by numerical values at discrete intervals of time, mathematical evaluation of the integral in Equation 8.2 is not possible and a numerical method must be used to solve the equation of motion. The numerical methods used to find the response of a linear single-degree-of-freedom system can be classified into the following categories:

1. Evaluation of the Duhamel's integral by a numerical method.
2. Direct numerical integration of the equation of motion.

A numerical evaluation of the Duhamel's integral provides the response of the system to an applied force. If the system starts with prescribed initial conditions, the response due to the specified initial displacement and velocity must be obtained from Equation 8.1 and superimposed on the response obtained by the numerical evaluation of the Duhamel's integral. On the other hand, when a direct numerical integration of the equation of motion is used, the combined response due to the initial conditions as well as the applied force can be directly obtained. For a nonlinear system, the Duhamel integral solution is no longer valid and recourse must be had to a direct numerical integration of the equation of motion.

In this chapter, we present several alternative methods for the numerical evaluation of the Duhamel's integral. This is followed by a detailed description of some of the methods used for the direct numerical integration of the equations of motion. It is observed that direct integration is a more general method since it is applicable to both linear and nonlinear systems. Also, as will be shown later, direct integration methods can be extended directly to the response analysis of a multi-degree-of-freedom system.

8.2 CONSERVATION OF ENERGY

Consider the undamped single-degree-of-freedom system shown in Figure 8.1a. The system is vibrating without the application of an external force. The equation of motion for such a system is given by

$$m\ddot{u} + ku = 0 \qquad (8.3)$$

The displacement and velocity responses of the system are shown in Figure 8.1b and c, respectively. On multiplying Equation 8.3 by velocity $\dot{u}$ and integrating, we obtain

$$\frac{1}{2}m(\dot{u})^2 + \frac{1}{2}ku^2 = C \qquad (8.4)$$

where C is a constant. The first term on the left-hand side of Equation 8.4 represents the kinetic energy of the system, while the second term represents the potential energy, which in this case is the strain energy of the spring. Equation 8.4 shows that for the system under consideration, the total energy, kinetic plus potential, is a constant. In general, for a system subjected only to conservative forces, the total energy is always conserved. A formal proof of the statement was provided in Section 4.9. In writing the energy equation, we must

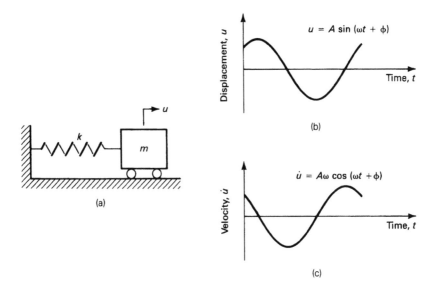

Figure 8.1. Undamped single-degree-of-freedom system.

include in the total potential energy the potential of the applied force as well. Equation 8.4 thus represents a special case when no external force is acting on the system and the potential energy consists only of the internal strain energy. We proved in Chapter 5 that a solution of Equation 8.3 is given by

$$u = A \sin(\omega t + \phi) \tag{8.5}$$

where ω is the natural frequency of the system and A and ϕ are arbitrary constants whose values are determined by the initial conditions. By using Equation 8.5, the kinetic energy T and the potential energy V can be shown to be

$$T = \frac{1}{2} m\omega^2 A^2 \cos^2(\omega t + \phi) \tag{8.6a}$$

$$V = \frac{1}{2} kA^2 \sin^2(\omega t + \phi) \tag{8.6b}$$

For the kinetic energy to be a maximum, $\cos(\omega t + \phi)$ must be equal to 1. When such is the case, $\sin(\omega t + \phi)$, and therefore the potential energy, is zero. At such an instance, the total energy is equal to the kinetic energy, given by

$$T_{\max} = \frac{1}{2} m\omega^2 A^2 \tag{8.7}$$

In a similar manner, the potential energy is a maximum when $\sin(\omega t + \phi)$ is 1, at which time $\cos(\omega t + \phi)$ and therefore, the kinetic energy is zero. At such an instant, the total energy is equal to the potential energy given by

$$V_{\max} = \frac{1}{2} kA^2 \tag{8.8}$$

Because the total energy is constant, using Equations 8.4, 8.7, and 8.8, we obtain

$$C = T_{max} = V_{max} \tag{8.9a}$$

or

$$\frac{1}{2} m\omega^2 A^2 = \frac{1}{2} kA^2 \tag{8.9b}$$

Equation 8.9 enables us to find an expression for the frequency ω:

$$\omega = \sqrt{\frac{k}{m}} \tag{8.10}$$

The procedure just outlined, which uses the principle of conservation of energy, is known as Rayleigh's method. Equation 8.10 is, in fact, identical to the one derived in Chapter 5 and gives the true vibration frequency of the system. It appears, then, that for the system considered, the Rayleigh method leads to the true frequency and seems to have no particular advantage over the direct method. This is in general true, although some problems are more easily solved by the energy method, as the following examples illustrate.

Example 8.1

A wheel and axle assembly of mass moment of inertia I_0 is inclined from the vertical by an angle α, as shown in Figure E8.1. If an unbalanced weight w is attached to the wheel at a distance a from the axis and the assembly undergoes small vibrations about the position of rest, determine the frequency of vibration.

Solution

If the air resistance and friction in the bearings are neglected, the only force acting on the system is that due to gravity. The gravity force being conservative, the total energy will be conserved.

As the wheel swings, the height of the unbalanced weight above the datum and hence its potential energy changes. The kinetic energy, of course, depends on the angular velocity of the wheel axle assembly. Denoting the angular velocity of the wheel by $\dot{\theta}$, the kinetic energy T is given by

$$T = \frac{1}{2}I_0(\dot{\theta})^2 + \frac{1}{2}\frac{w}{g}a^2(\dot{\theta})^2 \tag{a}$$

The height of the unbalanced weight changes from h_1 to h as the wheel swings through an angle θ. Taking the datum as the lowest position of weight w, the potential energy V is given by

$$V = w(h - h_1)$$
$$= wa \sin \alpha(1 - \cos \theta) \tag{b}$$

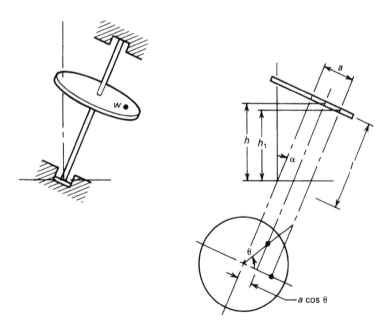

Figure E8.1. Vibration of a wheel axle assembly.

For small angle θ, Equation b becomes

$$V = wa \sin \alpha \frac{\theta^2}{2} \tag{c}$$

For simple harmonic motion

$$\theta = A \sin \omega t$$
$$\dot{\theta} = A\omega \cos \omega t \tag{d}$$

and

$$T_{max} = V_{max} \tag{e}$$

From Equations a, c, and d,

$$T_{max} = \frac{A^2 \omega^2}{2} \left(I_0 + \frac{w}{g} a^2 \right) \tag{f}$$

$$V_{max} = w \frac{aA^2}{2} \sin \alpha$$

Finally, from Equations e and f,

$$\omega^2 = \frac{wa \sin \alpha}{I_0 + (w/g)a^2} \tag{g}$$

Example 8.2

The system shown in Figure E8.2 is rotating about the vertical axis with a constant angular speed of Ω rads/s. The mass m is attached to the end of a light cantilever strip which has a moment of inertia I for flexure in the plane of the paper but is rigid in the plane of rotation. Obtain the natural frequency of vibration of mass m.

Solution

When the system is at rest, the gravity force on mass m is balanced by the spring force exerted by the cantilever strip. If the displacement of the mass m is measured from its position at rest, we may neglect the gravity force in our formulation. The only other force acting on mass m is the centripetal force given by $m\Omega^2 R$, where R is the radial distance of mass m from the axis of

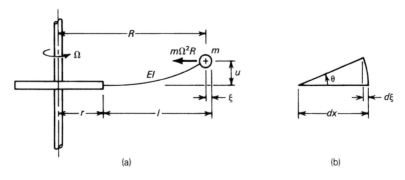

Figure E8.2. Wheel with an attached mass rotating at constant angular speed.

rotation. Since a central force, defined as a force that is always directed toward or away from a fixed point, is conservative, the total energy in the system is conserved. The kinetic energy of the system is given by

$$T = \frac{1}{2}m(\dot{u})^2 + C \tag{a}$$

where C is a constant that will account for the kinetic energy of rotation of the mass as well as the flywheel.

The potential energy consists of the strain energy of the spring and the work done by the centripetal force. To obtain the former, we determine an equivalent force P which when applied at the end of the cantilever will cause a tip deflection of u

$$\frac{Pl^3}{3EI} = u \tag{b}$$

The strain energy U is equal to the work done by the average force $\frac{1}{2}P$ in undergoing deflection u

$$U = \frac{1}{2}Pu = \frac{3EI}{2l^3}u^2 \tag{c}$$

To obtain the work done by the centripetal force, we first assume that for small displacements, the force is constant and is given by

$$F = m\Omega^2(l + r) \tag{d}$$

We also need the displacement of mass m in the horizontal direction. Referring to Figure E8.2, the horizontal displacement over a small distance dx is given by

$$d\xi = dx(1 - \cos\theta)$$

$$= \frac{1}{2}\theta^2\,dx \tag{e}$$

$$= \frac{1}{2}(u')^2\,dx$$

The total displacement, obtained by integrating Equation e, is

$$\xi = \int_0^l \frac{1}{2}(u')^2\,dx \tag{f}$$

Now, for a tip force P

$$u' = \frac{P}{EI}\left(lx - \frac{x^2}{2}\right) \tag{g}$$

Substitution of Equation g into Equation f gives

$$\xi = \frac{P^2 l^5}{15(EI)^2} \tag{h}$$

On using Equation b, ξ can be expressed in terms of u as

$$\xi = \frac{3}{5}\frac{u^2}{l} \tag{i}$$

The work done by the centripetal force is

$$W = m\Omega^2(l+r)\frac{3}{5}\frac{u^2}{l} \tag{j}$$

and the total potential energy becomes

$$V = \frac{3}{2}\frac{EI}{l^3}u^2 + \frac{3}{5}m\Omega^2\frac{u^2}{l}(l+r) \tag{k}$$

Noting that for simple harmonic motion energy is conserved, the following expression is obtained for frequency ω

$$\omega^2 = \frac{\frac{3}{2}(EI/l^3) + \frac{3}{5}(m\Omega^2/l)(l+r)}{m/2}$$

or

$$\omega^2 = \omega_0^2\left\{1 + \frac{6}{5}\left(\frac{\Omega}{\omega_0}\right)^2\left(1 + \frac{r}{l}\right)\right\} \tag{l}$$

where $\omega_0^2 = 3EI/ml^3$.

8.3 APPLICATION OF RAYLEIGH METHOD TO MULTI-DEGREE-OF-FREEDOM SYSTEMS

As noted earlier, the main application of the Rayleigh method is to the determination of an approximate value of the fundamental frequency of vibration of a multi-degree-of-freedom system. The method relies on the estimation of a vibration shape for the system so that the latter is converted to an equivalent single-degree-of-freedom system. The frequency of the equivalent system is then obtained by applying the principle of conservation of energy.

For an illustration of the procedure, consider the system shown in Figure 8.2, consisting of a mass M attached to a uniform spring having stiffness k and total mass m. Because the spring possesses a mass, the system has, in fact, an infinite number of degrees of freedom. However, if we make the assumption

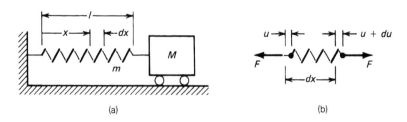

(a)   (b)

Figure 8.2. Vibration of mass M attached to a spring having mass m.

that the displacement of the spring–mass system is given by

$$u(x,t) = z(t)\psi(x) \tag{8.11}$$

where $z(t)$ is a generalized coordinate, considered as the unknown, and $\psi(x)$ is a shape function that must be selected, the system reduces to a single-degree-of-freedom system, because once $z(t)$ is determined, the displaced shape of the entire system is known.

By definition, the spring constant k is the force required to stretch the spring of length l by a unit distance. Thus, the force required to stretch an element of length dx by a unit distance will be lk/dx. Noting that the extension of the element is du, the force F acting on it is given by

$$F = \frac{l}{dx}k\,du \tag{8.12}$$

The strain energy stored in the element is

$$dU = \frac{1}{2}F\,du = \frac{1}{2}kl\frac{du}{dx}\,du$$
$$= \frac{1}{2}kl\left(\frac{du}{dx}\right)^2 dx \tag{8.13}$$

The total strain energy is given by

$$U = \frac{1}{2}kl\int_0^l\left(\frac{du}{dx}\right)^2 dx$$
$$= \frac{1}{2}klz^2\int_0^l(\psi')^2\,dx \tag{8.14}$$

The total kinetic energy of the spring and the attached mass is readily found to be

$$T = \int_0^l \frac{1}{2}\bar{m}(x)\{\dot{u}(x,t)\}^2\,dx + \frac{1}{2}M\{\dot{u}(l)\}^2$$
$$= \frac{1}{2}\frac{m}{l}(\dot{z})^2\int_0^l\{\psi(x)\}^2\,dx + \frac{1}{2}M(\dot{z})^2\{\psi(l)\}^2 \tag{8.15}$$

where $\bar{m}$ is the mass per unit length of the spring. Applying the principle of conservation of energy and assuming that $z = A\sin\omega t$, we obtain

$$\omega^2 = \frac{\frac{1}{2}kl\int_0^l\{\psi'(x)\}^2\,dx}{\frac{1}{2}(m/l)\int_0^l\{\psi(x)\}^2\,dx + \frac{1}{2}M\{\psi(l)\}^2} \tag{8.16}$$

If the displacement of the spring is assumed to vary linearly, $\psi(x)=x/l$. The selected shape function ψ satisfies the boundary condition of zero displacement at the fixed end, $x=0$. The potential and kinetic energies are obtained from Equations 8.14 and 8.15, respectively

$$U = \frac{1}{2}kz^2 \tag{8.17a}$$

and

$$T = \frac{1}{2}\left(M + \frac{m}{3}\right)\dot{z}^2 \tag{8.17b}$$

Assuming simple harmonic motion and equating the maximum values of U and T, we get

$$\omega = \sqrt{\frac{k}{M + m/3}} \tag{8.18}$$

Equation 8.18 shows that the frequency of the system of Figure 8.2 with a spring of mass m is approximately equal to that of an equivalent system with a massless spring and an attached mass that is equal to M plus one-third the mass of the spring. When m is zero, Equation 8.18 reduces to $\omega = \sqrt{k/M}$, which is the true frequency of a system with massless spring. This is because the linear displacement shape is, in fact, the true vibration shape of such a system.

As an alternative to the linear displacement shape, let us assume that $\psi = \sin(\pi x/2l)$. Note that, with this choice, too, $u=0$ at $x=0$ so that the boundary condition at the fixed end is satisfied. Substitution into Equations 8.14 and 8.15 gives

$$U = \frac{1}{2}klz^2\frac{\pi^2}{4l^2}\int_0^l \cos^2\frac{\pi x}{2l}\,dx$$

$$= \frac{1}{2}\frac{\pi^2}{8}kz^2 \tag{8.19}$$

and

$$T = \frac{1}{2}\frac{m}{l}(\dot{z})^2\int_0^l \sin^2\frac{\pi x}{2l}\,dx + \frac{1}{2}M(\dot{z})^2$$

$$= \frac{1}{2}\frac{m}{2}(\dot{z})^2 + \frac{1}{2}M(\dot{z})^2 \tag{8.20}$$

The frequency ω is now given by

$$\omega = \frac{\pi}{\sqrt{8}}\sqrt{\frac{k}{M + \frac{1}{2}m}} \tag{8.21}$$

If the spring–mass m is zero, Equation 8.21 will give $\omega = \pi/\sqrt{8}\sqrt{k/M}$. When compared to the true frequency of $\sqrt{k/M}$, this value is higher by about 11%. On the other hand, if the attached mass M is zero, the frequency works out to $\omega = \pi/2\sqrt{k/m}$, which is exactly equal to the true frequency of vibration of a uniform spring. This is because sine function $\psi = \sin(\pi x/2l)$ happens to be the true vibration shape for that case.

The foregoing example shows that the accuracy of the computed frequency depends on how closely the estimated vibration shape resembles the true vibration shape. Criteria for the selection of a vibration shape are discussed in a later section. The example also shows that even with a crude estimate of the vibration shape, the error in the calculated frequency is not too large. The application of the Rayleigh method to discrete multi-degree-of-freedom systems follows lines similar to those described in the foregoing paragraphs. This is illustrated in the following example.

Example 8.3
The three-story building frame shown in Figure E8.3 is modeled by assuming that all the mass is lumped at floor levels and that the columns are axially rigid. It is further assumed that the floor beams are infinitely rigid. This last assumption gives a model that is commonly referred to as a shear-type building model. The interstory stiffness of such a model can be easily obtained from known flexural rigidities of the columns. The story stiffness as well as the floor masses are shown in the figure. Determine the fundamental frequency of the building frame.

Solution
The building model is, in fact, a three-degree-of-freedom system because the three displacement coordinates u_1, u_2, and u_3 must be specified to determine the configuration of the frame. To convert the model to an equivalent single-degree-of-freedom system, we assume a suitable

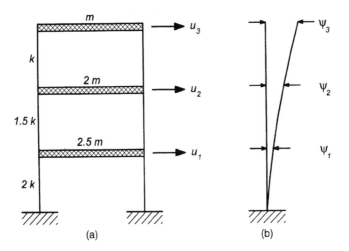

Figure E8.3. Lateral vibrations of a shear building: (a) building model; (b) displaced shape, ψ.

displacement shape ψ, where in this case, ψ is a vector with three elements. In matrix notation, we write

$$\mathbf{u} = z(t)\psi \tag{a}$$

in which $z(t)$ is the unknown generalized coordinate. The potential energy of the system is best expressed in matrix notation as

$$U = \frac{1}{2}\mathbf{u}^T\mathbf{K}\mathbf{u}$$

$$= \frac{1}{2}\{z(t)\}^2\psi^T\mathbf{K}\psi \tag{b}$$

where $\mathbf{K}$ is the stiffness matrix of the frame given by

$$\mathbf{K} = k\begin{bmatrix} 3.5 & -1.5 & 0 \\ -1.5 & 2.5 & -1 \\ 0 & -1 & 1 \end{bmatrix} \tag{c}$$

The kinetic energy of the frame is obtained from

$$T = \frac{1}{2}\dot{\mathbf{u}}^T\mathbf{M}\dot{\mathbf{u}}$$

$$= \frac{1}{2}\{\dot{z}(t)\}^2\psi^T\mathbf{M}\psi \tag{d}$$

where $\mathbf{M}$ is the mass matrix of the frame given by

$$\mathbf{M} = m\begin{bmatrix} 2.5 & & \\ & 2 & \\ & & 1 \end{bmatrix} \tag{e}$$

Let us now assume that the vibration shape is

$$\psi = \begin{bmatrix} 1 \\ 2 \\ 3 \end{bmatrix} \tag{f}$$

Substitution of Equations c and f into Equation b gives

$$U = \frac{1}{2}z^2 4.5k \tag{g}$$

Also, from Equations d, e, and f we get

$$T = \frac{1}{2}\dot{z}19.5m \tag{h}$$

Assuming that the motion is simple harmonic and that energy is conserved, we get the following value of frequency from Equations g and h

$$\omega = 0.48\sqrt{\frac{k}{m}} \tag{i}$$

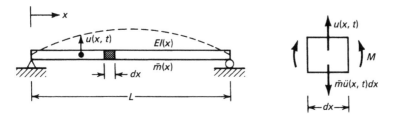

Figure 8.3. Flexural vibrations of a beam.

8.3.1   *Flexural vibrations of a beam*

Consider the lateral vibrations of a simply supported nonuniform beam. The beam shown in Figure 8.3 has a mass of $\bar{m}(x)$ per unit length and flexural rigidity $EI(x)$, both functions of coordinate x. The system represented by the beam has an infinite number of degrees of freedom. However, if we can make an estimate of the vibration shape, the system can be reduced to an equivalent single-degree-of-freedom system.

Let us assume that deflection $u(x,t)$ is given by

$$u(x,t) = z(t)\psi(x) \tag{8.22}$$

where $z(t)$ is the unknown coordinate and $\psi(x)$ is the selected deflection shape. Standard textbooks on the mechanics of materials show that strain energy of the beam due to flexural deformations is given by

$$U = \frac{1}{2}z^2 \int_0^L EI(x)\{\psi''(x)\}^2 \, dx \tag{8.23}$$

The kinetic energy is obtained from

$$T = \frac{1}{2}\dot{z}^2 \int_0^L \bar{m}(x)\{\psi(x)\}^2 \, dx \tag{8.24}$$

The free undamped vibrations of the beam can be assumed as harmonic, so that $z = A \sin \omega t$. By substituting this value of z in Equations 8.23 and 8.24, we obtain the following expressions for the maximum potential and kinetic energies

$$U_{\max} = \frac{1}{2}A^2 \int_0^L EI(x)\{\psi''(x)\}^2 \, dx \tag{8.25}$$

$$T_{\max} = \frac{1}{2}A^2\omega^2 \int_0^L \bar{m}(x)\{\psi(x)\}^2 \, dx \tag{8.26}$$

Applying the principle of conservation of energy, we get

$$T_{\max} = U_{\max} \tag{8.27}$$

or

$$\omega^2 = \frac{\int_0^L EI(x)\{\psi''(x)\}^2\, dx}{\int_0^L \bar{m}(x)\{\psi(x)\}^2\, dx} \tag{8.28}$$

Referring to Section 2.7.4, we note that the numerator in Equation 8.28 is identical to the generalized stiffness k^*, while the denominator is equal to the generalized mass m^*, so that

$$\omega^2 = \frac{k^*}{m^*} \tag{8.29}$$

It is observed that Rayleigh's method gives the same result for the frequency as a formulation based on the concept of generalized coordinates. This is to be expected because the formulation is similar in both cases; $z(t)$ in the Rayleigh method, being, in effect, a generalized coordinate. The only difference in the two approaches is that in the Rayleigh method, we have used the principle of conservation of energy, while in the generalized coordinate approach, the virtual work principle was used.

As a specific example, let us assume that the beam is uniform and that the shape function is $\psi = \sin(\pi x/L)$. Note that ψ is zero for both $x = 0$ and $x = L$. The assumed vibration shape therefore satisfies the two geometric boundary conditions. Substitution for ψ in Equations 8.23 and 8.24 gives

$$U = \frac{1}{4} z^2 EI \frac{\pi^4}{L^3} \tag{8.30}$$

$$T = \frac{1}{2} (\dot{z})^2 \bar{m} L \tag{8.31}$$

Assuming simple harmonic motion, and equating the maximum potential energy to the maximum kinetic energy, we get

$$\omega = \pi^2 \sqrt{\frac{EI}{mL^3}} \tag{8.32}$$

where $m = \bar{m}L$ is the total mass of the beam. The frequency estimate given by Equation 8.32 is, in fact, exact because the selected shape function happens to be the true vibration shape of a uniform beam.

Example 8.4

Find the fundamental frequency of the simply supported uniform beam shown in Figure E8.4a. Assume that the vibration shape of the beam is the same as the deflection shape assumed by the

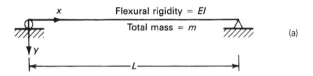

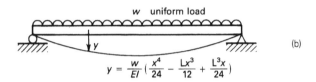

$$y = \frac{w}{EI}\left(\frac{x^4}{24} - \frac{Lx^3}{12} + \frac{L^3x}{24}\right)$$

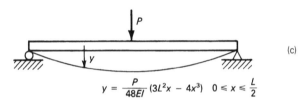

$$y = \frac{P}{48EI}(3L^2x - 4x^3) \quad 0 \le x \le \frac{L}{2}$$

Figure E8.4. Flexural vibrations of a uniform beam.

beam under (i) a uniformly distributed load (Fig. E8.4b), and (ii) a central concentrated load (Fig. E8.4c).

Solution

(i) The deflected shape of the beam under a uniformly distributed load of w per unit length is given by

$$y = \frac{w}{EI}\left(\frac{x^4}{24} - \frac{Lx^3}{12} + \frac{L^3x}{24}\right) \tag{a}$$

We now assume that the lateral deflection of the vibrating beam is given by

$$u(x,t) = z(t)\psi(x) \tag{b}$$

where

$$\psi(x) = \frac{1}{EI}\left(\frac{x^4}{24} - \frac{Lx^3}{12} + \frac{L^3x}{24}\right) \tag{c}$$

and the arbitrary load w has been included in z. Substitution into Equations 8.23 and 8.24 gives

$$U = \frac{1}{2}z^2\frac{1}{EI}\frac{L^5}{120} \tag{d}$$

$$T = \frac{1}{2}\dot{z}^2\left(\frac{1}{EI}\right)^2\frac{31L^9}{362,880} \tag{e}$$

As an alternative, the strain energy can be obtained by recognizing that it is equal to the work done by the applied force, so that

$$U = \frac{1}{2} \int_0^L z \cdot z\psi(x)\, dx$$

$$= \frac{1}{2}z^2 \frac{L^5}{120EI}$$

(f)

The vibration frequency is obtained from Equation 8.28

$$\omega = 9.8767 \sqrt{\frac{EI}{mL^3}}$$

(g)

where m is the total mass of the beam. In comparison with the exact frequency given by Equation 8.32, this is only 0.07% too high.

(ii) The deflected shape of the beam under a central concentrated load P is given by

$$y = \frac{P}{48EI}(3L^2 x - 4x^3) \quad \text{for } 0 \le x \le L/2$$

(h)

The deflections are symmetrical about the center line. Again, we assume that the lateral deflections are given by Equation b, where

$$\psi(x) = \frac{1}{48EI}(3L^2 x - 4x^3)$$

(i)

and the arbitrary load P has been included in z. The strain energy and the kinetic energy, respectively, are given by

$$U = \frac{1}{2}z^2 2 \int_0^{L/2} EI\{\psi''(x)\}^2\, dx$$

$$= \frac{1}{2}z^2 \frac{L^3}{48EI}$$

(j)

and

$$T = \frac{1}{2}z^2 2 \int_0^{L/2} \bar{m}\{\psi(x)\}^2\, dx$$

$$= \frac{1}{2}z^2 \frac{\bar{m}L^7}{(48EI)^2}\frac{17}{35}$$

(k)

Applying Rayleigh's principle, we obtain

$$\omega = 9.941 \sqrt{\frac{EI}{mL^3}}$$

(l)

The frequency given by Equation l is 0.72% too high.

Note that the strain energy could be obtained by computing the work done by the external force, so that

$$U = \frac{1}{2}z^2\psi\left(\frac{L}{2}\right)$$

$$= \frac{1}{2}\frac{z^2}{EI}\frac{L^3}{48}$$

(m)

8.4 IMPROVED RAYLEIGH METHOD

The examples presented in the previous sections show that the frequencies obtained by assuming a vibration shape which is not the same as the true shape are always higher than the exact frequency. On physical considerations, this can be explained by the fact that external constraints must be applied to the system if it has to be forced to vibrate in a shape that is different from its true vibration shape. These constraints make the system stiffer and hence increase its frequency. Later, we shall provide more rigorous mathematical proofs for the foregoing statement, both for discrete and continuous systems. The statement also implies that if by using the Rayleigh method, more than one estimate is obtained for the frequency of a system, the lowest of these estimated values is closest to the true frequency.

It is also obvious that in the application of Rayleigh method, the closer the assumed vibration shape is to the true vibration shape, the better is the estimate of the frequency. We can systematically improve the assumption of vibration shape by recognizing that the dynamic deflections result from the application of inertia forces. For example, consider the free vibration equation of a discrete multi-degree-of-freedom system which can be expressed in the alternative form

$$\mathbf{M\ddot{u} + Ku} = 0 \tag{8.33}$$

$$\mathbf{u} = -\mathbf{K}^{-1}\mathbf{M\ddot{u}} \tag{8.34}$$

Equation 8.34 expresses the fact that the displacement $\mathbf{u}$ results from the application of inertia force $\mathbf{M\ddot{u}}$. If $\mathbf{u}$ is assumed to be equal to $A\sin(\omega t + \phi)\boldsymbol{\psi}$, Equation 8.34 reduces to

$$\boldsymbol{\psi} = \omega^2\mathbf{K}^{-1}\mathbf{M\psi} \tag{8.35}$$

Equation 8.35 holds if $\boldsymbol{\psi}$ is the true vibration shape. However, if instead of $\boldsymbol{\psi}$, we use an approximate vibration shape $\boldsymbol{\psi}_0$, we cannot expect Equation 8.35 to hold and the product on the right-hand side would give a new vector $\bar{\boldsymbol{\psi}}_1$ which will be different from $\boldsymbol{\psi}_0$, the initial assumption of the vibration shape.

$$\bar{\boldsymbol{\psi}}_1 = \omega^2\mathbf{K}^{-1}\mathbf{M\psi}_0 \tag{8.36}$$

Intuitively, we can reason that $\bar{\boldsymbol{\psi}}_1$ should be closer to the true vibration shape than $\boldsymbol{\psi}_0$ is, so that if we use $\bar{\boldsymbol{\psi}}_1$ in the Rayleigh method, we should obtain a better estimate of the frequency. The foregoing reasoning forms the basis of an improvement in the Rayleigh method. In practice, since ω is not known, we use inertia forces given by $\mathbf{M\psi}_0$ rather than $\omega^2\mathbf{M\psi}_0$. The resulting vibration shape, $\boldsymbol{\psi}_1 = \mathbf{K}^{-1}\mathbf{M\psi}_0$, is proportional to $\bar{\boldsymbol{\psi}}_1$. Thus

$$\bar{\boldsymbol{\psi}}_1 = \omega^2\boldsymbol{\psi}_1 \tag{8.37}$$

Using $\bar{\boldsymbol{\psi}}_1$ to obtain the strain energy, we get

$$
\begin{aligned}
(U)_{\max} &= \frac{1}{2}A^2\bar{\boldsymbol{\psi}}_1^T\mathbf{K}\bar{\boldsymbol{\psi}}_1 \\
&= \frac{1}{2}A^2\omega^4\boldsymbol{\psi}_1^T\mathbf{K}\boldsymbol{\psi}_1 \\
&= \frac{1}{2}A^2\omega^4\boldsymbol{\psi}_1^T\mathbf{K}\mathbf{K}^{-1}\mathbf{M}\boldsymbol{\psi}_0 \\
&= \frac{1}{2}A^2\omega^4\boldsymbol{\psi}_1^T\mathbf{M}\boldsymbol{\psi}_0
\end{aligned}
\tag{8.38}
$$

Equating this to the maximum kinetic energy given by $T_{\max} = \frac{1}{2}A^2\omega^2\boldsymbol{\psi}_0^T\mathbf{M}\boldsymbol{\psi}_0$, we get

$$
\omega^2 = \frac{\boldsymbol{\psi}_0^T\mathbf{M}\boldsymbol{\psi}_0}{\boldsymbol{\psi}_1^T\mathbf{M}\boldsymbol{\psi}_0}
\tag{8.39}
$$

We may, however, get a further improvement in the frequency if we use $\bar{\boldsymbol{\psi}}_1$ instead of $\boldsymbol{\psi}_0$ in calculating the maximum kinetic energy. Thus

$$
\begin{aligned}
T_{\max} &= \frac{1}{2}A^2\omega^2\bar{\boldsymbol{\psi}}_1^T\mathbf{M}\bar{\boldsymbol{\psi}}_1 \\
&= \frac{1}{2}A^2\omega^6\boldsymbol{\psi}_1^T\mathbf{M}\boldsymbol{\psi}_1
\end{aligned}
\tag{8.40}
$$

Equations 8.38 and 8.40 give

$$
\omega^2 = \frac{\boldsymbol{\psi}_1^T\mathbf{M}\boldsymbol{\psi}_0}{\boldsymbol{\psi}_1^T\mathbf{M}\boldsymbol{\psi}_1}
\tag{8.41}
$$

The estimated value can be improved even further by using $\boldsymbol{\psi}_1$ to obtain a new vibration shape $\boldsymbol{\psi}_2$. If the process is repeated a sufficient number of times, $\boldsymbol{\psi}_n$ will eventually converge to the true vibration shape and the frequency obtained by using $\boldsymbol{\psi}_n$ will be the exact frequency. We will provide a formal proof of this statement in a later chapter. However, because the advantage of Rayleigh method lies in obtaining a quick but reasonably accurate estimate of the frequency, the extra work involved in carrying the improvement too far is not justified.

The reasoning of the foregoing paragraphs applies equally well to a continuous system. Assume that the vibration shape of the flexural beam, $\psi(x)$, is the same as the deflection shape produced by an external load $p(x)$. Deflection $u(x, t) = z\psi(x)$ will therefore be caused by a force $zp(x)$ and the strain energy stored in the beam will be given by

$$U = \frac{1}{2} \int_0^L z\,p(x)z\psi(x)\,dx$$
$$= \frac{1}{2}z^2 \int_0^L p(x)\psi(x)\,dx$$

(8.42)

Correspondingly, the maximum strain energy will be

$$(U)_{\max} = \frac{1}{2}A^2 \int_0^L p(x)\psi(x)\,dx$$

(8.43)

On equating this to the maximum kinetic energy given by Equation 8.26, we obtain

$$p(x) = \omega^2 \bar{m}(x)\psi(x)$$

(8.44)

Thus, an applied load $p(x)$ equal to the inertia force produces a deflected shape that is identical to the vibration shape of the beam. The statement will be true provided that $\psi(x)$ is the true vibration shape. On the other hand, if the assumed vibration shape denoted by $\psi_0(x)$ in not the same as the true vibration shape, the deflection produced by force $\omega^2 \bar{m}(x)\psi_0(x)$ will not be $\psi_0(x)$ but slightly different from it. Let us represent such a deflected shape by $\bar{\psi}_1(x)$. Again, intuitively we can state that $\bar{\psi}_1(x)$ will be closer to the true vibration shape than $\psi_0(x)$ is. We cannot yet derive $\bar{\psi}_1(x)$ from $\psi_0(x)$ because ω^2 in Equation 8.44 is unknown. We therefore use $\psi_1(x)$, the deflected shape resulting from the application of load $\bar{m}(x)\psi_0(x)$. Shapes $\psi_1(x)$ and $\bar{\psi}_1(x)$ are related by the equation

$$\bar{\psi}_1(x) = \omega^2 \psi_1(x)$$

(8.45)

and the strain energy expression obtained from Equation 8.43 becomes

$$(U)_{\max} = \frac{1}{2}A^2 \int_0^L \omega^2 \bar{m}(x)\psi_0(x)\bar{\psi}_1(x)\,dx$$
$$= \frac{1}{2}A^2\omega^4 \int_0^L \bar{m}(x)\psi_0(x)\psi_1(x)\,dx$$

(8.46)

Equating the maximum strain energy, Equation 8.46, to the maximum kinetic energy given by Equation 8.26 with $\psi_0(x)$ substituted in place of $\psi(x)$, we get

$$\omega^2 = \frac{\int_0^L \bar{m}(x)\{\psi_0(x)\}^2\,dx}{\int_0^L \bar{m}(x)\psi_0(x)\psi_1(x)\,dx}$$

(8.47)

We can, however, obtain a further improvement by using $\bar{\psi}_1(x)$ instead of $\psi_0(x)$ in the kinetic energy expression Equation 8.26, so that

$$T_{max} = \frac{1}{2}A^2\omega^6 \int_0^L \bar{m}(x)\psi_1^2(x)\,dx \tag{8.48}$$

On equating T_{max} from Equation 8.48 to $(U)_{max}$ from Equation 8.46, we get

$$\omega^2 = \frac{\int_0^L \bar{m}(x)\psi_0(x)\psi_1(x)\,dx}{\int_0^L \bar{m}(x)\{\psi_1(x)\}^2\,dx} \tag{8.49}$$

As in the case of a discrete system, the accuracy of the Rayleigh method can be further improved by using $\psi_1(x)$ to derive a new shape $\psi_2(x)$, which can be used to give a more accurate estimate of the frequency. However, the extra computations involved in the process are hardly justified.

Example 8.5
For the frame shown in Example 8.3, obtain improved estimates of the fundamental frequency.

Solution
The initial estimate of the displaced shape is given by

$$\psi_0^T = \begin{bmatrix} 1 & 2 & 3 \end{bmatrix} \tag{a}$$

The inertia forces $M\psi_0$ are applied to the frame as shown in Figure E8.5. The resulting displacements are easily computed and are also shown in the figure. We have

$$\psi_1^T = \frac{m}{k}\begin{bmatrix} 4.75 & 9.42 & 12.42 \end{bmatrix} \tag{b}$$

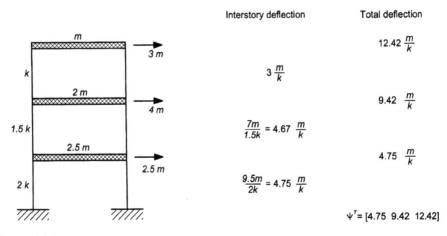

Figure E8.5. Frequency of vibration of a three-story shear frame.

Substituting for ψ_0 and ψ_1 in Equation 8.39, we obtain a more accurate estimate of the frequency

$$\omega^2 = 0.2246 \frac{k}{m}$$

$$\omega = 0.474 \sqrt{\frac{k}{m}}$$

A further improvement is obtained in the estimate of frequency, ω, by using Equation 8.41

$$\omega^2 = 0.2237 \frac{k}{m}$$

$$\omega = 0.473 \sqrt{\frac{k}{m}}$$

Example 8.6
Find the fundamental frequency of vibration of the cantilever beam shown in Figure E8.6a using the shape function $\psi_0(x) = (x/L)^2$. Obtain a more accurate estimate of the frequency by the improved Rayleigh method.

Solution
The strain energy of the cantilever is obtained from Equation 8.23

$$U = \frac{1}{2}z^2 \int_0^L EI \left(\frac{2}{L^2}\right)^2 dx$$

$$= \frac{1}{2}z^2 \frac{4EI}{L^3}$$

(a)

Equation 8.24 gives the kinetic energy

$$T = \frac{1}{2}\dot{z}^2 \int_0^L \bar{m} \left(\frac{x^2}{L^2}\right)^2 dx$$

$$= \frac{1}{2}\dot{z}^2 \frac{\bar{m}L}{5}$$

(b)

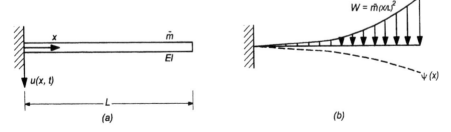

Figure E8.6. Vibration of a uniform cantilever beam.

Assuming simple harmonic motion and that the energy is conserved, we obtain the estimated frequency as

$$\omega^2 = 20 \frac{EI}{mL^3}$$

$$\omega = 4.472 \sqrt{\frac{EI}{mL^3}}$$

(c)

where m is the total mass of the beam. As compared to the exact frequency of $3.5159\sqrt{EI/mL^3}$ this is 27.2% too high.

To obtain a better estimate of the frequency, we apply inertia forces $\bar{m}\psi_0(x)$ on the cantilever beam as shown in Figure E8.6b and calculate the deflected shape under the load. Denoting the resulting deflection as $y = \psi_1(x)$, we have from the elementary beam theory

$$EI \frac{d^4 y}{dx^4} = \bar{m} \frac{x^2}{L^2}$$

(d)

Equation d is solved for y with the following two geometric and two natural boundary conditions

At $x = 0$: $y = 0$

At $x = 0$: $\dfrac{dy}{dx} = 0$

The two natural boundary conditions require that both the moment and the shear are zero at the free end. The condition that the moment equal to zero at the free end gives

$$x = L: \quad EI \frac{d^2 y}{dx^2} = 0$$

The condition that shear equal to zero at the free end gives

$$x = L: \quad EI \frac{d^3 y}{dx^3} = 0$$

On carrying out the necessary integration, we get the following value for y

$$y = \psi_1(x) = \frac{\bar{m}}{EI} \left(\frac{x^6}{360 L^2} - \frac{Lx^3}{18} + \frac{L^2 x^2}{8} \right)$$

(e)

Substitution of ψ_0 and ψ_1 in Equation 8.47 gives

$$\omega^2 = 12.46 \frac{EI}{mL^3}$$

$$\omega = 3.53 \sqrt{\frac{EI}{mL^3}}$$

(f)

The use of Equation 8.49 provides a further improvement in the value of ω, giving

$$\omega^2 = 12.365 \frac{EI}{mL^3}$$

$$\omega = 3.5164 \sqrt{\frac{EI}{mL^3}}$$

(g)

This is very close to the exact frequency of $3.5159\sqrt{\frac{EI}{mL^3}}$.

8.5 SELECTION OF AN APPROPRIATE VIBRATION SHAPE

Since the accuracy of the frequency estimate obtained by Rayleigh's method depends on the vibration shape selected, a discussion of the considerations that guide such selection is relevant. We have observed that the generalized coordinate method and the Rayleigh method are essentially the same; the criteria presented in Section 2.7.4 for the selection of a deflected shape therefore apply to the Rayleigh method as well.

Briefly, the selected displacement shape should satisfy all of the geometric or essential boundary conditions. Use of vibration shapes that violate one or more of such conditions may lead to estimates of frequency that are inaccurate and unreliable. Additional accuracy is obtained if the selected shapes satisfy some or all of the natural or force boundary conditions. Thus, referring to the vibration of a cantilever beam, the shape $\psi = (x/L)^2$ satisfies the two essential boundary conditions: namely, the deflection and the slope are both zero at the built in end, where $x = 0$. The deflection shape violates the condition that moment should be zero at the free end, but the condition that shear at the free end should be zero is satisfied. For the case of a simply supported beam, the deflection shapes

$$\psi = \frac{w}{EI}\left(\frac{x^4}{24} - \frac{Lx^3}{12} + \frac{L^3x}{24}\right)$$

and

$$\psi = \frac{P}{48EI}[3L^2x - 4x^3]$$

both satisfy all four boundary conditions: two on the deflections at supports and two on the moments at those locations.

The foregoing discussion suggests that a deflection shape produced by the application of a static load on the system should be an appropriate shape to use in the Rayleigh method, because such a shape will automatically satisfy all boundary conditions. The effectiveness of using a deflected shape produced by application of static load was demonstrated in Example 8.4. The choice has an additional advantage because the calculation of strain energy is simplified, such energy simply being equal to the work done by the static loads in riding through the deflections produced by them. This avoids the need to use the alternative energy expression of Equation 8.23, which involves the second derivative of the assumed shape function. Because the error involved in the use of derivatives of an approximate displacement shape is greater than in the use of the shape function itself, the direct method of calculating the strain energy is always more accurate than Equation 8.23. Although, as demonstrated in Example 8.5, any reasonable choice of a static load deflection shape is effective, usually the most appropriate choice is the deflection shape resulting from the application of a gravity load equal to that produced by the mass of the system. As an

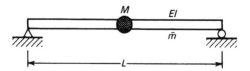

Figure 8.4. Vibrations of a beam with central mass.

illustration, consider the simply supported uniform beam with a central mass shown in Figure 8.4. It is required to find the frequency of lateral vibrations of the beam.

The deflected shape produced by uniform gravity load $\bar{m}g$ and a central load Mg is given by

$$\psi(x) = y_1(x) + y_2(x) \tag{8.50}$$

where $y_1(x)$ is the deflection produced by the uniform load $\bar{m}g$

$$y_1(x) = \frac{\bar{m}g}{24EI}(x^4 - 2Lx^3 + L^3x) \quad \text{for } 0 \leq x \leq L \tag{8.51}$$

and $y_2(x)$ is the deflection due to the central concentrated load Mg

$$y_2(x) = \frac{Mg}{48EI}(3L^2x - 4x^3) \quad \text{for } 0 \leq x \leq L/2 \tag{8.52}$$

The strain energy stored in the beam is given by

$$
\begin{aligned}
U_{\max} &= \frac{1}{2}\int_0^L \bar{m}g\,y_1(x)\,dx + \frac{1}{2}2\int_0^{L/2} \bar{m}g\,y_2(x)\,dx \\
&\quad + \frac{1}{2}Mg\{y_1(L/2) + y_2(L/2)\} \\
&= \frac{1}{2}\frac{M^2L^3g^2}{48EI}\left\{\frac{2}{5}\left(\frac{m}{M}\right)^2 + \frac{5}{4}\left(\frac{m}{M}\right) + 1\right\}
\end{aligned}
\tag{8.53}
$$

where $m = \bar{m}L$ is the total mass of the system. The maximum kinetic energy is given by

$$
\begin{aligned}
T_{\max} &= \frac{1}{2}\omega^2\left[2\int_0^{L/2}\bar{m}\{y_1(x) + y_2(x)\}^2\,dx + M\{y_1(L/2) + y_2(L/2)\}^2\right] \\
&= \frac{1}{2}\frac{\omega^2 M^3 L^6 g^2}{(48EI)^2}\left\{\frac{62}{315}\left(\frac{m}{M}\right)^3 + \frac{113}{112}\left(\frac{m}{M}\right)^2 + \frac{243}{140}\left(\frac{m}{M}\right) + 1\right\}
\end{aligned}
\tag{8.54}
$$

On equating U_{max} to T_{max}, we obtain the following expression for the frequency

$$\omega^2 = \frac{48EI}{ML^3} \frac{(2/5)\alpha^2 + (5/4)\alpha + 1}{(62/315)\alpha^3 + (113/112)\alpha^2 + (243/140)\alpha + 1} \qquad (8.55)$$

where $\alpha = m/M$.

Several points are worth noting. First, the gravity constant g drops out from the expression for ω and need not be included in the computations. This is to be expected, because a constant multiplier in the deflection shape does not affect the final results. Second, if the beam is massless, that is, if $\alpha = 0$, we get $\omega^2 = 48EI/ML^3$, which is the exact frequency for a uniform massless beam with a central mass M. The exact result is obtained because $y_2(x)$ is the true vibration shape in this case. Finally, if the central mass is 0, α tends to ∞ and Equation 8.55 gives in the limit

$$\omega^2 = 97.55 \frac{EI}{mL^3}$$
$$\omega = 9.8767 \sqrt{\frac{EI}{mL^3}} \qquad (8.56)$$

which, of course, is the same result as obtained in Example 8.4. It is of interest to note that the deflection shape obtained by applying gravity loading produced by the mass in the system is, in fact, the same as the shape ψ_1 obtained by the procedure described in the preceding section when ψ_0 is taken to be a constant equal to 1. This is because with $\psi_0 = 1$, the inertia force $m\psi_0$ is proportional to the gravity load mg.

When the equivalence between the procedure for obtaining improved shape ψ_1, given $\psi_0 = 1$, and that for obtaining a deflection shape from mass-produced gravity loads is recognized, it is easy to appreciate that the direction of gravity may have to be adjusted to match the deflected shape with the expected vibration shape. Consider, for example, the beam with overhangs shown in Figure 8.5. The overhangs are expected to move in a direction opposite to that of the span section of the beam. The first estimate of the vibration shape, ψ_0, should therefore be as indicated in Figure 8.5b. Similarly, the direction of gravity loads should be as shown in Figure 8.5c, where the loads in the overhanging portions are acting in a direction opposite to that of forces within the span.

As a final example of the application of mass-produced gravity load, consider a discrete system composed of masses $M_1, M_2, \ldots, M_n$. Let the deflections at the location of the point masses due to gravity loads of magnitudes $M_1, M_2, \ldots, M_n$ be equal to $y_1, y_2, \ldots, y_n$. The shape function ψ to be used in the Rayleigh method is then given by

$$\psi^T = [y_1 \quad y_2 \quad \cdots \quad y_n] \qquad (8.57)$$

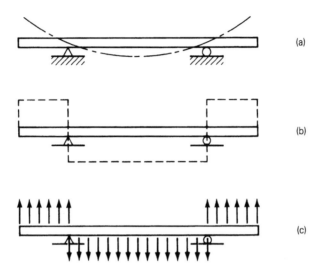

Figure 8.5. Vibrations of an overhanging beam: (a) anticipated vibration shape; (b) first estimate of vibration shape, ψ_0; (c) mass-produced gravity load.

The maximum strain energy is

$$U_{\max} = \frac{1}{2}[M_1 y_1 + M_2 y_2 + \cdots + M_n y_n] \tag{8.58}$$

The maximum kinetic energy is given by

$$T_{\max} = \frac{1}{2}\omega^2[M_1 y_1^2 + M_2 y_2^2 + \cdots + M_n y_n^2] \tag{8.59}$$

Equating $(U)_{\max}$ to $T_{\max}$ we obtain the frequency ω

$$\omega^2 = \frac{\sum M_i y_i}{\sum M_i y_i^2} \tag{8.60}$$

Note that we have ignored the gravity constant in our computations because it eventually cancels out.

Example 8.7
For the frame shown in Example 8.3, calculate the deflected shape resulting from the application of gravity forces equal to the story masses. Use this deflected shape to calculate the frequency of vibration.

Solution
The frame represents a discrete system with three concentrated masses and the corresponding three degrees of freedom. Since the vibrations are in the lateral direction, we apply the gravity forces so that they also act in the horizontal direction. The applied forces and the calculations of

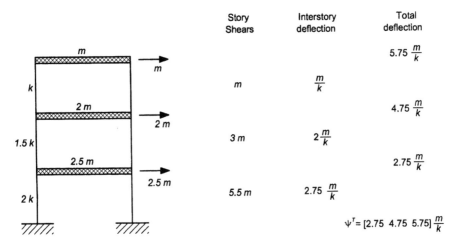

Figure E8.7. Lateral vibrations of a three-story shear frame.

resulting displacements are indicated in Figure E8.7. The assumed vibration shape then becomes

$$\boldsymbol{\psi}^T = \frac{m}{k}[2.75 \quad 4.75 \quad 5.75] \tag{a}$$

The maximum strain energy is given by

$$U_{max} = \frac{1}{2}\left(m \times 5.75\frac{m}{k} + 2m \times 4.75\frac{m}{k} + 2.5m \times 2.75\frac{m}{k}\right)$$

$$= \frac{1}{2} \times 22.125\frac{m^2}{k} \tag{b}$$

The kinetic energy is

$$T_{max} = \frac{1}{2}\omega^2\frac{m^3}{k^2} \times 97.094 \tag{c}$$

Equating U_{max} to T_{max} we get

$$\omega^2 = 0.2279\frac{k}{m}$$

$$\omega = 0.477\sqrt{\frac{k}{m}}$$

The gravity constant has not been included in the computations because it cancels out.

8.6 SYSTEMS WITH DISTRIBUTED MASS AND STIFFNESS: ANALYSIS OF INTERNAL FORCES

In Section 2.7.4, we discussed a method by which a system with distributed mass and stiffness can be represented by a single-degree-of-freedom system using generalized coordinate and an assumed shape function (Eq. 2.19). The resulting equation for forced damped vibrations works out to

$$m^* \ddot{z}(t) + c^* \dot{z}(t) + k^* z(t) = p^*(t) \tag{8.61}$$

where m^* is the generalized mass, c^* the generalized damping, k^* the generalized stiffness, and p^* the generalized force, all determined according to Section 2.7.4.

Consider for example the vibrations of the simply supported beam shown in Figure 8.3, and suppose that the equation of motion has been solved to obtain the value of generalized coordinate $z(t)$. The displacement of the beam can be obtained by using Equation 2.19. It is now required to determine the internal forces, namely the bending moments and shears. This can be achieved through a static analysis of the beam for a load that will produce the displaced shape given by $z(t)\psi(x)$. From ordinary beam theory, we obtain

$$w_1(x,t) = \frac{\delta^2}{\delta x^2}\left[EI(x)\frac{\delta^2 u}{\delta x^2}\right] = \frac{\delta^2}{\delta x^2}[EI(x)\psi''(x)]z(t) \tag{8.62}$$

The equivalent load involves fourth-order differential of the shape function. The shears and moments will, in turn, involve third and second order differentials of the shape function, respectively. The internal forces obtained from the equivalent load are thus less accurate than the displacement because derivatives of the approximate shape function provide poorer approximations than does the shape function itself. A better estimate of the internal forces can be obtained by using the following expression for the equivalent load.

$$w_2(x,t) = \omega^2 \bar{m}\psi(x)z(t) \tag{8.63}$$

Equation 8.63 does not involve derivatives of the shape function and is therefore expected to provide better estimates of the internal forces. The equivalent loads obtained from Equations 8.62 and 8.63 are identical only when the shape function $\psi(x)$ is one of the shapes in which the beam may execute free vibrations. A proof of this is provided in Chapter 15 (Eq. 15.18). A special shape of this kind is known as a mode shape of the beam. When $\psi(x)$ is not one of the mode shapes, loads w_1 and w_2 are not the same. However, the total strain energy stored in the beam as it deflects under the two different loads is the same.

An expression for the strain energy stored through deflection u produced by load w_1 is given by

$$U = \frac{1}{2}\int_0^L w_1(x,t)u(x,t)\,dx = \frac{1}{2}z(t)\int_0^L w_1(x,t)\psi(x)\,dx \tag{8.64}$$

The strain energy given by Equation 8.64 should be the same as that obtained by considering the work done by the flexural moments (Eq. 8.23). Thus

$$\frac{1}{2}z(t)\int_0^L w_1(x,t)\psi(x)\,dx = \frac{1}{2}z^2(t)\int_0^L EI(x)\{\psi''(x)\}^2\,dx \tag{8.65}$$

On substituting Equation 8.28 in Equation 8.65 we get

$$\frac{1}{2}z(t)\int_0^L w_1(x,t)\psi(x)\,dx = \frac{1}{2}z^2(t)\omega^2\int_0^L \bar{m}(x)\psi^2(x)\,dx$$

$$= \frac{1}{2}z(t)\int_0^L [\omega^2\bar{m}(x)\psi(x)z(t)]\psi(x)\,dx \tag{8.66}$$

On comparing the left- and right-hand sides of Equation 8.66, we conclude that an equivalent load $w_2(x,t) = \omega^2\bar{m}(x)\psi(x)z(t)$ produces the same strain energy as load $w_1(x,t)$.

Example 8.8

For the precast concrete T-beam of Example 6.1, calculate the maximum mid-span moment produced due to clapping and stamping by spectators during a sporting event. Use the shape function given in Example 6.1.

Solution

The following data has been supplied.

$$E = 3.7 \times 10^{10}\ \text{N/m}^2, \quad I = 6.327 \times 10^{-3}\ \text{m}^4, \quad \bar{m} = 567.3\ \text{kg/m}, \quad \xi = 0.03$$

$$k^* = 701.6\ \text{kN/m}, \quad m^* = 326.6\ \text{kg}, \quad p^* = 0.7137\sin\Omega t = p_0\sin\Omega t\ \text{kN}$$

$$\omega = 7.38\ \text{Hz}, \quad \Omega = 3.0\ \text{Hz}, \quad \beta = 0.4065$$

$$\psi(x) = \frac{x}{L} - 2\frac{x^3}{L^3} + \frac{x^4}{L^4}$$

Using the data supplied, we get

$$z(t) = \frac{p^*}{k^*}[(1 - \beta^2)^2 + (2\beta\xi)^2]^{-1/2}\sin(\Omega t - \phi)$$

$$= 1.218 \times 10^{-3}\sin(\Omega t - \phi)\ \text{m} \tag{a}$$

The equivalent load is given by

$$w_1(x,t) = EIz(t)\frac{d^4\psi}{dx^4}$$

$$= EIz(t)\frac{24}{L^4} = 0.365\sin(\Omega t - \phi)\ \text{kN/m} \tag{b}$$

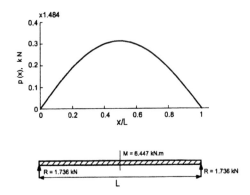

Figure E8.8. Dynamic distributed load on a precast concrete T-beam used to support stadium seating.

The maximum mid-span moment is given by

$$(M)_{L/2} = \frac{0.365L^2}{8} = 6.247 \text{ kN.m} \tag{c}$$

As an alternative, we may obtain the equivalent load from

$$w_2(x,t) = \omega^2 \bar{m} \psi(x) z(t)$$

$$= 46.35^2 \times 0.5673 \times 1.218 \times 10^{-3} \psi(x) \sin(\Omega t - \phi) \tag{d}$$

$$= 1.484 \psi(x) \sin(\Omega t - \phi) \text{ kN/m}$$

The maximum value of the distributed load represented by Equation d is shown in Figure E8.8. The end reaction works out to 1.736 kN and the maximum mid-span moment to 6.447 kN.m. In the present case, there is not a large difference between the two estimates of the mid-span moment because the assumed vibration shape is quite close to the first free-vibration mode shape of the beam.

8.7 NUMERICAL EVALUATION OF DUHAMEL'S INTEGRAL

As stated earlier, there may be instances when the response history of a dynamic system rather than its frequency is of interest. For a linear system, the response of the system to a specified exciting force is given by the Duhamel's equation (Eq. 8.2). Except when the exciting force is a simple mathematical function, a closed-form solution of Duhamel's integral is not possible, and a numerical method of evaluation must be used. In this section, we present three alternative procedures for the numerical evaluation of Duhamel's integral and illustrate the computations involved by several examples. For an undamped system, the

Duhamel's integral reduces to

$$u(t) = \frac{1}{m\omega} \int_0^t p(\tau) \sin \omega(t - \tau) \, d\tau$$

$$= \frac{1}{m\omega} \int_0^t p(\tau)(\sin \omega t \cos \omega \tau - \cos \omega t \sin \omega \tau) \, d\tau \qquad (8.67)$$

$$= A \sin \omega t - B \cos \omega t$$

where

$$A = \frac{1}{m\omega} \int_0^t p(\tau) \cos \omega \tau \, d\tau$$

$$\qquad (8.68)$$

$$B = \frac{1}{m\omega} \int_0^t p(\tau) \sin \omega \tau \, d\tau$$

The Duhamel's integral for a damped system can be reduced as follows

$$u(t) = \frac{1}{m\omega_d} e^{-\xi\omega t} \int_0^t e^{\xi\omega\tau} p(\tau)(\sin \omega_d t \cos \omega_d \tau - \cos \omega_d t \sin \omega_d \tau) \, d\tau$$

$$= A e^{-\xi\omega t} \sin \omega_d t - B e^{-\xi\omega t} \cos \omega_d t \qquad (8.69)$$

where

$$A = \frac{1}{m\omega_d} \int_0^t e^{\xi\omega\tau} p(\tau) \cos \omega_d \tau \, d\tau$$

$$\qquad (8.70)$$

$$B = \frac{1}{m\omega_d} \int_0^t e^{\xi\omega\tau} p(\tau) \sin \omega_d \tau \, d\tau$$

It is our objective to carry out a numerical evaluation of the integrals in Equations 8.68 and 8.70. To evaluate A or B, we denote the integrand by $f(\tau)$ and plot $y = f(\tau)$ as a function of τ as shown in Figure 8.6a. The required integral is equal to the area enclosed by the graph of $f(\tau)$ and the τ axis between 0 and t. Several alternative procedures are available for the numerical evaluation of the required areas; among them, the following are more commonly used:
1. Rectangular summation.
2. Trapezoidal method.
3. Simpson's method.

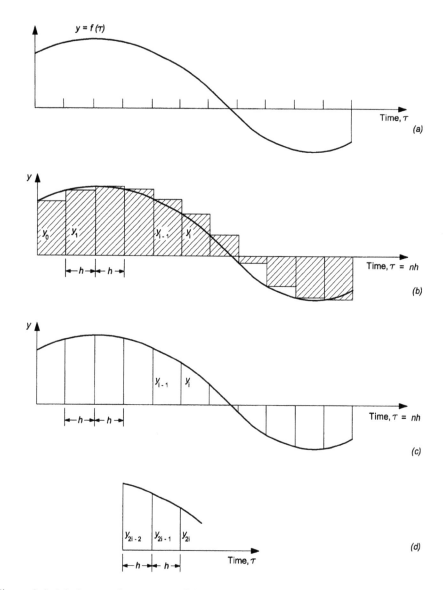

Figure 8.6. (a) Area under a curve; (b) rectangular summation (c) trapezoidal method; (d) Simpson's method.

8.7.1   *Rectangular summation*

To obtain the area enclosed by function $f(\tau)$, we divide the time axis from 0 to t into n small intervals of time, each of duration h. The area under the curve is now approximated by the sum of the areas of N rectangles shown shaded in Figure 8.6b. The ordinate of the first rectangle is y_0 and that of the *i*th is y_{i-1}. The area of the *i*th rectangle is $h y_{i-1}$. The total area is therefore given by

$$\int_0^t y \, dt = h(y_0 + y_1 + y_2 + \cdots + y_{N-1}) \tag{8.71}$$

Since the upper boundaries of the rectangles do not match the outline of the curve, the area obtained by the foregoing procedure is in error. In a general case, these errors are self-compensating, as is evident from Figure 8.6b, and provided that *h* is small enough and a large number of rectangles is used in the summation, reasonable accuracy will be obtained.

8.7.2 *Trapezoidal method*

As in the case of rectangular summation, the time axis is divided into *N* equal intervals each of duration *h*. The portion of the area bounded by the curve, ordinates y_{i-1} and y_i, and the τ axis is now approximated by a trapezium of base *h* and heights y_{i-1} and y_i. The area of the *i*th trapezium is $(h/2)(y_{i-1}+y_i)$. The total area is obtained by summing the *N* trapeziums

$$\int_0^t y \, d\tau = \frac{h}{2}(y_0 + 2y_1 + 2y_2 + \cdots + 2y_{N-1} + y_N) \tag{8.72}$$

It is obvious from a reference to Figure 8.6c that for the same value of *h* the trapezoidal method will, in general, give a more accurate estimate of the area under the curve $y = f(\tau)$ than rectangular summation would.

8.7.3 *Simpson's method*

The time axis is again divided into *N* equal intervals each of duration *h*. However, *N* in this case must be an even number. The portion of the curve connecting ordinates y_{2i-2}, y_{2i-1}, and y_{2i} is now approximated by a parabola, as shown in Figure 8.6d. It is easily seen that there are *N*/2 segments of the type shown in the figure and that the area of the *i*th segment is

$$\frac{h}{3}(y_{2i-2} + 4y_{2i-1} + y_{2i})$$

Summation of the areas of *N*/2 segments gives the total area

$$\int_0^t y \, d\tau = \frac{h}{3}(y_0 + 4y_1 + 2y_2 + 4y_3 + 2y_4 + \cdots$$

$$+ 2y_{N-2} + 4y_{N-1} + y_N) \tag{8.73}$$

Example 8.9

The water tower shown in Figure E8.9a is idealized as a single-degree-of-freedom system. It is subjected to the half-sine-wave loading shown in Figure E8.9b. Calculate the displacement history for the first 1.0 s using numerical evaluation of the Duhamel's integral and $h = 0.1$ s. Neglect damping and assume that the tower is initially at rest.

Solution

The mass m of the tower and the frequency ω are obtained first

$$m = \frac{978.8}{386.4} = 2.533 \text{ kip} \cdot \text{s}^2/\text{in.}$$

$$\omega = \sqrt{\frac{k}{m}} = \sqrt{\frac{100}{2.533}} = 6.283 \text{ rad/s}$$

The displacement response is obtained in terms of Duhamel's integral as

$$u(t) = A \sin \omega t - B \cos \omega t \tag{a}$$

where A and B have been defined in Equation 8.68. The response calculations for the first 0.6 s are shown in Tables E8.9a and E8.9b, where results are given for three alternative methods: rectangular summation, trapezoidal method, and the Simpson's method. After 0.6 s, A and B remain unchanged and the response for subsequent intervals of time is given by Equation a, with A and B equal to their values at 0.6 s. Because the forcing function is, in this case, a simple mathematical function, it is possible to obtain a closed-form solution. When the sine-wave function is expressed as

$$P(t) = p_0 \sin \Omega t \tag{b}$$

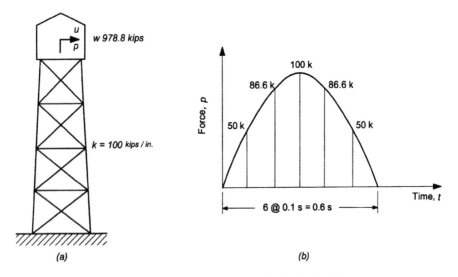

Figure E8.9. Forced vibration of a tower structure: (a) tower; (b) forcing function.

Table E8.9a. Numerical evaluation of Duhamel's equation, undamped system.

| τ (s) | $P(\tau)$ (kips) | $\cos \omega\tau$ | $p(\tau) \times \cos \omega\tau$ | $\bar{A}$ | | | $\sin \omega\tau$ | $p(\tau) \times \sin \omega\tau$ | $\bar{B}$ | | |
				Rectang sum	Trapezoidal method	Simpson's method			Rectang sum	Trapezoidal method	Simpson's method
(1)	(2)	(3)	(4)	(5)	(6)	(7)	(8)	(9)	(10)	(11)	(12)
0.0	0.0	1.000	0.00	0.00	0.0	0.0	0.000	0.00	0.0	0.0	0.0
0.1	50.0	0.809	40.45	0.00	40.5		0.588	29.39	0.0	29.4	
0.2	86.6	0.309	26.76	40.45	107.7	188.6	0.951	82.36	29.4	141.1	199.9
0.3	100.0	-0.309	-30.90	67.21	103.5		0.957	95.10	111.7	318.6	
0.4	86.6	-0.809	-70.06	36.32	2.6	21.7	0.588	50.92	206.8	464.6	713.6
0.5	50.0	-1.000	-50.00	-33.74	-117.5		0.000	0.00	257.8	515.5	
0.6	0.0	-0.809	0.00	-83.74	-167.5	-248.4	-0.588	0.00	257.8	515.5	764.5

$$\frac{\Delta\tau}{m\omega} = \frac{0.1}{2.533 \times 6.283} = 6.283 \times 10^{-3}$$

$$\frac{\Delta\tau}{2m\omega} = 3.142 \times 10^{-3}$$

$$\frac{\Delta\tau}{3m\omega} = 2.094 \times 10^{-3}$$

Table E8.9b. Numerical evaluation of Duhamel's equation, undamped system.

t	$\sin \omega t$	$\cos \omega t$	Rectangular sum			Trapezoidal method			Simpson's method		
			$A = \bar{A}\frac{\Delta\tau}{m\omega}$	$B = \bar{B}\frac{\Delta\tau}{m\omega}$	$A \sin \omega t - B \cos \omega t$	$A = \bar{A}\frac{\Delta\tau}{2m\omega}$	$B = \bar{B}\frac{\Delta\tau}{2m\omega}$	$A \sin \omega t - B \cos \omega t$	$A = \bar{A}\frac{\Delta\tau}{3m\omega}$	$B = \bar{B}\frac{\Delta\tau}{3m\omega}$	$A \sin \omega t - B \cos \omega t$
0.1	0.588	0.809	0.0000	0.0000	0.0000	0.1271	0.0923	0.0000			
0.2	0.951	0.309	0.2542	0.1847	0.1847	0.3382	0.4434	0.1846	0.3949	0.4187	0.2458
0.3	0.951	-0.309	0.4223	0.7019	0.6182	0.3252	1.0010	0.6184			
0.4	0.588	-0.809	0.2282	1.2994	1.1849	0.0080	1.4597	1.1856	0.0454	1.4946	1.2358
0.5	0.000	-1.000	-0.2434	1.6199	1.6199	-0.3692	1.6197	1.6197			
0.6	-0.588	-0.809	-0.5262	1.6199	1.6199	-0.5262	1.6197	1.6197	-0.5203	1.6013	1.6162

Table E8.9c. Comparison of numerical solution with theoretical results.

t	Duhamel integral $A \sin \omega t - B \cos \omega t$	Closed-form solution, Eq. d or e
0.1	0.0000	0.0333
0.2	0.1846	0.2411
0.3	0.6184	0.6788
0.4	1.1856	1.2290
0.5	1.6197	1.6334
0.6	1.6197	1.6000
0.7	1.0000	0.9880
0.8	0.0000	0.0000
0.9	−1.0000	−0.9900
1.0	−1.6197	−1.6006

the response for the first 0.6 s is given by

$$u(t) = \frac{p_0}{k} \frac{1}{1 - \beta^2} (\sin \Omega t - \beta \sin \omega t) \tag{c}$$

where $\beta = \Omega/\omega$. We have

$$\Omega = \frac{\pi}{0.6} = 5.236 \text{ rad/s}$$

$$\beta = \frac{\Omega}{\omega} = \frac{5.236}{6.283} = 0.833$$

$$p_0 = 100 \text{ kips}$$

so that

$$u(t) = 3.267(\sin 5.236t - 0.833 \sin 6.283t) \tag{d}$$

At $t = 0.6$ s

$$u(t) = 3.267(\sin 0.6 \times 5.236 - 0.833 \sin 0.6 \times 6.283)$$

$$= 1.6 \text{ in.}$$

$$\dot{u}(t) = 3.267(5.236 \cos 0.6 \times 5.236 - 0.833 \times 6.283 \cos 6.283 \times 0.6)$$

$$= -3.272 \text{ in./s}$$

The exact response subsequent to 0.6 s is given by

$$u(t) = 1.6 \cos \omega(t - 0.6) - \frac{3.272}{6.283} \sin(t - 0.6) \tag{e}$$

The theoretical response values for the first 1 s obtained from Equations d and e are compared with the results of the numerical evaluation of Duhamel's integral using the trapezoidal method (Table E8.9c).

Example 8.10

Repeat Example 8.9 with 10% damping in the system. Use the trapezoidal method.

Solution

The displacement response is now given by

$$u(t) = A \sin \omega_d t - B \cos \omega_d t \tag{a}$$

where A and B are as in Equation 8.70, and

$$\omega_d = \omega \sqrt{1 - \xi^2}$$

$$= 6.283 \sqrt{1 - 0.1^2}$$

$$= 6.2515 \text{ rad/s}$$

The response calculations for the first 0.6s are shown in Tables E8.10a and E8.10b. After 0.6s, A and B remain unchanged and the response for subsequent intervals of time is given by Equation a with A and B equal to their values at 0.6 s. The exact response to the sine-wave force for the first 0.6 s is given by

$$u(t) = \frac{p_0}{k} \frac{1}{(1 - \beta^2)^2 + (2\beta\xi)^2} \{(1 - \beta^2) \sin \Omega t - 2\beta\xi \cos \Omega t\}$$

Table E8.10a. Numerical evaluation of Duhamel's equation, damped system.

τ (s)	$p(\tau)$ (kips)	$e^{\xi\omega\tau}$	$\cos \omega_d \tau$	$p(\tau)e^{\xi\omega\tau} \cos \omega_d \tau$	$\bar{A}$ trapezoidal	$\sin \omega_d \tau$	$p(\tau)e^{\xi\omega\tau} \sin \omega_d \tau$	$\bar{B}$ trapezoidal
0.0	0.0	1.0000	1.0000	0.000	0.000	0.0000	0.000	0.000
0.1	50.0	1.0648	0.8109	43.172	43.172	0.5852	31.156	31.156
0.2	86.6	1.1339	0.3150	30.392	117.276	0.9491	93.198	155.510
0.3	100.0	1.2074	−0.3000	−36.222	111.986	0.9540	115.186	363.894
0.4	86.6	1.2857	−0.8015	−89.240	−13.476	0.5980	66.823	545.903
0.5	50.0	1.3691	−0.9999	−68.448	−171.164	0.0158	1.084	613.810
0.6	0.0	1.4579	−0.8200	0.000	−239.612	−0.5723	0.000	614.894

$$\frac{1}{m\omega_d} \frac{\Delta\tau}{2} = \frac{1 \times 0.1}{2.533 \times 6.2515 \times 2} = 3.1575 \times 10^{-3}$$

Table E8.10b. Numerical evaluation of Duhamel's equation, damped system.

t (1)	A (2)	B (3)	$\sin \omega_d t$ (4)	$\cos \omega_d t$ (5)	$e^{-\xi\omega t}$ (6)	$e^{-\xi\omega t}A \sin \omega_d t$ (7)	$e^{-\xi\omega t}B \cos \omega_d t$ (8)	7−8 (9)	Eq. b
0.1	0.1362	0.0983	0.5852	0.8109	0.9391	0.0749	0.0749	0.0000	0.0323
0.2	0.3700	0.4907	0.9491	0.3150	0.8819	0.3097	0.1363	0.1734	0.2254
0.3	0.3533	1.1481	0.9540	−0.3000	0.8282	0.2792	−0.2853	0.5645	0.6204
0.4	−0.0425	1.7224	0.5980	−0.8015	0.7778	−0.0198	−1.0737	1.0539	1.0961
0.5	−0.5401	1.9366	0.0158	−0.9999	0.7304	−0.0063	−1.4144	1.4081	1.4251
0.6	−0.7560	1.9400	−0.5723	−0.8200	0.6850	0.2968	−1.0912	1.3880	1.3772

$$+ \frac{p_0}{k} \frac{e^{-\omega \xi t}}{(1 - \beta^2)^2 + (2\beta\xi)^2} \left\{ 2\beta\xi \cos \omega_d t + \frac{\Omega}{\omega_d} \right. \tag{b}$$

$$\left. \left(1 + \beta^2 - 2\frac{\omega_d^2}{\omega^2} \right) \sin \omega_d t \right\}$$

where $p_0 = 100$ kips, $\Omega = 5.236$ rad/s, and $\beta = \Omega/\omega = 0.833$. The response values obtained from Equation b are also shown in Table E8.10b.

8.8   DIRECT INTEGRATION OF THE EQUATIONS OF MOTION

The direct integration of the equations of motion provides the response of the system at discrete intervals of time which are usually equally spaced. Determination of the response involves the computation of three basic parameters: displacement, velocity, and acceleration. The integration algorithms are based on appropriately selected expressions that relate the response parameters at a given interval of time to their values at one or more previous time points. In general, two independent expressions of this nature must be specified. The equation of motion written for the time interval under consideration provides the third expression necessary to determine the three unknown parameters.

Direct integration thus involves a marching along the time dimension. It is assumed that at the beginning of the integration, response parameter values have been specified or have been computed at one or more points preceding the time range of interest. These specified or computed values permit the marching scheme to be begun so that response can be computed at as many subsequent points as desired. The accuracy and stability of a scheme depend on the expressions selected for relating the response parameters at a time to their historic value, as well as on the magnitude of the time interval used in the computation.

In the following sections, we present several methods used in numerical integration of the equation of motion and discuss their relative accuracy as well as the conditions that must be met to guarantee stability of the solution. The following notations are used in the presentation: $h =$ time interval, $t_n = nh$, $u_n =$ displacement at time nh, $\dot{u}_n =$ velocity at time nh, $\ddot{u}_n =$ acceleration at time nh, and $p_n =$ the applied forces at time nh.

8.9   INTEGRATION BASED ON PIECE-WISE LINEAR REPRESENTATION OF THE EXCITATION

If the forcing function can reasonably be represented by a series of straight lines, it is possible to develop exact formulas for integration of the equation of motion of a linear system. Consider the time-dependent force $p(t)$ shown in

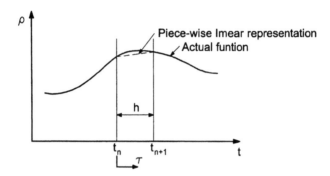

Figure 8.7. Piece-wise linear representation of force $p(t)$.

Figure 8.7. Over a sufficiently small interval of time h, the variation of force with time can be assumed to be linear as shown by the dashed line in Figure 8.7. With the origin located at time t_n, $p(\tau)$ is given by

$$p(\tau) = p_n + \frac{p_{n+1} - p_n}{h}\tau \qquad (8.74)$$

An exact solution of the equation of motion exists for the forcing function given by Equation 8.74.

Consider first an undamped system. The solution to Equation 8.74 consists of three components: 1. a free-vibration component with initial displacement u_n and initial velocity $\dot{u}_n$, 2. forced-vibrations induced by a constant force p_n, and 3. forced-vibrations induced by a linearly varying force $(p_{n+1} - p_n)\tau/h$. The free-vibration component is given by Equation 5.9

$$u_1(\tau) = u_n \cos \omega\tau + \frac{\dot{u}_n}{\omega}\sin \omega\tau \qquad (8.75)$$

The response to a constant force is obtained from Equation 7.14 with $\xi = 0$.

$$u_2(\tau) = \frac{p_n}{k}(1 - \cos \omega\tau) \qquad (8.76)$$

The response to linearly varying force is obtained from Equation 7.21.

$$u_3(\tau) = \frac{p_{n+1} - p_n}{kh}\left(\tau - \frac{\sin \omega\tau}{\omega}\right) \qquad (8.77)$$

The displacement at t_{n+1} is obtained by adding u_1, u_2, and u_3 and substituting h for τ.

$$u_{n+1} = Au_n + B\dot{u}_n + Cp_n + Dp_{n+1} \qquad (8.78)$$

where

$$A = \cos \omega h$$

$$B = \frac{\sin \omega h}{\omega}$$

$$C = \frac{1}{k} \left(\frac{\sin \omega h}{\omega h} - \cos \omega h \right)$$ (8.79)

$$D = \frac{1}{k} \left(1 - \frac{\sin \omega h}{\omega h} \right)$$

The velocity at t_{n+1} is obtained by adding the differentials $\dot{u}_1$, $\dot{u}_2$, and $\dot{u}_3$ and substituting h for τ.

$$\dot{u}_{n+1} = A_1 u_n + B_1 \dot{u}_n + C_1 p_n + D_1 p_{n+1}$$ (8.80)

where

$$A_1 = - \omega \sin \omega h$$

$$B_1 = \cos \omega h$$

$$C_1 = \frac{1}{k} \left(\omega \sin \omega h - \frac{1}{h} + \frac{\cos \omega h}{h} \right)$$ (8.81)

$$D_1 = \frac{1}{k} \left(\frac{1}{h} - \frac{\cos \omega h}{h} \right)$$

Recurrence formulas 8.78 and 8.80 permit step-by-step integration, provided the displacement and velocity at time $t = 0$ are given. If the acceleration history is also of interest, it can be obtained by satisfying the equation of motion at each time step. Thus

$$\ddot{u}_{n+1} = \frac{1}{m}(p_{n+1} - k u_{n+1})$$ (8.82)

Theoretically, the time interval h may be varied from step to step. However, a uniform time step is usually preferred to save time on the computations.

The method is readily extended to damped system. The free-vibration component is now given by Equation 5.37. The response to the constant force p_n is obtained from Equation 7.14, and the response to linearly varying force is obtained by solving Equation 7.20. On combining the three components and their derivatives, we obtain recurrence formulas that are similar to Equations

8.78 and 8.80 with the coefficients given by the following

$$A = e^{-\xi\omega h}\left(\frac{\xi}{\sqrt{1-\xi^2}}\sin\omega_d h + \cos\omega_d h\right)$$

$$B = e^{-\xi\omega h}\left(\frac{1}{\omega_d}\sin\omega_d h\right)$$

$$C = \frac{1}{k}\left[\frac{2\xi}{\omega h} + e^{-\xi\omega h}\left\{\left(\frac{1-2\xi^2}{\omega_d h} - \frac{\xi}{\sqrt{1-\xi^2}}\right)\right.\right.$$
$$\left.\left.\sin\omega_d h - \left(1+\frac{2\xi}{\omega h}\right)\cos\omega_d h\right\}\right]$$

$$D = \frac{1}{k}\left\{1 - \frac{2\xi}{\omega h} + e^{-\xi\omega h}\left(\frac{2\xi^2-1}{\omega_d h}\sin\omega_d h + \frac{2\xi}{\omega h}\cos\omega_d h\right)\right\}$$

$$A_1 = -e^{-\xi\omega h}\left(\frac{\omega}{\sqrt{1-\xi^2}}\sin\omega_d h\right)$$

$$B_1 = e^{-\xi\omega h}\left(\cos\omega_d h - \frac{\xi}{\sqrt{1-\xi^2}}\sin\omega_d h\right)$$

$$C_1 = \frac{1}{k}\left[-\frac{1}{h} + e^{-\xi\omega h}\left\{\left(\frac{\omega}{\sqrt{1-\xi^2}} + \frac{\xi}{h\sqrt{1-\xi^2}}\right)\sin\omega_d h + \frac{1}{h}\cos\omega_d h\right\}\right]$$

$$D_1 = \frac{1}{k}\left\{\frac{1}{h} - \frac{e^{-\xi\omega h}}{h}\left(\frac{\xi}{\sqrt{1-\xi^2}}\sin\omega_d h + \cos\omega_d h\right)\right\}$$

(8.83)

It may be noted that while the integration formulas are exact, approximation is involved in representing the forcing function by a series of straight lines. For a small h, the errors involved are expected to be small.

Example 8.11
Calculate the response of the tower in Example 8.9 to the loading shown there for the first 1.0 s using piece-wise linear representation of the force, as shown in Figure E8.11. Assume that damping in the system is 10% of critical and use $h = 0.1$ s.

Solution
The following properties have been obtained in Examples 8.9 and 8.10.

$k = 100$ kips/in.

$m = 2.533$ kip $\cdot$ s^2/in.

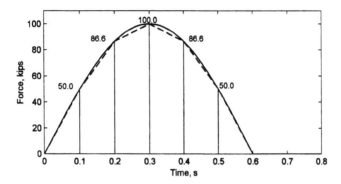

Figure E8.11. Piece-wise linear representation of a sinusoidal forcing function.

Table E8.11. Piece-wise linear representation of the excitation.

Time	u_n	$\dot{u}_n$	p_n	p_{n+1}	u_{n+1} (Eq. 8.73)	$\dot{u}_{n+1}$ (Eq. 8.75)	u_n (theoretical)
0.0	0.0000	0.0000	0.00	50.0	0.0313	0.9164	0.0000
0.1	0.0313	0.9164	50.0	86.6	0.2206	2.9448	0.0323
0.2	0.2206	2.9448	86.6	100.0	0.6062	4.5651	0.2254
0.3	0.6062	4.5651	100.0	86.6	1.0713	4.3453	0.6204
0.4	1.0713	4.3453	86.6	50.0	1.3928	1.6855	1.0961
0.5	1.3928	1.6855	50.0	0.0	1.3461	−2.8245	1.4251
0.6	1.3461	−2.8245	0.0	0.0	0.8510	−6.6664	1.3772
0.7	0.8510	−6.6664	0.0	0.0	0.1090	−7.6618	0.8683
0.8	0.1090	−7.6618	0.0	0.0	−0.5845	−5.7895	0.1105
0.9	−0.5845	−5.7895	0.0	0.0	−0.9864	−2.0601	−0.5974
1.0	−0.9864	−2.0601	0.0				−1.0073

$$\omega = 6.283 \text{ rad/s}$$

$$\omega_d = 6.2515 \text{ rad/s}$$

$$u_0 = \dot{u}_0 = \ddot{u}_0 = 0$$

The coefficients in the recurrence formulas are determined from Equation 8.83.

$$A = 0.81672, \quad B = 0.08791, \quad C = 0.0012073, \quad D = 0.00062547,$$

$$A_1 = -3.4706, \quad B_1 = 0.70625, \quad C_1 = 0.016378, \quad D_1 = 0.018328$$

The displacements and velocities obtained from Equations 8.78 and 8.80, respectively, are shown in Table E8.11. The theoretical response for a sine-wave loading is given by Equation b of Example 8.10. The equation is used to obtain the theoretical response values for the first 0.6 s shown in Tables E8.11. For finding the response at time intervals beyond the first 0.6 s, we need the displacement as well as the velocity at $t = 0.6$ s. The displacement has already been calculated; the velocity is obtained by differentiating Equation b of Example 8.10 and substituting

$t = 0.6$ s. Displacement response beyond 0.6 s is now given by

$$u(t) = e^{-\xi\omega(t-0.6)} \left\{ \frac{v_{0.6} + u_{0.6}\xi\omega}{\omega_d} \sin \omega_d(t - 0.6) + u_{0.6} \cos \omega_d(t - 0.6) \right\} \tag{j}$$

where $u_{0.6}$ and $v_{0.6}$ are the displacement and velocity, respectively, at $t = 0.6$ s.

Theoretical displacement values for $t \geq 0.6$ s are calculated from Equation j and are entered in Table E8.11. The difference between the calculated and theoretical values of the response can be attributed to the difference between the actual forcing function and its piece-wise linear representation. To improve the accuracy of computations, we need to use a smaller interval of time, so that the piece-wise linear representation more closely matches the actual function.

8.10   DERIVATION OF GENERAL FORMULAE

As mentioned in Section 8.8, time integration methods are based on appropriately selected expressions relating the response parameters at a given interval of time to their values at one or more previous points. A general expression of this type can be written as

$$u_{n+1} = \sum_{l=n-k}^{n} A_l u_l + \sum_{l=n-k}^{n+1} B_l \dot{u}_l + \sum_{l=n-k}^{n+1} C_l \ddot{u}_l + R \tag{8.84}$$

where R is a remainder term representing the error in the expression, and A_l, B_l, and C_l are constants, some of which may equal zero. Equation 8.84 relates the displacement, velocity, and acceleration at time point $n+1$, to their values at the previous points $n-k$, $n-k+1, \ldots, n$. The equation has $m = 5+3k$ undetermined constants A, B, and C. By a suitable choice for the values of these constants, the equation can be made exact in the special case where u is a polynomial of order $m - 1$. When the formula is exact, the remainder term R will equal zero. Now, if Equation 8.84 is exact for a polynomial of order $(m - 1)$, it will also be exact when u takes any one of the following values

$$1, \ t, \ t^2, \ \ldots, \ t^{m-1}$$

Therefore, to evaluate the m constants, we can successively set u equal to each of these values in turn, and in each case, make a substitution in Equation 8.84. This procedure yields m simultaneous equations involving the set of m coefficients A_l, B_l, and C_l. A solution of these equations provides the required values for the coefficients.

Now, if $u = t^m$ is substituted into Equation 8.84, R cannot be expected to equal zero automatically. We will designate the resulting value of R by E_m. It will be evident that the larger the number of terms in Equation 8.84, the smaller will be the magnitude of truncation error R. Equations having many terms may, however, encounter spurious roots and instability, characteristics that we shall examine in some detail later. It may not, therefore, be desirable to use all

m free constants to reduce the truncation error. An approach having practical usefulness is to employ Equation 8.84 to represent exactly a polynomial of order $p - 1$, p being an integer smaller than m. Then, $(m - p)$ constants become available which can be assigned arbitrary values chosen so as to improve the stability or convergence characteristics of the resulting formula.

A formula that is exact for polynomials up to an order $m - 1$ has a truncation error E_m when $u = t^m$. It is of interest to estimate the truncation error term when u is a polynomial of order higher than m, or in the more general case, when u is a function other than a polynomial. Standard textbooks on numerical analysis show that in most cases, the truncation error term R is given by

$$R = \frac{E_m}{m!} u^{(m)}(\xi), \quad (n - k)h \leq \xi \leq (n + 1)h \tag{8.85}$$

where $u^{(m)}$ is the value of the mth differential of u at $t = \xi$. Parameter ξ is, in general, not known, and it may not therefore be possible to obtain a precise value for R. Nevertheless, Equation 8.85 can often be used to provide an upper bound on R. An estimate of the truncation error term may be helpful in choosing one formula over the other, although such error is not the only criterion in the choice of a formula.

The procedure described in the foregoing paragraph is a very general method of finding formulas. In fact, all formulae of type Equation 8.84 that are used in structural dynamics for time integration can be derived using the procedure described. However, very often such formulae are obtained from physical considerations, such as, for example, an assumed variation in the acceleration $\ddot{u}$, or from the well-known finite difference approximations of the differentials. In the following sections, we develop, both from the underlying physical considerations, and by using the technique of this section, a number of more commonly used methods of numerical integration of the equation of motion. In each case, we also derive the truncation error term.

8.11   CONSTANT-ACCELERATION METHOD

In deriving the formulae for this method, we make the assumption that over a small interval of time h, the acceleration of the system is constant and is equal to its value at the beginning of the interval, as shown in Figure 8.8. For simplicity and without loss of generality, we shift the origin to the time point n. The constant-acceleration assumption gives

$$\ddot{u} = \ddot{u}_n \tag{8.86}$$

On integrating Equation 8.86, we get

$$\dot{u} = \ddot{u}_n t + A \tag{8.87}$$

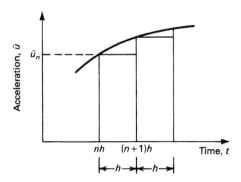

Figure 8.8. Constant-acceleration method.

where A is an arbitrary constant, whose value is obtained from boundary condition

$$\dot{u} = \dot{u}_n \quad \text{for } t = 0 \tag{8.88}$$

Substitution of Equation 8.88 in Equation 8.87 gives $A = \dot{u}_n$. Now, setting $t = h$ in Equation 8.87 and recognizing that for this value of t, $\dot{u} = \dot{u}_{n+1}$, we have

$$\dot{u}_{n+1} = \dot{u}_n + h\ddot{u}_n \tag{8.89}$$

Integrating Equation 8.87 and using the boundary condition $u = u_n$ at $t = 0$, we get

$$u = u_n + \dot{u}_n t + \frac{\ddot{u}_n t^2}{2} \tag{8.90}$$

For $t = h$, $u = u_{n+1}$, so that Equation 8.90 gives

$$u_{n+1} = u_n + h\dot{u}_n + \frac{h^2}{2}\ddot{u}_n \tag{8.91}$$

We next derive expressions similar to those of Equations 8.89 and 8.91 by the method of Section 8.10. The velocity expression can be written as

$$\dot{u}_{n+1} = a_1 \dot{u}_n + a_2 \ddot{u}_n \tag{8.92}$$

where a_1 and a_2 are constants to be determined. Since we have two free constants, we can expect to make Equation 8.92 exact for $u = 1$ and $u = t$. For $u = 1$, $\dot{u} = 0$, and $\ddot{u} = 0$, and Equation 8.92 is automatically satisfied because we get zeros on both sides. For $u = t$, $\dot{u} = 1$, and $\ddot{u} = 0$. Substitution in Equation 8.92 gives

$$1 = a_1 \tag{8.93}$$

We are still left with one free constant and can therefore make the formula exact for $u = t^2$ as well. In this case, we have $\dot{u} = 2t$ and $\ddot{u} = 2$, and substitution into Equation 8.92 gives

$$2h = 0 + 2a_2 \tag{8.94}$$

Equations 8.92, 8.93, and 8.94 lead to the expression

$$\dot{u}_{n+1} = \dot{u}_n + h\ddot{u}_n \tag{8.95}$$

which is the same as Equation 8.89. Now, if we substitute $u = t^3$, we cannot expect Equation 8.95 to be satisfied. In fact, a remainder term is obtained in this case. We denote this term by E_3, so that

$$3h^2 = E_3 \tag{8.96}$$

Finally, using Equation 8.85, we write Equation 8.95 with the error term included

$$\dot{u}_{n+1} = \dot{u}_n + h\ddot{u}_n + \frac{E_3}{3!}u^{(3)}(\xi)$$

$$= \dot{u}_n + h\ddot{u}_n + \frac{h^2}{2}u^{(3)}(\xi) \tag{8.97}$$

Next, we write the displacement expression as

$$u_{n+1} = b_1 u_n + b_2 \dot{u}_n + b_3 \ddot{u}_n + R \tag{8.98}$$

On making Equation 8.98 exact for $u = 1$, t, t^2, we get

$$\begin{aligned} u &= 1 & 1 &= b_1 \\ u &= t & h &= b_2 \\ u &= t^2 & h^2 &= 2b_3 \end{aligned} \tag{8.99}$$

Substitution of Equation 8.99 into Equation 8.98 gives

$$u_{n+1} = u_n + h\dot{u}_n + \frac{h^2}{2}\ddot{u}_n + R \tag{8.100}$$

To obtain the error term R, we try $u = t^3$. Substitution into Equation 8.98 gives

$$h^3 = E_3 \tag{8.101}$$

From Equation 8.85

$$R = \frac{E_3}{3!}u^{(3)}(\xi) = \frac{h^3}{6}u^{(3)}(\xi) \tag{8.102}$$

Equations 8.89 and 8.91 provide two of the three relations required for time integration. The third relationship is obtained by writing the equation of motion at time point $n + 1$

$$m\ddot{u}_{n+1} + c\dot{u}_{n+1} + ku_{n+1} = p_{n+1} \tag{8.103}$$

Together, Equations 8.89, 8.91, and 8.103 allow us to obtain the parameters u_{n+1}, $\dot{u}_{n+1}$, and $\ddot{u}_{n+1}$ in terms of u_n, $\dot{u}_n$ and $\ddot{u}_n$. Equation 8.89 gives velocity $\dot{u}_{n+1}$, while Equation 8.91 gives displacement u_{n+1}. Substituting for u_{n+1} and $\dot{u}_{n+1}$ in Equation 8.103, we get the acceleration $\ddot{u}_{n+1}$

$$\ddot{u}_{n+1} = \frac{1}{m} \left\{ p_{n+1} - ku_n - (c + kh)\dot{u}_n - \left(ch + \frac{kh^2}{2} \right) \ddot{u}_n \right\} \tag{8.104}$$

To begin the time integration, we need to know the values of u_0, $\dot{u}_0$, and $\ddot{u}_0$, the displacement, velocity, and acceleration, respectively, at time $t = 0$. Two of them must be specified; the third is obtained by using the equation of motion at $t = 0$.

8.12   NEWMARK'S β METHOD

In 1959, Newmark devised a series of numerical integration formulae which are collectively known as Newmark's β methods. We derive Newmark's formulae by the general method of Section 8.10. The velocity expression is of the form

$$\dot{u}_{n+1} = a_1 \dot{u}_n + a_2 \ddot{u}_n + a_3 \ddot{u}_{n+1} \tag{8.105}$$

Equation 8.105 is automatically satisfied with $u = 1$, so that we still have three free constants, a_1, a_2, and a_3. We use two of them to make the formula exact for $u = t$ and t^2, leaving one of the constants slack. Substitution into Equation 8.105 gives

$$\begin{aligned} u &= t \quad\quad 1 = a_1 \\ u &= t^2 \quad 2h = 2a_2 + 2a_3 \end{aligned} \tag{8.106}$$

We now have two equations and three unknowns. We assign an arbitrary value to one of the unknowns and determine the other two from Equation 8.106. Let us select $a_3 = \gamma h$, where γ is an arbitrary constant. We then get

$$\begin{aligned} a_1 &= 1 \\ a_2 &= h(1 - \gamma) \end{aligned} \tag{8.107}$$

Equation 8.105 now takes the form

$$\dot{u}_{n+1} = \dot{u}_n + h(1 - \gamma)\ddot{u}_n + h\gamma\ddot{u}_{n+1} + R \tag{8.108}$$

The error term is obtained by substituting $u = t^3$ in Equation 8.108, so that

$$3h^2 = 6h^2\gamma + E_3$$
$$E_3 = h^2(3 - 6\gamma) \tag{8.109a}$$

Using Equations 8.85 and 8.109a, we get

$$R = \frac{E_3}{3!}u^{(3)}(\xi)$$

$$= h^2\left(\frac{1}{2} - \gamma\right)u^{(3)}(\xi) \tag{8.109b}$$

Constant γ is usually selected to be $\frac{1}{2}$. With this value for γ, Equation 8.108 becomes

$$\dot{u}_{n+1} = \dot{u}_n + \frac{h}{2}\ddot{u}_n + \frac{h}{2}\ddot{u}_{n+1} + R \tag{8.110}$$

Furthermore the error term given by Equation 8.109b becomes zero. This implies that the formula is exact for polynomials of order up to 3. The new error term is obtained by substituting $u = t^4$ and can be shown to be

$$R = -\frac{h^3}{12}u^{(4)}(\xi) \tag{8.111}$$

The displacement expression in Newmark's β method is

$$u_{n+1} = b_1 u_n + b_2 \dot{u}_n + b_3 \ddot{u}_n + b_4 \ddot{u}_{n+1} \tag{8.112}$$

Equation 8.112 has four free constants. We use three of them to make the formula exact for $u = 1$, t, and t^2. Substituting these values of u, in turn, in Equation 8.112, we obtain

$$u = 1 \quad 1 = b_1$$
$$u = t \quad h = b_2 \tag{8.113}$$
$$u = t^2 \quad h^2 = 2b_3 + 2b_4$$

Equation 8.113 provides three relationships among four unknowns. We assign an arbitrary value βh^2 to b_4 so that $b_3 = h^2(\frac{1}{2} - \beta)$. Equation 8.112 now reduces to

$$u_{n+1} = u_n + h\dot{u}_n + h^2\left(\frac{1}{2} - \beta\right)\ddot{u}_n + h^2\beta\ddot{u}_{n+1} + R \tag{8.114}$$

The error term R is obtained by substituting $u = t^3$ in Equation 8.114, which gives

$$h^3 = 6\beta h^3 + E_3$$
$$E_3 = h^3(1 - 6\beta) \tag{8.115a}$$

and

$$R = \frac{E_3}{3!} u^{(3)}(\xi)$$

$$= h^3 \left(\frac{1}{6} - \beta \right) u^{(3)}(\xi)$$

(8.115b)

By assigning different values to γ and β, a series of integration formulas can be obtained. As an example, when $\gamma = 0$ and $\beta = 0$, Newmark's β method reduces to the constant-acceleration method, which we have already discussed. Other more commonly used versions of the Newmark's β method are:
1. The average acceleration method, $\gamma = \frac{1}{2}$, $\beta = \frac{1}{4}$.
2. The linear acceleration method, $\gamma = \frac{1}{2}$, $\beta = \frac{1}{6}$.
A description of these methods is presented in the following paragraphs.

8.12.1 *Average acceleration method*

When γ is set equal to $\frac{1}{2}$ and $\beta = \frac{1}{4}$, Equations 8.108 and 8.114 reduce to

$$\dot{u}_{n+1} = \dot{u}_n + \frac{h}{2}(\ddot{u}_n + \ddot{u}_{n+1}) - \frac{h^3}{12} u^{(4)}(\xi)$$

(8.116)

$$u_{n+1} = u_n + h\dot{u}_n + \frac{h^2}{4}(\ddot{u}_n + \ddot{u}_{n+1}) + 0.152 h^3 u^{(3)}(\xi_m)$$

(8.117)

The error term in Equation 8.116 has been obtained from Equation 8.111. For $\beta = \frac{1}{4}$, the error expression of Equation 8.115b is not applicable. The error term in Equation 8.117, obtained by a somewhat more complicated procedure [1], represents an upper bound. The coordinate value ξ_m lies between 0 and h and is selected to give the maximum value of differential $u^{(3)}$.

Equations 8.116 and 8.117 can also be derived by assuming that, as indicated in Figure 8.9, the acceleration of the system remains constant over the small interval h and its value is equal to the average of the values of accelerations at the beginning and end of the interval

$$\ddot{u} = \frac{1}{2}(\ddot{u}_n + \ddot{u}_{n+1})$$

(8.118)

For simplicity, we shift the origin on the time axis to point t_n. Then, by integrating Equation 8.118 and applying the boundary conditions $\dot{u} = \dot{u}_n$ at $t = 0$ and $\dot{u} = \dot{u}_{n+1}$ at $t = h$, we get

$$\dot{u} = \dot{u}_n + \frac{t}{2}(\ddot{u}_n + \ddot{u}_{n+1})$$

(8.119)

---

[1] R.W. Hamming. 1962. *Numerical Methods for Scientists and Engineers*. New York: McGraw Hill Book Co. Inc.

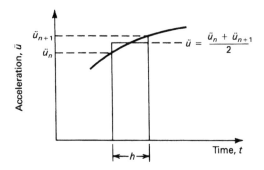

Figure 8.9. Average acceleration method.

and

$$\dot{u}_{n+1} = \dot{u}_n + \frac{h}{2}(\ddot{u}_n + \ddot{u}_{n+1}) \tag{8.120}$$

Integration of Equation 8.119 with the boundary conditions $u = u_n$ at $t = 0$ and $u = u_{n+1}$ at $t = h$ yields

$$u_{n+1} = u_n + \dot{u}_n h + \frac{h^2}{4}(\ddot{u}_n + \ddot{u}_{n+1}) \tag{8.121}$$

Equations 8.120 and 8.121 are, of course, the same as Equations 8.116 and 8.117, respectively. Equations 8.120 and 8.121 combined with Equation 8.103 enable us to solve for u_{n+1}, $\dot{u}_{n+1}$, and $\ddot{u}_{n+1}$ in terms of u_n, $\dot{u}_n$, and $\ddot{u}_n$. From Equation 8.121, we obtain

$$\ddot{u}_{n+1} = \frac{4}{h^2}(u_{n+1} - u_n - h\dot{u}_n) - \ddot{u}_n \tag{8.122}$$

Substitution of Equation 8.122 into Equation 8.120 gives

$$\dot{u}_{n+1} = -\dot{u}_n + \frac{2}{h}(u_{n+1} - u_n) \tag{8.123}$$

Finally, on substituting for $\ddot{u}_{n+1}$ and $\dot{u}_{n+1}$ from Equations 8.122 and 8.123 in Equation 8.103 and collecting terms, we get

$$\left(\frac{4m}{h^2} + \frac{2c}{h} + k\right)u_{n+1} = p_{n+1} + m\left(\frac{4}{h^2}u_n + \frac{4}{h}\dot{u}_n + \ddot{u}_n\right) + c\left(\frac{2}{h}u_n + \dot{u}_n\right) \tag{8.124}$$

Equation 8.124 can be solved for u_{n+1}. Substitution into Equations 8.122 and 8.123 then gives $\ddot{u}_{n+1}$ and $\dot{u}_{n+1}$, respectively. To begin the integration, u_0, $\dot{u}_0$, and $\ddot{u}_0$ must be known. Two of these, usually the initial displacement and the initial velocity, will be given; the third can be determined by using the equation of motion written at $t = 0$.

8.12.2   *Linear acceleration method*

With $\gamma = \frac{1}{2}$ and $\beta = \frac{1}{6}$, Equations 8.108 and 8.114 reduce to

$$\dot{u}_{n+1} = \dot{u}_n + \frac{h}{2}(\ddot{u}_n + \ddot{u}_{n+1}) - \frac{h^3}{12}u^{(4)}(\xi) \tag{8.125}$$

$$u_{n+1} = u_n + h\dot{u}_n + \frac{h^2}{3}\ddot{u}_n + \frac{h^2}{6}\ddot{u}_{n+1} - \frac{h^4}{24}u^{(4)}(\xi) \tag{8.126}$$

The error term in Equation 8.125 is obtained from Equation 8.111. However, when Equation 8.115 is used to obtain the error term for the displacement relation, a zero value is obtained. This is because Equation 8.126 is exact for a polynomial of order up to 3. The error term can, however, be obtained by standard procedure, first substituting $u = t^4$ in Equation 8.126 to obtain E_4 and then by using Equation 8.85.

Equations 8.125 and 8.126 can also be derived by assuming that, as indicated in Figure 8.10, the acceleration of the system varies linearly over a small interval of time h. With the origin centered on the time axis at t_n, we have

$$\ddot{u} = \ddot{u}_n + \frac{\ddot{u}_{n+1} - \ddot{u}_n}{h}t \tag{8.127}$$

Integration of Equation 8.127 and substitution of the boundary conditions at $t = 0$ and $t = h$ gives

$$\dot{u} = \dot{u}_n + \ddot{u}_n t + (\ddot{u}_{n+1} - \ddot{u}_n)\frac{t^2}{2h} \tag{8.128}$$

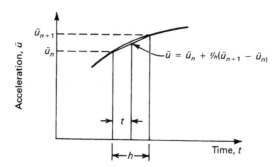

Figure 8.10. Linear acceleration method.

and

$$\dot{u}_{n+1} = \dot{u}_n + \frac{h}{2}(\ddot{u}_n + \ddot{u}_{n+1}) \tag{8.129}$$

On integrating Equation 8.128 and substituting the boundary conditions, we get

$$u_{n+1} = u_n + h\dot{u}_n + \frac{h^2}{3}\ddot{u}_n + \frac{h^2}{6}\ddot{u}_{n+1} \tag{8.130}$$

Together, Equations 8.129 and 8.130 and the equation of motion written at t_{n+1} (Eq. 8.103) enable us to solve for u_{n+1}, $\dot{u}_{n+1}$, and $\ddot{u}_{n+1}$ in terms of u_n, $\dot{u}_n$, and $\ddot{u}_n$. Thus, Equation 8.130 gives

$$\ddot{u}_{n+1} = \frac{6}{h^2}\left(u_{n+1} - u_n - h\dot{u}_n - \frac{h^2}{3}\ddot{u}_n\right) \tag{8.131}$$

Substitution of Equation 8.131 in Equation 8.129 yields

$$\dot{u}_{n+1} = \frac{3}{h}(u_{n+1} - u_n) - 2\dot{u}_n - \frac{h}{2}\ddot{u}_n \tag{8.132}$$

Finally, on substituting Equations 8.131 and 8.132 in Equation 8.103 and collecting terms, we get

$$\left(\frac{6m}{h^2} + \frac{3c}{h} + k\right)u_{n+1} = p_{n+1} + m\left(\frac{6}{h^2}u_n + \frac{6}{h}\dot{u}_n + 2\ddot{u}_n\right)$$

$$+ c\left(\frac{3}{h}u_n + 2\dot{u}_n + \frac{h}{2}\ddot{u}_n\right) \tag{8.133}$$

Equation 8.133 is solved for u_{n+1}. Equations 8.131 and 8.132 then give the acceleration and velocity, respectively, at time t_{n+1}. Again, to start the time integration, we need to know u_0, $\dot{u}_0$, and $\ddot{u}_0$. Two of these three parameters will be specified and the third can be obtained from the equation of motion written at $t = 0$.

Example 8.12

Calculate the response of the tower in Example 8.9 to the loading shown there for the first 1.0 s using a numerical integration technique. Assume that damping in the system is 10% of critical and use $h = 0.1$ s.

Solution

We will obtain the response using
 (i) the constant-acceleration method,
 (ii) the average acceleration method, and
 (iii) the linear acceleration method. For each case

$$k = 100 \text{ kips/in.}$$

$$m = 2.533 \text{ kip} \cdot \text{s}^2/\text{in.}$$

$$\omega = 6.283 \text{ rad/s}$$

$$c = 2\xi\omega m = 3.183 \text{ kip} \cdot \text{s/in.}$$

$$u_0 = \dot{u}_0 = \ddot{u}_0 = 0$$

(i) Constant-acceleration method. Using Equations 8.91, 8.89, and 8.104, we get

$$u_{n+1} = u_n + 0.1\dot{u}_n + 0.005\ddot{u}_n \tag{a}$$

$$\dot{u}_{n+1} = \dot{u}_n + 0.1\ddot{u}_n \tag{b}$$

$$\ddot{u}_{n+1} = \frac{1}{2.533}(p_{n+1} - 100u_n - 13.183\dot{u}_n - 0.8183\ddot{u}_n) \tag{c}$$

Step-by-step time integration is now carried out using Equations a, b, and c. Response calculations for the first 1.0 s are shown in Table E8.12a.

(ii) Average acceleration method. Substituting the values of m, c, k, and h in Equation 8.124, we get

$$1176.9u_{n+1} = p_{n+1} + 1076.9u_n + 104.5\dot{u}_n + 2.533\ddot{u}_n \tag{d}$$

Equations 8.123 and 8.122 give

$$\dot{u}_{n+1} = 20(u_{n+1} - u_n) - \dot{u}_n \tag{e}$$

$$\ddot{u}_{n+1} = 400(u_{n+1} - u_n) - 40\dot{u}_n - \ddot{u}_n \tag{f}$$

Time integration is carried out using Equations d, e, and f. Response calculations for the first 1.0 s are shown in Table E8.12b.

(iii) Linear acceleration method. Substitution for m, c, k, and h in Equation 8.133 gives

$$1715.3u_{n+1} = p_{n+1} + 1615.3u_n + 158.35\dot{u}_n + 5.225\ddot{u}_n \tag{g}$$

Equations 8.132 and 8.131 give

$$\dot{u}_{n+1} = 30(u_{n+1} - u_n) - 2\dot{u}_n - 0.05\ddot{u}_n \tag{h}$$

$$\ddot{u}_{n+1} = 600(u_{n+1} - u_n) - 60\dot{u}_n - 2\ddot{u}_n \tag{i}$$

Time integration is carried out using Equations g, h, and i. Response calculations for the first 1.0 s are shown in Table E8.12c.

The theoretical response for a sine-wave loading is given by Equation b of Example 8.10. The equation is used to obtain the theoretical response values for the first 0.6 s shown in Tables E8.12a, E8.12b, and E8.12c. Theoretical displacement values for $t \geq 0.6$ s are calculated from Equation j of Example 8.11 and are entered in Tables E8.12a, E8.12b, and E8.12c. An examination of the response values in these tables shows that the average acceleration and the linear acceleration methods give reasonable results. The time step h of 0.1 s, which is one-tenth the natural period of the system and one-twelfth the period of exciting force is probably the maximum one can use to ensure a reasonable accuracy; a smaller step size must be used if better accuracy is desired.

The results obtained from the constant-acceleration method are not satisfactory. A considerably smaller step size must be used in this case to obtain acceptable accuracy. In fact, the method is not very effective, for several reasons that will be discussed later.

Table E8.12a. Constant-acceleration method.

Time	u_n	$\dot{u}_n$	$\ddot{u}_n$	p_{n+1}	u_{n+1} (Eq. a)	$\dot{u}_{n+1}$ (Eq. b)	$\ddot{u}_{n+1}$ (Eq. c)	u_n (theoretical)
0.0	0.0000	0.000	0.00	50.0	0.0000	0.000	19.74	0.0000
0.1	0.0000	0.000	19.74	86.6	0.0987	1.974	27.81	0.0323
0.2	0.0987	1.974	27.81	100.0	0.4351	4.755	16.32	0.2254
0.3	0.4351	4.755	16.32	86.6	0.9922	6.387	−13.01	0.6204
0.4	0.9923	6.387	−13.01	50.0	1.5659	5.086	−48.47	1.0961
0.5	1.5660	5.086	−48.47	0.0	1.8322	0.239	−72.63	1.4251
0.6	1.8320	0.239	−72.63	0.0	1.4927	−7.024	−50.11	1.3772
0.7	1.4930	−7.024	−50.11	0.0	0.5401	−12.036	−6.19	0.8683
0.8	0.5400	−12.036	−6.19	0.0	−0.6949	−12.655	43.33	0.1105
0.9	−0.6950	−12.655	43.33	0.0	−1.7430	−8.322	79.29	−0.5974
1.0	−1.7430	−8.322	79.29					−1.0073

Table E8.12b. Average acceleration method.

Time	u_n	$\dot{u}_n$	$\ddot{u}_n$	p_{n+1}	u_{n+1} (Eq. d)	$\dot{u}_{n+1}$ (Eq. e)	$\ddot{u}_{n+1}$ (Eq. f)	u_n (theoretical)
0.0	0.0000	0.0000	0.000	50.0	0.0425	0.8497	16.994	0.0000
0.1	0.0425	0.8497	16.994	86.6	0.2245	2.7902	21.816	0.0323
0.2	0.2245	2.7902	21.816	100.0	0.5851	4.4218	10.815	0.2254
0.3	0.5851	4.4218	10.815	86.6	1.0248	4.3736	−11.779	0.6204
0.4	1.0248	4.3736	−11.779	50.0	1.3433	1.9943	−35.806	1.0961
0.5	1.3432	1.9943	−35.806	0.0	1.3291	−2.2767	−49.613	1.4251
0.6	1.3291	−2.2767	−49.613	0.0	0.9073	−6.1607	−28.066	1.3772
0.7	0.9073	−6.1607	−28.066	0.0	0.2227	−7.5297	0.686	0.8683
0.8	0.2227	−7.5297	0.686	0.0	−0.4633	−6.1909	26.089	0.1105
0.9	−0.4633	−6.1909	26.089	0.0	−0.9175	−2.8928	39.871	−0.5974
1.0	−0.9175	−2.8928	39.871					−1.0073

Table E8.12c. Linear acceleration method.

Time	u_n	$\dot{u}_n$	$\ddot{u}_n$	p_{n+1}	u_{n+1} (Eq. g)	$\dot{u}_{n+1}$ (Eq. h)	$\ddot{u}_{n+1}$ (Eq. k)	u_n theoretical
0.0	0.0000	0.0000	0.000	50.0	0.0291	0.8745	17.490	0.0000
0.1	0.0291	0.8745	17.490	86.6	0.2119	2.8603	22.227	0.0323
0.2	0.2119	2.8603	22.227	100.0	0.5896	4.4991	10.549	0.2254
0.3	0.5896	4.4991	10.549	86.6	1.0532	4.3819	−12.893	0.6204
0.4	1.0532	4.3819	−12.893	50.0	1.3862	1.8707	−37.331	1.0961
0.5	1.3862	1.8707	−37.331	0.0	1.3644	−2.5299	−50.680	1.4251
0.6	1.3644	−2.5299	−50.680	0.0	0.8969	−6.4303	−27.328	1.3772
0.7	0.8969	−6.4303	−27.328	0.0	0.1678	−7.6476	2.981	0.8683
0.8	0.1678	−7.6476	2.981	0.0	−0.5389	−6.0547	28.875	0.1105
0.9	−0.5389	−6.0547	28.875	0.0	−0.9785	−2.5210	41.789	−0.5974
1.0	−0.9785	−2.5210	41.789					−1.0073

8.13 WILSON-θ METHOD

This method proposed by E. L. Wilson is similar to the linear acceleration method. It is based on the assumption that the acceleration varies linearly over an extended interval θh as shown in Figure 8.11. Parameter θ, which is always greater than 1, is selected to give the desired characteristics of accuracy and stability. The basic relationships used in the method are similar to Equations 8.131, 8.132, and 8.133 with $n + 1$ replaced by $n + \theta$, and h replaced by θh.

$$\ddot{u}_{n+\theta} = \frac{6}{(\theta h)^2} \left\{ u_{n+\theta} - u_n - \theta h \dot{u}_n - \frac{(\theta h)^2}{3} \ddot{u}_n \right\} \tag{8.134}$$

$$\dot{u}_{n+\theta} = \frac{3}{\theta h}(u_{n+\theta} - u_n) - 2\dot{u}_n - \frac{\theta h}{2}\ddot{u}_n \tag{8.135}$$

$$\left\{ \frac{6m}{(\theta h)^2} + \frac{3c}{\theta h} + k \right\} u_{n+\theta} = p_{n+\theta} + m \left\{ \frac{6u_n}{(\theta h)^2} + \frac{6}{\theta h}\dot{u}_n + 2\ddot{u}_n \right\}$$

$$+ c \left\{ \frac{3}{\theta h}u_n + 2\dot{u}_n + \frac{\theta h}{2}\ddot{u}_n \right\} \tag{8.136}$$

Since the acceleration is assumed to vary linearly from time nh to $(n+\theta)h$, the exciting force is also assumed to vary linearly over the same interval of time and $p_{n+\theta}$ is therefore obtained by projecting the exciting force value to time $(n + \theta)h$. Thus

$$p_{n+\theta} = p_n + \frac{p_{n+1} - p_n}{h}\theta h$$

$$= p_n(1 - \theta) + p_{n+1}\theta \tag{8.137}$$

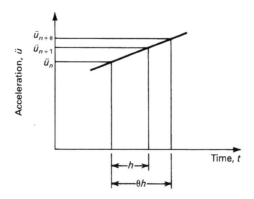

Figure 8.11. Wilson-θ method.

Equation 8.136 is used to calculate $u_{n+\theta}$. This value is then substituted in Equation 8.134 to obtain $\ddot{u}_{n+\theta}$. The acceleration at normal time increment h is now given by

$$\ddot{u}_{n+1} = \ddot{u}_n + \frac{\ddot{u}_{n+\theta} - \ddot{u}_n}{\theta} \tag{8.138}$$

This value of $\ddot{u}_{n+1}$ is substituted in Equations 8.129 and 8.130 to obtain the velocity and displacement, respectively, at normal increment of time.

Example 8.13
Solve Example 8.12 by the Wilson θ method using $\theta = 1.5$ and $h = 0.1$ s.

Solution
Substitution of m, c, k, h, and θ in Equation 8.136 gives

$$839.13u_{n+\theta} = p_{n+\theta} + 739.13u_n + 107.69\dot{u}_n + 5.305\ddot{u}_n \tag{a}$$

where $p_{n+\theta}$ is obtained from Equation 8.137. Equation 8.134 now leads to

$$\ddot{u}_{n+\theta} = 266.7(u_{n+\theta} - u_n) - 40\dot{u}_n - 2\ddot{u}_n \tag{b}$$

From Equation 8.138, we get

$$\ddot{u}_{n+1} = 0.6667\ddot{u}_{n+\theta} + 0.3333\ddot{u}_n \tag{c}$$

Finally, substitution of Equation c in Equations 8.129 and 8.130 gives

$$\dot{u}_{n+1} = \dot{u}_n + 0.05(\ddot{u}_n + \ddot{u}_{n+1}) \tag{d}$$

$$u_{n+1} = u_n + 0.1\dot{u}_n + 0.00333\ddot{u}_n + 0.00167\ddot{u}_{n+1} \tag{e}$$

Step-by-step time integration is carried out using Equations a through e. Response calculations for the first 1 s are shown in Table E8.13.

Table E8.13. Wilson-θ method.

Time	u_n	$\dot{u}_n$	$\ddot{u}_n$	$p_{n+\theta}$	$u_{n+\theta}$ (Eq. a)	$\ddot{u}_{n+\theta}$ (Eq. b)	$\ddot{u}_{n+1}$ (Eq. c)	$\dot{u}_{n+1}$ (Eq. d)	u_{n+1} (Eq. e)	u_n (theoretical)
0.0	0.0000	0.0000	0.00	75.0	0.0894	23.837	15.89	0.7946	0.0265	0.0000
0.1	0.0265	0.7946	15.89	104.9	0.3508	22.923	20.58	2.6188	0.1932	0.0323
0.2	0.1932	2.6181	20.58	106.7	0.7634	6.197	10.99	4.1966	0.5419	0.2254
0.3	0.5419	4.1966	10.99	79.9	1.1806	−19.507	−9.34	4.2791	0.9827	0.6204
0.4	0.9827	4.2791	−9.34	31.7	1.3934	−42.927	−31.73	2.2254	1.3265	1.0961
0.5	1.3265	2.2254	−31.73	−25.0	1.2237	−52.993	−45.91	−1.6564	1.3668	1.4251
0.6	1.3668	−1.6564	−45.91	0.0	0.7011	−19.467	−28.28	−5.3658	1.0010	1.3772
0.7	1.0010	−5.3658	−28.28	0.0	0.0143	8.039	−4.07	−6.9832	0.3634	0.8683
0.8	0.3634	−6.9832	−4.07	0.0	−0.6018	30.041	18.67	−6.2529	−0.3174	0.1105
0.9	−0.3174	−6.2529	18.67	0.0	−0.9640	40.323	33.11	−3.6641	−0.8252	−0.5974
1.0	−0.8252	−3.6641	33.11							−1.0073

8.14   METHODS BASED ON DIFFERENCE EXPRESSIONS

8.14.1   *Central difference method*

The method uses standard *central difference expressions* to relate the time derivatives of displacement, that is, velocity and acceleration, to the displacement values at selected intervals of time. The velocity expressions is given by

$$\dot{u}_n = \frac{1}{2h}(u_{n+1} - u_{n-1}) + R \tag{8.139}$$

A graphical representation of Equation 8.139 is shown in Figure 8.12a. Substitution of $u = 1$, t, t^2, in turn, in Equation 8.139 shows that the equation is exact for these values of u, so that the remainder term is zero. For $u = t^3$, we get $R = E_3 = -h^2$. In a general case, therefore

$$R = \frac{E_3}{3!} u^{(3)}(\xi) = -\frac{h^2}{6} u^{(3)}(\xi) \tag{8.140}$$

The central difference expression used for acceleration is

$$\ddot{u}_n = \frac{1}{h^2}(u_{n+1} - 2u_n + u_{n-1}) + R \tag{8.141}$$

A graphical representation of Equation 8.141 is shown in Figure 8.12b. By substitution, it can be shown that Equation 8.141 is exact for $u = 1$, t, t^2, t^3. For $u = t^4$, we get $R = E_4 = -2h^2$. In a general case, therefore

$$R = \frac{E_4}{4!} u^{(4)}(\xi) = -\frac{h^2}{24} u^{(4)}(\xi) \tag{8.142}$$

The equation of dynamic equilibrium is formed at time point n,

$$m\ddot{u}_n + c\dot{u}_n + ku_n = p_n \tag{8.143}$$

Substituting Equations 8.139 and 8.141 in Equation 8.143 and solving for u_{n+1}, we get

$$\left(\frac{m}{h^2} + \frac{c}{2h}\right)u_{n+1} = p_n + \left(-k + \frac{2m}{h^2}\right)u_n + \left(\frac{c}{2h} - \frac{m}{h^2}\right)u_{n-1} \tag{8.144}$$

Equations 8.139 and 8.141 are now used to find $\dot{u}_n$ and $\ddot{u}_n$. It will be evident from Equation 8.144 that to start the time integration, we need the values of both u_0 and u_{-1}. When Equations 8.139 and 8.141 are written at $t = 0$ and solved simultaneously, the following expression is obtained for u_{-1}

$$u_{-1} = u_0 + \frac{h^2}{2}\ddot{u}_0 - h\dot{u}_0 \tag{8.145}$$

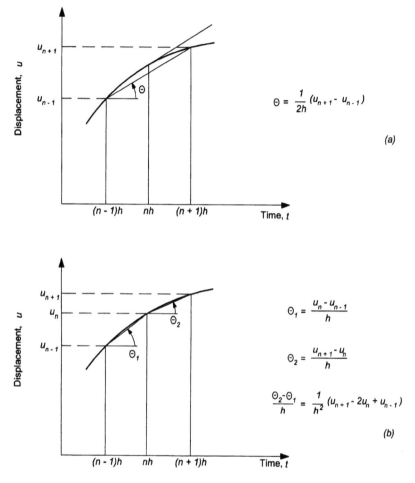

$$\Theta = \frac{1}{2h}(u_{n+1} - u_{n-1})$$

(a)

$$\Theta_1 = \frac{u_n - u_{n-1}}{h}$$

$$\Theta_2 = \frac{u_{n+1} - u_n}{h}$$

$$\frac{\Theta_2 - \Theta_1}{h} = \frac{1}{h^2}(u_{n+1} - 2u_n + u_{n-1})$$

(b)

Figure 8.12. Central difference method.

In a well-posed problem, two of the three parameters u_0, $\dot{u}_0$, and $\ddot{u}_0$ will be specified; the third can be obtained from the equation of motion (Eq. 8.143) written at $t=0$.

It will be observed that in all the methods discussed so far, except in the central difference method, one of the three expressions used in the solution procedure was the equation of motion written at time point $t=(n+1)h$. In contrast, the central difference method uses the equation of motion at $t=nh$. Methods that use the equation of motion at $n+1$ are called *implicit methods*, while those that use the equation at n are called *explicit methods*. As we shall see later, in the case of a multi-degree-of-freedom system, use of an implicit method may require much more computation than that needed for an explicit method. On the other hand, while explicit methods are only conditionally stable, some of the implicit methods are unconditionally stable. Conditional stability may be a serious restriction in the analysis of some multi-degree-of-freedom systems. Stability of integration methods is discussed in Section 8.16.

Example 8.14

Solve Example 8.12 by the central difference method with $h = 0.1$ s.

Table E8.14. Central difference method.

Time	p_n	u_{n-1}	u_n	u_{n+1} (Eq. a)	$\dot{u}_n$ (Eq. b)	$\ddot{u}_n$ (Eq. c)	u_n (theoretical)
0.0	0.0	0.0000	0.0000	0.0000	0.0000	0.000	0.0000
0.1	50.0	0.0000	0.0000	0.1857	0.9286	18.573	0.0323
0.2	86.6	0.0000	0.1857	0.6022	3.0110	23.074	0.2254
0.3	100.0	0.1857	0.6022	1.1172	4.6574	9.854	0.6204
0.4	86.6	0.6022	1.1172	1.4780	4.3792	−15.418	1.0961
0.5	50.0	1.1172	1.4780	1.4329	1.5785	−40.594	1.4251
0.6	0.0	1.4780	1.4329	0.8609	−3.0859	−52.693	1.3772
0.7	0.0	1.4329	0.8609	0.0366	−6.9813	−25.216	0.8683
0.8	0.0	0.8609	0.0366	−0.7038	−7.8231	8.381	0.1105
0.9	0.0	0.0366	−0.7038	−1.0953	−5.6593	34.893	−0.5974
1.0	0.0	−0.7038	−1.0953				−1.0073

Solution

On substituting for m, c, k, and h, Equation 8.144 becomes

$$269.21u_{n+1} = p_n + 406.6u_n - 237.39u_{n-1} \tag{a}$$

Equations 8.139 and 8.141 then give

$$\dot{u}_n = 5(u_{n+1} - u_{n-1}) \tag{b}$$

$$\ddot{u}_n = 100(u_{n+1} - 2u_n + u_{n-1}) \tag{c}$$

To start the integration, Equation 8.145 is used to obtain u_{-1}. In this case, because u_0, $\dot{u}_0$, and $\ddot{u}_0$ are all zero, u_{-1} is also zero. Equations a, b, and c are now used to carry out step-by-step integration. Response calculations are shown in Table E8.14 for the first 1 s. If the velocity and acceleration are not of interest, the corresponding columns may be omitted from the table.

8.14.2   Houbolt's method

Houbolt's method provides another example of a method that uses difference operators to represent the time derivatives of displacement. The method employs double backward difference operators to obtain expressions for velocity and acceleration. The velocity expression is of the form

$$\dot{u}_{n+1} = a_1 u_{n-2} + a_2 u_{n-1} + a_3 u_n + a_4 u_{n+1} \tag{8.146}$$

Equation 8.146 has four free constants and can be made exact for $u = 1$, t, t^2, and t^3. Substitution of these values in turn leads to the following relationships

$$0 = a_1 + a_2 + a_3 + a_4$$

$$1 = -2ha_1 - ha_2 + ha_4$$

$$2h = 4h^2 a_1 + h^2 a_2 + h^2 a_4 \tag{8.147}$$

$$3h^2 = -8h^3 a_1 - h^3 a_2 + h^3 a_4$$

Simultaneous solution of Equation 8.147 gives $a_1 = -1/(3h)$, $a_2 = 3/(2h)$, $a_3 = -3/h$, and $a_4 = 11/(6h)$. Equation 8.146 therefore becomes

$$\dot{u}_{n+1} = \frac{1}{6h}(-2u_{n-2} + 9u_{n-1} - 18u_n + 11u_{n+1}) \tag{8.148}$$

The acceleration expression is of the form

$$\ddot{u}_{n+1} = b_1 u_{n-2} + b_2 u_{n-1} + b_3 u_n + b_4 u_{n+1} \tag{8.149}$$

Equation 8.149 also has four free constants and can be made exact for $u = 1$, t, t^2, and t^3. Substitution in turn into Equation 8.149 leads to

$$0 = b_1 + b_2 + b_3 + b_4$$

$$0 = -2hb_1 - hb_2 + hb_4$$

$$2 = 4h^2 b_1 + h^2 b_2 + h^2 b_4 \tag{8.150}$$

$$6h = -8h^3 b_1 - h^3 b_2 + h^3 b_4$$

Solution of the set of Equations 8.150 gives $b_1 = -1/h^2$, $b_2 = 4/h^2$, $b_3 = -5/h^2$, and $b_4 = 2/h^2$. Equation 8.149 therefore becomes

$$\ddot{u}_{n+1} = \frac{1}{h^2}(-u_{n-2} + 4u_{n-1} - 5u_n + 2u_{n+1}) \tag{8.151}$$

The third expression used in Houbolt's method is the equation of motion written at $t = (n + 1)h$. Simultaneous solution of Equations 8.148 and 8.151 and the equation of motion gives

$$\left(\frac{2m}{h^2} + \frac{11c}{6h} + k\right) u_{n+1} = p_{n+1} + m\left(\frac{5u_n}{h^2} - \frac{4u_{n-1}}{h^2} + \frac{u_{n-2}}{h^2}\right)$$

$$+ c\left(\frac{3u_n}{h} - \frac{3u_{n-1}}{2h} + \frac{u_{n-2}}{3h}\right) \tag{8.152}$$

It will be observed from Equation 8.152 that u_0, u_1, and u_2 must be known or determined before time integration can be started. Displacement u_0 will be specified, while u_1 and u_2 are calculated by one of the other methods described earlier, such as, for example, the average acceleration method or the linear acceleration method. Houbolt's method belongs to the class of implicit methods and, as shown later, is unconditionally stable.

8.15   ERRORS INVOLVED IN NUMERICAL INTEGRATION

The errors involved in the numerical integration of differential equations, such as for example an equation of motion, can be classified into three types:
1. Round-off errors introduced by repeated computations using a small step size.
2. Truncation errors involved in representing u_{n+1} or $\dot{u}_{n+1}$ by a finite number of terms in the Taylor series expansion. This is the error term represented by R in the preceding sections.
3. Propagated error introduced by replacing the differential equation by a finite difference equivalent.

We will not consider round-off errors in detail here. They are random in nature and therefore must be treated by statistical methods. Error bounds obtained by statistical means often grossly overestimate the real errors. In computer calculations, it is often possible to reduce the round-off errors by using a higher precision in the computations.

Truncation errors are accumulated locally at each step. Unless the integration method being used is unstable, truncation error is a good indication of the accuracy of the numerical solution. Truncation error terms therefore provide a useful criterion for assessing the relative accuracy of the various methods of numerical integration. An estimate of the truncation error in a formula can be obtained by considering the equation of undamped free vibration

$$\ddot{u} + \omega^2 u = 0 \tag{8.153}$$

Equation 8.153 has a solution of the type

$$u = c_1 \sin \omega t + c_2 \cos \omega t \tag{8.154}$$

in which c_1 and c_2 are the constants of integration. In the integration formulas presented in this chapter, the truncation term is, in general, of the form

$$Ch^p u^{(p)}(\xi) \tag{8.155}$$

where C is a constant and p is the degree of the formula. From Equation 8.154, we note that the pth derivative of u contains a factor ω^p; the magnitude of the truncation error will therefore depend on the value of $(\omega h)^p$. It is evident that to keep the truncation error low, the time step size h should be chosen so that

ωh is less than 1. Propagated errors affect the stability of the computation and are discussed in detail in the next section.

8.16 STABILITY OF THE INTEGRATION METHOD

As stated in Section 8.15, errors are introduced into the numerical solution due to truncation. It is important to know the effect of the error introduced at one step on the computations at the next step. If the error has a tendency to grow, the solution soon becomes unbounded and meaningless. In such a situation, the computational method is said to be unstable. In a general discussion of the stability of integration methods, it is convenient to use certain matrix notations. We review here briefly those terms in matrix algebra that we shall find useful.

Let $\mathbf{A}$ be a square matrix of order N. The *trace* of $\mathbf{A}$ is defined as the sum of the elements on its leading diagonal. Denoting half-trace of $\mathbf{A}$ by α_1, we have

$$2\alpha_1 = a_{11} + a_{22} + \cdots + a_{NN} \tag{8.156}$$

The *minor* M_{ij} of matrix $\mathbf{A}$ is the determinant of the $N-1$ by $N-1$ matrix obtained by deleting row i and column j from the original matrix. A minor is called a *principal minor* when $i = j$.

The *eigenvalues* λ of a square matrix $\mathbf{A}$ are obtained by solving the *characteristic polynomial* equation

$$det\,[\mathbf{A} - \lambda\mathbf{I}] = 0 \tag{8.157}$$

In particular, when $\mathbf{A}$ is a 3×3 matrix, expansion of Equation 8.157 gives

$$\lambda^3 - 2\alpha_1\lambda^2 + a_2\lambda - a_3 = 0 \tag{8.158}$$

where α_2 is the sum of principal minors of $\mathbf{A}$ and α_3 is the determinant of $\mathbf{A}$. Solution of Equation 8.158 will give three different values of λ. The three eigenvalues are denoted by λ_1, λ_2, and λ_3. The *spectral radius*, $\rho(\mathbf{A})$, of matrix $\mathbf{A}$ is defined by

$$\rho(\mathbf{A}) = max\{|\lambda_1|, |\lambda_2|, |\lambda_3|\}$$

We examine the stability of numerical integration technique with reference to the solution of the equation of undamped free vibration (Eq. 8.153). The condition of stability will be affected by the presence of damping, but for small amounts of damping this effect can be ignored. In any case, the presence of damping will make the stability conditions less restrictive. If a solution method is unstable under conditions of free vibration, it is also likely to be unstable for the solution of the problems of forced vibration, because instability in the complementary solution will soon make the total solution meaningless.

In a majority of numerical integration techniques presented in this chapter, the three response parameters, displacement, velocity, and acceleration, at time point $n + 1$ can be expressed in terms of their values at point n as follows

$$\mathbf{r}_{n+1} = \mathbf{A}\mathbf{r}_n \qquad (8.159)$$

where $\mathbf{r}_n^T = [u_n \quad h\dot{u}_n \quad h^2\ddot{u}_n]$ and $\mathbf{A}$ is called the *amplification matrix*. Equation 8.159 assumes that the exciting force is zero. We write Equation 8.159 at two previous time points

$$\mathbf{r}_n = \mathbf{A}\mathbf{r}_{n-1}$$
$$\mathbf{r}_{n-1} = \mathbf{A}\mathbf{r}_{n-2} \qquad (8.160)$$

It can be shown that the use of Equations 8.159 and 8.160 to eliminate the velocities and accelerations gives an equation of the form

$$u_{n+1} - 2\alpha_1 u_n + \alpha_2 u_{n-1} - \alpha_3 u_{u-2} = 0 \qquad (8.161)$$

where, as defined earlier, α_1 is $\frac{1}{2}$ trace of $\mathbf{A}$, α_2 is the sum of principal minors of $\mathbf{A}$, and α_3 is the determinant of $\mathbf{A}$. Equation 8.161 is a *difference equation* which has a solution of the form

$$u_n = c\rho^n \qquad (8.162)$$

in which c is a constant. Substitution of Equation 8.162 in Equation 8.161 gives

$$\rho^3 - 2\alpha_1\rho^2 + \alpha_2\rho - \alpha_3 = 0 \qquad (8.163)$$

Equation 8.163 has three roots ρ_1, ρ_2, and ρ_3. By comparing Equations 8.163 and 8.158, we can conclude that these roots are, in fact, equal to the three eigenvalues of matrix $\mathbf{A}$. A general solution of Equation 8.161 is given by

$$u_n = \sum_{i=1}^{3} c_i\rho_i^n \qquad (8.164)$$

where c_i's are arbitrary constants to be determined from the initial conditions.

If the absolute value of any of the three roots of Equation 8.163 is greater than 1, the solution given by Equation 8.164 will become unbounded as time progresses, that is, as n increases. In such a situation, the solution procedure is said to be unstable. Stability thus requires that the absolute value of each of the three roots of Equation 8.163 be less than or equal to 1, or equivalently that the spectral radius of matrix $\mathbf{A}$ be less than or equal to 1.

Comparison between Equations 8.164 and 8.154 shows that the difference equation has three roots, while the differential equation has only two. The extra root, denoted by ρ_3, is called a *spurious root*, while the other two roots, ρ_1 and ρ_2, are the *principal roots*. If the contribution from the spurious root has to be small, it must be real and less than 1. Also, the differential equation has an oscillating solution; therefore, if the solution of the difference equation has to correspond to that of the differential equation, the two principal roots ρ_1 and ρ_2 must be complex conjugates of each other. In the following, we examine the stability of each of the integration methods described in this chapter.

8.16.1 *Newmark's β method*

The amplification matrix for Newmark's β method is obtained by simultaneous solution of Equations 8.108 and 8.114 and the equation of motion at time point $n+1$ for the three parameters u_{n+1}, $\dot{u}_{n+1}$, and $\ddot{u}_{n+1}$. After some algebraic manipulation, we obtain

$$\mathbf{A} = \frac{1}{D} \begin{bmatrix} 1 + 2\xi\gamma\omega h & 1 + 2\xi\omega h(\gamma - \beta) & (\frac{1}{2} - \beta) + 2\xi\omega h(\frac{1}{2}\gamma - \beta) \\ -\gamma\omega^2 h^2 & 1 + \omega^2 h^2(\beta - \gamma) & (1 - \gamma) + \omega^2 h^2(\beta - \frac{\gamma}{2}) \\ -\omega^2 h^2 & -(2\xi\omega h + \omega^2 h^2) & -2\xi\omega h(1 - \gamma) - \omega^2 h^2(\frac{1}{2} - \beta) \end{bmatrix}$$

(8.165)

where

$$D = 1 + 2\xi\gamma\omega h + \beta\omega^2 h^2$$

With $\xi = 0$ the invariants of matrix $\mathbf{A}$, α_1, α_2, and α_3 become

$$\alpha_1 = \frac{1 + \omega^2 h^2/2(-\frac{1}{2} + 2\beta - \gamma)}{1 + \beta\omega^2 h^2}$$

$$\alpha_2 = \frac{1 + \omega^2 h^2(\frac{1}{2} + \beta - \gamma)}{1 + \beta\omega^2 h^2}$$

(8.166)

$$\alpha_3 = 0$$

Equation 8.163 now reduces to

$$\rho^2 - 2\alpha_1\rho + \alpha_2 = 0$$

(8.167)

where α_1 and α_2 are given by Equation 8.166.

In fact, Equation 8.167 is, in this case, obtained more easily directly from Equations 8.108 and 8.114. We begin by rewriting the velocity equation, Equation 8.108, at time $t_n = nh$

$$\dot{u} - \dot{u}_{n-1} = h(1 - \gamma)\ddot{u}_{n-1} + \gamma h\ddot{u}_n$$

(8.168)

Next, we write the displacement equation (Eq. 8.114) at t_n

$$u_n = u_{n-1} + h\dot{u}_{n-1} + h^2\left(\frac{1}{2} - \beta\right)\ddot{u}_{n-1} + h^2\beta\ddot{u}_n \tag{8.169}$$

Subtraction of Equation 8.169 from Equation 8.114 without the remainder term gives

$$u_{n+1} - 2u_n + u_{n-1} = h(\dot{u}_n - \dot{u}_{n-1}) + h^2\left(\frac{1}{2} - \beta\right)(\ddot{u}_n - \ddot{u}_{n-1})$$

$$+ h^2\beta(\ddot{u}_{n+1} - \ddot{u}_n) \tag{8.170}$$

On substituting for $\dot{u} - \dot{u}_{n-1}$ from Equation 8.168 and using Equation 8.153 to express the accelerations in terms of displacements, we get

$$u_{n+1} - \frac{2 + \omega^2 h^2(-\frac{1}{2} + 2\beta - \gamma)}{1 + \beta\omega^2 h^2}u_n + \frac{1 + \omega^2 h^2(\frac{1}{2} + \beta - \gamma)}{1 + \beta\omega^2 h^2}u_{n-1} = 0 \tag{8.171}$$

which is the same as Equation 8.167. Equation 8.167 indicates that Newmark's β method provides only two roots. They are given by

$$\rho_{1,2} = \alpha_1 \pm \sqrt{\alpha_1^2 - \alpha_2} \tag{8.172}$$

For the difference equation to provide an oscillatory solution, α_1^2 must be less than α_2. Assuming that this condition is satisfied, Equation 8.172 can be expressed as

$$\rho_{1,2} = \alpha_2^{1/2}(\cos\phi \pm i\sin\phi) \tag{8.173}$$

where

$$\tan\phi = \pm\frac{\sqrt{\alpha_2 - \alpha_1^2}}{\alpha_1} \tag{8.174}$$

The general solution of the difference equation thus becomes

$$u_n = \alpha_2^{n/2}(c_1 \cos n\phi + c_2 \sin n\phi) \tag{8.175}$$

where c_1 and c_2 are arbitrary constants to be obtained from initial conditions.

The solution given by Equation 8.175 will remain bounded provided that α_2 is less than or equal to 1. In summary, then, the condition for stable oscillatory response is

$$\alpha_1^2 < \alpha_2 \leq 1 \tag{8.176}$$

To obtain the period of oscillation, we rewrite Equation 8.175 in the form

$$u_n = \alpha_2^{n/2}(c_1 \cos \bar{\omega}t_n + c_2 \sin \bar{\omega}t_n) \tag{8.177}$$

where $t_n = nh$ and $\bar{\omega} = \phi/h$. The period of oscillation is given by $\bar{T} = 2\pi/\bar{\omega}$. This will, in general, be slightly different from the true period $2\pi/\omega$.

Constant-acceleration method
In this case, we have $\gamma = 0$ and $\beta = 0$, so that

$$\alpha_1 = 1 - \frac{\omega^2 h^2}{4}$$

$$\alpha_2 = 1 + \frac{\omega^2 h^2}{2} \tag{8.178}$$

Parameter α_2 is always greater than 1 and hence the solution will diverge with time. The response will be oscillatory provided that

$$\alpha_1^2 < \alpha_2 \tag{8.179a}$$

or

$$\omega h < 4 \tag{8.179b}$$

Average acceleration method
In this case, $\gamma = \frac{1}{2}$ and $\beta = \frac{1}{4}$. Hence

$$\alpha_1 = \frac{1 - \omega^2 h^2/4}{1 + \omega^2 h^2/4}$$

$$\alpha_2 = 1 \tag{8.180}$$

Since α_2 is equal to 1, the amplitude of the numerical solution is exactly equal to that of the exact solution and does not decay or diverge with time. Also, α_1^2 is less than α_2 for all values of ωh. The method is therefore unconditionally stable.

Linear acceleration method
For this method, $\gamma = \frac{1}{2}$ and $\beta = \frac{1}{6}$, so that

$$\alpha_1 = \frac{1 - \omega^2 h^2/3}{1 + \omega^2 h^2/6}$$

$$\alpha_2 = 1 \tag{8.181}$$

Again, since $\alpha_2 = 1$, the amplitude of the solution is exact. The condition for oscillatory motion gives

$$\omega h < \sqrt{12} \tag{8.182a}$$

or

$$\frac{h}{T} < \frac{\sqrt{3}}{\pi} \equiv 0.55 \tag{8.182b}$$

The spectral radius of the amplification matrix, which represents the root of the difference equation having the largest absolute value, is plotted in Figure 8.14 as a function of h/T. The figure shows that for a linear acceleration method, the spectral radius is less than 1 as long as $h/T < \sqrt{3}/\pi$, which therefore is the condition for stability of the method.

8.16.2   *Wilson-θ method*

The relationships between displacement, velocity, and acceleration at time t_{n+1} and those at time t_n are obtained by the procedure described in Section 8.13. The resulting amplification matrix is given by

$$
\mathbf{A} = \frac{1}{D}
\begin{bmatrix}
\theta + \frac{\omega^2 h^2}{6}(\theta^3 - 1) & \theta + \frac{\theta \omega^2 h^2}{6}(\theta^2 - 1) & (\frac{\theta}{2} - \frac{1}{6}) + \frac{\theta^2 \omega^2 h^2}{12}(\theta - 1) \\
+\theta^2 \xi \omega h & +\xi \omega h(\theta^2 - \frac{1}{3}) & +\theta \xi \omega h(\frac{\theta}{2} - \frac{1}{3}) \\[2mm]
-\frac{\omega^2 h^2}{2} & \theta + \frac{\theta \omega^2 h^2}{6}(\theta^2 - 3) & (\theta - \frac{1}{2}) + \frac{\theta^2 \omega^2 h^2}{12}(2\theta - 3) \\
 & +\xi \omega h(\theta^2 - 1) & +\xi \omega h \theta(\theta - 1) \\[2mm]
-\omega^2 h^2 & -\theta \omega^2 h^2 - 2\xi \omega h & (\theta - 1) + \frac{\theta^2 \omega^2 h^2}{6}(\theta - 3) \\
 & & +\xi \omega h \theta(\theta - 2)
\end{bmatrix}
$$

$$(8.183)$$

where $D = \theta + \theta^2 \xi \omega h + \frac{1}{6}\theta^3 \omega^2 h^2$. With $\xi = 0$, the invariants of the amplification matrix are

$$
\alpha_1 = \frac{1}{\hat{D}}\left\{\frac{1}{2}(3\theta - 1) + \frac{\omega^2 h^2}{12}(3\theta^3 - 3\theta^2 - 3\theta - 1)\right\}
$$

$$
\alpha_2 = \frac{1}{\hat{D}^2}\left\{\theta(3\theta - 2) + \frac{\omega^2 h^2}{6}(5\theta^4 - 7\theta^3 + 5\theta - 1)\right.
$$

$$
\left. + \frac{\theta^3 \omega^4 h^4}{36}(3\theta^3 - 6\theta^2 + 4)\right\}
$$

$$
\alpha_3 = \frac{1}{\hat{D}^3}\left\{\theta^2(\theta - 1) + \frac{\omega^2 h^2}{6}(3\theta^5 - 5\theta^4 + 3\theta^3 + 2\theta^2 - 3\theta)\right.
$$

$$
+ \frac{\omega^4 h^4}{36}(3\theta^7 - 7\theta^6 + 6\theta^5 + \theta^4 - 6\theta^2 + 3\theta) \tag{8.184}
$$

$$
\left. + \frac{\omega^6 h^6}{216}(\theta^9 - 3\theta^8 + 3\theta^7 + 2\theta^6 - 3\theta^5)\right\}
$$

where $\hat{D} = \theta + \frac{1}{6}\theta^3 \omega^2 h^2$.

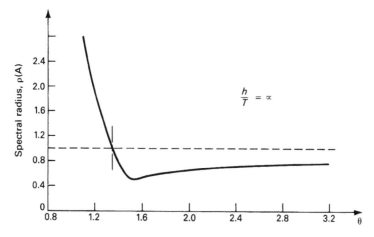

Figure 8.13. Wilson-θ method; spectral radius versus θ for $h/T = \infty$.

Because α_1, α_2, and α_3 may all be nonzero, it is evident that the difference equation in this case has three nonzero roots. If the method has to be stable, the absolute value of each of these must be smaller than 1. It may, however, be possible to chose a value of θ such that the requirement for stability is unconditionally satisfied. To see whether this is possible, we first obtain the limit of matrix $\mathbf{A}$ as ωh tends to ∞. This limit is easily shown to be given by

$$\mathbf{A} = \frac{1}{\theta^3} \begin{bmatrix} \theta^3 - 1 & \theta^3 - \theta & \frac{1}{2}(\theta^3 - \theta^2) \\ -3 & \theta^3 - 3\theta & \frac{1}{2}(2\theta^3 - 3\theta^2) \\ -6 & -6\theta & \theta^3 - 3\theta^2 \end{bmatrix} \tag{8.185}$$

The spectral radius $\rho(\mathbf{A})$ of matrix $\mathbf{A}$ in Equation 8.185 is calculated for a series of values of θ and is plotted against θ in Figure 8.13. It is observed that for $\theta > 1.37$, $\rho(\mathbf{A})$ is always less than 1. Wilson-θ method can therefore be made unconditionally stable by selecting $\theta > 1.37$. As we shall see later, the accuracy of the method is better for smaller θ and a value $\theta = 1.4$ is usually selected.

We now examine the three roots of the difference equation (Eq. 8.163) for two different values of θ: 1.4 and 2.0. The roots can be obtained by solving the cubic equation, or equivalently, by calculating the eigenvalues of amplification matrix $\mathbf{A}$ (Eq. 8.183) with $\xi = 0$. The absolute values of the three roots are plotted in Figure 8.14 as functions of $h/T = \omega h/2\pi$. It is found that one of the three roots is real. This root represents a spurious root in the solution. Its value is always less than 1, and hence its contribution dies out fairly rapidly. The other two roots, called the principal roots, are complex conjugates of each other, so that their absolute values are the same. This absolute value is found to

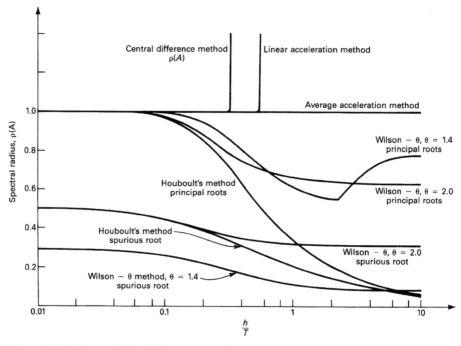

Figure 8.14. Roots of the difference equation.

be less than 1, which implies that the amplitude of calculated response decays with time and the numerical process has introduced an artificial damping.

For $\theta = 1.4$, the curve representing the principal roots dips to a minimum at $h/T \approx 2.29$. When $h/T > 2.29$, the principal roots take two distinct, real values and the solution of the difference equation is no longer oscillatory, although it is still bounded. For $h/T > 2.29$, the curve in Figure 8.14 shows the principal root having the higher absolute value.

8.16.3   Central difference method

In this case, the difference equation is obtained directly from Equation 8.144. With $c = 0$ and $p = 0$, we have

$$u_{n+1} - (2 - \omega^2 h^2)u_n + u_{n-1} = 0 \tag{8.186}$$

$$\alpha_1 = 1 - \frac{\omega^2 h^2}{2}$$
$$\alpha_2 = 1 \tag{8.187}$$

Since α_2 is equal to 1, the amplitude of the numerical solution is equal to that of the true solution. The condition for oscillatory motion gives

$$\omega h < 2 \tag{8.188a}$$

or

$$\frac{h}{T} < \frac{1}{\pi} \equiv 0.318 \tag{8.188b}$$

The spectral radius of the amplification matrix is plotted in Figure 8.14 as a function of h/T. The spectral radius is greater than 1 when $h/T > 1/\pi$, confirming that Equation 8.188b defines the condition of stability.

8.16.4 *Houbolt's method*

The difference equation for free undamped vibration is obtained from Equation 8.152 with p_{n+1} and c equal to zero

$$(2 + \omega^2 h^2)u_{n+1} - 5u_n + 4u_{n-1} - u_{n-2} = 0 \tag{8.189}$$

The characteristic equation is

$$(2 + \omega^2 h^2)\rho^3 - 5\rho^2 + 4\rho - 1 = 0 \tag{8.190}$$

From Equation 8.190, it is evident that, as in the case of the Wilson-θ method, Houbolt's method leads to three roots for the characteristic equation. Of these, the spurious root is found to be real, while the other two principal roots are complex conjugates of each other. The absolute values of the spurious root and the principal roots are plotted in Figure 8.14 as functions of h/T. These values are less than 1 for all h/T and the method is therefore unconditionally stable.

8.17 SELECTION OF A NUMERICAL INTEGRATION METHOD

A number of alternative methods of numerical integration have been presented in the preceding sections. It is observed that with the exception of constant-acceleration method, all of these methods are either unconditionally stable or stable when ωh or h/T is below a certain value. In all cases where there is a condition, the limiting value of ωh is seen to be larger than 1. On the other hand, as pointed out earlier, to ensure adequate accuracy in the numerical results, ωh must be selected to be considerably smaller than 1. It can therefore be concluded that for the response analysis of a single-degree-of-freedom system, the stability criteria are not restrictive and the selection of the method is not therefore governed by considerations of stability. As we shall see later, the situation is different for multi-degree-of-freedom systems where stability may be an important consideration.

Stability not being a criterion, the selection of numerical integration must be based on the relative accuracy of the results obtained. We measure the relative accuracy in reference to the solution of the undamped free-vibration equation (Eq. 8.153), which has a solution given by Equation 8.154. The numerical integration procedures, on the other hand, give a solution of the form

$$u_n = c_3 \rho_3^n + c_1 \rho_1^n + c_2 \rho_2^n \tag{8.191}$$

where ρ_3 is a spurious root, and ρ_1, ρ_2 are the principal roots which are of the form $a + bi$ and $a - bi$. Equation 8.191 can be expressed in the alternative form

$$u_n = c_3 \rho_3^n + \rho_p^n (\hat{c}_1 \cos \bar{\omega} t_n + \hat{c}_2 \sin \bar{\omega} t_n) \tag{8.192}$$

where $\rho_p = \sqrt{a^2 + b^2}$, $\tan \phi = b/a$, $t_n = nh$, $\bar{\omega} = \phi/h$, and $\hat{c}_1$ and $\hat{c}_2$ are new arbitrary constants.

In some of the integration methods presented here, the spurious solution does not exist. When it does exist, it rapidly dies out because ρ_3 is less than 1. We therefore restrict our comparison to the principal part of the solution. If the true solution (Eq. 8.154) and the numerical solution have to correspond to each other, ρ_p must be equal to 1 and $\bar{\omega}$ must be equal to ω. In general, these conditions are not satisfied. As measures of relative accuracy, we therefore define a period elongation (PE) and an amplitude decay (AD) as follows

$$\text{PE} = \frac{\bar{T} - T}{T} = \frac{\omega h - \phi}{\phi} \tag{8.193a}$$

$$\text{AD} = \frac{\rho_p^n - \rho_p^{n + \bar{T}/h}}{\rho_p^n} = 1 - \rho_p^{2\pi/\phi} \tag{8.193b}$$

where $\bar{T} = 2\pi/\bar{\omega}$. As an alternative, amplitude decay can also be represented by defining an equivalent viscous damping coefficient $\bar{\xi}$ such that

$$\begin{aligned}
\rho_p^n &= e^{-\bar{\xi} \bar{\omega} t_n} \\
&= e^{-\bar{\xi} \phi n}
\end{aligned} \tag{8.194}$$

Equation 8.194 gives

$$\bar{\xi} = -\frac{\ln \rho_p}{\phi} \tag{8.195}$$

The period elongation for several different methods is plotted in Figure 8.15 as a function of h/T. In Figure 8.16, we show amplitude decay as well as the equivalent viscous damping as a function of h/T for both the Wilson-θ

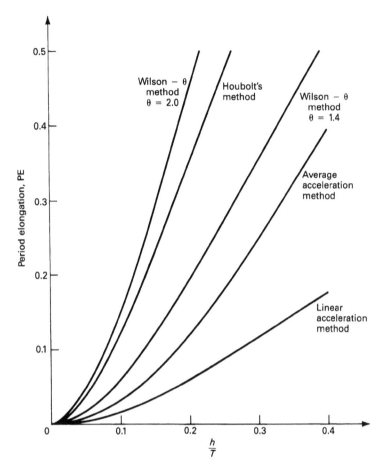

Figure 8.15. Period elongation versus h/T.

method and Houbolt's method. It should be noted that the average acceleration method and the linear acceleration method show no amplitude decay. From the data presented in Figures 8.15 and 8.16, it is evident that for the numerical integration of single-degree-of-freedom systems, the linear acceleration method, which gives no amplitude decay and the lowest period elongation, is the most suitable of the methods presented.

8.18   SELECTION OF TIME STEP

The selection of an appropriate time step for the numerical integration of the equation of motion is important in the success of the procedure. Too large a time step will give results that are unsatisfactory from the point of view

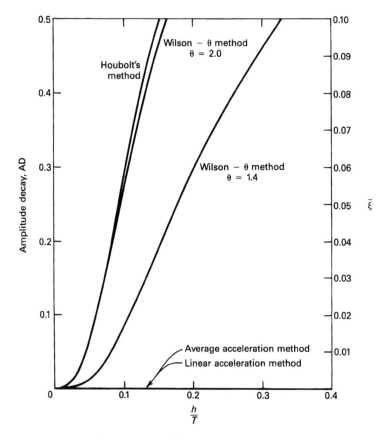

Figure 8.16. Amplitude decay versus h/T.

accuracy. A very small time step, on the other hand, will mean increased cost of computation. The natural period of the system as well as the characteristics of the exciting force are factors that govern the choice of a time step. An examination of the truncation error terms for the methods presented in the preceding sections shows that to keep the errors within limits, ωh must be less than 1. This is equivalent to h/T being less than 0.16. In general, acceptable accuracy can be achieved with a time step that is about one-tenth of the natural period of the system.

Numerical integration involves sampling of the exciting function at intervals equal to the selected time step. Such sampling should not lead to a distortion of the exciting function. If the function is viewed as a superposition of harmonic components, a time step that is about one-tenth the component with the smallest period should be quite satisfactory. It is not, however, necessary to decompose the forcing function into its harmonic components to determine a suitable value of h. In general, the time step size can be determined by an inspection of the exciting function. Further, the step size need not be much smaller than that governed by the natural period of the system, because high-frequency components of the exciting force which are not adequately represented by such a selection will not, in any case, excite a significant dynamic response of the system.

The accuracy of the numerical integration process may be verified by performing the response calculations with two different time steps that are close to each other. If the response obtained with the shorter time step is not too different from that obtained with the larger time step, the process may be taken to have converged to the true solution.

8.19   ANALYSIS OF NONLINEAR RESPONSE

The equation of motion of a vibrating system specifies the requirement of equilibrium between the applied force, the force of inertia, the damping force, and the spring force. The last three forces depend on the physical properties of the system: its mass, damping characteristics, and stiffness. When these properties do not vary with time, the system is said to be linear and the solution methods discussed in previous sections are applicable. However, if any of the physical characteristics of mass, damping, or stiffness vary with time, the system becomes nonlinear and special methods must be devised for its solution. In a majority of cases, the mass is time invariant. Since damping characteristics cannot, in any case, be defined with certainty, it is not unreasonable to assume that they also remain constant with time. We therefore restrict our discussion to those nonlinear systems in which the nonlinearity arises because of varying stiffness or a nonlinear force–displacement relationship.

Figure 8.17a shows the relationships between the inertia force f_I and the acceleration $\ddot{u}$. The slope of the tangent to the curve represents the mass at a specific instant of time. Since mass is considered to be time invariant, this slope is constant and the relationship between f_I and $\ddot{u}$ is a straight line. In a similar manner, the relationship between damping force f_D and velocity $\dot{u}$, shown in Figure 8.17b, is represented by a straight line whose slope is equal to the damping constant c. Finally, Figure 8.17c shows the relationship between the spring force f_S and the displacement u. In this case, the relationship is not a straight line since the spring stiffness, which is equal to the slope of the tangent to the curve, varies with the displacement and hence with time. The instantaneous stiffness, which is equal to the slope of the tangent at the specified instant of time, is represented by k_T. The equation of motion at time $t = nh$ can be expressed as

$$f_I(t) + f_D(t) + f_S(t) = p(t) \tag{8.196a}$$

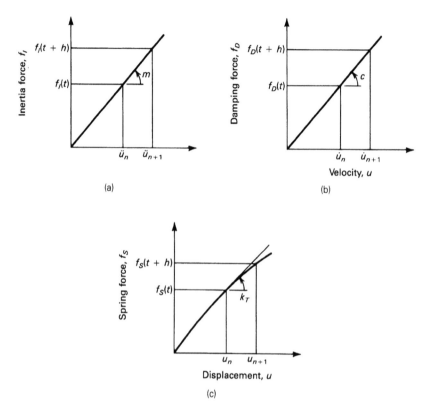

Figure 8.17. Variation of forces in a dynamic system: (a) inertia force versus acceleration; (b) damping force versus velocity; (c) spring force versus displacement.

or

$$m\ddot{u}_n + c\dot{u}_n + f_S(t) = p_n \qquad (8.196b)$$

In a similar manner, equilibrium at time $t + h$ or $(n + 1)h$ is represented by

$$m\ddot{u}_{n+1} + c\dot{u}_{n+1} + f_S(t + h) = p_{n+1} \qquad (8.197)$$

On subtracting Equation 8.196b from Equation 8.197, we obtain the equation of motion in its incremental form

$$m\Delta\ddot{u} + c\Delta\dot{u} + f_S(t + h) - f_S(t) = \Delta p \qquad (8.198)$$

If we make the assumption that over the small interval of time h, the stiffness is constant and is equal to the tangent stiffness at the beginning, we can write

$$f_S(t + h) - f_S(t) = k_T \Delta u \qquad (8.199)$$

A reference to Figure 8.17c will show that for a finite interval of time, Equation 8.199 is in error and the tangent stiffness k_T should, in fact, be replaced by the secant stiffness. The latter is, however, not known until the new displaced configuration is known, which is precisely what we are trying to determine. We must therefore use k_T, which can be determined from the present configuration and hope that for a sufficiently small time step, h, the error introduced by using the tangent stiffness rather than the secant stiffness will not be large.

Substitution of Equation 8.199 in Equation 8.198 gives an incremental equation that can be solved by one of the time integration schemes discussed in preceding sections. To illustrate the procedure, let us assume that Newmark's linear acceleration method is to be used in the response calculation. We first express Equation 8.130 in the alternative form

$$\Delta u = h\dot{u}_n + \frac{h^2}{2}\ddot{u}_n + \frac{h^2}{6}\Delta\ddot{u}$$

(8.200)

in which we have used $u_{n+1} = u_n + \Delta u$ and $\ddot{u}_{n+1} = \ddot{u}_n + \Delta\ddot{u}$. Equation 8.200 provides an expression for $\Delta\ddot{u}$ in terms of Δu

$$\Delta\ddot{u} = \frac{6}{h^2}\left(\Delta u - h\dot{u}_n - \frac{h^2}{2}\ddot{u}_n\right)$$

(8.201)

Next, we write Equation 8.129 in the incremental form

$$\Delta\dot{u} = h\ddot{u}_n + \frac{h}{2}\Delta\ddot{u}$$

(8.202)

where $\Delta\dot{u} = \dot{u}_{n+1} - \dot{u}_n$. Substitution of Equation 8.201 in Equation 8.202 gives

$$\Delta\dot{u} = \frac{3}{h}\Delta u - 3\dot{u}_n - \frac{h}{2}\ddot{u}_n$$

(8.203)

Now, if we substitute Equations 8.199, 8.201, and 8.203 in Equation 8.198, we get

$$\left(\frac{6m}{h^2} + \frac{3c}{h} + k_T\right)\Delta u = \Delta p + m\left(\frac{6\dot{u}_n}{h} + 3\ddot{u}_n\right) + c\left(3\dot{u}_n + \frac{h}{2}\ddot{u}_n\right)$$

(8.204a)

or

$$k_T^*\Delta u = \Delta p^*$$

(8.204b)

where

$$k_T^* = \frac{6m}{h^2} + \frac{3c}{h} + k_T$$

$$\Delta p^* = \Delta p + m\left(\frac{6\dot{u}_n}{h} + 3\ddot{u}_n\right) + c\left(3\dot{u}_n + \frac{h}{2}\ddot{u}_n\right)$$

Once Δu is obtained from Equation 8.204b, substitution in Equation 8.203 will give $\Delta \dot{u}$. These values of Δu and $\Delta \dot{u}$ can be added to u_n and $\dot{u}_n$ to obtain u_{n+1} and $\dot{u}_{n+1}$ respectively. The acceleration at time t_{n+1} can also be calculated similarly by using Equation 8.201 to find $\Delta \ddot{u}$. The three response parameters u_{n+1}, $\dot{u}_{n+1}$, and $\ddot{u}_{n+1}$ calculated in this manner will not quite satisfy the equilibrium equation at t_{n+1} because of the error involved in using tangent stiffness in place of the secant stiffness. To improve the accuracy of the procedure, one of the three parameters, usually the acceleration $\ddot{u}_{n+1}$, is obtained by enforcing total equilibrium at t_{n+1} rather than by adding an increment to the current value. Thus

$$\ddot{u}_{n+1} = \frac{1}{m}\{p_{n+1} - f_S(t+h) - c\dot{u}_{n+1}\} \tag{8.205}$$

Example 8.15

A tower supporting a tank of mass 0.1 kip · s²/in. is shown in Figure E8.15a. The columns are assumed to be massless, and damping in the system can be represented by a viscous damping coefficient $c = 0.2$ kip · s/in. The tower is subjected to the blast loading shown in Figure E8.15b. The force–displacement relationship for the tower structure is indicated in Figure E8.14c, in which the nonlinear part of the elastic force–displacement relation is given by $f_S = 12\{\frac{2}{3}u - \frac{1}{3}(2u/3)^3\}$. Obtain the response for first 0.8 s using the average acceleration method.

Solution

The incremental acceleration $\Delta \ddot{u}$ is obtained from Equation 8.122:

$$\Delta \ddot{u} = \frac{4}{h^2}\left(\Delta u - h\dot{u}_n - \frac{h^2}{2}\ddot{u}_n\right) \tag{a}$$

When this is substituted in Equation 8.120, we get

$$\Delta \dot{u} = \frac{2}{h}\Delta u - 2\dot{u}_n = 20\Delta u - 2\dot{u}_n \tag{b}$$

Substitution of Equations a and b in Equation 8.198 leads to

$$k^*\Delta u = \Delta p^* \tag{c}$$

where

$$k^* = \left(\frac{4m}{h^2} + \frac{2c}{h} + k_T\right) = 44 + k_T$$

$$\Delta p^* = \Delta p + m\left(\frac{4\dot{u}_n}{h} + 2\ddot{u}_n\right) + 2c\dot{u}_n \tag{d}$$

$$= \Delta p + 4.4\dot{u}_n + 0.2\ddot{u}_n$$

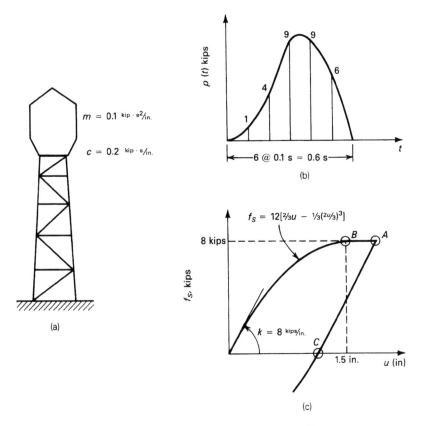

Figure E8.15. (a) Tower; (b) exciting force; (c) force–displacement relationship.

Equations c and b give Δu and $\Delta \dot{u}$ and hence u_{n+1} and $\dot{u}_{n+1}$. The equilibrium equation at t_{n+1} is then used to obtain $\ddot{u}_{n+1}$.

The smallest period of the system corresponds to the initial stiffness 8 k/in. and is given by

$$T = 2\pi \sqrt{\frac{m}{k_{\text{init}}}} = 0.702 \text{ s}$$

A time of step of 0.1 s, which is about one-seventh of the period, is used in the calculations. The response values for the first 0.8 s are shown in Table E8.15. The force–displacement relationship (Fig. E8.15c) shows that for $u > 1.5$ in., $f_S = 8$ kips. The displacement increases from 1.1128 in. to 1.7093 in. in the period 0.4 to 0.5 s. Evidently, the governing equation for f_S changes from $f_S = 12\{\frac{2}{3}u - \frac{1}{3}(2u/3)^3\}$ to $f_S = 8$ during this period. However, because of the use of a discrete time interval, this change is not detected until $t = 0.5$ s or $u = 1.7093$ in. Also, as soon as the velocity reverses, for example at A in the figure, the unloading path is followed and the stiffness changes from 0 to 8 kips/in. Theoretically, this change should take place when the velocity is zero. Because of the discrete time interval used in the response calculation, the change in the direction of velocity is detected at 0.6 s, where the velocity value has been calculated to be -0.4973 in./s. It is apparent that the velocity, in fact, changed direction between 0.5 and 0.6 s. Simultaneously with a change in the direction of velocity, the stiffness changed from 0 to

Table E8.15. Response of a nonlinear system.

t (s)	u_n (in.)	$\dot{u}_n$ (in./s)	$\ddot{u}_n$ (in./s)^2	p (kips)	K_T (kips/in.)	Δu (in.) Eq. c	$\Delta \dot{u}$ (in./s) Eq. b	u_{n+1} (in.) (2)+(7)	$\dot{u}_{n+1}$ (in./s) (3)+(8)	f_S (kips)	$\ddot{u}_{n+1}$ (in./s)^2
(1)	(2)	(3)	(4)	(5)	(6)	(7)	(8)	(9)	(10)	(11)	(12)
0	0.0000	0.0000	0.0000	0	8.0000	0.0192	0.3846	0.0192	0.3846	0.1538	7.6924
0.1	0.0192	0.3846	7.6924	1	7.9987	0.1198	1.6273	0.1391	2.0119	1.1093	24.8835
0.2	0.1391	2.0119	24.8835	4	7.9312	0.3626	3.2277	0.5016	5.2396	3.8635	40.8861
0.3	0.5016	5.2396	40.8861	9	7.1053	0.6111	1.7432	1.1128	6.9828	7.2691	3.3438
0.4	1.1128	6.9828	3.3438	9	3.5974	0.5965	−2.0350	1.7093	4.9477	8.0000	−29.8955
0.5	1.7093	4.9477	−29.8955	6	0.0000	0.2225	−5.4450	1.9318	−0.4973	8.0000	−79.0054
0.6	1.9318	−0.4973	−79.0054	0	8.0000	−0.3459	−5.9234	1.5859	−6.4207	5.2328	−39.4866
0.7	1.5859	−6.4207	−39.4866	0	8.0000	−0.6952	−1.0026	0.8907	−7.4833	−0.3288	18.2546
0.8	0.8907	−7.4833	18.2546	0							

8 kips/in. However, in the response calculation, $k_T = 0$ was assumed for the entire duration of time from 0.5 to 0.6 s. Evidently, the use of a finite time step in the integration gives rise to errors of this type whenever a transition occurs in the force–displacement relationship. Similar transition also occurs at point C in Figure E8.15c. It should be noted that in the iterations following 0.6 s, k_T has been changed to 8 kips/in., and the spring force follows the unloading path so that at 0.7 s, $f_S = 8 + 8 \times \Delta u = 8 - 8 \times 0.3459 = 5.2328$ kips.

8.20 ERRORS INVOLVED IN NUMERICAL INTEGRATION OF NONLINEAR SYSTEMS

The errors introduced into the numerical integration because the procedure in fact provides a solution of a finite difference approximation of the equation of motion rather than of the original difference equation, are common to both linear and nonlinear systems. Additional sources of error are, however, present in the latter case. These additional errors can be classified as those arising due to 1. use of tangent stiffness in place of secant stiffness, and 2. delay in detecting the transitions in the force–displacement relationship.

The errors due to the use of tangent stiffness can be minimized by using a process of iteration. Consider, for example, the graphic representation of Equation 8.204b shown in Figure 8.18a. The relationship shows a nonlinearity because the slope $k^* = 6m/h^2 + 3c/h + k_T$ is not constant, k_T being dependent on displacement and hence on time. In a static load case k^* will be equal to k_T and the nonlinearity will be even more severe. The presence of mass and damping terms in the expression for k^* smooth out the nonlinearity. In particular, when h is small, the constant terms $6m/h^2$ and $3c/h$ are much larger when compared to k_T.

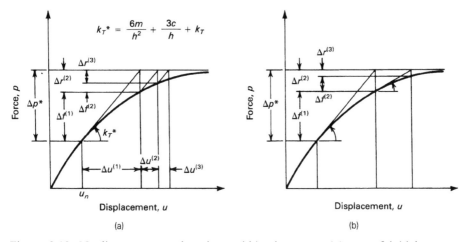

Figure 8.18. Nonlinear system: iterations within time step: (a) use of initial tangent stiffness; (b) use of current tangent stiffness.

The iteration is begun with equation

$$k_T^* \Delta u^{(1)} = \Delta p^* \tag{8.206}$$

For the softening type of nonlinearity shown in Figure 8.18a, where the spring stiffness decreases with an increase in displacement, the equivalent spring force due to a displacement of $\Delta u^{(1)}$ is less than that given by Equation 8.206. As a result, a residual force remains as shown in Figure 8.18a. Let this residual force be denoted by $\Delta r^{(2)}$. Additional displacement is caused by the residual force so that

$$\begin{aligned} k_T^* \Delta u^{(2)} &= \Delta r^{(2)} \\ &= \Delta p^* - \Delta f^{(1)} \end{aligned} \tag{8.207}$$

This additional displacement is used to find a new value of the residual force, and the process is continued until convergence is achieved. Iterations within a time step can be expressed as

$$k_T^* \Delta u^{(k)} = \Delta r^{(k)}$$

$$u_{n+1}^k = u_{n+1}^{(k-1)} + \Delta u^{(k)}$$

$$\Delta f^{(k)} = f_S^{(k)} - f_S^{(k-1)} + \frac{6m}{h^2} \Delta u^{(k)} + \frac{3c}{h} \Delta u^{(k)} \qquad k = 1, \ldots, \hat{k} \tag{8.208}$$

$$\Delta r^{(k+1)} = \Delta r^{(k)} - \Delta f^{(k)}$$

where f_S represents the spring force. Also, as shown by Equation 8.206, iteration commences with $\Delta r^{(1)} = \Delta p^*$. When the iterative process has converged, that is, when $\Delta r^{(k)}$ or $\Delta u^{(k)}$ is sufficiently small, the total incremental displacement is given by

$$\Delta u = \sum_{k=1}^{\hat{k}} \Delta u^{(k)} \tag{8.209}$$

The incremental acceleration and velocity can now be obtained from relationships similar to Equations 8.201 and 8.203.

The procedure outlined above, in which initial stiffness k_T is used in each iteration, is referred to as the modified Newton Raphson method. The iteration process will converge to the correct displacement, although the rate of convergence may be somewhat slow. A better rate of convergence is achieved by using the current tangent stiffness $k_T^{(k)}$ in place of the initial stiffness. This, however, requires that tangent stiffness be evaluated at each iteration, involving additional computations. The improved iteration process is graphically illustrated in Figure 8.18b and is called the full Newton Raphson method.

A convergence criterion must be established for the iteration process. One possible choice is to require that the ratio $\Delta r^{(k)}/\Delta p^*$ is smaller than a specified value. This criterion may not be satisfactory for a system that is elasto-plastic or nearly so. For such a system, even when the residual force is sufficiently small, the displacement may be significantly in error. An alternative is to require that $\Delta u^{(k)}/\Delta u$ is sufficiently small. Since the total increment Δu will not be known during the iteration process, the current estimate of Δu, that is, $\sum_k \Delta u^{(k)}$, may be used in its place. Another possible convergence criterion is to require that the work done in the incremental displacement during an iteration is small in relation to the work done on the total displacement during the time step, that is, $(\Delta r^{(k)} \Delta u^{(k)})/(\Delta p^* \Delta u)$ is small.

The errors involved in numerical integration during transitions in the force–displacement relationship can be minimized by using a subincrement of time to carry out the integration during an interval in which a transition is detected. For example, if it is found that the velocity changed sign during the interval from t_n to t_{n+1}, integration can be retraced in the interval t_n to t_{n+1} with a smaller time step, say, $h/5$. Alternatively, an iterative process may be used in which integration is resumed from t_n with a smaller step whose size is progressively adjusted so that at the end of such an adjusted time step, the velocity is close to zero.

The criteria for the selection of time step used in integration are similar to those for a linear system. The natural period of the system, however, varies as the stiffness changes. The limits imposed on ωh or h/T to minimize truncation errors and to ensure stability of the system should therefore be based on the largest value of ω or the smallest value of the period.

Example 8.16
Repeat Example 8.15 using an iterative procedure to take care of the nonlinearity and the transition phase in the force–displacement relationship.

Solution
A time step of 0.1 s is used except during the transition phase at A, when the time step is adjusted so that zero velocity condition occurs at the end of a time interval. The iteration procedure is given by

$$k_T^* \Delta u^{(k)} = \Delta r^{(k)}$$

$$u_{(n+1)}^{(k)} = u_{(n+1)}^{(k-1)} + \Delta u^{(k)}$$

$$\Delta f^{(k)} = f_S^{(k)} - f_S^{(k-1)} + \frac{4m}{h^2} \Delta u^{(k)} + \frac{2c}{h} \Delta u^{(k)}$$

$$\Delta r^{(k+1)} = \Delta r(k) - \Delta f^{(k)}$$

where

$$k_T^* = \frac{4m}{h^2} + \frac{2c}{h} + k_T$$

$$= 44 + k_T$$

and

$$\Delta p^* = \Delta p + m\left(\frac{4}{h}\dot{u}_n + 2\ddot{u}_n\right) + 2c\dot{u}_n$$

$$= \Delta p + 4.4\dot{u}_n + 0.2\ddot{u}_n$$

Also, it is assumed that convergence has taken place when Δu has no significant digit in the fourth decimal place. Response calculations for the first 0.8 s are shown in Table E8.16. During a time step when nonlinearity exists, more than one iteration is required for convergence. These iterations are all shown against the time step under consideration.

As an example, consider the progression from 0.3 s to 0.4 s. The first iteration is given by

$$\Delta u^{(1)} = \frac{1}{k_T^*}\Delta r^{(1)}$$

$$= \frac{1}{k_T^*}\Delta p^*$$

$$= \frac{31.3839}{51.0968} = 0.6142$$

Hence

$$u^{(1)}(0.4) = u^{(1)}(0.3) + 0.6142 = 1.1182$$

$$\Delta f^{(1)} = f_S(1.1182) - f_S(0.5040) + \frac{4m}{h^2}\Delta u^{(1)} + \frac{2c}{h}\Delta u^{(1)}$$

$$= 7.2885 - 3.8803 + 44 \times 0.6142 = 30.4330$$

$$\Delta r^{(2)} = 31.3839 - 30.4330 = 0.9509$$

$$\Delta u^{(2)} = \frac{0.9509}{51.0968} = 0.0186$$

$$u^{(2)}(0.4) = 1.1182 + 0.0186 = 1.1368$$

$$\Delta f^{(2)} = f_S(1.1368) - f_S(1.1182) + 44 \times 0.0186$$

$$= 7.3532 - 7.2885 + 44 \times 0.0186 = 0.8831$$

$$\Delta r^{(3)} = \Delta r^{(2)} - \Delta f^{(2)}$$

$$= 0.9509 - 0.8831 = 0.0678$$

$$\Delta u^{(3)} = \frac{0.0678}{51.0968} = 0.0013$$

and so on.

Table E8.16. Response of nonlinear system: iteration within time step.

t	u_n	$\dot{u}_n$	$\ddot{u}_n$	p	K_T	K^*	$\Delta p^*/\Delta r$	Δu	u_n
0.0000	0.0000	0.0000	0.0000	0	8.0000	52.0000	1.0000	0.0192	0.0192
							0.84×10^{-5}	0.02×10^{-5}	0.0192
0.1000	0.0192	0.3846	7.6923	1	7.9987	51.9987	6.2307	0.1198	0.1391
							0.0037	0.72×10^{-5}	0.1391
0.2000	0.1391	2.0119	24.8801	4	7.9312	51.9312	18.8284	0.3626	0.5017
							0.1197	0.0023	0.5040
							0.0021	0.0000	0.5040
0.3000	0.5040	5.2861	40.6251	9	7.0968	51.0968	31.3839	0.6142	1.1182
							0.9509	0.0186	1.1368
							0.0678	0.0013	1.1381
							0.0061	0.0001	1.1382
0.4000	1.1382	7.3959	1.6316	9	3.3949	47.3949	29.8683	0.6302	1.7683
							1.4972	0.0316	1.7999
							0.1068	0.0023	1.8021
							0.0056	0.0001	1.8022
0.5000	1.8022	5.8881	-31.7762	6	0.0000	44.0000	13.5524	0.3080	2.1102
0.6000	2.1102	0.2719	-80.5438	0	0.0000	34924.51	16.0826	0.00046	2.1107
0.6034	2.1107	-0.0002	-80.0004	0	8.0000	54.9968	-16.0010	-0.2909	1.8198
0.7000	1.8198	-6.0218	-44.6844	0	8.0000	52.0000	-36.2780	-0.6977	1.1221
0.8000	1.1221	-7.9322	14.9524						

Between 0.6 and 0.7 s, the velocity is found to cross over from positive to negative. By a process of iteration, the time at which $\dot{u}_n = 0$ is found to be 0.60339s. After this time, the stiffness reverts to 8 kips/in. The calculations between 0.6 and 0.60339 s are carried out with a time step $h = 0.00339$ s. Similarly, the calculations between 0.60339 and 0.7 s use $h = 0.09661$ s. Beyond $t = 0.60339$ s, the spring force–displacement relationship follows the unloading path so that the spring force reduces by $8\Delta u$ during each time step. Comparison of results in Table E8.15 and E8.16 shows that for nonlinear systems, response computations in which tangent stiffness is used without further iteration within the time step may lead to significant errors.

SELECTED READINGS

Argyris, J.H., Dunne, P.C. & Angelopoulos, T. 1973. Dynamic Response by Large Step Integration. *Earthquake Engineering and Structural Dynamics*, Vol. 2: 185–203.

Bathe, K.J. 1996. *Finite Element Procedures*. Englewood Cliffs: Prentice Hall.

Bathe, K.J. & Wilson, E.L. 1973. Stability and Accuracy Analysis of Direct Integration Methods. *Earthquake Engineering and Structural Dynamics*, Vol. 1: 283–291.

Belytschko, T. & Scoeberle, D.F. 1975. On the Conditional Stability of an Implicit Algorithm for Nonlinear Structural Dynamics. *Journal of Applied Mechanics*, Vol. 42: 865–869.

Clough, R.W. & Penzien, J. 1993. *Dynamics of Structures*. New York: McGraw-Hill. 2nd Edition.

Goudreau, G.L. & Taylor, R.L. 1972. Evaluation of Numerical Integration Methods in Elastodynamics. *Computer Methods in Applied Mechanics and Engineering*, Vol. 2: 69–97.

Hamming, R.W. 1962. *Numerical Methods for Scientists and Engineers*. New York: McGraw-Hill.

Hilber, H.M. & Hughes, T.J.R. 1978. Collocation, Dissipation, and Overshoot for Time Integration Schemes in Structural Dynamics. *Earthquake Engineering and Structural Dynamics*, Vol. 6: 99–117.

Hilber, H.M., Hughes, T.J.R. & Taylor, R.L. 1977. Improved Numerical Dissipation for Time Integration Algorithms in Structural Dynamics. *Earthquake Engineering and Structural Dynamics*, Vol. 5: 283–292.

Houbolt, J.C. 1950. A Recurrence Matrix Solution for the Dynamic Response of Elastic Aircraft. *Journal of the Aeronautical Sciences*, Vol. 17: 540–550.

Hughes, T.J.R. 1976. Stability, Convergence and Growth and Decay of Energy of the Average Acceleration Method in Nonlinear Structural Dynamics. *Computers and Structures*, Vol. 6: 313–324.

Hughes, T.J.R. 1977. A Note on the Stability of Newmark's Algorithm in Nonlinear Structural Dynamics. *International Journal for Numerical Methods in Engineering*, Vol. 11: 383–386.

Humar, J.L. & Wright, E.W. 1974. Numerical Methods in Structural Dynamics. *Canadian Journal of Civil Engineering*, Vol. 1: 179–193.

Hurty, W.C. & Rubinstein, M.F. 1964. *Dynamics of Structures*. Englewood Cliffs: Prentice Hall.

Krieg, R.D. 1973. Unconditional Stability in Numerical Time Integration Methods. *Journal of Applied Mechanics*, Vol. 40: 417–421.

Mewmark, N.M. 1959. A Method of Computation for Structural Dynamics. *Journal of Engineering Mechanics Division*, Vol. 85: 67–94. ASCE.

Park, K.C. 1975. Evaluating Time Integration Methods for Nonlinear Dynamic Analysis. In T. Belytschko, J.R. Osias & P.V. Marcal (eds) *Finite Element Analysis of Transient Nonlinear Behavior*. AMD-Vol. 14. New York: ASME.

Rayleigh, Lord. 1945. *The Theory of Sound*. New York: Dover.

Temple, G. & Bickley, W.G. 1956. *Rayleigh Principle and its Application to Engineering*. New York: Dover.

Thomson, W.T. 1964. *Theory of Vibration with Applications*. Englewood Cliffs, Prentice Hall. 2nd Edition.

Wilson, E.L., Farhoomand, I. & Bathe, K.J. 1973. Nonlinear Dynamic Analysis of Complex Structures. *Earthquake Engineering and Structural Dynamics*, Vol. 1: 241–252.

PROBLEMS

8.1 Obtain the frequency of vibration of the system shown in Problem 2.1 by using Rayleigh's method.

8.2 Obtain the frequency of vibration for small oscillations of the system shown in Problem 2.3. Neglect damping and use Rayleigh's method.

8.3 Using Rayleigh's method, obtain the frequency of vibration for small oscillations of the inverted pendulum shown in Figure P8.3.

8.4 A steel beam is simply supported over a span of 4.0 m carries weights of 20 kN at the center and 12 kN at 1.0 m from each end (Fig. P8.4). Calculate the frequency of transverse vibration assuming that the deflection shape is similar to the static deflection shape and that the mass of the beam can be neglected in comparison to the mass of the concentrated weights. The static deflection is 1.2 mm at the 20-kN load and 0.85 mm at each of the 12-kN loads.

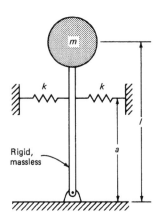

Figure P8.3.

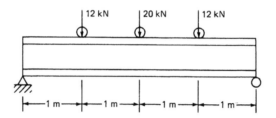

Figure P8.4.

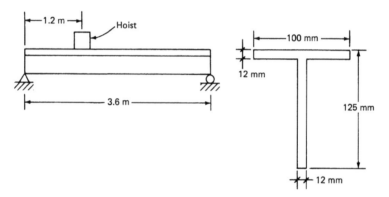

Figure P8.5.

8.5 The aluminum beam shown in Figure P8.5 carries a small hoist weighing 450 N. Calculate the frequency of vibration when the hoist is at a distance of 1.2 m from one support. What are the lowest and highest frequencies and the corresponding hoist locations? $E = 69,000\,\text{MPa}$ and $\rho = 2770\,\text{kg/m}^3$ for aluminum.

8.6 Obtain the frequency of vibration of the structure shown in Figure P8.6 by using Rayleigh's method. Neglect deformations due to shear and axial forces and assume that the vibration shape is that produced by a vertical downward load acting at point C.

8.7 Determine the frequency of vibration of the beam shown in Figure P8.7 by using Rayleigh's method. Assume that the vibration shape is similar to that produced by the deflection of the beam under self weight gravity load applied in the appropriate direction.

8.8 Obtain an estimate of the fundamental frequency of the two degree-of-freedom system shown in Figure P8.8 using Rayleigh's method and the following deflection shape.

$$\psi_0 = \begin{bmatrix} u_{10} \\ u_{20} \end{bmatrix} = \begin{bmatrix} 1 \\ 1 \end{bmatrix}$$

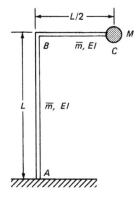

Figure P8.6.

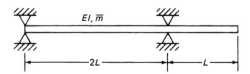

Figure P8.7.

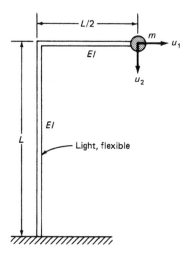

Figure P8.8.

Refine your frequency estimate by using the improved Rayleigh's method.

8.9 The single story building shown in Figure P8.9 is subjected to a blast load whose effect can be represented by a story level lateral force as

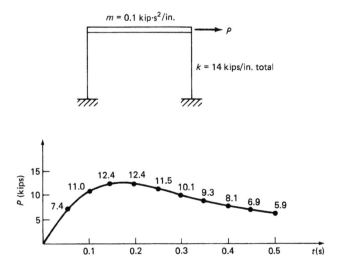

Figure P8.9.

indicated. The total lateral stiffness of the building is 14 kips/in and the floor deck mass is $0.1 \text{kip.s}^2/\text{in}$. Obtain the displacement response history for the first 0.5 s by a numerical integration of the Duhamel's integral. Use a time step of 0.05 s and the trapezoidal method of integration.

8.10   Obtain the response of the structure in Problem 8.9 if damping is 10% of critical. As in Problem 8.9, evaluate the Duhamel's integral by the trapezoidal method of integration and use a time step of 0.05 s.

8.11   Obtain the response of the structure in Problem 8.9 by the average acceleration method. Assume that damping is 10% of critical and use a time step of 0.05 s.

8.12   Repeat Problem 8.11 using the linear acceleration method.

8.13   Repeat Problem 8.11 using the central difference method.

8.14   Solve Problem 8.11 by Houbolt's method. To start the numerical integration, use the displacement values at 0.05 s and 0.10 s obtained in Problem 8.11.

8.15   An eccentric mass shaker installed on the top floor of the three-story shear frame building shown in Figure P8.15 produces a harmonic force $p_0 \sin \Omega t$, where $p_0 = 45 \text{kN}$ and $\Omega = 5\pi/3 \text{rad/s}$. The frame is modeled as a single-degree-of-freedom system by assuming that the vibration shape is $\psi^T = [1 \; 1.75 \; 2.5]$. Find the history of the top-floor displacement for the first 0.6 s using a piece-wise linear representation of the forcing function and a time step of 0.1 s. Find the maximum base shear produced during this time. Neglect damping.

8.16   Repeat Problem 8.15 assuming that damping is 10% of critical.

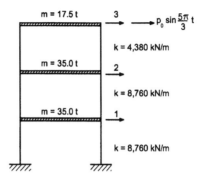

Figure P8.15.

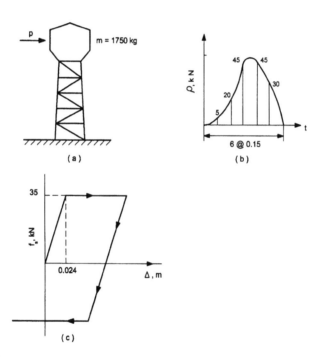

Figure P8.17.

8.17   A water tank modeled as a single-degree-of-freedom system has a mass of 17500 kg (Fig. P8.17). The force–displacement relationship for the supporting structure is elasto-plastic in nature with a yield strength of 35 kN and yield displacement of 0.024 m. The tank is subjected to the blast load shown in the figure. Obtain the response of the tank for the first 0.8 s using the linear acceleration method and a time step of 0.1 s. Neglect damping. Identify the source of errors in your results.

CHAPTER 9

Analysis of response in the frequency domain

9.1 TRANSFORM METHODS OF ANALYSIS

The methods of time-domain analysis outlined in Chapters 5 through 8 can be effectively used to obtain the response of any single-degree-of-freedom system. It is, however, possible in certain situations to greatly improve the efficiency of the analytical procedure by employing the transform method of mathematics. Transform methods have proved to be very useful in many engineering applications. The simplest of these methods is the logarithm. Logarithms allow the transformation of a problem of algebraic multiplication to a simpler problem of addition. The success of the method, however, depends on the availability of the tables of logarithms and antilogarithms of numbers. Logarithm is a transform that applies to discrete value of a variable. There are other transforms that apply to continuous functions; the *Laplace transform* is one of them. It allows the conversion of a differential equation to an algebraic equation which is much simpler to solve. Again, the success of the method depends on the availability of the tables of Laplace transforms and inverse Laplace transforms.

In the analysis of dynamic response of linear structures, Fourier transform can play an important role. Briefly, this transform converts the task of evaluating the convolution integral of Equation 7.9 into that of finding the product of the Fourier transforms of the two functions involved in the convolution. The inverse Fourier transform of the product then gives the desired response function of time. The method of analysis is commonly referred to as *dynamic analysis in the frequency domain*. It would appear that like other transforms, the success of the method should depend on the ease with which the Fourier transform and the inverse Fourier transform of the functions involved in the analysis can be evaluated. Unfortunately, except for simple cases, where the direct method of solving the convolution integral is as effective, the application of Fourier transform involves evaluation of integrals that are not always easy to solve. In most practical cases, therefore, Fourier transform is used in its discrete form, which permits the replacement of the integrals by equivalent numerical computations of areas. In fact, the efficiency of Fourier methods is most apparent in cases where the excitation is specified in terms of numerical values at regular intervals of time rather than as a mathematical function. In such a situation, even the time-domain analysis, which eventually depends on the evaluation of a convolution integral, must rely on numerical methods, and the alternative method of analysis through discrete Fourier transform proves to be most effective.

Ordinarily, the computations involved in obtaining the discrete Fourier transforms of the functions being convolved, taking the product of these transforms and then evaluating the discrete inverse Fourier transform are no less than those in a direct evaluation of the discrete convolution. However, the development of a special algorithm called fast Fourier transform (FFT) has completely altered this position. The FFT algorithm, which derives its efficiency from exploiting the harmonic property of a discrete transform, cuts down the computations by several orders of magnitude and makes frequency-domain analysis highly efficient. The FFT has found widespread application in several areas of engineering analysis. Its use in dynamic analysis of structures is, however, more recent. Textbooks on structural dynamics therefore contain at best a very cursory treatment of the topic. The subject is, however, important enough to justify a more in-depth coverage.

Besides computational efficiency, analysis in the frequency domain possesses several other advantages. By providing a clearer representation of the frequency content of the forcing function, it enables one to evaluate the potential for it to excite a given structure. In addition, for problems involving infinite domains such as those of wave motion in reservoirs of large extent or in unbounded soil media, and in other situations where the physical characteristics of the system are dependent on the vibration frequency, frequency-domain analysis may prove to be more effective than a time-domain analysis. It should, however, be noted that frequency-domain analysis applies only to time-invariant, linear systems for which the principle of superposition holds.

This chapter begins by presenting the response analysis for periodic excitation, which, in fact, is an example of the simplest type of frequency-domain analysis. Analytical methods for evaluating the response of nonperiodic excitation are dealt with next. The concept of Fourier transforms is developed at this stage. This is followed by the introduction of discrete Fourier transforms and their application in the evaluation of dynamic response.

9.2   FOURIER SERIES REPRESENTATION OF A PERIODIC FUNCTION

A function $g(t)$ of time t is said to be a *periodic function* of t with a period equal to T_0 if it satisfies the following relationship

$$g(t + nT_0) = g(t) \tag{9.1}$$

where n is an integer with negative or positive values.

Figure 9.1 presents several examples of periodic functions. A periodic function with a finite number of discontinuities and a finite number of maximas or

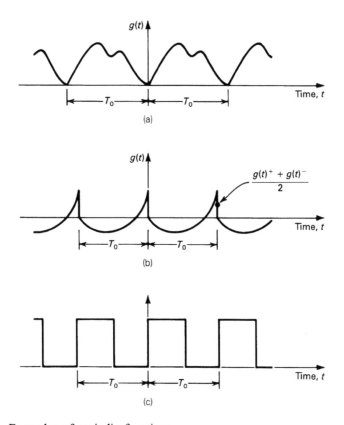

Figure 9.1. Examples of periodic functions.

minimas within a range of time equal to its period T_0 can be represented by an infinite trigonometric series as follows

$$g(t) = a_0 + \sum_{n=1}^{\infty} a_n \cos 2\pi n f_0 t + \sum_{n=1}^{\infty} b_n \sin 2\pi n f_0 t \qquad (9.2)$$

where, a_n and b_n are constants to be determined and $f_0 = 1/T_0$ is the frequency in cycles per second. The trigonometric series of Equation 9.2 is known as the *Fourier series*. Setting $\omega_0 = 2\pi f_0$, where ω_0 is the frequency in radians per second, the series can be expressed in the alternative form

$$g(t) = a_0 + \sum_{n=1}^{\infty} a_n \cos n\omega_0 t + \sum_{n=1}^{\infty} b_n \sin n\omega_0 t \qquad (9.3)$$

To obtain the coefficients a_n, we multiply the left- and the right-hand sides of Equation 9.3 by $\cos m\omega_0 t$ and integrate over a period. Then noting that

$$\int_{-T_0/2}^{T_0/2} \cos n\omega_0 t \cos m\omega_0 t \, dt = \begin{cases} 0 & \text{for } n \neq m \\ \dfrac{T_0}{2} & \text{for } n = m \end{cases} \qquad (9.4a)$$

$$\int_{-T_0/2}^{T_0/2} \cos m\omega_0 t \, dt = \begin{cases} 0 & \text{for } m \neq 0 \\ T_0 & \text{for } m = 0 \end{cases} \tag{9.4b}$$

and

$$\int_{-T_0/2}^{T_0/2} \sin n\omega_0 t \, \cos m\omega_0 t \, dt = 0 \quad \text{for all } m \text{ and } n \tag{9.4c}$$

we obtain the following expression for coefficients a

$$a_0 = \frac{1}{T_0} \int_{-T_0/2}^{T_0/2} g(t) \, dt \tag{9.5a}$$

$$a_n = \frac{2}{T_0} \int_{-T_0/2}^{T_0/2} g(t) \cos n\omega_0 t \, dt \tag{9.5b}$$

Coefficients b_n are obtained in a similar manner by multiplying the left- and the right-hand sides of Equation 9.3 by $\sin m\omega_0 t$ and integrating over a period. Then, because of the relationships

$$\int_{-T_0/2}^{T_0/2} \sin n\omega_0 t \, \sin m\omega_0 t \, dt = \begin{cases} 0 & \text{for } n \neq m \\ T_0/2 & \text{for } n = m \end{cases} \tag{9.6}$$

and that given by Equation 9.4c, the coefficient b_n is obtained as

$$b_n = \frac{2}{T_0} \int_{-T_0/2}^{T_0/2} g(t) \sin n\omega_0 t \, dt \tag{9.7}$$

It can be shown that at a discontinuity such as the one shown in Figure 9.1b, the Fourier series of Equations 9.2 or 9.3 converges to a value that is the average of the values of the function $g(t)$ immediately to the left and the right of the discontinuity.

9.3   RESPONSE TO A PERIODICALLY APPLIED LOAD

From the foregoing discussion, we note that any periodic load can be expressed in terms of the sum of a series of harmonic components. Theoretically speaking, an infinite number of such components must be included to get a precise representation of the load. In practice, the contribution of the successive terms becomes smaller and smaller, and only the first few terms need to be included in the series to obtain a reasonable level of accuracy. Because each component in the series represents a harmonic load, the response to which we already know how to evaluate, the total response of a single-degree-of-freedom system to the periodic load can be obtained simply by summing the responses to the individual components provided that the system is linear and the principle of superposition holds. It should be noted, however, that our periodic representation of the load necessarily implies that the load has been in existence for an indefinite period of time even before time zero. The response that we can hope to evaluate from such a periodic representation will therefore be the steady-state response to the applied load and will not include the transient term.

In Chapter 6, we developed expressions for the steady-state response of a system to several different types of applied loads. When the applied load is constant and has a magnitude p_0, the steady-state response is given by

$$u(t) = \frac{p_0}{k} \tag{9.8}$$

For a sine function load, $p_0 \sin \Omega t$, the steady-state response of an undamped system is

$$u(t) = \frac{p_0}{k} \frac{1}{1 - \beta^2} \sin \Omega t \tag{9.9}$$

where, $\beta = \Omega/\omega$ and ω is the natural frequency of the system. The response of a damped system is given by

$$u(t) = \frac{p_0}{k} \frac{1}{(1 - \beta^2)^2 + (2\xi\beta)^2} \{(1 - \beta^2) \sin \Omega t - 2\xi\beta \cos \Omega t\} \tag{9.10}$$

For a cosine function load, $p_0 \cos \Omega t$, the undamped response is

$$u(t) = \frac{p_0}{k} \frac{1}{1 - \beta^2} \cos \Omega t \tag{9.11}$$

while the damped response is

$$u(t) = \frac{p_0}{k} \frac{1}{(1 - \beta^2)^2 + (2\xi\beta)^2} \{2\xi\beta \sin \Omega t + (1 - \beta^2) \cos \Omega t\} \tag{9.12}$$

Using expressions 9.8, 9.9, and 9.11, the undamped steady-state response of a system to a periodic force represented by the Fourier series of Equation 9.3 is obtained as

$$u(t) = \frac{a_0}{k} + \sum_{n=1}^{\infty} \frac{a_n}{k} \frac{1}{1 - \beta_n^2} \cos n\omega_0 t + \sum_{n=1}^{\infty} \frac{b_n}{k} \frac{1}{1 - \beta_n^2} \sin n\omega_0 t \tag{9.13}$$

where $\beta_n = n\omega_0/\omega$. In a similar manner, the damped response of the system can be shown to be

$$u(t) = \frac{a_0}{k} + \sum_{n=1}^{\infty} \frac{a_n}{k} \frac{1}{(1 - \beta_n^2)^2 + (2\xi\beta_n)^2}$$

$$\times \{2\xi\beta_n \sin n\omega_0 t + (1 - \beta_n^2) \cos n\omega_0 t\}$$

$$+ \sum_{n=1}^{\infty} \frac{b_n}{k} \frac{1}{(1 - \beta_n^2)^2 + (2\xi\beta_n)^2} \{(1 - \beta_n^2) \sin n\omega_0 t - 2\xi\beta_n \cos n\omega_0 t\}$$

$$= \frac{a_0}{k} + \sum_{n=1}^{\infty} \frac{1}{k} \frac{1}{(1 - \beta_n^2)^2 + (2\xi\beta_n)^2} [\{a_n 2\xi\beta_n + b_n(1 - \beta_n^2)\} \sin n\omega_0 t$$

$$+ \{a_n(1 - \beta_n)^2 - b_n 2\xi\beta_n\} \cos n\omega_0 t] \qquad (9.14)$$

Equations 9.13 and 9.14 can be recognized as Fourier series representations of the response, which must therefore be periodic with a period equal to T_0.

Example 9.1
Obtain the steady-state response of the system shown in Figure E9.1a to the loading shown in Figure E9.1b.

Solution
To obtain a periodic representation of the excitation force, we extend the function on the negative direction of the time axis as shown in Figure E9.1b. The period T_0 is seen to be 4 s

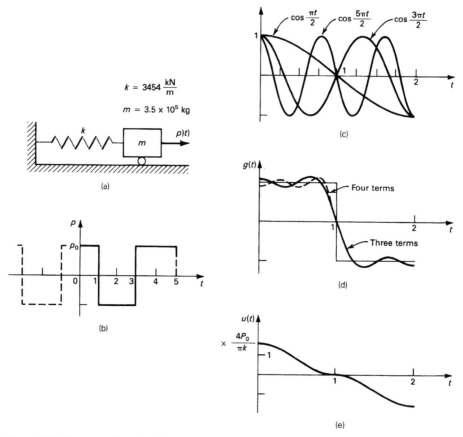

Figure E9.1. Response of a single-degree-of-freedom system to a periodic load: (a) system model; (b) periodic load; (c) first three harmonics of the loading function; (d) Fourier series representation of the loading function; (e) response to periodic load.

and the coefficients of the Fourier series are therefore given by

$$a_0 = \frac{1}{4} \int_{-2}^{2} g(t) \, dt \tag{a}$$

$$a_n = \frac{2}{4} \int_{-2}^{2} g(t) \cos \frac{2n\pi t}{4} \, dt \tag{b}$$

$$b_n = \frac{2}{4} \int_{-2}^{2} g(t) \sin \frac{2n\pi t}{4} \, dt \tag{c}$$

Because $g(t)$ is even, Equations a, b, and c give

$$a_0 = \frac{1}{2} \int_{0}^{2} g(t) \, dt$$

$$= \frac{1}{2} p_0 \int_{0}^{1} dt - \frac{1}{2} p_0 \int_{1}^{2} dt = 0 \tag{d}$$

$$a_n = \int_{0}^{2} g(t) \cos \frac{2n\pi t}{4} \, dt$$

$$= p_0 \int_{0}^{1} \cos \frac{2n\pi t}{4} \, dt - p_0 \int_{1}^{2} \cos \frac{2n\pi t}{4} \, dt$$

$$= \frac{4 p_0}{n\pi} \sin \frac{n\pi}{2} \tag{e}$$

$$b_n = 0 \tag{f}$$

The Fourier series representation of the periodic forcing function is, therefore

$$g(t) = \frac{4 p_0}{\pi} \left(\cos \frac{\pi t}{2} - \frac{1}{3} \cos \frac{3\pi t}{2} + \frac{1}{5} \cos \frac{5\pi t}{2} - \frac{1}{7} \cos \frac{7\pi t}{2} + \cdots \right) \tag{g}$$

The first three harmonics and their sum are shown in Figure E9.1c and d. It is seen that the first harmonic term, which has a period of 4 s or a frequency of $\frac{1}{4}$ Hz, has an amplitude of $4p_0/\pi$. The amplitude decreases rapidly as the frequency increases, so that the second harmonic with a frequency of $\frac{3}{4}$ Hz has an amplitude of only $4p_0/3\pi$, while the third harmonic has a frequency of $\frac{5}{4}$ Hz and an amplitude of $4p_0/5\pi$. It is also seen that even when just three terms are included, the series provides a reasonable representation of the rectangular forcing function.

At $t = 1$ s where the forcing function is discontinuous, the Fourier series converges to the average value of the function, which in this case is zero. The natural frequency of the system is $\omega = \sqrt{k/m} = \pi$ rad/s. The response of the system, which is undamped, is now obtained by substituting the value of a_n in Equation 9.13

$$u(t) = \frac{4 p_0}{\pi k} \left(\frac{1}{1 - \beta_1^2} \cos \frac{\pi t}{2} - \frac{1}{3} \frac{1}{1 - \beta_3^2} \cos \frac{3\pi t}{2} \right.$$

$$\left. + \frac{1}{5} \frac{1}{1 - \beta_5^2} \cos \frac{5\pi t}{2} + \cdots \right) \tag{h}$$

where $\beta_n = n\omega_0/\omega = n/2$. Figure E9.1e shows the response over half a period, obtained by summing the first three terms of the series in Equation h.

9.4   EXPONENTIAL FORM OF FOURIER SERIES

In developing the frequency-domain analysis procedure, it is convenient first to obtain an exponential form for the Fourier series of Equation 9.3. Using de Moivre's theorem, the sine and cosine functions in Equation 9.3 can be expressed in terms of complex exponentials as follows

$$\sin n\omega_0 t = \frac{e^{in\omega_0 t} - e^{-in\omega_0 t}}{2i} \tag{9.15}$$

$$\cos n\omega_0 t = \frac{e^{in\omega_0 t} + e^{-in\omega_0 t}}{2} \tag{9.16}$$

Substitution of Equation 9.16 in Equation 9.5b yields

$$a_n = \frac{1}{T_0} \int_{-T_0/2}^{T_0/2} g(t)(e^{in\omega_0 t} + e^{-in\omega_0 t})\, dt \tag{9.17a}$$

If we replace n by $-n$ in Equation 9.17a, we get

$$a_{-n} = \frac{1}{T_0} \int_{-T_0/2}^{T_0/2} g(t)(e^{-in\omega_0 t} + e^{in\omega_0 t})\, dt = a_n \tag{9.17b}$$

In a similar manner, b_n is obtained by substituting Equation 9.15 in Equation 9.7

$$b_n = \frac{1}{T_0} \int_{-T_0/2}^{T_0/2} g(t) \frac{e^{in\omega_0 t} - e^{-in\omega_0 t}}{i}\, dt \tag{9.18a}$$

$$b_{-n} = \frac{1}{T_0} \int_{-T_0/2}^{T_0/2} g(t) \frac{e^{-in\omega_0 t} - e^{in\omega_0 t}}{i}\, dt = -b_n \tag{9.18b}$$

The Fourier series of Equation 9.3 can be expressed as

$$g(t) = a_0 + \sum_{n=1}^{\infty} a_n \frac{e^{in\omega_0 t} + e^{-in\omega_0 t}}{2} + \sum_{n=1}^{\infty} b_n \frac{e^{i\omega_0 n t} - e^{-in\omega_0 t}}{2i}$$

$$= a_0 + \sum_{n=1}^{\infty} e^{in\omega_0 t} \left(\frac{a_n}{2} + \frac{b_n}{2i} \right) + \sum_{n=1}^{\infty} e^{-in\omega_0 t} \left(\frac{a_n}{2} - \frac{b_n}{2i} \right) \tag{9.19}$$

Using relationships 9.17b and 9.18b, Equation 9.19 becomes

$$g(t) = a_0 + \sum_{n=1}^{\infty} e^{in\omega_0 t} \left(\frac{a_n}{2} + \frac{b_n}{2i} \right) + \sum_{n=-\infty}^{-1} e^{in\omega_0 t} \left(\frac{a_n}{2} + \frac{b_n}{2i} \right)$$

$$= \sum_{n=-\infty}^{\infty} c_n e^{in\omega_0 t} \tag{9.20}$$

in which coefficient c_n is given by

$$c_n = \frac{1}{2}(a_n - ib_n)$$

$$= \frac{1}{2T_0} \int_{-T_0/2}^{T_0/2} g(t)\{(e^{in\omega_0 t} + e^{-in\omega_0 t}) - (e^{in\omega_0 t} - e^{-in\omega_0 t})\} \, dt$$

$$= \frac{1}{T_0} \int_{-T_0/2}^{T_0/2} g(t)e^{-in\omega_0 t} \, dt \tag{9.21a}$$

and

$$c_0 = \frac{1}{T_0} \int_{-T_0/2}^{T_0/2} g(t) \, dt = a_0 \tag{9.21b}$$

Equations 9.20 and 9.21 together represent the *exponential form* of the Fourier series, which is seen to be much more compact than the alternative form given by Equations 9.3, 9.5, and 9.7.

9.5 COMPLEX FREQUENCY RESPONSE FUNCTION

In Section 9.3, we developed a method for obtaining the response of a system to a periodically applied load. The procedure we used was first to express the applied load as the sum of a series of harmonic components. Knowing the steady-state response to a harmonic load, we could obtain the response to each harmonic component of the periodic load. The sum of these component responses gave us the total response. We can use a similar procedure when the Fourier series representing the applied load is given in the exponential form, but we must first obtain the steady-state response of the system to an excitation given by the complex function $p = p_0 e^{i\omega_0 t}$. This can be achieved by solving the following differential equation for u.

$$m\ddot{u} + c\dot{u} + ku = p_0 e^{i\omega_0 t} \tag{9.22}$$

The complementary solution to Equation 9.22 is given by Equation 5.36.

$$u = e^{-\xi\omega t}(A \cos \omega_d t + B \sin \omega_d t) \tag{5.36}$$

The particular integral solution of Equation 9.22 is of the form

$$u = Ge^{i\omega_0 t} \tag{9.23}$$

Substituting u and its derivatives from Equation 9.23 into Equation 9.22, we get

$$(-m\omega_0^2 + ic\omega_0 + k)Ge^{i\omega_0 t} = p_0 e^{i\omega_0 t} \tag{9.24}$$

The complete solution of Equation 9.22 is then obtained by summing Equations 5.36 and 9.23 and substituting the value of G from Equation 9.24

$$u = e^{-\xi\omega t}(A\cos\omega_d t + B\sin\omega_d t) + \frac{p_0}{-m\omega_0^2 + ic\omega_0 + k}e^{i\omega_0 t} \tag{9.25}$$

The first term on the right-hand side of Equation 9.25 rapidly dies out with time and represents the *transient response*. The second term in Equation 9.25 is the *steady-state response*. In summary, the steady-state response of a system to an applied load $p_0 e^{i\omega_0 t}$ is given by $H(\omega_0)p_0 e^{i\omega_0 t}$, where

$$H(\omega_0) = \frac{1}{-m\omega_0^2 + ic\omega_0 + k} \tag{9.26a}$$

$$= \frac{1}{k(-\beta^2 + 2i\xi\beta + 1)} \tag{9.26b}$$

Function $H(\omega_0)$ is called the *complex frequency function*. When damping is zero, Equation 9.26 reduces to

$$H(\omega_0) = \frac{1}{m(\omega^2 - \omega_0^2)} \tag{9.27}$$

The response to a periodic load represented by Equation 9.20 is now easily obtained as

$$u(t) = \sum_{n=-\infty}^{\infty} c_n H(n\omega_0)e^{in\omega_0 t} \tag{9.28a}$$

where

$$c_n = \frac{1}{T_0} \int_{-T_0/2}^{T_0/2} g(t)e^{-in\omega_0 t}\,dt \tag{9.28b}$$

9.6   FOURIER INTEGRAL REPRESENTATION OF A NONPERIODIC LOAD

To obtain a Fourier representation of a nonperiodic load, we first construct a periodic version of the given load. The first step is to select a value for the period. The selected period should, of course, be larger than the duration of the applied load. Within each period, the periodic version has a magnitude equal to the specified load for the duration of the latter but is zero otherwise. A periodic version constructed as above is shown in Figure 9.2.

It is evident that the curves shown by dashed lines represent fictitious loads that, in fact, do not exist. If T_0 is now increased to a very large value, the fictitious loads will move to infinity and in the limit, we will get a true representation of the applied load. We apply this reasoning to derive a Fourier

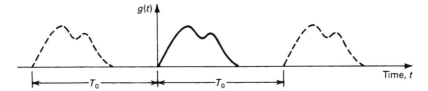

Figure 9.2. Periodic version of a loading function.

representation of the applied load from Equations 9.20 and 9.21. First, since T_0 is very large, we set

$$\omega_0 = \frac{2\pi}{T_0} = \Delta\Omega \tag{9.29a}$$

and

$$n\omega_0 = \Omega_n \tag{9.29b}$$

With this notation, Equation 9.21 gives

$$c_n T_0 = \int_{-T_0/2}^{T_0/2} g(t)e^{-i\Omega_n t}\, dt \tag{9.30}$$

From Equation 9.20

$$g(t) = \frac{1}{T_0} \sum_{-\infty}^{\infty} c_n T_0 e^{i\Omega_n t}$$

$$= \frac{1}{2\pi} \sum_{-\infty}^{\infty} c_n T_0 e^{i\Omega_n t} \Delta\Omega \tag{9.31}$$

In the limit, as T_0 approaches infinity, Ω_n becomes a continuous function and Equation 9.30 becomes

$$c_n T_0 = G(\Omega) = \int_{-\infty}^{\infty} g(t)e^{-i\Omega t}\, dt \tag{9.32}$$

while Equation 9.31 takes the form

$$g(t) = \frac{1}{2\pi} \int_{-\infty}^{\infty} G(\Omega)e^{i\Omega t}\, d\Omega \tag{9.33}$$

where we have replaced the discrete coordinate $\Omega_n = n\Delta\Omega$ by its continuous version Ω.

Equation 9.32 represents the *Fourier transform* of the time function $g(t)$, while Equation 9.33 is the *inverse Fourier transform*. The two together are called Fourier transform pair. They are central to the analysis through frequency domain. In the discussion that follows, we will denote the time function by a lowercase letter, and the Fourier transform of the function by the same uppercase letter. The relationship between a transform pair will be expressed by the following notation

$$g(t) \Leftrightarrow G(\Omega) \tag{9.34}$$

9.7   RESPONSE TO A NONPERIODIC LOAD

Having obtained the Fourier representation of a nonperiodic load, it is straightforward to obtain the response of a linear system to an applied load that is nonperiodic. Thus, when T_0 is large, Equations 9.28 and 9.29 give

$$u(t) = \frac{1}{T_0} \sum_{-\infty}^{\infty} (c_n T_0) H(\Omega_n) e^{i\Omega_n t}$$

$$= \frac{1}{2\pi} \sum_{-\infty}^{\infty} (c_n T_0) H(\Omega_n) e^{i\Omega_n t} \Delta\Omega \tag{9.35}$$

In the limit, as T_0 approaches ∞, Equation 9.35 becomes

$$u(t) = \frac{1}{2\pi} \int_{-\infty}^{\infty} G(\Omega) H(\Omega) e^{i\Omega t} \, d\Omega \tag{9.36}$$

where $G(\Omega) = c_n T_0$ is given by Equation 9.32. A nonperiodic load of special interest is a unit impulse applied at, say, $t = 0$.

The *unit impulse load* is most conveniently expressed in terms of the delta function. Thus

$$g(t) = \delta(t) \tag{9.37}$$

In Chapter 7, we had derived expressions for the response of a system to a unit impulse load. For an undamped and a damped system, respectively, these expressions are given by Equations 7.4 and 7.5.

$$h(t) = \frac{1}{m\omega} \sin \omega t \tag{7.4}$$

$$h(t) = \frac{1}{m\omega_d} e^{-\xi\omega t} \sin \omega_d t \tag{7.5}$$

The Fourier transform of the unit impulse is

$$G(\Omega) = \int_{-\infty}^{\infty} \delta(t) e^{-i\Omega t} \, dt$$

$$= 1 \tag{9.38}$$

and the response to a unit impulse is therefore given by

$$h(t) = \frac{1}{2\pi} \int_{-\infty}^{\infty} G(\Omega)H(\Omega)e^{i\Omega t}\,d\Omega$$

$$= \frac{1}{2\pi} \int_{-\infty}^{\infty} H(\Omega)e^{i\Omega t}\,d\Omega$$

(9.39)

Equation 9.39 shows that $h(t)$ is the inverse transform of $H(\Omega)$ and we have the important relationship

$$h(t) \Leftrightarrow H(\Omega)$$

(9.40)

The choice of symbol h to represent the unit impulse function and the uppercase version H to represent the complex frequency response function is thus in conformity with our selected notation for a Fourier transform pair.

9.8   CONVOLUTION INTEGRAL AND CONVOLUTION THEOREM

The convolution integral and the convolution theorem which enables us to obtain the Fourier transform of the convolution integral play a central role in the frequency-domain analysis of the dynamic response of structures. In Chapter 7, we indicated the form of convolution integral and proved that it gives the forced response of a linear time invariant system starting from zero initial conditions. In this section, we present a formal definition of the convolution integral, demonstrate its physical interpretation, and show how it is possible to obtain its value through Fourier transform analysis. The convolution of two functions $g(t)$ and $h(t)$, denoted as $g(t) * h(t)$, is given by

$$u(t) = g(t) * h(t) = \int_{-\infty}^{\infty} g(\tau)h(t - \tau)\,d\tau$$

(9.41)

A physical interpretation of the convolution integral can be provided by considering an example. Let it be required to convolve the two functions of time $g(\tau)$ and $h(\tau)$ shown in Figure 9.3a and b, respectively. Function $g(\tau)$ may, for example, represent an exciting force function while $h(\tau)$ may represent a unit impulse response function. For the purposes of demonstration, we have kept the two functions simple. Also note that in keeping with our description of $g(\tau)$ and $h(\tau)$ as possible exciting force and unit impulse response, respectively, we have chosen our functions so that they have a null value for τ less than 0.

As shown in Figure 9.3c, function $h(-\tau)$ is the mirror image of $h(\tau)$ about the y axis. The process of obtaining $h(-\tau)$ is also referred to as folding of the original function about the y axis. Function $h(t - \tau)$ is simply the function $h(-\tau)$ shifted to the right by time t. As an example, Figure 9.3d shows $h(t - \tau)$

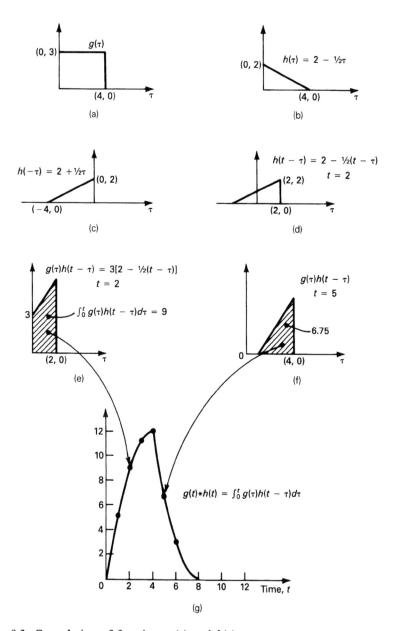

Figure 9.3. Convolution of functions $g(t)$ and $h(t)$.

for $t = 2$. The integrals of the product function $g(\tau)h(t-\tau)$ between limits $-\infty$ and ∞ are shown in Figure 9.3g for different values of t. We note that since $g(\tau)$ is 0 for $\tau < 0$, the lower limit of integration is effectively zero. In a similar manner, because $h(t - \tau)$ is 0 for $\tau > t$, the upper limit of integration becomes

t. The convolution integral of Equation 9.41 can therefore be expressed as

$$u(t) = g(t) * h(t) = \int_0^t g(\tau)h(t - \tau)\, d\tau \tag{9.42}$$

Equation 9.42 is precisely the same as the Duhamel's equation. Therefore, when $g(\tau)$ is the exciting force and $h(\tau)$ the unit impulse response function of a linear time invariant system, the convolution integral $g(t) * h(t)$ gives the response of that system at time t.

The frequency-domain analysis procedure depends on an evaluation of the Fourier transform of the convolution integral (Eq. 9.41 or 9.42). The required transform is given by

$$U(\Omega) = \int_{-\infty}^{\infty} \int_{-\infty}^{\infty} g(\tau)h(t - \tau)\, d\tau\, e^{-i\Omega t}\, dt \tag{9.43}$$

By interchanging the order of integration in Equation 9.43, we obtain

$$U(\Omega) = \int_{-\infty}^{\infty} \int_{-\infty}^{\infty} h(t - \tau)e^{-i\Omega t}\, dt\, g(\tau)\, d\tau \tag{9.44}$$

If we set $t - \tau = \alpha$, so that $t = \alpha + \tau$, Equation 9.44 becomes

$$U(\Omega) = \int_{-\infty}^{\infty} \int_{-\infty}^{\infty} h(\alpha)e^{-i\Omega\alpha}\, d\alpha\, e^{-i\Omega\tau}g(\tau)\, d\tau$$

$$= \int_{-\infty}^{\infty} H(\Omega)e^{-i\Omega\tau}g(\tau)\, d\tau \tag{9.45}$$

$$= H(\Omega)G(\Omega)$$

Equation 9.45 proves a theorem of great significance in the frequency-domain analysis. It shows that the Fourier transform of the convolution integral is equal to the product of the Fourier transforms of the functions being convolved.

As stated earlier, convolution between the exciting force function and the unit impulse response function gives the resultant response of a linear time invariant system to the exciting force. The Fourier transform of the convolution integral in that case is the product of function $G(\Omega)$, the transform of the exciting force, and $H(\Omega)$, the transform of the unit impulse function. It was shown that the Fourier transform of the unit impulse function is, in fact, equal to the complex frequency response function given by Equation 9.26. In summary, the following relationships hold

$$g(t) \Leftrightarrow G(\Omega)$$

$$h(t) \Leftrightarrow H(\Omega) \tag{9.46}$$

$$u(t) = g(t) * h(t) \Leftrightarrow G(\Omega)H(\Omega)$$

It also follows that $u(t)$ can be obtained by taking the inverse Fourier transform of $G(\Omega)H(\Omega)$. Thus

$$u(t) = \frac{1}{2\pi} \int_{-\infty}^{\infty} G(\Omega)H(\Omega)e^{i\Omega t}\, d\Omega \tag{9.47}$$

This is the same relationship that we obtained directly in Section 9.7.

9.9   DISCRETE FOURIER TRANSFORM

In the foregoing sections, we developed the essential steps in the frequency domain-analysis of the dynamic response of structures. These steps can be summarized as follows:
1. Obtain the Fourier transforms of the excitation function and the unit impulse response function.
2. Take the product of the two transforms obtained in step 1.
3. Take the inverse Fourier transform of the product to obtain the desired response.

The success of the procedure therefore depends on being able to obtain the direct and inverse Fourier transforms of the desired functions. However, as noted earlier, the transform expressions involve complex integrals whose evaluation is often quite tedious. In fact, in many problems, a numerical evaluation of the transforms may be the only practical method of obtaining a solution. Quite often, the forcing function or the unit impulse response function, or both, are available only in the form of discrete values at specified intervals of time. There are also cases where the Fourier transform of the impulse function is available directly but only in a discrete form. In all such situations, recourse must be had to a numerical evaluation of the transforms involved. In view of these considerations, it is useful to develop the concept of discrete Fourier transform and to establish a relationship between continuous and discrete transforms.

Consider the time function shown in Figure 9.4. Let it be sampled at N equal intervals from $n=0$ to $N-1$, implying that N discrete values of the function are known at time steps $0, \Delta t, 2\Delta t, \dots, (N-1)\Delta t$. It can then be shown that with a suitable choice for the coefficients a_n, the curve represented by the

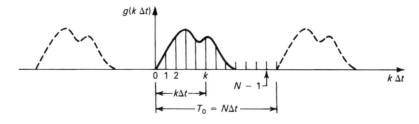

Figure 9.4. Sampled periodic function.

following series can be made to pass through the given points where $T_0 = N\Delta t$ and $\Delta\Omega = 2\pi/T_0$.

$$g(k\Delta t) = \frac{1}{2\pi} \sum_{n=0}^{N-1} a_n e^{2\pi i k \Delta t (n/T_0)} \Delta\Omega$$
$$= \frac{1}{T_0} \sum_{n=0}^{N-1} a_n e^{2\pi i k n/N} \tag{9.48}$$

To obtain the coefficient a_n, we multiply both sides of Equation 9.48 by $e^{-2\pi i k m/N}$ and take a sum as k varies from 0 to $N-1$.

$$\sum_{k=0}^{N-1} g(k\Delta t) e^{-2\pi i k m/N} = \frac{1}{T_0} \sum_{n=0}^{N-1} \sum_{k=0}^{N-1} a_n e^{2\pi i k n/N} e^{-2\pi i k m/N}$$
$$= \frac{1}{T_0} \sum_{n=0}^{N-1} \sum_{k=0}^{N-1} a_n e^{2\pi i k(n-m)/N} \tag{9.49}$$

The nth term in the outer sum is given by

$$p_n = \frac{1}{T_0} \sum_{k=0}^{N-1} a_n e^{2\pi i k(n-m)/N} \tag{9.50}$$

For the case when $n = m$, Equation 9.50 gives

$$p_m = \frac{1}{T_0} a_m N = \frac{a_m}{\Delta t} \tag{9.51}$$

When $n \neq m$, Equation 9.50 represents a geometric progression with common ratio $r = e^{2\pi i(n-m)/N}$ whose sum is given by

$$p_n = \frac{1}{T_0} a_n \frac{1 - r^N}{1 - r}$$
$$= \frac{1}{T_0} a_n \frac{1 - e^{2\pi i(n-m)}}{1 - e^{2\pi i(n-m)/N}}$$
$$= 0 \tag{9.52}$$

where the last result is obtained by expanding $e^{2\pi i(n-m)}$ in the numerator by de Moivre's theorem and noting that $\cos 2\pi(n-m) = 1$ and $\sin 2\pi(n-m) = 0$.
   Equation 9.49 thus reduces to

$$\sum_{k=0}^{N-1} g(k\Delta t) e^{-2\pi i k m/N} = \frac{a_m}{\Delta t} \tag{9.53a}$$

or

$$a_m = \sum_{k=0}^{N-1} g(k\Delta t) e^{-2\pi i k m/N} \Delta t \tag{9.53b}$$

From Equations 9.48 and 9.53 we obtain the following *discrete Fourier trans-form* pair

$$G(n\Delta\Omega) = \sum_{k=0}^{N-1} g(k\Delta t)e^{-2\pi ikn/N}\Delta t$$

$$= \sum_{k=0}^{N-1} g(k\Delta t)e^{-ik\Delta tn\Delta\Omega}\Delta t \qquad (9.54\text{a})$$

$$g(k\Delta t) = \frac{1}{2\pi}\sum_{n=0}^{N-1} G(n\Delta\Omega)e^{2\pi ikn/N}\Delta\Omega \qquad (9.54\text{b})$$

where $G(n\Delta\Omega)$ is equivalent to a_n.

From Equation 9.54b

$$g\{(N+k)\Delta t\} = \frac{1}{2\pi}\sum_{n=0}^{N-1} G(n\Delta\Omega)e^{2\pi in(k+N)/N}\Delta\Omega$$

$$= \frac{1}{2\pi}\sum_{n=0}^{N-1} G(n\Delta\Omega)e^{2\pi ink/N}e^{2\pi in}\Delta\Omega$$

$$= \frac{1}{2\pi}\sum_{n=0}^{N-1} G(n\Delta\Omega)e^{2\pi ink/N}\Delta\Omega \qquad (9.55)$$

$$= g(k\Delta t)$$

The function represented by Equation 9.54b is therefore a periodic extension of the original function we were trying to represent, the period being $T_0 = N\Delta t$. In a similar manner, we can show that the discrete Fourier transform given by Equation 9.54a is periodic with a period $(N\Delta\Omega)$.

On comparing Equation 9.54a and its continuous counterpart (Eq. 9.32), we note that the two are equivalent. The integral in the continuous version has been replaced by a rectangular sum over a finite time period $N\Delta t$ in the discrete version. Obviously, if the discrete version has to parallel the continuous version closely, the time step Δt should be sufficiently small. In a similar manner, a comparison of Equations 9.54b and 9.33 shows that the two are equivalent. Again, integration in the continuous version has been replaced by rectangular sum over a finite frequency band $N\Delta\Omega$ in the discrete version, and if the two versions have to match each other closely, the frequency step $\Delta\Omega$ should be small.

An important difference exists between the continuous and discrete Fourier transforms. While the continuous transform provides a true representation of the given function, the discrete transform represents only a periodic version of the function. Within a period $T_0 = N\Delta t$, the periodic version is similar to the continuous version; outside the period the two are quite different, unless the original function also happens to be periodic with a period T_0.

9.10 DISCRETE CONVOLUTION AND DISCRETE CONVOLUTION THEOREM

We define the *discrete convolution* of two functions as

$$u(k\Delta t) = \sum_{m=0}^{N-1} g(m\Delta t)h\{(k-m)\Delta t\}\Delta t \tag{9.56}$$

where both $g(t)$ and $h(t)$ are periodic functions with a period $N\Delta t$, so that

$$\begin{aligned} g(m+rN)\Delta t &= g(m\Delta t) \\ h(m+rN)\Delta t &= h(m\Delta t) \end{aligned} \quad r = 0, \pm 1, \pm 2, \dots \tag{9.57}$$

On comparing Equations 9.41 and 9.56, we note that the discrete convolution is entirely analogous to continuous convolution. Two important differences, however, exist. First, the discrete convolution is carried out on periodic extensions of the actual functions. This is necessary because the use of discrete Fourier transforms is possible only if the functions being convolved are periodic. Depending on the type of functions being convolved, the convolution of the extended functions may or may not be different from that of the original functions. The second difference is that the integration in a continuous convolution is replaced by rectangular summation in the discrete convolution. If the time interval Δt is small, the difference between the two is expected to be negligible.

Analogous to the case of continuous functions, a relationship exists between the discrete convolution of two periodic functions of time and the product of their discrete Fourier transforms. Mathematically, this relationship can be expressed as

$$g(k\Delta t) * h(k\Delta t) \Leftrightarrow G(n\Delta\Omega)H(n\Delta\Omega) \tag{9.58}$$

A discrete Fourier transform of the convolution exists only when both $g(k\Delta t)$ and $h(k\Delta t)$ are periodic. It is then obtained by using Equation 9.54a

$$\begin{aligned} U(n\Delta\Omega) &= \sum_{k=0}^{N-1} u(k\Delta t)e^{-2\pi i kn/N}\Delta t \\ &= \sum_{k=0}^{N-1}\left[\sum_{m=0}^{N-1} g(m\Delta t)h\{(k-m)\Delta t\}\Delta t\right]e^{-2\pi i kn/N}\Delta t \end{aligned} \tag{9.59}$$

Changing the order of summation in Equation 9.59, we get

$$U(n\Delta\Omega) = \sum_{m=0}^{N-1}\left[\sum_{k=0}^{N-1} h\{(k-m)\Delta t\}e^{-2\pi i kn/N}\Delta t\right]g(m\Delta t)\Delta t \tag{9.60}$$

Substitution of $k - m = \alpha$ in Equation 9.60 gives

$$U(n\Delta\Omega) = \sum_{m=0}^{N-1} \left\{ \sum_{\alpha=-m}^{N-1-m} h(\alpha\Delta t)e^{-2\pi i\alpha n/N}\Delta t \right\} g(m\Delta t)e^{-2\pi imn/N}\Delta t \quad (9.61)$$

Because $h(\alpha\Delta t)$ is periodic and the summation from $\alpha = -m$ to $(N - 1 - m)$ extends over a period

$$\sum_{\alpha=-m}^{N-1-m} h(\alpha\Delta t)e^{-2\pi i\alpha n/N}\Delta t = \sum_{\alpha=0}^{N-1} h(\alpha\Delta t)e^{-2\pi i\alpha n/N}\Delta t$$

$$= H(n\Delta\Omega) \tag{9.62}$$

where the last relationship follows from the definition of discrete Fourier transform (Eq. 9.54a). Substitution of Equation 9.62 into Equation 9.61 gives

$$U(n\Delta\Omega) = \sum_{m=0}^{N-1} H(n\Delta\Omega)g(m\Delta t)e^{-2\pi imn/N}\Delta t = H(n\Delta\Omega)G(n\Delta\Omega) \quad (9.63)$$

which proves Equation 9.58.

The application of discrete transforms in the evaluation of the response of a linear system can now be summarized as follows.

1. Obtain the discrete transform of a periodic extension of the applied load $g(t)$.

$$G(n\Delta\Omega) = \sum_{k=0}^{N-1} g(k\Delta t)e^{-2\pi ikn/N}\Delta t \tag{9.64}$$

2. Obtain the discrete transform of a periodic extension of the unit impulse function $h(t)$. This will usually require a truncation of $h(t)$ at a time equal to or less than the selected period T_0.

$$H(n\Delta\Omega) = \sum_{k=0}^{N-1} h(k\Delta t)e^{-2\pi ikn/N}\Delta t \tag{9.65}$$

3. Obtain the inverse transform of the product of $G(n\Delta\Omega)$ and $H(n\Delta\Omega)$.

$$U(n\Delta\Omega) = G(n\Delta\Omega)H(n\Delta\Omega) \tag{9.66a}$$

$$u(k\Delta t) = \frac{1}{2\pi} \sum_{n=0}^{N-1} U(n\Delta\Omega)e^{2\pi ikn/N}\Delta\Omega \tag{9.66b}$$

The ability of the discrete response $u(k\Delta t)$ to reproduce the continuous response depends on how well the discrete convolution represents the continuous convolution.

It is of interest to compare the foregoing expressions for the evaluation of discrete response with the expressions for the response to a periodic load that we derived directly in Section 9.5. On comparing Equations 9.28b and 9.64 and noting that $c_n T_0$ is the same as $G(n \Delta \Omega)$, we find that the two expressions are similar except that the integral over a period in Equation 9.28b is replaced by a summation in Equation 9.64. The complex frequency function $H(n \Delta \Omega)$ given by Equation 9.65 is a discrete version of H obtained from Equations 9.26 or 9.27. Finally, Equation 9.66b for discrete response is equivalent to Equation 9.28a, except that rather than including an infinite number of frequencies, we have restricted the summation to over a finite number of frequencies.

9.11   COMPARISON OF CONTINUOUS AND DISCRETE FOURIER TRANSFORMS

In application to structural dynamics, the Fourier transform method is used primarily to obtain the convolution of two time functions, one of them being the forcing function and the other the unit impulse response function. As outlined in Section 9.10, the first step in the analysis is computation of the Fourier transforms of the two functions. In most practical applications, the transforms used in the analysis are discrete rather than continuous. A product of the two transforms is obtained next. The inverse transform of this product gives the desired response. It is evident that the evaluation of the transforms represents an intermediate step in the computation and the transforms themselves are of no direct interest. Thus, there would be no reason to examine how closely a discrete transform represents a continuous Fourier transform. Occasionally however, it is not the unit impulse response function but its continuous Fourier transform, the complex response function, which is specified. In such a situation, the given continuous transform must be converted into a discrete form so that it can be used in forming a product with the discrete transform of the forcing function. The inverse transform of the product gives the required response. It therefore becomes necessary to examine the relationship between the discrete and continuous transforms of a function, and to study the adjustments that must be made to the continuous transform so that it resembles the corresponding discrete transform.

Consider the rectangular pulse function shown in Figure 9.5a. We first construct a periodic extension of the function with a period $T_0 = 1.6$ s, and then sample the function at N time points within a period, N being equal to 16. The sampled time intervals are given by $k \Delta t$, where k varies from 0 to $N - 1$ and $\Delta t = 0.1$ s. At a point of discontinuity, the continuous Fourier transform converges to a value that is an average of the values on either side of the discontinuity. Therefore, for a proper comparison of the continuous and the discrete Fourier transform, we define the sampled value at a discontinuity to be equal to such an average. Figure 9.5b indicates the sampled values at $k = 0$ and $k = 4$, where the function is discontinuous.

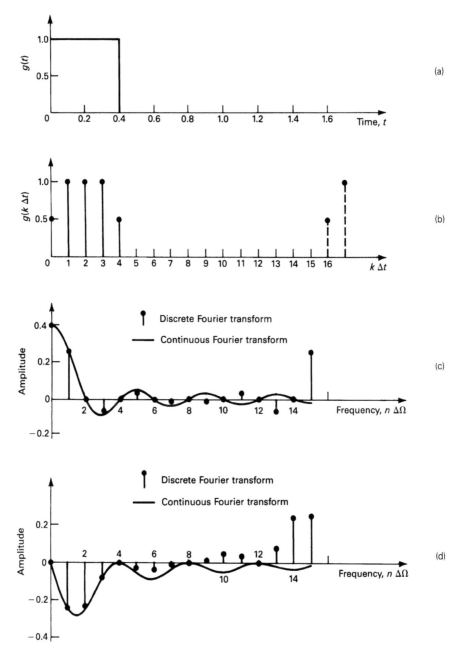

Figure 9.5. (a) Rectangular pulse function; (b) sampled time function; (c) real part of DFT; (d) imaginary part of DFT.

We now use Equation 9.54a to obtain the discrete Fourier transform of the sampled function. Figure 9.5c shows the real part of the discrete transform over one period, while Figure 9.5d shows the imaginary part. The horizontal axes in these diagrams represent the frequency scale given by $n\Delta\Omega$ or $n\Delta f$, where n varies from 0 to $N-1$ and Δf is the frequency increment in hertz. We have

$$\Delta f = \frac{\Delta\Omega}{2\pi} = \frac{1}{T_0} \tag{9.67}$$

giving $\Delta\Omega = 1.25\pi$ rad/s or $\Delta f = 0.625$ Hz. The discrete Fourier transform is periodic with a period of $N\Delta\Omega = 20\pi$ rad/s or $N\Delta f = 10$ Hz. Figure 9.5c and d show that the real part of the discrete transform is symmetric about $n = N/2$ while the imaginary part is antisymmetric about that point. We prove below that this is to be expected.

For a frequency of $m\Delta\Omega$, the discrete Fourier transform is obtained from Equation 9.54a by substituting m for n

$$G(m\Delta\Omega) = \sum_{k=0}^{N-1} g(k\Delta t)e^{-2\pi ikm/N}\Delta t \tag{9.68}$$

A frequency that is symmetric to $m\Delta\Omega$ about $N\Delta\Omega/2$ is $l\Delta\Omega$, where

$$l = N/2 + (N/2 - m)$$

$$= N - m \tag{9.69}$$

The Fourier transform at the symmetric frequency is obtained by setting $n = N - m$ in Equation 9.54a

$$G\{(N-m)\Delta\Omega\} = \sum_{k=0}^{N-1} g(k\Delta t)e^{-2\pi ik(N-m)/N}\Delta t$$

$$= \sum_{k=0}^{N-1} g(k\Delta t)e^{2\pi ikm/N}\Delta t \tag{9.70}$$

It is easily seen that Equation 9.70 is the complex conjugate of Equation 9.68, so that the real parts of the two are identical, while the imaginary parts are equal in magnitude but have opposite signs.

It is also of interest to examine the discrete transform value for $n = -m$. Equation 9.54a gives

$$G\{(-m)\Delta\Omega)\} = \sum_{k=0}^{N-1} g(k\Delta t)e^{2\pi ikm/N} \tag{9.71}$$

Equations 9.70 and 9.71 are identical, indicating that the transform values for frequencies higher than $N\Delta\Omega/2$ are the same as those for the corresponding negative frequency values. The frequency $N\Delta\Omega/2$ is referred to as the *folding frequency* or the *Nyquist frequency* and is the highest frequency represented in the transform. The folding frequency should be chosen to be sufficiently large so that the highest frequency of significance in the time function is represented. Since $N\Delta\Omega/2$ is equal to $\pi/\Delta t$, this implies that Δt should be chosen to be sufficiently small. From a physical point of view, this means that if the time function shows rapid variation with time, implying significant high frequency content, it should be sampled more frequently.

The continuous Fourier transform of the rectangular pulse function is obtained from Equation 9.32. If t_1 denotes the duration of the rectangular pulse

$$
\begin{aligned}
G(\Omega) &= \int_0^{t_1} e^{-i\Omega t}\, dt \\[2mm]
&= \int_0^{t_1} \cos \Omega t\, dt - i \int_0^{t_1} \sin \Omega t\, dt \\[2mm]
&= \frac{\sin \Omega t_1}{\Omega} + i\frac{\cos \Omega t_1 - 1}{\Omega}
\end{aligned}
\tag{9.72}
$$

For purposes of comparison, the real and imaginary parts of the continuous Fourier transform are also plotted in Figure 9.5c and d respectively.

It is noted that although the magnitude of continuous Fourier transform decreases as the frequency increases, it is never quite zero for all values of frequency higher than any selected value. On the other hand, because the discrete Fourier transform is periodic with a period of $N\Delta\Omega$, transform values for frequencies higher than $N\Delta\Omega/2$ cannot be represented by the discrete Fourier transform. Within the range of frequencies from 0 to $N\Delta\Omega/2$, the continuous and discrete transforms are quite close. The difference between the two becomes more significant as the frequency approaches the folding frequency. This is because of the *aliasing* effect explained below.

Figure 9.6a shows the real part of the continuous Fourier transform of the rectangular pulse function of Figure 9.5a. Let the pulse function be sampled at sample interval $\Delta t = 0.2$ s. The Fourier transform of the sampled waveform is different from that of the continuous version, but can be obtained from that of the latter as follows. It can be shown that the Fourier transform of the sampled waveform is periodic with a period equal to $2\pi/\Delta t$ rad/s or $1/\Delta t = 5$ Hz. To obtain the transform of sampled time function, we first construct a series of frequency functions of the same shape as that of the continuous transform but shifted on the frequency axis by $2\pi m/\Delta t$ rad/s, where m is an integer that may range from $-\infty$ to $+\infty$. Superposition of these frequency functions gives the desired transform.

Figure 9.6b shows two versions of the continuous transform function, one shifted by $+2\pi/\Delta t$ and the other by $-2\pi/\Delta t$. A superposition of Figure 9.6a and b is shown in Figure 9.6c. The resulting transform is different from the

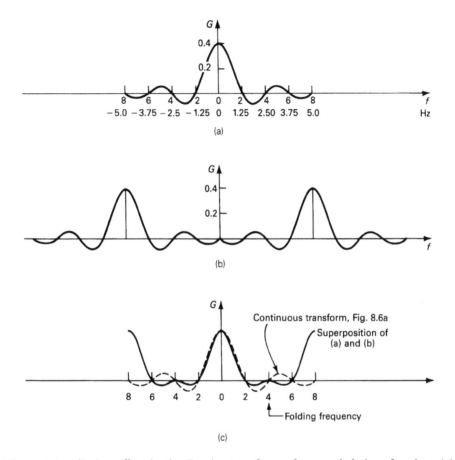

Figure 9.6. Aliasing effect in the Fourier transform of a sampled time function: (a) continuous transform of rectangular pulse; (b) two shifted images of the continuous transform; (c) superposition of (a) and (b).

continuous transform because of the overlapping effect. It is obvious that such overlapping will always occur unless the continuous transform is zero for frequency values greater than $\pi/\Delta t$. Frequency functions that have zero values outside a certain range of frequencies are called *bandlimited*. For such transform functions, overlapping, also known as aliasing, can be avoided if Δt is chosen such that $\pi/\Delta t$, the folding frequency, is higher than the highest frequency in the transform. This leads to the important conclusion that if the transform of the sample time function is required to closely approximate that of the continuous time function, the sampling intervals should be chosen sufficiently small so as to avoid aliasing. For transform functions that are not bandlimited, aliasing will occur whenever a time function is sampled. However, if the transform value decreases with frequency, the aliasing effect can be reduced by making Δt sufficiently small. It will also be noted that for transform functions that decrease in value with frequency, the effect of aliasing is most significant near the folding frequency.

We have seen that sampling in the time domain leads to a transform function that is periodic, the period being equal to $2\pi/\Delta t$ rad/s. In a similar manner, sampling of the frequency function will lead to a time-domain function that is periodic with a period given by $T_0 = 2\pi/\Delta\Omega$. Also, just as sampling in the time domain resulted in aliasing in the frequency domain, sampling in the frequency domain will result in aliasing in the time domain, unless the function is *time limited* (is zero outside a certain time range), and $\Delta\Omega$ is sufficiently small to avoid time function overlap. In the example under consideration, the value of $T_0 = 1.6$ s is sufficiently large so that time-domain aliasing does not occur.

It is of interest to note the effect of choosing $T_0 = 0.8$, which gives $\Delta\Omega = 2.5$ π rad/s or 1.25 Hz. The discrete Fourier transform values will be those in Figure 9.5c corresponding to $n = 0, 2, 4, 6$, and 9. The first of these values is equal to 0.4; all others are zero. When these results are compared with continuous Fourier transform values, it is obvious that the sampling interval for the frequency function is so coarse that the function is not adequately represented. In summary, if the discrete transform has to closely approximate the continuous transform, Δt should be significantly small so that the time function is adequately represented and T_0 should be sufficiently large so that the frequency function is adequately represented.

Example 9.2
Compare the continuous and discrete Fourier transforms of a unit impulse function. The undamped natural frequency of the system is 2π rad/s, the sampling interval $\Delta t = 0.1$ s and $m = 0.25$ lb · s^2/in. Assume that the damping ratio is zero.

Solution
The unit impulse function $h(t)$ given by

$$h(t) = \begin{cases} \frac{1}{m\omega} \sin \omega t & t > 0 \\ 0 & t < 0 \end{cases} \tag{a}$$

is shown in Figure E9.2a. The continuous Fourier transform of $h(t)$ is the complex frequency response function $H(\Omega)$, given by

$$H(\Omega) = \frac{1}{m(\omega^2 - \Omega^2)} \tag{b}$$

The unit impulse function is a sine wave with a period of 1 s. To obtain a discrete transform, the function is truncated at 1 s and sampled at 0.1 s, giving 10 sample points, as shown in Figure E9.2b. Discrete transform of the sampled function is obtained by using Equation 9.54a. The discrete and continuous transforms, both shown in Figure E9.2c, are quite different. The discrete transform is, in fact, the transform of a periodic version of the truncated impulse function. As seen from Figure 9.2b, this periodic version is a continuous sine wave in contrast to the unit impulse function, which is zero for $t < 0$ and follows a sine wave only for $t \geq 0$. The continuous transform of a sine wave is comprised of two pulses and is identical to the discrete transform obtained here.

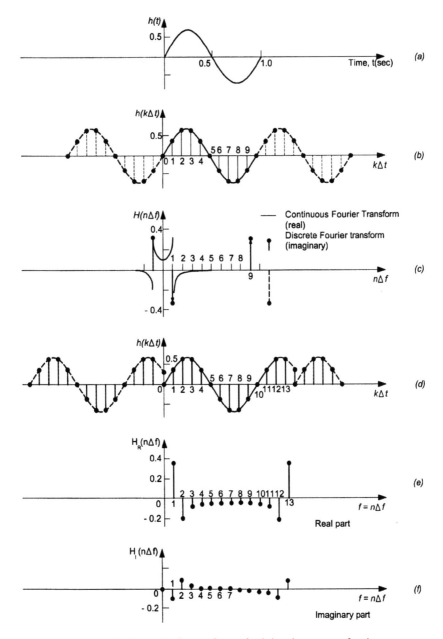

Figure E9.2. Continuous and discrete Fourier transforms of unit impulse response function.

It is of interest to note the effect of truncating the impulse function at a time interval that is not a multiple of the period of the sine wave. Let the impulse function be truncated at 1.4 s and sampled as before at 0.1 s. The discrete Fourier transform shown in Figure E9.2e and f is now different from that obtained earlier. This is because the transform represents the periodic function of Figure E9.2d, which is not the same as that shown in Figure E9.2b.

Example 9.3
Repeat Example 9.2 with a damping ratio $\xi = 0.05$.

Solution
The unit impulse function is now given by

$$h(t) = \begin{cases} \dfrac{1}{m\omega_d} e^{-\xi\omega t} \sin \omega_d t & t > 0 \\ 0 & t < 0 \end{cases} \tag{a}$$

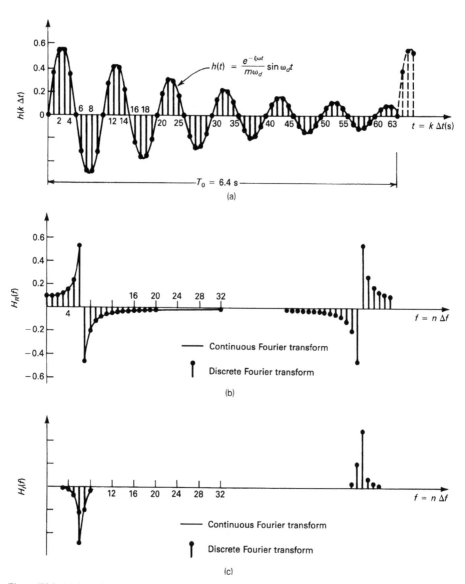

Figure E9.3. (a) Sampled time function; (b) Fourier transform, real part; (c) Fourier transform, imaginary part.

The complex frequency response function $H(\Omega)$, which is the continuous Fourier transform of $h(t)$, is given by

$$H(\Omega) = \frac{1}{m\{-\Omega^2 + 2i\xi\omega\Omega + \omega^2\}} \tag{b}$$

To obtain the discrete transform of $h(t)$, the function must be truncated at a certain time and then sampled. The truncation interval is chosen as 6.4 s and the sampling interval as 0.1 s. The discrete transform will represent a periodic function of time with a period of 6.4 s as shown in Figure E9.3a.

The real and imaginary parts of the discrete transform obtained from Equation 9.54a are shown in Figure E9.3b and c, respectively. For the purpose of comparison, the two components of continuous Fourier transform obtained from Equation b are also shown. The effect of truncation and aliasing are small in this case, so that the discrete and continuous transforms are fairly close. The sampling interval in the frequency domain is $1/T_0 = 0.1563$ Hz. The maximum frequency represented in the discrete transform, known as the folding frequency, is $32 \times 0.1563 = 5$ Hz. Also, as expected, the real part of the discrete transform is symmetric about the folding frequency, while the imaginary part is antisymmetric about it.

9.12   APPLICATION OF DISCRETE INVERSE TRANSFORM

As stated in Section 9.11, in many problems of dynamics, the continuous Fourier transform of unit impulse function, which is the same as the complex frequency response function, is specified directly. In such situations, a discrete version of the specified frequency response function is multiplied by the discrete Fourier transform of the forcing function. The inverse discrete transform of the product gives the response of the system in the time domain. For the method to be successful, the discrete version of the frequency response function must truly represent the unit impulse function of the system. In other words, the discrete inverse transform of the sampled frequency response function must closely match the impulse function. This will require careful sampling and reconstruction of the continuous function.

As an example of such sampling and reconstruction, consider the continuous Fourier transform of a rectangular pulse function given by Equation 9.72. Let this function be discretized with a sampling interval of $\Delta\Omega$ and the total number of sample points equal to N. The discretization requires that the frequency function be converted to its periodic version with a period equal to $N\Delta\Omega$. Furthermore, it is evident from the discussion in the preceding paragraph that the real part of the discrete transform is symmetric about the folding frequency of $N\Delta\Omega/2$, while the imaginary part is antisymmetric about that frequency. Thus, to construct a proper discrete transform, we take $N/2$ samples of the real part and fold them about a frequency of $N\Delta\Omega/2$. In a similar manner, we take $N/2$ samples of the imaginary part, but instead of just folding them, we fold and flip them about the frequency $N\Delta\Omega/2$. Sample values for frequencies greater than $N\Delta\Omega/2$ are thus complex conjugates of those at symmetric frequencies below $N\Delta\Omega/2$.

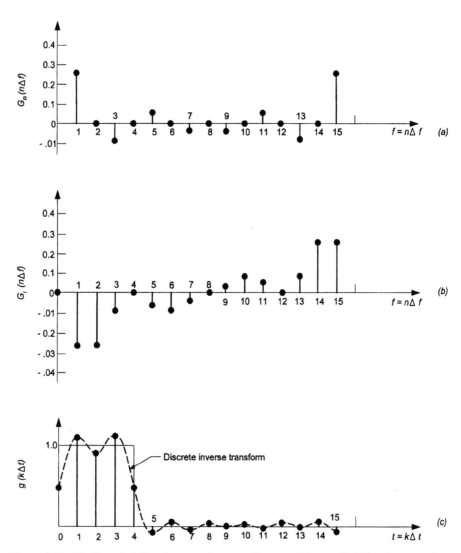

Figure 9.7. (a) Sampled Fourier transform, real part; (b) sampled Fourier transform, imaginary part; (c) time function and discrete inverse transform.

Figure 9.5c and d show, respectively, the real and imaginary parts of the continuous Fourier transform of the rectangular pulse function of Figure 9.5b. Taking $N = 16$ and $\Delta\Omega = 1.25\pi\,\mathrm{rad/s}$ or $\Delta f = 0.625\,\mathrm{Hz}$, discrete versions of the real and imaginary parts of the continuous transforms are constructed as shown in Figure 9.7a and b, respectively. Discrete inverse transform is now obtained by using Equation 9.54b and results in the periodic time function shown in Figure 9.7c. For purposes of comparison, the original rectangular pulse function is also shown there.

The time function obtained by taking the inverse transform oscillates about the true value. These oscillations are present because we truncated the frequency function at a frequency of $N\Delta\Omega/2$ and thus ignored the higher-frequency content of the time function reflected in nonzero amplitudes in the continuous Fourier transform at such higher frequencies. The value of N should be increased to reduce the oscillations. Similar oscillations were observed in the Fourier series representation of the periodic version of rectangular pulse function (Example 9.1).

As another example of the use of discrete inverse transform, consider the continuous Fourier transform of the unit impulse function for an undamped system given by

$$H(\Omega) = \frac{1}{m(\omega^2 - \Omega^2)} \tag{9.73}$$

Let $m = 0.25$ lb·s²/in. and the natural frequency of the system $\omega = 2\pi$ rad/s. Suppose that we want to obtain the time function $h(t)$ by taking the discrete inverse transform of $H(\Omega)$. To do this, we must convert the frequency function into a periodic function of frequency and sample it at regular intervals. We choose the sampling interval as 0.15625 Hz and the period as equal to 64 such intervals, that is, 10 Hz. The folding frequency is thus 5 Hz or 32 sampling intervals. The discrete frequency function values at the first 32 samples are obtained by using Equation 9.73. The remaining 32 samples are simply the complex conjugates of the first 32 samples. We note that in our case, the frequency function has no imaginary part, hence the second set of 32 samples are obtained by folding the first 32 about a frequency of 5 Hz. The discrete frequency function constructed as above is shown in Figure 9.8a.

The discrete inverse transform of the function of Figure 9.8a is obtained by using Equation 9.54b and is shown in Figure 9.8b. The continuous inverse transfer of $H(\Omega)$ is the unit impulse function $h(t)$ given by

$$h(t) = \begin{cases} \frac{1}{m\omega} \sin \omega t & t > 0 \\ 0 & t < 0 \end{cases} \tag{9.74}$$

For purposes of comparison, the function $h(t)$ is also shown in Figure 9.8b and is seen to be quite different from the inverse discrete transform value.

The large difference between the continuous and discrete inverse Fourier transform values would have been expected because of the results obtained in Example 9.2, where we found the discrete transform of unit impulse function to be quite different from the continuous transform. It is, however, of some interest to study this difference more carefully. We do this for the general case of a damped system. The special case of an undamped system will then follow by setting $\xi = 0$.

Figure 9.9a shows the unit impulse function, $h(t)$, for a damped system with damping ratio ξ. Figure 9.9b shows the amplitude of the continuous Fourier

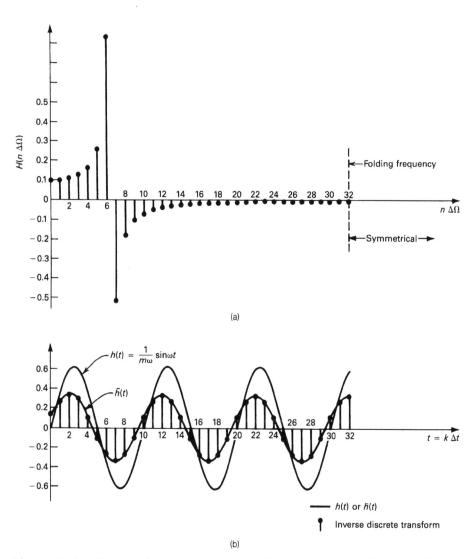

(a)

(b)

Figure 9.8. (a) Discretized version of continuous Fourier transform; (b) inverse Fourier
transform of discrete frequency function.

transform of $h(t)$. Let us now convert the continuous frequency function of
Figure 9.9b into a discrete form by taking an infinite number of samples spaced
at a frequency of $\Delta\Omega$ as shown in Figure 9.9e. The inverse transform of this
sampled function is not the same as that shown in Figure 9.9a but can be
obtained from the latter. The inverse Fourier transform of the sampled waveform
is periodic with a period $T_0 = 2\pi/\Delta\Omega$. To obtain the inverse transform of the
sampled frequency function, we first construct a series of time function of the
same shape as the continuous inverse transform $h(t)$ but shifted on the time

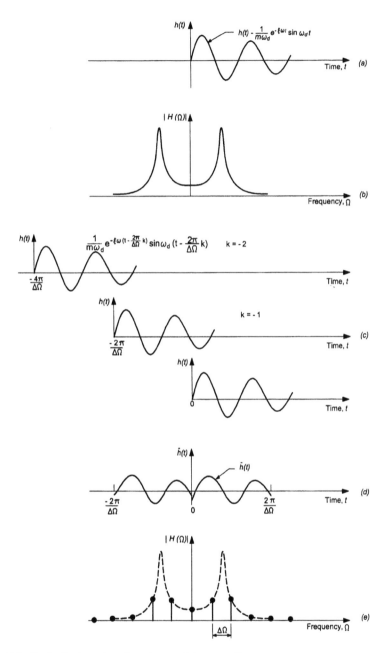

Figure 9.9. Periodic impulse function and its transform.

axis by an amount $2\pi r/\Delta\Omega$ s, where r is an integer ranging from $-\infty$ to ∞. Three such time functions are shown in Figure 9.9c. Superposition of the time functions gives the desired inverse transform of the sampled frequency function. We will denote the inverse transform so obtained by $\bar{h}(t)$.

Referring to Figure 9.9d, which shows the inverse transform obtained by superposition, the value of $\bar{h}(t)$ in any one period is given by

$$\bar{h}(t) = \frac{1}{m\omega_d} \sum_{r=-\infty}^{0} e^{-\xi\omega(t-rT_0)} \sin \omega_d(t - rT_0) \tag{9.75}$$

Using the relation $\sin\theta = \frac{1}{2}(e^{i\theta} - e^{-i\theta})$, Equation 9.75 becomes

$$\bar{h}(t) = \frac{1}{m\omega_d} e^{-\xi\omega t} \sum_{r=-\infty}^{0} e^{\xi\omega r T_0} \frac{e^{i\omega_d(t-rT_0)} - e^{-i\omega_d(t-rT_0)}}{2i}$$

$$= \frac{1}{m\omega_d} \frac{e^{-\xi\omega t}}{2i}$$

$$\left\{ e^{i\omega_d t} \sum_{r=-\infty}^{0} e^{-rT_0(-\xi\omega+i\omega_d)} - e^{-i\omega_d t} \sum_{r=-\infty}^{0} e^{-rT_0(-\xi\omega-i\omega_d)} \right\} \tag{9.76}$$

On replacing $-r$ by r in Equation 9.76, we get

$$\bar{h}(t) = \frac{1}{m\omega_d} \frac{e^{-\xi\omega t}}{2i} \left\{ e^{i\omega_d t} \sum_{r=0}^{\infty} e^{rT_0(-\xi\omega+i\omega_d)} - e^{-i\omega_d t} \sum_{r=0}^{\infty} e^{rT_0(-\xi\omega-i\omega_d)} \right\} \tag{9.77}$$

Using the expression for the sum of a geometric progression, Equation 9.77 can be written as

$$\bar{h}(t) = \frac{1}{m\omega_d} \frac{e^{-\xi\omega t}}{2i} \left\{ \frac{e^{i\omega_d t}}{1 - e^{T_0(-\xi\omega+i\omega_d)}} - \frac{e^{-i\omega_d t}}{1 - e^{T_0(-\xi\omega-i\omega_d)}} \right\} \tag{9.78}$$

which after simplification becomes

$$\bar{h}(t) = \frac{e^{-\xi\omega t}}{m\omega_d} \left\{ \frac{\sin \omega_d t - e^{-\xi\omega T_0} \sin \omega_d(t - T_0)}{1 - 2e^{-\xi\omega T_0} \cos \omega_d T_0 + e^{-2\xi\omega T_0}} \right\} \tag{9.79}$$

The time function $\bar{h}(t)$ is different from $h(t)$ because of the overlapping or aliasing. If T_0 is large, the effect of the overlap reduces, and in the limit, as T_0 approaches ∞, $\bar{h}(t)$ should approach $h(t)$. This can be verified by setting $T_0 = \infty$ in Equation 9.79.

For an undamped system, $\xi = 0$, and Equation 9.79 reduces to

$$\bar{h}(t) = \frac{1}{2m\omega} \left(\sin \omega t + \frac{\cos \omega t \sin \omega T_0}{1 - \cos \omega T_0} \right) \tag{9.80}$$

The value of $\bar{h}(t)$ has also been plotted in Figure 9.8b and is seen to match the discrete inverse transform quite closely. The small difference between the two exists because the frequency function must be truncated before calculating its inverse Fourier transform, while function $\bar{h}(t)$ is obtained by assuming an infinite number of samples. Finally, it should be noted that while for a damped system, $\bar{h}(t)$ approaches $h(t)$ as T_0 becomes large, such is not the case for an undamped system.

9.13   COMPARISON BETWEEN CONTINUOUS AND DISCRETE CONVOLUTION

In Section 9.10, it was mentioned that important differences exist between continuous and discrete convolutions. Because practical frequency-domain analysis depends on the possibility of approximating the continuous convolution of exciting force and the unit impulse response by a discrete convolution of the two functions, we must examine the differences between the two types of convolutions more carefully.

Consider again the two functions of time shown in Figure 9.3a. The continuous convolution of these two functions was obtained in Section 9.8 and is shown in Figure 9.3g. Let it be required to obtain a discrete convolution of the two functions. To achieve this objective, we construct a periodic version of the two functions and sample them at a suitable interval. We choose the sampling interval as $\Delta t = 1$ s and assume that the period is equal to $N\Delta t$, N being equal to 6. The reconstructed functions are shown in Figure 9.10a and b. Our choice of the period has led to the inclusion of a number of zero-amplitude samples in a period.

Next, we fold one of the two functions, say $h(m\Delta t)$, about the y axis. The folded function is shown in Figure 9.10c. To get the discrete convolution value at time $k\Delta t$, we first construct the shifted function $h\{(k - m)\Delta t\}$. Figure 9.10d shows the shifted function for $k = 1$. In accordance with the definition given in Equation 9.56, the convolution, $u(k\Delta t)$, is now obtained by taking the products of the corresponding sample points of $g(m\Delta t)$ and $h\{(k - m)\Delta t\}$ for all values of m lying in a period, that is, ranging from 0 to $N - 1$, and summing the products. It is evident from Figure 9.10d that the discrete convolution value will be different from the continuous convolution value. This is because in the case of discrete convolution, sample points from the previous period of $h(m\Delta t)$ or the succeeding period of $h(-m\Delta t)$ are interfering in the convolution. The overlapping portion is shown shaded in Figure 9.10d.

The discrete convolution for all values of k from 0 to $N-1$ is shown in Figure 9.10f. It represents a very poor approximation to the continuous convolution because of period overlap. The errors introduced by the overlap can be avoided provided that we make the period of sampled time functions sufficiently large by introducing appropriate number of samples with zero function values. If one of the time functions is of duration $p\Delta t$ and the other of duration $q\Delta t$, the minimum length of period required to avoid overlap is easily proved to be equal to $(p + q + 1)\Delta t$. For the example being considered, this period is equal to $9\Delta t = 9$ s. This is readily seen to be true by reference to Figure 9.11.

Figure 9.11b shows the shifted function $h\{(k - m)\Delta t\}$ for $k = 8$. The significant portion of h is now just overlapping the significant portion of g at $m = 4$. The previous period of $h(-m\Delta t)$ should at this time be clear of the first $(p+q)\Delta t$ seconds, giving $N \geq p+q+1$. Figure 9.11c shows the shifted function

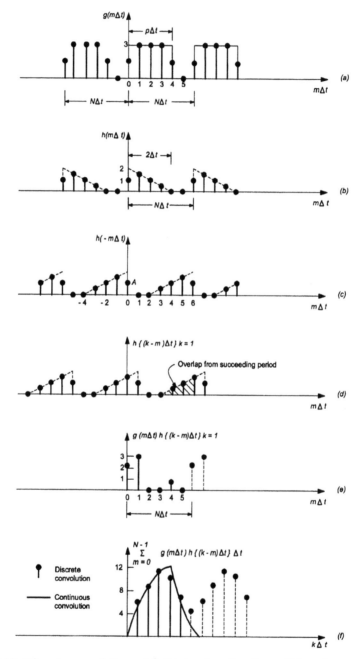

Figure 9.10. Discrete convolution of finite-duration functions. N less than $p + q + 1$.

$h\{(k - m)\Delta t\}$ for $k = 0$. The significant portion of h from one period is just overlapping the significant portion of g at $m = 0$. At this time, the significant portion for the succeeding period of $h(-m\Delta t)$ should be clear of the significant

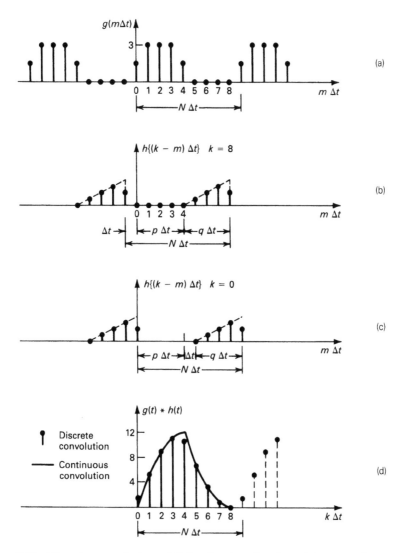

Figure 9.11. Discrete convolution of finite-duration functions N greater than or equal to $p+q+1$.

portion of g, again leading to $N \geq p+q+1$. The discrete convolution for $N=9$ is shown in Figure 9.11d, where the continuous convolution is also shown for the purpose of comparison. The two now match more closely; the small difference between them arises from the fact that integration in the continuous convolution has been replaced by rectangular summation in the discrete convolution.

In the foregoing discussion, we have assumed that both g and h are of finite duration; that is, p and q are finite. In such a situation, N can be chosen to be equal to or greater than $p+q+1$ to avoid overlap. If either one of g and h is not of finite duration, N will theoretically be infinite and the suggested method will not work. Procedures that can be applied in such cases are discussed in succeeding paragraphs.

Example 9.4

Find the response of a damped single-degree-of-freedom system to an exciting force which consists of the rectangular pulse shown in Figure E9.4a. The system has a mass of $0.25\,\text{lb}\cdot\text{in./s}^2$, a natural frequency of $2\pi\,\text{rad/s}$, and a damping fraction $\xi = 0.05$.

Solution

The unit impulse response function $h(t)$ is given by

$$h(t) = \frac{1}{m\omega_d} e^{-\xi\omega t} \sin \omega_d t \tag{a}$$

where $\omega_d = \omega\sqrt{1 - \xi^2}$.

Function $h(t)$ is shown in Figure E9.4b. It is noted that although $g(t)$, the forcing function, is of finite duration, $h(t)$ is not. However, the magnitude of $h(t)$ decreases fairly rapidly with time, so that we can truncate the function at a sufficiently large value of t and assume that it is zero for all greater t.

Let the sampling interval be chosen as 0.1 s and $h(t)$ be truncated at $19 \times 0.1 = 1.9$ s. We have $p = 4$ and $q = 19$, so that the period should be selected as $N = p + q + 1 = 24$. The sampled functions with a period of $24 \times 0.1 = 2.4$ s are shown in Figure E9.4c and d, respectively. The discrete convolution of the two sampled functions is obtained through the frequency domain by using the convolution theorem of Equation 9.58. An appropriate computer program is used to obtain the direct and inverse Fourier transforms involved. The resulting convolution values are shown in Figure E9.4e and Table E9.4. The exact response of the system to the rectangular pulse function of amplitude p_0 is given by the following expressions

$$u(t) = 0 \qquad t < 0 \tag{b}$$

$$u(t) = \frac{p_0}{m\omega^2}\left\{ 1 - e^{-\xi\omega t}\left(\cos \omega_d t + \frac{\xi}{\sqrt{1 - \xi^2}} \sin \omega_d t \right) \right\} \qquad 0 \le t \le t_1 \tag{c}$$

$$u(t) = \frac{p_0}{m\omega^2} e^{-\xi\omega(t - t_1)}$$

$$\left[\left\{ e^{-\xi\omega t_1}\left(\sin \omega_d t_1 - \frac{\xi}{\sqrt{1 - \xi^2}} \cos \omega_d t_1 \right) + \frac{\xi}{\sqrt{1 - \xi^2}} \right\} \right.$$

$$\times \sin \omega_d(t - t_1)$$

$$+ \left\{ 1 - e^{-\xi\omega t_1}\left(\cos \omega_d t_1 + \frac{\xi}{\sqrt{1 - \xi^2}} \sin \omega_d t_1 \right) \right\}$$

$$\left. \times \cos \omega_d(t - t_1) \right] \qquad t > t_1 \tag{d}$$

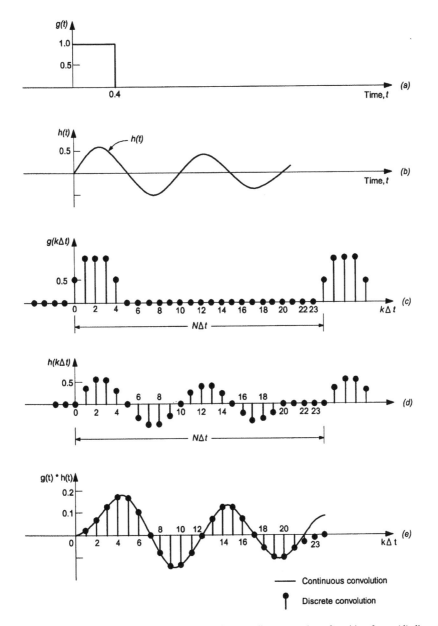

Figure E9.4. (a) Exciting force; (b) unit impulse function; (c) discrete version of exciting force; (d) discrete version of unit impulse function; (e) response function.

The exact response, which is the same as the response obtained from a continuous convolution, is also shown in Figure E9.4e and Table E9.4. The discrete convolution results closely match the exact values. The difference is because of the rectangular summation used in discrete convolution. The results can therefore be improved by using a smaller value for Δt, say 0.05 s.

Table E9.4. Response of a single-degree-of-freedom system to a rectangular pulse function.

k	$t(s)$	$u(k\,\Delta t)$ Exact	$u(k\,\Delta t)$ Frequency domain
0	0.0	0.0000	0.0000
1	0.1	0.0190	0.0183
2	0.2	0.0673	0.0647
3	0.3	0.1252	0.1208
4	0.4	0.1708	0.1650
5	0.5	0.1689	0.1635
6	0.6	0.1047	0.1017
7	0.7	0.0055	0.0059
8	0.8	−0.0896	−0.0862
9	0.9	−0.1458	−0.1408
10	1.0	−0.1446	−0.1399
11	1.1	−0.0899	−0.0873
12	1.2	−0.0053	−0.0056
13	1.3	0.0761	0.0733
14	1.4	0.1244	0.1202
15	1.5	0.1237	0.1197
16	1.6	0.0772	0.0750
17	1.7	0.0050	0.0052
18	1.8	−0.0647	−0.0622
19	1.9	−0.1062	−0.1025
20	2.0	−0.1058	−0.1021
21	2.1	−0.0663	−0.0733
22	2.2	−0.0047	−0.0383
23	2.3	0.0550	−0.0105

It will be noted that the match between the exact and discrete convolution values is not as good on samples 20 through 23. This is because of the truncation applied to the unit impulse function. In a later section, we discuss a method of applying suitable correction to improve these values.

9.14   DISCRETE CONVOLUTION OF AN INFINITE- AND A FINITE-DURATION WAVEFORM

In Section 9.13, we discussed a procedure by which a discrete convolution between two finite-duration functions can be made to approximate their continuous convolution. Very often, however, only one of the two functions being convolved is of finite duration, while the other continues for an infinite time. A practical example of such a case is the response analysis of a system subjected to a dynamic force that acts for a limited duration. The unit impulse function, representing the second function involved in the convolution, is in this case, of infinite duration. However, if the interval of interest during which the response is required is not too large, the problem can be converted to that of the convolution of two finite-duration functions.

As an example, consider the response analysis of a system subjected to a forcing function, g, of duration T_p shown in Figure 9.12a. The unit impulse response function, h, for the system is shown in Figure 9.12b. If the interval

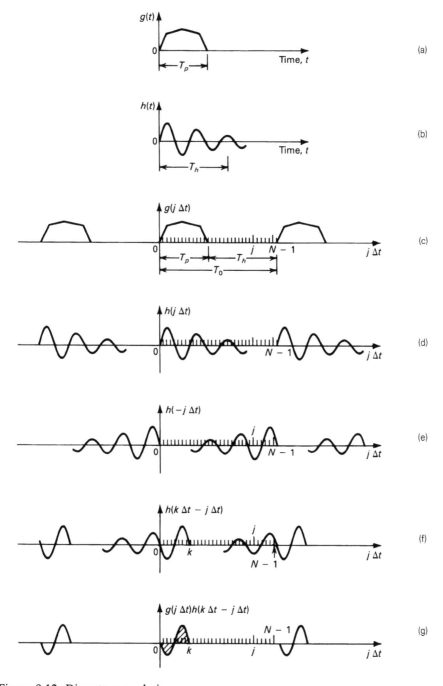

Figure 9.12. Discrete convolution.

of interest during which the response is sought is T_h, the unit impulse function is truncated at T_h. Functions g and h are now discretized and converted into periodic forms. If $T_p = p\Delta t$ and $T_h = q\Delta t$, the period T_0 is selected so that $T_0 = N\Delta t = (p + q + 1)\Delta t$. Figure 9.12c and d show the discretized periodic versions of the two functions. Both functions have been augmented by a sufficient number of zeros to make up a period. The discretized functions are now convolved. Figure 9.12e through g show the steps involved in convolution, the shaded area in Figure 9.12g representing the response value at time $k\,\Delta t$.

The results of the discrete convolution for t from 0 to T_h are similar to those for a continuous convolution and give the required response. This method of using discrete convolution to obtain the response for a desired interval of time is sometimes referred to as the *fast convolution technique*. Its main disadvantage is that, in general, convolution needs to be carried out over an extended time period of $T_p + T_h$ to obtain useful results over a time T_h.

The convolution values beyond T_h are incorrect because of the truncation applied to h. However, if damping has caused the unit impulse response to decay to a small magnitude after the elapse of time T_h, the error caused by the truncation of h would be small and the discrete convolution would closely approximate the continuous convolution over the entire period T_0. In effect, we are then convolving two functions of finite durations T_p and T_h, respectively, and the convolution procedure is identical to that described in Section 9.13.

Next, suppose that although T_h is sufficiently large for the damping to reduce the unit impulse function to a negligible value, the duration of the excitation function is so long that convolution over $T_p + T_h$ is not possible within the memory capacity of the computer. Evaluation of the discrete convolution in such a case will require that the excitation function be sectioned. Let each section of g be of length t_p, as shown in Figure 9.13a. Discrete convolution is now carried out over the first section of g using the fast convolution technique, as shown in Figure 9.13b. The convolution results are correct over the first t_p seconds but are in error over the remaining time interval of T_h seconds because of the truncation applied to g. The second section of g is now convolved with the unit impulse response function. The results of this convolution must be superimposed on those of the previous convolution in such a manner that the last T_h seconds of the first convolution overlap the first T_h seconds of the second convolution, as shown in Figure 9.13c. If there are more sections remaining in g, the process is repeated. Superposition of the convolution results from individual sections in the manner described above will provide an approximation to the true response of the system.

The procedure described in the foregoing paragraph using superposition of convolution values is referred to as *overlap-add sectioning*. It is effective whenever one of the two functions being convolved is of finite duration. In particular, if the unit impulse response function is not of finite duration, but the excitation function is, the impulse response function could be divided into sections of t_h

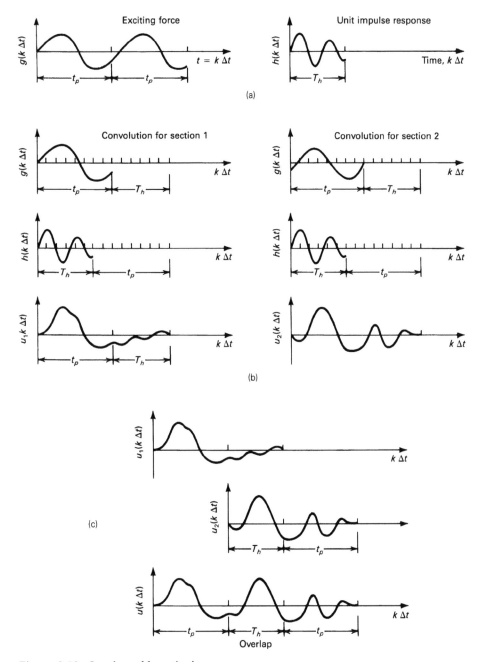

Figure 9.13. Overlap-add sectioning.

seconds. The convolution results of one section must then be superposed on that of the preceding section so that the first T_p seconds of the former overlap the last T_p seconds of the latter.

496   *Humar*

Example 9.5

In Example 9.4 the response values obtained at the last four time points, 20 through 23, were in error because of the truncation applied to $h(t)$. Use overlap-add sectioning method to obtain a better estimate of the response at these points.

Solution

In this case, the excitation function is of finite duration, but the unit impulse response function lasts for a long time and must be sectioned. The first section of $h(t)$ consisting of sample points

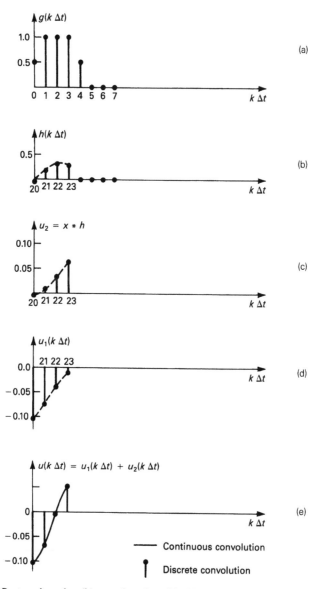

Figure E9.5. (a) Rectangular pulse; (b) second section of h; (c) convolution of (a) and (b); (d) convolution results from Example 9.4; (e) resultant response = (c) + (d).

Table E9.5. Use of overlap-add sectioning to obtain corrected response.

k	$t(s)$	Exact response $\bar{u}(k\,\Delta t)$	Response from Example 9.4	Corrective response	Resultant response
20	2.0	−0.1058	−0.1021	−0.0003	−0.1024
21	2.1	−0.0663	−0.0733	0.0089	0.0644
22	2.2	−0.0047	−0.0383	0.0335	−0.0049
23	2.3	0.0550	−0.0105	0.0634	0.0529

0 to 19, was convolved with the rectangular pulse function in Example 9.4 after augmenting both $p(t)$ and the truncated function $h(t)$ by a sufficient number of zeros to give a period of $N\Delta t = (p + q + 1)\Delta t = 24\Delta t$. Results of that convolution were correct for points 0 through 19, but were in error for points 20 through 23. The first four convolution results from the next section of $h(t)$ should be superposed on the values already obtained for points 20 through 23 to get a better estimate of the true response.

We construct the second section of $h(t)$ by taking its values from sample points 20 through 23, and then convolve this section with the rectangular pulse function. The two functions to be convolved are shown in Figure E9.5a and b. Both functions have been shifted to the origin and augmented by the sufficient number of zeros to avoid aliasing.

The convolution results for points 20 through 23, shown in Figure E9.5c, are now superposed on those obtained in Example 9.4. The resulting values are close to the true response, as seen in Figure E9.5e and Table E9.5.

9.15   CORRECTIVE RESPONSE SUPERPOSITION METHODS

The fast convolution method and the overlap-add method are both effective in adapting the discrete convolution so that it becomes representative of the corresponding continuous convolution. However, both methods depend on augmenting the forcing function as well as the truncated unit impulse function by a band of zeros. The response must be calculated over the extended time period, but useful results are restricted to the duration of the truncated impulse function, the remaining part of the response being inaccurate. The need to compute the response over an extended period seriously affects the computational efficiency of the procedure.

Veletsos and others have suggested a series of methods that do not rely on the addition of a long grace band of zeros to the functions being convolved and therefore significantly improve the efficiency of discrete convolution. Conceptually, these methods rely on the superposition of an appropriate corrective response on the response of the system to a periodic extension of the prescribed excitation. Consider, for example, a system subjected to the forcing function of duration T_p shown in Figure 9.14a. The maximum response of the system to the given excitation may be achieved either within the duration T_p or in a free-vibration era following the termination of excitation. The length of free-vibration era during which a maximum may be reached will be no longer than

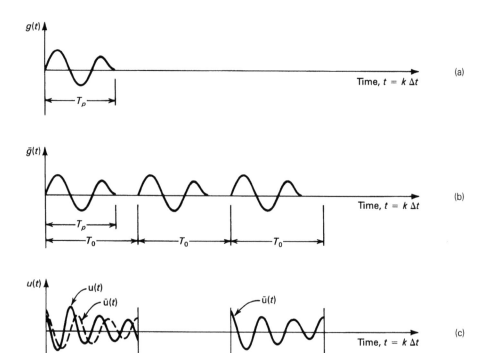

Figure 9.14. (a) Forcing function; (b) periodic version of forcing function; (c) comparison of transient and periodic response.

half a cycle of motion at the natural frequency of the system, or in other words, $T/2$, where T is the system period. The total length of time over which the response needs to be calculated is thus $T_0 = T_p + T/2$.

The first step in the *corrective response methods* is to construct a periodic extension of the prescribed excitation with a period T_0. The extended function is shown in Figure 9.14b. The exact response of the system to the periodic extension of the prescribed excitation is obtained next. This requires the determination of $\bar{h}(t)$, the periodic impulse response of the system, defined as the response at time t $(0 < t < T_0)$ due to a sequence of unit impulses applied at $t = 0, -T_0, -2T_0, \ldots$. The periodic impulse response $\bar{h}(t)$ was obtained in Section 9.12 and is given by Equation 9.79. In a discrete form, the periodic response of the system, $\bar{u}(k\Delta t)$, to a periodic extension of the exciting force $g(t)$ is given by

$$\bar{u}(k\,\Delta t) = \sum_{j=0}^{N-1} g(j\Delta t)\bar{h}\{(k-j)\Delta t\}\Delta t \tag{9.81}$$

Equation 9.81 for $\bar{u}(k\Delta t)$ can be quite effectively evaluated in terms of discrete Fourier transform as

$$\bar{u}(k\,\Delta t) = \frac{1}{2\pi} \sum_{n=0}^{N-1} G(n\,\Delta\Omega)\bar{H}(n\,\Delta\Omega)e^{2\pi ikn/N}\,\Delta\Omega \tag{9.82}$$

where $G(n\,\Delta\Omega)$ is the discrete Fourier transform of the excitation $g(k\,\Delta t)$, and $\bar{H}(n\,\Delta\Omega)$ is the discrete Fourier transform of $\bar{h}(k\,\Delta t)$.

The periodic response $\bar{u}(k\,\Delta t)$ and the true transient response $u(k\,\Delta t)$ are both shown in Figure 9.14c, where, for the sake of generality, it has been assumed that the system starts from nonzero initial conditions so that the transient response is not zero at $t=0$. In general, $\bar{u}(k\,\Delta t)$ is not the same as $u(k\,\Delta t)$. Since the specified excitation and its periodic extension are identical over $0<t<T_0$, the difference in the two responses must arise because the initial conditions at the start of the time period T_0 are not the same in the two cases. This difference in the initial conditions is a result of the fact that the periodic response $\bar{u}(k\,\Delta t)$ is affected by excitation in the preceding periods, while the transient response starts from the specified initial conditions. This suggests that if we could evaluate a corrective response that when superposed on the periodic response $\bar{u}(k\,\Delta t)$ will make the initial conditions in the corrected response similar to those in $u(k\,\Delta t)$, then the corrected response will be identical to the true transient response. All the methods proposed by Veletsos et al. attempt to do this, although the details of the procedure used are different. In the following paragraphs we describe three different approaches that have been proposed.

9.15.1 *Corrective transient response based on initial conditions*

Let the specified initial conditions at $t=0$ be $u(0)$ and $\dot{u}(0)$. The displacement at the start of a period in the periodic response $\bar{u}(k\,\Delta t)$ is $\bar{u}(0)$, while the velocity at that time is $\dot{\bar{u}}(0)$, one or both of which are different from the corresponding initial conditions specified. The displacement at $t=0$ in the periodic response, $\bar{u}(0)$, is obtained directly from Equation 9.82 by substituting $k=0$. To obtain the initial velocity, we proceed as follows.

Figure 9.15a and b show, respectively, the real and the imaginary part of the frequency function $U(n\,\Delta\Omega) = G(n\,\Delta\Omega)\bar{H}(n\,\Delta\Omega)$. Since $U(n\,\Delta\Omega)$ is periodic, the summation over a period can be taken from $-N/2$ to $N/2$, and Equation 9.82 can be expressed as

$$\bar{u}(k\,\Delta t) = \frac{1}{T_0} \sum_{n=-N/2}^{N/2} U(n\,\Delta\Omega)e^{2\pi int/T_0} \tag{9.83}$$

in which use has been made of the relations $t=k\,\Delta t$, $T_0=N\Delta t$, and $\Delta\Omega=2\pi/T_0$. The velocity at time $t=0$ is obtained by differentiating Equation 9.83 and

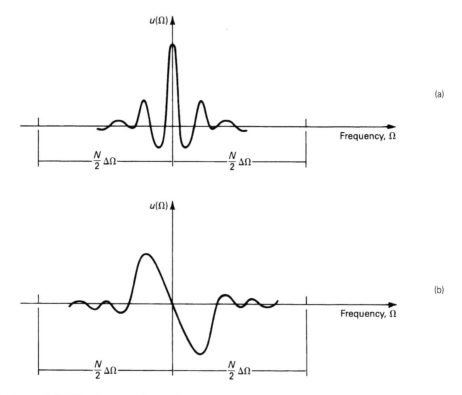

Figure 9.15. Fourier transform of response: (a) real part; (b) imaginary part.

substituting for t:

$$\dot{u}(0) = \frac{2\pi i}{(T_0)^2} \sum_{n=-N/2}^{N/2} nU(n\,\Delta\Omega)$$

$$= \frac{2\pi i}{(T_0)^2} \sum_{n=-N/2}^{N/2} n\operatorname{Re}\{U(n\,\Delta\Omega)\} + in\operatorname{Im}\{U(n\,\Delta\Omega)\} \qquad (9.84)$$

where $\operatorname{Re}\{U(n\,\Delta\Omega)\}$ represents the real part and $\operatorname{Im}\{U(n\,\Delta\Omega)\}$ the imaginary part of $U(n\,\Delta\Omega)$.

Since $\operatorname{Re}\{U(n\,\Delta\Omega)\}$ is symmetric about the origin as shown in Figure 9.15a, the product $n\operatorname{Re}\{U(n\,\Delta\Omega)\}$ is antisymmetric and the first term in the summation in Equation 9.84 vanishes. Also, since $\operatorname{Im}\{U(n\,\Delta\Omega)\}$ is antisymmetric, the product $n\operatorname{Im}\{U(n\,\Delta\Omega)\}$ is symmetric. Equation 9.84 therefore reduces to

$$\dot{u}(0) = -\frac{4\pi}{(T_0)^2} \sum_{n=0}^{N/2} n\operatorname{Im}\{U(n\,\Delta\Omega)\} \qquad (9.85)$$

The difference between the initial conditions in the transient and periodic responses can now be evaluated as

$$\Delta u(0) = u(0) - \bar{u}(0)$$
$$\Delta \dot{u}(0) = \dot{u}(0) - \dot{\bar{u}}(0)$$

(9.86)

The corrective response to be superposed on $\bar{u}(k\,\Delta t)$ is the transient free-vibration response of the system due to initial conditions $\Delta u(0)$ and $\Delta \dot{u}(0)$. The transient response due to a unit displacement change at $t=0$ is obtained from Equation 5.37 and is given by

$$r(t) = e^{-\xi\omega t}\left(\cos\omega_d t + \frac{\xi\omega}{\omega_d}\sin\omega_d t\right)$$

(9.87)

The response due to a unit velocity change is also obtained from Equation 5.37 and is

$$s(t) = \frac{e^{-\xi\omega t}}{\omega_d}\sin\omega_d t$$

(9.88)

The corrective responses are therefore given by

$$\eta_1(t) = \Delta u(0)r(t)$$

(9.89a)

$$\eta_2(t) = \Delta \dot{u}(0)s(t)$$

(9.89b)

and the corrected response is

$$u(k\,\Delta t) = \bar{u}(k\,\Delta t) + \eta_1(k\,\Delta t) + \eta_2 k(\Delta t)$$

(9.90)

Example 9.6
Find the maximum response of an undamped single-degree-of-freedom system to an exciting force which consists of a rectangular pulse of duration 1.6 s shown in Figure E9.6a. The system has a mass of 0.25 lb · in./s^2 and a natural frequency of 2π rad/s.

Solution
The natural period of the system is $T = 2\pi/\omega = 1$ s. The maximum response may be attained either within the duration of the forcing function or in a free-vibration era after the force has ceased to act. If the maximum does occur within the free-vibration era, it will be attained within a half-cycle of free motion. We should, therefore, compute the response of the system for a length of time equal to at least $1.6 + 0.5T = 2.1$ s.

We select a time duration of 2.4 s for the calculation of response. This also becomes the time period T_0 for a frequency-domain analysis using discrete Fourier transform. The forcing function is extended by adding a sufficient number of zeros to make up the period. It is then sampled at 0.1 s intervals, as shown in Figure E9.6a. The number of samples N is equal to 24.

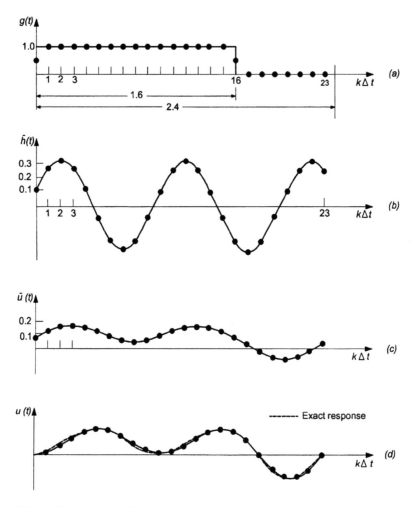

Figure E9.6. (a) Forcing function; (b) periodic unit impulse response; (c) periodic response; (d) transient response.

The periodic impulse response function $\bar{h}(t)$, obtained from Equation 9.80, is shown in Figure E9.6b. The response of the system to a periodic version of the forcing function with a period of T_0 is obtained by convolving the sampled forcing function and $\bar{h}(t)$ through the frequency domain using Equation 9.82. The resulting periodic response $\bar{u}(t)$ is shown in Figure E9.6c and Table E9.6. The periodic response is expected to be different from the transient response because in the former the displacement and velocity at the beginning of a period are not zero. The displacement at $t = 0$ in the periodic response is found to be 0.0793 in., while the velocity at the same time calculated from Equation 9.85 works out to 0.4105 in./s.

The corrective responses are obtained from Equations 9.89a and 9.89b with $\Delta u(0) = -0.0793$ and $\Delta \dot{u}(0) = -0.4105$ in./s. The estimate of transient response obtained by the superposition of periodic response $\bar{u}(t)$ and the corrective responses is shown in Figure E9.6d and Table E9.6. For the purpose of comparison, the exact response obtained from Equations b, c, and d of Example 9.4 is also shown.

Table E9.6. Response calculation through superposition of corrective responses based on initial conditions.

k	$t(s)$	$\bar{u}(k\,\Delta t)$ (in.) Equation 9.82	$u(k\,\Delta t)$ (in.) Equation 9.90	Exact response
0	0.0	0.0793	0.0000	0.0000
1	0.1	0.1167	0.0142	0.0194
2	0.2	0.1470	0.0603	0.0700
3	0.3	0.1585	0.1209	0.1326
4	0.4	0.1470	0.1727	0.1833
5	0.5	0.1167	0.1959	0.2026
6	0.6	0.0793	0.1817	0.1833
7	0.7	0.0489	0.1356	0.1326
8	0.8	0.0374	0.0751	0.0700
9	0.9	0.0489	0.0233	0.0194
10	1.0	0.0793	0.0000	0.0000
11	1.1	0.1167	0.0142	0.0194
12	1.2	0.1470	0.0603	0.0700
13	1.3	0.1585	0.1209	0.1326
14	1.4	0.1470	0.1727	0.1833
15	1.5	0.1167	0.1959	0.2026
16	1.6	0.0793	0.1817	0.1833
17	1.7	0.0303	0.1169	0.1133
18	1.8	−0.0303	0.0074	0.0000
19	1.9	−0.0793	−0.1050	−0.1133
20	2.0	−0.0980	−0.1772	−0.1833
21	2.1	−0.0793	−0.1818	−0.1833
22	2.2	−0.0303	−0.1169	−0.1133
23	2.3	0.0793	−0.0074	0.0000

9.15.2 *Corrective periodic response based on initial conditions*

In the preceding section, we used a corrective response that was equal to the transient response obtained by imposing appropriate displacement and velocity changes at $t=0$. The resultant response obtained by superposing $\bar{u}(t)$ and the corrective responses gave the true response $u(t)$ for $0<t<T_0$. As an alternative, the corrective response may be obtained by imposing a series of periodic displacement and velocity changes at $t=0, \pm T_0, \pm 2T_0, \ldots$. The magnitude of these changes must be selected so that the response obtained by a superposition of $\bar{u}(t)$ and the corrective responses has initial displacement and velocity that are identical to the specified initial conditions. To determine the magnitude of displacement and velocity changes, we first express these changes in terms of appropriate periodic force pulses. As will be shown later, the required force pulses consist of unit impulses $\delta(t)$ and unit doublets $\dot{\delta}(t)$.

The transient response of a system to pulse $\delta(t)$ is obtained from the following Duhamel integral

$$u_1(t) = \frac{1}{m\omega_d} \int_0^t e^{-\xi\omega(t-\tau)} \delta(\tau) \sin \omega_d(t+\tau)\, d\tau$$

$$= \frac{1}{m\omega_d} e^{-\xi\omega t} \sin \omega_d t$$

$$(9.91)$$

On substituting $t=0$ in Equation 9.91, we get

$$u_1(0) = 0 \qquad (9.92)$$

Differentiation of Equation 9.91 gives

$$\dot{u}_1(t) = -\frac{\xi\omega}{m\omega_d}e^{-\xi\omega t}\sin\omega_d t + \frac{1}{m}e^{-\xi\omega t}\cos\omega_d t \qquad (9.93)$$

so that

$$\dot{u}_1(0) = \frac{1}{m} \qquad (9.94)$$

Equations 9.92 and 9.94 show that a unit impulse force $\delta(t)$ introduces a velocity change of $1/m$ and a displacement change of zero at $t=0$. The force pulse required to produce a unit velocity change and no displacement change is therefore given by

$$P_1(t) = m\delta(t) \qquad (9.95)$$

The transient response to a unit doublet is given by

$$u_2(t) = \frac{1}{m\omega_d}\int_0^t e^{-\xi\omega(t-\tau)}\dot{\delta}(\tau)\sin\omega_d(t-\tau)\,d\tau \qquad (9.96)$$

Integration by parts of Equation 9.96 gives

$$u_2(t) = \frac{1}{m\omega_d}[\delta(\tau)e^{-\xi\omega(t-\tau)}\sin\omega_d(t-\tau)|_0^t$$

$$-\int_0^t \delta(\tau)\xi\omega e^{-\xi\omega(t-\tau)}\sin\omega_d(t-\tau)\,d\tau$$

$$+\int_0^t \delta(\tau)\omega_d e^{-\xi\omega(t-\tau)}\cos\omega_d(t-\tau)\,d\tau] \qquad (9.97)$$

The first term in Equation 9.97 is zero since $\delta(\tau)=0$ for $0<\tau<t$. Equation 9.97 therefore reduces to

$$u_2(t) = \frac{e^{-\xi\omega t}}{m\omega_d}(-\xi\omega\sin\omega_d t + \omega_d\cos\omega_d t) \qquad (9.98)$$

Substitution of $t=0$ in Equation 9.98 yields

$$u_2(0) = \frac{1}{m} \qquad (9.99)$$

Differentiation of Equation 9.98 gives

$$\dot{u}_2(t) = \frac{e^{-\xi\omega t}}{m\omega_d}\{(2\xi^2\omega^2 - \omega^2)\sin\omega_d t - 2\xi\omega\omega_d\cos\omega_d t\} \qquad (9.100)$$

so that

$$\dot{u}_2(0) = \frac{-2\xi\omega}{m} \tag{9.101}$$

From Equations 9.99 and 9.101 we conclude that a doublet $m\dot{\delta}(u)$ introduces a unit displacement change as well as a velocity change of $-2\xi\omega$. The force pulse required to introduce only a unit displacement change and no velocity change is therefore given by

$$P_2(t) = m\dot{\delta}(t) + 2\xi\omega m\delta(t) \tag{9.102}$$

The second term on the right-hand side of Equation 9.102 produces a velocity change of $2\xi\omega$ which exactly cancels the velocity change of $-2\xi\omega$ introduced by the pulse $m\dot{\delta}(t)$, so that the net result of the simultaneous action of the two pulses is a unit displacement change alone.

Next, let us evaluate the periodic response due to a set of unit impulses applied at $t = 0, \pm T_0, \pm 2T_0, \ldots$. The response is again given by the Duhamel integral, except that we must now use the periodic impulse function $\bar{h}(t)$ in place of $h(t)$. Thus

$$\begin{aligned}
\bar{u}_1(t) &= \int_0^t \delta(\tau)\bar{h}(t - \tau)\,d\tau \\
&= \frac{1}{m\omega_d\Delta} \int_0^t e^{-\xi\omega(t-\tau)}\delta(\tau)\{\sin\omega_d(t - \tau) \\
&\quad - e^{-\xi\omega T_0}\sin\omega_d(t - \tau - T_0)\}\,d\tau \\
&= \frac{e^{-\xi\omega t}}{m\omega_d\Delta}\{\sin\omega_d t - e^{-\xi\omega T_0}\sin\omega_d(t - T_0)\}
\end{aligned} \tag{9.103}$$

where

$$\Delta = 1 - 2e^{-\xi\omega T_0}\cos\omega_d T_0 + e^{-2\xi\omega T_0} \tag{9.104}$$

In a similar manner, the periodic response due to a set of unit doublets applied at $t = 0, \pm T_0, \pm 2T_0 \ldots$ is given by

$$\begin{aligned}
\bar{u}_2(t) &= \int_0^t \dot{\delta}(\tau)\bar{h}(t - \tau)\,d\tau \\
&= \frac{1}{m\omega_d\Delta} \int_0^t \dot{\delta}(\tau)e^{-\xi\omega(t-\tau)}\{\sin\omega_d(t - \tau) \\
&\quad - e^{-\xi\omega T_0}\sin\omega_d(t - \tau - T_0)\}\,d\tau
\end{aligned} \tag{9.105}$$

Integration of Equation 9.105 by parts leads to

$$\bar{u}_2(t) = -\frac{e^{-\xi\omega t}}{m\omega_d\Delta}[\xi\omega\{\sin\omega_d t - e^{-\xi\omega T_0}\sin\omega_d(t-T_0)\}$$
$$-\omega_d\{\cos\omega_d t - e^{-\xi\omega T_0}\cos\omega_d(t-T_0)\}]$$

(9.106)

The magnitude of force pulse required to introduce a unit displacement but no velocity change is given by Equation 9.102. When such a pulse is applied periodically, the resultant response, $\bar{r}(t)$ is obtained by using Equations 9.103 and 9.106

$$\bar{r}(t) = m\bar{u}_2(t) + 2\xi\omega m\bar{u}_1(t)$$

$$= \frac{e^{-\xi\omega t}}{\Delta}\left\{\cos\omega_d t - e^{-\xi\omega T_0}\cos\omega_d(t-T_0) + \frac{\xi\omega}{\omega_d}\sin\omega_d t\right.$$
$$\left. - \frac{\omega\xi e^{-\xi\omega T_0}}{\omega_d}\sin\omega_d(t-T_0)\right\}$$

(9.107)

In a similar manner, the periodic pulses required to introduce a unit velocity change will generate a response given by

$$\bar{s}(t) = m\bar{u}_1(t)$$

$$= \frac{1}{\Delta\omega_d}e^{-\xi\omega t}\{\sin\omega_d t - e^{-\xi\omega T_0}\sin\omega_d(t-T_0)\}$$

(9.108)

Functions $\bar{r}(t)$ and $\bar{s}(t)$ obtained from Equations 9.107 and 9.108 for specified values of ω, ξ, and T_0 are shown in Figure 9.16a and b, respectively. As would be expected, function $\bar{r}(t)$ exhibits a unit discontinuity at $t=0$. Differentiation of the expressions for $\bar{r}(t)$ and $\bar{s}(t)$ yields the following additional conditions

$$\dot{\bar{r}}(t) = -\omega^2\bar{s}(t)$$

(9.109)

$$\dot{\bar{s}}(t) = \bar{r}(t) - 2\xi\omega\bar{s}(t)$$

(9.110)

Equations 9.109 and 9.110 will be useful later in completing the formulation.

Now let the corrective responses be given by

$$\bar{\eta}_1(t) = a\bar{r}(t)$$

(9.111a)

$$\bar{\eta}_2(t) = b\bar{s}(t)$$

(9.111b)

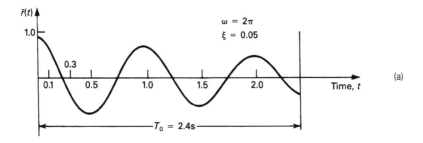

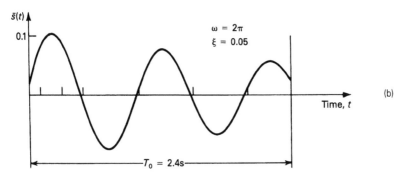

Figure 9.16. (a) Periodic response due to unit displacement change; (b) periodic response due to unit velocity change.

where a and b are unknown weighting factors that must be applied to the basic response functions $\bar{r}(t)$ and $\bar{s}(t)$. Since the displacement and velocity at $t=0$ in the corrected response should be identical to the respective initial conditions in the true transient response, we have

$$a\bar{r}(0) + b\bar{s}(0) = u(0) - \bar{u}(0) = \Delta u(0) \tag{9.112a}$$

$$a\dot{\bar{r}}(0) + b\dot{\bar{s}}(0) = \dot{u}(0) - \dot{\bar{u}}(0) = \Delta \dot{u}(0) \tag{9.112b}$$

Equations 9.112 can be expressed in the matrix form as

$$\begin{bmatrix} \bar{r}(0) & \bar{s}(0) \\ \dot{\bar{r}}(0) & \dot{\bar{s}}(0) \end{bmatrix} \begin{bmatrix} a \\ b \end{bmatrix} = \begin{bmatrix} \Delta u(0) \\ \Delta \dot{u}(0) \end{bmatrix} \tag{9.113}$$

Periodic response values $\bar{r}(0)$ and $\bar{s}(0)$ are obtained from Equations 9.107 and 9.108, respectively.

$$\bar{r}(0) = \frac{1}{\Delta}(1 - e^{-\xi\omega T_0}\cos\omega_d T_0 + \frac{\omega\xi}{\omega_d}e^{-\xi\omega T_0}\sin\omega_d T_0) \tag{9.114a}$$

$$\bar{s}(0) = \frac{1}{\Delta \omega_d} e^{-\xi \omega T_0} \sin \omega_d T_0 \qquad (9.114b)$$

Derivatives $\dot{\bar{r}}(0)$ and $\dot{\bar{s}}(0)$ are obtained by substituting $\bar{r}(0)$ and $\bar{s}(0)$ in Equations 9.109 and 9.110. The procedure for computing $\Delta u(0)$ and $\Delta \dot{u}(0)$ is identical to that described in the preceding section. Equations 9.113 can now be solved for a and b. Substitution of the resulting values in Equations 9.111 will give $\bar{\eta}_1(t)$ and $\bar{\eta}_2(t)$. The corrected response is then obtained by superposing $\eta_1(t)$ and $\eta_2(t)$ on the periodic response $\bar{u}(t)$

$$u(k \,\Delta t) = \bar{u}(k \,\Delta t) + \bar{\eta}_1(k \,\Delta t) + \bar{\eta}_2(k \,\Delta t) \qquad (9.115)$$

The application of corrective periodic responses $\bar{r}(t)$ and $\bar{s}(t)$ given, respectively, by Equations 9.107 and 9.108 serves as an alternative to the use of corrective transient responses $r(t)$ and $s(t)$ (Eqs. 9.87 and 9.88). Since the expressions for $\bar{r}(t)$ and $\bar{s}(t)$ are more involved than those for $r(t)$ and $s(t)$, there does not appear to be any particular advantage in using the former. However, because $\bar{r}(t)$ and $\bar{s}(t)$ are periodic, it is possible to evaluate them by using a frequency-domain analysis. This may lead to significant computational efficiency.

To obtain $\bar{r}(t)$ and $\bar{s}(t)$ through a frequency analysis, we make use of Equations 9.28a and 9.28b. Recognizing that $\bar{r}(t)$ is produced by periodic force pulses $m\dot{\delta}(t) + 2\xi \omega m \delta(t)$ and using Equation 9.28b, we get

$$R(n\omega_0) = c_n T_0 = \int_{-T_0/2}^{T_0/2} \{m\dot{\delta}(t) + 2\xi \omega m \delta(t)\} e^{-in\omega_0 t}\, dt$$

$$= m\delta(t)e^{-in\omega_0 t}\Big|_{-T_0/2}^{T_0/2} + in\omega_0 \int_{-T_0/2}^{T_0/2} m\delta(t)e^{-in\omega_0 t}\, dt + 2\xi \omega m$$

$$= in\omega_0 m + 2\xi \omega m \qquad (9.116)$$

Substitution into Equation 9.28a gives

$$\bar{r}(t) = \frac{1}{T_0} \sum_{n=-\infty}^{\infty} (in\omega_0 m + 2\xi \omega m) H(n\omega_0) e^{in\omega_0 t} \qquad (9.117)$$

Restricting the summation in Equation 9.112 to over a range of significant frequencies $(-N/2)\omega_0$ to $(N/2)\omega_0$, we get

$$\bar{r}(t) = \frac{m\omega}{T_0} \sum_{n=-N/2}^{N/2} (i\beta_n + 2\xi) H_n e^{i\Omega_n t} \qquad (9.118)$$

where $\Omega_n = n\omega_0$, $\beta_n = \Omega_n/\omega$, and $H_n = H(n\omega_0)$.

In a similar manner, recognizing that $\bar{s}(t)$ is produced by periodic force pulses $m\delta(t)$, we get from Equation 9.28b

$$S(n\omega_0) = \int_{-T_0/2}^{T_0/2} m\delta(t)e^{-in\omega_0 t}\,dt$$

$$= m \tag{9.119}$$

Substitution in Equation 9.28a gives

$$\bar{s}(t) = \frac{m}{T_0} \sum_{n=-N/2}^{N/2} H_n e^{i\Omega_n t} \tag{9.120}$$

Before Equations 9.118 and 9.120 are used for obtaining $\bar{r}(t)$ and $\bar{s}(t)$, several points need to be given consideration. It is evident from Equations 9.111 and 9.113 that for evaluating the corrective responses $\bar{\eta}_1(t)$ and $\bar{\eta}_2(t)$, we need to determine $\bar{r}(0), \bar{s}(0), \dot{\bar{r}}(0)$, and $\dot{\bar{s}}(0)$. The derivatives $\dot{\bar{r}}(0)$ and $\dot{\bar{s}}(0)$ are, however, related to $\bar{r}(0)$ and $\bar{s}(0)$ through Equations 9.109 and 9.110. The magnitude of $\bar{\eta}_1(t)$ and $\bar{\eta}_2(t)$ is thus dependent on the terminal values of $\bar{r}$ and $\bar{s}$, namely $\bar{r}(0)$ and $\bar{s}(0)$.

Figure 9.16 shows that $\bar{r}(0)$ is discontinuous at $t=0$. As a consequence, when frequency analysis is used to compute $\bar{r}(0)$, the result will converge to the average value at the discontinuity and not to the value just to the right of $t=0$ needed in the computation. Further, because of the truncation applied in the frequency summation, aliasing will occur in the time domain and the terminal value $\bar{r}(0)$ may be significantly affected by such aliasing. Finally, the presence of term β_n in the numerator of Equation 9.118 will slow down the convergence, since β_n increases with n. These difficulties are obviated by using the following procedure.

By taking out the term corresponding to $n=0$ from the series in Equation 9.118, $\bar{r}(t)$ can be expressed as

$$\bar{r}(t) = \frac{m\omega}{T_0} 2\xi H_0 + \frac{m\omega}{T_0} \sum_{\substack{n=-N/2 \\ n\neq 0}}^{N/2} (i\beta_n + 2\xi) H_n e^{i\Omega_n t} \tag{9.121}$$

Substituting for H_n from Equation 9.26 and noting that $H_0 = 1/k$, Equation 9.121 can be written as

$$\bar{r}(t) = \frac{2\xi}{T_0\omega} + \frac{1}{T_0\omega} \sum_{n=-N/2}^{N/2} (i\beta_n + 2\xi) \frac{e^{i\Omega_n t}}{-\beta_n^2 + 2i\xi\beta_n + 1}$$

$$= \frac{2\xi}{T_0\omega} + \frac{1}{T_0\omega} \sum_{n=-N/2}^{N/2} \frac{1}{i\beta_n}(-\beta_n^2 + 2i\xi\beta_n + 1 - 1)\frac{e^{i\Omega_n t}}{-\beta_n^2 + 2i\xi\beta_n + 1}$$

$$= \frac{2\xi}{T_0\omega} + \frac{1}{T_0\omega} \sum_{n=-N/2}^{N/2} \frac{-e^{i\Omega_n t}}{i\beta_n(-\beta_n^2 + 2i\xi\beta_n + 1)} + \frac{e^{i\Omega_n t}}{i\beta_n} \tag{9.122}$$

It can be shown that the series $\{1/(T_0\omega)\} \sum\limits_{n=-N/2}^{N/2} \{e^{i\Omega_n t}/(i\beta_n)\}$ converges to $(\frac{1}{2} - t/T_0)$. Equation 9.122 therefore reduces to

$$\bar{r}(t) = \left(\frac{1}{2} - \frac{t}{T_0}\right) + \frac{1}{T_0\omega}\left\{2\xi - \sum_{\substack{n=-N/2 \\ n\neq 0}}^{N/2} \frac{e^{i\Omega_n t}}{i\beta_n(-\beta_n^2 + 2i\xi\beta_n + 1)}\right\} \quad (9.123)$$

The discontinuity in $\bar{r}(t)$ at $t = 0$ is incorporated in the term $\frac{1}{2} - t/T_0$ in Equation 9.123; the remaining terms in that equation are continuous. Further, the series in Equation 9.123 converges rapidly because β_n now appears in the denominator unlike in Equation 9.121, where it appeared in the numerator.

It is also possible to improve the accuracy of $\bar{s}(0)$ obtained from Equation 9.120. The approximation in the expression for $\bar{s}(0)$ arises because the frequency summation has been restricted to the finite range $n = -N/2$ to $N/2$. An improved value for $\bar{s}(0)$ can be expressed as

$$\bar{s}(0) = \frac{m}{T_0} \sum_{n=-N/2}^{N/2} H_n + \Delta\bar{s}(0) \quad (9.124)$$

where

$$\Delta\bar{s}(0) = \frac{m}{T_0} \sum_{n=-\infty}^{-N/2-1} H_n + \frac{m}{T_0} \sum_{n=N/2+1}^{\infty} H_n \quad (9.125)$$

Since the real part of H_n is symmetric about the y axis, while the imaginary part is antisymmetric, Equation 9.125 can be expressed as

$$\Delta\bar{s}(0) = \frac{2m}{T_0} \sum_{n=N/2+1}^{\infty} \mathrm{Re}\,(H_n) \quad (9.126)$$

For large values of n, H_n can be approximated as

$$H_n = -\frac{1}{k\beta_n^2}$$

$$= -\frac{1}{mn^2\omega_0^2}$$

$$= -\frac{T_0^2}{4\pi^2 mn^2} \quad (9.127)$$

where we have used the relationship $\omega_0 = 2\pi/T_0$. Substitution of Equation 9.127 in Equation 9.126 gives

$$\Delta\bar{s}(0) = -\frac{T_0}{2\pi^2} \sum_{n=N/2+1}^{\infty} \frac{1}{n^2}$$

$$= -\frac{T_0}{2\pi^2} \left(\sum_{n=1}^{\infty} \frac{1}{n^2} - \sum_{n=1}^{N/2} \frac{1}{n^2}\right) \quad (9.128)$$

The first series in Equation 9.128 converges to $\pi^2/6$. The expression for $\Delta \bar{s}(0)$ therefore reduces to

$$\Delta \bar{s}(0) = -T_0 \left(\frac{1}{12} - \frac{1}{2\pi^2} \sum_{n=1}^{N/2} \frac{1}{n^2} \right) \tag{9.129}$$

Functions $\bar{r}(0)$ and $\bar{s}(0)$ are obtained from Equations 9.123 and 9.124. These values are then substituted in Equations 9.109 and 9.110 to obtain $\dot{\bar{r}}(0)$ and $\dot{\bar{s}}(0)$.

Example 9.7
Solve the problem of Example 9.6 by superposition of corrective periodic responses.

Solution
As in Example 9.6, the response is calculated over a total time of 2.4 s, which also becomes the period T_0. The convolution of the sampled forcing function and $\bar{h}(t)$ gives the periodic response $\bar{u}(t)$, which is, of course, identical to that obtained in Example 9.6.

The periodic corrective response corresponding to a unit displacement change, $\bar{r}(t)$, obtained from Equation 9.107 is shown in Figure E9.7a and also in Table E9.7. The periodic corrective response corresponding to a unit velocity change, $\bar{s}(t)$, obtained from Equation 9.108 is shown

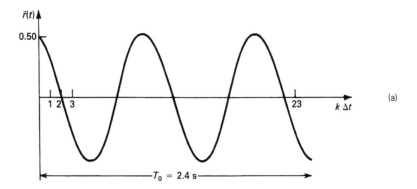

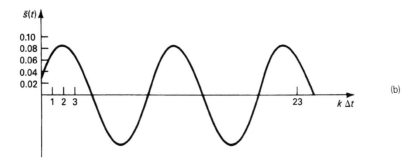

Figure E9.7. (a) Periodic response due to unit displacement change; (b) periodic response due to unit velocity change.

Table E9.7. Corrective periodic responses for unit displacement and unit velocity change.

k	$t(s)$	Exact values		Frequency domain values	
		$\bar{r}(t)$	$\bar{s}(t)$	$\bar{r}(t)$	$\bar{s}(t)$
0	0.0	0.5000	0.0259	0.5000	0.0260
1	0.1	0.3090	0.0677	0.3114	0.0669
2	0.2	0.0000	0.8367	−0.0016	0.0837
3	0.3	−0.3090	0.0677	−0.3078	0.0679
4	0.4	−0.5000	0.0259	−0.5000	0.0256
5	0.5	−0.5000	−0.0259	−0.4993	−0.0255
6	0.6	−0.3090	−0.0677	−0.3096	−0.0681
7	0.7	0.0000	−0.0836	0.0004	−0.0833
8	0.8	0.3090	−0.0677	0.3087	−0.0681
9	0.9	0.5000	−0.0259	0.5002	−0.0255
10	0.0	0.5000	0.0259	0.4999	0.0255
11	1.1	0.3090	0.0677	0.3091	0.0681
12	1.2	0.0000	0.0837	0.0000	0.0833
13	1.3	−0.3090	0.0677	−0.3091	0.0681
14	1.4	−0.5000	0.0259	−0.4999	0.0255
15	1.5	−0.5000	−0.0259	−0.5002	−0.0255
16	1.6	−0.3090	−0.0677	−0.3087	−0.0681
17	1.7	0.0000	−0.0837	−0.0004	−0.0833
18	1.8	0.3090	−0.0677	0.3096	−0.0681
19	1.9	0.5000	−0.0259	0.4993	−0.0256
20	2.0	0.5000	0.0259	0.5000	0.0256
21	2.1	0.3090	0.0677	0.3078	0.0679
22	2.2	0.0000	0.0837	0.0016	0.0837
23	2.3	−0.3090	0.0677	−0.3114	0.0669
24	2.4	−0.5000	0.0259	−0.5000	0.0260

in Figure E9.7b and Table E9.7. The weighting factors a and b, to be applied to $\bar{r}(t)$ and $\bar{s}(t)$, are calculated from Equation 9.113 with $\Delta u(0) = -0.0793$ in. and $\Delta \dot{u} = -0.4105$ in./s and are given by

$$a = -0.1050$$

$$b = -1.0353$$

The corrected response is now calculated from Equation 9.115. As would be expected, the results are identical to those obtained in Example 9.6. As an alternative, the periodic impulse responses $\bar{r}(t)$ and $\bar{s}(t)$ can be obtained through a frequency-domain analysis by using Equations 9.123, 9.120 and 9.124. The resulting values are also shown in Table E9.7 and closely match the exact values.

9.15.3 *Corrective responses obtained from a pair of force pulses*

In this method, the corrective responses are obtained by applying two periodic force impulses of appropriate magnitude at the end of the time period for which response is to be determined. As shown in Figure 9.17, the time duration for which the response calculations are desired is augmented by two zero force points $N - 2$ and $N - 1$. A periodic unit impulse $\delta(t)$ is now applied at $N - 2$ and the response $\bar{u}_1$ of the system to this impulse is determined. In a similar

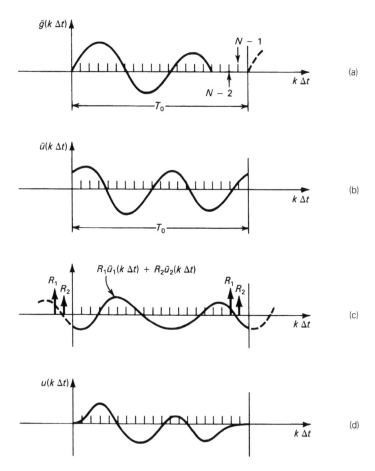

Figure 9.17. Use of corrective force pulses in the calculation of transient response: (a) periodic version of exciting force; (b) periodic response; (c) corrective response; (d) transient response.

manner, a unit impulse is applied at $N - 1$ and the resulting response $\bar{u}_2$ is determined. It should be noted that $\bar{u}_1$ is the same as $\bar{h}$ shifted on the time axis by $(N - 2)\,\Delta t$, while $\bar{u}_2$ is equal to $\bar{h}$ shifted by $(N - 1)\,\Delta t$. If the magnitudes of correcting force impulses are R_1 and R_2, as shown in Figure 9.17, the total response due to the periodic version of the exciting force and the correcting impulses is given by

$$u(k\,\Delta t) = \bar{u}(k\,\Delta t) + R_1\bar{u}_1(k\,\Delta t) + R_2\bar{u}_2(k\,\Delta t) \tag{9.130}$$

The combined response should satisfy the initial conditions in the transient response. This requirement leads to the following two equations

$$R_1\bar{u}_1(0) + R_2\bar{u}_2(0) = u(0) - \bar{u}(0) \tag{9.131a}$$

$$R_1 \ddot{\bar{u}}_1(0) + R_2 \ddot{\bar{u}}_2(0) = \dot{u}(0) - \dot{\bar{u}}(0) \tag{9.131b}$$

In Equations 9.131, $u(0)$ and $\dot{u}(0)$ are specified quantities, $\bar{u}(0)$ is determined from Equation 9.81 or 9.82, while $\dot{\bar{u}}(0)$ is obtained from Equation 9.85. To calculate $\ddot{\bar{u}}_1(0)$ and $\ddot{\bar{u}}_2(0)$, we need the derivative of $\bar{h}(t)$. This is readily obtained by differentiating Equation 9.79

$$\dot{\bar{h}}(t) = \frac{1}{m\omega_d\Delta} e^{-\xi\omega t} \{\omega_d \cos \omega_d t - \omega_d e^{-\xi\omega T_0} \cos \omega_d (t - T_0)\} - \xi\omega\bar{h}(t)$$

$$\tag{9.132}$$

where $\Delta = 1 - 2e^{-\xi\omega T_0} \cos \omega_d T_0 + e^{-2\xi\omega T_0}$. Derivative $\dot{\bar{u}}_1(t)$ can be obtained by noting that it is equal to $\bar{h}(t)$ shifted on the time axis by $(N-2)\Delta t$. In a similar manner, $\ddot{\bar{u}}_2(t)$ is the same as $\bar{h}(t)$ shifted on the time axis by $(N-1)\Delta t$.

Based on the foregoing observations, we get

$$\bar{u}_1(0) = \bar{h}(2\Delta t) \quad \dot{\bar{u}}_1(0) = \dot{\bar{h}}(2\Delta t)$$

$$\bar{u}_2(0) = \bar{h}(\Delta t) \quad \dot{\bar{u}}_2(0) = \dot{\bar{h}}(\Delta t). \tag{9.133}$$

Equations 9.131 can now be solved for R_1 and R_2. Substitution in Equation 9.130 then gives the corrected transient response $u(k\,\Delta t)$. Figure 9.17b shows the periodic response $\bar{u}(k\,\Delta t)$, Figure 9.17c shows the two periodic force pulses as well as the response produced by them, and Figure 9.17d shows the true transient response obtained by superposing Figure 9.17b and c.

Example 9.8
Solve the problem of Example 9.6 by superposition of corrective responses obtained by applying two force pulses of appropriate magnitude.

Solution
As in Example 9.6, response is obtained over a period of 2.4 s, so that $N=24$. A force pulse of magnitude R_1 is applied at time point $N-2$, and a pulse of magnitude R_2 is applied at $N-1$. The unknown magnitudes R_1 and R_2 are determined by solving Equations 9.131, which can be expressed as

$$\begin{bmatrix} \bar{h}(2\Delta t) & \bar{h}(\Delta t) \\ \dot{\bar{h}}(2\Delta t) & \dot{\bar{h}}(\Delta t) \end{bmatrix} \begin{bmatrix} R_1 \\ R_2 \end{bmatrix} = \begin{bmatrix} \Delta u(0) \\ \Delta \dot{u}(0) \end{bmatrix} \tag{a}$$

Terms $\bar{h}(2\Delta t)$ and $\bar{h}(\Delta t)$ are obtained from Equation 9.79 with $\Delta t = 0.1$ s. Derivatives $\dot{\bar{h}}(2\Delta t)$ and $\dot{\bar{h}}(\Delta t)$ are obtained from Equation 9.132. Also, as in Example 9.6, $\Delta u(0) = -0.0793$ in. and $\Delta \dot{u} = -0.4105$ in./s. Equation a thus reduces to

$$\begin{bmatrix} 0.3347 & 0.2708 \\ 0.0000 & 1.2361 \end{bmatrix} \begin{bmatrix} R_1 \\ R_2 \end{bmatrix} = - \begin{bmatrix} 0.0793 \\ 0.4105 \end{bmatrix} \tag{b}$$

On solving Equations b, we get $R_1 = 0.03188$ and $R_2 = -0.33212$. The corrected response is now obtained by using the superposition equation

$$u(k\,\Delta t) = \bar{u}(k\,\Delta t) + R_1 \bar{h}\{(k+2)\,\Delta t\} + R_2 \bar{h}\{(k+1)\,\Delta t\}$$

$$k = 1, 2, \ldots, N-3 \tag{c}$$

For $k = 1$ to $N - 3$, the corrected response should provide a good estimate of the true transient response. The results obtained are identical to those in Examples 9.6 and 9.7.

9.16 EXPONENTIAL WINDOW METHOD

While the corrective response superposition methods overcome some of the limitations associated with the fast convolution technique, they are not readily extended to the frequency domain analysis of general multi-degree-of-freedom systems or continuous systems. Recently a new and powerful method has been suggested that can be applied to the analysis of all types of systems, and that is particularly effective in the analysis of systems with no damping or low damping. Named as the exponential window method or the method of artificial damping, the method relies on the scaling of both the forcing function and the unit impulse response function by an exponentially decaying function. Such scaling ensures that the two functions die rapidly toward the end of a selected period, so that time domain aliasing is avoided in their convolution. The scaled functions are convolved through the frequency domain. The response to the original time function is now recovered from the result obtained from the convolution of the scaled function. The method works equally well when the frequency response function $H(\Omega)$, rather than the unit impulse function $h(t)$, has been specified. For low damping values $h(t)$ cannot be recovered from $H(\Omega)$ because of aliasing involved in the use of discrete transforms, and the methods that rely on a knowledge of $h(t)$ cannot be used.

Rigorous theoretical background to the exponential window method is available in the literature. Here, we provide a less rigorous but simpler explanation of the theory. Given the forcing function $g(t)$ and the unit impulse response function $h(t)$, define scaled functions $\hat{g}(t)$ and $\hat{h}(t)$ as follows

$$\hat{g}(t) = e^{-at} g(t)$$

$$\hat{h}(t) = e^{-at} h(t) \tag{9.134}$$

where a is an arbitrary constant.
Convolution of $\hat{g}(t)$ and $\hat{h}(t)$ gives

$$\hat{u}(t) = \int_0^t \hat{g}(\tau)\hat{h}(t - \tau)\,d\tau$$

$$= \int_0^t e^{-a\tau} g(\tau) e^{-a(t-\tau)} h(t - \tau)\,d\tau$$

$$= e^{-at} \int_0^t g(\tau)h(t-\tau)\,d\tau$$

$$= e^{-at} u(t) \tag{9.135}$$

Equation 9.135 shows that once $\hat{u}(t)$ has been obtained through the convolution of $\hat{g}(t)$ and $\hat{h}(t)$, the response $u(t)$ can be obtained from the relationship

$$u(t) = e^{at}\hat{u}(t) \tag{9.136}$$

Due to exponential scaling, both $\hat{g}(t)$ and $\hat{h}(t)$ die rapidly with time and are expected to approach a very small value toward the end of a selected period of time over which the response is to be calculated. This implies that their convolution may be carried out in the frequency domain using discrete Fourier transform without the need for adding trailing zeros or the superposition of a corrective response. Frequency domain convolution, however, requires that we evaluate the Fourier transforms of both $\hat{g}(t)$ and $\hat{h}(t)$. The required transforms are obtained as follows

$$\hat{G}(\Omega) = \int_{-\infty}^{\infty} \hat{g}(t)e^{-i\Omega t}\,dt$$

$$= \int_{-\infty}^{\infty} g(t)e^{-i(\Omega - ia)t}\,dt$$

$$= G(\Omega - ia) \tag{9.137}$$

In a similar manner, $\hat{H}(\Omega) = H(\Omega - ia)$. The relationship derived in Equation 9.137, in fact, follows from the well-known frequency shifting theorem for Fourier transforms. It implies that $\hat{G}(\Omega)$ and $\hat{H}(\Omega)$ required in the convolution of $\hat{g}(t)$ and $\hat{h}(t)$ through the frequency domain are obtained from $G(\Omega)$ and $H(\Omega)$, respectively, by substituting $\Omega - ia$ for Ω.

The question that still remains is related to what is an appropriate value for the exponential decay constant a. This parameter should be sufficiently large to minimize aliasing. However, a very large value of a may cause numerical difficulties, and the values obtained for the response $u(t)$ may become quite inaccurate toward the end of the period T_0. This is because in recovering the value of $u(t)$ from $\hat{u}(t)$, according to Equation 9.136, the factor e^{at} also amplifies any noise that may be present in the computations. Such amplification increases with t, so that the recovered values of $u(t)$ become seriously contaminated as t approaches T_0. A good criterion for selecting a is to ensure that the value of factor e^{-at} at the end of the period T_0 is not less than 10^{-m}, where m is a constant whose value depends on the precision used in the computation.

In general, good results are obtained by keeping m between 2 and 3. Thus

$$e^{-aT_0} = 10^{-m}$$

$$a = \frac{m \ln 10}{T_0} \qquad (9.138)$$

The essential steps in the frequency domain analysis by the exponential window method can now be summarized as follows:

1. Select a value for the decay parameter a by using Equation 9.138.
2. Obtain the discrete Fourier transform of the scaled forcing function $\hat{g}(t) = e^{-at}g(t)$. As shown in Equation 9.137, transforms G and $\hat{G}$ are related, so that $\hat{G}(\Omega) = G(\Omega - ia)$.
3. Obtain the discrete Fourier transform of the scaled unit impulse function $\hat{h}(t) = e^{-at}h(t)$. Again, transforms H and $\hat{H}$ are related, so that $\hat{H}(\Omega) = H(\Omega - ia)$. If the complex frequency response function $H(\Omega)$, rather than the unit impulse function, has been specified, construct a discrete version as explained in Section 9.11. Now obtain the discrete Fourier transform of the scaled unit impulse function by noting that $\hat{H}(\Omega) = H(\Omega - ia)$.
4. Obtain the scaled response $\hat{u}(t)$ by taking the inverse discrete Fourier transform of the product of $\hat{G}(\Omega)$ and $\hat{H}(\Omega)$. The true response $u(t)$ is now given by $u(t) = \hat{u}(t)e^{at}$.

Example 9.9
Solve the problem of Example 9.6 by the exponential window method.

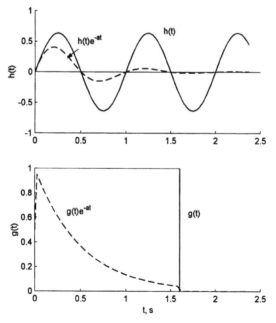

Figure E9.9. Excitation functions $g(t)$ and $g(t)e^{-at}$; unit impulse response functions $h(t)$ and $h(t)e^{-at}$.

Table E9.9. Response obtained by exponential window method.

k	t (s)	$u(k\,\Delta t)$ (in.) Equation 9.90	Exact response
	0.0	0.0007	0.0000
1	0.1	0.0199	0.0194
2	0.2	0.0690	0.0700
3	0.3	0.1294	0.1326
4	0.4	0.1780	0.1833
5	0.5	0.1962	0.2026
6	0.6	0.1771	0.1833
7	0.7	0.1279	0.1326
8	0.8	0.0675	0.0700
9	0.9	0.0190	0.0194
10	1.0	0.0007	0.0000
11	1.1	0.0199	0.0194
12	1.2	0.0690	0.0700
13	1.3	0.1294	0.1326
14	1.4	0.1780	0.1833
15	1.5	0.1962	0.2026
16	1.6	0.1772	0.1833
17	1.7	0.1095	0.1133
18	1.8	0.0000	0.0000
19	1.9	−0.1095	−0.1133
20	2.0	−0.1772	−0.1833
21	2.1	−0.1772	−0.1833
22	2.2	−0.1095	−0.1133
23	2.3	−0.0000	0.0000

Solution
The response is calculated over a total time of $T_0 = 2.4$ s. To obtain a we use

$$a = \frac{2\ln 10}{T_0} = 1.92$$

We select $a = 2$ and apply exponential scaling to both $g(t)$ and $h(t) = 1/m(\omega^2 - \Omega^2)$. The scaled functions are shown in Figure E9.9. As expected, they die rapidly as t approaches T_0. The two functions are sampled at 0.1 s. Discrete Fourier transforms of the sampled functions, calculated by using a standard computer program, provide $\hat{G}(\Omega)$ and $\hat{H}(\Omega)$. Response $\hat{u}(t)$ is obtained by taking the inverse discrete Fourier transform of the product of $\hat{G}(\Omega)$ and $\hat{H}(\Omega)$. The true response $u(t)$ is now recovered from $\hat{u}(t)$ by using Equation 9.136. The computed displacements are compared with the exact values in Table E9.9; the match between the two is quite good.

Example 9.10
The gravity dam and the impounded reservoir shown in Figure E9.10a are excited by an earthquake motion along the y axis, that is, in the vertical direction. Assuming that the ground motion does not vary along the length of the dam and the dam is long in comparison to its cross-sectional dimension, the system can be represented by the two-dimensional model shown in Figure E9.10a. It can be shown that when ground motion is $e^{i\Omega t}$, the total pressure on the face of the dam, expressed as a fraction of the hydrostatic force $F_s = \frac{1}{2}wH^2$, is given by

$$F(\Omega) = \frac{\tilde{F}(\Omega)}{F_s} = \frac{2c^2}{g\Omega^2 H^2} \frac{1 - \cos(\Omega H/c)}{\cos(\Omega H/c)} \qquad (a)$$

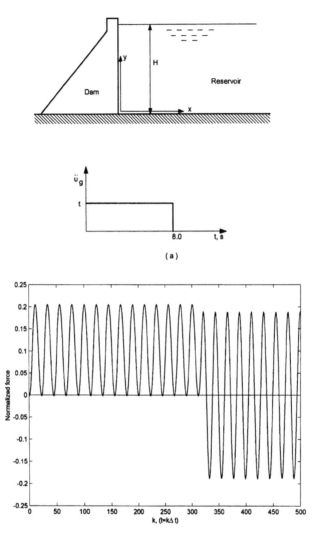

Figure E9.10. Analysis of a rigid gravity dam in frequency domain; (a) dam cross-section and ground acceleration; (b) frequency domain response obtained by the exponential window method.

where w is the unit weight of water, H the height of the reservoir $= 200$ m, c the velocity of sound in water $= 1440$ m/s, and g is the acceleration due to gravity $= 9.81$ m/s^2. The variation of ground acceleration is also shown in Figure E9.9a. Calculate the variation of total hydrodynamic force on the dam face with time for a duration of 25 s.

Solution
Since the reservoir bottom is rigid and there is no loss of energy by wave radiation, the system is completely undamped. Also, it is the frequency response function rather than the unit impulse response function that is specified. The standard frequency domain analysis procedures using discrete Fourier transforms do not, therefore, work. However, the exponential window method is quite effective. In carrying out the analysis the following data is used: time step

$\Delta t = 0.025$ s; $T_0 = 25$ s; frequency step $\Delta\Omega = 2\pi/T_0 = 0.2513$ rad/s. The exponential decay factor *a* is obtained as follows

$$a = 2\frac{\ln 10}{T_0} = 0.184 \tag{b}$$

The exciting function $g(t)$ is sampled at intervals of 0.025, giving $N = 1000$ samples and scaled by the exponential decay function $e^{-ak\Delta t}$. Discrete Fourier transform of the scaled function $\hat{g}(t)$ is obtained using a standard computer program. A discrete version of the frequency response function $\hat{F}(\Omega)$ must be obtained next. Values of $\hat{F}(\Omega)$ are obtained from Equation a at frequency intervals of $\Delta\Omega = 0.2513$ rad/s by substituting $\Omega = (n\Delta\Omega - ia)$ instead of $n\Delta\Omega$. Discretization is carried out upto a frequency of $\pi/\Delta t = 125.7$ corresponding to $n = 500$. The sample values of $\hat{F}(\Omega)$ for $\Omega > \pi/\Delta t$ are taken as being the complex conjugates of values for frequencies located symmetrically about $\pi/\Delta t$. Thus $\hat{F}(N/2 + m) = \hat{F}^*(N/2 - m)$, where * denotes a complex conjugate. The scaled response $\hat{p}(t)$ is obtained by taking the inverse discrete Fourier transform of the product of $\hat{G}(\Omega)$ and $\hat{F}(\Omega)$. The desired response, that is the normalized total force on the dam face, is now calculated from $p(k\Delta t) = e^{ak\Delta t}\hat{p}(k\Delta t)$. The results are presented for the first 500 sample points in Figure E9.10b.

9.17   THE FAST FOURIER TRANSFORM

As stated earlier, the dynamic response analysis of a linear system involves evaluation of the convolution integral of two functions of time, or equivalently, of the discrete convolution given by Equation 9.56

$$u(k\,\Delta t) = \sum_{m=0}^{N-1} g(m\,\Delta t)h\{(k-m)\,\Delta t\}\,\Delta t \tag{9.56}$$

In the frequency domain, discrete convolution is replaced by the computations involved in the following steps. Compute the discrete transforms from Equations 9.64 and 9.65

$$G(n\,\Delta\Omega) = \sum_{k=0}^{N-1} g(k\,\Delta t)e^{-2\pi ink/N}\,\Delta t \tag{9.64}$$

$$H(n\,\Delta\Omega) = \sum_{k=0}^{N-1} h(k\,\Delta t)e^{-2\pi ink/N}\,\Delta t \tag{9.65}$$

followed by the product

$$U(n\,\Delta\Omega) = G(n\,\Delta\Omega)H(n\,\Delta\Omega) \tag{9.66a}$$

and finally the inverse discrete transform

$$u(k\,\Delta t) = \frac{1}{2\pi}\sum_{n=0}^{N-1} U(n\,\Delta\Omega)e^{2\pi ink/N}\,\Delta\Omega \tag{9.66b}$$

For the evolution of response at N sample points, Equation 9.56 involves N^2 multiplications of real quantities. On the other hand, Equations 9.64 and 9.65

each require N^2 multiplications of a real and a complex quantity, Equation 9.66b involves N^2 multiplications of two complex quantities, and Equation 9.66a involves N multiplications of two complex quantities.

From the foregoing discussion it is obvious that the frequency-domain analysis will require much more computation than a time-domain analysis. However, if advantage is taken of the harmonic properties of some of the functions involved in Equations 9.64, 9.65, and 9.66, the computation time can be drastically reduced. The technique that utilizes this harmonic property for efficient computation of a Fourier transform is commonly known as *fast Fourier transform*(FFT). An FFT algorithm was first developed by J.W. Cooley and J.W. Tukey in 1965. Since then a number of different versions of the algorithm have been devised. The advent of the fast Fourier transform has made frequency-domain analysis so efficient that in many situations it provides a better alternative to a time-domain analysis.

9.18   THEORETICAL BACKGROUND TO FAST FOURIER TRANSFORM

It should be realized that fast Fourier transform is simply a particular technique of calculating discrete Fourier transform. Although the efficiency of the technique depends to some extent on the details of the specific algorithm used, the underlying theory is similar and merits discussion. It is presented here with relation to the original method devised by Cooley and Tukey, which relied on N being chosen to be to be equal to 2^γ, where γ is an integer. In more recent algorithms, this restriction has been removed.

Consider the discrete Fourier transform

$$X(n) = \sum_{k=0}^{N-1} x(k)e^{-2\pi i n k/N} \tag{9.139}$$

where we have temporarily dispensed with the scaling factor Δt. For the purpose of illustration we will assume that N has been chosen as 8, so that $\gamma = 3$. Also, for ease of reference we will use the notation

$$W = e^{-2\pi i/N} \tag{9.140}$$

Integers n and k can be expressed in decimal equivalent binary form as

$$n = 4n_2 + 2n_1 + n_0$$
$$k = 4k_2 + 2k_1 + k_0 \tag{9.141}$$

where n_0, n_1, n_2 and k_0, k_1, k_2 are integers with possible values 0 and 1.

Substitution of Equations 9.140 and 9.141 into Equation 9.139 gives

$$X(n_2, n_1, n_0) = \sum_{k_0=0}^{1} \sum_{k_1=0}^{1} \sum_{k_2=0}^{1} x(k_2, k_1, k_0)W^{nk} \tag{9.142}$$

It should be noted that Equation 9.142 is, in fact, a set of eight equations corresponding to two different values of each of the integers n_0, n_1 and n_2, the subscript of X ranging from 0 to 7.

Function W^{nk} in Equation 9.142 can be rewritten as

$$W^{nk} = W^{(4n_2+2n_1+n_0)(4k_2+2k_1+k_0)}$$

$$= W^{(4n_2+2n_1+n_0)4k_2} W^{(4n_2+2n_1+n_0)2k_1} W^{(4n_2+2n_1+n_0)k_0} \qquad (9.143)$$

The harmonic property of function W is now used to simplify Equation 9.143. We note that

$$W^{mN} = e^{-2\pi i m N/N}$$

$$= \cos 2m\pi - i \sin 2m\pi$$

$$= 1 \qquad (9.144)$$

Thus, $W^{16n_2k_2} = W^{8n_1k_2} = W^{8n_2k_1} = 1$ since the exponent in each of these terms is an integer multiple of $N=8$. Equation 9.143 therefore reduces to

$$W^{nk} = W^{4n_0k_2} W^{(2n_1+n_0)2k_1} W^{(4n_2+2n_1+n_0)k_0} \qquad (9.145)$$

Substitution into Equation 9.142 gives

$$X(n_2, n_1, n_0) = \sum_{k_0=0}^{1} \sum_{k_1=0}^{1} \left\{ \sum_{k_2=0}^{1} x(k_2, k_1, k_0) W^{4n_0k_2} \right\}$$

$$\times W^{(2n_1+n_0)2k_1} W^{(4n_2+2n_1+n_0)k_0} \qquad (9.146)$$

Let us examine the inner sum given by

$$x_1(n_0, k_1, k_0) = \sum_{k_2=0}^{1} x(k_2, k_1, k_0) W^{4n_0k_2} \qquad (9.147)$$

Equation 9.142 again represents a set of eight equation for a combination of two possible values of each of the integers k_1, k_0, and n_0. These individual equations are of the form

$$x_1(n_0, k_1, k_0) = x(0, k_1, k_0) W^0 + x(1, k_1, k_0) W^{4n_0} \qquad (9.148)$$

The resulting values are arranged as shown in the *signal flow graph* of Figure 9.18. Each value of x_1 originates from two different values of x, listed in the first column of the graph. As an example, for $n_0 = 1$, $k_1 = 0$, and $k_0 = 1$

$$x_1(1, 0, 1) = x_1(5)$$

$$= x(0, 0, 1) W^{4\times1\times0} + x(1, 0, 1) W^{4\times1\times1}$$

$$= x(1) + x(5) W^4 \qquad (9.149)$$

The paths leading from values of x to x_1 are shown on the signal flow graph. The multiplier applicable to each originating term is placed next to the path leading from it, except that when the multiplier is 1, it is omitted. Referring to Equation 9.149, we note that the multiplier on $x(1)$ is 1, while that on $x(5)$ is W^4. The latter is indicated on the path leading from $x(5)$ to $x_1(5)$.

The power of W, which is $4n_0k_2$, will be zero for $k_2=0$ and $4n_0$ when $k_2=1$. For the former case, the multiplier is 1; for the latter it is W^{4n_0}. Also, in the special case when $n_0=0$, the latter multiplier also becomes 1, but for the purpose of generality, we have indicated it as W^0. From Equation 9.148 it is apparent that in going from x to x_1, we need to carry out N complex multiplications followed by $(N-1)$ complex additions.

The number of complex multiplications involved can be further reduced if we recognize that series x_1 can be divided into sets of two terms each, in which both terms are calculated from the same originating terms in x. For example, both $x_1(0)$ and $x_1(4)$ are calculated from the same two values of x. Thus

$$x_1(0)=x(0)+x(4)W^0$$
$$x_1(4)=x(0)+x(4)W^4 \tag{9.150}$$

Terms in x_1 that are members of such a set are called *dual nodes*. It is seen from the signal flow graph that dual nodes have identical values of k_1 and k_0. However, the first of the two nodes has $n_0=k_2=0$, while the second node has $n_0=k_2=1$. The nodes are separated by a distance of $N/2=4$, and if the multiplier for the first node is W^m, that for the second node is $W^{m+(N/2)}$. Noting that

$$W^{m+(N/2)} = e^{(2i\pi/N)(N/2+m)}$$
$$= e^{i\pi}e^{2i\pi m/N} \tag{9.151}$$
$$= -W^m$$

the dual-node computations can be written as

$$x_1(j)=x(j)+x(N/2+j)W^m$$
$$x_1(N/2+j)=x(j)-x(N/2+j)W^m \tag{9.152}$$

From Equation 9.152 it is seen that only one complex multiplication is required in evaluating a pair of dual nodes. The total number of complex multiplications required in going from x to x_1 is thus $N/2$. The second sum in Equation 9.146 can now be expressed as

$$x_2(n_0, n_1, k_0)= \sum_{k_1=0}^{1} x_1(n_0, k_1, k_0)W^{(2n_1+n_0)2k_1} \tag{9.153}$$

Again, Equation 9.153 represents a set of eight equations for a combination of each of the integers n_0, n_1, and k_0. In each case, the result is denoted by

$x_2(n_0, n_1, k_0)$. The arrangement of results is shown on the signal flow graph. As before, each value of x_2 originates from two different values of x_1. The paths from the originating terms as well as the applicable multipliers, $W^{(2n_1+n_0)2k_1}$, are shown on the graph. By recognizing the presence of dual nodes, it can easily be shown that the computations involved in going from x_1 to x_2 are $N/2$ complex multiplications followed by $N - 1$ complex additions.

Finally, the outermost sum in Equation 9.146 can be expressed as

$$x_3(n_0, n_1, n_2) = \sum_{k_0=0}^{1} x(n_0, n_1, k_0)W^{(4n_2+2n_1+n_0)k_0} \tag{9.154}$$

Equation 9.154 represents a set of eight equations for a combination of two possible values of each of the integers n_0, n_1, and n_2. The results are denoted by $x_3(n_0, n_1, n_2)$ and are shown on the signal flow graph. The computations involved in going from x_2 to x_3 consist of $N/2$ complex multiplications and $N - 1$ complex additions. The relationship between X and x_3 is easily derived from the last two columns of the signal flow graph.

$$X(n_2, n_1, n_0) = x_3(n_0, n_1, n_2) \tag{9.155}$$

The relationship is again demonstrated in Figure 9.18b, where X has been arranged in normal order.

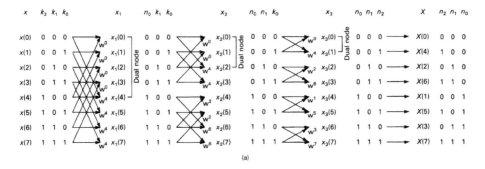

(a)

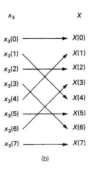

(b)

Figure 9.18. Signal flow graphs.

The number of columns of intermediate results x_1, x_2, x_3, in the example just cited, is 3 or $\gamma = \log_2 N$. The total number of complex multiplications in the evaluation of discrete Fourier transform thus reduces from N^2 to $(N/2)\log_2 N$. For large values of N, this implies a drastic reduction in the computation time. The reasoning presented in the foregoing paragraphs with reference to $\gamma = 3$ can easily be extended to any other value of γ.

9.19 COMPUTING SPEED OF FFT CONVOLUTION

On the basis of discussion presented in the foregoing paragraph, we can now make a comparison between the computing speed of an FFT convolution as represented by Equations 9.63 through 9.66 and a direct convolution (Eq. 9.56). If the computing speed is assumed to be proportional to the number of multiplications involved, what is required is a comparison of such multiplications. Direct convolution by Equation 9.56 requires N^2 real multiplications. By comparison, FFT convolution using Equations 9.64 through 9.66 will require $(3/2)N\log_2 N + N$ complex multiplications, when no advantage is taken of the fact that in Equations 9.64 and 9.65 only one of the two quantities involved in the product is complex, the other being real. For large values of N, FFT convolution will be significantly faster. The exact speed will, of course, depend on the details of the computer algorithm used. Assuming that a complex multiplication is equivalent to four real multiplications, the number of real multiplications involved in a direct and an FFT convolution are shown in Table 9.1 as a function of N. It is seen that even with N as low as 64, FFT convolution is already substantially faster than a direct convolution.

Table 9.1. Number of real multiplications in discrete convolution.

N	α	Time domain, N^2	Frequency domain $4(\frac{3N}{2}\log_2 N + N)$	Ratio, time Domain to frequency domain
2	1	4	20	0.20
4	2	16	64	0.25
8	3	64	176	0.36
16	4	256	448	0.57
32	5	1,024	1,088	0.94
64	6	4,096	2,560	1.60
128	7	16,384	5,888	2.78
256	8	65,536	13,312	4.92
512	9	262,144	29,696	8.83
1,024	10	1,048,576	65,536	16.00

SELECTED READINGS

Bergland, G.D. 1969. A Guided Tour of the Fast Fourier Transform. *Institute of Electrical and Electronics Engineers Spectrum*: 41–52.

Brigham, E.O. 1974. *The Fast Fourier Transform.* Englewood Cliffs: Prentice Hall.

Bracewell, R. 1978. *The Fourier Transform and its Applications.* New York: 232–236, McGraw-Hill. 2nd Edition.

Cooley, J.W. & Tukey, J.W. 1965. An Algorithm for the Machine Calculation of Complex Fourier Series. *Mathematics of Computation*, Vol. 19: 297–301.

Hall, J.F. 1982. An FFT Algorithm for Structural Dynamics. *Earthquake Engineering and Structural Dynamics*, Vol. 10: 797–811.

Humar, J.L. & Xia, H. 1993. Dynamic Response Analysis in the Frequency Domain. *Earthquake Engineering and Structural Analysis*, Vol. 22: 1–12.

Kausel, E. & Roesset, J.M. 1992. Frequency Domain Analysis of Undamped Systems. *Journal of Engineering Mechanics*, Vol. 118: 721–734. ASCE.

Liu, S.C. & Fagel, L.W. 1971. Earthquake Interaction by Fast Fourier Transform. *Journal of Engineering Mechanics Division*, Vol. 97: 1223–1237. ASCE.

Meek, J.W. & Veletsos, A.S. 1972. Dynamic Analysis by Extra Fast Fourier Transform. *Journal of Engineering Mechanics Division*, Vol. 98: 367–384. ASCE.

Veletsos, A.S. & Kumar, A. 1983. Steady State and Transient Response of Linear Structures. *Journal of Engineering Mechanics Division*, Vol. 109: 1215–1230. ASCE.

Veletsos, A.S. & Ventura, C.E. 1984. Efficient Analysis of Dynamic Response of Linear Systems. *Earthquake Engineering and Structural Dynamics*, Vol. 12: 521–536.

PROBLEMS

9.1 A single degree of freedom system with frequency ω, stiffness k, and negligible damping is subjected to the periodic load shown in Figure P9.1. Obtain the response of the system to the applied load. Show that resonance takes place not only when $\Omega = \omega$ but also when Ω takes any of the values in the series $\Omega = \omega/3,\ \omega/5,\ \omega/7,\ldots$.

9.2 Obtain the response of a system with an undamped frequency ω to the periodic load shown in Figure P9.2. The stiffness of the system is k and damping is 10% of critical. The exciting frequency $\Omega = 1.5\omega$.

9.3 The response of the system in Problem 9.1 consists of a constant term that is equal to the static displacement under an average load $p_0/2$ and

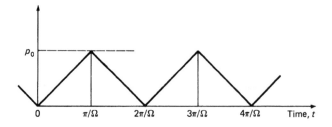

Figure P9.1.

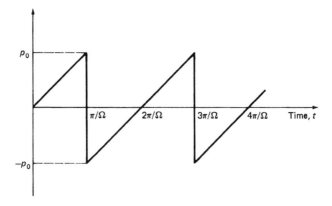

Figure P9.2.

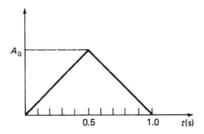

Figure P9.4.

a series of harmonic terms. Plot the response of the system $\frac{u}{(p_0/k)}$ as a function of time using the constant term and the first three harmonics. Given $\Omega = 1.5\omega$.

9.4 Using an appropriate computer program, obtain the discrete Fourier transform of the function shown in Figure P9.4. Use a sampling interval of 0.1 s and a period of 1 s. Now obtain the inverse Fourier transform of the frequency function obtained by you. Compare this inverse transform with the original time function and explain the differences, if any, between the two.

9.5 Show that the continuous Fourier transform of $g(t) = e^{-a|t|}$ where $a > 0$ is given by $G(\Omega) = 2a/(a^2 + \Omega^2)$. Now with $a = 1$, a sampling interval of 0.125 Hz and a period of 1.25 cycles construct a frequency function whose inverse discrete transform will closely resemble the original time function. Using an appropriate computer program, obtain the inverse discrete transform of the frequency function constructed by you. Compare this inverse transform with the original time function. Note the differences between the two and explain how these differences arise.

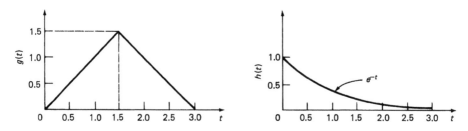

Figure P9.6.

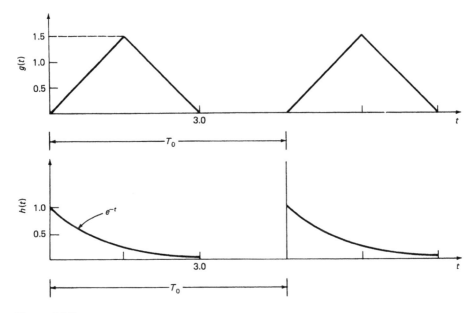

Figure P9.7.

9.6 Obtain the continuous convolution of the two time function shown in Figure P9.6. Next, with a sampling interval of 0.5 s, obtain the discrete convolution of the two functions. Compare the two sets of results and note the errors introduced in the discrete convolution. What is the cause of these errors and what can be done to improve the accuracy of the results.

9.7 Figure P9.7 shows the periodic versions of the time functions shown in Figure P9.6. The selected period is T_0. If the discrete convolution of the functions shown in Figure P9.7 has to be identical to that of the functions in Figure P9.6 over the time interval 0 to T_0 what should be the minimum value of T_0. Using a period equal to or greater than the minimum value obtain the discrete convolution of the two function

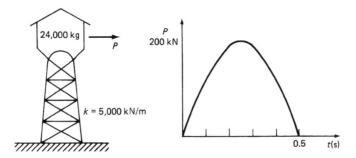

Figure P9.8.

through a frequency domain analysis. Compare the results obtained by you with the discrete convolution values calculated in Problem 9.6. Now carry out the frequency analysis with a period that is 1 s less than the minimum value of T_0. Compare the new results obtained by you with those obtained earlier, note the differences and explain the cause(s) of these differences.

9.8 The water tower shown in Figure P9.8 has a mass of 24,000 kg which can be assumed as being lumped at the center of the tank. The lateral stiffness of the supporting frame is 5,000 kN/m. The tank is subjected to a lateral load which varies as a half sine wave of amplitude 200 kN and period 1.0 s as shown in the figure. Using a frequency domain analysis and Fast convolution technique obtain the history of response for the first second of motion. Use a sampling interval of 0.05 s. Compare your results with a closed form solution. Neglect damping.

9.9 Repeat Problem 9.8 with a damping equal to 10% of critical.

9.10 A single degree of freedom system with stiffness 20 N/mm and period 1.0 s is subjected to a force of duration 0.6 s which varies as shown in Figure P9.10. The response of the system is to be computed for the first 1.6 s using a frequency domain analysis and the Fast Fourier technique. The time function sampling rate is 0.1 s. Limitation on computer memory dictates that the duration of time function to be discretized at one time not exceed 1.6 s. Use the overlap-add-sectioning method to determine the required response. Damping is 10% of critical.

9.11 Solve Problems 9.10 using a corrective transient response based on initial conditions.

9.12 Solve Problem 9.10 using a corrective periodic response based on initial conditions.

9.13 Solve Problem 9.10 using corrective responses obtained from a pair of force pulses of appropriate magnitude.

9.14 The initial velocity resulting from a forcing function which is a periodic extension of the one shown in Figure P9.10 must be determined in

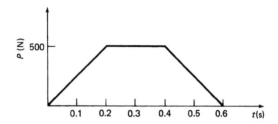

Figure P9.10.

order to proceed with the solution of Problem 9.11. This velocity can be calculated by using Equation 9.85. As a possible alternative $\dot{\bar{u}}(0)$ can be obtained by fitting a parabola to $\bar{u}(0)$, $\bar{u}(\Delta t)$ and $\bar{u}(2\Delta t)$ and taking the slope of this parabola at $t = u$. Repeat Problem 9.11 using this alternative procedure for obtaining an initial velocity estimate.

9.15 Suppose that the system of Problem 9.10 starts with an initial displacement of 20 mm and an initial velocity of 100 mm/s. Obtain the response of the system to the applied force shown in Figure P9.10 using the corrective transient response method.

9.16 Solve Problem 9.8 using the exponential window method. Neglect damping.

9.17 Solve Problem 9.8 using the exponential window method. Assume that damping is 10% of critical.

PART 3

CHAPTER 10

Free-vibration response: Multi-degree-of-freedom systems

10.1 INTRODUCTION

As in the case of a single-degree-of-freedom system, a system with multiple degrees of freedom will vibrate even without the presence of an external force whenever it is subjected to disturbances in the form of initial displacements, or initial velocities, or both, along one or more of its degrees of freedom. The free vibrations of a multi-degree-of-freedom system are governed by the following equation of motion

$$\mathbf{M\ddot{u} + C\dot{u} + Ku = 0} \tag{10.1}$$

where, as usual, $\mathbf{M}$ is the mass matrix, $\mathbf{C}$ the viscous damping matrix, $\mathbf{K}$ the stiffness matrix, and $\mathbf{u}$ the displacement vector. When damping resistances are absent or are negligible, the free vibrations of the system are said to be un-damped. Analysis of undamped free-vibration response provides the displacements and stresses in such a system. But apart from this direct application, free-vibration analysis is also useful in obtaining methods for the solution of both damped free-vibration and forced-vibration problems. A study of the free-vibration response is, therefore, of considerable interest. The equations of un-damped free-vibration are obtained by setting $\mathbf{C=0}$ in Equation 10.1

$$\mathbf{M\ddot{u} + K\dot{u} = 0} \tag{10.2}$$

Equation 10.2 has a solution of the form $\mathbf{u = q}\sin(\omega t + \theta)$, where $\mathbf{q}$ is an arbitrary vector, ω is referred to as the frequency of vibration, and θ is a phase angle, all yet to be determined. To test the validity of our proposed solution, we substitute from it values of $\mathbf{u}$ and $\mathbf{\ddot{u}}$ in Equation 10.2. This gives

$$(\mathbf{K} - \omega^2\mathbf{M})\mathbf{q}\sin(\omega t + \theta) = \mathbf{0} \tag{10.3}$$

Since Equation 10.3 must be true for all values of t, we may cancel $\sin(\omega t + \theta)$ and obtain the equation

$$\mathbf{Kq} = \omega^2\mathbf{Mq} \tag{10.4}$$

whose solution should allow us to determine $\mathbf{q}$, and ω.

Equation 10.4 represents a problem that is commonly known as a *linearized eigenvalue problem*. Apart from vibrations, eigenvalue problem occurs in several other areas of engineering analysis, including, for example, buckling analysis, and the analysis of heat conduction. For a large system, its solution presents a substantial challenge as well as a formidable computational task. A large volume of literature dealing with eigenvalue problems is now in existence, and several different methods have been devised for solving such problems.

In this chapter, we discuss the nature and the properties of the eigenvalue problem, primarily from the point of view of vibration and structural dynamics. We then describe the application of eigenvalue solution to the determination of free vibration response. The numerical procedures for solving the eigenvalue problem are presented in Chapter 11. The application to the solution of forced vibration problem is covered in a subsequent chapter.

10.2   STANDARD EIGENVALUE PROBLEM

The *eigenvector* of a matrix is defined as a vector which has the property that when premultiplied by the matrix, it yields another vector that is proportional to the original vector, the constant of proportionality being called an *eigenvalue*. Mathematically, this is expressed as

$$\mathbf{A}\mathbf{q} = \lambda\mathbf{q} \tag{10.5}$$

in which $\mathbf{A}$ is a square matrix of order N, $\mathbf{q}$ is a right eigenvector, also of order N, and λ is the corresponding eigenvalue. For Equation 10.5 to be conformable, the product on the left hand side must be a vector of size N and hence the matrix $\mathbf{A}$ must be of size $N \times N$. Thus only square matrices can have eigenvalues and eigenvectors.

In general, when Equation 10.5 is to be solved to obtain the eigenvalues λ, it is referred to as an *eigenvalue problem*. When both the eigenvalues λ and the eigenvectors $\mathbf{q}$ have to be determined, the problem is more commonly referred to as an *eigenproblem*. Equation 10.5 can be expressed in the alternative form

$$(\mathbf{A} - \lambda\mathbf{I})\mathbf{q} = 0 \tag{10.6}$$

where $\mathbf{I}$ is an identity matrix of the same order as $\mathbf{A}$. In eigenproblems of interest to us, $\mathbf{A}$ will generally be real. Equation 10.6 therefore represents a set of homogeneous simultaneous equations in the N unknown elements of $\mathbf{q}$, with the coefficients of all the elements being real. In general, a set of homogeneous equations of the form of Equation 10.6 leads to zero values for all of the unknowns. Such a solution is called a trivial solution and is of no interest to us. There is one exception under which Equation 10.6 gives a nontrivial solution in which at least one of the solution vectors $\mathbf{q}$ is nonzero. This exception occurs when the determinant of the coefficient matrix $\mathbf{A} - \lambda\mathbf{I}$ is zero. The foregoing condition for a nontrivial solution can be used to obtain λ, which is as yet undetermined. Application of the condition for a nontrivial solution leads to an Nth order algebraic equation in λ, called the *characteristic equation* of matrix A.

The solution of the characteristic equation of an Nth order matrix gives N values of λ, all of which need not be distinct. These N values of λ are called the eigenvalues of matrix $\mathbf{A}$. Corresponding to each value of λ, Equation 10.6 gives one nontrivial solution for $\mathbf{q}$. However, because Equation 10.6 represents a set of homogeneous equations, the solution for $\mathbf{q}$ is not unique and the elements of $\mathbf{q}$ are determined only in terms of their relative values. In other words, if $\mathbf{q}_i$ represents a solution of Equation 10.6, $r\mathbf{q}_i$ is also a solution, r being a constant.

In summary, Equation 10.6 gives N values for λ, which are called the eigenvalues of matrix $\mathbf{A}$. Associated with each eigenvalue λ_i, there is a vector $\mathbf{q}_i$ called an eigenvector which may be scaled in any arbitrary manner and would still retain its characteristic properties. In our subsequent discussion, we will denote the N eigenvalues by $\lambda_1, \lambda_2, \ldots, \lambda_N$, and the corresponding eigenvectors by $\mathbf{q}_1, \mathbf{q}_2, \ldots, \mathbf{q}_N$. Also, we will assume that the eigenvalues are ordered so that $\lambda_1 < \lambda_2 < \lambda_3 < \cdots < \lambda_N$.

The eigenvalues and eigenvectors of a matrix possess certain characteristic properties which depend on the nature of the matrix. An eigenproblem of special interest in vibration analysis is the linearized problem given by Equation 10.4. The linearized eigenproblem is discussed in some detail in the following sections of this chapter.

10.3 LINEARIZED EIGENVALUE PROBLEM AND ITS PROPERTIES

The free vibration equation (Eq. 10.4) can now be identified as an eigenvalue problem of special form given by

$$\mathbf{Kq} = \lambda \mathbf{Mq} \tag{10.7}$$

in which $\lambda = \omega^2$. The problem represented by Equation 10.7 is known as a linearized eigenproblem; at times, the term "generalized eigenproblem" is also used. The linearized eigenproblem can readily be converted into the standard form of Equation 10.5 by using one of the methods described in Section 11.3.2. In general, the eigenvalues of Equation 10.7 are both real and positive. However, when matrix $\mathbf{M}$ is singular, Equation 10.7 possesses one or more infinite eigenvalues. On the other hand, when $\mathbf{K}$ is singular, the equation has one or more zero eigenvalues.

The eigenvectors obtained from Equation 10.7 possess a special characteristic, called the orthogonality property. For an illustration of this property, let $\mathbf{q}_i$

and $\mathbf{q}_j$ be any two eigenvectors and λ_i and λ_j the corresponding eigenvalues, then

$$\mathbf{Kq}_i = \lambda_i \mathbf{Mq}_i \tag{10.8a}$$

$$\mathbf{Kq}_j = \lambda_j \mathbf{Mq}_j \tag{10.8b}$$

Premultiplication of both sides of Equation 10.8a by $\mathbf{q}_j^T$ yields

$$\mathbf{q}_j^T \mathbf{Kq}_i = \lambda_i \mathbf{q}_j^T \mathbf{Mq}_i \tag{10.9}$$

Transposing the two sides of Equation 10.9 and using the symmetry property of $\mathbf{K}$ and $\mathbf{M}$ matrices by which $\mathbf{K}^T = \mathbf{K}$ and $\mathbf{M}^T = \mathbf{M}$, we get

$$\mathbf{q}_i^T \mathbf{Kq}_j = \lambda_i \mathbf{q}_i^T \mathbf{Mq}_j \tag{10.10}$$

Premultiplication of both sides of Equation 10.8b by $\mathbf{q}_i^T$ gives

$$\mathbf{q}_i^T \mathbf{Kq}_j = \lambda_j \mathbf{q}_i^T \mathbf{Mq}_j \tag{10.11}$$

Finally, on subtracting Equation 10.11 from Equation 10.10, we get

$$(\lambda_i - \lambda_j)\mathbf{q}_i^T \mathbf{Mq}_j = 0 \tag{10.12}$$

or

$$\mathbf{q}_i^T \mathbf{Mq}_j = 0 \quad \lambda_i \neq \lambda_j \tag{10.13}$$

Equation 10.13 states that eigenvectors corresponding to distinct eigenvalues are orthogonal with respect to the mass matrix. This is not necessarily so when an eigenvalue is repeated, that is, when $\lambda_i = \lambda_j$. In that case, the eigenvectors in a group having a common eigenvalue are orthogonal only to the eigenvectors outside the group. However, with an appropriate adjustment, even the eigenvectors within the group can be made orthogonal to each other provided that the matrices involved in the eigenproblem are symmetric, which is true of both the stiffness and the mass matrix. In general, therefore, the orthogonality relationship can be written as

$$\mathbf{q}_i^T \mathbf{Mq}_j = 0 \quad i \neq j \tag{10.14}$$

Substitution of Equation 10.14 in Equation 10.10 gives

$$\mathbf{q}_i^T \mathbf{Kq}_j = 0 \quad i \neq j \tag{10.15}$$

Further, on premultiplying both sides of Equation 10.8a by $\mathbf{q}_i^T$, we obtain

$$\lambda_i = \frac{\mathbf{q}_i^T \mathbf{K} \mathbf{q}_i}{\mathbf{q}_i^T \mathbf{M} \mathbf{q}_i} \tag{10.16}$$

As pointed out earlier, the elements of an eigenvector are known only in a relative sense, their absolute magnitudes being indeterminate. In view of this, it is useful to define some standard procedure for scaling the eigenvectors. The process of scaling is called *normalization.* One possible way of normalizing is to make the absolute value of the largest element in the vector equal to 1. Another method of scaling, often used for the eigenvectors of Equation 10.7, is what is commonly referred to as *mass orthonormalization,* in which the eigenvectors are scaled so that they satisfy the following relationship

$$\boldsymbol{\phi}_i^T \mathbf{M} \boldsymbol{\phi}_i = 1 \tag{10.17}$$

in which $\boldsymbol{\phi}_i$ denotes the ith normalized eigenvector. Henceforth, we will use the symbol $\boldsymbol{\phi}$ to denote a mass orthonormal eigenvector. The orthogonality property of eigenvectors $\boldsymbol{\phi}$ can now be stated as

$$\boldsymbol{\phi}_i^T \mathbf{M} \boldsymbol{\phi}_j = \delta_{ij} \tag{10.18}$$

$$\boldsymbol{\phi}_i^T \mathbf{K} \boldsymbol{\phi}_j = \lambda_i \delta_{ij} \tag{10.19}$$

where δ_{ij} is the Kronecker delta, having the property $\delta_{ij} = 0$, for $i \neq j$, and $\delta_{ij} = 1$ for $i = j$.

The eigenvectors of a linearized eigenproblem are also referred to as *mode shapes,* while the square roots of the eigenvalues are referred to as the frequencies. It is customary to arrange the frequencies in the ascending order of magnitude. The first frequency in the sequence is then referred to as the lowest frequency or the *fundamental frequency* and the corresponding mode shape as the *fundamental mode shape.* As we shall see later, the mode shapes and frequencies play an important role in the free and forced vibration analyses of a multi-degree-of-freedom system.

The $N \times N$ matrix, formed by arranging side by side the N mode shapes of a system, is referred to as the *modal matrix* and has several useful applications. When mass orthonormal mode shapes are used in forming the modal matrix, the latter is denoted by $\boldsymbol{\Phi}$ and is given by

$$\boldsymbol{\Phi} = [\boldsymbol{\phi}_1 \quad \boldsymbol{\phi}_2 \quad \cdots \quad \boldsymbol{\phi}_N] \tag{10.20}$$

By using the orthogonality property represented by Equations 10.18 and 10.19, it can easily be shown that

$$\boldsymbol{\Phi}^T \mathbf{M} \boldsymbol{\Phi} = \mathbf{I} \tag{10.21}$$

Table 10.1. Properties of a linearized eigenvalue problem.

1. The eigenvalues of a linearized eigenvalue problem in which $\mathbf{K}$ and $\mathbf{M}$ are both symmetric are all real.
2. When matrices $\mathbf{K}$ and $\mathbf{M}$ are both positive definite, the eigenvalues are all positive.
3. When $\mathbf{K}$ is singular, at least one of the eigenvalues must be zero. When $\mathbf{M}$ is singular, at least one of the eigenvalues must be infinite.
4. The eigenvectors are orthogonal with respect to both $\mathbf{K}$ and $\mathbf{M}$. For symmetric $\mathbf{K}$ and $\mathbf{M}$ matrices

$$\mathbf{q}_i^T \mathbf{M} \mathbf{q}_j = 0 \quad i \neq j$$
$$\mathbf{q}_i^T \mathbf{K} \mathbf{q}_j = 0 \quad i \neq j$$

5. The eigenvectors of a linearized eigenvalue problem in which the matrices involved are symmetric, including those corresponding to repeated eigenvalues, are all independent.
6. Any arbitrary vector of order N can be expressed as a superposition of the eigenvectors of an $N \times N$ symmetrical eigenvalue problem. Thus

$$\mathbf{x} = \boldsymbol{\Phi} \mathbf{c}$$

in which $\mathbf{c} = \boldsymbol{\Phi}^T \mathbf{M} \mathbf{x}$ and $\boldsymbol{\Phi}$ is the matrix of mass orthonormal eigenvectors.

and

$$\boldsymbol{\Phi}^T \mathbf{K} \boldsymbol{\Phi} = \boldsymbol{\Lambda} \tag{10.22}$$

where $\boldsymbol{\Lambda}$, called the *spectral matrix*, is the diagonal matrix of eigenvalues λ. The properties of the eigenvalues and eigenvectors obtained from a linearized eigenproblem are summarized in Table 10.1.

Example 10.1
A two-story shear building frame is shown in Figure E10.1a. The mass of the frame is assumed to be lumped at the floor levels and the floors are considered rigid. The floor masses and the story stiffnesses are indicated in the figure. Obtain the mode shapes and frequencies of the frame for vibrations in the plane of the paper.

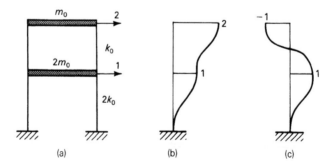

Figure E10.1. Mode shapes of a two-story shear frame: (a) frame; (b) first mode shape; (c) second mode shape.

Solution
Corresponding to the two degrees of freedom shown in the figure, the stiffness matrix is given by

$$\mathbf{K} = k_0 \begin{bmatrix} 3 & -1 \\ -1 & 1 \end{bmatrix} \tag{a}$$

The mass matrix is

$$\mathbf{M} = m_0 \begin{bmatrix} 2 & 0 \\ 0 & 1 \end{bmatrix} \tag{b}$$

The linearized eigenproblem is given by

$$[\mathbf{K} - \lambda \mathbf{M}]\mathbf{q} = 0 \tag{c}$$

or

$$\begin{bmatrix} 3k_0 - 2m_0\lambda & -k_0 \\ -k_0 & k_0 - m_0\lambda \end{bmatrix} \mathbf{q} = 0 \tag{d}$$

A nontrivial solution for $\mathbf{q}$ will be obtained from Equation d provided that the determinant of the coefficient matrix is zero. This leads to the characteristic equation

$$\begin{vmatrix} 3k_0 - 2m_0\lambda & -k_0 \\ -k_0 & k_0 - m_0\lambda \end{vmatrix} = 0 \tag{e}$$

or

$$\lambda^2 - \frac{5}{2}\frac{k_0}{m_0}\lambda + \frac{k_0^2}{m_0^2} = 0 \tag{f}$$

On solving Equation f, we get

$$\lambda_1 = \frac{k_0}{2m_0}$$

$$\lambda_2 = \frac{2k_0}{m} \tag{g}$$

The mode shapes are obtained by substituting for λ in the equation $[\mathbf{K} - \lambda \mathbf{M}]\mathbf{q} = 0$. For $\lambda = k_0/2m_0$, Equation d reduces to

$$\begin{bmatrix} 2k_0 & -k_0 \\ -k_0 & \frac{k_0}{2} \end{bmatrix} \mathbf{q}_1 = 0 \tag{h}$$

Since the coefficient matrix in Equation h is singular, there are infinitely many solutions for $\mathbf{q}$. However, if we arbitrarily select $q_{11} = 1$, both of the equations in Equation h give $q_{21} = 2$. The first mode shape is thus given by

$$\mathbf{q}_1 = \begin{bmatrix} 1 \\ 2 \end{bmatrix} \tag{i}$$

The second mode is obtained in a similar manner by substituting $\lambda = 2k_0/m_0$ in Equation d and solving for $\mathbf{q}$, which gives

$$\mathbf{q}_2 = \begin{bmatrix} 1 \\ -1 \end{bmatrix} \tag{j}$$

We may wish to normalize the mode shapes so that they are orthonormal with respect to the mass matrix. The normalized mode shapes are given by

$$\boldsymbol{\phi}_1 = \frac{1}{\sqrt{6m_0}} \begin{bmatrix} 1 \\ 2 \end{bmatrix}$$

$$\boldsymbol{\phi}_2 = \frac{1}{\sqrt{3m_0}} \begin{bmatrix} 1 \\ -1 \end{bmatrix} \tag{km}$$

These mode shapes have been plotted in Figure E10.1b and c, respectively.

10.4   EXPANSION THEOREM

It is often useful to recognize that a vector $\mathbf{x}$ of order N can be represented by the superposition of N independent vectors also of order N. Such a set of vectors is said to span an N-dimensional space and form a complete basis for representing any vector in that space. Since the eigenvectors of a symmetric matrix of order N are independent, they constitute a valid set of basis vectors in an N-dimensional space and can be used to represent any Nth order vector. The superposition can be expressed as

$$\mathbf{x} = \sum_{i=1}^{N} \mathbf{q}_i c_i \tag{10.23}$$

or as

$$\mathbf{x} = \mathbf{Qc} \tag{10.24}$$

in which $\mathbf{Q}$ is the modal matrix of eigenvectors $\mathbf{q}_i$ and $\mathbf{c}$ is a vector of weighting factors given by

$$\mathbf{c} = \mathbf{Q}^{-1}\mathbf{x} \tag{10.25}$$

It is important to note that $\mathbf{Q}^{-1}$ always exists because vectors $\mathbf{q}_i$ are independent. In fact, for symmetric $\mathbf{A}$, $\mathbf{Q}^{-1} = \mathbf{Q}^{T}$.

Since the mode shapes obtained from a linearized eigenproblem are also independent, they can be equally effectively used in the representation of vector $\mathbf{x}$. Denoting the mode shapes by $\mathbf{q}_i$, we have from Equation 10.23

$$\mathbf{x} = c_1\mathbf{q}_1 + c_2\mathbf{q}_2 + \cdots + c_N\mathbf{q}_N \tag{10.26}$$

To obtain c_i, we premultiply both sides of Equation 10.26 by $\mathbf{q}_i^T \mathbf{M}$. Then, since all terms on the right-hand side except $\mathbf{q}_i^T \mathbf{M} \mathbf{q}_i$ vanish because of the mass orthogonality of the eigenvectors, we have

$$c_i = \frac{\mathbf{q}_i^T \mathbf{M} \mathbf{x}}{\mathbf{q}_i^T \mathbf{M} \mathbf{q}_i} \tag{10.27}$$

If mass orthonormal mode shapes are used in the expansion of $\mathbf{x}$, so that

$$\mathbf{x} = c_1 \boldsymbol{\phi}_1 + c_2 \boldsymbol{\phi}_2 + \cdots + c_N \boldsymbol{\phi}_N \tag{10.28}$$

then because $\boldsymbol{\phi}_i^T \mathbf{M} \boldsymbol{\phi}_i = 1$, coefficient c_i is given by

$$c_i = \boldsymbol{\phi}_i^T \mathbf{M} \mathbf{x} \quad \text{for } i = 1, 2, \ldots, N \tag{10.29a}$$

or

$$\mathbf{c} = \boldsymbol{\Phi}^T \mathbf{M} \mathbf{x} \tag{10.29b}$$

The representation of a vector as a superposition of the eigenvectors of a matrix as in Equations 10.24 or 10.26 is, at times, referred to as *expansion theorem*. In a more general sense, it is a Ritz vector representation in which the eigenvectors have been used as the Ritz vectors.

10.5 RAYLEIGH QUOTIENT

Let $\mathbf{K}$ and $\mathbf{M}$ represent, respectively, the stiffness and mass matrices of order N, and let $\mathbf{u}$ be any arbitrary vector also of order N. Now define a scalar ρ given by

$$\rho = \frac{\mathbf{u}^T \mathbf{K} \mathbf{u}}{\mathbf{u}^T \mathbf{M} \mathbf{u}} \tag{10.30}$$

The scalar ρ is known as the Rayleigh quotient and has several important properties. To derive these properties, we express the arbitrary vector $\mathbf{u}$ as a superposition of the eigenvectors of Equation 10.7 using Equation 10.28. Thus

$$\mathbf{u} = \boldsymbol{\Phi} \mathbf{c} \tag{10.31}$$

The denominator in Equation 10.30 can now be expressed as

$$\mathbf{u}^T \mathbf{M} \mathbf{u} = \mathbf{c}^T \boldsymbol{\Phi}^T \mathbf{M} \boldsymbol{\Phi} \mathbf{c}$$

$$= \sum_{i=1}^{N} c_i^2 \tag{10.32}$$

while the numerator is obtained from

$$\mathbf{u}^T\mathbf{K}\mathbf{u} = \mathbf{c}^T\mathbf{\Phi}^T\mathbf{K}\mathbf{\Phi}\mathbf{c}$$

$$= \mathbf{c}^T\mathbf{\Lambda}\mathbf{c}$$

$$= \sum_{i=1}^{N} \lambda_i c_i^2 \tag{10.33}$$

The Rayleigh quotient is given by

$$\rho = \frac{\lambda_1 c_1^2 + \lambda_2 c_2^2 + \cdots + \lambda_N c_N^2}{c_1^2 + c_2^2 + \cdots + c_N^2} \tag{10.34a}$$

$$= \lambda_1 \frac{c_1^2 + (\lambda_2/\lambda_1)c_2^2 + \cdots + (\lambda_N/\lambda_1)c_N^2}{c_1^2 + c_2^2 + \cdots + c_N^2} \tag{10.34b}$$

Since $\lambda_1 < \lambda_2 < \lambda_3 < \cdots < \lambda_N$, the numerator in Equation 10.34b is always greater than the denominator, so that

$$\rho \geq \lambda_1 \tag{10.35}$$

The equality is obtained when $c_2, c_3, \ldots, c_N$ are zero and the arbitrary vector $\mathbf{u}$ coincides with the lowest eigenvector. The minimum value of Rayleigh quotient obtained by varying $\mathbf{u}$ is thus equal to the lowest eigenvalue. Alternatively, Equation 10.34a can be expressed as

$$\rho = \lambda_N \frac{(\lambda_1/\lambda_N)c_1^2 + (\lambda_2/\lambda_N)c_2^2 + \cdots + c_N^2}{c_1^2 + c_2^2 + \cdots + c_N^2} \tag{10.36}$$

The numerator in Equation 10.36 is always smaller than the denominator and hence

$$\rho \leq \lambda_N \tag{10.37}$$

Combining Equations 10.35 and 10.37, we have

$$\lambda_1 \leq \rho \leq \lambda_N \tag{10.38}$$

If the arbitrary vector $\mathbf{u}$ is selected from a subset of vectors that are orthogonal to the first $s - 1$ eigenvectors, we have

$$\mathbf{u}^T\mathbf{M}\phi_j = 0 \quad j = 1, 2, \ldots, s - 1 \tag{10.39}$$

and in the expansion of $\mathbf{u}$

$$c_j = 0 \quad j = 1, 2, \ldots, s - 1 \tag{10.40}$$

The Rayleigh quotient is now given by

$$\rho = \lambda_s \frac{c_s^2 + (\lambda_{s+1}/\lambda_s)c_{s+1}^2 + \cdots + (\lambda_N/\lambda_s)c_N^2}{c_s^2 + c_{s+1}^2 + \cdots + c_N^2} \tag{10.41}$$

Thus $\rho \geq \lambda_s$ and the minimum value of ρ obtained by varying $\mathbf{u}$ is λ_s. If the arbitrary vector $\mathbf{u}$ used in forming the Rayleigh quotient approximates the ith eigenvector, the deviation from the true eigenvector being represented by the error $\varepsilon\mathbf{v}$, where ε is a small quantity, then the Rayleigh quotient approximates λ_i with an error of order ε^2. To prove this, let

$$\mathbf{u} = \boldsymbol{\phi}_i + \varepsilon\mathbf{v} \tag{10.42}$$

in which $\boldsymbol{\phi}_i$ is the ith eigenvector and $\varepsilon\mathbf{v}$ represents the contribution of all other eigenvectors to $\mathbf{u}$. The numerator of the Rayleigh quotient is then given by

$$\mathbf{u}^T\mathbf{K}\mathbf{u} = (\boldsymbol{\phi}_i^T + \varepsilon\mathbf{v}^T)\mathbf{K}(\boldsymbol{\phi}_i + \varepsilon\mathbf{v})$$

$$= \boldsymbol{\phi}_i^T\mathbf{K}\boldsymbol{\phi}_i + \varepsilon\boldsymbol{\phi}_i^T\mathbf{K}\mathbf{v} + \varepsilon\mathbf{v}^T\mathbf{K}\boldsymbol{\phi}_i + \varepsilon^2\mathbf{v}^T\mathbf{K}\mathbf{v} \tag{10.43}$$

Noting that $\mathbf{v}$ can be expressed as

$$\mathbf{v} = \sum_{\substack{j=1 \\ j\neq i}}^{N} \alpha_j\boldsymbol{\phi}_j \tag{10.44}$$

and using the orthonormality property of the eigenvectors, we have

$$\boldsymbol{\phi}_i^T\mathbf{M}\mathbf{v} = \mathbf{v}^T\mathbf{M}\boldsymbol{\phi}_i = 0$$

$$\boldsymbol{\phi}_i^T\mathbf{K}\mathbf{v} = \mathbf{v}^T\mathbf{K}\boldsymbol{\phi}_i = 0$$

$$\boldsymbol{\phi}_i^T\mathbf{K}\boldsymbol{\phi}_i = \lambda_i\boldsymbol{\phi}_i^T\mathbf{M}\boldsymbol{\phi}_i = \lambda_i \tag{10.45}$$

$$\mathbf{v}^T\mathbf{M}\mathbf{v} = \sum_{\substack{j=1 \\ j\neq i}}^{N} \alpha_j^2$$

$$\mathbf{v}^T\mathbf{K}\mathbf{v} = \sum_{\substack{j=1 \\ j\neq i}}^{N} \alpha_j^2\lambda_j \tag{10.46}$$

Equation 10.43 thus reduces to

$$\mathbf{u}^T\mathbf{K}\mathbf{u} = \lambda_i + \varepsilon^2(\alpha_1^2\lambda_1 + \alpha_2^2\lambda_2 + \cdots + \alpha_{i-1}^2\lambda_{i-1} + \alpha_{i+1}^2\lambda_{i+1}$$

$$+ \cdots + \alpha_N^2\lambda_N) \tag{10.47}$$

In a similar manner, the denominator of the Rayleigh quotient can be shown to be given by

$$\mathbf{u}^T\mathbf{M}\mathbf{u} = \boldsymbol{\phi}_i^T\mathbf{M}\boldsymbol{\phi}_i + \varepsilon\boldsymbol{\phi}_i^T\mathbf{M}\mathbf{v} + \varepsilon\mathbf{v}^T\mathbf{M}\boldsymbol{\phi}_i + \varepsilon^2\mathbf{v}^T\mathbf{M}\mathbf{v}$$

$$= 1 + \varepsilon^2(\alpha_1^2 + \alpha_2^2 + \alpha_3^2 + \cdots + \alpha_{i-1}^2 + \alpha_{i+1}^2 + \cdots + \alpha_N^2)$$

$$(10.48)$$

Finally, the Rayleigh quotient becomes

$$\rho = \frac{\lambda_i + \varepsilon^2(\alpha_1^2\lambda_1 + \alpha_2^2\lambda_2 + \cdots + \alpha_{i-1}^2\lambda_{i-1} + \alpha_{i+1}^2\lambda_{i+1} + \cdots + \alpha_N^2\lambda_N)}{1 + \varepsilon^2(\alpha_1^2 + \alpha_2^2 + \cdots + \alpha_{i-1}^2 + \alpha_{i+1}^2 + \cdots + \alpha_N^2)}$$

$$(10.49)$$

Considering that ε is small, and neglecting terms of order higher than ε^2, we can reduce Equation 10.49 to

$$\rho = \lambda_i + \varepsilon^2\{\alpha_1^2(\lambda_1 - \lambda_i) + \alpha_2^2(\lambda_2 - \lambda_i) + \cdots + \alpha_{i-1}^2(\lambda_{i-1} - \lambda_i)$$

$$+ \alpha_{i+1}^2(\lambda_{i+1} - \lambda_i) + \cdots + \alpha_N^2(\lambda_N - \lambda_i)\}$$

$$(10.50)$$

Thus, the Rayleigh quotient approximates λ_i with an error of order ε^2.

Vector $\mathbf{u}$ in Equation 10.30 can be interpreted as a vibration shape of the system. The numerator in Equation 10.30 will then be recognized as the generalized stiffness defined in Section 2.8, while the denominator represents the generalized mass. With appropriate choice of $\mathbf{u}$, the Rayleigh quotient can be used to provide an estimate of the lowest or the fundamental frequency. As proved here, this estimate always represents an upper bound to the fundamental frequency. When $\mathbf{u}$ coincides with the first vibration mode shape of the system, the Rayleigh quotient becomes exactly equal to the fundamental frequency.

Example 10.2
Obtain the fundamental frequency of the system shown in Figure E10.2 by minimizing the Rayleigh quotient.

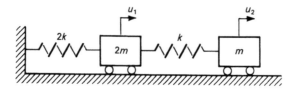

Figure E10.2. Two-degree-of-freedom system.

Solution
The stiffness matrix **K** and the mass matrix **M** of the system are given by

$$\mathbf{K} = \begin{bmatrix} 3k & -k \\ -k & k \end{bmatrix} \tag{a}$$

$$\mathbf{M} = \begin{bmatrix} 2m & 0 \\ 0 & m \end{bmatrix} \tag{b}$$

For the displacement shape of the system, assume a vector

$$\begin{bmatrix} u_1 \\ u_2 \end{bmatrix} = \begin{bmatrix} d \\ \alpha d \end{bmatrix} \tag{c}$$

where d is any arbitrary scalar and α is a parameter that is yet to be determined. We now form the Rayleigh quotient

$$\rho = \frac{\mathbf{u}^T \mathbf{K} \mathbf{u}}{\mathbf{u}^T \mathbf{M} \mathbf{u}}$$

$$= \frac{d^2 k (3 - 2\alpha + \alpha^2)}{d^2 m (2 + \alpha^2)} \tag{d}$$

To obtain the value of α for which ρ will be a minimum, we set $d\rho/d\alpha = 0$. This gives

$$\frac{d\rho}{d\alpha} = \frac{-4 + 2\alpha^2 - 2\alpha}{(2 + \alpha^2)^2}$$

$$= 0 \tag{e}$$

Equation e gives $\alpha = 2$ or $\alpha = -1$. One of these two values will give a minimum for ρ, while the other will give a maximum. For $\alpha = 2$, we get $\rho = k/2m$, while for $\alpha = -1$, $\rho = 2k/m$. Thus $\alpha = 2$ minimizes ρ and the fundamental frequency ω_1 is given by

$$\omega_1 = \sqrt{\frac{k}{2m}} \tag{f}$$

The fundamental mode shape is $\mathbf{q}_1^T = [1 \ \ 2]$. The maximum value of ρ leads to the second frequency

$$\omega_2 = \sqrt{\frac{2k}{m}} \tag{g}$$

The corresponding mode shape is $\mathbf{q}_2^T = [1 \ \ -1]$. Note that the mode shapes have not been properly normalized.

10.6   SOLUTION OF THE UNDAMPED FREE-VIBRATION PROBLEM

As stated earlier in this chapter, the free-vibration equation (Equation 10.2) has a solution $\mathbf{u} = \mathbf{q} \sin(\omega t + \theta)$ in which $\mathbf{q}$ and ω satisfy the linearized eigenvalue problem of Equation 10.7:

$$\mathbf{K} \mathbf{q} = \lambda \mathbf{M} \mathbf{q} \tag{10.7}$$

where $\lambda = \omega^2$.

For an N-degree-of-freedom system, Equation 10.7 gives N values of λ, all of which are real and positive but may not be distinct. Corresponding to each eigenvalue there is an associated eigenvector. These vectors are determined within a scalar multiple and form an orthogonal set. In vibration analysis, the eigenvalues, or rather their square roots, are referred to as the frequencies and the eigenvectors as the mode shapes. Procedures for determining the mode shapes and frequencies of a multi-degree-of-freedom system are presented in Chapter 11.

The most general solution of Equation 10.2 is obtained as a superposition of the N mode shapes and can be written as

$$\mathbf{u} = \sum_{n=1}^{N} a_n \boldsymbol{\phi}_n \sin(\omega_n t + \theta_n) \tag{10.51a}$$

or

$$\mathbf{u} = \sum_{n=1}^{N} (c_n \boldsymbol{\phi}_n \sin \omega_n t + d_n \boldsymbol{\phi}_n \cos \omega_n t) \tag{10.51b}$$

where c_n and d_n are arbitrary constants and $\boldsymbol{\phi}_n$ are the normalized mode shapes. The unknown arbitrary constants can be determined by using the $2N$ initial conditions which must be specified. These initial conditions consist of N initial displacements and N initial velocities, one corresponding to each of the N degrees of freedom.

To determine constants d, we note that at $t = 0$, $\mathbf{u} = \mathbf{u}_0$, the vector of initial displacements. Thus setting $t = 0$ in Equation 10.51b

$$\mathbf{u}_0 = \sum_{n=1}^{N} d_n \boldsymbol{\phi}_n \tag{10.52a}$$

or

$$\mathbf{u}_0 = \boldsymbol{\Phi} \mathbf{d} \tag{10.52b}$$

where $\mathbf{d}$ is the vector of constants d_n. Premultiplying both sides of Equation 10.52b by $\boldsymbol{\Phi}^T \mathbf{M}$ and assuming that the mode shapes have been properly normalized so that $\boldsymbol{\Phi}^T \mathbf{M} \boldsymbol{\Phi} = \mathbf{I}$, we obtain

$$\mathbf{d} = \boldsymbol{\Phi}^T \mathbf{M} \mathbf{u}_0 \tag{10.53a}$$

or

$$d_n = \boldsymbol{\phi}_n^T \mathbf{M} \mathbf{u}_0 \quad n = 1, 2, \ldots, N \tag{10.53b}$$

In a similar manner, if the vector of initial velocities is $\mathbf{v}_0$,

$$\mathbf{v}_0 = \sum_{n=1}^{N} c_n \omega_n \boldsymbol{\phi}_n \tag{10.54}$$

and

$$c_n = \frac{1}{\omega_n} \boldsymbol{\phi}_n^T \mathbf{M} \mathbf{v}_0 \tag{10.55}$$

10.7 MODE SUPERPOSITION ANALYSIS OF FREE-VIBRATION RESPONSE

The solution procedure outlined in the preceding section can also be expressed in terms of a transformation of coordinates. This transformation is represented as

$$\mathbf{u} = \boldsymbol{\Phi} \mathbf{y}$$

$$= \sum_{n=1}^{N} \boldsymbol{\phi}_n y_n \tag{10.56}$$

where $\boldsymbol{\Phi}$ is the modal matrix and $\mathbf{y}$ is a vector of transformed coordinates called *modal coordinates* or *normal coordinates*. The transformed equations of undamped free vibration become

$$\boldsymbol{\Phi}^T \mathbf{M} \boldsymbol{\Phi} \ddot{\mathbf{y}} + \boldsymbol{\Phi}^T \mathbf{K} \boldsymbol{\Phi} \mathbf{y} = 0 \tag{10.57}$$

Because of the orthogonality relationships given by Equations 10.18 and 10.19, Equation 10.57 reduces to

$$\ddot{\mathbf{y}} + \boldsymbol{\Lambda} \mathbf{y} = 0 \tag{10.58}$$

or

$$\ddot{y}_n + \omega_n^2 y_n = 0 \quad n = 1, 2, \dots, N \tag{10.59}$$

The transformation to modal coordinates thus reduces the N simultaneous equations of motion to N uncoupled equations. The solution to the nth equation is given by

$$y_n = y_{0n} \cos \omega_n t + \frac{\dot{y}_{0n}}{\omega_n} \sin \omega_n t \tag{10.60}$$

in which y_{0n} is the value of y_n at $t = 0$ and $\dot{y}_{0n}$ is the value of $\dot{y}_n$ at $t = 0$. Vectors $\mathbf{y}_0$ and $\dot{\mathbf{y}}_0$ can be obtained from the given initial conditions and the orthogonality condition (Equation 10.18) as follows

$$\mathbf{u}_0 = \boldsymbol{\Phi} \mathbf{y}_0$$
$$\mathbf{y}_0 = \boldsymbol{\Phi}^T \mathbf{M} \mathbf{u}_0 \tag{10.61}$$

and

$$\mathbf{v}_0 = \boldsymbol{\Phi} \dot{\mathbf{y}}_0$$
$$\dot{\mathbf{y}}_0 = \boldsymbol{\Phi}^T \mathbf{M} \mathbf{v}_0 \tag{10.62}$$

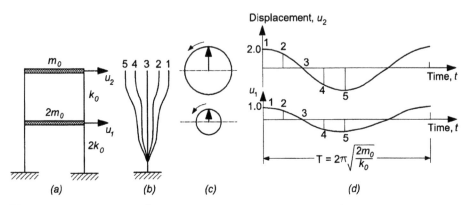

Figure 10.1. Response of a two-story shear frame in the first mode: (a) frame; (b) displaced shape of a column; (c) rotating vector representation; (d) displacement-time history.

Once the response along the modal coordinates has been determined by substitution of y_{0n}, $\dot{y}_{0n}$, and ω_n in Equation 10.60, response in the physical coordinates, u, can be obtained from the transformation equation (Eq. 10.56).

The foregoing procedure is referred to as the *mode superposition method*. It can be used to solve not only the undamped free-vibration problem but also the damped free-vibration problem as well as the more general forced vibration problem. The mode superposition method thus plays a central role in the vibration analysis of multi-degree-of-freedom systems and will be discussed further in subsequent chapters.

To gain some insight into the physical meaning of the vibration frequencies and mode shapes, consider the simple two-story shear frame shown in Figure 10.1a. The mass is assumed to be concentrated at the floors, which are also considered rigid. The value of the floor masses and story stiffnesses have been indicated. The stiffness and mass matrices as well as the frequencies and mode shapes for this simple structure were obtained in Example 10.1. Let the building frame be given certain initial disturbances and then allowed to vibrate. Its displaced shape at any instance of time will be given by

$$\mathbf{u} = a_1 \boldsymbol{\phi}_1 \sin(\omega_1 t + \theta_1) + a_2 \boldsymbol{\phi}_2 \sin(\omega_2 t + \theta_2) \tag{10.63a}$$

or

$$\mathbf{u} = c_1 \boldsymbol{\phi}_1 \sin \omega_1 t + d_1 \boldsymbol{\phi}_1 \cos \omega_1 t$$
$$+ c_2 \boldsymbol{\phi}_2 \sin \omega_2 t + d_2 \boldsymbol{\phi}_2 \cos \omega_2 t \tag{10.63b}$$

where c_1, c_2, d_1 and d_2 are arbitrary constants.

To determine these arbitrary constants, four initial conditions – two displacements and two velocities – must be specified. Let the specified conditions be

$$u_{01} = 1$$
$$u_{02} = 2$$
$$v_{01} = 0$$
$$v_{02} = 0$$

(10.64)

Substitution of the displacement conditions in Equation 10.63 and the velocity conditions in the derivative of that equation will give

$$d_1 = \sqrt{6m_0} \qquad d_2 = 0 \qquad c_1 = 0 \qquad c_2 = 0 \qquad (10.65)$$

Equation 10.63b now reduces to

$$\mathbf{u} = \begin{bmatrix} 1 \\ 2 \end{bmatrix} \cos \sqrt{\frac{k_0}{2m_0}} t \qquad (10.66)$$

The displacement response given by Equation 10.66 is shown in Figure 10.1b, c, and d. Figure 10.1b shows the displaced position of one of the columns at several different instants of time. The numbers 1 through 5 refer to the instants of time identified on Figure 10.1d, which shows the displacement–time history at each floor level. Figure 10.1c shows a rotating vector representation for the displacement at each story level. The rotating vector for story level 2 has a length of 2 units, starts off from a vertical position, and rotates counterclockwise at $\omega_1 = \sqrt{k_0/2m_0}$ rad/s. The vector for level 1 also starts off from a vertical position and is rotating at the same angular frequency, that is, ω_1, but its length is equal to 1 unit. The projection of a rotating vector on the vertical axis gives the displacement response.

The displacement at the second-story level is always twice that at the first-story level. The maximum positive values of the displacements at these levels are 2 and 1, respectively, the maximum negative values are -2 and -1. Both levels attain their maximum and minimum displacements at the same instant of time. Both levels pass through zero displacement simultaneously. In fact, they are vibrating exactly in phase with each other with a period of $2\pi\sqrt{2m_0/k_0}$ seconds. The frame is said to be vibrating in its first mode.

Now, let the specified initial conditions be

$$u_{01} = -1$$
$$u_{02} = 1$$
$$v_{01} = 0$$
$$v_{02} = 0$$

(10.67)

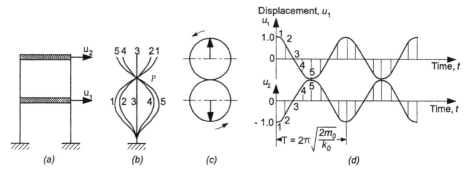

Figure 10.2. Response of a two-story frame in the second mode: (a) frame; (b) displaced shape of a column; (c) rotating vector representation; (d) displacement-time history.

instead of those given in Equation 10.64. It is easily shown that the vibration shape will, in this case, be given by

$$\mathbf{u} = \begin{bmatrix} -1 \\ 1 \end{bmatrix} \cos \sqrt{\frac{2k_0}{m_0}} t \tag{10.68}$$

The displacement response (Equation 10.68) is shown in Figure 10.2, which also indicates the displaced shapes, the displacement–time history for both levels, as well as the two rotating vector representations. The displacement at the second-story level is, in this case, always equal but opposite to that at the first-story level. The two levels are vibrating exactly 180° out of phase. As in the first mode, all points along the height of the frame attain their maximum displacements at the same instant of time, although in this case, these displacements may be in the opposite directions. Also, all the points pass through zero displacements at the same time. The frame is, in fact, vibrating in the second mode with a vibration period of $2\pi\sqrt{m_0/2k_0}$ seconds. It is of interest to note that point P remains stationary. A stationary point such as P is referred to as a *node* and there is one node in this case.

In the preceding example, the building frame was vibrating either in its first mode or in its second mode. This was so because we adjusted the initial conditions to excite only one of the two mode shapes. In a general case, the vibration shape will be a linear combination of the two mode shapes and it will be difficult to visualize a pattern in the vibration. As an example, suppose that the initial conditions are

$$u_{01} = \frac{3}{2}$$

$$u_{02} = \frac{3}{2}$$

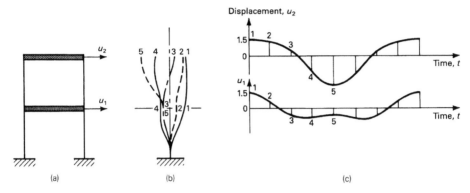

Figure 10.3. Response of a two-story frame subject to arbitrary initial conditions: (a) frame; (b) displaced shape of a column; (c) displacement-time history.

$$v_{01} = 0$$

$$v_{02} = 0 \tag{10.69}$$

Substitution into Equation 10.63 and its derivative gives

$$d_1 = \sqrt{6m_0}$$

$$d_2 = \frac{1}{2}\sqrt{3m_0} \tag{10.70}$$

$$c_1 = 0$$

$$c_2 = 0$$

The vibration shape is now obtained from

$$\mathbf{u} = \begin{bmatrix} 1 \\ 2 \end{bmatrix} \cos\sqrt{\frac{k_0}{2m_0}}t + \begin{bmatrix} \frac{1}{2} \\ -\frac{1}{2} \end{bmatrix} \cos\sqrt{\frac{2k_0}{m_0}}t \tag{10.71}$$

The displacement response given by Equation 10.71 is shown in Figure 10.3, which also shows the displacement–time histories for each level. It is not possible, in this case, to represent the response by a single rotating vector. However, the response can be viewed as a superposition of two vectors rotating at different angular speeds.

The same results will be obtained when the alternative mode superposition method is used in the analysis. The transformation to modal coordinates is expressed as

$$\mathbf{u} = \begin{bmatrix} \frac{1}{\sqrt{6m_0}} & \frac{1}{\sqrt{3m_0}} \\ \frac{2}{\sqrt{6m_0}} & \frac{-1}{\sqrt{3m_0}} \end{bmatrix} \begin{bmatrix} y_1 \\ y_2 \end{bmatrix} \tag{10.72}$$

The transformed equations are obtained from Equation 10.59

$$\ddot{y}_1 + \frac{k_0}{2m_0} y_1 = 0$$

$$\ddot{y}_2 + \frac{2k_0}{m_0} y_2 = 0$$

(10.73)

The solutions to Equations 10.73 are

$$y_1 = y_{01} \cos \sqrt{\frac{k_0}{2m_0}} t + \dot{y}_{01} \sqrt{\frac{2m_0}{k_0}} \sin \sqrt{\frac{k_0}{2m_0}} t$$

$$y_2 = y_{02} \cos \sqrt{\frac{2k_0}{m_0}} t + \dot{y}_{02} \sqrt{\frac{m_0}{2k_0}} \sin \sqrt{\frac{2k_0}{m_0}} t$$

(10.74)

The initial conditions in the modal coordinates are obtained from Equations 10.61 and 10.62. If the specified conditions in the physical coordinates are those in Equation 10.64, then

$$\mathbf{y}_0 = \begin{bmatrix} \frac{1}{\sqrt{6m_0}} & \frac{2}{\sqrt{6m_0}} \\ \frac{1}{\sqrt{3m_0}} & \frac{-1}{\sqrt{3m_0}} \end{bmatrix} \begin{bmatrix} 2m_0 & 0 \\ 0 & m_0 \end{bmatrix} \begin{bmatrix} 1 \\ 2 \end{bmatrix}$$

$$= \begin{bmatrix} \sqrt{6m_0} \\ 0 \end{bmatrix}$$

(10.75a)

$$\dot{\mathbf{y}}_0 = 0$$

(10.75b)

Substitution in Equations 10.74 gives

$$y_1 = \sqrt{6m_0} \cos \sqrt{\frac{k_0}{2m_0}} t$$

$$y_2 = 0$$

(10.76)

The displacements in the physical coordinates are obtained by using the transformation Equation 10.72.

$$\mathbf{u} = \begin{bmatrix} \frac{1}{\sqrt{6m_0}} & \frac{1}{\sqrt{3m_0}} \\ \frac{2}{\sqrt{6m_0}} & \frac{-1}{\sqrt{3m_0}} \end{bmatrix} \begin{bmatrix} \sqrt{6m_0} \cos \sqrt{\frac{k_0}{2m_0}} t \\ 0 \end{bmatrix}$$

$$= \begin{bmatrix} 1 \\ 2 \end{bmatrix} \cos \sqrt{\frac{k_0}{2m_0}} t$$

(10.77)

These displacements are identical to those given by Equation 10.66.

As another example of vibration mode shapes, consider the three-story building frame shown in Figure 10.4a. The building has three different mode shapes,

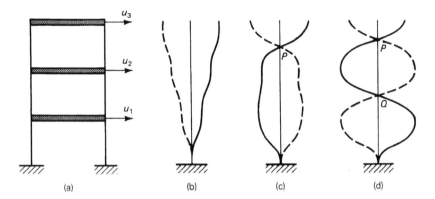

Figure 10.4. Mode shapes of a three-story shear frame: (a) frame; (b) first mode; (c) second mode; (d) third mode.

also shown in the figure. The second mode shape exhibits one node point that remains stationary. The third mode shape has two nodes. The vibration response of the building either due to initial excitation or due to applied time-varying forces or both can be represented as a weighted linear superposition of the three modes shown in Figure 10.4b, c, and d.

10.8 SOLUTION OF THE DAMPED FREE-VIBRATION PROBLEM

When the damping term in Equation 10.1 cannot be neglected, the solution procedure must be altered. For a general type of damping, the procedure is, in fact, fairly complex. However, as we shall observe later, when damping is of a special nature, considerable simplification is possible. The damped free-vibration equation (Eq. 10.1) has a solution of the form

$$\mathbf{u} = \mathbf{q}e^{st} \tag{10.78}$$

Substitution for $\mathbf{u}$ and its derivatives from Equation 10.78 into Equation 10.1 gives

$$(s^2\mathbf{M} + s\mathbf{C} + \mathbf{K})\mathbf{q} = 0 \tag{10.79}$$

Equation 10.79 will give a nontrivial solution for $\mathbf{q}$ only, provided that the matrix within the parentheses on the left-hand side is singular. This leads to the condition

$$\det(s^2\mathbf{M} + s\mathbf{C} + \mathbf{K}) = 0 \tag{10.80}$$

Equation 10.80 gives a characteristic equation which is a real polynomial of order $2N$ in s. It has $2N$ solutions which are either real or in complex conjugate pairs.

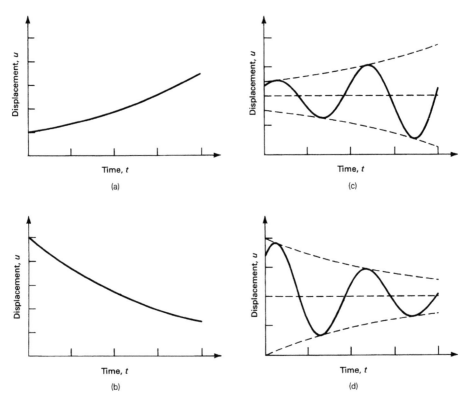

Figure 10.5. Damped free-vibration response: (a) steady exponential divergence; (b) steady exponential convergence; (c) oscillating divergence; (d) oscillating convergence.

Equation 10.79, in fact, represents what is commonly referred to as a *quadratic eigenvalue problem*. Its solution gives $2N$ eigenvalues, s, and corresponding eigenvectors, $\mathbf{q}$. The general solution to the free vibration problem is obtained by a superposition of the $2N$ solutions obtained by substituting for $\mathbf{q}$ and s in Equation 10.78

$$\mathbf{u} = \sum_{n=1}^{2N} \alpha_n \mathbf{q}_n e^{s_n t} \tag{10.81}$$

where α_n is an arbitrary constant. When a root s_n is real and positive, the corresponding solution shows steady exponential divergence (Fig. 10.5a); if it is real but negative, the solution exhibits steady exponential convergence, as shown in Figure 10.5b. For complex conjugate pairs of the form $s_m = \mu_m + i v_m$ and $s_m^* = \mu_m - i v_m$, the corresponding eigenvectors are also complex conjugates and can be represented as $\mathbf{q}_m = \mathbf{a}_m + i \mathbf{b}_m$ and $\mathbf{q}_m^* = \mathbf{a}_m - i \mathbf{b}_m$. The component of solution for $\mathbf{u}$ corresponding to the pair s_m, s_m^* is given by

$$\mathbf{u}_m = (c_m + id_m)e^{(\mu_m + iv_m)t}(\mathbf{a}_m + i\mathbf{b}_m)$$

$$+ (c_m - id_m)e^{(\mu_m - iv_m)t}(\mathbf{a}_m - i\mathbf{b}_m) \tag{10.82}$$

in which c_m and d_m are arbitrary constants. If μ_m is positive, Equation 10.82 represents a solution which shows oscillating divergence. On the other hand, if μ_m is negative, the solution exhibits oscillating convergence. These cases are shown in Figure 10.5c and d, respectively. The last case represents the most practical situation for mechanical or structural systems.

If all roots of the eigenproblem are complex conjugates, there will be N solutions of the form of Equation 10.82 and the total displacement response will be given by

$$\mathbf{u} = \sum_{m=1}^{N} \mathbf{u}_m \tag{10.83}$$

Equation 10.82 can be expressed in the alternative form

$$\mathbf{u}_m = e^{\mu_m t}(\mathbf{v}_m \sin \omega_m t + \mathbf{w}_m \cos \omega_m t) \tag{10.84}$$

where

$$\omega_m = v_m$$

$$\mathbf{v}_m = -2(d_m \mathbf{a}_m + c_m \mathbf{b}_m) \tag{10.85}$$

$$\mathbf{w}_m = 2(c_m \mathbf{a}_m - d_m \mathbf{b}_m)$$

and c_m and d_m are arbitrary constants to be determined from the $2N$ initial conditions that must be specified. As would be expected, the final solution is real.

The quadratic eigenproblem can also be transformed to either of a standard and a linearized form. The latter is of particular interest, since it allows for its solution the use of methods that apply to the eigenvalue solution of an undamped case. To obtain the linearized form, we first define a $2N \times 1$ vector of unknown velocities and displacements

$$\mathbf{x} = \begin{bmatrix} \dot{\mathbf{u}} \\ \mathbf{u} \end{bmatrix} \tag{10.86}$$

The equations of motion are now written in an augmented form as follows

$$\mathbf{M}\dot{\mathbf{u}} - \mathbf{M}\dot{\mathbf{u}} = 0 \tag{10.87a}$$

$$\mathbf{M}\ddot{\mathbf{u}} + \mathbf{C}\dot{\mathbf{u}} + \mathbf{K}\mathbf{u} = 0 \tag{10.87b}$$

Equations 10.87a and 10.87b can be combined into the following matrix equation

$$\begin{bmatrix} 0 & \mathbf{M} \\ \mathbf{M} & \mathbf{C} \end{bmatrix} \begin{bmatrix} \ddot{\mathbf{u}} \\ \dot{\mathbf{u}} \end{bmatrix} + \begin{bmatrix} -\mathbf{M} & 0 \\ 0 & \mathbf{K} \end{bmatrix} \begin{bmatrix} \dot{\mathbf{u}} \\ \mathbf{u} \end{bmatrix} = \begin{bmatrix} 0 \\ 0 \end{bmatrix}$$

(10.88)

or

$$\mathbf{A}\dot{\mathbf{x}} = \mathbf{B}\mathbf{x}$$

(10.89)

where

$$A = \begin{bmatrix} 0 & -\mathbf{M} \\ -\mathbf{M} & -\mathbf{C} \end{bmatrix}$$

(10.90a)

and

$$B = \begin{bmatrix} -\mathbf{M} & 0 \\ 0 & \mathbf{K} \end{bmatrix}$$

(10.90b)

Equation 10.89 represents a set of equations referred to as the reduced equations of motion. It is of interest to note that matrices $\mathbf{A}$ and $\mathbf{B}$ are both symmetrical and of order $2N$. Equation 10.89 has a solution of the form $\mathbf{x} = \mathbf{v}e^{st}$. Substitution of this solution gives

$$\mathbf{B}\mathbf{v} = s\mathbf{A}\mathbf{v}$$

(10.91)

Equation 10.91 will be recognized as a linearized eigenvalue problem. If desired, it can be converted into a standard form by premultiplying the two sides of Equation 10.91 by $\mathbf{A}^{-1}$. The inverse of $\mathbf{A}$ is given by

$$\mathbf{A}^{-1} = \begin{bmatrix} \mathbf{M}^{-1}\mathbf{C}\mathbf{M}^{-1} & -\mathbf{M}^{-1} \\ -\mathbf{M}^{-1} & 0 \end{bmatrix}$$

(10.92)

where it is assumed that $\mathbf{M}$ is nonsingular. The standard form of the eigenproblem thus becomes

$$\mathbf{A}^{-1}\mathbf{B}\mathbf{v} = s\mathbf{v}$$

(10.93)

or

$$\begin{bmatrix} -\mathbf{M}^{-1}\mathbf{C} & -\mathbf{M}^{-1}\mathbf{K} \\ \mathbf{I} & 0 \end{bmatrix} \mathbf{v} = s\mathbf{v}$$

(10.94)

Equation 10.94 represents a standard eigenproblem of order $2N$. Because the coefficient matrix is real, the eigenvalues must either be real or in complex conjugate pairs. The eigenvectors corresponding to the complex eigenvalues are also complex and occur in conjugate pairs. A majority of mechanical or structural systems are underdamped, in which case all of the eigenvalues are in complex conjugate pairs with a negative real part and the solution of the equations of motion represent oscillating exponential convergence.

As is the case of undamped free vibration, eigenvectors $\mathbf{v}_n$ are orthogonal with respect to both $\mathbf{A}$ and $\mathbf{B}$

$$\mathbf{v}_m^T \mathbf{A} \mathbf{v}_n = 0 \quad s_m \neq s_n \tag{10.95}$$

$$\mathbf{v}_m^T \mathbf{B} \mathbf{v}_n = 0 \quad s_m \neq s_n \tag{10.96}$$

The proof of the orthogonality conditions follows lines similar to those for the case of undamped vibrations and need not be repeated. The difference is that the eigenvectors are now in general complex and occur in conjugate pairs.

It is useful to normalize the eigenvectors so that they are orthonormal to matrix $\mathbf{A}$, giving

$$\mathbf{v}_m^T \mathbf{A} \mathbf{v}_m = 1 \quad m = 1, 2, \ldots, 2N \tag{10.97}$$

Thus, if we define a modal matrix $\mathbf{V}$ of the $2N$ eigenvectors, the conditions of orthogonality can be expressed as

$$\mathbf{V}^T \mathbf{A} \mathbf{V} = \mathbf{I} \tag{10.98}$$

$$\mathbf{V}^T \mathbf{B} \mathbf{V} = \mathbf{S} \tag{10.99}$$

where $\mathbf{S}$, called the spectral matrix, is a diagonal matrix of eigenvalues s. The solution of the damped free-vibration equation can now be achieved by transformation into modal coordinates. The transformation is expressed as

$$\mathbf{x} = \sum_{n=1}^{2N} \mathbf{v}_n y_n$$

$$= \mathbf{V}\mathbf{y} \tag{10.100}$$

where $\mathbf{y}$ is the vector of $2N$ modal coordinates. Substitution in Equation 10.89 and premultiplication by $\mathbf{V}^T$ gives

$$\dot{\mathbf{y}} = \mathbf{S}\mathbf{y} \tag{10.101}$$

Because $\mathbf{S}$ is a diagonal matrix, Equation 10.101, in fact, represents $2N$ uncoupled equations of the form

$$\dot{y}_n = s_n y_n \quad n = 1, 2, \ldots, 2N \tag{10.102}$$

Equation 10.102 has the solution

$$y_n = d_n e^{s_n t} \tag{10.103}$$

where d_n are arbitrary constants to be determined from initial conditions. Thus using Equation 10.100, we obtain

$$\mathbf{x}_0 = \mathbf{V}\mathbf{y}_0$$
$$= \mathbf{V}\mathbf{d} \tag{10.104}$$

where $\mathbf{d}$ is a vector of the $2N$ constants, d_n, and $\mathbf{x}_0$ is the vector of the $2N$ given initial conditions: in order, the N initial velocities and the N initial displacements. Premultiplying both sides of Equation 10.104 by $\mathbf{V}^T\mathbf{A}$, we get

$$\mathbf{d} = \mathbf{V}^T\mathbf{A}\mathbf{x}_0 \tag{10.105}$$

The modal coordinates in Equation 10.103 are thus fully determined. Solution in terms of the physical coordinates is obtained by using the transformation equation (Eq. 10.100).

Although, theoretically, the procedure described in the foregoing paragraphs can be used to solve a damped free-vibration problem, the method becomes impractical for the solution of real problems, because of the very large volume of computations that must be performed. Roughly speaking, the computations required in the solution of an eigenproblem are proportional to the cube of the order of the matrices involved. For an N-degree-of-freedom system, the solution of a quadratic eigenproblem will thus require about eight times as much effort as that in the solution of the linearized eigenproblem of the associated undamped system. In addition, because the eigenvalues and eigenvectors are complex, the numerical procedure for obtaining them are rendered more complicated. The quadratic eigenproblem solution is therefore seldom employed; instead, a mode superposition method based on the eigenvalues and eigenvectors of the associated undamped system is used.

As a first step in the solution of the damped free-vibration equation, a normal coordinate transformation is carried out using the transformation given by Equation 10.56. The transformed equations become

$$\mathbf{\Phi}^T\mathbf{M}\mathbf{\Phi}\ddot{\mathbf{y}} + \mathbf{\Phi}^T\mathbf{C}\mathbf{\Phi}\dot{\mathbf{y}} + \mathbf{\Phi}^T\mathbf{K}\mathbf{\Phi}\mathbf{y} = 0 \tag{10.106}$$

where $\mathbf{\Phi}$ is the modal matrix of the associated undamped system. It should be noted that the vectors in $\mathbf{\Phi}$ are not the true mode shapes of the damped system. Matrix $\mathbf{\Phi}$ is simply a transformation matrix which is convenient to use because it diagonalizes both the mass matrix and the stiffness matrix. The transformed matrix $\mathbf{\Phi}^T\mathbf{C}\mathbf{\Phi}$ is not, however, diagonal unless $\mathbf{C}$ is of a special form. For the present, we will assume that $\mathbf{C}$ is indeed of such a special form, so that $\mathbf{\Phi}^T\mathbf{C}\mathbf{\Phi}$ is diagonal and the nth diagonal term is c_n. Equations 10.106 can then be written as

$$m_n\ddot{y}_n + c_n\dot{y}_n + k_n y_n = 0 \quad n = 1, 2, \ldots, N \tag{10.107a}$$

or

$$\ddot{y}_n + c_n \dot{y}_n + \omega_n^2 y_n = 0 \quad n = 1, 2, \ldots, N \tag{10.107b}$$

Equations 10.107 represent N uncoupled equations in the modal coordinates y_n, in which m_n is 1 because of the mass orthonormality of $\mathbf{\Phi}$, $k_n = \omega_n^2$, and c_n is the damping coefficient, which, in analogy with a single-degree-of-freedom system, can be expressed as $c_n = 2\xi_n \omega_n m_n = 2\xi_n \omega_n$, where ξ_n is interpreted as the damping ratio in the nth mode.

The solution to Equations 10.107 is given by

$$y_n = e^{-\xi_n \omega_n t} \left(\frac{\dot{y}_{0n} + y_{0n}\xi_n \omega_n}{\omega_{Dn}} \sin \omega_{Dn}t + y_{0n} \cos \omega_{Dn}t \right) \tag{10.108}$$

in which $\omega_{Dn} = \omega_n \sqrt{1 - \xi_n^2}$, and y_{0n} and $\dot{y}_{0n}$ are the nth elements of the vectors $\mathbf{y}_0$ and $\dot{\mathbf{y}}_0$, respectively. The latter two vectors are obtained from the given initial conditions by using Equations. 10.61 and 10.62. Once the modal coordinates have been computed from Equation 10.108, the response in physical coordinates can be obtained by using the transformation equation (Eq. 10.56).

It must be emphasized that the method described in the foregoing paragraphs is based on the assumption that the damping matrix becomes diagonal under a modal coordinate transformation in which the undamped mode shapes of the system are used as the basis vectors. When the damping matrix possesses such a property, damping in the system is said to be *proportional*. As a corollary, when damping is nonproportional, a transformation that uses the undamped mode shapes will not diagonalize the damping matrix and will not, therefore, uncouple the equations of motion. The conditions for damping orthogonality are discussed in the following section.

Example 10.3
For the two-degree-of-freedom system shown in Figure E10.3, the mass stiffness and damping matrices are as indicated below:

$$\mathbf{M} = \begin{bmatrix} 2 & 0 \\ 0 & 1 \end{bmatrix}$$

$$\mathbf{K} = \begin{bmatrix} 3 & -1 \\ -1 & 1 \end{bmatrix}$$

$$\mathbf{C} = \begin{bmatrix} 0.4 & -0.05 \\ -0.05 & 0.2 \end{bmatrix}$$

The damped mode shapes and frequencies have been derived as follows:

$$s_1 = -0.08334 - 0.70221i \quad \mathbf{v}_1 = \begin{bmatrix} 0.1924 + 0.1492i \\ 0.3817 + 0.3031i \\ -0.2415 + 0.2454i \\ -0.4893 + 0.4855i \end{bmatrix}$$

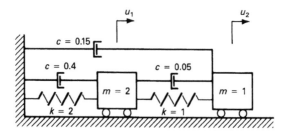

Figure E10.3. Free vibrations of a system with damping coupling.

$$s_2 = -0.08334 + 0.70221i \quad \mathbf{v}_2 = \begin{bmatrix} 0.1924 - 0.1492i \\ 0.3817 - 0.3031i \\ -0.2415 - 0.2454i \\ -0.4893 - 0.4855i \end{bmatrix}$$

$$s_3 = -0.11667 - 1.40933i \quad \mathbf{v}_3 = \begin{bmatrix} 0.3687 + 0.3173i \\ -0.3759 - 0.3084i \\ -0.2451 + 0.2413i \\ 0.2393 - 0.2469i \end{bmatrix}$$

$$s_4 = -0.11667 + 1.40933i \quad \mathbf{v}_4 = \begin{bmatrix} 0.3687 - 0.3173i \\ -0.3759 + 0.3084i \\ -0.2451 - 0.2413i \\ 0.2393 + 0.2469i \end{bmatrix}$$

The mode shapes have been normalized so that $\mathbf{V}^T\mathbf{A}\mathbf{V}=\mathbf{I}$, where $\mathbf{A}$ is defined by Equation 10.90a. The system is given initial displacements $u_{10}=1$, $u_{20}=2$ and set vibrating. The initial velocities are zero. Obtain expressions for the response of the system.

Solution
The equations of motion are uncoupled by using the transformation given by Equation 10.100. The uncoupled equations have a solution of the form of Equation 10.103, in which arbitrary constants d_n are obtained from the initial condition

$$\mathbf{d} = \mathbf{y}_0 = \mathbf{V}^T\mathbf{A}\mathbf{x}_0 \tag{a}$$

The initial conditions are

$$\mathbf{x}_0 = \begin{bmatrix} 0 \\ 0 \\ 1 \\ 2 \end{bmatrix}$$

and the matrix $\mathbf{A}$ is given by

$$
\mathbf{A} = \begin{bmatrix} 0 & 0 & -2 & 0 \\ 0 & 0 & 0 & -1 \\ -2 & 0 & -0.4 & 0.05 \\ 0 & -1 & 0.05 & -0.2 \end{bmatrix}
$$

On substituting for $\mathbf{x}_0$ and $\mathbf{A}$ in Equation a, we get

$$
\mathbf{d} = \begin{bmatrix} -0.90449 - 1.14808i \\ -0.90449 + 1.14808i \\ 0.004315 - 0.0038015i \\ 0.004315 + 0.0038015i \end{bmatrix} \tag{b}
$$

The response in terms of the physical coordinates are obtained by using the transformation equation (Eq. 10.100)

$$
\mathbf{x} = \mathbf{V}\mathbf{y} \tag{c}
$$

where $y_n = d_n e^{s_n t}$. The eigenvalues s_n are complex and can be expressed in the form $\alpha_n + i\beta_n$. Hence

$$
e^{s_n t} = e^{\alpha_n t}(\cos \beta_n t + i \sin \beta_n t) \tag{d}
$$

Equation c can now be expressed as

$$
\mathbf{x} = \sum_{n=1}^{4} d_n e^{\alpha_n t}(\cos \beta_n t + i \sin \beta_n t)\mathbf{v}_n \tag{e}
$$

Substitution of the appropriate quantities in Equation e gives the resultant response, which is, of course, real.

$$
\dot{u}_1 = e^{-0.08334t}(-0.00558 \cos 0.70221t - 0.71164 \sin 0.70221t)
$$

$$
+ e^{-0.11667t}(0.00559 \cos 1.40933t - 0.000066 \sin 1.40933t)
$$

$$
\dot{u}_2 = e^{-0.08334t}(-0.00558 \cos 0.70221t - 1.42474 \sin 0.70221t)
$$

$$
+ e^{-0.11667t}(0.00559 \cos 1.40933t + 0.000193 \sin 1.40933t)
$$

$$
u_1 = e^{-0.08334t}(1.00028 \cos 0.70221t + 0.11076 \sin 0.70221t) \tag{f}
$$

$$
+ e^{-0.11667t}(-0.00028 \cos 1.40933t + 0.00395 \sin 1.40933t)
$$

$$
u_2 = e^{-0.08334t}(1.99983 \cos 0.70221t + 0.24528 \sin 0.70221t)
$$

$$
+ e^{-0.11667t}(0.00019 \cos 1.40933t - 0.00395 \sin 1.40933t)
$$

As a check, the velocities can also be obtained by taking time differentials of the expressions for displacements. The resulting values should be the same as those obtained above directly.

The initial conditions are such that the response in the second mode with a frequency of 1.40933 rad/s is minimal. For simplicity, if we neglect these responses, the displacements in Equations f can be expressed as

$$u_1 = 1.0064e^{-0.08334t}\sin(0.70221t + \theta_1)$$

$$u_2 = 2.0148e^{-0.08334t}\sin(0.70221t + \theta_2)$$

(g)

where

$$\theta_1 = \tan^{-1}9.031 = 83.68°$$

$$\theta_2 = \tan^{-1}8.153 = 83.01°$$

Thus even within a mode the responses along the two degrees of freedom are not entirely in phase. This is unlike the case of undamped vibration, where individual responses within a mode are all in phase. This phase difference will increase as the damping coupling becomes stronger.

Example 10.4

In Example 10.3, the equations of motion are transformed by using the undamped mode shapes. This will not uncouple the equations of motion, because the damping matrix does not satisfy the condition of orthogonality. Uncouple the equations by neglecting the off-diagonal elements of the transformed damping matrix. Now obtain the response of the system for the initial conditions given in Example 10.3.

Solution

The undamped frequencies and mode shapes were derived in Example 10.2.

$$\omega_1 = \frac{1}{\sqrt{2}} \quad \mathbf{q}_1 = \begin{bmatrix} 1 \\ 2 \end{bmatrix}$$

$$\omega_2 = \sqrt{2} \quad \mathbf{q}_2 = \begin{bmatrix} 1 \\ -1 \end{bmatrix}$$

We normalize the eigenvectors, so that

$$\boldsymbol{\phi}_n^T \mathbf{M} \boldsymbol{\phi}_n = 1$$

(a)

$$\boldsymbol{\phi}_1 = \frac{\mathbf{q}_1}{\sqrt{\mathbf{q}_1^T \mathbf{M} \mathbf{q}_1}} = \begin{bmatrix} \frac{1}{\sqrt{6}} \\ \frac{2}{\sqrt{6}} \end{bmatrix}$$

$$\boldsymbol{\phi}_2 = \frac{\mathbf{q}_2}{\sqrt{\mathbf{q}_2^T \mathbf{M} \mathbf{q}_2}} = \begin{bmatrix} \frac{1}{\sqrt{3}} \\ \frac{-1}{\sqrt{3}} \end{bmatrix}$$

(b)

The modal matrix is given by

$$\boldsymbol{\Phi} = \begin{bmatrix} \frac{1}{\sqrt{6}} & \frac{1}{\sqrt{3}} \\ \frac{2}{\sqrt{6}} & -\frac{1}{\sqrt{6}} \end{bmatrix}$$

(c)

Using the transformation $\mathbf{u} = \boldsymbol{\Phi}\mathbf{y}$, the transformed damping matrix $\hat{\mathbf{C}}$ is given by

$$\hat{\mathbf{C}} = \boldsymbol{\Phi}^T \mathbf{C} \boldsymbol{\Phi}$$

$$= \begin{bmatrix} 0.16667 & -0.01178 \\ -0.01178 & 0.23333 \end{bmatrix} \tag{d}$$

When the off-diagonal terms of the damping matrix are neglected, the uncoupled equations of motion become

$$\ddot{y}_1 + 0.16667\dot{y}_1 + 0.5y_1 = 0$$
$$\ddot{y}_2 + 0.23333\dot{y}_2 + 2y_2 = 0 \tag{e}$$

The damping ratios in the two modes are derived as follows

$$2\xi_1\omega_1 = 0.16667$$
$$\xi_1 = 0.11785$$
$$2\xi_2\omega_2 = 0.23333 \tag{f}$$
$$\xi_2 = 0.08249$$

To solve Equations e, we need the initial conditions $\mathbf{y}_0$ and $\dot{\mathbf{y}}_0$. From the transformation relationship between the physical and the modal coordinates, we have

$$\mathbf{u}_0 = \boldsymbol{\Phi}\mathbf{y}_0 \tag{g}$$

Premultiplying both sides of Equation g by $\boldsymbol{\Phi}^T\mathbf{M}$, we get

$$\mathbf{y}_0 = \boldsymbol{\Phi}^T\mathbf{M}\mathbf{u}_0$$

$$= \begin{bmatrix} \sqrt{6} \\ 0 \end{bmatrix} \tag{h}$$

and

$$\dot{\mathbf{y}}_0 = \boldsymbol{\Phi}^T\mathbf{M}\dot{\mathbf{u}}_0$$
$$= 0 \tag{i}$$

Equations e have solutions of the form

$$y_n = e^{-\xi_n\omega_n t}\left(\frac{\dot{y}_{0n} + y_{0n}\xi_n\omega_n}{\omega_{Dn}}\sin\omega_{Dn}t + y_{0n}\cos\omega_{Dn}t\right) \tag{j}$$

where $\omega_{Dn} = \omega_n\sqrt{1 - \xi_n^2}$.

Using the values of damping ratio and initial conditions derived above, we get

$$y_1 = e^{-0.08333t}(0.2907\sin 0.70218t + 2.4495\cos 0.70218t])$$
$$y_2 = 0 \tag{k}$$

Transformation to the physical coordinates gives

$$u_1 = e^{-0.08333t}(0.11868\sin 0.70218t + \cos 0.70218t)$$
$$u_2 = e^{-0.08333t}(0.23736\sin 0.70218t + 2\cos 0.70218t) \tag{l}$$

The initial conditions are such that no response is excited in the second mode. The response is thus entirely in the first mode. The displacement along degree-of-freedom 2 is always twice that along degree-of-freedom 1 and the two displacements are always in phase. To fully appreciate this last statement, we may rewrite Equations 1 in the following alternative form

$$u_1 = 1.007e^{-0.08333t} \sin(0.70218t + \theta_1)$$

$$u_2 = 2.014e^{-0.08333t} \sin(0.70218t + \theta_2)$$

(m)

where θ_1 and θ_2 are the phase angles given by

$$\theta_1 = \theta_2 = \tan^{-1} 8.426 = 83.23°$$

10.9   ADDITIONAL ORTHOGONALITY CONDITIONS

It is clear from the previous discussion that if a normal coordinate transformation using the undamped free vibration mode shapes diagonalizes the damping matrix, the vibration analysis is considerably simplified. It is useful to know what type of damping matrix will be diagonalized under normal coordinate transformation, not only for determining whether such a transformation will be effective in the analysis, but also when it is necessary to construct a damping matrix that will provide specified values of damping in one or more of the vibration modes. Before establishing the conditions for damping orthogonality, we need to develop a series of additional mode shape orthogonality conditions similar to those given by Equations 10.18 and 10.19.

As a first step, we rewrite Equation 10.7 in terms of $\boldsymbol{\phi}_i$ and λ_i

$$\mathbf{K}\boldsymbol{\phi}_i = \lambda_i \mathbf{M}\boldsymbol{\phi}_i \tag{10.109}$$

On premultiplying both sides of Equation 10.109 by $\boldsymbol{\phi}_j^T \mathbf{K}\mathbf{M}^{-1}$, we get

$$\begin{aligned} \boldsymbol{\phi}_j^T \mathbf{K}\mathbf{M}^{-1}\mathbf{K}\boldsymbol{\phi}_i &= \lambda_i \boldsymbol{\phi}_j^T \mathbf{K}\boldsymbol{\phi}_i \\ &= 0 \end{aligned} \qquad i \neq j \tag{10.110}$$

Also

$$\begin{aligned} \boldsymbol{\phi}_i^T \mathbf{K}\mathbf{M}^{-1}\mathbf{K}\boldsymbol{\phi}_i &= \lambda_i \boldsymbol{\phi}_i^T \mathbf{K}\boldsymbol{\phi}_i \\ &= \lambda_i^2 \end{aligned} \tag{10.111}$$

In a similar manner, if we premultiply both sides of Equation 10.109 by $\boldsymbol{\phi}_j^T \mathbf{K}\mathbf{M}^{-1}\mathbf{K}\mathbf{M}^{-1}$, we have

$$\begin{aligned} \boldsymbol{\phi}_j^T \mathbf{K}\mathbf{M}^{-1}\mathbf{K}\mathbf{M}^{-1}\mathbf{K}\boldsymbol{\phi}_i &= \lambda_i \boldsymbol{\phi}_j^T \mathbf{K}\mathbf{M}^{-1}\mathbf{K}\boldsymbol{\phi}_i \\ &= 0 \end{aligned} \tag{10.112}$$

where the last relationship follows from Equation 10.110. Also

$$\boldsymbol{\phi}_i^T \mathbf{K} \mathbf{M}^{-1} \mathbf{K} \mathbf{M}^{-1} \mathbf{K} \boldsymbol{\phi}_i = \lambda_i \boldsymbol{\phi}_i^T \mathbf{K} \mathbf{M}^{-1} \mathbf{K} \boldsymbol{\phi}_i$$
$$= \lambda_i^3 \tag{10.113}$$

By repeated application of the procedure above, we obtain a family of relationships

$$\boldsymbol{\phi}_j^T (\mathbf{K} \mathbf{M}^{-1})^a \mathbf{K} \boldsymbol{\phi}_i = 0 \quad \begin{cases} i \neq j \\ a = 0, 1, 2, \ldots, \infty \end{cases} \tag{10.114}$$

$$\boldsymbol{\phi}_i^T (\mathbf{K} \mathbf{M}^{-1})^a \mathbf{K} \boldsymbol{\phi}_i = \lambda^{a+1} \quad a = 0, 1, 2, \ldots, \infty \tag{10.115}$$

Equations 10.110 and 10.111 can be written in the following alternative forms

$$\boldsymbol{\phi}_j^T \mathbf{M} \mathbf{M}^{-1} \mathbf{K} \mathbf{M}^{-1} \mathbf{K} \boldsymbol{\phi}_i = 0 \quad i \neq j \tag{10.116a}$$

or

$$\boldsymbol{\phi}_j^T \mathbf{M} (\mathbf{M}^{-1} \mathbf{K})^2 \boldsymbol{\phi}_i = 0 \quad i \neq j \tag{10.116b}$$

and

$$\boldsymbol{\phi}_i^T \mathbf{M} \mathbf{M}^{-1} \mathbf{K} \mathbf{M}^{-1} \mathbf{K} \boldsymbol{\phi}_i = \lambda_i^2 \tag{10.117a}$$

or

$$\boldsymbol{\phi}_i^T \mathbf{M} (\mathbf{M}^{-1} \mathbf{K})^2 \boldsymbol{\phi}_i = \lambda_i^2 \tag{10.117b}$$

The general form can now be deduced as

$$\boldsymbol{\phi}_j^T \mathbf{M} (\mathbf{M}^{-1} \mathbf{K})^b \boldsymbol{\phi}_i = 0 \quad \begin{cases} i \neq j \\ b = 0, 1, 2, \ldots, \infty \end{cases} \tag{10.118}$$

$$\boldsymbol{\phi}_i^T \mathbf{M} (\mathbf{M}^{-1} \mathbf{K})^b \boldsymbol{\phi}_i = \lambda_i^b \quad b = 0, 1, 2, \ldots, \infty \tag{10.119}$$

Note that $b = 0$ corresponds to the basic case (Eq. 10.18).

Now, if we premultiply both sides of Equation 10.109 by $\boldsymbol{\phi}_j^T \mathbf{M} \mathbf{K}^{-1}$, we have

$$\boldsymbol{\phi}_j^T \mathbf{M} \boldsymbol{\phi}_i = \lambda_i \boldsymbol{\phi}_j^T \mathbf{M} \mathbf{K}^{-1} \mathbf{M} \boldsymbol{\phi}_i \tag{10.120}$$

Using Equation 10.18 and assuming that $\lambda_i \neq 0$ gives

$$\boldsymbol{\phi}_j^T \mathbf{M} \mathbf{K}^{-1} \mathbf{M} \boldsymbol{\phi}_i = 0 \tag{10.121a}$$

or

$$\boldsymbol{\phi}_j^T \mathbf{M} (\mathbf{M}^{-1} \mathbf{K})^{-1} \boldsymbol{\phi}_i = 0 \quad i \neq j \ \lambda_i \neq 0 \tag{10.121b}$$

Premultiplication of Equation 10.109 by $\boldsymbol{\phi}_i^T \mathbf{M} \mathbf{K}^{-1}$ gives

$$\boldsymbol{\phi}_i^T \mathbf{M} \boldsymbol{\phi}_i = \lambda_i \boldsymbol{\phi}_i^T \mathbf{M} \mathbf{K}^{-1} \mathbf{M} \boldsymbol{\phi}_i \tag{10.122a}$$

or

$$\boldsymbol{\phi}_i^T \mathbf{M}(\mathbf{M}^{-1}\mathbf{K})^{-1}\boldsymbol{\phi}_i = \lambda_i^{-1} \quad \lambda_i \neq 0 \tag{10.122b}$$

In a similar manner, on premultiplying both sides of Equation 10.109 by $\boldsymbol{\phi}_j^T \mathbf{M}\mathbf{K}^{-1}\mathbf{M}\mathbf{K}^{-1}$, we get

$$\boldsymbol{\phi}_j^T \mathbf{M}\,\mathbf{K}^{-1}\mathbf{M}\boldsymbol{\phi}_i = \lambda_i \boldsymbol{\phi}_j^T \mathbf{M}\mathbf{K}^{-1}\mathbf{M}\mathbf{K}^{-1}\mathbf{M}\boldsymbol{\phi}_i \tag{10.123}$$

But the left-hand side of Equation 10.123 is zero because of Equation 10.121. Hence

$$\boldsymbol{\phi}_j^T \mathbf{M}\mathbf{K}^{-1}\mathbf{M}\mathbf{K}^{-1}\mathbf{M}\boldsymbol{\phi}_i = 0 \tag{10.124a}$$

or

$$\boldsymbol{\phi}_j^T \mathbf{M}(\mathbf{M}^{-1}\mathbf{K})^{-2}\boldsymbol{\phi}_i = 0 \quad i \neq j, \ \lambda_i \neq 0 \tag{10.124b}$$

Also, premultiplication of both sides of Equation 10.109 by $\boldsymbol{\phi}_i^T \mathbf{M}\mathbf{K}^{-1}\mathbf{M}\mathbf{K}^{-1}\mathbf{M}\boldsymbol{\phi}_i$ and use of Equation 10.122 gives

$$\boldsymbol{\phi}_i^T \mathbf{M}(\mathbf{M}^{-1}\mathbf{K})^{-2}\boldsymbol{\phi}_i = \lambda_i^{-2} \quad \lambda_i \neq 0 \tag{10.125}$$

By repeated application of the procedure above, we get the family of relationships

$$\boldsymbol{\phi}_j^T \mathbf{M}(\mathbf{M}^{-1}\mathbf{K})^b\boldsymbol{\phi}_i = 0 \quad \begin{cases} i \neq j, \quad \lambda_i \neq 0 \\ b = -1, -2, \ldots, -\infty \end{cases} \tag{10.126}$$

$$\boldsymbol{\phi}_i^T \mathbf{M}(\mathbf{M}^{-1}\mathbf{K})^b\boldsymbol{\phi}_i = \lambda_i^b \quad \begin{cases} \lambda_i \neq 0 \\ b = -1, -2, \ldots, -\infty \end{cases} \tag{10.127}$$

Equations 10.118, 10.119, 10.126, and 10.127 can be combined into the following general relationships

$$\boldsymbol{\phi}_j^T \mathbf{M}(\mathbf{M}^{-1}\mathbf{K})^b\boldsymbol{\phi}_i = \begin{cases} 0 & i \neq j \\ \lambda_i^b & i = j \end{cases} \quad -\infty < b < \infty \tag{10.128}$$

10.10   DAMPING ORTHOGONALITY

As stated earlier, it is useful to be able to determine the condition under which a damping matrix would be diagonalized by a normal coordinate transformation. It should be noted that when a system is assumed to possess proportional damping, the mode superposition method can be used in the analysis and the damping matrix is not required, provided that a damping ratio can be specified for each mode that has been included. In fact, because it is difficult to relate damping resistance arising from internal friction to a measurable physical characteristics, it is easier and more meaningful to specify a modal damping ratio than to construct a damping matrix. It would therefore appear that there is no need to specify a damping matrix explicitly. There are, however, situations when a damping matrix is required. For example, when a direct integration of the equations of motion is used instead of a mode superposition analysis, the damping matrix must be known. In fact, if the system is nonlinear, mode superposition method is no longer valid and a numerical time integration must be used and the damping matrix has to be specified. When a damping matrix is required, it should be constructed in such a manner that it would lead to specified values for the damping ratio in some or all of the modes. The construction of such a matrix requires a study of the conditions of damping orthogonality.

We can use the orthogonality relationships given by Equation 10.128 to develop a damping matrix that will satisfy the condition of orthogonality. A damping matrix that is proportional to the mass matrix evidently meets the desired condition. For example, let

$$\mathbf{C} = \alpha_0 \mathbf{M} \tag{10.129}$$

where α_0 is the constant of proportionality. If we premultiply both sides of Equation 10.129 by ϕ_j^T and postmultiply them by ϕ_i, we get

$$\phi_j^T \mathbf{C} \phi_i = \alpha_0 \phi_j^T \mathbf{M} \phi_i \quad i \neq j \tag{10.130}$$
$$= 0$$

In the form given by Equation 10.129, there is one free parameter α_0 which we can select so as to obtain a desired value of damping in any one mode shape. Thus, if we want the damping ratio for the ith mode shape to be ξ_i, we have

$$\phi_i^T \mathbf{C} \phi_i = c_i = 2\xi_i \omega_i \tag{10.131}$$

Equation 10.131 gives

$$\alpha_0 \phi_i^T \mathbf{M} \phi_i = 2\xi_i \omega_i$$
$$\alpha_0 = 2\xi_i \omega_i \tag{10.132}$$

Once a value of α_0 has been selected as above, the damping in any other mode will be given by

$$\xi_j = \frac{\alpha_0}{2\omega_j}$$
$$= \xi_i \frac{\omega_i}{\omega_j} \tag{10.133}$$

The damping ratio is thus seen to be inversely proportional to the frequency. The variation of ξ with ω for a *mass proportional* damping is plotted in Figure 10.6a.

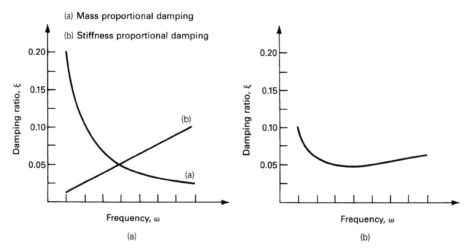

Figure 10.6. (a) Mass and stiffness proportional damping; (b) Rayleigh damping.

Instead of choosing a mass proportional damping, we may choose a *stiffness proportional* damping, in which case the damping matrix is given by

$$\mathbf{C} = \alpha_1 \mathbf{K} \tag{10.134}$$

It is easily verified that the damping matrix given by Equation 10.134 possesses the orthogonality property. To determine α_1, the constant of proportionality, we specify a damping ratio ξ_i for the ith mode. Then, from Equation 10.134

$$\boldsymbol{\phi}_i^T \mathbf{C} \boldsymbol{\phi}_i = \alpha_1 \omega_i^2 \tag{10.135a}$$

or

$$c_i = 2\xi_i \omega_i = \alpha_1 \omega_i^2 \tag{10.135b}$$

which gives the following value for α_1.

$$\alpha_1 = \frac{2\xi_i}{\omega_i} \tag{10.136}$$

Having selected a value of α_1 as above, the damping ratio for any other mode can be determined

$$\xi_j = \frac{1}{2}\omega_j \alpha_1$$
$$= \xi_i \frac{\omega_j}{\omega_i} \tag{10.137}$$

Equation 10.137 and curve b in Figure 10.6a show that for stiffness proportional damping, the damping ratio increases with the frequency.

As a more general case, we can select the damping matrix to be a linear combination of the mass and the stiffness matrices, so that

$$\mathbf{C} = \alpha_0 \mathbf{M} + \alpha_1 \mathbf{K} \tag{10.138}$$

Equation 10.138 has two free parameters, α_0 and α_1. We can thus specify the damping ratio for any two modes, say ith and jth. Then, from Equation 10.138

$$\begin{aligned} \boldsymbol{\phi}_i^T \mathbf{C} \boldsymbol{\phi}_i &= 2\xi_i \omega_i = \alpha_0 + \alpha_1 \omega_i^2 \\ \boldsymbol{\phi}_j^T \mathbf{C} \boldsymbol{\phi}_j &= 2\xi_j \omega_j = \alpha_0 + \alpha_1 \omega_j^2 \end{aligned} \tag{10.139}$$

Equation 10.139 can be expressed in a matrix form as

$$\frac{1}{2} \begin{bmatrix} \frac{1}{\omega_i} & \omega_i \\ \frac{1}{\omega_j} & \omega_j \end{bmatrix} \begin{bmatrix} \alpha_0 \\ \alpha_1 \end{bmatrix} = \begin{bmatrix} \xi_i \\ \xi_j \end{bmatrix} \tag{10.140}$$

Equation 10.140 can be solved for α_0 and α_1. Once these parameters have been determined in terms of the given values of ξ_i and ξ_j, the damping ratio in any other mode can be determined. Damping ratio ξ now varies with ω according to the curve shown in Figure 10.6b.

Damping that leads to a matrix satisfying Equation 10.138 is called *Rayleigh damping*, which is a special case of proportional damping. With Rayleigh damping, which also covers mass proportional and stiffness proportional damping, damping ratios can be assigned specified values for at most two modes. In the practical application of Rayleigh damping, damping ratios are usually specified for the fundamental mode and one other mode that is close to the highest mode of interest. Also, the two specified damping ratios are commonly selected to be equal to each other. In that case, Equation 10.140 provides the following values for the unknown coefficients α_0 and α_1

$$\begin{aligned} \alpha_0 &= \frac{2\xi_0 \omega_i \omega_j}{\omega_i + \omega_j} \\ \alpha_1 &= \frac{2\xi_0}{\omega_i + \omega_j} \end{aligned} \tag{10.141}$$

where ξ_0 is the specified damping ratio. Modes that lie between the two selected modes will have a damping ratio slightly smaller than ξ_0. Modes higher than the highest mode for which damping ratio has been specified will have a damping ratio greater than ξ_0 and this damping ratio will increase with the mode number.

When more than two damping ratios have to be specified, use must be made of additional orthogonality conditions given by Equation 10.128. The damping matrix is then represented by

$$\mathbf{C} = \sum_b \alpha_b \mathbf{M} (\mathbf{M}^{-1} \mathbf{K})^b \tag{10.142}$$

The number of terms to be included in the summation must be equal to M, the number of modes for which damping ratio has to be specified. The damping orthogonality condition of Equation 10.128 yields

$$2\xi_i\omega_i = \boldsymbol{\phi}_i^T\mathbf{C}\boldsymbol{\phi}_i$$

$$= \boldsymbol{\phi}_i^T\left(\sum_b \alpha_b\mathbf{M}(\mathbf{M}^{-1}\mathbf{K})^b\right)\boldsymbol{\phi}_i \tag{10.143}$$

$$= \sum_b \alpha_b\omega_i^{2b}$$

A equation similar to Equation 10.143 is written for each of the M selected modes. Together, these equations lead to a set of simultaneous equations that can be solved for the coefficients α_b. As an example, suppose that damping ratios ξ_i, ξ_j, and ξ_k, respectively, have to be specified for modes i, j, and k. We may then choose a damping matrix of the form

$$\mathbf{C} = \alpha_0\mathbf{M} + \alpha_1\mathbf{K} + \alpha_2\mathbf{K}\mathbf{M}^{-1}\mathbf{K} \tag{10.144}$$

where the three values selected for b are 0, 1, and 2
By using the orthogonality conditions (Eq. 10.128) we get

$$\boldsymbol{\phi}_i^T\mathbf{C}\boldsymbol{\phi}_i = 2\xi_i\omega_i = \alpha_0 + \alpha_1\omega_i^2 + \alpha_2\omega_i^4$$

$$\boldsymbol{\phi}_j^T\mathbf{C}\boldsymbol{\phi}_j = 2\xi_j\omega_j = \alpha_0 + \alpha_1\omega_j^2 + \alpha_2\omega_j^4 \tag{10.145}$$

$$\boldsymbol{\phi}_k^T\mathbf{C}\boldsymbol{\phi}_k = 2\xi_k\omega_k = \alpha_0 + \alpha_1\omega_k^2 + \alpha_2\omega_k^4$$

or in matrix form

$$\frac{1}{2}\begin{bmatrix} \frac{1}{\omega_i} & \omega_i & \omega_i^3 \\ \frac{1}{\omega_j} & \omega_j & \omega_j^3 \\ \frac{1}{\omega_k} & \omega_k & \omega_k^3 \end{bmatrix}\begin{bmatrix} \alpha_0 \\ \alpha_1 \\ \alpha_2 \end{bmatrix} = \begin{bmatrix} \xi_i \\ \xi_j \\ \xi_k \end{bmatrix} \tag{10.146}$$

Equation 10.146 can be solved to obtain α_0, α_1, and α_2.

The damping matrix obtained from Equation 10.142 is known as the *Caughey damping matrix*. It may be noted that unlike mass and stiffness matrices, Caughey damping matrix is fully populated. Hence its use in a step-by-step integration of the equations of motion increases the computation time. In addition, the coefficient matrix in the simultaneous equations for the unknowns α_b, such as the one in Equation 10.146, is ill-conditions, because the terms in this matrix differ by several orders of magnitude. Finally, care should be taken in the use of Caughey matrix to ensure that modes for which a damping ratio has not been specified do not end up with an unreasonable amount of damping. For particular combination of modes for which damping has been specified and the selected values of b, some of the modes with unspecified damping may end

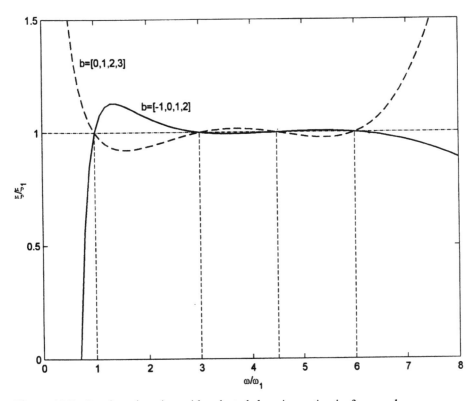

Figure 10.7. Caughey damping with selected damping ratios in four modes.

up having negative damping ratios. In such a case, the response in the affected modes is likely to grow as the step-by-step integration proceeds, making the solution unreliable. As an example, consider a case in which a damping ratio of ξ_0 has been specified for four selected modes, say $1, i, j$, and n. Let the relative frequencies for these modes be in the ratio $1, 3, 4.5$, and 6. The variation of ξ with mode number can now be calculated for a given set of values of the parameter b. Curves corresponding to two different set of values for b, namely $[-1, 0, 1, 2]$ and $[0, 1, 2, 3]$, have been plotted in Figure 10.7. It is evident from the data presented that for $b = [-1, 0, 1, 2]$, the higher modes will end up having negative damping.

In practical application of Caughey damping, damping ratios will be specified for a small number of modes so as to avoid serious ill-conditioning. The list of modes for which a damping value is specified should normally include the fundamental mode and the highest mode of interest. The other modes in the list could be evenly spaced between the first and the highest mode.

An alternative method exists for obtaining an orthogonal damping matrix. Thus, if the desired damping ratios in the N modes are $\xi_1, \xi_2, \dots, \xi_N$, respectively, the transformed damping matrix can be written as

$$\mathbf{\Phi}^T\mathbf{C}\mathbf{\Phi} = \begin{bmatrix} 2\xi_1\omega_1 & 0 & . & 0 & . & 0 \\ 0 & 2\xi_2\omega_2 & . & 0 & . & 0 \\ . & . & . & . & . & . \\ . & . & . & 2\xi_n\omega_n & . & . \\ . & . & . & . & . & . \\ 0 & 0 & . & 0 & . & 2\xi_N\omega_N \end{bmatrix}$$

$$= \hat{\mathbf{C}} \tag{10.147}$$

where $\hat{\mathbf{C}}$ represents the transformed damping matrix, which is diagonal. Equation 10.147 leads to

$$\mathbf{C} = (\mathbf{\Phi}^T)^{-1}\hat{\mathbf{C}}\mathbf{\Phi}^{-1} \tag{10.148}$$

The inverse of the modal matrix $\mathbf{\Phi}$ is obtained from the orthogonality relationship (Eq. 10.18)

$$\mathbf{\Phi}^T\mathbf{M}\mathbf{\Phi} = \mathbf{I} \tag{10.18}$$

so that

$$(\mathbf{\Phi}^T)^{-1} = \mathbf{M}\mathbf{\Phi} \tag{10.149}$$

and

$$\mathbf{\Phi}^{-1} = \mathbf{\Phi}^T\mathbf{M} \tag{10.150}$$

Substitution of Equations 10.149 and 10.150 in Equation 10.148 gives

$$\mathbf{C} = \mathbf{M}\mathbf{\Phi}\hat{\mathbf{C}}\mathbf{\Phi}^T\mathbf{M} \tag{10.151}$$

Equation 10.151 can be expressed in the following alternative form

$$\mathbf{C} = \mathbf{M}\left(\sum_{n=1}^{N} 2\xi_n\omega_n\phi_n\phi_n^T\right)\mathbf{M} \tag{10.152}$$

which identifies more clearly the contribution from each mode.

When any particular mode is not included in the summation in Equation 10.152, it is equivalent to specifying a damping ratio equal to zero for that mode; in other words, that mode would be undamped. To ensure that modes not included in Equation 10.152 still have a reasonable amount of damping, the damping specified in by Equation 10.152 should normally be used in association with stiffness proportional damping. Let ξ_m be the desired damping in mode m, which is the highest mode of interest. From Equations 10.134 and 10.136 we know that such a damping can be produced by a stiffness proportional damping matrix given by

$$\tilde{C} = \frac{2\xi_m}{\omega_m}K \tag{10.153}$$

The corresponding damping in other modes is obtained from

$$\tilde{\xi}_i = \xi_m \frac{\omega_i}{\omega_m} \tag{10.154}$$

If the desired damping in mode i is ξ_i, the remaining amount of damping $\bar{\xi}_i = \xi_i - \tilde{\xi}_i$ should be produced by the matrix constructed as in Equation 10.152. Combining Equations 10.153 and 10.152 we get

$$C = \frac{2\xi_m}{\omega_m}K + M\left(\sum_{n=1}^{N} 2\bar{\xi}_n \omega_n \phi_n \phi_n^T\right)M \tag{10.155}$$

Finally, it should be noted that the damping matrix obtained from either Equation 10.152 or Equation 10.155 is fully populated.

Example 10.5
The mass and stiffness matrices of the three-degree-of-freedom system shown in Figure E10.5 are as indicated below. The three natural frequencies of the system are also shown. Derive a Rayleigh damping matrix such that the damping in the first and second modes is, respectively, 5% and 3% of critical. What is the resulting damping ratio in the third mode?

$$M = \begin{bmatrix} 2 & & \\ & 2 & \\ & & 2 \end{bmatrix}$$

$$K = \begin{bmatrix} 600 & -600 & 0 \\ -600 & 1800 & -1200 \\ 0 & -1200 & 3600 \end{bmatrix}$$

$$\omega = \begin{bmatrix} 11.62 \\ 27.50 \\ 45.90 \end{bmatrix} \text{rad/s}$$

Solution
The Rayleigh damping matrix is given by

$$C = \alpha_0 M + \alpha_1 K \tag{a}$$

where a_0 and a_1 are obtained by solving the following equations

$$\frac{1}{2}\begin{bmatrix} \frac{1}{\omega_1} & \omega_1 \\ \frac{1}{\omega_2} & \omega_2 \end{bmatrix}\begin{bmatrix} \alpha_0 \\ \alpha_1 \end{bmatrix} = \begin{bmatrix} \xi_1 \\ \xi_2 \end{bmatrix} \tag{b}$$

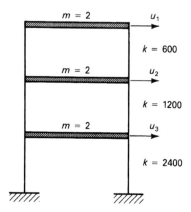

Figure E10.5. Three-degree-of-freedom system.

Substitution of the given values of $\omega_1, \omega_2, \xi_1$ and ξ_2 in Equation b gives

$$\frac{1}{2}\begin{bmatrix} \frac{1}{11.62} & 11.62 \\ \frac{1}{27.50} & 27.5 \end{bmatrix}\begin{bmatrix} \alpha_0 \\ \alpha_1 \end{bmatrix} = \begin{bmatrix} 0.05 \\ 0.03 \end{bmatrix} \tag{c}$$

On solving Equation c, we get $\alpha_0 = 1.056$ and $\alpha_1 = 0.0078$. The damping matrix is now obtained from Equation a.

$$\mathbf{C} = \alpha_0\mathbf{M} + \alpha_1\mathbf{K} = \begin{bmatrix} 2.583 & -0.471 & 0 \\ -0.471 & 3.526 & -0.943 \\ 0 & -0.943 & 4.940 \end{bmatrix}$$

The damping ratio in the third mode is given by

$$\xi_3 = \frac{1}{2}\left(\frac{\alpha_0}{\omega_3} + \alpha_1\omega_3\right)$$

$$= \frac{1}{2}\left(\frac{1.056}{45.90} + 0.0078 \times 45.90\right)$$

$$= 0.029$$

Example 10.6
For the system of Example 10.5, derive a damping matrix that will give a damping ratio of 5% in each of the three modes.

Solution
The required damping matrix is given by

$$\mathbf{C} = \alpha_0\mathbf{M}(\mathbf{M}^{-1}\mathbf{K})^0 + \alpha_1\mathbf{M}(\mathbf{M}^{-1}\mathbf{K})^1 + \alpha_2\mathbf{M}(\mathbf{M}^{-1}\mathbf{K})^2$$

$$= \alpha_0\mathbf{M} + \alpha_1\mathbf{K} + \alpha_2\mathbf{K}\mathbf{M}^{-1}\mathbf{K} \tag{a}$$

where the coefficients α_0, α_1, and α_3 are determined from

$$\frac{1}{2}\begin{bmatrix} \frac{1}{\omega_1} & \omega_1 & \omega_1^3 \\ \frac{1}{\omega_2} & \omega_2 & \omega_2^3 \\ \frac{1}{\omega_3} & \omega_3 & \omega_3^3 \end{bmatrix}\begin{bmatrix} \alpha_0 \\ \alpha_1 \\ \alpha_2 \end{bmatrix} = \begin{bmatrix} 0.05 \\ 0.05 \\ 0.05 \end{bmatrix} \tag{b}$$

On substituting for ω_1, ω_2, and ω_3 in Equation b and solving for α_0, α_1, and α_2, we get $\alpha_0 = 0.755$, $\alpha_1 = 0.310 \times 10^{-2}$, and $\alpha_2 = -0.606 \times 10^{-6}$. The damping matrix from Equation a is then

$$\mathbf{C} = \begin{bmatrix} 3.15 & -1.42 & -0.22 \\ -1.42 & 5.56 & -1.75 \\ -0.22 & -1.75 & 8.30 \end{bmatrix} \tag{c}$$

Example 10.7
The mode shapes for the system of Example 10.5 are as given below. Using the given mode shapes, obtain a damping matrix that will give 5% damping in each of the three modes.

$$\mathbf{Q} = \begin{bmatrix} 1.000 & 1.000 & 1.000 \\ 0.548 & -1.522 & -6.260 \\ 0.198 & -0.872 & 12.10 \end{bmatrix}$$

Solution
We first normalize the mode shapes so that $\mathbf{\Phi}^T\mathbf{M}\mathbf{\Phi} = \mathbf{I}$. Thus for the first mode shape

$$M_{11} = \mathbf{q}_1^T\mathbf{M}\mathbf{q}_1$$

$$= [1.000 \quad 0.548 \quad 0.198]\begin{bmatrix} 2 & & \\ & 2 & \\ & & 2 \end{bmatrix}\begin{bmatrix} 1.000 \\ 0.548 \\ 0.198 \end{bmatrix}$$

$$= 2.679 \tag{a}$$

The normalized value of the first mode is given by

$$\phi_1 = \frac{1}{\sqrt{2.679}}\mathbf{q}_1 = \begin{bmatrix} 0.611 \\ 0.335 \\ 0.121 \end{bmatrix} \tag{b}$$

The other mode shapes are normalized in a similar manner, so that

$$\mathbf{\Phi} = \begin{bmatrix} 0.611 & 0.350 & 0.052 \\ 0.335 & -0.533 & -0.324 \\ 0.121 & -0.305 & 0.626 \end{bmatrix} \tag{c}$$

The required damping matrix is given by

$$\mathbf{C} = \mathbf{M}\mathbf{\Phi}\hat{\mathbf{C}}\mathbf{\Phi}^T\mathbf{M} \tag{d}$$

where

$$\hat{C} = \begin{bmatrix} 2\xi_1\omega_1 & & \\ & 2\xi_2\omega_2 & \\ & & 2\xi_1\omega_1 \end{bmatrix} = \begin{bmatrix} 1.162 & 0 & 0 \\ 0 & 2.75 & 0 \\ 0 & 0 & 4.59 \end{bmatrix} \qquad \text{(e)}$$

Substitution into Equation d gives

$$C = \begin{bmatrix} 3.13 & -1.41 & -0.24 \\ -1.41 & 5.57 & -1.75 \\ -0.24 & -1.75 & 8.30 \end{bmatrix} \qquad \text{(f)}$$

which is equal to that derived in Example 10.6 within round-off errors.

SELECTED READINGS

Biggs, J.M. 1964. *Introduction to Structural Dynamics*. New York: McGraw-Hill.

Caughey, T.K. 1960. Classical Normal Modes in Damped Linear Dynamic Systems. *Journal of Applied Mechanics*, Vol. 27: 269–271. ASME.

Clough, R.W. & Penzien, J. 1993. *Dynamics of Structures*. New York: McGraw-Hill. 2nd Edition.

Crandall, S.H. & McCalley, R.B. Jr. 1996. Matrix Methods of Analysis. In C.M. Harris (ed.) *Shock and Vibration Handbook*. New York: McGraw-Hill. 4th Edition.

Hurty, W.C. & Rubinstein, M.F. 1964. *Dynamics of Structures*. Englewood Cliffs: Prentice Hall.

Meirovitch, L. 1967. *Analytical Methods in Vibrations*. London: Macmillan.

Wilson, E.L. & Penzien, J. 1972. Evaluation of Orthogonal Damping Matrices. *International Journal for Numerical Methods in Engineering*, Vol. 4: 5–10.

PROBLEMS

10.1   Obtain the equations of free vibrations of the system shown in Problem 8.8. Determine the frequencies of vibration and the mode shapes of the system. Suppose that the system is given an initial displacement of 1 in. along coordinate 1 and released from that position. Obtain expressions that will determine the subsequent response. The following data are supplied.

$$E = 30,000 \text{ ksi} \quad I = 500 \text{ in.}^4 \quad L = 144 \text{ in.} \quad m = 14.6 \text{ lb.s}^2/\text{in.}$$

10.2   A simplified model of an automobile is shown in Figure P10.2. The characteristics of the system are also identified in the figure. Obtain the frequencies and mode shapes of vibration of the vehicle. The automobile is traveling across a rough road. Uneven joints located every 16 m on the road pavement cause vibration of the vehicle in the second mode. Calculate the speed of the automobile. If a passenger in the vehicle is not to experience any acceleration, where should he be seated?

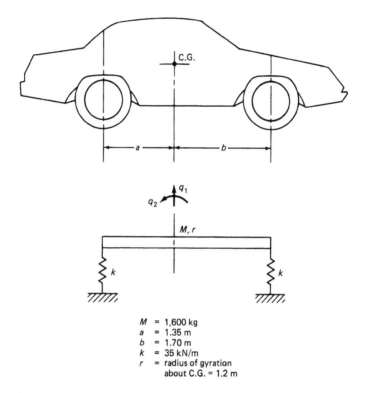

$$
\begin{aligned}
M &= 1{,}600 \text{ kg} \\
a &= 1.35 \text{ m} \\
b &= 1.70 \text{ m} \\
k &= 35 \text{ kN/m} \\
r &= \text{radius of gyration} \\
  &\quad \text{about C.G.} = 1.2 \text{ m}
\end{aligned}
$$

Figure P10.2.

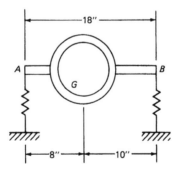

Figure P10.3.

10.3 A motor weighing 100 lb is mounted on two springs as indicated in Figure P10.3. The springs are deflected 2.00 in. by the weight of the motor. The radius of gyration of the motor with respect to a horizontal axis through the center of mass G and perpendicular to the plane of the paper is 4.00 in. Two 25 lb forces are applied and suddenly removed as follows: (a) 25 lb

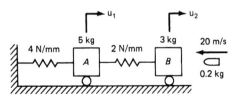

Figure P10.4.

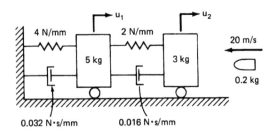

Figure P10.5.

down at each A and B; (b) 25 lb down at A, up at B; 50 lb down at B only. Determine the resulting motion of the motor for each case.

10.4 Two wooden blocks of mass 5 kg and 3 kg, respectively, are restrained by a system of springs, as shown in Figure P10.4. The restraining springs have stiffnesses 4 N/mm and 2 N/mm respectively as indicated. A bullet of mass 0.2 kg is fired at a speed of 20 m/s and embeds itself into block B. Determine the resulting motion of the system.

10.5 Obtain the response of the system shown in Problem 10.4 assuming that the system possesses viscous damping represented by dash pots of damping coefficients 0.032N.s/mm and 0.016N.s/mm as shown in Figure P10.5.

10.6 The mode shapes and frequencies of the three-story frame of Figure P10.6 are shown on the figure. The story masses are also indicated. The second floor of the frame is pulled out to the right a distance of 0.75 in. and then suddenly released. Determine the subsequent motion of the frame. Neglect damping.

10.7 Repeat Problem 10.6 assuming that the damping in each mode is 10% of critical.

10.8 The frame of Problem 10.6 is set vibrating after being imparted a velocity of 5 in./s at floor level 3, 10 in./s at floor level 2, and 5 in./s at floor level 1. Determine the subsequent motion of the frame. Neglect damping.

10.9 For the frame shown in Figure P10.6, obtain a Rayleigh damping matrix that will give 5% of critical damping in each of the 1st and 3rd modes. What is the resulting damping ratio in the second mode?

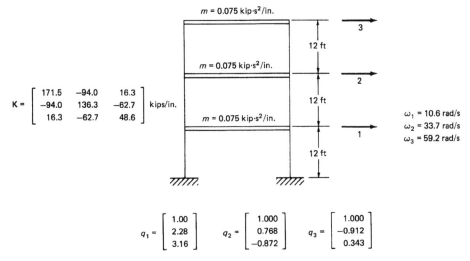

$$K = \begin{bmatrix} 171.5 & -94.0 & 16.3 \\ -94.0 & 136.3 & -62.7 \\ 16.3 & -62.7 & 48.6 \end{bmatrix} \text{kips/in.}$$

$m = 0.075 \ \text{kip·s}^2/\text{in.}$

$m = 0.075 \ \text{kip·s}^2/\text{in.}$

$m = 0.075 \ \text{kip·s}^2/\text{in.}$

12 ft

12 ft

12 ft

3

2

1

$\omega_1 = 10.6 \ \text{rad/s}$
$\omega_2 = 33.7 \ \text{rad/s}$
$\omega_3 = 59.2 \ \text{rad/s}$

$$q_1 = \begin{bmatrix} 1.00 \\ 2.28 \\ 3.16 \end{bmatrix} \qquad q_2 = \begin{bmatrix} 1.000 \\ 0.768 \\ -0.872 \end{bmatrix} \qquad q_3 = \begin{bmatrix} 1.000 \\ -0.912 \\ 0.343 \end{bmatrix}$$

Figure P10.6.

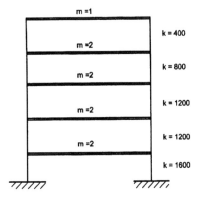

m = 1

k = 400

m = 2

k = 800

m = 2

k = 1200

m = 2

k = 1200

m = 2

k = 1600

Figure P10.11.

10.10 Obtain a damping matrix for the frame of Figure P10.6 using Equation 10.152. Include the 1st and the 3rd modes in your formulation and assume that the damping ratios in each of these modes is 5%. Verify that the resulting damping matrix leads to zero damping in the 2nd mode.

10.11 The frequencies in rad/s and mass-orthonormal mode shapes of the frame shown in Figure P10.11 are as given below. Using Equation 10.152 construct a damping matrix that will give 5% damping in each of the first and second modes. Verify that such a damping matrix gives zero damping in modes 3, 4, and 5.

$$[\Lambda] = \begin{bmatrix} 7.77 \\ 18.98 \\ 27.66 \\ 36.39 \\ 45.71 \end{bmatrix}$$

$$\Phi = \begin{bmatrix} 0.1029 & 0.2098 & 0.2688 & 0.4454 & 0.4176 \\ 0.2296 & 0.3635 & 0.2840 & 0.0552 & -0.4810 \\ 0.3333 & 0.2988 & -0.0631 & -0.4566 & 0.2955 \\ 0.4385 & -0.0675 & -0.4628 & 0.2871 & -0.0822 \\ 0.5161 & -0.6783 & 0.5077 & -0.1246 & 0.0195 \end{bmatrix}$$

10.12 Construct a damping matrix for the frame of Problem 10.11 using Equation 10.149 in association with a stiffness proportional damping, so as to provide 5% damping in the first and second modes and 1% damping in mode No. 5. What is the resulting damping in modes 3 and 4?

CHAPTER 11

Numerical solution of the eigenproblem

11.1 INTRODUCTION

In Chapter 10, we discussed a method for the solution of undamped and damped free-vibration problems which involved a transformation of the equations to a set of coordinates called modal coordinates or normal coordinates. Modal coordinate transformation uncouples the equations of undamped free vibrations so that the problem reduces to the solution of a set of N independent second-order differential equations. In the case when damping is present, modal transformation will uncouple the equations of motion only provided that the damping matrix satisfies the orthogonality condition. In practice, because the nature of physical characteristics that determine damping are difficult to define or to measure, damping is usually best specified in terms of modal damping ratios, an approach which while implying that the damping matrix does satisfy the orthogonality conditions, does not require an explicit determination of the matrix.

Modal coordinate transformation is, in fact, a special case of Ritz vector method of solution, in which the mode shapes of the system are used as the Ritz vectors. It can be applied to the solution of both free- and forced-vibration problems. Furthermore, in many problems of vibration, the response can be represented by a superposition of the responses in only a few modes, usually those having the lower frequencies of vibration. In such cases, it is not necessary to determine all the mode shapes of the system and it is usually sufficient to calculate only the first few such shapes, along with the associated frequencies of vibration.

It is apparent from the discussion above that mode shapes and frequencies sometimes play a central role in the solution of a free- or a forced-vibration problem. As described in Chapter 10, the eigenvalue or frequencies of a system can be obtained by finding the roots of the characteristic equation. This requires an explicit evaluation of the coefficients $c_0, c_1, \ldots, c_N$ in the characteristic polynomial.

$$p(\lambda) = c_N \lambda^N + c_{N-1} \lambda^{N-1} + \cdots + c_1 \lambda + c_0 \qquad (11.1)$$

Once the polynomial in Equation 11.1 has been obtained, a standard numerical procedure can be used to find the roots of the equation. We used this method in Chapter 10 for solving several simple problems. The direct evaluation of the coefficients in the polynomial (Eq. 11.1), however, requires a very large number of numerical operations. Further, the roots of the characteristic equation are very sensitive to the coefficient values and small errors in the evaluation of the latter, inevitable because of round-off, lead to significant loss of precision in the values of the roots. The method is not therefore practicable for the solution of large systems. Indeed, the numerical solution of an eigenproblem to determine the mode shapes and frequencies is by no means a trivial exercise. Considerable effort has been directed toward developing efficient computational procedures for solving the problem. The advent of the digital computer gave a very significant impetus to the work in this area. As a result, since 1954, a number of new methods have been devised for eigenproblem solution. These methods can be classified into the following broad categories:

1. Transformation methods.
2. Iterative methods.
3. Determinant search methods.

The selection of a numerical procedure in a specific case depends on the nature of the problem, which affects the size of the matrices involved and their properties, such as symmetry, positive definiteness, bandedness, and sparsity. The suitability of solution procedure also depends on the type of solution desired, for example, whether all the mode shapes and frequencies are required or only a few need to be found, and whether the few modes to be found are those with the lowest or the highest frequencies. Very often, the procedure best suited to the problem at hand involves a combination of two or more of the methods that may belong to the same or to different categories. In the following sections, we describe several different methods in each of the categories listed above. Our discussion is related primarily to the eigenproblem encountered in vibrations, but the methods discussed may be equally applicable to problems in other fields of engineering analysis.

In engineering analysis it can often be the case that not all the eigenvalues and eigenvectors are of interest. For certain class of dynamic loads, a major portion of the response of the system is contained in only some of the modes. The remaining modes contribute very little to the response and the mode shapes, and frequencies for such modes need therefore not be evaluated. For example, earthquake forces typically excite only the mode shapes with lower frequencies. In view of the foregoing, it is useful to order the eigenvalue on the basis of their magnitudes. We will adopt an ordering in which the eigenvalues or frequencies are arranged in order of increasing magnitude; the lowest eigenvalue will be denoted by λ_1 and the highest by λ_N, while the corresponding frequencies will be denoted by ω_1 and ω_N respectively.

11.2 PROPERTIES OF STANDARD EIGENVALUES AND EIGENVECTORS

In the following paragraph we discuss some useful properties of the eigenvalues and eigenvectors obtained from a solution of the eigenvalue problem defined by Equations 10.5 or 10.6. First, consider the eigenvalue problem for transposed matrix $\mathbf{A}^T$

$$\mathbf{A}^T \mathbf{p} = \bar{\lambda} \mathbf{p} \tag{11.2}$$

The eigenvalues $\bar{\lambda}$ are obtained from the characteristic equation

$$\det(\mathbf{A}^T - \bar{\lambda} \mathbf{I}) = 0 \tag{11.3a}$$

or

$$\det(\mathbf{A} - \bar{\lambda} \mathbf{I})^T = 0 \tag{11.3b}$$

Since the determinant of the transpose of a matrix is the same as the determinant of the original matrix, Equation 11.3b is entirely equivalent to Equation 10.6, and hence $\bar{\lambda}$ must be equal to λ. The eigenvectors $\mathbf{p}$ are, however, not necessarily equal to $\mathbf{q}$, except when $\mathbf{A}$ is symmetric.

Equation 11.2 can be expressed in the alternative form

$$\mathbf{p}^T \mathbf{A} = \lambda \mathbf{p}^T \tag{11.4}$$

In view of the relationship given by Equation 11.4, vectors $\mathbf{p}$ are referred to as the *left eigenvectors* of $\mathbf{A}$. The left and right eigenvectors of a matrix bear a special relationship to each other. To obtain this relationship, consider a left eigenvector $\mathbf{p}_i$ and a right eigenvector $\mathbf{q}_j$. They satisfy the following equations

$$\mathbf{p}_i^T \mathbf{A} = \lambda_i \mathbf{p}_i^T \tag{11.5a}$$

$$\mathbf{A} \mathbf{q}_j = \lambda_j \mathbf{q}_j \tag{11.5b}$$

Postmultiplication of both sides of Equation 11.5a by $\mathbf{q}_j$ and premultiplication of the two sides of Equation 11.5b by $\mathbf{p}_i^T$ gives

$$\mathbf{p}_i^T \mathbf{A} \mathbf{q}_j = \lambda_i \mathbf{p}_i^T \mathbf{q}_j \tag{11.6a}$$

$$\mathbf{p}_i^T \mathbf{A} \mathbf{q}_j = \lambda_j \mathbf{p}_i^T \mathbf{q}_j \tag{11.6b}$$

Subtraction of Equation 11.6b from Equation 11.6a yields

$$(\lambda_i - \lambda_j) \mathbf{p}_i^T \mathbf{q}_j = 0 \tag{11.7}$$

Equation 11.7 gives the relationship

$$\mathbf{p}^T\mathbf{q}_j = 0 \qquad \lambda_i \neq \lambda_j \tag{11.8}$$

The relationship expressed in Equation 11.8 states that the left and right eigenvectors of a real matrix corresponding to two distinct eigenvalues are orthogonal to each other.

It can be shown that any arbitrary vector $\mathbf{u}$ of dimension N may be expressed as a linear combination of the N eigenvectors of $\mathbf{A}$. Thus

$$\mathbf{u} = c_1\mathbf{q}_1 + c_2\mathbf{q}_2 + \cdots + c_N\mathbf{q}_N \tag{11.9}$$

To obtain coefficient c_i in Equation 11.9, we premultiply both sides by the left eigenvector $\mathbf{p}_i^T$. This gives

$$c_i = \frac{\mathbf{p}_i^T\mathbf{u}}{\mathbf{p}_i^T\mathbf{q}_i} \tag{11.10}$$

When $\mathbf{A}$ is symmetric, Equation 11.10 reduces to

$$c_i = \frac{\mathbf{q}_i^T\mathbf{u}}{\mathbf{q}_i^T\mathbf{q}_i} \tag{11.11}$$

11.3   TRANSFORMATION OF A LINEARIZED EIGENVALUE PROBLEM TO THE STANDARD FORM

The standard eigenvalue problem appears in a variety of problems in science and engineering. Many different procedures have therefore been developed for its solution. In the following sections, we describe some of these procedures. In order for these procedure to be used in the solution of linearized eigenvalue problem encountered in structural dynamics, we need to transform the linearized eigenvalue problem to a standard form. This can be accomplished in several different ways. On multiplying both sides of Equation 10.7 by $\mathbf{M}^{-1}$, we get

$$\mathbf{M}^{-1}\mathbf{K}\mathbf{q} = \lambda\mathbf{q} \tag{11.12a}$$

or

$$\mathbf{E}\mathbf{q} = \lambda\mathbf{q} \tag{11.12b}$$

where $\mathbf{E} = \mathbf{M}^{-1}\mathbf{K}$.

As an alternative, we may multiply both sides of Equation 10.7 by $\mathbf{K}^{-1}$ to get

$$\mathbf{K}^{-1}\mathbf{M}\mathbf{q} = \frac{1}{\lambda}\mathbf{q} \tag{11.13a}$$

or

$$\mathbf{D}\mathbf{q} = \gamma\mathbf{q} \tag{11.13b}$$

where $\mathbf{D} = \mathbf{K}^{-1}\mathbf{M}$ and $\gamma = 1/\lambda$. Each of Equations 11.12b and 11.13b represent a standard eigenvalue problem. It should be noted that matrices $\mathbf{E}$ and $\mathbf{D}$ are both unsymmetric. It is, however, possible to convert the linearized eigenvalue problem into a standard form in which the matrix involved is symmetric. Provided that $\mathbf{K}$ is positive definite, this can be achieved by factorizing the stiffness matrix $\mathbf{K}$ into a lower triangular matrix $\mathbf{L}_K$ and its transpose, so that

$$\mathbf{K} = \mathbf{L}_K \mathbf{L}_K^T \tag{11.14}$$

The transformation of $\mathbf{K}$ into the form of Equation 11.14 is called *Choleski decomposition*. Equation 10.7 can now be expressed as

$$\mathbf{M}\mathbf{q} = \frac{1}{\lambda} \mathbf{L}_K \mathbf{L}_K^T \mathbf{q} \tag{11.15a}$$

$$\mathbf{L}_K^{-1} \mathbf{M} (\mathbf{L}_K^T)^{-1} \mathbf{L}_K^T \mathbf{q} = \gamma \mathbf{L}_K^T \mathbf{q}$$

$$\tilde{\mathbf{D}}\tilde{\mathbf{q}} = \gamma \tilde{\mathbf{q}} \tag{11.15b}$$

in which $\gamma = 1/\lambda$, $\tilde{\mathbf{D}} = \mathbf{L}_K^{-1} \mathbf{M} (\mathbf{L}_K^T)^{-1}$ is a symmetric matrix, and the modified eigenvector $\tilde{\mathbf{q}}$ is related to $\mathbf{q}$, such that $\mathbf{L}_K^T \mathbf{q} = \tilde{\mathbf{q}}$.

As an alternative, if the mass matrix is positive definite, Choleski decomposition can be applied to it leading to the modified eigenvalue problem

$$\mathbf{L}_M^{-1} \mathbf{K} (\mathbf{L}_M^T)^{-1} \mathbf{L}_M^T \mathbf{q} = \lambda \mathbf{L}_M^T \mathbf{q}$$

$$\tilde{\mathbf{E}}\tilde{\mathbf{q}} = \lambda \tilde{\mathbf{q}} \tag{11.16}$$

in which $\tilde{\mathbf{E}} = \mathbf{L}_M^{-1} \mathbf{M} (\mathbf{L}_M^T)^{-1}$ is a symmetric matrix, and the modified eigenvector $\tilde{\mathbf{q}}$ is related to $\mathbf{q}$, such that $\mathbf{L}_M^T \mathbf{q} = \tilde{\mathbf{q}}$.

When the mass matrix is diagonal, Choleski decomposition is quite straightforward because matrices $\mathbf{L}_M$ and $\mathbf{L}_M^T$ are given by

$$\mathbf{L}_M = \mathbf{L}_M^T = \mathbf{M}^{1/2} \tag{11.17}$$

Matrix $\mathbf{M}^{1/2}$ is a diagonal matrix whose elements are square roots of the respective elements in $\mathbf{M}$, while the inverse matrix $\mathbf{L}_M^{-1}$ is a diagonal matrix whose elements are the reciprocals of the elements of $\mathbf{M}^{1/2}$. Obviously, the method will not work when there are one or more zeros on the diagonal of $\mathbf{M}$, that is, when $\mathbf{M}$ is positive semidefinite. In that case, the degrees of freedom corresponding to zero mass should first be eliminated from the equations of motion by the process of static condensation described in Section 3.7.

11.4   TRANSFORMATION METHODS

The *transformation methods* are, in general, useful when the matrices involved are of comparatively small order and are more or less fully populated or have a large bandwidth. These methods lead to simultaneous evaluation of all eigenvalues. The eigenvectors can then be obtained by a process of inverse transformation or by using an appropriate iterative process described later in this chapter. These methods are not particularly appropriate when the matrices involved are of a large order and only a few eigenvalues need to be determined. Also, because transformations destroy the sparsity of a matrix, transformation methods are not efficient in dealing with sparse matrices. Transformation methods rely on the fact that under certain types of transformations, the eigenvalues of a matrix remain unchanged. A series of such transformations are carried out on the matrix being analyzed, until it is reduced to such a form that the eigenvalues can either be determined by inspection or are easy to compute. A majority of transformation methods are applicable to an eigenproblem in the standard form, where the matrix involved is symmetric. A linearized eigenproblem must therefore first be converted into a standard symmetric form before one of the methods described in this section can be used for its solution. In some cases, however, a variant of the transformation method described is available that can be applied directly to the solution of a linearized eigenproblem.

The most general type of transformation that leaves the eigenvalues unaltered is called a *similarity transformation*. Thus, if $\mathbf{N}$ is an arbitrary nonsingular transformation matrix of the same order as $\mathbf{A}$, the standard eigenproblem Equation 10.5 can be expressed as

$$\mathbf{N}^{-1}\mathbf{A}\mathbf{N}\mathbf{N}^{-1}\mathbf{q} = \lambda \mathbf{N}^{-1}\mathbf{q} \tag{11.18a}$$

or

$$\tilde{\mathbf{A}}\tilde{\mathbf{q}} = \lambda\tilde{\mathbf{q}} \tag{11.18b}$$

in which the transformed matrix $\tilde{\mathbf{A}} = \mathbf{N}^{-1}\mathbf{A}\mathbf{N}$, and the modified eigenvector $\tilde{\mathbf{q}} = \mathbf{N}^{-1}\mathbf{q}$. Obviously, the eigenvalues remain unchanged under such a transformation.

When $\mathbf{A}$ is symmetric, an *orthogonal transformation* is most effective. The transformation matrix $\mathbf{N}$ is chosen so that $\mathbf{N}^{-1} = \mathbf{N}^T$, and Equation 11.18 reduces to

$$\mathbf{N}^T\mathbf{A}\mathbf{N}\mathbf{N}^T\mathbf{q} = \lambda\mathbf{N}^T\mathbf{q} \tag{11.19}$$

The transformed matrix $\tilde{\mathbf{A}} = \mathbf{N}^T\mathbf{A}\mathbf{N}$ is, in this case, symmetrical. After a series of orthogonal transformations involving matrices $\mathbf{N}_1, \mathbf{N}_2, \ldots, \mathbf{N}_m$, we have

$$\tilde{\mathbf{A}} = \mathbf{N}_m^T \cdots \mathbf{N}_2^T\mathbf{N}_1^T\mathbf{A}\mathbf{N}_1\mathbf{N}_2 \cdots \mathbf{N}_m \tag{11.20}$$

and

$$\tilde{\mathbf{q}} = \mathbf{N}_m^T \cdots \mathbf{N}_2^T \mathbf{N}_1^T \mathbf{q} \qquad (11.21a)$$

so that

$$\mathbf{q} = \mathbf{N}_1 \mathbf{N}_2 \cdots \mathbf{N}_m \tilde{\mathbf{q}} \qquad (11.21b)$$

The transformation methods available for the solution of an eigenvalue problem are listed below:
1. Jacobi diagonalization.
2. Givens tridiagonalization.
3. Householder's transformation.
4. The LR transformation.
5. The QR transformation.
Some of the transformation methods of interest to us are described below.

11.4.1 *Jacobi diagonalization*

Although *Jacobi diagonalization* is the oldest transformation method available, its conceptual simplicity and the resulting ease with which it can be programmed on a computer makes it still attractive for the solution of eigenproblems of small size. It is often used for obtaining the eigenvalues of the secondary matrices encountered in the subspace iteration technique described later.

In its standard form, the method is used for the solution of a standard symmetric eigenvalue problem and involves repeated application of orthogonal transformations to the symmetric matrix until the latter becomes diagonal. Each transformation is designed to eliminate two symmetric off-diagonal elements. The transformation matrix for reducing the elements a_{ij} and a_{ji} to zero is given by

$$\mathbf{N} = \begin{array}{c} \\ \\ \\ i \\ \\ j \\ \\ \end{array}
\begin{pmatrix}
1 & 0 & . & 0 & . & 0 & . & 0 \\
0 & 1 & . & 0 & . & 0 & . & 0 \\
. & . & . & . & . & . & . & . \\
0 & 0 & . & \cos\alpha & . & -\sin\alpha & . & 0 \\
. & . & . & . & . & . & . & . \\
0 & 0 & . & \sin\alpha & . & \cos\alpha & . & 0 \\
0 & 0 & . & 0 & . & 0 & . & 1
\end{pmatrix} \qquad (11.22)$$

Matrix $\mathbf{N}$ is readily seen to be an orthogonal matrix for which $\mathbf{N}^{-1} = \mathbf{N}^T$. Further, the transformation can be interpreted as representing a rotation of the i, jth degrees of freedom through an angle α. For example, let a vector $\mathbf{v}$ in

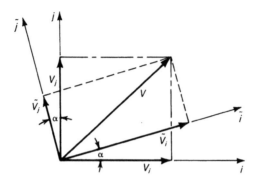

Figure 11.1. Rotational transformation of coordinates.

the *ij* plane have components v_i and v_j along the coordinate directions *i* and *j*. Let $\tilde{i}$ and $\tilde{j}$ represent new coordinate directions inclined at an angle α to the original coordinates as shown in Figure 11.1,   and let $\tilde{v}_i$ and $\tilde{v}_j$ represent the components of the vector in the inclined coordinates; then

$$\begin{bmatrix} v_i \\ v_j \end{bmatrix} = \begin{bmatrix} \cos\alpha & -\sin\alpha \\ \sin\alpha & \cos\alpha \end{bmatrix} \begin{bmatrix} \tilde{v}_i \\ \tilde{v}_j \end{bmatrix} \tag{11.23a}$$

$$\mathbf{v} = \mathbf{C}\tilde{\mathbf{v}} \tag{11.23b}$$

where **C** is the transformation matrix that represents the rotation. Comparison of Equations 11.22 and 11.23 shows the validity of interpreting matrix **N** as representing a rotation of the *i*, *j* degree of freedom.

When the orthogonal transformation $\tilde{\mathbf{A}} = \mathbf{N}^T \mathbf{A} \mathbf{N}$ is evaluated, the elements on the rows and columns *i* and *j* of matrix $\tilde{\mathbf{A}}$ are given by

$$\tilde{a}_{ii} = a_{ii}\cos^2\alpha + a_{jj}\sin^2\alpha + 2a_{ij}\sin\alpha\cos\alpha$$

$$j > i \tag{11.24}$$

$$\tilde{a}_{jj} = a_{ii}\sin^2\alpha + a_{jj}\cos^2\alpha - 2a_{ij}\sin\alpha\cos\alpha$$

and

$$\tilde{a}_{ij} = \tilde{a}_{ji}$$

$$= (a_{jj} - a_{ii})\sin\alpha\cos\alpha + a_{ij}(\cos^2\alpha - \sin^2\alpha) \tag{11.25}$$

By equating the off-diagonal term to zero, we obtain

$$\tan 2\alpha = \frac{2a_{ij}}{a_{ii} - a_{jj}} \quad a_{ii} \neq a_{jj} \tag{11.26a}$$

$$\alpha = \frac{\pi}{4} \quad a_{ii} = a_{jj} \tag{11.26b}$$

Each transformation affects all elements in rows and columns i and j. Apparently, therefore, an off-diagonal element reduced to zero during a transformation will become nonzero during subsequent transformations. The process is, in fact, iterative. With repeated iterations, the off-diagonal elements reduce in magnitude until they are negligible in comparison to the diagonal values. After a sufficient number of iterations, the transformed matrix becomes diagonal. The eigenvalues of the diagonal matrix are equal to the elements on the diagonal. Since the transformations do not alter the eigenvalues, the eigenvalues of the diagonal matrix are also those of the original. The eigenvectors of the diagonal matrix are unit vectors, hence $\tilde{\mathbf{Q}} = \mathbf{I}$, the identity matrix. The eigenvectors of the original matrix are obtained by using the relationship

$$\mathbf{Q} = \mathbf{N}_1 \mathbf{N}_2 \cdots \mathbf{N}_m \tilde{\mathbf{Q}} \tag{11.27}$$

In the standard Jacobi diagonalization procedure, the off-diagonal elements with the largest absolute value are chosen for elimination at each step. In a computer implementation, the process of finding the largest element is likely to be quite time consuming, adding to the cost of computation. A variation of the process in which the algorithm does not involve such a search is therefore generally preferred. One possible alternative is to eliminate off-diagonal elements in a systematic order, for example, $(i, j) = (1, 2)$, $(1, 3), \ldots, (1, N)$ followed by $(2, 3)$, $(2, 4), \ldots, (2, N)$, and so on. This process, called *serial Jacobi*, still converges to the right solution, although the convergence may be slower.

In another variation, a threshold value is established, and although the elements are eliminated in a serial order, elimination is skipped if the modulus of the off-diagonal element under consideration lies below the threshold. After a complete pass through the matrix, the threshold value is lowered and the process repeated until the transformed matrix is nearly diagonal. Since the objective is to reduce the magnitude of off-diagonal elements in comparison to those of the diagonal, comparison of element (i, j) with a threshold is usually accomplished by using the following inequality

$$\left[\frac{(a_{ij})^2}{a_{ii} a_{jj}} \right]^{1/2} < 10^{-s} \tag{11.28}$$

where s is a measure of the desired convergence.

For its solution by the standard Jacobi diagonalization process, the eigenvalue problem of Equation 10.7 must first be reduced to the standard symmetric form given by Equation 10.5. However, a variation of the method, called the *generalized Jacobi method*, can be applied directly to Equation 10.7. The procedure consists of repeated simultaneous transformations of $\mathbf{K}$ and $\mathbf{M}$ matrices until they are both reduced to a diagonal form. Mathematically, the transformation can be expressed as

$$\mathbf{N}^T\mathbf{KNN}^{-1}\mathbf{q} = \lambda\mathbf{N}^T\mathbf{MNN}^{-1}\mathbf{q} \tag{11.29a}$$

or

$$\tilde{\mathbf{K}}\tilde{\mathbf{q}} = \lambda\tilde{\mathbf{M}}\tilde{\mathbf{q}} \tag{11.29b}$$

in which $\tilde{\mathbf{K}} = \mathbf{N}^T\mathbf{KN}$, $\tilde{\mathbf{M}} = \mathbf{N}^T\mathbf{MN}$, and $\mathbf{q} = \mathbf{N}\tilde{\mathbf{q}}$.

The transformation matrix $\mathbf{N}$ is given by

$$
\mathbf{N} = \begin{array}{c} \\ \\ i \\ \\ j \\ \\ \\ \end{array}
\begin{array}{cc} i & j \\ & \\ \left(\begin{array}{ccccccc} 1 & . & 0 & . & 0 & . & 0 \\ . & . & . & . & . & . & . \\ 0 & . & 1 & . & \gamma & . & 0 \\ . & . & . & . & . & . & . \\ 0 & . & \alpha & . & 1 & . & 0 \\ 0 & . & 0 & . & 0 & . & 1 \end{array}\right) \end{array} \tag{11.30}
$$

On carrying out the transformations and equating the off-diagonal elements $\tilde{k}_{ij}$ and $\tilde{m}_{ij}$ of the transformed stiffness and mass matrices to zero, we get

$$
\begin{aligned}
k_{ii}\gamma + k_{jj}\alpha + k_{ij}(1 + \alpha\gamma) &= 0 \\
m_{ii}\gamma + m_{jj}\alpha + m_{ij}(1 + \alpha\gamma) &= 0
\end{aligned} \tag{11.31}
$$

Solution of Equation 11.31 gives

$$
\begin{aligned}
a &= k_{jj}m_{ij} - k_{ij}m_{jj} \\
b &= k_{ii}m_{jj} - k_{jj}m_{ii} \\
c &= k_{ii}m_{ij} - k_{ij}m_{ii} \\
\alpha &= \frac{1}{2a}[b + \text{sign}(b)\sqrt{b^2 + 4ac}] \qquad a \neq 0 \\
\alpha &= -\frac{c}{b} \qquad\qquad\qquad\qquad\qquad a = 0 \\
\gamma &= -\alpha\frac{a}{c}
\end{aligned} \tag{11.32}
$$

where k_{ij} and m_{ij} represent elements of $\mathbf{K}$ and $\mathbf{M}$, respectively, and $i < j$. If a and b are simultaneously zero, α is taken as zero and $\gamma = -k_{ij}/k_{ii}$. In the expression for α, the square-root term is taken with the sign of b to avoid having to take the difference of two quantities that may be of similar magnitude. This improves the numerical stability of the method.

Use of the generalized Jacobi method avoids the need for transforming the linear eigenvalue problem to a standard form which may involve a substantial amount of computations and, in addition, may introduce numerical inaccuracies when the matrices are ill conditioned. In particular, when the off-diagonal elements of the matrices are small, conversion to a standard form will give a matrix with a combination of very large and very small numbers. Application of the standard Jacobi technique is, in such cases, likely to result in numerical inaccuracies. Finally, when the stiffness and mass matrices are banded so that the off-diagonal element are already zero, use of the generalized Jacobi method for reduction of the matrices to a diagonal form is very efficient. Conversion to a standard form on the other hand, destroys the bandedness of the matrices so that no advantage can be taken of their sparsity.

Example 11.1
Use Jacobi's method to determine the eigenvalue of the symmetric matrix **A** given below.

$$
\mathbf{A} = \begin{bmatrix}
2640 & 230 & -1330 & 0 \\
230 & 770 & 0 & 0 \\
-1330 & 0 & 2640 & -230 \\
0 & 0 & -230 & 770
\end{bmatrix}
$$

Solution
The standard Jacobi transformation given by Equations 11.22 and 11.26 is used. In forming the transformation matrix (Eq. 11.22), we need $\sin\alpha$ and $\cos\alpha$. These are obtained directly from Equation 11.26 by using the following expressions

$$\cos^2\alpha = 0.5 + \frac{a_{ii} - a_{jj}}{2R}$$

$$\sin^2\alpha = 0.5 - \frac{a_{ii} - a_{jj}}{2R} \tag{a}$$

$$\sin\alpha\cos\alpha = \frac{a_{ij}}{R}$$

in which

$$R = [(a_{ii} - a_{jj})^2 + 4a_{ij}^2]^{1/2}$$

Function $\sin\alpha$ is chosen as positive so that $\cos\alpha$ takes the sign of a_{ij}. At each iteration step, the off-diagonal element with the largest absolute value is reduced to zero. The iteration steps are listed below.

Iteration no.	Row i	Col j	$\cos\alpha$	$\sin\alpha$	Transformed matrix $N^T A N$			
1	1	3	−0.7071	0.7071	3970.00	−162.63	0	−162.63
					−162.63	770.00	−162.63	0
					0	−162.63	1310.00	162.63
					−162.63	0	162.63	770.00
2	1	4	−0.99872	0.05063	3978.20	162.40	8.20	0
					162.4	770.00	−162.63	8.20
					8.20	−162.63	1310.00	−162.40
					0	8.20	−162.40	761.80
3	2	3	−0.26777	0.96348	3978.20	−35.56	−158.70	0
					−35.56	1355.20	0	−158.70
					−158.70	0	724.80	35.56
					0	−158.70	35.56	761.80
⋮	⋮	⋮	⋮	⋮	⋮			
11	2	3	−0.999994	0.011111	3986.40	0.06	0	0
					0.06	1394.70	0	0
					0	0	753.55	−0.06
					0	0	−0.06	685.32

The largest off-diagonal element in the original matrix is in row 1 and column 3, its absolute magnitude being 1330. In the first iteration step, element a_{13} and a_{31} are reduced to zero. In the second iteration step, a_{14} and a_{41} are reduced to zero. It should be noted that following the latter transformation, element a_{13} and a_{31} are restored to nonzero values. At the end of 11 iteration cycles, the off-diagonal elements are all zero up to the first decimal digit. The diagonal elements at this stage give the desired eigenvalues. Thus

$$\lambda_1 = 685.32$$

$$\lambda_2 = 753.55$$

$$\lambda_3 = 1394.7$$

$$\lambda_4 = 3986.4$$

If desired, the eigenvectors can be obtained from Equation 11.11. Substitution of $N_1, N_2, \ldots, N_{11}$ and $\tilde{Q} = I$ gives

$$Q = \begin{bmatrix} -0.2443 & -0.0504 & 0.6636 & 0.7053 \\ 0.6636 & 0.7053 & 0.2443 & 0.0504 \\ -0.2443 & 0.0504 & 0.6636 & -0.7053 \\ -0.6636 & 0.7053 & -0.2443 & 0.0504 \end{bmatrix}$$

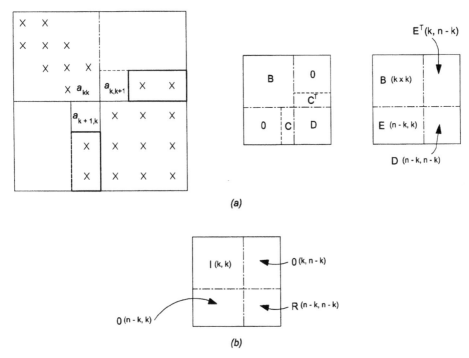

Figure 11.2. Householder's transformation: (a) matrix A; (b) transformation matrix N.

11.4.2 *Householder's transformation*

Householder's method reduces a symmetric matrix to tridiagonal form by the application of a series of orthogonal transformation. The eigenvalues and eigenvectors of the tridiagonal matrix are then computed by either the LR or the QR method, to be described later. The eigenvalues so computed are the same as those of the original matrix, while the eigenvectors of the tridiagonal matrix are related to those of the original by a relationship of the form of Equation 11.21b. Since Householder's method can be applied to a symmetric matrix only, the linearized eigenvalue problem must first be converted to a standard symmetric form.

To illustrate the method, consider Figure 11.2, which shows the transformed matrix **A** after reduction has been completed to row $k - 1$. At this stage, the matrix is tridiagonal from row 1 to $k - 1$. In the kth transformation, we wish to reduce to zero all nonzero elements in row k as well as column k except element $\bar{a}_{k,k-1}$, $\bar{a}_{k,k}$, $\bar{a}_{k,k+1}$, $\bar{a}_{k-1,k}$, and $\bar{a}_{k+1,k}$, where the bar on a indicates that the current value of the element is taken. The elements to be reduced to zero are enclosed in boxes in Figure 11.2. The kth transformation is given by

$$\mathbf{A}_{k+1} = \mathbf{N}_k^T \mathbf{A}_k \mathbf{N}_k \tag{11.33}$$

The transformation matrix N_k is shown in a partitioned form in Figure 11.2b. It consists of submatrices $\mathbf{I}$ (the identity matrix) and $\mathbf{R}$ and two null matrices. The submatrix $\mathbf{R}$ is selected to be of the form

$$\mathbf{R} = \mathbf{I} - 2\mathbf{w}\mathbf{w}^T \tag{11.34}$$

where $\mathbf{w}$ is a column vector of size $N - k$ with Euclidean norm equal to unity, so that

$$\mathbf{w}^T\mathbf{w} = 1 \tag{11.35}$$

By taking the transpose of Equation 11.34, it is seen that $\mathbf{R}^T = \mathbf{R}$. It follows that $\mathbf{N}^T = \mathbf{N}$. Thus, both $\mathbf{R}$ and $\mathbf{N}$ are symmetric. Further

$$\begin{aligned} \mathbf{R}\mathbf{R}^T &= (\mathbf{I} - 2\mathbf{w}\mathbf{w}^T)(\mathbf{I} - 2\mathbf{w}\mathbf{w}^T) \\ &= \mathbf{I} - 4\mathbf{w}\mathbf{w}^T + 4\mathbf{w}\mathbf{w}^T\mathbf{w}\mathbf{w}^T \end{aligned} \tag{11.36}$$

Substitution of Equation 11.35 into Equation 11.36 gives

$$\mathbf{R}\mathbf{R}^T = \mathbf{I} \tag{11.37}$$

By using the relationship in Equation 11.37, it can be proved that $\mathbf{N}\mathbf{N}^T = \mathbf{I}$. This implies that the transformation specified by Equation 11.33 is orthogonal.

Expansion of the basic transformation Equation 11.33 using the partitioned forms of $\mathbf{A}$ and $\mathbf{N}$ shown in Figure 11.2 gives

$$\mathbf{A}_{k+1} = \left[\begin{array}{c|cc} \mathbf{B} & \multicolumn{2}{c}{\mathbf{0}} \\ & \multicolumn{2}{c}{\mathbf{c}^T\mathbf{R}} \\ \hline \mathbf{0} & \mathbf{R}\mathbf{c} & \mathbf{R}\mathbf{D}\mathbf{R} \end{array} \right] \tag{11.38}$$

We note from Equation 11.38 that the transformation leaves the submatrix $\mathbf{B}$ unaltered. Thus, the computations in the kth step do not change that portion of $\mathbf{A}$ which has already been tridiagonalized.

To determine $\mathbf{w}$ and hence $\mathbf{R}$, we note that the matrix product $\mathbf{R}\mathbf{c}$ should give a vector all of whose elements except the first are zero. Thus

$$\mathbf{R}\mathbf{c} = \mathbf{g} \tag{11.39a}$$

or

$$[\mathbf{I} - 2\mathbf{w}\mathbf{w}^T]\mathbf{c} = \mathbf{g} \tag{11.39b}$$

where $\mathbf{g}^T = [r \quad 0 \quad 0 \quad \cdots \quad 0], r$ is as yet undermined and

$$\mathbf{c}^T = [\bar{a}_{k,k+1} \quad \bar{a}_{k,k+2} \quad \cdots \quad \bar{a}_{k,N}] \tag{11.40}$$

From Equation 11.39a, we have

$$\mathbf{g}^T\mathbf{g} = \mathbf{c}^T\mathbf{R}^T\mathbf{R}\mathbf{c} \tag{11.41a}$$

or

$$r^2 = \mathbf{c}^T\mathbf{c} \tag{11.41b}$$

Now premultiplying both sides of Equation 11.39b by $\mathbf{c}^T$ and denoting the scalar quantity $\mathbf{w}^T\mathbf{c} = \mathbf{c}^T\mathbf{w}$ by h, we get

$$\mathbf{c}^T\mathbf{c} - 2h^2 = \mathbf{c}^T\mathbf{g} \tag{11.42}$$

Noting that $\mathbf{c}^T\mathbf{g} = \bar{a}_{k,k+1}r$ and using Equation 11.41b, Equation 11.42 reduces to

$$2h^2 = r^2 - \bar{a}_{k,k+1}r \tag{11.43}$$

Equations 11.41b and 11.43 provide the values of r and h. The sign of r is chosen so that the product $\bar{a}_{k,k+1}r$ in Equation 11.43 is negative. Now using Equation 11.39b and that $\mathbf{w}^T\mathbf{c} = h$

$$2h\mathbf{w} = \mathbf{c} - \mathbf{g} \equiv \mathbf{v} \tag{11.44}$$

Finally, Equation 11.34 leads to the transformation matrix $\mathbf{R}$

$$\mathbf{R} = \mathbf{I} - \frac{1}{2h^2}\mathbf{v}\mathbf{v}^T \tag{11.45}$$

in which $\mathbf{v} = \mathbf{c} - \mathbf{g}$ and $2h^2$ is given by Equation 11.43. The Householder transformation begins with $\mathbf{A}_1 = \mathbf{A}$ and the vector $\mathbf{c}$ given by

$$\mathbf{c}^T = [a_{12} \quad a_{13} \quad \cdots \quad a_{1N}] \tag{11.46}$$

It should be noted that at each step, the transformed matrix $\mathbf{A}_k$ remains symmetric and hence only half of the matrix need to be kept in the computer storage. However, in the case of a banded matrix, the bandwidth of the unreduced part during an intermediate steps may be larger than that of the original matrix.

Example 11.2
Tridiagonalize the matrix in Example 11.1 by using Householder's method.

Solution
In the first step of transformation, elements a_{31}, a_{41} and a_{13}, a_{14} are reduced to zero. The steps involved in the computations are as follows

$$\mathbf{c}_1 = \begin{bmatrix} 230.0 \\ -1330.0 \\ 0.0 \end{bmatrix} \tag{a}$$

$$r_1^2 = \mathbf{c}_1^T \mathbf{c}_1$$

$$= 1,821,800 \tag{b}$$

$$2h_1^2 = r_1^2 - a_{12}r_1$$

$$= 1,821,800 - 230 \times (-1349.74)$$

$$= 2,132,240 \tag{c}$$

$$\mathbf{g}_1 = \begin{bmatrix} -1349.74 \\ 0 \\ 0 \end{bmatrix} \tag{d}$$

$$\mathbf{v}_1 = \mathbf{c}_1 - \mathbf{g}_1$$

$$= \begin{bmatrix} 1579.74 \\ -1330.00 \\ 0 \end{bmatrix} \tag{e}$$

$$\frac{\mathbf{v}_1 \mathbf{v}_1^T}{2h_1^2} = \begin{bmatrix} 1.17040 & -0.98537 & 0 \\ -0.098537 & 0.82960 & 0 \\ 0 & 0 & 0 \end{bmatrix} \tag{f}$$

$$\mathbf{R}_1 = \mathbf{I} - \frac{1}{2h_1^2}\mathbf{v}_1 \mathbf{v}_1^T$$

$$= \begin{bmatrix} -0.170400 & -0.98537 & 0 \\ -0.098537 & 0.17040 & 0 \\ 0 & 0 & 1 \end{bmatrix} \tag{g}$$

The transformed matrix obtained from Equation 11.38 becomes

$$\mathbf{A}_2 = \begin{bmatrix} 2640.0 & -1349.7 & 0 & 0 \\ -1349.7 & 2585.7 & 314.0 & -226.6 \\ 0 & 314.0 & 824.3 & -39.2 \\ 0 & -226.6 & -39.2 & 770.0 \end{bmatrix}$$

In the second step

$$\mathbf{c}_2 = \begin{bmatrix} 314.0 \\ -226.6 \end{bmatrix}$$

$$r_2^2 = 149,937$$

$$2h_2^2 = 271,547$$

$$\mathbf{v}_2 = \begin{bmatrix} 701.23 \\ -226.64 \end{bmatrix}$$

$$\mathbf{R}_2 = \begin{bmatrix} -0.81085 & 0.58526 \\ 0.58526 & 0.81085 \end{bmatrix}$$

At the end of second step, the transformed matrix, which is now tridiagonal, is given by

$$\mathbf{A}_3 = \begin{bmatrix} 2640.0 & -1349.7 & 0 & 0 \\ -1349.7 & 2585.7 & -387.24 & 0 \\ 0 & -387.24 & 842.90 & -13.43 \\ 0 & 0 & -13.42 & 751.40 \end{bmatrix}$$

A major portion of the computations involved in Householder's method are those required for the evaluation of matrix product **RDR**. For general matrices, the triple matrix product will need $2m^3$ multiplications, where m is the order of each of the matrices. However, because of the special nature of **R**, the number of multiplications can be reduced substantially by using the following steps in the calculations

$$\mathbf{s} = \mathbf{Dw}$$

$$\mathbf{p}^T = 2\mathbf{s}^T$$

$$\beta = \mathbf{p}^T \mathbf{w} \qquad\qquad (11.47)$$

$$\mathbf{q} = \mathbf{p} - 2\beta\mathbf{w}$$

$$\mathbf{RDR} = \mathbf{D} - \mathbf{wp}^T - \mathbf{qw}^T$$

The validity of the series of steps in Equation 11.47 can be verified by direct substitution. The number of multiplications is now $3m^2 + 3m$. For large m, this is substantially smaller than $2m^3$.

11.4.3   *QR transformation*

The *QR transformation*, which derives its name from the notations used in the transformation, can be applied to any symmetric matrix to reduce the latter to a diagonal form. However, it is usually more efficient first to reduce the matrix to a tridiagonal form by using Householder's method and then to apply the QR transformation to reduce the tridiagonal matrix to a diagonal form. As described earlier, the eigenvalues and eigenvectors of the diagonal matrix are obtained by inspection. In the following paragraph, we describe the application of QR method to the reduction of a tridiagonal matrix. Extension to the case of a more general symmetric matrix is straightforward.

The QR transformation is essentially iterative in nature. In the kth iteration step, matrix $\mathbf{A}_k$ is transformed to $\mathbf{A}_{k+1}$. This is done in two steps. Matrix $\mathbf{A}_k$ is first transformed to an upper triangular form. This is achieved by premultiplying $\mathbf{A}_k$ by a series of orthogonal matrices $\mathbf{N}_1^T, \mathbf{N}_2^T, \ldots, \mathbf{N}_i^T, \ldots, \mathbf{N}_{N-1}^T$. Matrix $\mathbf{N}_i$ is given by

$$
\mathbf{N}_i^T = \begin{array}{c} \\ i \\ i+1 \\ \\ \end{array}
\begin{pmatrix}
1 & 0 & . & 0 & 0 & . & 0 \\
 & . & . & . & . & . & . \\
0 & 0 & . & \cos\alpha & \sin\alpha & . & 0 \\
0 & 0 & . & -\sin\alpha & \cos\alpha & . & 0 \\
 & . & . & . & . & . & . \\
0 & 0 & . & 0 & 0 & . & 1
\end{pmatrix}
\qquad (11.48)
$$

With α chosen appropriately, premultiplication by $\mathbf{N}_i^T$ reduces the subdiagonal term $a_{i+1,i}$ in the current version of $\mathbf{A}$ to zero. Also, the subdiagonal terms already reduced to zero during earlier operations are unaltered. For ease of reference, we represent $\mathbf{N}_{i-1}^T, \ldots, \mathbf{N}_2^T \mathbf{N}_1^T \mathbf{A}_k$ by $\mathbf{A}_k^{(i)}$. The subdiagonal terms up to row i in matrix $\mathbf{A}_k^{(i)}$, have already been reduced to zero. On forming the product $\mathbf{N}_i^T \mathbf{A}_k^{(i)}$ and equating the term $a_{i+1,i}$ to zero, we get

$$
-\bar{a}_{ii} \sin\alpha + \bar{a}_{i+1,i} \cos\alpha = 0 \qquad (11.49)
$$

so that

$$
\tan\alpha = \frac{\bar{a}_{i+1,i}}{\bar{a}_{ii}} \qquad (11.50)
$$

where the bar on an element a indicates that its value from $\mathbf{A}_k^{(i)}$ has been used.

On completing the full series of transformations using $\mathbf{N}_1^T, \mathbf{N}_2^T, \ldots, \mathbf{N}_{N-1}^T$, we get

$$
\mathbf{N}_{N-1}^T \cdots \mathbf{N}_2^T \mathbf{N}_1^T \mathbf{A}_k = \mathbf{R}_k \qquad (11.51)
$$

or

$$\mathbf{Q}_k^T \mathbf{A}_k = \mathbf{R}_k \tag{11.52}$$

where $\mathbf{R}_k$ is an upper triangular matrix and

$$\mathbf{Q}_k^T = \mathbf{N}_{N-1}^T \cdots \mathbf{N}_2^T \mathbf{N}_1^T \tag{11.53}$$

Since each of the matrices $\mathbf{N}_1^T$, $\mathbf{N}_2^T, \ldots, \mathbf{N}_{N-1}^T$ are orthogonal, $\mathbf{Q}_k^T$ is also orthogonal, that is, $\mathbf{Q}_k^T = \mathbf{Q}_k^{-1}$.

The transformation to $\mathbf{A}_{k+1}$ is now completed by using the expression

$$\mathbf{A}_{k+1} = \mathbf{R}_k \mathbf{Q}_k \tag{11.54}$$

where

$$\mathbf{Q}_k = \mathbf{N}_1, \mathbf{N}_2, \ldots, \mathbf{N}_{N-1} \tag{11.55}$$

Substitution of Equation 11.52 into Equation 11.54 gives

$$\mathbf{A}_{k+1} = \mathbf{Q}_k^T \mathbf{A}_k \mathbf{Q}_k \tag{11.56}$$

It is clear from Equation 11.56 that the transformation from $\mathbf{A}_k$ to $\mathbf{A}_{k+1}$ is orthogonal under which the eigenvalues remain unchanged.

The matrix product on the left-hand side of Equation 11.51 strips all subdiagonal terms in $\mathbf{A}_k$ so that the product matrix $\mathbf{R}_k$ is an upper tridiagonal matrix. However, when the kth iteration is completed by carrying out the matrix operations indicated in Equation 11.54, the subdiagonal elements will, in general, be restored to nonzero values, but their magnitudes would have reduced. Eventually, after a sufficient number of iterations, matrix $\mathbf{A}$ would become almost diagonal, the elements on the diagonal being equal to the eigenvalues of the original tridiagonal matrix. If required, the eigenvectors of the tridiagonal matrix can be obtained by using an expression similar to Equation 11.21b

$$\mathbf{T} = \mathbf{Q}_1 \mathbf{Q}_2 \cdots \mathbf{Q}_k \mathbf{I} \tag{11.57}$$

In Equation 11.57, $\mathbf{I}$, the identity matrix, is the matrix of the eigenvectors of a diagonal matrix, $\mathbf{T}$ is the matrix of eigenvectors of the tridiagonal matrix, $\mathbf{Q}_1$, $\mathbf{Q}_2, \ldots, \mathbf{Q}_k$ are the orthogonal transformation matrices, and it is assumed that convergence has been achieved after k iterations. In practice, it is more efficient to evaluate the eigenvectors by using the inverse iteration method described later. In that case, only those eigenvectors that are of interest need to be evaluated. When a combination of the Householder reduction, QR transformation and inverse iteration is used for the evaluation of eigenvalues and eigenvectors, the method is commonly referred to as the *HQRI method*.

In the QR iteration procedure described above, eigenvalues are obtained in increasing order of magnitude. The lowest eigenvalue, λ_1, is obtained first and occupies the position $\bar{a}_{NN}$. When this element does not change significantly with successive iterations, $\bar{a}_{NN}$ has converged to λ_1. At this stage, if desired, the matrix can be deflated by removing the Nth row and Nth column and the iterations continued on the resulting $N - 1$ by $N - 1$ matrix until the next eigenvalue λ_2 is obtained. This process is continued, the ith eigenvalue being obtained by iterations on a deflated matrix of size $N - i + 1$.

Assuming that eigenvalues $\lambda_1, \lambda_2, \ldots, \lambda_{i-1}$ have already been obtained, convergence to λ_i depends on the ratio $|\lambda_i/\lambda_{i+1}|$; the smaller this ratio, the faster the convergence. This suggests that convergence can be improved by using a modified procedure called QR iteration with shifted origin. The procedure is based on the fact that the eigenvalues $\tilde{\lambda}$ of the matrix $\tilde{\mathbf{A}} = \mathbf{A} - \mu\mathbf{I}$, where μ is any scalar value, are related to the eigenvalues of $\mathbf{A}$ by the expression

$$\tilde{\lambda}_j = \lambda_j - \mu \tag{11.58}$$

Equation 11.58 can easily be proved by noting that

$$\tilde{\mathbf{A}}\mathbf{q} = \mathbf{A}\mathbf{q} - \mu\mathbf{q}$$

$$= (\lambda - \mu)\mathbf{q} \tag{11.59}$$

In the kth iteration of the QR method for obtaining λ_i, an estimate, even though approximate, has already been obtained for the desired eigenvalue. The estimated value denoted by μ_k may be taken as equal to the element a_{ii} of the current version of $\mathbf{A}$. The next transformation is now carried out on matrix $(\mathbf{A}_k - \mu_k\mathbf{I})$, so that

$$\tilde{\mathbf{Q}}_k^T(\mathbf{A}_k - \mu_k\mathbf{I}) = \tilde{\mathbf{R}}_k \tag{11.60}$$

$$\tilde{\mathbf{A}}_{k+1} = \tilde{\mathbf{R}}_k\tilde{\mathbf{Q}}_k \tag{11.61}$$

The transferred matrix is then restored by removing the shift. Thus

$$\mathbf{A}_{k+1} = \tilde{\mathbf{A}}_{k+1} + \mu_k\mathbf{I} \tag{11.62}$$

A new estimate is now obtained for λ_i. If convergence has not been achieved, the iteration process (Eqs. 11.60, 11.61, and 11.62) is repeated. It is evident that if the shift is close to the value of λ_i, the ratio $|\tilde{\lambda}_i/\tilde{\lambda}_{i+1}|$ will be small and the convergence will be speeded up significantly.

Example 11.3

Reduce the following tridiagonal matrix to a diagonal form by using QR iteration

$$A = \begin{bmatrix} 3 & -2 & 0 \\ -2 & 8 & -4 \\ 0 & -4 & 5 \end{bmatrix}$$

Solution

The first tridiagonal matrix Q_1 in the series of orthogonal transformation to be performed on A is itself the product of two other matrices and is given by

$$Q_1 = N_1 N_2 \tag{a}$$

in which N_1 is selected so that in $N_1^T A$, the subdiagonal element a_{21} is zero. Matrix N_2 eliminates element a_{32} without affecting a_{21}, so that in the product $N_2^T N_1^T A$ both a_{21} and a_{32} are zero. Computations for obtaining N_1 and N_2 are tabulated below. In each case, matrix N is given by Equation 11.48 with α determined from Equation 11.50. Using the later equation, the following expressions are obtained for $\cos \alpha$ and $\sin \alpha$

$$\cos \alpha = \frac{\bar{a}_{ii}}{\sqrt{\bar{a}_{ii}^2 + \bar{a}_{i+1,i}^2}}$$

$$\tag{b}$$

$$\sin \alpha = \frac{\bar{a}_{i+1,i}}{\sqrt{\bar{a}_{ii}^2 + \bar{a}_{i+1,i}^2}}$$

i	$\cos \alpha$	$\sin \alpha$	Transformed matrix, $A^{(i+1)} = N_i^T A^{(i)}$
1	0.83205	-0.55470	$\begin{bmatrix} 3.606 & -6.102 & 2.219 \\ 0 & 5.547 & -3.328 \\ 0 & -4.000 & 5.000 \end{bmatrix}$
2	0.81111	-0.58490	$\begin{bmatrix} 3.606 & -6.102 & 2.219 \\ 0 & 6.839 & -5.624 \\ 0 & 0 & 2.109 \end{bmatrix}$

Note that in the above, $A^{(1)} = A$ and the matrix $R_1 = A^{(3)}$ obtained at the end of second step is an upper triangular matrix. The first transformation matrix Q_1 is now given by

$$Q_1 = \bar{N}_1 N_2$$

$$= \begin{bmatrix} 0.83205 & 0.44992 & 0.32444 \\ -0.55470 & 0.67468 & 0.48666 \\ 0 & -0.58490 & 0.81111 \end{bmatrix}$$

The first transformation is completed by carrying out the matrix multiplications in

$$\mathbf{A}_2 = \mathbf{Q}_1^T \mathbf{A} \mathbf{Q}_1$$

$$= \mathbf{R}_1 \mathbf{Q}_1 \tag{c}$$

which gives

$$\mathbf{A}_2 = \begin{bmatrix} 6.385 & -3.794 & 0 \\ -3.794 & 7.905 & -1.235 \\ 0 & -1.234 & 1.711 \end{bmatrix}$$

At the end of eight transformations, we will get

$$\mathbf{A}_9 = \begin{bmatrix} 11.1110 & -0.0033 & 0 \\ -0.0033 & 3.5820 & -0.0008 \\ 0 & -0.0008 & 1.3060 \end{bmatrix}$$

At this stage, we have nearly converged to the lowest eigenvalue of 1.306, and if accuracy to about three decimal digits is considered adequate, the matrix can be deflated by omitting the last row and the last column. However, in this case, we will continue our iteration on the complete matrix. Four figure accuracy is achieved at the end of 12 transformation, giving

$$\mathbf{A}_{13} = \begin{bmatrix} 11.111 & 0 & 0 \\ 0 & 3.582 & 0 \\ 0 & 0 & 1.306 \end{bmatrix}$$

The eigenvalues of the given matrix are thus

$$\lambda_1 = 1.306$$

$$\lambda_2 = 3.582$$

$$\lambda_3 = 11.111$$

11.5   ITERATION METHODS

Iteration methods are ideally suited for eigenvalue problems of large size with sparse or banded matrices, particularly when only some of the eigenvalues, either the lowest few or the highest few, are required. The following iteration methods have been effectively applied to practical solutions of eigenvalue problems in the vibration of engineering structures:
1. Vector iteration.
2. Vector iteration with shift.
3. Subspace iteration.
4. Lanczos method.

A brief description of each of the methods is provided in the following paragraphs.

11.5.1 *Vector iteration*

Direct vector iteration or simply vector iteration is also known as *the power method.* It allows successive evaluation of as many eigenvalues and eigenvectors as required, starting with the eigenvalue with the largest absolute value. Considering the standard eigenvalue problem, assume that the eigenvalues have been arranged so that $|\lambda_1| < |\lambda_2| < |\lambda_3| < \cdots < |\lambda_N|$. Begin with an arbitrary vector $\mathbf{u}$ and let it be expressed in the form of Equation 11.9. Premultiplication of Equation 11.9 by $\mathbf{A}$ gives

$$\begin{aligned} \mathbf{Au} &= c_1 \mathbf{Aq}_1 + c_2 \mathbf{Aq}_2 + \cdots + c_N \mathbf{Aq}_N \\ &= c_1 \lambda_1 \mathbf{q}_1 + c_2 \lambda_2 \mathbf{q}_2 + \cdots + c_N \lambda_N \mathbf{q}_N \end{aligned} \tag{11.63}$$

It is easily seen that after k multiplications by $\mathbf{A}$ Equation 11.9 will give

$$\begin{aligned} \mathbf{A}^k \mathbf{u} &= c_1 \lambda_1^k \mathbf{q}_1 + c_2 \lambda_2^k \mathbf{q}_2 + \cdots + c_N \lambda_N^k \mathbf{q}_N \\ &= \lambda_N^k \left[c_1 \left(\frac{\lambda_1}{\lambda_N} \right)^k \mathbf{q}_1 + c_2 \left(\frac{\lambda_2}{\lambda_N} \right)^k \mathbf{q}_2 + \cdots \right. \\ &\qquad \left. + c_{N-1} \left(\frac{\lambda_{N-1}}{\lambda_N} \right)^k \mathbf{q}_{N-1} + c_N \mathbf{q}_N \right] \end{aligned} \tag{11.64}$$

Now because $|\lambda_i/\lambda_N| < 1$ for $i = 1, 2, 3, \ldots, N-1$, in the limit as k becomes large

$$\mathbf{u}_k = \mathbf{A}^k \mathbf{u} \simeq c_N \lambda_N^k \mathbf{q}_N \tag{11.65a}$$

and

$$\mathbf{u}_{k+1} = \mathbf{A}^{k+1} \mathbf{u} \simeq c_N \lambda_N^{k+1} \mathbf{q}_N \tag{11.65b}$$

The rate of convergence depends on the ratio λ_{N-1}/λ_N; the smaller this ratio, the faster the rate of convergence.

The product vector obtained from either Equations 11.65a or 11.65b is proportional to the eigenvector that corresponds to the eigenvalue with the largest magnitude. Also, the ratio of the value of an element in the $(k+1)$th iteration to the value of the same element in the kth iteration approaches λ_N as k becomes large. Rather than using the ratio of the two successive values of an arbitrarily selected element of the product vector, the following alternative expression may be used to obtain a better estimate of the eigenvalue after $k+1$ iterations:

$$\lambda_N \approx \rho_{k+1} = \frac{\mathbf{u}_{k+1}^T \mathbf{u}_{k+1}}{\mathbf{u}_{k+1}^T \mathbf{u}_k} \tag{11.66}$$

Equation 11.66 can easily be shown to be true by substituting for $\mathbf{u}_{k+1}$ and $\mathbf{u}_k$ from Equations 11.65b and 11.65a respectively.

In practice, because the magnitude of an eigenvector is arbitrary, it is convenient to normalize the product vector at the end of each iteration. In fact, with each multiplication by $\mathbf{A}$, the elements of the product vector grow or decrease in magnitude, and this growth or decrease may cause numerical problems in computations. Scaling of the vector after each iteration keeps element values within reasonable bounds. Scaling can be achieved either through a division by the largest element of the vector or by normalizing the vector so that its Euclidean norm is unity. When the latter method is adopted, the procedure is represented by the following steps

$$\bar{\mathbf{u}}_{k+1} = \mathbf{A}\mathbf{u}_k \tag{11.67}$$

$$\rho_{k+1} = \frac{\bar{\mathbf{u}}_{k+1}^T \bar{\mathbf{u}}_{k+1}}{\bar{\mathbf{u}}_{k+1}^T \mathbf{u}_k} \tag{11.68}$$

$$\mathbf{u}_{k+1} = \frac{\bar{\mathbf{u}}_{k+1}}{(\bar{\mathbf{u}}_{k+1}^T \bar{\mathbf{u}}_{k+1})^{1/2}} \tag{11.69}$$

Equation 11.68 is a Rayleigh quotient estimate of the eigenvalue, and Equation 11.69 represents normalization of the vector $\bar{\mathbf{u}}_{k+1}$, so that the normalized vector $\mathbf{u}_{k+1}$ has a Euclidean norm of unity.

Example 11.4

Using direct vector iteration, determine the eigenvalue with the largest absolute magnitude and the corresponding eigenvector of matrix $\mathbf{A}$ given by

$$\mathbf{A} = \begin{bmatrix} -64 & 48 & 36 \\ -136 & 102 & 76 \\ 72 & -54 & -40 \end{bmatrix}$$

Solution

We use the iteration procedure defined by Equations 11.67, 11.68, and 11.69, using for our initial trial vector a vector all of whose elements are equal to unity. The computations are tabulated as follows.

k	$\mathbf{u}_k = \dfrac{\bar{\mathbf{u}}_k}{\sqrt{\bar{\mathbf{u}}_k^T \bar{\mathbf{u}}_k}}$	$\bar{\mathbf{u}}_{k+1}^T = \mathbf{A}\mathbf{u}_k$	$\bar{\mathbf{u}}_{k+1}^T \bar{\mathbf{u}}_{k+1}$	$\bar{\mathbf{u}}_{k+1}^T \mathbf{u}_k$	$\rho = \dfrac{\bar{\mathbf{u}}_{k+1}^T \bar{\mathbf{u}}_{k+1}}{\bar{\mathbf{u}}_{k+1}^T \mathbf{u}_k}$
1	$\begin{bmatrix} 1 \\ 1 \\ 1 \end{bmatrix}$	$\begin{bmatrix} 20 \\ 42 \\ -22 \end{bmatrix}$	2648.0	40.0	66.2
2	$\begin{bmatrix} 0.3887 \\ 0.8162 \\ 0.4275 \end{bmatrix}$	$\begin{bmatrix} 29.69 \\ 62.88 \\ -33.19 \end{bmatrix}$	5936.8	48.68	77.05
3	$\begin{bmatrix} 0.3853 \\ 0.8161 \\ -0.4307 \end{bmatrix}$	$\begin{bmatrix} -0.9916 \\ -1.8918 \\ 0.9002 \end{bmatrix}$	5.373	-2.314	-2.322
4	$\begin{bmatrix} -0.4278 \\ -0.8162 \\ 0.3884 \end{bmatrix}$	$\begin{bmatrix} 2.184 \\ 4.447 \\ -2.263 \end{bmatrix}$	29.67	-5.443	-5.450
5	$\begin{bmatrix} 0.4010 \\ 0.8165 \\ -0.4155 \end{bmatrix}$	$\begin{bmatrix} -1.430 \\ 2.831 \\ 1.401 \end{bmatrix}$	12.02	-3.467	-3.467
6	$\begin{bmatrix} -0.4124 \\ -0.8165 \\ 0.4041 \end{bmatrix}$	$\begin{bmatrix} 1.748 \\ 3.515 \\ -1.766 \end{bmatrix}$	18.53	-4.305	-4.305
7	$\begin{bmatrix} 0.4061 \\ 0.8166 \\ -0.4103 \end{bmatrix}$	$\begin{bmatrix} -1.564 \\ -3.119 \\ 1.555 \end{bmatrix}$	14.59	-3.820	-3.820
8	$\begin{bmatrix} -0.4095 \\ -0.8164 \\ 0.4070 \end{bmatrix}$	$\begin{bmatrix} 1.673 \\ 3.351 \\ -1.678 \end{bmatrix}$	16.85	-4.104	-4.104
9	$\begin{bmatrix} 0.4076 \\ 0.8165 \\ -0.4089 \end{bmatrix}$	$\begin{bmatrix} -1.615 \\ -3.227 \\ 1.612 \end{bmatrix}$	15.62	-3.952	-3.952
10	$\begin{bmatrix} -0.4086 \\ -0.8165 \\ 0.4079 \end{bmatrix}$	$\begin{bmatrix} 1.643 \\ 3.287 \\ -1.644 \end{bmatrix}$	16.21	-4.026	-4.026

k	$\mathbf{u}_k = \frac{\bar{\mathbf{u}}_k}{\sqrt{\bar{\mathbf{u}}_k^T \bar{\mathbf{u}}_k}}$	$\bar{\mathbf{u}}_{k+1}^T = \mathbf{A}\mathbf{u}_k$	$\bar{\mathbf{u}}_{k+1}^T \bar{\mathbf{u}}_{k+1}$	$\bar{\mathbf{u}}_{k+1}^T \mathbf{u}_k$	$\rho = \frac{\bar{\mathbf{u}}_{k+1}^T \bar{\mathbf{u}}_{k+1}}{\bar{\mathbf{u}}_{k+1}^T \mathbf{u}_k}$
11	$\begin{bmatrix} 0.4081 \\ 0.8165 \\ -0.4084 \end{bmatrix}$	$\begin{bmatrix} -1.629 \\ -3.257 \\ 1.628 \end{bmatrix}$	15.912	−3.989	−3.989
12	$\begin{bmatrix} -0.4084 \\ -0.8165 \\ 0.4081 \end{bmatrix}$				

At this stage, convergence may be assumed to have been achieved. The eigenvalue is −3.99 and the corresponding eigenvector is given by

$$\mathbf{q}_1 = \begin{bmatrix} -0.4084 \\ -0.8165 \\ 0.4081 \end{bmatrix}$$

For comparison, the exact eigenvalue is $\lambda_1 = -4$ and the exact eigenvector is

$$\mathbf{q}_1 = \begin{bmatrix} -0.4082 \\ -0.8165 \\ 0.4082 \end{bmatrix}$$

11.5.2 *Inverse vector iteration*

The vector iteration method can be applied to the linearized eigenvalue problem by reducing to a standard form given by either Equation 11.12b or Equation 11.13b. When applied to the form of Equation 11.12b, iterations will converge to the most dominant value of λ, that is, to the highest frequency ω. When applied to Equation 11.13b, convergence will be to the largest value of γ, that is, to the lowest value of λ or ω.

In practical vibration problems, the lower frequencies and mode shapes are of greater interest, and therefore the form given by Equation 11.13b is preferred. Iteration on Equation 11.13b, which leads to the lowest value of ω, is usually referred to as *inverse vector iteration*. Inverse iteration involves repeated multiplication by matrix D, usually referred to as the *dynamic matrix*. In all other respects, the method is similar to the one described in Section 11.5.1. The algorithm can thus be summarized as

$$\bar{\mathbf{u}}_{k+1} = \mathbf{D}\mathbf{u}_k \tag{11.70}$$

$$\rho_{k+1} = \frac{\bar{\mathbf{u}}_{k+1}^T \mathbf{u}_k}{\bar{\mathbf{u}}_{k+1}^T \bar{\mathbf{u}}_{k+1}} \tag{11.71}$$

$$\mathbf{u}_{k+1} = \frac{\bar{\mathbf{u}}_{k+1}}{(\bar{\mathbf{u}}_{k+1}^T \bar{\mathbf{u}}_{k+1})^{1/2}} \tag{11.72}$$

Equation 11.71, which is the inverse of Equation 11.68, gives the reciprocal of the eigenvalue, so that $\rho_{k+1} = 1/\gamma = \lambda$.

For the purpose of numerical computations, it usually is more efficient to organize the inverse iteration on a linearized eigenvalue problem as follows

$$\mathbf{K}\bar{\mathbf{u}}_{k+1} = \mathbf{x}_k \tag{11.73}$$

$$\bar{\mathbf{x}}_{k+1} = \mathbf{M}\bar{\mathbf{u}}_{k+1} \tag{11.74}$$

$$\rho_{k+1} = \frac{\bar{\mathbf{u}}_{k+1}^T \mathbf{x}_k}{\bar{\mathbf{u}}_{k+1}^T \bar{\mathbf{x}}_{k+1}} \tag{11.75}$$

$$\mathbf{x}_{k+1} = \frac{\bar{\mathbf{x}}_{k+1}}{(\bar{\mathbf{u}}_{k+1}^T \bar{\mathbf{x}}_{k+1})^{1/2}} = \frac{\mathbf{M}\bar{\mathbf{u}}_{k+1}}{(\bar{\mathbf{u}}_{k+1}\mathbf{M}\bar{\mathbf{u}}_{k+1})^{1/2}} \tag{11.76}$$

Iteration is commenced with $\mathbf{x} = \mathbf{Mu}$, where $\mathbf{u}$ is an arbitrary trial vector. Equation 11.73 is solved for $\bar{\mathbf{u}}_{k+1}$, the solution being equivalent to multiplying the current version of the trial vector by $\mathbf{D} = \mathbf{K}^{-1}\mathbf{M}$. Equation 11.75 is a Rayleigh quotient estimate of the eigenvalue. Equation 11.76 normalizes the current value of the trial vector so that $\mathbf{x}_{k+1} = \mathbf{Mu}_{k+1}$ and $\mathbf{u}_{k+1}^T \mathbf{Mu}_{k+1} = 1$. When k is sufficiently large, ρ_{k+1} converges to $\lambda_1 = \omega_1^2$ and $\mathbf{u}_{k+1}$ converges to $\boldsymbol{\phi}_1$, so that

$$\lambda_1 = \rho_{k+1} \tag{11.77a}$$

$$\boldsymbol{\phi}_1 = \frac{\bar{\mathbf{u}}_{k+1}}{(\bar{\mathbf{u}}_{k+1}^T \bar{\mathbf{x}}_{k+1})^{1/2}} \tag{11.77b}$$

Convergence can be measured by comparing two successive values of λ. Thus

$$\frac{|\lambda_1^{(k+1)} - \lambda_1^{(k)}|}{\lambda_1^{(k+1)}} < \varepsilon \tag{11.78}$$

where ε is the specified tolerance. If $\varepsilon = 10^{-2s}$, the eigenvalue will be accurate to $2s$ digits, while the eigenvectors will be accurate to s digits.

It is evident that, when applied as described in the foregoing paragraphs, the inverse vector iteration will always converge to the least dominant eigenvalue and the corresponding eigenvector. Convergence to the second least dominant eigenvalue can, however, be achieved, provided that the trial vector is such that in its representation as a superposition of modes, the contribution of the first mode is nil. When such is the case, coefficient c_1 in Equation 11.9 is zero and the trial vector is said to be orthogonal to the first mode. Now, in general, an arbitrary trial vector cannot be expected to be orthogonal to the first mode. However, provided that the first mode shape has already been determined, the trial vector can be modified to sweep away the first mode shape from it. The procedure used in such modification is called *Gram–Schmidt orthogonalization*. Thus, suppose that the first mode shape is required to be swept off from an arbitrary trial vector $\mathbf{u}$, and that the purified vector which is free of the first mode is denoted by $\tilde{\mathbf{u}}$; then

$$\tilde{\mathbf{u}} = \mathbf{u} - c_1 \mathbf{q}_1$$

$$= \mathbf{u} - \frac{\mathbf{q}_1^T \mathbf{M} \mathbf{u}}{\mathbf{q}_1^T \mathbf{M} \mathbf{q}_1} \mathbf{q}_1 \tag{11.79}$$

in which Equation 10.27 with $i = 1$ has been used to substitute for c_1. Equation 11.79 can be expressed in the alternative form

$$\tilde{\mathbf{u}} = \left(\mathbf{I} - \frac{\mathbf{q}_1 \mathbf{q}_1^T \mathbf{M}}{\mathbf{q}_1^T \mathbf{M} \mathbf{q}_1} \right) \mathbf{u}$$

$$= \mathbf{S}_1 \mathbf{u} \tag{11.80}$$

where $\mathbf{S}_1$, called the first mode *sweeping matrix*, is given by

$$\mathbf{S}_1 = \mathbf{I} - \frac{\mathbf{q}_1 \mathbf{q}_1^T \mathbf{M}}{\mathbf{q}_1^T \mathbf{M} \mathbf{q}_1} \tag{11.81}$$

Theoretically, once the iterations have been commenced with a purified vector $\tilde{\mathbf{u}}$, the process should converge to the next most dominant eigenvalue. In practice, small errors in numeric computation during iterations, which are inevitable on account of finite-precision arithmetic, will result in the trial vector becoming contaminated by the first eigenvector, and purification must therefore be repeated at the end of each iteration. When repeated multiplication by the dynamic matrix $\mathbf{D} = \mathbf{K}^{-1} \mathbf{M}$ is used in the iteration, the purification at each iteration is conveniently achieved by modifying $\mathbf{D}$ as follows

$$\mathbf{D}_2 = \mathbf{D} \mathbf{S}_1 \tag{11.82}$$

and using $\mathbf{D}_2$ in place of $\mathbf{D}$ in the iterations.

The Gram–Schmidt process can, in fact, be used to orthogonalize a trial vector with respect to all the eigenvector or mode shapes that have already been determined, so that iteration on the purified vectors will converge to the next least dominant eigenvalue. To orthogonalize a vector **u** with respect to eigenvectors 1 to n, we carry out the following computations

$$\tilde{\mathbf{u}} = \mathbf{u} - \sum_{j=1}^{n} c_j \mathbf{q}_j$$

$$= \mathbf{u} - \sum_{j=1}^{n} \frac{\mathbf{q}_j^T \mathbf{M} \mathbf{u}}{\mathbf{q}_j^T \mathbf{M} \mathbf{q}_j} \mathbf{q}_j$$

$$= \mathbf{u} - \sum_{j=1}^{n} (\boldsymbol{\phi}_j^T \mathbf{M} \mathbf{u}) \boldsymbol{\phi}_j \tag{11.83}$$

where $\boldsymbol{\phi}_j$ are M-orthonormal mode shapes. Similar to Equation 11.81, we can construct a matrix, $\mathbf{S}_n$, that will sweep off modes 1 to n

$$\mathbf{S}_n = \mathbf{I} - \sum_{j=1}^{n} \frac{\mathbf{q}_j \mathbf{q}_j^T \mathbf{M}}{\mathbf{q}_j^T \mathbf{M} \mathbf{q}_j} \tag{11.84}$$

When matrix $\mathbf{D}_{n+1} = \mathbf{D} \mathbf{S}_n$ is used in inverse iteration, the process will converge to eigenvalue λ_{n+1} and the corresponding eigenvector.

The sweeping matrices can be obtained by an alternative procedure that is more convenient for hand computations. To illustrate the method, consider a three-degree-of-freedom system in which the first mode $\boldsymbol{\phi}_1$ has already been calculated. The purified trial vector for obtaining the second mode shape should be orthogonal to $\boldsymbol{\phi}_1$, so that

$$\boldsymbol{\phi}_1^T \mathbf{M} \tilde{\mathbf{u}} = 0 \tag{11.85}$$

Expansion of matrix Equation 11.85 gives

$$\phi_{11} \sum_{i=1}^{3} m_{1i} \tilde{u}_i + \phi_{21} \sum_{i=1}^{3} m_{2i} \tilde{u}_i + \phi_{31} \sum_{i=1}^{3} m_{3i} \tilde{u}_i = 0 \tag{11.86}$$

The single constraint equation can be solved to give one of the three elements of $\tilde{\mathbf{u}}$ in terms of the other two. If we choose to express $\tilde{u}_1$ in terms of $\tilde{u}_2$ and $\tilde{u}_3$, we get

$$\tilde{u}_1 = -\frac{\sum_{j=1}^{3} \phi_{j1} m_{j2}}{\sum_{j=1}^{3} \phi_{j1} m_{j1}} \tilde{u}_2 - \frac{\sum_{j=1}^{3} \phi_{j1} m_{j3}}{\sum_{j=1}^{3} \phi_{j1} m_{j1}} \tilde{u}_3 \tag{11.87}$$

Thus, if the initial trial vector for the second mode shape is given by

$$\mathbf{u} = \begin{bmatrix} u_1 \\ u_2 \\ u_3 \end{bmatrix} \tag{11.88}$$

then the purified trial vector, that is, one which is orthogonal to the first mode shape, can be constructed by taking $\tilde{u}_2 = u_2$, $\tilde{u}_3 = u_3$, and $\tilde{u}_1$ selected to satisfy Equation 11.87. In matrix form, these requirements can be stated as

$$
\begin{bmatrix} \tilde{u}_1 \\ \tilde{u}_2 \\ \tilde{u}_3 \end{bmatrix} = \begin{bmatrix} 0 & -\dfrac{\sum_{j=1}^{3} \phi_{j1} m_{j2}}{\sum_{j=1}^{3} \phi_{j1} m_{j1}} & -\dfrac{\sum_{j=1}^{3} \phi_{j1} m_{j3}}{\sum_{j=1}^{3} \phi_{j1} m_{j1}} \\ 0 & 1 & 0 \\ 0 & 0 & 1 \end{bmatrix} \begin{bmatrix} u_1 \\ u_2 \\ u_3 \end{bmatrix}
\tag{11.89a}
$$

or

$$\tilde{\mathbf{u}} = \mathbf{S}_1 \mathbf{u} \tag{11.89b}$$

where $\mathbf{S}_1$ is the required sweeping matrix. It is interesting to note that the sweeping matrix for the second mode has one column of zeros corresponding to the single constraint on the elements of the trial vector. It is easily shown that the sweeping matrix to obtain the third mode will have two columns of zeros, there being two constraint equations, so that $\tilde{u}_3 = u_3$ is arbitrary while $\tilde{u}_1$ and $\tilde{u}_2$ are related to $\tilde{u}_3$.

The procedure demonstrated above can be generalized as follows. Let s be the number of mode shapes already determined, and let the number of remaining mode shapes be γ. The constraint equations that the trial vector for $(s + 1)$th mode should satisfy are given by

$$\mathbf{\Phi}_s^T \mathbf{M} \tilde{\mathbf{u}} = 0 \tag{11.90}$$

Where, $\mathbf{\Phi}_s^T$ is a matrix of the first s mode shapes. Now, let $\mathbf{u}$ be partitioned so that $\mathbf{u}_s$ is a vector of the first s elements of $\mathbf{u}$ and $\mathbf{u}_r$ a vector of the remaining r elements. Also, let $\mathbf{M}$ be partitioned so that $\mathbf{M}_s$ is a matrix of the first s columns of $\mathbf{M}$ and $\mathbf{M}_r$ a matrix of the remaining r columns. Equation 11.90 can now be expressed as

$$
\underset{(s \times N)}{[\mathbf{\Phi}_s^T]} \quad \underset{(N \times s)}{[\mathbf{M}_s} \quad \underset{(N \times r)}{\mathbf{M}_r]} \begin{bmatrix} \tilde{\mathbf{u}}_s \\ (s \times 1) \\ \tilde{\mathbf{u}}_r \\ (r \times 1) \end{bmatrix} = 0
\tag{11.91}
$$

or

$$\tilde{\mathbf{u}}_s = -[\mathbf{\Phi}_s^T \mathbf{M}_s]^{-1}[\mathbf{\Phi}_s^T \mathbf{M}_r] \tilde{\mathbf{u}}_r \tag{11.92}$$

If the arbitrary trial vector is

$$\mathbf{u} = \begin{bmatrix} \mathbf{u}_s \\ \mathbf{u}_r \end{bmatrix} \tag{11.93}$$

then the purified trial vector is given by

$$
\{\tilde{\mathbf{u}}\} =
\begin{bmatrix}
\underset{(s \times s)}{\mathbf{0}} & \underset{(s \times r)}{-[\boldsymbol{\Phi}_s^T \mathbf{M}_s]^{-1}[\boldsymbol{\Phi}_s^T \mathbf{M}_r]} \\
\underset{(r \times s)}{\mathbf{0}} & \underset{(r \times r)}{\mathbf{I}}
\end{bmatrix}
\begin{bmatrix}
\underset{(s \times 1)}{\mathbf{u}_s} \\
\underset{(r \times 1)}{\mathbf{u}_r}
\end{bmatrix}
\tag{11.94}
$$

The sweeping matrix is thus obtained from

$$
\mathbf{S}_s =
\begin{bmatrix}
\mathbf{0} & -[\boldsymbol{\Phi}_s^T \mathbf{M}_s]^{-1}[\boldsymbol{\Phi}_s^T \mathbf{M}_r] \\
\mathbf{0} & \mathbf{I}
\end{bmatrix}
\tag{11.95}
$$

and has s columns of zero elements. The modified dynamic matrix to be used in the iteration is obtained on multiplying $\mathbf{D}$ by $\mathbf{S}_s$, so that

$$
\mathbf{D}_{s+1} = \mathbf{D}\mathbf{S}_s
\tag{11.96}
$$

and the iterations carried out with $\mathbf{D}_{s+1}$ will converge to the $(s+1)$th eigenpair.

When the algorithm given by Equations 11.73 through 11.76 is being used, Equation 11.74 is modified as follows

$$
\bar{\mathbf{x}}_{k+1} = \mathbf{M}\mathbf{S}_s\bar{\mathbf{u}}_{k+1}
\tag{11.97}
$$

Iterations for mode $s+1$ commence with $\mathbf{x}_1 = \mathbf{M}\mathbf{S}_s\mathbf{u}$, where $\mathbf{u}$ is the initial trial vector, and Equation 11.97 is used in place of Equation 11.74. The sweeping matrix technique can be used with the direct iteration method, in which case, the eigenvalues and eigenvectors will be obtained, in order, from the most dominant to the least dominant.

It is obvious from the foregoing discussion that, in order to use the vector iteration process to obtain the eigenvalues and eigenvectors other than the least or the most dominant ones, the trial vector must be purified of the eigenvectors already determined. The process of purification uses these eigenvectors; hence, to maintain numerical stability, they should be calculated with high numerical precision. In practical terms, each successive eigenvector determined by the iteration process becomes less and less accurate, and after a certain stage, the results become unreliable. It should be noted that for the use of inverse iteration technique, the stiffness matrix $\mathbf{K}$ should be nonsingular, and therefore positive definite; otherwise, $\mathbf{K}^{-1}$ will not exist and the product $\mathbf{K}^{-1}\mathbf{M}$ cannot be formed. A semidefinite stiffness matrix is singular and cannot be inverted. In such cases, iteration with shifts, as described in the next section, can be used.

Example 11.5

Use inverse iteration to obtain the eigenvalues and eigenvectors of the following matrix

$$A = \begin{bmatrix} 10 & 2 & -2 \\ 2 & 13 & -4 \\ -2 & -4 & 13 \end{bmatrix}$$

Solution

The eigenvalue problem is written as

$$Aq = \lambda q$$

or

$$A^{-1}q = \frac{1}{\lambda}q$$

$$= \gamma q \tag{a}$$

$$Dq = \gamma q$$

Matrix **D** is obtained from

$$D = A^{-1}$$

$$= \frac{1}{1458} \begin{bmatrix} 153 & -8 & 18 \\ -18 & 126 & 36 \\ 18 & 36 & 126 \end{bmatrix}$$

We now select a trial vector and repeatedly multiply it by the dynamic matrix **D**. At each stage of iteration, we normalize the product vector by dividing all its elements by the first element of the vector. Beginning with an initial trial vector $u_0^T = [1 \ 1 \ 1]$, the results obtained in each stage of iteration are as listed below. We have omitted the multiplier 1/1458 from the computations but reintroduce it when convergence has taken place.

k	u_{k-1}	$\bar{u}_k = Du_{k-1}$	u_k
1	$\begin{bmatrix} 1 \\ 1 \\ 1 \end{bmatrix}$	$\begin{bmatrix} 153 \\ 144 \\ 180 \end{bmatrix}$	$\begin{bmatrix} 1.000 \\ 0.941 \\ 1.176 \end{bmatrix}$
2	$\begin{bmatrix} 1.000 \\ 0.941 \\ 1.176 \end{bmatrix}$	$\begin{bmatrix} 157.23 \\ 142.90 \\ 200.05 \end{bmatrix}$	$\begin{bmatrix} 1.000 \\ 0.909 \\ 1.272 \end{bmatrix}$
3	$\begin{bmatrix} 1.000 \\ 0.909 \\ 1.272 \end{bmatrix}$	$\begin{bmatrix} 159.54 \\ 142.30 \\ 210.99 \end{bmatrix}$	$\begin{bmatrix} 1.000 \\ 0.892 \\ 1.322 \end{bmatrix}$

$$4 \quad \begin{bmatrix} 1.000 \\ 0.892 \\ 1.322 \end{bmatrix} \quad \begin{bmatrix} 160.74 \\ 141.97 \\ 216.68 \end{bmatrix} \quad \begin{bmatrix} 1.000 \\ 0.883 \\ 1.348 \end{bmatrix}$$

$$5 \quad \begin{bmatrix} 1.000 \\ 0.883 \\ 1.348 \end{bmatrix} \quad \begin{bmatrix} 161.37 \\ 141.79 \\ 219.64 \end{bmatrix} \quad \begin{bmatrix} 1.000 \\ 0.879 \\ 1.361 \end{bmatrix}$$

$$6 \quad \begin{bmatrix} 1.000 \\ 0.879 \\ 1.361 \end{bmatrix} \quad \begin{bmatrix} 161.69 \\ 141.70 \\ 221.13 \end{bmatrix} \quad \begin{bmatrix} 1.0000 \\ 0.8764 \\ 1.3617 \end{bmatrix}$$

$$7 \quad \begin{bmatrix} 1.0000 \\ 0.8764 \\ 1.3677 \end{bmatrix} \quad \begin{bmatrix} 161.84 \\ 141.66 \\ 221.88 \end{bmatrix} \quad \begin{bmatrix} 1.0000 \\ 0.8753 \\ 1.3710 \end{bmatrix}$$

$$8 \quad \begin{bmatrix} 1.0000 \\ 0.8753 \\ 1.3710 \end{bmatrix} \quad \begin{bmatrix} 161.92 \\ 141.64 \\ 222.25 \end{bmatrix} \quad \begin{bmatrix} 1.0000 \\ 0.8748 \\ 1.3727 \end{bmatrix}$$

$$9 \quad \begin{bmatrix} 1.0000 \\ 0.8748 \\ 1.3727 \end{bmatrix} \quad \begin{bmatrix} 161.96 \\ 141.64 \\ 222.45 \end{bmatrix} \quad \begin{bmatrix} 1.0000 \\ 0.8745 \\ 1.3735 \end{bmatrix}$$

At this stage, convergence could be assumed to have been achieved. The eigenvalue is the ratio of any two corresponding elements in $\tilde{\mathbf{u}}_9$ and $\mathbf{u}_8$. Introducing the multiplier $1/1458$, we have

$$\gamma_1 = \frac{161.96}{1458} = \frac{1}{9}$$

and hence, $\lambda_1 = 1/\gamma_1 = 9.00$. The corresponding eigenvector is $\mathbf{q}_1^T = [1.000 \ 0.8745 \ 1.3735]$.

To obtain the second eigenvector, we must construct a sweeping matrix which will remove the first mode component from the trial vector. Denoting the purified trial vector by $\tilde{\mathbf{u}}$, we have

$$\tilde{u}_1 + 0.8745\tilde{u}_2 + 1.3735\tilde{u}_3 = 0 \tag{c}$$

or

$$\begin{bmatrix} \tilde{u}_1 \\ \tilde{u}_2 \\ \tilde{u}_3 \end{bmatrix} = \begin{bmatrix} 0 & -0.8745 & -1.3735 \\ 0 & 1 & 1 \\ 0 & 0 & 1 \end{bmatrix} \begin{bmatrix} u_1 \\ u_2 \\ u_3 \end{bmatrix} \tag{d}$$

or

$$\tilde{\mathbf{u}} = \mathbf{S}_1 \mathbf{u}$$

The dynamic matrix $\mathbf{D}_2$ is now obtained as

$$\mathbf{D}_2 = \mathbf{DS}_1$$

$$= \frac{1}{1458} \begin{bmatrix} 0 & -151.80 & -190.15 \\ 0 & 141.74 & 60.72 \\ 0 & 20.26 & 101.28 \end{bmatrix} \tag{e}$$

Iteration are again begun with $\mathbf{u}_0^T = [1 \ 1 \ 1]$. Because the first column of $\mathbf{D}_2$ is zero, the first element in the iteration vector does not affect the computations. It is not therefore calculated until convergence has been achieved. Normalization is carried out by dividing the iteration vector by its second element.

k	$\mathbf{u}_{k-1}$	$\bar{\mathbf{u}}_k = \mathbf{D}_2\mathbf{u}_{k-1}$	$\mathbf{u}_k$
1	$\begin{bmatrix} 1 \\ 1 \\ 1 \end{bmatrix}$	$\begin{bmatrix} - \\ 202.46 \\ 121.54 \end{bmatrix}$	$\begin{bmatrix} - \\ 1.00 \\ 0.60 \end{bmatrix}$
2	$\begin{bmatrix} - \\ 1.00 \\ 0.60 \end{bmatrix}$	$\begin{bmatrix} - \\ 178.17 \\ 81.63 \end{bmatrix}$	$\begin{bmatrix} - \\ 1.000 \\ 0.455 \end{bmatrix}$
3	$\begin{bmatrix} - \\ 1.000 \\ 0.455 \end{bmatrix}$	$\begin{bmatrix} - \\ 169.35 \\ 66.32 \end{bmatrix}$	$\begin{bmatrix} - \\ 1.000 \\ 0.392 \end{bmatrix}$
4	$\begin{bmatrix} - \\ 1.000 \\ 0.392 \end{bmatrix}$	$\begin{bmatrix} - \\ 165.52 \\ 59.92 \end{bmatrix}$	$\begin{bmatrix} - \\ 1.000 \\ 0.362 \end{bmatrix}$
5	$\begin{bmatrix} - \\ 1.000 \\ 0.362 \end{bmatrix}$	$\begin{bmatrix} - \\ 163.72 \\ 56.92 \end{bmatrix}$	$\begin{bmatrix} - \\ 1.000 \\ 0.348 \end{bmatrix}$
6	$\begin{bmatrix} - \\ 1.000 \\ 0.348 \end{bmatrix}$	$\begin{bmatrix} - \\ 162.85 \\ 55.47 \end{bmatrix}$	$\begin{bmatrix} - \\ 1.000 \\ 0.341 \end{bmatrix}$
7	$\begin{bmatrix} - \\ 1.000 \\ 0.341 \end{bmatrix}$	$\begin{bmatrix} - \\ 162.42 \\ 54.75 \end{bmatrix}$	$\begin{bmatrix} - \\ 1.000 \\ 0.337 \end{bmatrix}$
8	$\begin{bmatrix} - \\ 1.000 \\ 0.337 \end{bmatrix}$	$\begin{bmatrix} - \\ 162.21 \\ 54.40 \end{bmatrix}$	$\begin{bmatrix} - \\ 1.000 \\ 0.335 \end{bmatrix}$

$$
9 \quad \begin{bmatrix} - \\ 1.000 \\ 0.335 \end{bmatrix} \quad \begin{bmatrix} - \\ 162.10 \\ 54.232 \end{bmatrix} \quad \begin{bmatrix} - \\ 1.0000 \\ 0.3345 \end{bmatrix}
$$

$$
10 \quad \begin{bmatrix} - \\ 1.0000 \\ 0.3345 \end{bmatrix} \quad \begin{bmatrix} - \\ 162.05 \\ 54.14 \end{bmatrix} \quad \begin{bmatrix} - \\ 1.0000 \\ 0.3341 \end{bmatrix}
$$

$$
11 \quad \begin{bmatrix} - \\ 1.0000 \\ 0.3341 \end{bmatrix} \quad \begin{bmatrix} -216.00 \\ 162.03 \\ 54.10 \end{bmatrix}
$$

At this stage, convergence has been achieved. The eigenvalue is obtained by taking the ratio of, say, the second element in $\bar{\mathbf{u}}_{11}$ and $\mathbf{u}_{10}$.

$$
\gamma_2 = \frac{162.03}{1458}
$$

Hence

$$
\lambda_2 = \frac{1}{\gamma_2} = 9.00
$$

The corresponding eigenvector is equal to $\bar{\mathbf{u}}_{11}$ normalized in any appropriate fashion. Using a division by the first element in the vector to normalize it, we get

$$
\mathbf{q}_2 = \begin{bmatrix} 1.00 \\ -0.75 \\ -0.25 \end{bmatrix}
$$

It is of interest to note that two eigenvalues obtained above are identical. It can be shown that the eigenvectors corresponding to repeated eigenvalues are not unique, and any linear combination of the individual eigenvectors is also an eigenvector. Thus, the vector $\mathbf{u} = \alpha_1 \mathbf{q}_1 + \alpha_2 \mathbf{q}_2$, where α_1 and α_2 are arbitrary constants is also an eigenvector of $\mathbf{A}$ as may be verified by substitution in $\mathbf{Au} = \lambda \mathbf{u}$, λ being the repeated eigenvalue.

The third eigenvalue can be obtained by using the following two orthogonality relationships

$$
\tilde{\mathbf{u}}^T \mathbf{q}_1 = 0
$$

$$
\tilde{\mathbf{u}}^T \mathbf{q}_2 = 0 \tag{f}
$$

Expansion of Equations f gives

$$
\tilde{u}_1 + 0.8745\tilde{u}_2 + 1.373\tilde{u}_3 = 0
$$
$$
\tilde{u}_1 - 0.75\tilde{u}_2 - 0.25\tilde{u}_3 = 0 \tag{g}
$$

Equations g can be solved for, say, $\tilde{u}_2$ and $\tilde{u}_3$ in terms of $\tilde{u}_1$. Arbitrarily taking $\tilde{u}_1 = 1$, we get

$$
\tilde{u}_2 = 2.0
$$

$$
\tilde{u}_3 = -2.0
$$

The complete vector is

$$\tilde{\mathbf{u}} = \begin{bmatrix} 1 \\ 2 \\ -2 \end{bmatrix} \tag{h}$$

The vector given by Equation h is, indeed, the third eigenvector $\mathbf{q}_3$. The third eigenvalue is obtained from

$$\lambda_3 = \frac{\mathbf{q}_3^T \mathbf{A} \mathbf{q}_3}{\mathbf{q}_3^T \mathbf{q}_3}$$

$$= 18 \tag{i}$$

Although, as demonstrated above, iteration are not necessary for the third eigenvector, we will show them for the purpose of illustration. The second sweeping matrix is obtained from Equations g.

$$\begin{bmatrix} \tilde{u}_1 \\ \tilde{u}_2 \\ \tilde{u}_3 \end{bmatrix} = \begin{bmatrix} 0 & 0 & -0.5 \\ 0 & 0 & -1.0 \\ 0 & 0 & 1.0 \end{bmatrix} \begin{bmatrix} u_1 \\ u_2 \\ u_3 \end{bmatrix}$$

or

$$\tilde{\mathbf{u}} = \mathbf{S}_2 \mathbf{u} \tag{j}$$

The dynamic matrix $\mathbf{D}_3$ is given by

$$\mathbf{D}_3 = \mathbf{D} \mathbf{S}_2$$

$$= \frac{1}{1458} \begin{bmatrix} 0 & 0 & -40.5 \\ 0 & 0 & -81.0 \\ 0 & 0 & 81.0 \end{bmatrix}$$

Iteration are commenced with $\mathbf{u}_0^T = \begin{bmatrix} 1 & 1 & 1 \end{bmatrix}$ and are tabulated as follows.

k	$\mathbf{u}_{k-1}$	$\bar{\mathbf{u}}_k = \mathbf{D}_3 \mathbf{u}_{k-1}$	$\mathbf{u}_k$
1	$\begin{bmatrix} 1 \\ 1 \\ 1 \end{bmatrix}$	$\begin{bmatrix} -40.5 \\ -81.0 \\ 81.0 \end{bmatrix}$	$\begin{bmatrix} 1 \\ 2 \\ -2 \end{bmatrix}$
	$\begin{bmatrix} 1 \\ \ddots \\ -2 \end{bmatrix}$	$\begin{bmatrix} 81.0 \\ \ddots \\ -162.0 \end{bmatrix}$	$\begin{bmatrix} 1 \\ \ddots \\ -2 \end{bmatrix}$

Convergence has been achieved after two steps. The third eigenvalue is given by

$$\gamma_3 = \frac{81}{1458}$$

Hence

$$\lambda_3 = \frac{1458}{81} = 18$$

11.5.3 *Vector iteration with shifts*

As discussed in the preceding section, the iteration procedure may become quite inaccurate after a number of modes have been determined and a new one is to be found by sweeping from the trial vector eigenvectors already found. An iteration procedure with shifts can be employed to obviate this difficulty. *Iteration with shifts* is most effective in the case of inverse iteration method, where the least dominant eigenvalue is determined first. We therefore describe the method in detail with respect to the linearized eigenvalue problem in which the lowest eigenvalues and eigenvectors are to be determined. Introduction of a shift μ in the origin of the eigenvalue axis (Fig. 11.3) gives

$$\lambda = \mu + \delta \tag{11.98}$$

where δ is the modified eigenvalue measured from the shifted origin. Substitution of Equation 11.98 into Equation 10.7 gives

$$(\mathbf{K} - \mu\mathbf{M})\mathbf{q} = \delta\mathbf{M}\mathbf{q} \tag{11.99a}$$

or

$$(\mathbf{K} - \mu\mathbf{M})^{-1}\mathbf{M}\mathbf{q} = \frac{1}{\delta}\mathbf{q} \tag{11.99b}$$

$$\hat{\mathbf{D}}\mathbf{q} = \hat{\gamma}\mathbf{q} \tag{11.99c}$$

where $\hat{\mathbf{D}} = (\mathbf{K} - \mu\mathbf{M})^{-1}\mathbf{M}$.

Iteration with Equation 11.99c will lead to the eigenvalue corresponding to the largest $\hat{\gamma}$, that is, to the lowest δ. As already pointed out in Section 11.5.1,

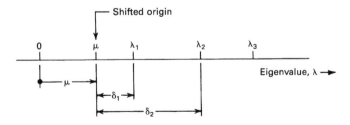

Figure 11.3. Measurement of eigenvalues from a shifted origin.

the rate of convergence depends on the ratio of the eigenvalue being sought to the next larger eigenvalue; the smaller this ratio, the faster the convergence. It is thus evident that by choosing a shift point close to the desired eigenvalue, a more accurate estimate of such eigenvalue as well as of the corresponding eigenvector can be obtained after a relatively small number of iterations. Thus, if the shift point is located between eigenvalues λ_n and λ_{n+1}, and $\mu - \lambda_n$ is smaller than $\lambda_{n+1} - \mu$, iteration will converge to λ_n and the rate of convergence will depend on $(\mu - \lambda_n)/(\lambda_{n+1} - \mu)$; the smaller this ratio, the faster the convergence. On the other hand, if $\lambda_{n+1} - \mu$ is smaller than $\mu - \lambda_n$, iteration will converge to λ_{n+1} and the rate of convergence will depend on the ratio $(\lambda_{n+1} - \mu)/(\mu - \lambda_n)$. Obviously, rapid convergence can be achieved if μ is located close to the desired eigenvalue.

Eigenvalue shift can be utilized with direct iteration. In this case, the eigenvalue equation becomes

$$\mathbf{M}^{-1}(\mathbf{K} - \mu\mathbf{M})\mathbf{q} = \delta\mathbf{q} \tag{11.100a}$$

or

$$\tilde{\mathbf{E}}\mathbf{q} = \delta\mathbf{q} \tag{11.100b}$$

where $\tilde{\mathbf{E}} = \mathbf{M}^{-1}(\mathbf{K} - \mu\mathbf{M})$. Iteration with $\tilde{\mathbf{E}}$ will converge to the eigenvalue that corresponds to the largest value of δ. Obviously, convergence can be achieved to either λ_1 or λ_N but not to any other intermediate frequency. If it is desired to converge to λ_N, for the most rapid convergence the shift point should be located as shown in Figure 11.4a, so that

$$\lambda_{N-1} - \mu = \mu - \lambda_1 \tag{11.101a}$$

or

$$\mu = \frac{\lambda_{N-1} + \lambda_1}{2} \tag{11.101b}$$

The convergence rate now depends on the ratio

$$\frac{\lambda_{N-1} - \mu}{\lambda_N - \mu} = \frac{\lambda_{N-1} - \lambda_1}{(\lambda_N - \lambda_1) + (\lambda_N - \lambda_{N-1})} \tag{11.102}$$

and cannot be improved any further. Similarly, if λ_1 is required, for the most rapid convergence, the shift point should be located as shown in Figure 11.4b, so that

$$\mu - \lambda_2 = \lambda_N - \mu \tag{11.103a}$$

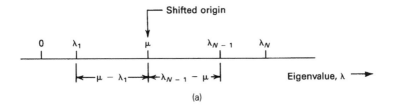

(a)

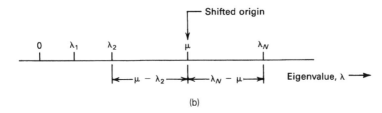

(b)

Figure 11.4. Direct iteration with shifts: (a) optimum location of shift point for convergence to λ_N; (b) optimum location of shift point for convergence to λ_1.

or

$$\mu = \frac{\lambda_2 + \lambda_N}{2} \qquad (11.103b)$$

In this case, convergence depends on the ratio

$$\frac{\mu - \lambda_2}{\mu - \lambda_1} = \frac{\lambda_N - \lambda_2}{(\lambda_N - \lambda_1) + (\lambda_2 - \lambda_1)} \qquad (11.104)$$

It is apparent that shifting of the eigenvalue origin is less effective when direct iteration is employed: first, the method can give only the lowest and the highest eigenpairs, and second, the rate of convergence cannot match that achieved with inverse iteration. The success of iteration with shift depends on the selection of an appropriate shift point. If the requirement is to obtain the eigenvalues nearest a specified value, the origin is shifted to such a specified value and the shift selection poses no difficulties.

As an example of a practical situation in which an eigenvalue nearest a specified value must be determined, consider the case of a structural or mechanical system subjected to an exciting force of given frequency. If the system has a natural frequency that is too close to the exciting frequency, resonance may take place and the system may require redesign. The possibility of resonance, and hence the need for redesign, may in some cases be detected simply by determining the system frequency closest to the exciting frequency.

Except for problems of the type described in the foregoing paragraph, the selection of the shift pint is not as straightforward. Very often, the requirement is to determine the lowest few eigenpairs. In such cases, inverse iteration with shift must be combined with iteration without shift. The first few, say five, frequencies and mode shapes are calculated by the standard inverse iteration technique along with Gram–Schmidt orthogonalization. The same process is then employed for the sixth eigenvalue, but instead of completing the iteration, it is discontinued after a few passes, at which time an estimate of the sixth frequency is obtained by use of the Rayleigh quotient (Equation 11.75). This estimated frequency is now used as the new shift point. The next five frequencies and mode shapes are then calculated using inverse iteration with the shifted origin.

Example 11.6

Construct the stiffness and mass matrices of the system shown in Figure E11.6. Obtain the eigenvalues and eigenvectors of the system by using inverse iteration. Given: $k_1 = 2$, $k_2 = 4$, $m_1 = m_3 = 2$, $m_2 = 4$.

Solution

Corresponding to the three degrees of freedom shown in the figure, the stiffness and mass matrices are given by

$$\mathbf{K} = \begin{bmatrix} 2 & -2 & 0 \\ -2 & 6 & -4 \\ 0 & -4 & 4 \end{bmatrix} \qquad \mathbf{M} = \begin{bmatrix} 2 & 0 & 0 \\ 0 & 4 & 0 \\ 0 & 0 & 2 \end{bmatrix}$$

The system shown in Figure E11.6 is not restrained against rigid-body motion and is said to be semidefinite. Its stiffness matrix is singular. We cannot therefore form the dynamic matrix $\mathbf{D} = \mathbf{K}^{-1}\mathbf{M}$, and inverse iteration will not work. By introducing a shift μ, we can modify the problem to the form given by Equation 11.99a, in which $(\mathbf{K} - \mu\mathbf{M})$ is not singular. Let us select $\mu = \frac{1}{2}$. The modified eigenvalue problem then becomes

$$\left(\mathbf{K} - \frac{1}{2}\mathbf{M} \right)^{-1} \mathbf{Mq} = \frac{1}{\delta}\mathbf{q}$$

$$\hat{\mathbf{D}}\mathbf{q} = \hat{\gamma}\mathbf{q}$$

(a)

in which $\hat{\gamma} = 1/\delta$ and $\lambda = \mu + \delta = \frac{1}{2} + \delta$.

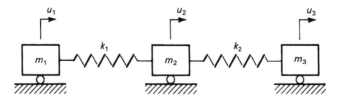

Figure E11.6. Three-degree-of-freedom semidefinite system.

Substituting for $\mathbf{K}$ and $\mathbf{M}$, we get

$$\mathbf{K} - \frac{1}{2}\mathbf{M} = \begin{bmatrix} 1 & -2 & 0 \\ -2 & 4 & -4 \\ 0 & -4 & 3 \end{bmatrix}$$

$$\hat{\mathbf{D}} = \left(\mathbf{K} - \frac{1}{2}\mathbf{M}\right)^{-1}\mathbf{M}$$

$$= -\frac{1}{16}\begin{bmatrix} -4 & 6 & 8 \\ 6 & 3 & 4 \\ 8 & 4 & 0 \end{bmatrix}\begin{bmatrix} 2 & & \\ & 4 & \\ & & 2 \end{bmatrix}$$

$$= -\frac{1}{4}\begin{bmatrix} -2 & 6 & 4 \\ 3 & 3 & 2 \\ 4 & 4 & 0 \end{bmatrix}$$

We now carry out iterations for the eigenproblem of Equation a. Again, for our trial vector, we choose $\mathbf{u}_0^T = [1\ 1\ 1]$. The computations are tabulated as follows.

k	$\mathbf{u}_{k-1}$	$\bar{\mathbf{u}}_k = \mathbf{D}\mathbf{u}_{k-1}$	$\mathbf{u}_k$
1	$\begin{bmatrix} 1 \\ 1 \\ 1 \end{bmatrix}$	$\begin{bmatrix} 8 \\ 8 \\ 8 \end{bmatrix}$	$\begin{bmatrix} 1 \\ 1 \\ 1 \end{bmatrix}$

Convergence takes place in one step. In fact, by coincidence, we have picked a trial vector that is exactly equal to an eigenvector of the system. The corresponding eigenvalue is given by

$$\hat{\gamma}_1 = -2$$

Thus

$$\delta_1 = -\frac{1}{2}$$

$$\lambda_1 = \mu + \delta_1$$

$$= 0$$

The eigenvector $\mathbf{q}_1^T = [1\ 1\ 1]$ represents a rigid-body motion of the system, and as would be expected, the corresponding eigenvalue is zero.

To obtain the second eigenvalue, we construct a sweeping matrix that will remove the eigenvector already determined. The orthogonality relationship gives the following constraint equation

$$\mathbf{q}_1^T\mathbf{M}\bar{\mathbf{u}} = 0$$

$$2\tilde{u}_1 + 4\tilde{u}_2 + 2\tilde{u}_3 = 0$$

In matrix form, we can write

$$\begin{bmatrix} \tilde{u}_1 \\ \tilde{u}_2 \\ \tilde{u}_3 \end{bmatrix} = \begin{bmatrix} 0 & -2 & -1 \\ 0 & 1 & 0 \\ 0 & 0 & 1 \end{bmatrix} \begin{bmatrix} u_1 \\ u_2 \\ u_3 \end{bmatrix}$$

or

$$\tilde{\mathbf{u}} = \mathbf{S}_1 \mathbf{u} \tag{b}$$

The new dynamic matrix is given by

$$\hat{\mathbf{D}}_2 = \hat{\mathbf{D}} \mathbf{S}_1$$

$$= \frac{1}{4} \begin{bmatrix} 0 & -10 & -6 \\ 0 & 3 & 1 \\ 0 & 4 & 4 \end{bmatrix} \tag{c}$$

We now carry out vector iterations with $\hat{\mathbf{D}}_2$ obtained from Equation c. Starting with $\mathbf{u}_0^T = [1 \ 1 \ 1]$, the computations proceed as shown below. The multiplier $\frac{1}{4}$ has been omitted from each stage. Also, the first element is not evaluated until convergence has been achieved.

k	$\mathbf{u}_{k-1}$	$\bar{\mathbf{u}}_k = \hat{\mathbf{D}}_2 \mathbf{u}_{k-1}$	$\mathbf{u}_k$
1	$\begin{bmatrix} 1 \\ 1 \\ 1 \end{bmatrix}$	$\begin{bmatrix} - \\ 4 \\ 8 \end{bmatrix}$	$\begin{bmatrix} - \\ 1 \\ 2 \end{bmatrix}$
2	$\begin{bmatrix} - \\ 1 \\ 2 \end{bmatrix}$	$\begin{bmatrix} - \\ 5 \\ 12 \end{bmatrix}$	$\begin{bmatrix} - \\ 1.0 \\ 2.4 \end{bmatrix}$
3	$\begin{bmatrix} - \\ 1 \\ 1 \end{bmatrix}$	$\begin{bmatrix} - \\ 5.4 \\ 13.6 \end{bmatrix}$	$\begin{bmatrix} - \\ 1.000 \\ 2.519 \end{bmatrix}$
4	$\begin{bmatrix} - \\ 1.000 \\ 2.519 \end{bmatrix}$	$\begin{bmatrix} - \\ 5.519 \\ 14.074 \end{bmatrix}$	$\begin{bmatrix} - \\ 1.000 \\ 2.550 \end{bmatrix}$
5	$\begin{bmatrix} - \\ 1.000 \\ 2.550 \end{bmatrix}$	$\begin{bmatrix} - \\ 5.55 \\ 14.20 \end{bmatrix}$	$\begin{bmatrix} - \\ 1.000 \\ 2.559 \end{bmatrix}$
6	$\begin{bmatrix} - \\ 1.000 \\ 2.559 \end{bmatrix}$	$\begin{bmatrix} - \\ 5.559 \\ 14.236 \end{bmatrix}$	$\begin{bmatrix} - \\ 1.000 \\ 2.561 \end{bmatrix}$
7	$\begin{bmatrix} - \\ 1.000 \\ 2.561 \end{bmatrix}$	$\begin{bmatrix} -25.366 \\ 5.561 \\ 14.244 \end{bmatrix}$	$\begin{bmatrix} -4.561 \\ 1.000 \\ 2.561 \end{bmatrix}$

At this stage, convergence has been achieved. The eigenvalue is given by

$$\hat{\gamma}_2 = \frac{1}{4} \times 5.561 = 1.39$$

$$\delta_2 = \frac{1}{\hat{\gamma}_2} = 0.7193$$

$$\lambda_2 = \delta_2 + \mu = 1.2193$$

The eigenvector, normalized so that the first element is equal to 1, is given by

$$\mathbf{q}_2 = \begin{bmatrix} 1.000 \\ -0.2192 \\ -0.5616 \end{bmatrix}$$

To obtain the third eigenvector, we write the two orthogonality relationships

$$\mathbf{q}_1^T \mathbf{M} \tilde{\mathbf{u}} = 0$$

$$\mathbf{q}_2^T \mathbf{M} \tilde{\mathbf{u}} = 0 \qquad (d)$$

Expansion of Equation d gives

$$2\tilde{u}_1 + 4\tilde{u}_2 + 2\tilde{u}_3 = 0$$

$$2\tilde{u}_1 - 0.8768\tilde{u}_2 - 1.1232\tilde{u}_3 = 0 \qquad (e)$$

Simultaneous solution of Equation e with $\tilde{u}_1$ arbitrarily set at 1 gives

$$\mathbf{q}_3 = \begin{bmatrix} \tilde{u}_1 \\ \tilde{u}_2 \\ \tilde{u}_3 \end{bmatrix} = \begin{bmatrix} 1.000 \\ -2.280 \\ 3.561 \end{bmatrix} \qquad (f)$$

The vector given by Equation f is the third eigenvector of the system. The corresponding eigenvalue can be obtained directly from

$$\lambda_3 = \frac{\mathbf{q}_3^T \mathbf{K} \mathbf{q}_3}{\mathbf{q}_3^T \mathbf{M} \mathbf{q}_3}$$

$$= 3.281$$

This example illustrates a method of determining the eigenvalues and eigenvectors of a semi-definite system.

11.5.4   *Subspace iteration*

The *subspace iteration method* developed by Bathe is an efficient method for the eigensolution of large systems when only the lower modes are of interest.

Conceptually, it is similar to the inverse iteration method, except that iteration is performed simultaneously on a number of trial vectors, and an eigenproblem in a reduced subspace is solved at the end of each iteration. The procedure ensures that the trial vectors will converge to different mode shapes and not all to the lowest one. The number of trial vectors m is usually much less than N, the order of the matrices involved, but is larger than p, the number of modes to be determined. In practice, the method is found to be most efficient when m is chosen to be the smaller of $2p$ and $p+8$. The linearized eigenvalue problem can be expressed as

$$\mathbf{KQ} = \mathbf{MQ\Lambda} \tag{11.105}$$

in which $\mathbf{Q}$ is an $N \times m$ matrix of the first m eigenvectors, and $\mathbf{\Lambda}$ is a diagonal matrix, the elements on the diagonal representing the first m eigenvalues. As in the case of inverse iteration method, the first step in the iteration is

$$\tilde{\mathbf{V}}_1 = \mathbf{K}^{-1}\mathbf{MV}_0 \tag{11.106}$$

where $\mathbf{V}_0$ is a matrix of m trial shapes.

Before proceeding with the next iteration step, the vectors in $\tilde{\mathbf{V}}_1$ must be orthogonalized so that they will converge to different eigenvectors or mode shapes and not all to the lowest one. Further, as in the case of iteration methods, the eigenvectors should be normalized in some way so that the numbers remain within reasonable bounds. The two requirements are satisfied by solving a new eigenvalue problem formed as follows

$$\mathbf{K}_1^* = \tilde{\mathbf{V}}_1^T \mathbf{K} \tilde{\mathbf{V}}_1$$

$$\mathbf{M}_1^* = \tilde{\mathbf{V}}_1^T \mathbf{M} \tilde{\mathbf{V}}_1 \tag{11.107}$$

$$\mathbf{K}_1^* \tilde{\mathbf{Z}}_1 = \mathbf{M}_1^* \tilde{\mathbf{Z}}_1 \mathbf{\Gamma}$$

where $\tilde{\mathbf{Z}}_1$ is a matrix of m eigenvectors and $\mathbf{\Gamma}$ is a diagonal matrix of m eigenvalues.

The eigenvalue problem of Equation 11.107 is of a comparatively small size, being of the order $m \times m$ and may conveniently be solved by any one of the transformation methods. The modal vector $\tilde{\mathbf{Z}}_1$ is then normalized so that

$$\mathbf{Z}_1^T \mathbf{M}_1^* \mathbf{Z}_1 = \mathbf{I} \tag{11.108}$$

The improved trial vectors for the next iteration are now given by

$$\mathbf{V}_1 = \tilde{\mathbf{V}}_1 \mathbf{Z}_1 \tag{11.109}$$

The improved trial vectors are orthogonal with respect to the stiffness and mass matrices of the original system and will therefore converge to different mode shapes. To prove the orthogonality, we note that

$$\mathbf{V}_1^T \mathbf{M} \mathbf{V}_1 = \mathbf{Z}_1^T \tilde{\mathbf{V}}_1^T \mathbf{M} \tilde{\mathbf{V}}_1 \mathbf{Z}_1$$

$$= \mathbf{Z}_1^T \mathbf{M}^* \mathbf{Z}_1$$

$$= \mathbf{I} \tag{11.110}$$

and

$$\mathbf{V}_1^T \mathbf{K} \mathbf{V}_1 = \mathbf{Z}_1^T \tilde{\mathbf{V}}_1^T \mathbf{K} \tilde{\mathbf{V}}_1 \mathbf{Z}_1$$

$$= \mathbf{Z}_1^T \mathbf{K}^* \mathbf{Z}_1$$

$$= \mathbf{\Gamma} \tag{11.111}$$

If the iteration procedure described above is repeated a sufficient number of times, $\mathbf{V}$ will converge to the first m mode shapes and $\mathbf{\Gamma}$ to the first m frequencies. Iteration may, in fact, be terminated when the first p frequencies do not change between iterations by more than a specified tolerance. The iteration procedure can be summarized as follows: Given $\mathbf{V}_k$ of size $N \times m$, find the next iteration vector $\tilde{\mathbf{V}}_{k+1}$ by solving

$$\mathbf{K} \tilde{\mathbf{V}}_{k+1} = \mathbf{M} \mathbf{V}_k \tag{11.112}$$

Find the projection of matrices $\mathbf{K}$ and $\mathbf{M}$ in an $m \times m$ space

$$\mathbf{K}_{k+1}^* = \tilde{\mathbf{V}}_{k+1}^T \mathbf{K} \tilde{\mathbf{V}}_{k+1}$$
$$\mathbf{M}_{k+1}^* = \tilde{\mathbf{V}}_{k+1}^T \mathbf{M} \tilde{\mathbf{V}}_{k+1} \tag{11.113}$$

Solve the following eigenvalue problem in the reduced space

$$\mathbf{K}_{k+1}^* \tilde{\mathbf{Z}}_{k+1} = \mathbf{M}_{k+1}^* \tilde{\mathbf{Z}}_{k+1} \mathbf{\Gamma}_{k+1} \tag{11.114}$$

Normalize the vectors in $\tilde{\mathbf{Z}}_{k+1}$ so that

$$\mathbf{Z}_{k+1}^T \mathbf{M} \mathbf{Z}_{k+1} = \mathbf{I} \tag{11.115}$$

Find an improved estimate of the eigenvectors

$$\mathbf{V}_{k+1} = \tilde{\mathbf{V}}_{k+1} \mathbf{Z}_{k+1} \tag{11.116}$$

When k becomes large, the vectors in $\mathbf{V}_{k+1}$ will converge to the first m eigenvectors and the diagonal elements of $\mathbf{\Gamma}_{k+1}$ will converge to the first m eigenvalues.

The selection of starting iteration vectors is important for the efficiency of a subspace iteration. If all vectors in the set of initialized vectors are orthogonal to a specific eigenvectors, that eigenvector will be completely missed. This is an unlikely event. However, the starting vector is still important from the point of view of achieving rapid convergence. The following procedure suggested by Bathe has been shown on the basis of experience to be quite effective. Select the first column in $\mathbf{MV}_0$ as equal to the diagonal of the mass matrix $\mathbf{M}$. This will ensure that all mass degrees of freedom are excited. For other columns of $\mathbf{MV}_0$, select unit vectors with entries $+1$ at the coordinates with smallest ratios of k_{ii}/m_{ii} and all other elements equal to zero.

11.5.5 *Lanczos iteration*

When first developed in 1950, the *Lanczos method* was looked upon as an orthogonal transformation that reduced a symmetric matrix to a tridiagonal form. With the development of more efficient methods of tridiagonalization, interest in the Lanczos method waned. Research in recent years has, however, shown that the method has great potential in the partial eigensolution of large systems when the matrices involved are sparse. Interest in the method has therefore revived, and currently, it is considered to be as effective a tool as the subspace iteration method for the solution of large vibration problems.

The Lanczos transformation matrix $\mathbf{X}$ consists of a series of mutually orthogonal vectors $\mathbf{x}_1, \mathbf{x}_2, \ldots, \mathbf{x}_N$, such that

$$\mathbf{X}^T\mathbf{X} = \mathbf{I} \tag{11.117}$$

The transformation can be expressed as

$$\mathbf{X}^T\mathbf{A}\mathbf{X} = \mathbf{T} \tag{11.118a}$$

or

$$\mathbf{A}\mathbf{X} = \mathbf{X}\mathbf{T} \tag{11.118b}$$

where $\mathbf{A}$ is the symmetric matrix to be tridiagonalized and $\mathbf{T}$ is the tridiagonal matrix. When written in its expanded form, Equation 11.118 becomes

$$
\mathbf{A}\begin{bmatrix} \mathbf{x}_1 & \mathbf{x}_2 & \cdots & \mathbf{x}_N \end{bmatrix}
$$

$$
= \begin{bmatrix} \mathbf{x}_1 & \mathbf{x}_2 & \cdots & \mathbf{x}_N \end{bmatrix}
\begin{bmatrix}
\alpha_1 & \beta_2 & 0 & . & 0 & 0 & 0 \\
\beta_2 & \alpha_2 & \beta_3 & . & 0 & 0 & 0 \\
. & . & . & . & . & . & . \\
0 & 0 & 0 & . & \beta_{N-1} & \alpha_{N-1} & \beta_N \\
0 & 0 & 0 & . & 0 & \beta_N & \alpha_N
\end{bmatrix}
\tag{11.119}
$$

or

$$\mathbf{Ax}_1 = \alpha_1\mathbf{x}_1 + \beta_2\mathbf{x}_2$$

$$\mathbf{Ax}_2 = \beta_2\mathbf{x}_1 + \alpha_2\mathbf{x}_2 + \beta_3\mathbf{x}_3$$

$$\cdots \qquad \cdots$$

$$\mathbf{Ax}_j = \beta_j\mathbf{x}_{j-1} + \alpha_j\mathbf{x}_j + \beta_{j+1}\mathbf{x}_{j+1} \qquad (11.120)$$

$$\cdots \qquad \cdots$$

$$\mathbf{Ax}_N = \beta_N\mathbf{x}_{N-1} + \alpha_N\mathbf{x}_N$$

The process of finding the Lanczos vectors begins with an arbitrary selection for $\tilde{\mathbf{x}}_1$, where

$$\tilde{\mathbf{x}}_1 = \beta_1\mathbf{x}_1 \qquad (11.121)$$

The unknown parameter β_1 and hence $\mathbf{x}_1$ is determined by considering that vector $\mathbf{x}_1$ should be orthonormal. Thus

$$\tilde{\mathbf{x}}_1^T\tilde{\mathbf{x}}_1 = \beta^2\mathbf{x}_1^T\mathbf{x}_1$$

$$= \beta_1^2 \qquad (11.122a)$$

and

$$\mathbf{x}_1 = \frac{\tilde{\mathbf{x}}_1}{\beta_1} \qquad (11.122b)$$

Multiplication of the first of Equation 11.120 by $\mathbf{x}_1^T$ gives

$$\mathbf{x}_1^T\mathbf{Ax}_1 = \alpha_1\mathbf{x}_1^T\mathbf{x}_1 + \beta_2\mathbf{x}_1^T\mathbf{x}_2 \qquad (11.123)$$

If we now select α_1 so that

$$\alpha_1 = \mathbf{x}_1^T\mathbf{Ax}_1 \qquad (11.124)$$

then substituting Equation 11.124 in Equation 11.123 and using the relationship $\mathbf{x}_1^T\mathbf{x}_1 = 1$, we have

$$\mathbf{x}_1^T\mathbf{x}_2 = 0 \qquad (11.125)$$

implying that vectors $\mathbf{x}_1$ and $\mathbf{x}_2$ are mutually orthogonal. The first of Equation 11.120 now gives

$$\tilde{\mathbf{x}}_2 = \beta_2\mathbf{x}_2 = \mathbf{Ax}_1 - \alpha_1\mathbf{x}_1 \qquad (11.126)$$

and

$$\tilde{\mathbf{x}}_2^T\tilde{\mathbf{x}}_2 = \beta_2^2\mathbf{x}_2^T\mathbf{x}_2 \qquad (11.127)$$

Since $x_2^T x_2 = 1$, β_2 is obtained from Equation 11.127. Thus

$$\beta_2 = (\tilde{x}_2^T \tilde{x}_2)^{1/2} \tag{11.128a}$$

and

$$x_2 = \frac{\tilde{x}_2}{\beta_2} \tag{11.128b}$$

Premultiplication of the second of Equation 11.120 by x_2^T gives

$$x_2^T A x_2 = \beta_2 x_2^T x_1 + \alpha_2 x_2^T x_2 + \beta_3 x_2^T x_3 \tag{11.129}$$

If α_2 is selected so that

$$\alpha_2 = x_2^T A x_2 \tag{11.130}$$

then, because $x_2^T x_1 = 0$ and $x_2^T x_2 = 1$, Equation 11.129 gives

$$x_2^T x_3 = 0 \tag{11.131}$$

Vector x_3 and constant β are now obtained from expressions similar to Equations 11.126, 11.127, and 11.128. The complete iteration process can be summarized as

$$\beta_j^2 = \tilde{x}_j^T \tilde{x}_j$$

$$x_j = \frac{\tilde{x}_j}{\beta_j} \tag{11.132}$$

$$\alpha_j = x_j^T A x_j$$

$$\tilde{x}_{j+1} = A x_j - \beta_j x_{j-1} - \alpha_j x_j$$

The iteration begins with $j = 1$ and $x_0 = 0$; it ends when x_N has been obtained. The standard eigenvalue problem is given by Equation 10.5:

$$Aq = \lambda q \tag{10.5}$$

Premultiplying both sides of Equation 10.5 by X^T and using the orthogonality relationship $XX^T = X^T X = I$, we have

$$X^T A X X^T q = \lambda X^T q \tag{11.133a}$$

or

$$T\tilde{q} = \lambda \tilde{q} \tag{11.133b}$$

where

$$\tilde{q} = X^T q \tag{11.134}$$

Equation 11.133 implies that the eigenvalues of matrix T are the same as those of matrix A, while the eigenvectors of the two matrices are related by

Equation 11.134. The eigenvalue of the tridiagonal matrix can be evaluated by any of the standard methods, such as the LR or the QR method. Theoretically, the Lanczos method vectors $\mathbf{x}_1, \mathbf{x}_2, \ldots, \mathbf{x}_N$ are mutually orthogonal to each other. However, because of round-off errors during computations, orthogonality relationship breaks down when vectors are sufficiently separated from each other. For large systems, this source of error makes Lanczos method unstable. It is therefore necessary to reorthogonalize a newly determined Lanczos vector by sweeping off any contribution from vectors determined previously. Purification is carried out by using the Gram–Schmidt process. Denoting the purified vector by $\bar{\mathbf{x}}_j$, we have

$$\bar{\mathbf{x}}_j = \mathbf{x}_j - \sum_{k=1}^{j-1} (\bar{\mathbf{x}}_k^T \mathbf{x}_j) \bar{\mathbf{x}}_k \tag{11.135}$$

The computations involved in the reorthogonalization procedure are very substantial. This makes the Lanczos procedure much less efficient than say, Householder's method for the reduction of symmetric matrices. The Lanczos method is, however, most effective when used to obtain the first few eigenvalues of large symmetric matrices. Thus, if only p eigenvalues are required where $p \ll N$, the tridiagonalization may by terminated after the first m Lanczos vectors have been found, m being sufficiently larger than p. The $N \times m$ transformation matrix $\mathbf{X}_m$ can now be used to give a tridiagonal matrix.

$$\mathbf{T}_m = \mathbf{X}_m^T \mathbf{A} \mathbf{X}_m \tag{11.136}$$

An eigensolution of $\mathbf{T}_m$ will give a good estimate of p of the lower eigenvalues.

To verify whether convergence has been achieved for the eigenvalues, it may be necessary to solve the eigenvalues for $\mathbf{T}_{m-1}$ as well, and to compare the magnitude of the desired eigenvalues obtained from $\mathbf{T}_{m-1}$ and $\mathbf{T}_m$. It should be noted that if the starting vector does not contain contributions from a specific mode, that mode will be missed completely. A procedure in which the individual elements of the starting vector are selected by a random number generator will minimize the likelihood of missed eigenvectors. However, when the system has multiple eigenvalues, the Lanczos method will give only one of the corresponding eigenvectors.

The Lanczos method can be applied to the solution of a linearized eigenvalue problems by converting the latter to the standard symmetric form, for example, that given by Equation 11.5. During Lanczos iterations, matrix $\tilde{\mathbf{D}}$ in Equation 11.15b plays the same role as matrix $\mathbf{A}$ in the iterations on standard form. The following alternative procedure dispenses with the need for transformation of linearized eigenproblem into a standard symmetric form. The Lanczos vector are selected to be orthonormal with respect to the mass matrix, so that

$$\mathbf{X}^T \mathbf{M} \mathbf{X} = \mathbf{I} \tag{11.137}$$

The eigenvalue problem is now expressed as

$$\mathbf{K}\phi = \lambda \mathbf{M}\phi \tag{11.138a}$$

$$\mathbf{K}^{-1}\mathbf{M}\phi = \frac{1}{\lambda}\phi \tag{11.138b}$$

Premultiplying Equation 11.138b by $\mathbf{X}^T\mathbf{M}$, we get

$$\mathbf{X}^T(\mathbf{M}\mathbf{K}^{-1}\mathbf{M})\mathbf{X}\tilde{\phi} = \frac{1}{\lambda}\mathbf{X}^T\mathbf{M}\mathbf{X}\tilde{\phi} \tag{11.139a}$$

or

$$\mathbf{T}\tilde{\phi} = \frac{1}{\lambda}\tilde{\phi} \tag{11.139b}$$

where

$$\phi = \mathbf{X}\tilde{\phi} \tag{11.140}$$

and $\mathbf{X}$ is selected so that the product $\mathbf{X}^T(\mathbf{M}\mathbf{K}^{-1}\mathbf{M})\mathbf{X}$ gives a tridiagonal matrix $\mathbf{T}$. Thus

$$\mathbf{X}^T\mathbf{M}\mathbf{K}^{-1}\mathbf{M}\mathbf{X} = \mathbf{T} \tag{11.141}$$

or because $\mathbf{X}^T\mathbf{M} = \mathbf{X}^{-1}$,

$$\mathbf{K}^{-1}\mathbf{M}\mathbf{X} = \mathbf{X}\mathbf{T} \tag{11.142}$$

As in the case of a standard eigenvalue problem, Equation 11.142 is written as

$$[\mathbf{K}^{-1}\mathbf{M}][\mathbf{x}_1, \mathbf{x}_2, \dots, \mathbf{x}_N] = [\mathbf{x}_1, \mathbf{x}_2, \dots, \mathbf{x}_N]\begin{bmatrix} \alpha_1 & \beta_2 & & & \\ \beta_2 & \alpha_2 & \beta_3 & & \\ & & & \ddots & \\ & & & & \beta_N & \alpha_N \end{bmatrix} \tag{11.143}$$

Expansion of Equation 11.143 gives, as before

$$\mathbf{K}^{-1}\mathbf{M}\mathbf{x}_1 = \alpha_1\mathbf{x}_1 + \beta_2\mathbf{x}_2$$

$$\mathbf{K}^{-1}\mathbf{M}\mathbf{x}_2 = \beta_2\mathbf{x}_1 + \alpha_2\mathbf{x}_2 + \beta_3\mathbf{x}_3$$

$$\cdots \qquad \cdots$$

$$\mathbf{K}^{-1}\mathbf{M}\mathbf{x}_j = \beta_j\mathbf{x}_{j-1} + \alpha_j\mathbf{x}_j + \beta_{j+1}\mathbf{x}_{j+1}$$

$$\cdots \qquad \cdots$$

$$\mathbf{K}^{-1}\mathbf{M}\mathbf{x}_N = \beta_N\mathbf{x}_{N-1} + \alpha_N\mathbf{x}_N \tag{11.144}$$

To begin the iteration, an arbitrary vector $\tilde{\mathbf{x}}_1$ is selected and is normalized with respect to the mass matrix, so that $\mathbf{x}_1^T \mathbf{M} \mathbf{x}_1 = 1$. Thus

$$\beta_1^2 = \tilde{\mathbf{x}}_1^T \mathbf{M} \tilde{\mathbf{x}}_1$$

$$\mathbf{x}_1 = \frac{\tilde{\mathbf{x}}_1}{\beta_1} \qquad (11.145)$$

Now premultiplying the first of Equations 11.144 by $\mathbf{x}_1^T \mathbf{M}$ and selecting $\alpha_1 = \mathbf{x}_1^T \mathbf{M} \mathbf{K}^{-1} \mathbf{M} \mathbf{x}_1$, we get $\mathbf{x}_1^T \mathbf{M} \mathbf{x}_2 = 0$. The first of Equations 11.144 then gives

$$\tilde{\mathbf{x}}_2 = \beta_2 \mathbf{x}_2$$

$$= \mathbf{K}^{-1} \mathbf{M} \mathbf{x}_1 - \alpha_1 \mathbf{x}_1 \qquad (11.146)$$

$$\beta_2^2 = \tilde{\mathbf{x}}_2^T \mathbf{M} \tilde{\mathbf{x}}_2$$

$$\mathbf{x}_2 = \frac{\tilde{\mathbf{x}}_2}{\beta_2}$$

The complete iteration process is easily deduced as

$$\beta_j^2 = \tilde{\mathbf{x}}_j^T \mathbf{M} \tilde{\mathbf{x}}_j$$

$$\mathbf{x}_j = \frac{\tilde{\mathbf{x}}_j}{\beta_j}$$

$$\mathbf{K} \tilde{\mathbf{x}}_{j+1} = \mathbf{M} \mathbf{x}_j \qquad (11.147)$$

$$\alpha_j = \mathbf{x}_j^T \mathbf{M} \tilde{\mathbf{x}}_{j+1}$$

$$\tilde{\mathbf{x}}_{j+1} = \tilde{\mathbf{x}}_{j+1} - \beta_j \mathbf{x}_{j-1} - \alpha_j \mathbf{x}_j$$

A considerable amount of research has been directed toward the Lanczos method in recent years. Many of the problems associated with the initial version of the method have been solved, so that currently, the method provides a very efficient tool for the partial eigensolution of large banded systems.

Example 11.7

Tridiagonalize the following matrix using Lanczos vectors. The starting vector may be taken as $\mathbf{x}^T = [0.5 \quad 0.5 \quad 0.5 \quad 0.5]$.

$$\mathbf{A} = \begin{bmatrix} 2640 & 230 & -1330 & 0 \\ 230 & 770 & 0 & 0 \\ -1330 & 0 & 2640 & -230 \\ 0 & 0 & -230 & 770 \end{bmatrix}$$

Solution

The Lanczos vectors are obtained by using Equation 11.132. The computations involved are tabulated below.

j	$\tilde{\mathbf{x}}_j$	β_j	$\mathbf{x}_j = \tilde{\mathbf{x}}_j/\beta_j$	$\alpha_j = \mathbf{x}_j^T \mathbf{A} \mathbf{x}_j$	$\tilde{\mathbf{x}}_{x+1} = \mathbf{A}\mathbf{x}_j - \beta_j \mathbf{x}_{j-1} - \alpha_j \mathbf{x}_j$
1	$\begin{bmatrix} 0.5 \\ 0.5 \\ 0.5 \\ 0.5 \end{bmatrix}$	1	$\begin{bmatrix} 0.5 \\ 0.5 \\ 0.5 \\ 0.5 \end{bmatrix}$	1040	$\begin{bmatrix} 250 \\ -20 \\ 20 \\ -250 \end{bmatrix}$
2	$\begin{bmatrix} 250 \\ -20 \\ 20 \\ -250 \end{bmatrix}$	354.68	$\begin{bmatrix} 0.7048 \\ -0.0564 \\ 0.0564 \\ -0.7048 \end{bmatrix}$	1599.28	$\begin{bmatrix} 468.25 \\ 31.54 \\ -894.00 \\ 394.21 \end{bmatrix}$
3	$\begin{bmatrix} 468.25 \\ 31.54 \\ -894.00 \\ 394.21 \end{bmatrix}$	1083.92	$\begin{bmatrix} 0.4320 \\ 0.0291 \\ -0.8248 \\ 0.3637 \end{bmatrix}$	3482.61	$\begin{bmatrix} -24.361 \\ 81.557 \\ -24.361 \\ -32.835 \end{bmatrix}$
4	$\begin{bmatrix} -24.361 \\ 81.557 \\ -24.361 \\ -32.835 \end{bmatrix}$	94.43	$\begin{bmatrix} -0.2580 \\ 0.8637 \\ -0.2580 \\ -0.3477 \end{bmatrix}$	698.12	

The Lanczos vectors are thus given by

$$\mathbf{X} = \begin{bmatrix} 0.5 & 0.7048 & 0.4320 & -0.2580 \\ 0.5 & -0.0564 & 0.0291 & 0.8637 \\ 0.5 & 0.0564 & -0.8248 & -0.2580 \\ 0.5 & -0.7048 & 0.3637 & -0.3477 \end{bmatrix}$$

The tridiagonal matrix is obtained from

$$\mathbf{T} = \mathbf{X}^T \mathbf{A} \mathbf{X}$$

$$= \begin{bmatrix} 1040.00 & 354.68 & & \\ 354.68 & 1599.28 & 1083.92 & \\ & 1083.92 & 3482.60 & 94.43 \\ & & 94.43 & 698.12 \end{bmatrix}$$

11.6 DETERMINANT SEARCH METHOD

When the matrices involved in the eigenvalue problem have a comparatively small bandwidth and only a few of the eigenvalues are to be determined, the *determinant search method* can be very effective in solving the eigenvalue problem. Consider the linearized eigenvalue problem

$$(\mathbf{K} - \lambda\mathbf{M})\mathbf{q} = 0 \tag{11.148}$$

As already stated, the determinant of $(\mathbf{K} - \lambda\mathbf{M})$ gives an Nth-order polynomial whose roots represent the N eigenvalues. The determinant search method uses an iterative procedure to locate as many zeros of the polynomial as are desired. Suppose that the least dominant eigenvalue λ_1 is to be calculated and that two rough estimates of the eigenvalue are available. Let these estimates be μ_{k-2} and μ_{k-1}, where $\mu_{k-2} < \mu_{k-1} < \lambda_1$. The procedure requires that the determinant $|\mathbf{K} - \lambda\mathbf{M}|$ be evaluated for these two estimated values of λ_1. Let the corresponding two values of the polynomial be denoted by p_{k-2} and p_{k-1}. Referring to Figure 11.5a, a new estimate, μ_k, for the zero of the polynomial

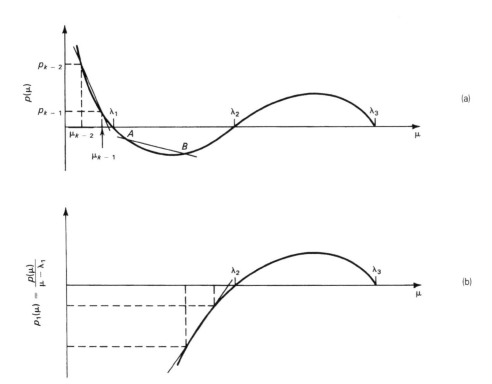

Figure 11.5. Determinant search method of determining eigenvalues: (a) determinant search for λ_1; (b) determinant search for λ_2.

can be obtained by fitting a straight line to the two points $(\mu_{k-2}, \; p_{k-2})$ and $(\mu_{k-1}, \; p_{k-1})$. The new eigenvalue estimate is given by

$$\mu_k = \mu_{k-1} - \eta \frac{\mu_{k-1} - \mu_{k-2}}{p_{k-1} - p_{k-2}} p_{k-1} \tag{11.149}$$

in which $\eta = 1$.

The polynomial is now evaluated for the new estimate μ_k, giving p_k. If p_k is sufficiently close to zero, the iteration process has converged and $\mu_k \simeq \lambda_1$. If convergence has not been achieved, the entire process is repeated with the two most recent values, p_{k-1}, p_k and μ_{k-1}, μ_k.

The procedure described above and represented by Equation 11.149 with $\eta = 1$ is the standard linear interpolation or secant iteration scheme. In this case, the root of the polynomial is approached only from one side, p does not change sign, and there is no chance of jumping over the root. However, the convergence may be slow at times. To accelerate the convergence, a value of 2 or more may be used for η. When $\eta > 2$, there is a possibility of jumping over one or more roots. Fortunately, this can be detected by using the Sturm sequence property so that if necessary, the iteration can be retraced.

A major computation effort is required in evaluating the determinant for a given value $\lambda = \mu$. The evaluation is carried out by obtaining an $\mathbf{LDL}^T$ factorization of matrix $(\mathbf{K} - \mu \mathbf{M})$. The determinant of the matrix is then obtained by taking the product of the diagonal elements of $\mathbf{D}$. According to the Sturm sequence property, the number of negative elements in $\mathbf{D}$ is equal to the number of eigenvalues less than μ. Skipped eigenvalues, if any, can thus be detected by counting the number of negative elements in $\mathbf{D}$.

To begin the iteration for λ_1, two estimates, μ_1 and μ_2, that are both smaller than λ_1 are required. The first estimate, μ_1, can be taken as zero. Because all eigenvalues are positive, $\mu_1 = 0$ is obviously less than or equal to λ_1. To obtain the second estimate, an inverse integration process without shift is used. The Rayleigh quotient estimate of λ_1, obtained after a few iterations, is then used to select a value for μ_2.

The determinant search method can also be employed to converge to higher roots. An additional complication arises this time. For the first root, if the iteration began with two values of μ, both less than λ_1, it should converge toward λ_1. However, for a higher root, this is not guaranteed. For example, iteration with points A and B in Figure 11.5a will converge back to λ_1. To overcome this difficulty, a *deflated polynomial* $p_j(\mu)$ is used instead of $p(\mu)$, where

$$p_j(\mu) = \frac{p(\mu)}{\coprod_{i=1}^{j}(\mu - \lambda_i)} \tag{11.150}$$

and λ_i ($i = 1$ to j) are the eigenvalues already determined. Iteration with $p_j(\mu)$ will converge to the $(j+1)$st eigenvalue, λ_{j+1}. The deflated polynomial for the second root is shown in Figure 11.5b. In practice, the determinant search method is used to obtain only a close estimate of the eigenvalue desired. More accurate estimates of the eigenvalue and the corresponding eigenvector are then obtained by using inverse vector iteration with shifts is described in Section 11.5.3.

It is apparent from the foregoing discussion that determinant search is most effective when used to locate a shift close to the eigenvalue desired. Thus, each eigenpair is calculated independently of others and the precision obtained in determination of previous eigenvalues has little bearing on the accuracy of the eigenpair currently being calculated. Also, since inverse iteration is used with shifts, orthogonalization of the vector need not be carried out the respect to all previous eigenvectors – just a few.

Example 11.8
For a system with the stiffness and mass matrices given below, obtain the eigenvalues and eigenvectors using a combination of determinant search method and inverse iteration with shift.

$$\mathbf{K} = \begin{bmatrix} 3 & -2 & 0 \\ -2 & 8 & -4 \\ 0 & -4 & 5 \end{bmatrix} \qquad \mathbf{M} = \begin{bmatrix} 2 & & \\ & 4 & \\ & & 2 \end{bmatrix}$$

Solution
Since the eigenvalues of a stable system are either equal to zero or positive, we begin the determinant search for the first eigenvalue with two initial estimates, $\mu_{k-2} = -1$ and $\mu_{k-1} = 0$. Both of these values must be less than the true eigenvalue. A new estimate for μ is obtained by using Equation 11.149 with $\eta = 1$. The computations are tabulated as follows.

k	μ_{k-2}	p_{k-2}	μ_{k-1}	p_{k-1}	$\mu_k = \mu_{k-1} - \dfrac{\mu_{k-1} - \mu_{k-2}}{p_{k-1} - p_{k-2}} p_{k-1}$	p_k
1	−1	312	0.0	52	0.2	26.1
2	0	52	0.2	26.1	0.4	7.1
3	0.2	26.1	0.4	7.1	0.475	1.65

It is evident that we are approaching the zero of the polynomial p. At this stage, we use inverse iteration with shift to obtain a more accurate estimate of the eigenvalue as well as the corresponding eigenvector. With a shift of $\mu = 0.475$

$$\hat{\mathbf{K}} = [\mathbf{K} - \mu\mathbf{M}]$$

$$= \begin{bmatrix} 2.05 & -2.00 & 0 \\ -2.00 & 6.10 & -4.00 \\ 0 & -4.00 & 4.05 \end{bmatrix}$$

The new eigenvalue problem is

$$\hat{\mathbf{K}}\mathbf{q} = \delta\mathbf{M}\mathbf{q}$$

Inverse iteration is carried out by using Equations 11.73 through 11.76. The initial trial vector is taken as

$$
y_1 = M \begin{bmatrix} 1 \\ 1 \\ 1 \end{bmatrix} = \begin{bmatrix} 2 \\ 4 \\ 2 \end{bmatrix}
$$

k	$\mathbf{x}_k$	$\bar{\mathbf{u}}_{k+1}$	$\bar{\mathbf{x}}_{k+1}$	$\bar{\mathbf{u}}_{k+1}^T \mathbf{x}_k$	$\bar{\mathbf{u}}_{k+1}^T \bar{\mathbf{x}}_{k+1}$	ρ_{k+1}	$\mathbf{x}_{k+1}$
1	$\begin{bmatrix} 2 \\ 4 \\ 2 \end{bmatrix}$	$\begin{bmatrix} 40 \\ 40 \\ 40 \end{bmatrix}$	$\begin{bmatrix} 80 \\ 160 \\ 80 \end{bmatrix}$	320	12,800	0.025	$\begin{bmatrix} 0.707 \\ 1.414 \\ 0.707 \end{bmatrix}$
2	$\begin{bmatrix} 0.707 \\ 1.414 \\ 0.707 \end{bmatrix}$	$\begin{bmatrix} 14.14 \\ 14.14 \\ 14.14 \end{bmatrix}$	$\begin{bmatrix} 28.28 \\ 56.57 \\ 28.28 \end{bmatrix}$	39.98	1,599.7	0.025	

Convergence has been achieved at this stage. We therefore have

$$
\delta = \rho_3
$$
$$
= 0.025
$$
$$
\lambda_1 = \mu + \delta
$$
$$
= 0.475 + 0.025 = 0.50
$$
$$
\phi_1 = \frac{\bar{\mathbf{u}}_3}{[\bar{\mathbf{u}}_3^T \bar{\mathbf{x}}_3]^{1/2}}
$$
$$
= \begin{bmatrix} 0.3535 \\ 0.3535 \\ 0.3535 \end{bmatrix}
$$

Determinant search for the second eigenvalue is carried out on a deflated polynomial given by

$$
p^{(2)} = \frac{p}{\mu - 0.5}
$$

To carry out the iterations, we need two estimates for the next eigenvalues. We will take these estimates as 0.6 and 1.0, respectively. It must be ensured that in doing so, we have not skipped the second eigenvalue; that is, none of the two estimates is larger than the second eigenvalue. This could be verified by using a Sturm sequence check. If $\mathbf{LDL}^T$ decomposition of $[\mathbf{K} - \mu\mathbf{M}]$ shows two negatives on $\mathbf{D}$, it is evident that the value of μ selected is larger than the first two eigenvalues and that a lower value should be selected for μ. The computations for the second eigenvalue are tabulated as follows.

k	μ_{k-2}	$p_{k-2}^{(2)}$	μ_{k-1}	$p_{k-1}^{(2)}$	μ_k	$p_k^{(2)}$
1	0.6	-56	1	-32	1.53	-6.68
1	1.0	-32	1.53	-6.68	1.67	-0.21

As this stage, we use inverse iteration with a shift of 1.67 to determine the second eigenvalue and the second eigenvector. If the shift point is closer to the second eigenvalue than to the first one, iteration should converge to the second eigenvalue. However, when eigenvalues are coincident or lie close together, the eigenvectors already determined must be swept from the trial vector. In practical applications, it is reasonable to sweep off, say, the six previous eigenvectors. For the present case, for simplicity, we will assume that it is not necessary to sweep off the previous eigenvector. The new stiffness matrix is

$$\hat{\mathbf{K}} = \begin{bmatrix} 3 & -2 & 0 \\ -2 & 8 & -4 \\ 0 & -4 & 5 \end{bmatrix} - 1.67 \begin{bmatrix} 2 & & \\ & 4 & \\ & & 2 \end{bmatrix}$$

$$= \begin{bmatrix} -0.34 & -2 & 0 \\ -2 & 1.32 & -4 \\ 0 & -4 & 1.66 \end{bmatrix}$$

The computations involved in the inverse iteration are tabulated below. We take care that the initial trial vector is not of the form $\mathbf{u}_1^T = [1 \quad 1 \quad 1]$, because that being precisely equal to the first mode shape, the iteration is likely to converge to the first eigenvalue. We start with $\mathbf{u}_1^T = [1 \quad 0 \quad 0]$, so that

$$x_1 = \mathbf{M}\mathbf{u}_1$$

$$= \begin{bmatrix} 2 \\ 0 \\ 0 \end{bmatrix}$$

The computations involved in the inverse iteration are tabulated below.

k	$\mathbf{x}_k$	$\bar{\mathbf{u}}_{k+1}$	$\bar{\mathbf{x}}_{k+1}$	$\bar{\mathbf{u}}_{k+1}^T \mathbf{x}_k$	$\bar{\mathbf{u}}_{k+1}^T \bar{\mathbf{x}}_{k+1}$	ρ_{k+1}	$\mathbf{x}_{k+1}$
1	$\begin{bmatrix} 2 \\ 0 \\ 0 \end{bmatrix}$	$\begin{bmatrix} 14.20 \\ -3.41 \\ -8.23 \end{bmatrix}$	$\begin{bmatrix} 28.40 \\ -13.65 \\ -16.45 \end{bmatrix}$	28.4	585.2	0.0485	$\begin{bmatrix} 1.17 \\ -0.56 \\ -0.68 \end{bmatrix}$
2	$\begin{bmatrix} 1.17 \\ -0.56 \\ -0.68 \end{bmatrix}$	$\begin{bmatrix} 12.10 \\ -2.64 \\ -6.78 \end{bmatrix}$	$\begin{bmatrix} 24.19 \\ -10.57 \\ -13.56 \end{bmatrix}$	20.25	412.5	0.0491	$\begin{bmatrix} 1.190 \\ -0.521 \\ -0.667 \end{bmatrix}$
3	$\begin{bmatrix} 1.190 \\ -0.521 \\ -0.667 \end{bmatrix}$	$\begin{bmatrix} 12.09 \\ -2.65 \\ -6.79 \end{bmatrix}$	$\begin{bmatrix} 24.18 \\ -10.60 \\ -13.58 \end{bmatrix}$	20.29	412.3	0.0492	

Convergence has been achieved at this stage. We therefore have

$$\delta = 0.0492$$

$$\lambda = \mu + \delta$$

$$= 1.67 + 0.0491$$

$$= 1.719$$

$$\phi_2 = \frac{\bar{\mathbf{u}}_4}{[\bar{\mathbf{u}}_4^T \bar{\mathbf{x}}_4]^{1/2}}$$

$$= \begin{bmatrix} 0.5952 \\ -0.1305 \\ -0.3342 \end{bmatrix}$$

In carrying out the determinant search for the third eigenvalue, we use the deflated polynomial

$$p^{(3)} = \frac{p}{(\mu - 0.5)(\mu - 1.719)}$$

For our two initial estimates of λ_3, we choose 2.0 and 2.5, respectively. The remaining computations are tabulated as follows

k	μ_{k-2}	$p_{k-2}^{(3)}$	μ_{k-1}	$p_{k-1}^{(3)}$	μ_k	$p_k^{(3)}$
1	2.0	14.24	2.5	10.24	3.781	0

We have converged exactly on the third eigenvalue. This is to be expected because $p^{(3)}$ is a linear function of μ, and secant iteration should therefore lead to the true zero in one step. To obtain the eigenvector, we may carry out an inverse iteration with a shift of 3.75. The final result are

$$\delta = 0.031$$

$$\lambda = \mu + \delta$$

$$= 3.75 + 0.031$$

$$= 3.781$$

$$\phi_3 = \begin{bmatrix} 0.1441 \\ -0.3286 \\ 0.5131 \end{bmatrix}$$

11.7 NUMERICAL SOLUTION OF COMPLEX EIGENVALUE PROBLEM

11.7.1 *Eigenvalue problem and the orthogonality relationship*

A majority of eigenvalue problems in engineering analysis involve matrices that are symmetric and positive definite. Occasionally, cases arise where eigenvalues of a real unsymmetric or even a complex matrix may be required. Systems with nonproportional damping discussed in Section 10.8 belong to the former category. The eigenvalue problem is, in such a case, given by

$$\mathbf{B}\mathbf{v} = s\mathbf{A}\mathbf{v} \qquad\qquad (11.151)$$

where **A** and **B** are the following matrices, each of order $2N$

$$\mathbf{A} = \begin{bmatrix} \mathbf{0} & -\mathbf{M} \\ -\mathbf{M} & -\mathbf{C} \end{bmatrix} \qquad \mathbf{B} = \begin{bmatrix} -\mathbf{M} & \mathbf{0} \\ \mathbf{0} & \mathbf{K} \end{bmatrix} \tag{11.152a}$$

and vector **v** is given by

$$\mathbf{v} = \begin{bmatrix} \dot{\mathbf{u}} \\ \mathbf{u} \end{bmatrix} \tag{11.152b}$$

It should be noted that although both **A** and **B** are symmetric and real, they are not positive definite.

The eigenvalue problem of Equation 11.151 can be expressed in the alternative form

$$\mathbf{Ev} = s\mathbf{v} \tag{11.153}$$

where $\mathbf{E} = \mathbf{A}^{-1}\mathbf{B}$ is given by

$$\mathbf{E} = \begin{bmatrix} -\mathbf{M}^{-1}\mathbf{C} & -\mathbf{M}^{-1}\mathbf{K} \\ \mathbf{I} & \mathbf{0} \end{bmatrix} \tag{11.154}$$

Matrix **E** is real but unsymmetric. For an underdamped system the $2N$ eigenvalues, s, and eigenvectors, **v**, are both complex. Furthermore, they both appear in complex conjugate pairs, and for a system that exhibits exponentially decaying free vibrations, the real part of each eigenvalue is negative.

Equation 11.151 can also be expressed in the form

$$\mathbf{B}^{-1}\mathbf{Av} = \frac{1}{s}\mathbf{v} \tag{11.155}$$

or ·

$$\mathbf{Dv} = \gamma \mathbf{v} \tag{11.156}$$

where $\mathbf{D} = \mathbf{B}^{-1}\mathbf{A}$ and $\gamma = 1/s$. Matrix **D**, called the dynamic matrix, is given by

$$\mathbf{D} = \begin{bmatrix} \mathbf{0} & \mathbf{I} \\ -\mathbf{K}^{-1}\mathbf{M} & -\mathbf{K}^{-1}\mathbf{C} \end{bmatrix} \tag{11.157}$$

in which submatrix $\mathbf{K}^{-1}\mathbf{M}$ would be recognized as the dynamic matrix for the undamped case. We denote the jth eigenvalue s_j as

$$s_j = \alpha_j + i\beta_j \tag{11.158}$$

and its complex conjugate $\bar{s}_j$ as

$$\bar{s}_j = \alpha_j - i\beta_j \tag{11.159}$$

where α_j and β_j are both real numbers. The corresponding eigenvectors are denoted by

$$\mathbf{v}_j = \boldsymbol{\xi}_j + i\boldsymbol{\eta}_j \tag{11.160a}$$

$$\bar{\mathbf{v}}_j = \boldsymbol{\xi}_j - i\boldsymbol{\eta}_j \tag{11.160b}$$

As proved in Section 10.8, the eigenvectors are orthogonal with respect to both **A** and **B**, so that

$$\mathbf{v}_j^T \mathbf{A} \mathbf{v}_k = 0 \qquad j \neq k \tag{11.161a}$$

$$\mathbf{v}_j^T \mathbf{B} \mathbf{v}_k = 0 \qquad j \neq k \tag{11.161b}$$

The orthogonality relationships hold even between two eigenvectors that are complex conjugated of each other, since their eigenvalues are different.

It is useful to express the orthogonality relationships in terms of the real and imaginary parts. Thus, Equation 11.161a is written as

$$(\boldsymbol{\xi}_j^T + i\boldsymbol{\eta}_j^T)\mathbf{A}(\boldsymbol{\xi}_k + i\boldsymbol{\eta}_k) = 0 \tag{11.162}$$

and since eigenvector $\mathbf{v}_k$ is orthogonal also to the conjugate of eigenvector $\mathbf{v}_j$

$$(\boldsymbol{\xi}_j^T - i\boldsymbol{\eta}_j^T)\mathbf{A}(\boldsymbol{\xi}_k + i\boldsymbol{\eta}_k) = 0 \tag{11.163}$$

Expansion of Equations 11.162 and 11.163 gives

$$\boldsymbol{\xi}_j^T \mathbf{A} \boldsymbol{\xi}_k - \boldsymbol{\eta}_j^T \mathbf{A} \boldsymbol{\eta}_k = 0$$

$$\boldsymbol{\eta}_j^T \mathbf{A} \boldsymbol{\xi}_k + \boldsymbol{\xi}_j^T \mathbf{A} \boldsymbol{\eta}_k = 0$$

$$\boldsymbol{\xi}_j^T \mathbf{A} \boldsymbol{\xi}_k + \boldsymbol{\eta}_j^T \mathbf{A} \boldsymbol{\eta}_k = 0 \tag{11.164}$$

$$\boldsymbol{\eta}_j^T \mathbf{A} \boldsymbol{\xi}_k - \boldsymbol{\xi}_j^T \mathbf{A} \boldsymbol{\eta}_k = 0$$

Combining the first and third, and second and fourth of Equations 11.164, we get the family of orthogonality relationships

$$\begin{aligned} \boldsymbol{\xi}_j^T \mathbf{A} \boldsymbol{\xi}_k &= 0 \\ \boldsymbol{\eta}_j^T \mathbf{A} \boldsymbol{\eta}_k &= 0 \\ \boldsymbol{\xi}_j^T \mathbf{A} \boldsymbol{\eta}_k &= 0 \\ \boldsymbol{\eta}_j^T \mathbf{A} \boldsymbol{\xi}_k &= 0 \end{aligned} \qquad j \neq k \tag{11.165}$$

A similar set of results can be expressed with respect to matrix **B**.

Now, since an eigenvector is also orthogonal to its own conjugate, we have

$$(\boldsymbol{\xi}_j^T + i\boldsymbol{\eta}_j^T)\mathbf{A}(\boldsymbol{\xi}_j - i\boldsymbol{\eta}_j) = 0 \tag{11.166}$$

and therefore

$$\boldsymbol{\xi}_j^T \mathbf{A} \boldsymbol{\xi}_j = -\boldsymbol{\eta}_j^T \mathbf{A} \boldsymbol{\eta}_j$$

$$\boldsymbol{\eta}_j^T \mathbf{A} \boldsymbol{\xi}_j = \boldsymbol{\xi}_j^T \mathbf{A} \boldsymbol{\eta}_j \tag{11.167}$$

11.7.2 *Matrix iteration for determining the complex eigenvalues*

The orthogonality relationships proved above are useful in developing a numerical technique for the solution of the aforesaid complex eigenvalue problem by a process of inverse iteration similar to that used in Section 11.5.2. It is in fact possible to start with an initial trial vector that is real and work entirely with real numbers to compute both the real and the imaginary parts of eigenvalues and eigenvectors. To develop the procedure, let us begin the iteration with a complex trial vector $\mathbf{p}_1^{(0)} + i\mathbf{q}_1^{(0)}$. The superscript (0) indicates an initial trial vector and the subscript 1 refers to the fact that the iteration are for the first eigenvector. For simplicity, we omit the subscript in the following discussion. Also, we will express the inverse eigenvalue γ in terms of its real and imaginary components

$$\gamma = \mu + iv \tag{11.168}$$

The iteration involves repeated multiplication of the trial vector by matrix $\mathbf{D}$. A single step in this iteration process is given by

$$\mathbf{D}(\mathbf{p}^{(n)} + i\mathbf{q}^{(n)}) = (\mu + iv)(\mathbf{p}^{(n)} + i\mathbf{q}^{(n)})$$

$$= \mathbf{p}^{(n+1)} + i\mathbf{q}^{(n+1)} \tag{11.169}$$

Expansion of Equation 11.169 gives

$$\mathbf{D}\mathbf{p}^{(n)} = \mu\mathbf{p}^{(n)} - v\mathbf{q}^{(n)}$$

$$= \mathbf{p}^{(n+1)} \tag{11.170a}$$

$$\mathbf{D}\mathbf{q}^{(n)} = \mu\mathbf{q}^{(n)} + v\mathbf{p}^{(n)}$$

$$= \mathbf{q}^{(n+1)} \tag{11.170b}$$

By eliminating $\mathbf{q}^{(n)}$ from Equations 11.170a and 11.170b, we get an expression for $\mathbf{q}^{(n+1)}$ in terms of $\mathbf{p}^{(n)}$ and $\mathbf{p}^{(n+1)}$

$$v\mathbf{q}^{(n+1)} = (\mu^2 + v^2)\mathbf{p}^{(n)} - \mu\mathbf{p}^{(n+1)} \tag{11.171}$$

Next, we write Equation 11.170a at step $n + 1$, so that

$$\mathbf{D}\mathbf{p}^{(n+1)} = \mu\mathbf{p}^{(n+1)} - \nu\mathbf{q}^{(n+1)}$$

$$= \mathbf{p}^{(n+2)} \tag{11.172}$$

or

$$\nu\mathbf{q}^{(n+1)} = \mu\mathbf{p}^{(n+1)} - \mathbf{p}^{(n+2)} \tag{11.173}$$

Substitution of Equation 11.173 in Equation 11.171 gives

$$\mathbf{p}^{(n+2)} - 2\mu\mathbf{p}^{(n+1)} + (\mu^2 + \nu^2)\mathbf{p}^{(n)} = 0 \tag{11.174}$$

Equation 11.174 can be viewed as a recursion equation that relates vector $\mathbf{p}$ at three successive iteration steps. Thus, the next equation in the sequence will be

$$\mathbf{p}^{(n+3)} - 2\mu\mathbf{p}^{(n+2)} + (\mu^2 + \nu^2)\mathbf{p}^{(n+1)} = 0 \tag{11.175}$$

Equations 11.174 and 11.175 hold for every element of vector $\mathbf{p}$. By solving these equations for any one of the elements of $\mathbf{p}$, we obtain the real part of the eigenvalue

$$\mu = \frac{1}{2} \frac{p^{(n)} p^{(n+3)} - p^{(n+1)} p^{(n+2)}}{p^{(n)} p^{(n+2)} - p^{(n+1)} p^{(n+1)}} \tag{11.176}$$

in which p denotes one of the elements of vector $\mathbf{p}$.

The same two equations, Equations 11.174 and 11.175, also give the squared amplitude $(\mu^2 + \nu^2)$ of the complex eigenvalue

$$\mu^2 + \nu^2 = \frac{p^{(n+1)} p^{(n+3)} - p^{(n+2)} p^{(n+2)}}{p^{(n)} p^{(n+2)} - p^{(n+1)} p^{(n+1)}} \tag{11.177}$$

With a sufficient number of iterations, μ will converge to the real part of the inverse eigenvalue, γ, having the highest amplitude and $\mu^2 + \nu^2$ will converge to the squared amplitude of that eigenvalue. Simultaneously, $\mathbf{p}^{(n)}$ will converge to the real part of the eigenvector. The imaginary part of the eigenvector can be obtained from Equation 11.170a

$$\mathbf{q}^{(n)} = \frac{\mu}{\nu}\mathbf{p}^{(n)} - \frac{1}{\nu}\mathbf{p}^{(n+1)} \tag{11.178}$$

To obtain the eigenvalue $s = \alpha + i\beta$, we use the relationships

$$\alpha = \frac{\mu}{\mu^2 + \nu^2} \tag{11.179a}$$

$$\beta = -\frac{\nu}{\mu^2 + \nu^2} \tag{11.179b}$$

Convergence cannot, in this case, be measured by comparing the elements of two successive values of the trial vector. Instead, comparison must be made between two successive values of μ or of the squared amplitude $\mu^2 + \nu^2$.

The procedure just described gives the eigenvalue with the lowest amplitude, or the inverse eigenvalue with highest amplitude, as well as the corresponding eigenvector. To obtain a higher eigenvalue, the eigenvalues already determined must be swept off from the trial vector. This is achieved by writing the orthogonality relationships between a trial vector for determining the nth eigenvalue and the $n-1$ eigenvectors determined previously. Let us assume that the eigenvector $(\xi_1 + i\eta_1)$ has already been obtained and the second eigenvalue and eigenvector are to be determined. Denoting the trial vector for the second eigenvector by $\mathbf{p}_2 + i\mathbf{q}_2$, we have

$$\begin{aligned}
\boldsymbol{\xi}_1^T \mathbf{A}\mathbf{p}_2 &= 0 \\
\boldsymbol{\eta}_1^T \mathbf{A}\mathbf{q}_2 &= 0 \\
\boldsymbol{\xi}_1^T \mathbf{A}\mathbf{q}_2 &= 0 \\
\boldsymbol{\eta}_1^T \mathbf{A}\mathbf{p}_2 &= 0
\end{aligned} \tag{11.180}$$

Since we carry out our iterations on a real vector, we need only the first and fourth of Equations 11.180. Observing that there are two constraint equations, we can assign arbitrary value to all elements of the trial vector except two, say the first two. The first two elements are then selected so that the complete trial vector satisfies the two constraint equations. The procedure is entirely similar to that described in Section 11.5.2, in which the constraint conditions were used to obtain a sweeping matrix. Premultiplication of the trial vector by this matrix resulted in the first eigenvector being swept off. In the present case, the sweeping matrix obtained from the two constraint equations will have two columns of zeros. This is to be expected because in calculating the first eigenvector, we have, in fact, obtained a complex conjugate pair rather than a single vector.

The use of the fourth equation of Equations 11.180 requires the computation of $\boldsymbol{\eta}_1$, the imaginary part of the first eigenvector. Vector $\boldsymbol{\eta}_1 = \mathbf{q}^{(n)}$ was probably calculated from Equation 11.178 during determination of the first eigenvector. However, as an alternative, we can use the following procedure, which avoids the use of $\boldsymbol{\eta}_1$. Noting that $\boldsymbol{\xi}_1 = \mathbf{p}_1^{(n)}$, we express the first of Equations 11.180 as

$$\{\mathbf{p}_1^{(n)}\}^T \mathbf{A}\mathbf{p}_2 = 0 \tag{11.181}$$

In a similar manner, noting that $\boldsymbol{\eta}_1 = \mathbf{q}_1^{(n)}$, we write the fourth of Equations 11.180 as

$$\{\mathbf{q}_1^{(n)}\}^T \mathbf{A}\mathbf{p}_2 = 0 \tag{11.182}$$

Substitution for $\mathbf{q}_1^{(n)}$ from Equation 11.178 into Equation 11.182 gives

$$\frac{\mu}{\nu}\{\mathbf{p}_1^{(n)}\}^T \mathbf{A}\mathbf{p}_2 - \frac{1}{\nu}\{\mathbf{p}_1^{(n+1)}\}^T \mathbf{A}\mathbf{p}_2 = 0 \tag{11.183}$$

The first term in Equation 11.183 is zero because of the relationship in Equation 11.181. Hence

$$\{\mathbf{p}_1^{(n+1)}\}^T \mathbf{A}\mathbf{p}_2 = 0 \tag{11.184}$$

Equations 11.181 and 11.184 represent the two constraints that $\mathbf{p}_2$ must satisfy.

Example 11.9

For the two-degree-of-freedom system described in Example 10.3, obtain the complex mode shapes and frequencies by the iteration method.

Solution

The mass, stiffness, and damping matrices are given in Example 10.3. The eigenvalue problem is defined by Equation 11.151

$$\mathbf{B}\mathbf{v} = s\mathbf{A}\mathbf{v}$$

where

$$\mathbf{A} = \begin{bmatrix} 0 & 0 & -2.0 & 0 \\ 0 & 0 & 0 & -1.0 \\ -2.0 & 0 & -0.4 & 0.05 \\ 0 & -1.0 & 0.05 & -0.2 \end{bmatrix} \tag{a}$$

and

$$\mathbf{B} = \begin{bmatrix} -2.0 & 0 & 0 & 0 \\ 0 & -1.0 & 0 & 0 \\ 0 & 0 & 3.0 & -1.0 \\ 0 & 0 & -1.0 & 1.0 \end{bmatrix} \tag{b}$$

For solution by the inverse iteration method, we express the problem in the form

$$\mathbf{D}\mathbf{v} = \gamma\mathbf{v} \tag{c}$$

where the dynamic matrix $\mathbf{D} = \mathbf{B}^{-1}\mathbf{A}$ is given by

$$\mathbf{D} = \begin{bmatrix} 0 & 0 & 1.0 & 0 \\ 0 & 0 & 0 & 1.0 \\ -1.0 & -0.5 & -0.175 & -0.075 \\ -1.0 & -1.5 & -0.125 & -0.274 \end{bmatrix} \tag{d}$$

Iteration is begun with a trial vector given by $\mathbf{p}^T = [1 \quad 1 \quad 1 \quad 1]$. Successive vectors obtained in the iteration process are tabulated as follows

$n=0$	$n=1$	$n=2$	$n=3$	$n=4$	$n=5$	$n=6$	$n=7$	$n=8$	$n=9$
1.0000	1.0000	−1.7500	−0.9763	3.4821	0.6115	−6.9749	1.1697	13.4550	−6.8456
1.0000	1.0000	−2.9000	−1.4837	6.6301	0.9433	−13.7630	2.6302	26.7500	−14.1530
1.0000	−1.7500	−0.9763	3.4821	0.6115	−6.9749	1.1697	13.4550	−6.8456	−24.5700
1.0000	−2.9000	−1.4837	6.6301	0.9433	−13.7630	2.6302	26.7500	−14.1530	−48.8320

After the first three new vectors have been calculated, Equations 11.176 and 11.177 are used to obtain estimates for μ and $\mu^2 + v^2$, respectively. The two equations can be evaluated using any one of the four elements of the iteration vector. In the following, we tabulate the values of μ and $\mu^2 + v^2$ obtained by using the third elements of the iteration vectors.

n	$p_3^{(n)}$	μ (Eq. 11.176)	$\mu^2 + v^2$ (Eq. 11.177)
0	1.0000	−0.21958	1.7448
1	−1.7500	−0.16527	1.8054
2	−0.9763	−0.18393	1.9385
3	3.4821	−0.16905	1.9437
4	0.6115	−0.17092	1.9863
5	−6.9749	−0.16810	1.9854
6	1.1697	−0.16761	1.9967
7	13.4550		
8	−6.8456		
9	−24.5700		

For the purpose of this example, we accept the convergence achieved at the sixth iteration ($n=6$). We can now use Equations 11.179a and 11.179b to obtain the real and imaginary parts of the eigenvalue s. Thus

$$\mu = -0.16761$$
$$v = \pm 1.4031 \tag{e}$$

$$\alpha = \frac{\mu}{\mu^2 + v^2} = -0.08394$$

$$\beta = \frac{-v}{\mu^2 + v^2} = \mp 0.70269 \tag{f}$$

The real part of the eigenvector is taken from the first table above, corresponding to $n = 6$

$$\xi = \begin{bmatrix} -6.9749 \\ -13.7630 \\ 1.1697 \\ 2.6302 \end{bmatrix} \tag{g}$$

The imaginary part of the eigenvector can be calculated from Equation 11.178

$$\boldsymbol{\eta} = \mathbf{q}^{(6)} = \frac{\mu}{\nu}\mathbf{p}^{(6)} - \frac{1}{\nu}\mathbf{p}^{(7)} \tag{h}$$

which gives

$$\boldsymbol{\eta} = \pm \begin{bmatrix} -0.00046122 \\ -0.23053 \\ -9.7290 \\ -19.380 \end{bmatrix} \tag{i}$$

Vectors $\boldsymbol{\xi}$ and $\boldsymbol{\eta}$ can be normalized with respect to matrix $\mathbf{A}$ so that

$$(\boldsymbol{\xi} + i\boldsymbol{\eta})^T \mathbf{A}(\boldsymbol{\xi} + i\boldsymbol{\eta}) = 1 \tag{j}$$

The normalized vectors are

$$\boldsymbol{\xi} = \begin{bmatrix} 0.19471 \\ 0.37928 \\ -0.24165 \\ -0.48973 \end{bmatrix} \tag{k}$$

$$\boldsymbol{\eta} = \pm \begin{bmatrix} 0.14985 \\ 0.30209 \\ 0.24648 \\ 0.48453 \end{bmatrix} \tag{l}$$

To obtain the second eigenvectors, we note that its real part must satisfy the two constraints, Equations 11.181 and 11.184. Substituting for $p^{(6)}$ and $p^{(7)}$ and $\mathbf{A}$ in these equations, we get

$$2.3393\,p_{1,2} + 2.6302\,p_{2,2} - 13.613\,p_{3,2} - 13.296\,p_{4,2} = 0$$
$$26.909\,p_{1,2} + 26.750\,p_{2,2} + 6.3836\,p_{3,2} + 7.3075\,p_{4,2} = 0 \tag{m}$$

In the constraint equations above, the first subscript on p denotes the element number in vector $\mathbf{p}$, and the second subscript signifies that the iteration vector is for the second mode.

With two constraint equations, we are free to assign arbitrary values to any two of the four elements of $\mathbf{p}_2$; the other two are then selected so that the constraint equations are satisfied. Let us choose to express elements p_1 and p_3 in terms of p_2 and p_4. Then, denoting the purified vector by $\tilde{\mathbf{p}}_2$, we can express the constrain relationship as

$$\tilde{\mathbf{p}}_2 = \mathbf{S}\mathbf{p}_2 \tag{n}$$

where $\mathbf{S}$ is the sweeping matrix given by

$$\mathbf{S} = \begin{bmatrix} 0 & -0.99919 & 0 & -0.38311 \\ 0 & 1 & 0 & 0 \\ 0 & -0.021504 & 0 & -0.98323 \\ 0 & 0 & 0 & 1 \end{bmatrix} \tag{o}$$

The dynamic matrix for calculating the second set of eigenvalues and eigenvectors is obtained from

$$\mathbf{D}_2 = \mathbf{DS}$$

$$= \begin{bmatrix} 0 & 0.021504 & 0 & -0.98323 \\ 0 & 0 & 0 & 1 \\ 0 & 0.49543 & 0 & 0.13538 \\ 0 & -0.50359 & 0 & -0.11379 \end{bmatrix} \tag{p}$$

Iterations are again begun with a trial vector given by $\mathbf{p}_2^T = [1 \quad 1 \quad 1 \quad 1]$. Successive vectors obtained in the iteration process are tabulated as follows

$n=0$	$n=1$	$n=2$	$n=3$	$n=4$
1.0000	-0.96172	0.62843	0.41272	-0.36337
1.0000	1.00000	-0.61728	-0.43326	0.36010
1.0000	0.63081	0.41137	-0.36447	-0.16590
1.0000	-0.61728	-0.43326	0.36010	0.17717

Estimates of μ and $\mu^2 + \nu^2$ are now obtained from Equations 11.176 and 11.177, respectively, and are tabulated as follows. Again, we use the third element of the iteration vector in our calculations.

n	$\mathbf{p}_{3,2}^{(n)}$	μ_2 (Eq. 11.176)	$\mu^2 + \nu^2$ (Eq. 11.177)
0	1.00000	-0.056893	0.503495
1	-0.61728	-0.056893	0.503495
2	-0.43326		
3	0.36010		
4	0.17717		

Convergence has been achieved at $n=1$. The corresponding values of μ, ν, α, and β are obtained as follows

$$\mu_2 = -0.056893$$

$$\nu_2 = \pm 0.70729$$

$$\alpha_2 = -0.11300 \tag{q}$$

$$\beta_2 = \mp 1.4048$$

The real part of the eigenvectors is equal to the iteration vector for $n=1$

$$\xi_2 = \begin{bmatrix} -0.96172 \\ 1.00000 \\ 0.63081 \\ -0.61728 \end{bmatrix} \tag{r}$$

The imaginary part is obtained from Equation 11.178

$$\boldsymbol{\eta}_2 = \pm \begin{bmatrix} -0.81115 \\ 0.79230 \\ -0.63306 \\ 0.66221 \end{bmatrix} \tag{s}$$

The normalized eigenvectors are given by

$$\boldsymbol{\xi}_2 = \begin{bmatrix} 0.36533 \\ -0.38012 \\ -0.24526 \\ 0.24021 \end{bmatrix} \tag{t}$$

$$\boldsymbol{\eta}_2 = \pm \begin{bmatrix} 0.31610 \\ -0.30905 \\ 0.24001 \\ -0.25126 \end{bmatrix} \tag{u}$$

The results obtained in this example compare well with the more exact values given in Example 10.3. The small differences exist because, in the present case, iterations for the first set of vectors were terminated early at $n = 6$.

11.8 SEMI-DEFINITE OR UNRESTRAINED SYSTEMS

11.8.1 *Characteristics of an unrestrained system*

It can be shown that for a stable system that is adequately restrained against rigid-body motion, the stiffness matrix is *positive definite*. This follows from the fact that when such a system is subjected to arbitrary displacement, the resulting strain energy stored in the system is always positive. If the arbitrary displacement vector is **u**, the forces required to cause such displacements are $\mathbf{f} = \mathbf{Ku}$. The work done by these forces as they gradually increase from zero to their full value is stored as strain energy V, which is given by

$$V = \frac{1}{2}\mathbf{u}^T\mathbf{f}$$

$$= \frac{1}{2}\mathbf{u}^T\mathbf{Ku} \tag{11.185}$$

Since V is positive for any arbitrary **u**, the term $\mathbf{u}^T\mathbf{Ku}$ is also positive. By definition, therefore, **K** is a positive definite matrix. In systems that are not properly restrained, rigid-body displacements can take place without the application of any force. Thus, denoting a possible rigid-body displacement by $\mathbf{u}_r$, we have

$$\mathbf{f}_r = \mathbf{K}\mathbf{u}_r = 0 \tag{11.186}$$

For nonzero $\mathbf{u}_r$, Equation 11.186 can be satisfied provided only that $\mathbf{K}$ is singular. It follows from Equation 11.186 that the strain energy stored in a rigid-body displacement is also zero

$$
\begin{aligned}
V_r &= \frac{1}{2}\mathbf{u}_r^T\mathbf{f}_r \\
&= \frac{1}{2}\mathbf{u}_r^T\mathbf{K}\mathbf{u}_r \\
&= 0
\end{aligned}
\tag{11.187}
$$

A matrix $\mathbf{K}$ which has the property that for arbitrary $\mathbf{u}$, the quadratic form $\mathbf{u}^T\mathbf{K}\mathbf{u}$ is either positive or zero is called a *semi-definite matrix*. Therefore, an unrestrained system that permits rigid-body displacement has a semi-definite stiffness matrix. Such a matrix is also singular. To illustrate these concepts, consider the system of Example 11.6. The stiffness matrix of the system is given by

$$\mathbf{K} = \begin{bmatrix} k_1 & -k_1 & 0 \\ -k_1 & k_1 + k_2 & -k_2 \\ 0 & -k_2 & k_2 \end{bmatrix} \tag{11.188}$$

Suppose that we wish to impose arbitrary displacements u_1, u_2, and u_3 along coordinates 1, 2, and 3, respectively. External forces must be applied to the system to maintain the specified displacements. These forces will perform work as they undergo their corresponding displacements, and the work done will be stored in the system as stain energy. Now, consider a special displacement pattern given by $u_1 = u_2 = u_3$. The forces required to cause such a displacement can be found by substitution in Equation 11.186 and are easily seen to be zero. The corresponding strain energy is also therefore zero. The special displacement pattern $u_1 = u_2 = u_3$ will be recognized as a rigid-body motion of the system. In fact, since the system is constrained to move in the plane of the paper, this is the only possible rigid-body motion. As would be expected, the stiffness matrix given by Equation 11.188 is singular, the second row of the matrix being a linear combination of the first and the last rows. The degree of singularity of the matrix is 1, corresponding to the one possible rigid-body displacement pattern.

It is of interest to note that a rigid-body displacement shape is also an eigenvector of the unrestrained system. For this statement to be true, the displacement vector should satisfy Equation 10.4 or equivalently, Equation 10.7. Thus

$$\mathbf{K}\mathbf{u}_r = \omega^2 \mathbf{M}\mathbf{u}_r \tag{11.189}$$

Since $\mathbf{Ku}_r$ is zero, Equation 11.189 can be satisfied by choosing $\omega = 0$. This proves the statement that $\mathbf{u}_r$ is an eigenvector or a mode shape, the corresponding eigenvalue or frequency being zero. The number of zero eigenvalues of Equation 11.189 is equal to the number of rigid-body modes, that is, equal to the degree of singularity of matrix $\mathbf{K}$. Rigid-body displacement shapes are also referred to as rigid-body modes. They have the same properties as other mode shapes. In particular, they satisfy the orthogonality relationships given by Equations 10.14 and 10.15.

A system can, of course, have more than one rigid-body mode. In the most general case, up to six rigid-body modes are possible. For example, a spacecraft or an aeroplane in flight has all six possible rigid-body modes, three translations and three rotations, one along each of the three axes.

11.8.2 *Eigenvalue solution of a semi-definite system*

As pointed out in the foregoing paragraphs, an unrestrained or semi-definite system has a singular stiffness matrix. This implies that in evaluating the eigenvalue of such a system, all those methods that rely on obtaining an inverse of the stiffness matrix cannot be used. Thus, the linearized eigenvalue problem cannot be reduced to a symmetric form by using Equation 11.15b. Also, it is not possible to use inverse iteration, the subspace iteration, or the Lanczos iteration methods to obtain an eigenvalue solution.

As noted in earlier sections of this chapter, in many cases, an iteration method may be the most efficient method of finding the mode shapes and frequencies of a system. Such is the case, for example, when the system is large and only the first few mode shapes and frequencies are required. Iteration methods cannot, however, be applied directly to the eigensolution of unrestrained system and special procedures must therefore be used in such cases. Three alternative procedures exist and are described in the following paragraphs.

Addition of a small fictitious stiffness
The singular stiffness matrix of the unrestrained system can be made nonsingular by adding a small but fictitious stiffness along an adequate number of degrees of freedom. This is equivalent to adding restraining springs along one or more degrees of freedom, so that the rigid-body motions are inhibited. To account for the presence of such springs, appropriate stiffness terms must be added to the diagonal of the original stiffness matrix. Addition of restraining springs while making the stiffness matrix nonsingular also modifies the system. However, if the restraining springs have a very low stiffness, the frequencies and mode shapes of the modified system will be very nearly equal to those of the original system. Example 11.10 illustrates the application of this method to the eigensolution of system shown in Figure E11.6.

Example 11.10

Obtain the eigenvalues and eigenvectors of the system shown in Figure E11.6, by adding a fictitious stiffness to degree-of-freedom 1.

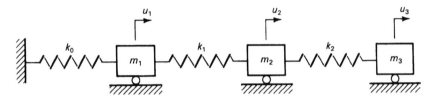

Figure E11.10. Semidefinite system restrained by a fictitious spring.

Solution

The semi-definite system of Figure E11.6 is modified by adding a spring of stiffness k_0 along degree-of-freedom 1. The modified system is shown in Figure E11.10. If we select $k_0 = 0.01$, the stiffness matrix of the system becomes

$$\mathbf{K} = \begin{bmatrix} 2.01 & -2 & 0 \\ -2 & 6 & -4 \\ 0 & -4 & 4 \end{bmatrix}$$

The mass matrix remains unchanged. An inverse iteration method as described by Equations 11.73 through 11.76 is used to obtain the eigenvalues and eigenvectors. The steps involved are presented below in brief. For each mode, the first trial vector u is selected so that $\mathbf{u}^T = [1 \quad 1 \quad 1]$.

First mode

Iteration begins with $\mathbf{x}_1 = \mathbf{Mu}$. The computations are tabulated as follows.

k	$\mathbf{x}_k$	$\bar{\mathbf{u}}_{k+1}$	$\bar{\mathbf{x}}_{k+1}$	$\mathbf{x}_{k+1}$	ρ_{k+1}
1	$\begin{bmatrix} 2 \\ 4 \\ 2 \end{bmatrix}$	$\begin{bmatrix} 800.00 \\ 803.0 \\ 803.50 \end{bmatrix}$	$\begin{bmatrix} 1600.0 \\ 3212.0 \\ 1607.0 \end{bmatrix}$	$\begin{bmatrix} 0.70501 \\ 1.4153 \\ 0.70810 \end{bmatrix}$	0.0012463
2	$\begin{bmatrix} 0.70501 \\ 1.41530 \\ 0.70810 \end{bmatrix}$	$\begin{bmatrix} 282.84 \\ 283.90 \\ 284.08 \end{bmatrix}$	$\begin{bmatrix} 565.68 \\ 1135.60 \\ 568.16 \end{bmatrix}$	$\begin{bmatrix} 0.70501 \\ 1.4153 \\ 0.70810 \end{bmatrix}$	0.0012463

Convergence has been achieved at this stage. The eigenvalue is 0.0012463, as compared to the exact value of 0.0. The eigenvector is given by $\bar{\mathbf{u}}_3$; when normalized so that the first element is equal to 1, it becomes

$$\mathbf{q}_1 = \begin{bmatrix} 1 \\ 1.0037 \\ 1.0044 \end{bmatrix}$$

compared to the rigid-body mode given by $\mathbf{q}_1^T = [1 \quad 1 \quad 1]$. As an alternative, the eigenvector can be mass orthonormalized using Equation 11.76. This gives

$$\boldsymbol{\phi}_1 = \begin{bmatrix} 0.35251 \\ 0.35383 \\ 0.35405 \end{bmatrix}$$

Second mode
To obtain the second eigenpair, a sweeping matrix is constructed using

$$\mathbf{S}_1 = \mathbf{I} - \boldsymbol{\phi}_1 \boldsymbol{\phi}_1^T \mathbf{M}$$

$$= \begin{bmatrix} 0.75148 & 0.49891 & -0.24961 \\ -0.24945 & 0.49922 & -0.25055 \\ -0.24961 & -0.50109 & 0.74930 \end{bmatrix}$$

Iteration for the second mode begin with $\mathbf{x}_1 = \mathbf{M}\mathbf{S}_1\mathbf{u}$. Also, Equation 11.97 is used in place of Equation 11.74. The computations are tabulated as follows

k	$\mathbf{x}_k$	$\bar{\mathbf{u}}_{k+1}$	$\bar{\mathbf{x}}_{k+1}$	$\mathbf{x}_{k+1}$	ρ_{k+1}
1	$\begin{bmatrix} 0.0059318 \\ -0.0031066 \\ -0.0028013 \end{bmatrix}$	$\begin{bmatrix} 0.0023930 \\ -0.00056097 \\ -0.0025226 \end{bmatrix}$	$\begin{bmatrix} 0.0047060 \\ -0.0022439 \\ -0.0025226 \end{bmatrix}$	$\begin{bmatrix} 1.2005 \\ -0.56285 \\ -0.63277 \end{bmatrix}$	1.2251
2	$\begin{bmatrix} 1.20050 \\ -0.56285 \\ -0.63277 \end{bmatrix}$	$\begin{bmatrix} 0.48840 \\ -0.10941 \\ -0.26760 \end{bmatrix}$	$\begin{bmatrix} 0.97681 \\ -0.43762 \\ -0.53520 \end{bmatrix}$	$\begin{bmatrix} 1.1950 \\ -0.53537 \\ -0.65474 \end{bmatrix}$	1.2231
3	$\begin{bmatrix} 1.1950 \\ -0.53537 \\ -0.65474 \end{bmatrix}$	$\begin{bmatrix} 0.48771 \\ -0.10735 \\ -0.27103 \end{bmatrix}$	$\begin{bmatrix} 0.97542 \\ -0.42938 \\ -0.54206 \end{bmatrix}$	$\begin{bmatrix} 1.19280 \\ -0.52507 \\ -0.66286 \end{bmatrix}$	1.2228
4	$\begin{bmatrix} 1.19280 \\ -0.52507 \\ -0.66286 \end{bmatrix}$	$\begin{bmatrix} 0.48740 \\ -0.10656 \\ -27228 \end{bmatrix}$	$\begin{bmatrix} 0.97480 \\ -0.42626 \\ -0.54456 \end{bmatrix}$	$\begin{bmatrix} 1.19200 \\ -0.52122 \\ -0.66587 \end{bmatrix}$	1.2228

Convergence has been achieved at this stage. The second eigenvalue is 1.2228; the corresponding eigenvector is given by $\bar{\mathbf{u}}_5$. When normalized so that the first element is equal to 1, the eigenvector becomes

$$\mathbf{q}_2 = \begin{bmatrix} 1.00000 \\ -0.21863 \\ -0.55863 \end{bmatrix}$$

In comparison, the more precise value obtained in Example 11.6 was

$$\mathbf{q}_2 = \begin{bmatrix} 1.0000 \\ -0.2192 \\ -0.5616 \end{bmatrix}$$

When mass orthonormalized, the second eigenvector becomes

$$\phi_2 = \begin{bmatrix} 0.59577 \\ -0.12981 \\ -0.33371 \end{bmatrix}$$

It is of interest to note that the selection of $\mathbf{u}^T = [1 \quad 1 \quad 1]$ as the initial trial vector was not a very good choice in this case, because the trial is very close to the first mode shape. As a result, when the first mode shape is swept off from it, the elements of the resulting vectors are very small, as shown by the small values in $\mathbf{x}_1 = \mathbf{MS}_1\mathbf{u}$. This does not, however, pose any special problem with the precision available on a computer.

Third mode
The sweeping matrix needed to obtain the third mode is given by

$$\mathbf{S}_2 = \mathbf{I} - \boldsymbol{\phi}_1\boldsymbol{\phi}_1^T\mathbf{M} - \boldsymbol{\phi}_2\boldsymbol{\phi}_2^T\mathbf{M}$$

$$= \mathbf{S}_1 - \boldsymbol{\phi}_2\boldsymbol{\phi}_2^T\mathbf{M}$$

$$= \begin{bmatrix} 0.041607 & -0.18956 & 0.14801 \\ -0.094778 & 0.43182 & -0.33718 \\ 0.14801 & -0.67437 & 0.52658 \end{bmatrix}$$

Iterations are begun with $\mathbf{x}_1 = \mathbf{MS}_2\mathbf{u}$. The computations are tabulated as follows

k	$\mathbf{x}_k$	$\bar{\mathbf{u}}_{k+1}$	$\bar{\mathbf{x}}_{k+1}$	$\mathbf{x}_{k+1}$	ρ_{k+1}
1	$\begin{bmatrix} 0.13251 \times 10^{-3} \\ -0.57933 \times 10^{-3} \\ 0.44706 \times 10^{-3} \end{bmatrix}$	$\begin{bmatrix} 0.24091 \times 10^{-4} \\ -0.42044 \times 10^{-4} \\ 0.69722 \times 10^{-4} \end{bmatrix}$	$\begin{bmatrix} 0.38554 \times 10^{-4} \\ -0.17579 \times 10^{-3} \\ -0.13727 \times 10^{-3} \end{bmatrix}$	$\begin{bmatrix} 0.28848 \\ -1.31143 \\ 1.02620 \end{bmatrix}$	3.2840
2	$\begin{bmatrix} 0.28848 \\ -1.31430 \\ 1.02620 \end{bmatrix}$	$\begin{bmatrix} 0.043901 \\ -0.10011 \\ 0.15645 \end{bmatrix}$	$\begin{bmatrix} 0.037919 \\ -0.40057 \\ 0.31278 \end{bmatrix}$	$\begin{bmatrix} 0.28846 \\ -1.13143 \\ 1.0262 \end{bmatrix}$	3.2810
3	$\begin{bmatrix} 2.8846 \\ -1.13143 \\ 1.0262 \end{bmatrix}$	$\begin{bmatrix} 0.043885 \\ -0.10013 \\ -0.15643 \end{bmatrix}$	$\begin{bmatrix} 0.087919 \\ -0.40057 \\ 0.31278 \end{bmatrix}$	$\begin{bmatrix} 0.28846 \\ -1.13143 \\ 1.0262 \end{bmatrix}$	3.2810

Iterations have converged; the eigenvalue is 3.2810 as compared to the more precise value of 3.271. The eigenvector is given by $\bar{\mathbf{u}}_4$. When normalized so that the first element is 1, it becomes

$$\mathbf{q}_3 = \begin{bmatrix} 1.0000 \\ -2.2816 \\ 3.5645 \end{bmatrix}$$

as compared to the more precise value in Example 11.6

$$\mathbf{q}_3 = \begin{bmatrix} 1.000 \\ -2.280 \\ 3.561 \end{bmatrix}$$

On mass orthonormalizing the eigenvector, we obtain

$$\boldsymbol{\phi}_3 = \begin{bmatrix} 0.14399 \\ -0.32851 \\ 0.31325 \end{bmatrix}$$

It is clear from the solution in Example 11.10 that the fictitious spring could have been added to any one or more of the three degrees of freedom. It is also apparent that a certain approximation is involved in the solution because of the addition of one or more fictitious springs. However, with the numeric precision that can be obtained in modern computers, the stiffness of the fictitious springs can be kept very small, so that the eigenvalues and eigenvectors of the modified system will, for all practical purposes, be equal to those of the original system. Also, the computer algorithm can be designed so that whenever during reduction of the stiffness matrix a zero is encountered on the diagonal, it is replaced by a small positive value and the reduction resumed from that point. This automates the procedure of adding springs to restrain rigid-body motions.

Vector iteration with shift
As explained in Section 11.5.3, introduction of a shift μ in the origin of the eigenvalue axis modifies the eigenvalue problem to that given by Equation 11.99a

$$(\mathbf{K} - \mu\mathbf{M})\mathbf{q} = \delta\mathbf{Mq} \tag{11.99a}$$

where λ, the eigenvalue of the original problem, is related to δ by Equation 11.98, $\lambda = \mu + \delta$. The modified stiffness matrix $(\mathbf{K} - \mu\mathbf{M})$ is, in general, nonsingular (if it is not, another value can be selected for μ), so that the vector iteration method can be applied to the solution of a modified problem. The eigenvectors remain unchanged under the shift; the eigenvalue λ can be obtained from δ by using Equation 11.98.

Iteration with shifts does not involve any approximation; both the eigenvalues and eigenvectors are obtained accurately. It may also be noted that when the mass matrix is diagonal, a shift is equivalent to adding springs along all degrees of freedom, the stiffness of a spring being equal to $-\mu$ times the corresponding term on the diagonal of the mass matrix. The application of iteration with shift to the eigensolution of a semi-definite system is illustrated by Example 11.6.

Sweeping of rigid-body modes

The rigid-body modes of an unrestrained or partially restrained system can usually be obtained by inspection. The condition that the remaining modes of the system be orthogonal to the rigid-body modes can therefore be used to set up as many constraint equations as there are rigid-body modes. Thus, if of a total of N mode shapes, s is the number of rigid-body modes and r the number of remaining modes, there exist s constraint equations. For an eigensolution of the unrestrained system, these equations are used to obtain a coordinate transformation in which the $s + r$ coordinates are expressed in terms of the r independent coordinates. A reduced eigenvalue problem is then formed in the transformed coordinates. Since the rigid-body modes have been removed in the process of transformation, the reduced system is positive definite and its eigenvalues and eigenvectors can be obtained by any iteration process.

The procedure is entirely equivalent to that described in Section 11.5.2 for sweeping off the mode shapes already determined. The r remaining modes satisfy the constraint equations expressed by Equations 11.90, 11.91, and 11.92. If one of the remaining mode shape $\mathbf{u}$ is partitioned as shown in Equation 11.93, the constrain condition can be expressed as

$$\mathbf{u} = \begin{bmatrix} (\mathbf{\Phi}_s^T \mathbf{M}_s)^{-1}(\mathbf{\Phi}_s^T \mathbf{M}_r) \\ \mathbf{I} \end{bmatrix} \mathbf{u}_r \tag{11.190a}$$

$$= \mathbf{T}\mathbf{u}_r \tag{11.190b}$$

Equation 11.190 can be viewed as a coordinate transformation in which transformation matrix $\mathbf{T}$ is of size N by r and vector $\mathbf{u}_r$ is of size r. The transformation is used to form an eigenvalue problem in the reduced space of size r. This gives

$$\tilde{\mathbf{K}}\mathbf{u}_r = \lambda \tilde{\mathbf{M}}\mathbf{u}_r \tag{11.191}$$

where

$$\tilde{\mathbf{K}} = \mathbf{T}^T \mathbf{K} \mathbf{T}$$
$$\tilde{\mathbf{M}} = \mathbf{T}^T \mathbf{M} \mathbf{T} \tag{11.192}$$

and both $\tilde{\mathbf{K}}$ and $\tilde{\mathbf{M}}$ are of size r by r. Solution of the eigenvalue Equation 11.191 will provide the remaining r eigenvalues and the corresponding eigenvectors. It should be noted that an eigenvector of Equation 11.191 will have only r elements. The corresponding full eigenvector of the original problem of size N can, however, be obtained by using Equation 11.190.

For a system of significant size, the number of rigid-body modes is small compared to the total number of degrees of freedom. As a result, the size of reduced stiffness matrix $\tilde{\mathbf{K}}$ or of the reduced mass matrix $\tilde{\mathbf{M}}$ is not much smaller than that of the original matrices. On the other hand, the transformation given by Equation 11.192 destroys the bandedness of the stiffness and mass matrices. Because of this, the method of sweeping off the rigid-body modes is computationally inefficient in comparison to the other two methods described in this section.

Example 11.11

Obtain the eigenvalues and eigenvectors of the unrestrained system of Example 11.6 by sweeping off the rigid-body modes.

Solution

It is easily seen that the system has one rigid-body mode, corresponding to a horizontal translation in the plane of the paper. This mode is given by

$$\mathbf{q}_1 = \begin{bmatrix} 1 \\ 1 \\ 1 \end{bmatrix} \tag{a}$$

The corresponding eigenvalue is $\lambda_1 = 0$.

Let one of the two remaining modes be represented by $\mathbf{u}$, where $\mathbf{u}^T = [u_1 \; u_2 \; u_3]$. Since this mode should be orthogonal to the rigid-body mode already determined, we have

$$\mathbf{q}_1^T \mathbf{M} \mathbf{u} = 0$$

or

$$2u_1 + 4u_2 + 2u_3 = 0 \tag{b}$$

Equation b can be expressed in the form of a coordinate transformation as follows

$$\begin{bmatrix} u_1 \\ u_2 \\ u_3 \end{bmatrix} = \begin{bmatrix} -2 & -1 \\ 1 & 0 \\ 0 & 1 \end{bmatrix} \begin{bmatrix} u_2 \\ u_3 \end{bmatrix} \tag{c}$$

or as

$$\mathbf{u} = \mathbf{T}\mathbf{u}_r \tag{d}$$

where

$$\mathbf{T} = \begin{bmatrix} -2 & -1 \\ 1 & 0 \\ 0 & 1 \end{bmatrix} \tag{e}$$

We now form a reduced eigenvalue problem as follows

$$\tilde{\mathbf{K}} = \mathbf{T}^T \mathbf{K} \mathbf{T}$$

$$= \begin{bmatrix} 22 & 2 \\ 2 & 6 \end{bmatrix} \tag{f}$$

$$\tilde{\mathbf{M}} = \mathbf{T}^T \mathbf{M} \mathbf{T}$$

$$= \begin{bmatrix} 12 & 4 \\ 4 & 4 \end{bmatrix} \tag{g}$$

$$\tilde{\mathbf{K}} \mathbf{u}_r = \lambda \tilde{\mathbf{M}} \mathbf{u}_r \tag{h}$$

We note that although $\mathbf{K}$ was singular, $\tilde{\mathbf{K}}$ is not. The eigenproblem of Equation h can therefore be solved by inverse iteration. The lower of the two eigenvalues obtained from Equation h is 1.2192. This is the second eigenvalue of the original system. Thus

$$\lambda_2 = 1.2192 \tag{i}$$

The corresponding eigenvector in the reduced space is

$$\mathbf{u}_2 = \begin{bmatrix} 1.0000 \\ 2.5617 \end{bmatrix} \tag{j}$$

We now use the coordinate transformation given by Equation d to obtain the full eigenvector.

$$\tilde{\mathbf{q}}_2 = \begin{bmatrix} -2 & -1 \\ 1 & 0 \\ 0 & 1 \end{bmatrix} \begin{bmatrix} 1.0000 \\ 2.5617 \end{bmatrix}$$

$$= \begin{bmatrix} -4.5617 \\ 1.0000 \\ 2.5617 \end{bmatrix} \tag{k}$$

On dividing through by the first element, we obtain the normalized vector

$$\mathbf{q}_2 = \begin{bmatrix} 1.0000 \\ -0.2192 \\ -0.5615 \end{bmatrix}$$

which is identical to that obtained in Example 11.6. On continuing the inverse iteration procedure, we get

$$\lambda_3 = 3.2807$$
$$\mathbf{u}_3 = \begin{bmatrix} 1.0000 \\ -1.5615 \end{bmatrix} \tag{l}$$

$$\tilde{\mathbf{q}}_3 = \begin{bmatrix} -0.4387 \\ 1.0000 \\ -1.5615 \end{bmatrix} \tag{m}$$

and

$$\mathbf{q}_3 = \begin{bmatrix} 1.0000 \\ -2.2805 \\ -3.5610 \end{bmatrix} \tag{n}$$

11.9   SELECTION OF A METHOD FOR THE DETERMINATION OF EIGENVALUES

The discussion in previous sections of this chapter shows that several alternative methods are available for the solution of an eigenproblem. Which method is most effective in a particular case depends on the type of solution desired and the characteristics of the matrices involved. The advantages and limitations of the methods presented were pointed out as each method was discussed. It is instructive to summarize these and to show how they affect the selection of a particular technique for the solution of the problem in hand. Often, the procedure that is most effective for the problem to be solved will involve a combination of two or more different methods.

The transformation methods are most useful when the matrices are of comparatively small order and are more or less fully populated or have a large bandwidth. They are not particularly appropriate when the matrices involved are of a large order and only a few eigenvalues need to be determined. In all cases, the eigenvector corresponding to a computed eigenvalue is best determined by a process of inverse iteration with shift equal to the eigenvalue.

The Jacobi transformation method is simple and stable. It is appropriate for calculating the eigenvalues of a standard symmetric matrix. The matrix need not be positive definite and the eigenvalues being sought may be negative, zero, or positive. All eigenvalues are obtained simultaneously. The generalized Jacobi method is suitable for the solution of a linearized eigenvalue problem of small order, particularly when the matrices involved are banded.

The Householder transformation matrix operates on a standard symmetric matrix. It must be combined with the QR method for the determination of eigenvalues. The eigenvectors are then obtained by using inverse iteration with shifts. For standard symmetric matrices, the combination of Householder and QR method is, in general, more efficient than the Jacobi method. As in the standard Jacobi method, the matrix need not be positive definite, and negative, zero, or positive eigenvalues may be obtained. Also, all eigenvalues are obtained simultaneously.

Iteration methods are most effective when the matrices involved are of a large order and only a few eigenvalues are desired. They may be applied directly to linearized eigenvalue problems without the need of transforming the latter to a standard form. Direct iteration gives the eigenvalues with the largest magnitude. For the solution of a linearized eigenproblem by direct iteration, the mass matrix must be positive definite. Inverse iteration leads to the lowest eigenvalue. For solution of a linearized eigenproblem by inverse iteration, the stiffness matrix must be positive definite.

When using an iteration method to determine an eigenvalue other than the most dominant or the least dominant, the eigenvectors already determined must be swept off from the trial vector by using Gram–Schmidt orthogonalization.

The accuracy of subsequent eigenvectors depends on the precision with which the previous eigenvectors have been determined. Therefore, unless high numerical precision is maintained, the methods may not be suitable for determining eigenpairs other than the few most dominant or least dominant. When more than a few lower-order eigenpairs are required, inverse iteration with shifts is suitable. The shifting procedure is, however, not very effective with forward or direct iteration.

For problems of large size where only a few eigenpairs are required, Lanczos iteration, the determinant search method, and subspace iteration are most effective. The determinant search method is used in conjunction with inverse iteration with shifts and Sturm sequence check. Subspace iteration is used together with Jacobi transformation and Sturm sequence check.

SELECTED READINGS

Bathe, K.J. 1996. *Finite Element Procedures*. Englewood Cliffs, New Jersey: Prentice Hall.

Bathe, K.J. & Wilson, E.L. 1973. Solution Methods for Eigenvalue Problems in Structural Mechanics. *International Journal for Numerical Methods in Engineering*, Vol. 6: 213–226.

Bathe, K.J. & Wilson, E.L. 1973. Eigensolutions of Large Structural Systems with Small Bandwidth. *Journal of Engineering Mechanics Division*, Vol. 99: 467–479. ASCE.

Bathe, K.J. & Wilson, E.L. 1976. *Numerical Methods in Finite Element Analysis*. Englewood Cliffs: Prentice Hall.

Francis, J.G.F. 1961–62. The QR Transformation Parts I and II. *Computer Journal*, Vol. 4: 265–371 & 332–345.

Goldstine, H.H., Murray, F.J. & Von Neumann, J. 1959. The Jacobi Method for Real Symmetric Matrices. *Journal of the Association for Computing Machinery*, Vol. 6: 59–96.

Gupta, K.K. 1972. Solutions of Eigenvalue Problems by the Sturm Sequence Method. *International Journal for Numerical Methods in Engineering*, Vol. 4: 379–404.

Householder, A.S. 1963. *Principles of Numerical Analysis*. New York: McGraw-Hill.

Jennings, A. 1992. *Matrix Computation for Engineers and Scientists*. Chichester, U.K.: John Wiley. 2nd Edition.

Jennings, A. 1981. Eigenvalue Methods and the Analysis of Structural Vibration. In I.S. Duff (ed.) *Sparse Matrices and Their Uses*: 109–138. London: Academic Press.

Ortega, J.M. & Kaiser, H.F. 1963–64. The LLT and QR Methods for Symmetric Tridiagonal Matrices. *Computer Journal*, Vol. 6: 99–101.

Parlett, B.N. 1980. *The Symmetric Eigenvalue Problem*. Englewood Cliffs: Prentice Hall.

Parlett, B.N. & Scott, S.N. 1979. The Lanczos Algorithm with Selective Orthogonalization. *Mathematics of Computation*. Vol. 33: 217–218.

Scott, D.S. 1981. The Lanczos Algorithm. In I.S. Duff (ed.) *Sparse Matrices and Their Uses*: 139–159. London: Academic Press.

Weaver, W. Jr. & Yoshida, D.M. 1971. The Eigenvalue Problem for Bounded Matrices. *Computers and Structures*, Vol. 1: 651–664.

Wilkinson, J.H. 1965. *The Algebraic Eigenvalue Problem.* Oxford, U.K.: Clarendon Press.

PROBLEMS

11.1   Find the mass and the flexibility matrices of the simply supported uniform beam shown in Figure P11.1 corresponding to the 3 coordinate direction indicated.

   The eigenvalue problem for the flexural vibrations of the beam can be expressed as

$$\mathbf{K}^{-1}\mathbf{Mq} = \frac{1}{\lambda}\mathbf{q}$$

$$\mathbf{Dq} = \gamma\mathbf{q}$$

   Show that for the given system, **D** is a symmetric matrix. Using standard Jacobi method, obtain the 3 natural frequencies of the system.

11.2   By inverting the flexibility matrix determined in Problem 11.1, obtain the stiffness matrix of the beam. Compute the three frequencies of the beam by using the generalized Jacobi method.

11.3   Determine the frequencies and mode shapes of the beam in Figure P11.1 by the inverse iteration method.

11.4   Figure P11.4 shows a three-story shear frame. The story stiffnesses and floor masses are as indicated. Form the stiffness and the mass matrices of the frame. Obtain an estimate of the fundamental frequency by the Rayleigh method and an assumed vibration shape given by

$$\mathbf{q}_1^T = [1 \ 2 \ 3]$$

   Using inverse iteration with a shift, obtain a more precise value of the fundamental frequency and the corresponding mode shape of the building.

11.5   The stiffness and mass matrices of the three-story building frame of Figure P11.5 are shown on the figure. There is a concern for possible resonant vibrations due to a blower fan mounted at the second-story level. The operating speed of the motor is 250 rpm. Obtain the natural frequency of the building frame that is closest to the blower frequency.

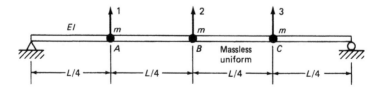

Figure P11.1.

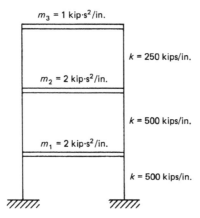

Figure P11.4.

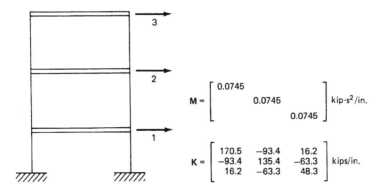

Figure P11.5.

11.6 Using Lanczos vectors, reduce the eigenvalue problem for the frame of Figure P11.5 to a tridiagonal form.

11.7 From the tridiagonal matrix obtained in Problem 11.6, compute the three natural frequencies of the frame by using the QR method.

11.8 Obtain estimates of the first two frequencies of the frame of Figure P11.5 by using the determinant search method.

11.9 A train made up of 3 cars of weight 60,000 lb each connected by coupling springs of stiffness 15,000 lb/in. is released by a shunting engine and rolls freely along a track (Fig. P11.9). Determine the mode shapes and frequencies of vibration of the system of cars by sweeping the rigid body mode.

11.10 Determine the mode shapes and frequencies of the system of cars in Problem 11.9 by adding a fictitious spring to restrain the rigid body

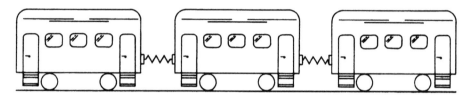

Figure P11.9.

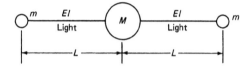

Figure P11.12.

mode.

11.11 Solve Problem 11.11 by using vector iteration with shift.

11.12 A small two engined plane in flight is modeled as shown in Figure P11.12. By sweeping off the rigid body modes of the plane, obtain a reduced model and hence determine the remaining frequency or frequencies of the plane. Neglect the rotatory inertia of the engine masses and the fuselage.

11.13 The stiffness and mass matrices of a discrete parameter system are as given below.

$$
\mathbf{M} = \begin{bmatrix} 2 & & & & \\ & 2 & & & \\ & & 2 & & \\ & & & 2 & \\ & & & & 1 \end{bmatrix}
$$

$$
\mathbf{K} = \begin{bmatrix} 2800 & -1200 & 0 & 0 & 0 \\ -1200 & 2400 & -1200 & 0 & 0 \\ 0 & -1200 & 2000 & -800 & 0 \\ 0 & 0 & -800 & 1200 & -400 \\ 0 & 0 & 0 & -400 & 400 \end{bmatrix}
$$

Using subspace iteration and the following trial vectors, obtain the first two mode shapes and frequencies of the system. Assume that acceptable

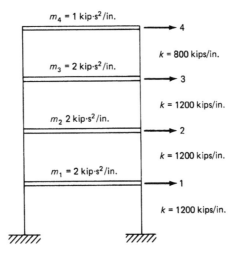

Figure P11.14.

accuracy has been achieved after three iterations.

$$\Psi^0 = \begin{bmatrix} 0.2 & -0.5 \\ 0.4 & -1.0 \\ 0.6 & -0.5 \\ 0.8 & 0.5 \\ 1.0 & 1.0 \end{bmatrix}$$

11.14 The mass and stiffness properties of a four-story shear frame are shown in Figure P11.14. Using the following trial vectors and subspace iteration, obtain the first two mode shapes and frequencies of the frame.

$$\begin{bmatrix} 0.50 & 0.50 \\ 1.00 & 1.00 \\ 1.19 & -0.05 \\ 1.48 & -1.62 \end{bmatrix}$$

11.15 Transform the eigenvalue equation in Problem 11.14 to standard symmetric form by using Equations 11.16 and 11.17. Now, reduce the resulting problem to a tridiagonal form by using Householder's transformation.

11.16 From the tridiagonal matrix obtained in Problem 11.15, compute the four frequencies and mode shapes by using the QR method.

CHAPTER 12

Forced dynamic response: Multi-degree-of-freedom systems

12.1 INTRODUCTION

The dynamic response of a multi-degree-of-freedom system is governed by the following equations of motion

$$\mathbf{M\ddot{u} + C\dot{u} + Ku = p}(t) \tag{12.1}$$

in which $\mathbf{p}(t)$ is the vector of time-dependent forcing functions. If the system is undamped, the damping term $\mathbf{C\dot{u}}$ will be absent from the equations of motion.

Equation 12.1 represents N simultaneous equations in the displacement vector $\mathbf{u}$ and its time derivatives. These equations can, in general, be solved only by a numerical method of integration. Because such methods of integration are based on certain mathematical idealizations, usually related to the manner in which the accelerations vary over a short period, the results obtained are only approximate. An exact mathematical solution may be possible when the time-varying forces are simple mathematical functions that would permit integration of the equations. Such exact solutions involve a transformation of the equations to a special set of coordinates called normal coordinates. The transformation uncouples the equations of motion, so that they are reduced to N independent equations whose mathematical solutions are relatively easily obtained.

The method based on normal coordinate transformation, called the *mode superposition method*, is similar to that described in Chapter 10 for the solution of free-vibration problems of multi-degree-of-freedom systems. Application of the mode superposition method to the solution of forced dynamic response is described in this chapter. Direct integration of the equations of motion by numerical methods is discussed in Chapter 13.

12.2 NORMAL COORDINATE TRANSFORMATION

The application of normal coordinate transformation to obtain the solution to free-vibration problems of a multi-degree-of-freedom system was described in

Section 10.7. The transformation is given by Equation 10.56, restated here for ease of reference

$$\mathbf{u} = \sum_{n=1}^{N} \boldsymbol{\phi}_n y_n \qquad (12.2a)$$

or

$$\mathbf{u} = \boldsymbol{\Phi} \mathbf{y} \qquad (12.2b)$$

where $\boldsymbol{\Phi}$ is the matrix of undamped mode shapes and $\mathbf{y}$ is the vector of normal coordinates. On applying this transformation to Equation 12.1 and using the procedure of Section 10.7, we get the following set of transformed equations

$$\mathbf{M}^* \ddot{\mathbf{y}} + \mathbf{C}^* \dot{\mathbf{y}} + \mathbf{K}^* \mathbf{y} = \mathbf{p}^* \qquad (12.3)$$

where

$$\mathbf{M}^* = \boldsymbol{\Phi}^T \mathbf{M} \boldsymbol{\Phi}$$
$$\mathbf{C}^* = \boldsymbol{\Phi}^T \mathbf{C} \boldsymbol{\Phi} \qquad (12.4)$$
$$\mathbf{K}^* = \boldsymbol{\Phi}^T \mathbf{K} \boldsymbol{\Phi}$$

As explained in Chapter 10, because of the orthogonality property of the mode shapes, both the mass matrix and the stiffness matrix are diagonal. In fact, if the mode shapes have been normalized to be mass orthonormal, $\mathbf{M}^*$ is an identity matrix. Further, if the damping matrix possesses the orthogonality property, $\mathbf{C}^*$ is also diagonal. In such a case, Equation 12.3 reduces to a set of N uncoupled equations. If the nth diagonal term in the transformed mass matrix is denoted by M_n, that in the transformed damping matrix by C_n, and that in the transformed stiffness matrix by K_n, the nth equation in the set of transformed equations becomes

$$M_n \ddot{y}_n + C_n \dot{y}_n + K_n y_n = p_n \qquad (12.5)$$

where $p_n = \boldsymbol{\phi}_n^T \mathbf{p}$. For the sake of generality, we have retained the term M_n, even though with mass orthonormal modes it will be equal to unity.

Equation 12.5 is the equation of motion of a single-degree-of-freedom system. On dividing through by M_n, it reduces to

$$\ddot{y}_n + \frac{C_n}{M_n} \dot{y}_n + \frac{K_n}{M_n} y_n = \frac{p_n}{M_n} \qquad (12.6)$$

As for a single-degree-of-freedom system, $C_n/M_n = 2\xi_n \omega_n$ and $K_n/M_n = \omega_n^2$, where ω_n is the nth natural frequency and ξ_n is the damping ratio in the nth mode. Equation 12.6 therefore becomes

$$\ddot{y}_n + 2\xi_n \omega_n \dot{y}_n + \omega_n^2 y_n = \frac{p_n}{M_n} \qquad (12.7)$$

It is of interest to note that in Equation 12.2a, $\phi_n y_n$ represents the contribution of the *n*th mode of vibration to the total response. Further, y_n, the *n*th normal coordinate, is completely determined by solving Equation 12.7, which involves only those parameters that are related to the *n*th mode. The total response in Equation 12.2 can thus be viewed as a superposition of the responses in individual modes. It is because of this fact that the solution procedure being discussed here is called the mode superposition method.

In deriving Equation 12.7, it is only necessary to know the damping ratio ξ_n, not the damping term C_n, which can only be determined provided that the damping matrix **C** is specified. As already mentioned, it is easier to assign a reasonable value to ξ_n than to determine the damping matrix **C**. It is, therefore, clearly an advantage of the mode superposition method that it does not require an explicit specification of the damping matrix. It remains to solve the single-degree-of-freedom equation, Equation 12.7, to determine the generalized coordinate y_n. Methods of solution of such an equation were discussed in detail in Chapters 5 through 9.

Complete solution of Equation 12.7 consists of a free- and a forced-vibration component. The former is given by Equation 5.37, while the latter can be obtained from Duhamel's integral. The complete solution is thus given by the following equation, which is identical to Equation 7.10

$$y_n = e^{-\xi_n \omega_n t} \left(\frac{\dot{y}_{0n} + y_{0n} \xi_n \omega_n}{\omega_{Dn}} \sin \omega_{Dn} t + y_{0n} \cos \omega_{Dn} t \right)$$

$$+ \frac{1}{M_n \omega_{Dn}} \int_0^t p_n(\tau) e^{-\omega \xi_n (t-\tau)} \sin \omega_{Dn}(t - \tau) \, d\tau \qquad (12.8)$$

As noted in Chapter 10, y_{0n} and $\dot{y}_{0n}$ in Equation 12.8 are the initial values of y_n and $\dot{y}_n$, respectively, determined from the given initial conditions by using Equations 10.61 and 10.62. The damped frequency $\omega_{Dn} = \omega_n \sqrt{1 - \xi_n^2}$. When $p_n(t)$ is a simple mathematical function of time t, Equation 12.8 can be evaluated exactly. In other cases, some form of numerical integration technique must be used.

For a complete solution of the original problem, Equation 12.8 must be solved for each of the N mode shapes. The response in the physical coordinates is obtained by superposing the modal responses by using Equation 12.2.

In many practical problems, a major portion of the response is contained in only a few of the mode shapes, usually those corresponding to the lowest frequencies. In such cases, modal response need be obtained for only the first few of the N modes, and therefore only such mode shapes and frequencies that are required need to be calculated. For large problems, this may mean a significant saving in the computational time.

In the mode superposition method described above, the undamped mode shapes were used to carry out a transformation of coordinates. The transformation uncouples the equations of motion provided that the damping matrix $\mathbf{C}$ satisfies the orthogonality condition. In a general case, where $\mathbf{C}$ does not satisfy such a condition, uncoupling can still be achieved provided that complex eigenvectors of the damped system are used in the transformation. The procedure is similar to that described in Section 10.8 for the solution of a damped free-vibration problem. However, as noted in that section, the determination of complex eigenvalues and eigenvectors involves large computational effort. Besides, the representation of damping in the system by a damping matrix is only an idealization. It is, therefore, more expedient and equally reasonable to use the undamped mode shapes in the transformation and to select a $\mathbf{C}$ that would become diagonal under such a transformation.

12.3   SUMMARY OF MODE SUPERPOSITION METHOD

We can now summarize the steps involved in mode superposition analysis of the forced vibration of a multi-degree-of-freedom system.

1. From the specified physical characteristics of the system, obtain the mass, stiffness, and damping matrices. Formulate the equations of motion using the physical property matrices and the given force vector

$$\mathbf{M\ddot{u}} + \mathbf{C\dot{u}} + \mathbf{Ku} = \mathbf{p} \tag{12.9}$$

For the mode superposition method to be successful, the damping matrix $\mathbf{C}$ must possess the orthogonality property. In most cases, the physical damping characteristics are not known. Damping is then specified at the modal level and the damping matrix $\mathbf{C}$ need not be determined.

2. Obtain the free-vibration mode shapes and frequencies of the undamped system by solving the following linearized eigenvalue problem by one of the methods described in Chapter 11

$$\mathbf{M\ddot{u}} + \mathbf{Ku} = 0 \tag{12.10a}$$

or

$$\mathbf{K\phi} = \omega^2 \mathbf{M\phi} \tag{12.10b}$$

in which ω is a frequency and ϕ is the corresponding mode shape.

3. As stated earlier, in the mode superposition method, it is not necessary to use all N mode shapes, N being the number of degrees of freedom. Reasonable accuracy can be achieved by using only a few modes. How many modes need to be used in a specific case is a question that we deal with more fully in Chapter 13; for the present discussion, let us assume that the first M modes will give reasonable accuracy in the determination of response. It will then be necessary to determine only the first M mode shapes and frequencies.

4. For each of the M mode shapes, determine the modal mass M_n, the modal damping C_n and the modal force p_n, given by

$$M_n = \boldsymbol{\phi}_n^T \mathbf{M} \boldsymbol{\phi}_n \tag{12.11a}$$

$$C_n = \boldsymbol{\phi}_n^T \mathbf{C} \boldsymbol{\phi}_n \tag{12.11b}$$

$$p_n = \boldsymbol{\phi}_n^T \mathbf{p} \tag{12.11c}$$

Note that since the mode shapes used in Equations 12.11 are mass orthonormal M_n is equal to unity. Also, if damping is to be specified at the mode level, it would be given in the form of a damping ratio ξ_n and the damping constant would then be given by $C_n = 2\xi_n \omega_n M_n$.

5. The equation of motion corresponding to the nth mode now becomes

$$\ddot{y}_n + \frac{C_n}{M_n}\dot{y} + \frac{K_n}{M_n}y = \frac{p_n}{M_n} \tag{12.12a}$$

or

$$\ddot{y}_n + 2\xi_n\omega_n\dot{y}_n + \omega_n^2 y_n = \frac{p_n}{M_n} \tag{12.12b}$$

6. From the given initial conditions in the physical coordinates, obtain the values of $\mathbf{y}_0$ and $\dot{\mathbf{y}}_0$, where $\mathbf{y}_0$ is the vector of modal displacements at time $t = 0$, and $\dot{\mathbf{y}}_0$ is the vector of the time derivatives of the modal displacements, again at time $t = 0$. It is possible to obtain these directly from Equation 12.2b as follows

$$\mathbf{y}_0 = \boldsymbol{\Phi}^{-1}\mathbf{u}_0 \tag{12.13a}$$

$$\dot{\mathbf{y}}_0 = \boldsymbol{\Phi}^{-1}\mathbf{v}_0 \tag{12.13b}$$

It is not, however, necessary to use Equations 12.13, which involve the inverse of $\boldsymbol{\Phi}$; instead, $\mathbf{y}_0$ and $\dot{\mathbf{y}}_0$ can be obtained more simply by using the mass orthonormality property of the mode shapes as in Equations 10.61 and 10.62

$$\mathbf{y}_0 = \boldsymbol{\Phi}^T \mathbf{M} \mathbf{u}_0 \tag{12.14a}$$

$$\dot{\mathbf{y}}_0 = \boldsymbol{\Phi}^T \mathbf{M} \mathbf{v}_0 \tag{12.14b}$$

7. Obtain solutions to the M single-degree-of-freedom equations given by Equation 12.12b for $n = 1$ to M. The Duhamel integral solution is given by

$$y_n = e^{-\xi_n\omega_n t}\left(\frac{\dot{y}_{0n} + y_{0n}\xi_n\omega_n}{\omega_{Dn}}\sin\omega_{Dn}t + y_{0n}\cos\omega_{Dn}t\right)$$

$$+ \frac{1}{M_n\omega_{Dn}}\int_0^t p_n(\tau)e^{-\omega\xi_n(t-\tau)}\sin\omega_{Dn}(t-\tau)\,d\tau \tag{12.15}$$

The integral in Equation 12.15 can be evaluated mathematically if $p_n(\tau)$ is a simple function; otherwise, a method of numerical integration must be used.

8. The response in physical coordinates can now be obtained by superposing the modal responses

$$\mathbf{u} = \sum_{n=1}^{M} \boldsymbol{\phi}_n y_n \qquad (12.16)$$

9. If desired, the spring forces can be obtained by using the relationship

$$\mathbf{f}_S = \mathbf{K}\mathbf{u}$$

$$= \sum_{n=1}^{M} \mathbf{K}\boldsymbol{\phi}_n y_n \qquad (12.17)$$

Since the mode shapes $\boldsymbol{\phi}_n$ satisfy the relationship $\mathbf{K}\boldsymbol{\phi}_n = \omega_n^2 \mathbf{M}\boldsymbol{\phi}_n$, Equation 12.17 can be expressed as

$$\mathbf{f}_S = \sum_{n=1}^{M} \omega_n^2 \mathbf{M}\boldsymbol{\phi}_n y_n \qquad (12.18)$$

Both Equations 12.17 and 12.18 show that the spring forces are also obtained as a superposition of the modal contributions.

Example 12.1
Derive the equations of motion for the system shown in Figure E12.1 and calculate the mode shapes and frequencies. Obtain the response of the system when it is subjected to a suddenly applied constant force F along coordinate 3. Given $m = 1$, $k = 1$, and $F = 1$.

Solution
We use an energy method to derive the equations of motion. The kinetic energy, T, of the system is given by

$$T = \frac{1}{2}m\dot{u}_1^2 + \frac{1}{2}2m\dot{u}_2^2 + \frac{1}{2}I\dot{\theta}^2 + \frac{1}{2}2m\dot{u}_3^2 \qquad (a)$$

where $\dot{\theta}$ is the angular velocity and I is the mass moment of inertia of the roller.

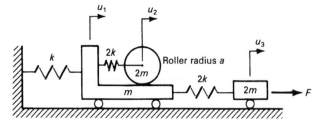

Figure E12.1. Three-degree-of-freedom system, forced vibrations.

Since the roller is assumed to roll without slipping

$$u_2 - u_1 = a\theta \tag{b}$$

Substituting $\dot{\theta}$ from Equation b and the value $I = ma^2$ in Equation a, we get

$$T = \frac{1}{2}m\dot{u}_1^2 + m\dot{u}_2^2 + \frac{1}{2}m(\dot{u}_2 - \dot{u}_1)^2 + m\dot{u}_3^2 \tag{c}$$

The potential energy stored in the springs is given by

$$V = \frac{1}{2}ku_1^2 + \frac{1}{2}2k(u_2 - u_1)^2 + \frac{1}{2}2k(u_3 - u_1)^2 \tag{d}$$

Using Lagrange's equations (Eq. 4.91), the equations of motion are obtained as follows

$$2m\ddot{u}_1 - m\ddot{u}_2 + 5ku_1 - 2ku_2 - 2ku_3 = 0$$

$$-m\ddot{u}_1 + 3m\ddot{u}_2 - 2ku_1 + 2ku_2 = 0 \tag{e}$$

$$2m\ddot{u}_3 - 2ku_1 + 2ku_3 - F = 0$$

In matrix form, Equations e can be expressed as

$$\begin{bmatrix} 2m & -m & 0 \\ -m & 3m & 0 \\ 0 & 0 & 2m \end{bmatrix} \begin{bmatrix} \ddot{u}_1 \\ \ddot{u}_2 \\ \ddot{u}_3 \end{bmatrix} + \begin{bmatrix} 5k & -2k & -2k \\ -2k & 2k & 0 \\ -2k & 0 & 2k \end{bmatrix} \begin{bmatrix} u_1 \\ u_2 \\ u_3 \end{bmatrix} = \begin{bmatrix} 0 \\ 0 \\ F \end{bmatrix} \tag{f}$$

or

$$\mathbf{M\ddot{u} + Ku = p}$$

The eigenvalue problem is given by Equation f with the right-hand side equal to zero. Its solution can be obtained from one of the standard method of eigensolution described in Chapter 11. The details of calculation are not shown, but the resulting eigenvalues and eigenvectors are as follows

$$\lambda_1 = 0.1703 \quad \omega_1 = 0.4127$$

$$\lambda_2 = 0.7663 \quad \omega_2 = 0.8754 \tag{g}$$

$$\lambda_3 = 3.063 \quad \omega_3 = 1.7503$$

$$\mathbf{q}_1 = \begin{bmatrix} 1.0000 \\ 1.2289 \\ 1.2054 \end{bmatrix}$$

$$\mathbf{q}_2 = \begin{bmatrix} 1.0000 \\ -4.1268 \\ 4.2792 \end{bmatrix} \tag{h}$$

$$\mathbf{q}_3 = \begin{bmatrix} 1.0000 \\ 0.1479 \\ -0.4847 \end{bmatrix}$$

The modal matrix is given by

$$\mathbf{Q} = [\mathbf{q}_1 \quad \mathbf{q}_2 \quad \mathbf{q}_3] \tag{i}$$

A normal coordinate transformation is now carried out using the modal matrix $\mathbf{Q}$, so that

$$\mathbf{u} = \mathbf{Q}\mathbf{y}$$
$$\tilde{\mathbf{M}} = \mathbf{Q}^T \mathbf{M} \mathbf{Q}$$
$$\tilde{\mathbf{K}} = \mathbf{Q}^T \mathbf{K} \mathbf{Q} \tag{j}$$
$$\tilde{\mathbf{p}} = \mathbf{Q}^T \mathbf{p}$$

and the transformed equations are

$$\tilde{\mathbf{M}}\ddot{\mathbf{y}} + \tilde{\mathbf{K}}\mathbf{y} = \tilde{\mathbf{p}} \tag{k}$$

On carrying out the computations, we will find that the set of equations given by Equation k are uncoupled. The uncoupled equations are

$$6.9788 \ddot{y}_1 + 1.1892 y_1 = 1.2054F$$

$$97.969 \ddot{y}_2 + 75.073 y_2 = 4.2792F \tag{l}$$

$$2.2396 \ddot{y}_3 + 6.8603 y_3 = -0.4847F$$

The solutions to Equations l can be obtained from Equation 7.14 with the damping ratio $\xi = 0$

$$y_1 = \frac{1.2054F}{1.1892}(1 - \cos 0.4127t)$$

$$y_2 = \frac{4.2792F}{75.073}(1 - \cos 0.8754t) \tag{m}$$

$$y_3 = \frac{-0.4847F}{6.8603}(1 - \cos 1.7503t)$$

The response in the physical coordinates is now obtained by using the transformation $\mathbf{u} = \mathbf{Q}\mathbf{y}$, giving

$$u_1 = 0.9999 - 1.0136 \cos 0.4127t - 0.0570 \cos 0.8754t$$
$$\qquad + 0.0707 \cos 1.7503t$$

$$u_2 = 0.9999 - 1.2456 \cos 0.4127t + 0.2352 \cos 0.8754t \tag{n}$$
$$\qquad + 0.0105 \cos 1.7503$$

$$u_3 = 1.500 - 1.2218 \cos 0.4127t - 0.2439 \cos 0.8754t$$
$$\qquad - 0.0343 \cos 1.7503t$$

12.4 COMPLEX FREQUENCY RESPONSE

As in the case of a single-degree-of-freedom system, the response of a multi-degree-of-freedom system to a harmonic excitation can conveniently be expressed in the notation of complex algebra. Assume that the forcing function is given by

$$\mathbf{p} = \mathbf{f}e^{i\Omega t} \tag{12.19}$$

where $\mathbf{f}$ represents the spatial variation of force and is independent of time. The equation of motion of a damped multi-degree-of freedom system subjected to the exciting force given by Equation 12.19 can be expressed as

$$\mathbf{M}\ddot{\mathbf{u}} + \mathbf{C}\dot{\mathbf{u}} + \mathbf{K}\mathbf{u} = \mathbf{f}e^{i\Omega t} \tag{12.20}$$

The steady-state solution to Equation 12.20 is

$$\mathbf{u} = \mathbf{U}e^{i\Omega t} \tag{12.21}$$

Substitution of Equation 12.21 in Equation 12.20 gives

$$[-\Omega^2\mathbf{M} + i\Omega\mathbf{C} + \mathbf{K}]\mathbf{U}e^{i\Omega t} = \mathbf{f}e^{i\Omega t} \tag{12.22}$$

Equation 12.22 can be solved for $\mathbf{U}$ to yield

$$\mathbf{U} = [-\Omega^2\mathbf{M} + i\Omega\mathbf{C} + \mathbf{K}]^{-1}\mathbf{f}$$
$$= \mathbf{R}_d\mathbf{f} \tag{12.23}$$

As in the case of a single-degree-of-freedom system $\mathbf{R}_d$ is referred to as the *receptance matrix*. It can be obtained by taking the inverse of matrix $[-\Omega^2\mathbf{M} + i\Omega\mathbf{C} + \mathbf{K}]$. This, however, involves considerable amount of computations, particularly because the matrix is complex-valued. For a system with proportional damping, a simpler and more instructive procedure is to obtain a solution by transforming the equations of motion to normal coordinates. The transformed equations are obtained from either Equation 12.5 or Equation 12.7.

$$\ddot{y}_n + 2\xi_n\omega_n\dot{y} + \omega^2 y_n = \boldsymbol{\phi}_n^T\mathbf{f}e^{i\Omega t} \quad n = 1, 2, \dots, N \tag{12.24}$$

The steady-state solution of Equation 12.24 is given by

$$y_n = Y_n e^{i\Omega t} \tag{12.25}$$

where

$$Y_n = \frac{\boldsymbol{\phi}_n\mathbf{f}}{(\omega_n^2 - \Omega^2) + 2i\xi_n\omega_n\Omega} \tag{12.26}$$

The solution in physical coordinates is given by

$$\mathbf{U} = \sum_{n=1}^{N} \boldsymbol{\phi}_n Y_n$$

$$\mathbf{R}_d\mathbf{f} = \sum_n \frac{\boldsymbol{\phi}_n\boldsymbol{\phi}_n^T\mathbf{f}}{(\omega_n^2 - \Omega^2) + 2i\xi_n\omega_n\Omega}$$

(12.27)

The receptance matrix $\mathbf{R}_d$ is an $N \times N$ matrix, in which the term $(R_d)_{lk}$ represents the displacement at coordinate l produced by a unit harmonic load applied at coordinate k. For $l = k$, $(R_d)_{kk}$ is referred to as a *point receptance*. When $l \neq k$, $(R_d)_{lk}$ is called *cross receptance*. The velocity response to harmonic excitation is obtained by differentiating Equation 12.21

$$\dot{\mathbf{u}} = i\Omega\mathbf{U}e^{i\Omega t}$$

$$= i\Omega\mathbf{R}_d\mathbf{f}e^{i\Omega t}$$

$$= \mathbf{R}_v\mathbf{f}e^{i\Omega t}$$

(12.28)

where $\mathbf{R}_v = i\Omega\mathbf{R}_d$ is referred to as the *mobility matrix*. The acceleration response is given by

$$\ddot{\mathbf{u}} = -\Omega^2\mathbf{R}_d\mathbf{f}e^{i\Omega t}$$

$$= \mathbf{R}_a\mathbf{f}e^{i\Omega t}$$

(12.29)

in which $\mathbf{R}_a = -\Omega^2\mathbf{R}_d$ is called *inertance matrix*. It is instructive to examine an individual element of matrix $\mathbf{R}_d$, given by

$$(R_d)_{lk} = \sum_n \frac{\phi_{ln}\phi_{kn}}{(\omega_n^2 - \Omega^2) + 2i\xi_n\omega_n\Omega}$$

(12.30)

in which ϕ_{ln} is the lth term of $\boldsymbol{\phi}_n$, ϕ_{kn} is the kth term of $\boldsymbol{\phi}_n$ and summation extends over all N modes. It is evident from Equation 12.30 that $(R_d)_{lk} = (R_d)_{kl}$ and matrix $\mathbf{R}_d$ is symmetric. Also, an individual term in the summation depends only on the parameters related to a single mode and, of course, the excitation frequency.

Graphical representation of the complex frequency response function (FRF) given by Equation 12.30 may consist of a plot of the amplitude of FRF versus the excitation frequency Ω and a plot of the phase angel versus Ω. Alternatively, a Nyquist graph, in which the imaginary part of the FRF is plotted against the real part, may be used. As an illustration, consider the undamped two-degree-of-freedom system shown in Figure 12.1. Assume that $k_2 = 0.5k_1$ and $m_2 = 0.5m_1$.

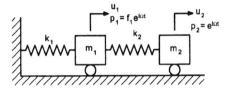

Figure 12.1. Harmonic force excitation of a two-degree-of-freedom system.

The two mode shapes and frequencies of the system are

$$\omega_1 = \sqrt{\frac{k_1}{2m_1}} \qquad \omega_2 = \sqrt{\frac{2k_1}{m_1}}$$

$$\phi_1 = \begin{bmatrix} \frac{1}{\sqrt{3m_1}} \\ \frac{2}{\sqrt{3m_1}} \end{bmatrix} \qquad \phi_2 = \begin{bmatrix} \frac{1}{\sqrt{1.5m_1}} \\ -\frac{1}{\sqrt{1.5m_1}} \end{bmatrix}$$

The elements of the receptance matrix are obtained on substituting appropriate values in Equation 12.30

$$(R_d)_{11} = \frac{1/3k_1}{1/2 - (\Omega/\omega_1^*)^2} + \frac{1/1.5k_1}{2 - (\Omega/\omega_1^*)^2} \tag{12.31a}$$

$$(R_d)_{12} = (R_d)_{21} = \frac{2/3k_1}{1/2 - (\Omega/\omega_1^*)^2} - \frac{2/3k_1}{2 - (\Omega/\omega_1^*)^2} \tag{12.31b}$$

$$(R_d)_{22} = \frac{4/3k_1}{1/2 - (\Omega/\omega_1^*)^2} + \frac{2/3k_1}{2 - (\Omega/\omega_1^*)^2} \tag{12.31c}$$

where $\omega_1^* = \sqrt{k_1/m_1}$. The amplitudes of the two terms in the point receptance given by Equation 12.31a are plotted in Figure 12.2. Also shown in that figure is the resultant value $(R_d)_{11}$. It may be noted that the amplitudes of the individual terms in Equation 12.31a represent the absolute values of these terms, the sign being reflected in the phase angle, which is either $0°$ or $180°$. However, in taking the sum of the two terms to get the resultant, account must be taken of the signs of the individual terms.

Several interesting points may be noted by observing the plots shown in Figure 12.2. When the excitation frequency is close to one of the modal frequencies, the term related to that mode dominates in the expression for FRF. In general, it can be stated that for well-separated modes the following provides

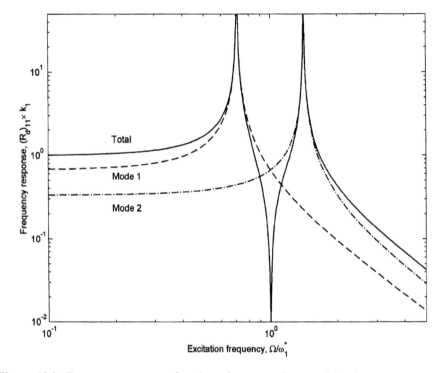

Figure 12.2. Frequency response functions for a two-degree-of-freedom system, point receptance plots.

a reasonable estimate of the frequency response.

$$(R_d)_{lk} \approx \frac{\phi_{lm}\phi_{km}}{(\omega_m^2 - \Omega^2) + 2\xi_m\omega_m\Omega} \qquad \text{for } \Omega \approx \omega_m \qquad (12.32)$$

When Ω coincides with one of the natural frequencies, the response for the undamped system becomes infinite. This phenomenon is referred to as resonance.

For excitation frequencies lying between two consecutive natural frequencies, that is $\omega_1 < \Omega < \omega_2$, the two terms in the point receptance, for example those in Equation 12.31a, are of opposite sign. This is because the numerator ϕ_{1n}^2 is always positive; on the other hand, while the denominator $\omega_1^2 - \Omega^2$ is negative, $\omega_2^2 - \Omega^2$ is positive. Consequently, at some value of Ω between ω_1 and ω_2, the two terms cancel each other and the resultant response is zero. In the present case, this happens when $\Omega/\omega_1^* = 1$. A point in the FRF where the response becomes very small is called *antiresonance*.

Resonances are also evident in the cross-receptance plot $(R_d)_{12}$ shown in Figure 12.3. However, antiresonance does not exist between ω_1 and ω_2, although a minimum still exists. In the present case, the two terms in the expression for $(R_d)_{12}$ are both positive and do not, therefore, cancel each other.

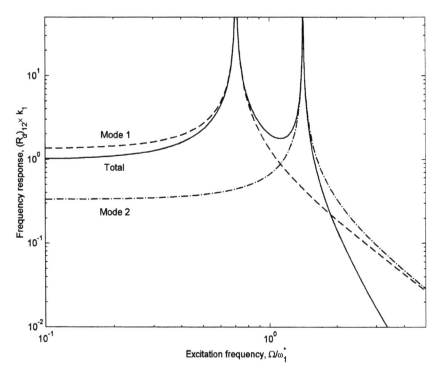

Figure 12.3. Frequency response functions for a two-degree-of-freedom system, cross-receptance plots.

Example 12.2

Assume that viscous damping exists in the two-degree-of-freedom system shown in Figure 2.1 and is represented by damping ratios $\xi_1 = \xi_2 = 0.1$ in each of the two modes. Also, $k_2 = 0.5k_1$ and $m_2 = 0.5m_1$. Plot the amplitude and phase angle against the exciting frequency for the receptance $(R_d)_{11}$. Also plot the Nyquist diagrams for both $(R_d)_{11}$ and $(R_d)_{12}$.

Solution

The undamped frequencies and mode shapes of the system were derived earlier and are given by

$$\omega_1 = \sqrt{\frac{k_1}{2m_1}} \qquad \omega_2 = \sqrt{\frac{2k_1}{m_1}}$$

$$\phi_1 = \begin{bmatrix} \frac{1}{\sqrt{3m_1}} \\ \frac{2}{\sqrt{3m_1}} \end{bmatrix} \qquad \phi_2 = \begin{bmatrix} \frac{1}{\sqrt{1.5m_1}} \\ -\frac{1}{\sqrt{1.5m_1}} \end{bmatrix}$$

The receptances are obtained by substituting the above values in Equation 12.30

$$(R_d)_{11} = \frac{1}{3k_1(0.5 - \bar{\Omega}^2 + i\sqrt{2}\xi_1\bar{\Omega})} + \frac{2}{3k_1(2 - \bar{\Omega}^2 + i2\sqrt{2}\xi_2\bar{\Omega})} \qquad (a)$$

$$(R_d)_{12} = \frac{2}{3k_1(0.5 - \bar{\Omega}^2 + i\sqrt{2}\xi_1\bar{\Omega})} - \frac{2}{3k_1(2 - \bar{\Omega}^2 + i2\sqrt{2}\xi_2\bar{\Omega})} \qquad (b)$$

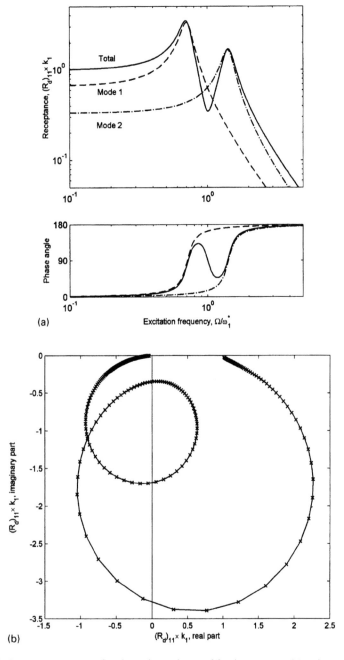

(a)

(b)

Figure E12.2. Frequency response of a damped two-degree-of-freedom system; (a) point receptance and phase angle; (b) Nyquist plot for point receptance; (c) Nyquist plot for cross-receptance.

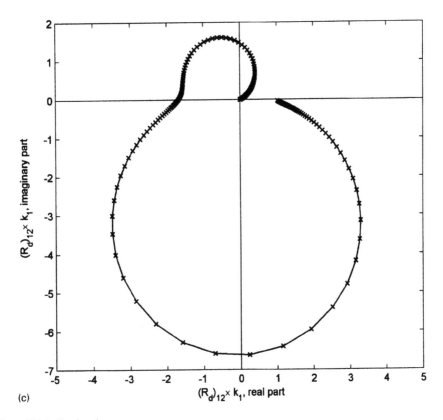

(c)

Figure E12.2. Continued.

where $\bar{\Omega} = \Omega/\omega_1^*$ and $\omega_1^* = \sqrt{k_1/m_1}$. The amplitude and phase angle for $(R_d)_{11}$ are plotted in Figure E12.2a. Both resonance and antiresonance exist in the amplitude plot; however, damping has caused a smoothening of the curves in each case. The Nyquist diagram for $(R_d)_{11}$ is obtained by plotting the imaginary part against the real part in Equation a and is shown in Figure E12.2b. The Nyquist plot for $(R_d)_{12}$ obtained from Equation b is shown in Figure E12.2c.

12.5 VIBRATION ABSORBERS

Vibration absorbers are devices that are used to kill or reduce unwanted vibrations in mechanical or structural systems. Although the detailed design of a vibration absorber must consider many factors, the conceptual background to the design is quite straightforward and relies on the phenomenon of antiresonance discussed in the preceeding section. Referring to Figure 12.1, let the primary system be represented by a single-degree-of-freedom system having a mass m_1 and spring stiffness k_1. Vibrations in the system are caused by a harmonic force $p_0 e^{i\Omega t}$ acting on the mass m_1. To reduce these vibrations, a mass m_2 is attached to the primary system through a spring of stiffness k_2. The response of the combined system can be obtained from Equations 12.21 and 12.27. It can be shown that the displacement amplitudes U_1 and U_2 are given by

$$U_1 = p_0 \frac{k_2 - m_2 \Omega^2}{D}$$

$$U_2 = p_0 \frac{k_2}{D}$$

(12.33)

where D is obtained from

$$D = (k_1 + k_2 - m_1 \Omega^2)(k_2 - m_2 \Omega^2) - k_2^2$$

(12.34)

Equation 12.33 shows that the vibrations of the primary system will be completely killed if the characteristics of the vibration absorbers are selected so that its frequency is equal to the operating frequency Ω, that is

$$\omega_2^* = \sqrt{\frac{k_2}{m_2}} = \Omega$$

(12.35)

In fact, the FRF $(R_d)_{11}$ reaches an antiresonance when this condition is satisfied, and for operating frequencies in the vicinity of antiresonance, vibrations of the primary system are reduced quite considerably. Because the characteristics of the vibration absorber are matched to a specific requirement, it is also referred to as a *tuned mass damper*. Obviously, a tuned mass damper is only effective if the operating frequencies are known to lie within a narrow range. The size and mass required for the absorber may also pose practical difficulty in its design, particularly when the primary structure is massive. At antiresonance, that is when $\sqrt{k_2/m_2} = \Omega$, the denominator D in Equation 12.33 becomes $-k_2^2$, so that the displacement amplitude for mass m_2 becomes

$$U_2 = -\frac{p_0}{k_2}$$

(12.36)

Equation 12.36 shows that the force exerted by the vibration absorber exactly balances the applied force. For a massive structure p_0 is expected to be large, and to keep the displacement of absorber mass within a reasonable bound, k_2 must be large. At the same time, a large k_2 should be accompanied by a large value of m_2, so as to satisfy Equation 12.35.

As an example of the effectiveness of a tuned mass damper, consider the case when $m_2 = 0.1 m_1$ and $k_2 = 0.1 k_1$. The FRF for the response of primary system after the attachment of tuned mass damper is shown in Figure 12.4, in which $\omega_1^* = \sqrt{(k_1/m_1)}$ is the uncoupled frequency of the primary system. Antiresonance is achieved when $\Omega = \omega_2^*$. Note that in the present example, $\omega_2^* = \omega_1^*$. Suppose that the design calls for the dynamic displacements not to exceed the static displacement under p_0. The range of operating frequencies

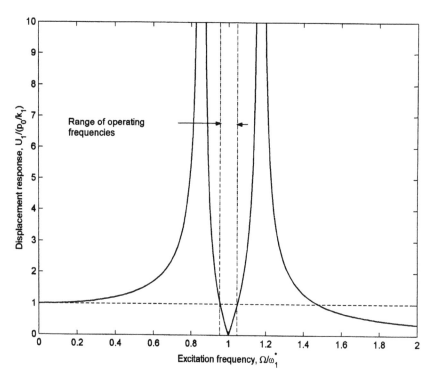

Figure 12.4. Frequency response function for a system equipped with vibration absorber.

over which the tuned mass damper will be effective are then as shown in Figure 12.4.

12.6 EFFECT OF SUPPORT EXCITATION

In many situations, a system may be made to vibrate by the motion of supports to which it is attached. Support-induced vibrations cause deformations and stresses in the structural or mechanical system, just as vibrations due to applied dynamic forces do. Obviously, such deformations and stresses must be taken into account in the design of the system. Many different examples can be cited of systems excited by support movement. Dynamic vibrations of an instrument attached to a moving frame is one such example. Earthquake induced vibrations of a building is another example.

In calculating the dynamic response of a system excited by support motions, we will, as usual, need the mass, stiffness, and damping matrices of the system. There are, however, no external forces except those acting at the attachments of the system to the supports. For a general treatment of the response, we include in our coordinates the displacement degrees of freedom at the supports as well.

The mass matrix of the system can then be partitioned as follows

$$\mathbf{M} = \begin{bmatrix} \mathbf{M}_{aa} & \mathbf{M}_{ab} \\ \mathbf{M}_{ba} & \mathbf{M}_{bb} \end{bmatrix} \tag{12.37}$$

where suffix a denotes superstructure degrees of freedom and suffix b denotes support degrees of freedom. Matrices $\mathbf{M}_{ab}$ and $\mathbf{M}_{ba}$ denote coupling mass matrices between the superstructure and the supports. Note that in a lumped mass idealization, both $\mathbf{M}_{ab}$ and $\mathbf{M}_{ba}$ will be null matrices while $\mathbf{M}_{aa}$ and $\mathbf{M}_{bb}$ will be diagonal. Coupling terms will exist, for example, when a consistent mass formulation is used (Section 3.6.2). The stiffness and damping matrices are partitioned in a similar manner.

$$\mathbf{K} = \begin{bmatrix} \mathbf{K}_{aa} & \mathbf{K}_{ab} \\ \mathbf{K}_{ba} & \mathbf{K}_{bb} \end{bmatrix} \tag{12.38}$$

$$\mathbf{C} = \begin{bmatrix} \mathbf{C}_{aa} & \mathbf{C}_{ab} \\ \mathbf{C}_{ba} & \mathbf{C}_{bb} \end{bmatrix} \tag{12.39}$$

The force vector is given by

$$\mathbf{f} = \begin{bmatrix} 0 \\ \mathbf{f}_b \end{bmatrix} \tag{12.40}$$

Note that there are no external forces along the superstructure degrees of freedom.

We now denote the displacements along the superstructure degrees of freedom by $\mathbf{u}'_a$ and those along the base degrees of freedom by $\mathbf{u}_b$. The superscript t on the superstructure displacements signifies that these displacements are absolute and is used to differentiate the absolute displacements from the relative displacements, which we use later in the formulation. With the notations described above, the equations of motion can be written as

$$\begin{bmatrix} \mathbf{M}_{aa} & \mathbf{M}_{ab} \\ \mathbf{M}_{ba} & \mathbf{M}_{bb} \end{bmatrix} \begin{bmatrix} \ddot{\mathbf{u}}'_a \\ \ddot{\mathbf{u}}_b \end{bmatrix} + \begin{bmatrix} \mathbf{C}_{aa} & \mathbf{C}_{ab} \\ \mathbf{C}_{ba} & \mathbf{C}_{bb} \end{bmatrix} \begin{bmatrix} \dot{\mathbf{u}}'_a \\ \dot{\mathbf{u}}_b \end{bmatrix}$$
$$+ \begin{bmatrix} \mathbf{K}_{aa} & \mathbf{K}_{ab} \\ \mathbf{K}_{ba} & \mathbf{K}_{bb} \end{bmatrix} \begin{bmatrix} \mathbf{u}'_a \\ \mathbf{u}_b \end{bmatrix} = \begin{bmatrix} 0 \\ \mathbf{f}_b \end{bmatrix} \tag{12.41}$$

Equations 12.41 can be expressed in expanded form as

$$\mathbf{M}_{aa}\ddot{\mathbf{u}}'_a + \mathbf{C}_{aa}\dot{\mathbf{u}}'_a + \mathbf{K}_{aa}\mathbf{u}'_a + \mathbf{M}_{ab}\ddot{\mathbf{u}}_b + \mathbf{C}_{ab}\dot{\mathbf{u}}_b + \mathbf{K}_{ab}\mathbf{u}_b = 0 \tag{12.42a}$$

$$\mathbf{M}_{bb}\ddot{\mathbf{u}}_b + \mathbf{C}_{bb}\dot{\mathbf{u}}_b + \mathbf{K}_{bb}\mathbf{u}_b + \mathbf{M}_{ba}\ddot{\mathbf{u}}'_a + \mathbf{C}_{ba}\dot{\mathbf{u}}'_a + \mathbf{K}_{ba}\mathbf{u}'_a = \mathbf{f}_b \tag{12.42b}$$

In Equations 12.42, the mass, damping, and stiffness matrices can be determined from the known physical characteristics of the system, while the support motions $\mathbf{u}_b$, $\dot{\mathbf{u}}_b$, and $\ddot{\mathbf{u}}_b$ must be specified. The unknowns to be determined are the displacements along the superstructure degrees of freedom, $\mathbf{u}_a$, and the support forces, $\mathbf{f}_b$. To reduce Equations 12.42 to a more useful form, let us examine the response of the system when the support displacements $\mathbf{u}_b$ are applied in a static manner. This is possible, provided we assume that $\mathbf{u}_b$ can be expressed as the product of a vector corresponding to the spatial degrees of freedom multiplied by a single time function, so that

$$\mathbf{u}_b = \tilde{\mathbf{u}}_b f(t) \tag{12.43}$$

where $f(t)$ denotes a function of time.

Static application of $\mathbf{u}_b$ means that the time variation is absent and the applied displacement vector is therefore equal to $\tilde{\mathbf{u}}_b$. The response caused by the application of $\tilde{\mathbf{u}}_b$ can be obtained from Equation 12.41 by noting that the velocities and accelerations will be zero. Thus

$$\begin{bmatrix} \mathbf{K}_{aa} & \mathbf{K}_{ab} \\ \mathbf{K}_{ba} & \mathbf{K}_{bb} \end{bmatrix} \begin{bmatrix} \tilde{\mathbf{u}}_a^s \\ \tilde{\mathbf{u}}_b \end{bmatrix} = \begin{bmatrix} 0 \\ \tilde{\mathbf{f}}_b^s \end{bmatrix} \tag{12.44}$$

in which $\tilde{\mathbf{u}}_a^s$ represents displacement of the superstructure and $\tilde{\mathbf{f}}_b^s$ represents support forces, both obtained from a static application of $\tilde{\mathbf{u}}_b$.
Equation 12.44 gives

$$\mathbf{K}_{aa} \tilde{\mathbf{u}}_a^s + \mathbf{K}_{ab} \tilde{\mathbf{u}}_b = 0 \tag{12.45a}$$

or

$$\tilde{\mathbf{u}}_a^s = -\mathbf{K}_{aa}^{-1} \mathbf{K}_{ab} \tilde{\mathbf{u}}_b$$
$$= \mathbf{R} \tilde{\mathbf{u}}_b \tag{12.45b}$$

where

$$\mathbf{R} = -\mathbf{K}_{aa}^{-1} \mathbf{K}_{ab} \tag{12.46}$$

On multiplying both sides of Equation 12.45b by $f(t)$, we get

$$\tilde{\mathbf{u}}_a^s f(t) = \mathbf{R} \tilde{\mathbf{u}}_b f(t) \tag{12.47a}$$

or

$$\mathbf{u}_a^s = \mathbf{R} \mathbf{u}_b \tag{12.47b}$$

where $\mathbf{u}_a^s = \tilde{\mathbf{u}}_a^s f(t)$ can be defined as a pseudostatic response. We now express the total displacements $\mathbf{u}_a^t$ as a superposition of the pseudostatic displacements $\mathbf{u}_a^s$ and relative displacements $\mathbf{u}_a$, so that

$$\mathbf{u}_a^t = \mathbf{u}_a^s + \mathbf{u}_a$$
$$= \mathbf{R} \mathbf{u}_b + \mathbf{u}_a \tag{12.48}$$

Substitution in Equation 12.42a gives

$$\mathbf{M}_{aa}\ddot{\mathbf{u}}_a + \mathbf{C}_{aa}\dot{\mathbf{u}}_a + \mathbf{K}_{aa}\mathbf{u}_a + \mathbf{M}_{aa}\mathbf{R}\ddot{\mathbf{u}}_b + \mathbf{M}_{ab}\ddot{\mathbf{u}}_b$$
$$+ \mathbf{C}_{aa}\mathbf{R}\dot{\mathbf{u}}_b + \mathbf{C}_{ab}\dot{\mathbf{u}}_b + \mathbf{K}_{aa}\mathbf{R}\mathbf{u}_b + \mathbf{K}_{ab}\mathbf{u}_b = 0 \tag{12.49}$$

Now, because $(\mathbf{K}_{aa}\mathbf{R}\mathbf{u}_b + \mathbf{K}_{ab}\mathbf{u}_b) = 0$ according to Equation 12.45a, Equation 12.49 reduces to

$$\mathbf{M}_{aa}\ddot{\mathbf{u}}_a + \mathbf{C}_{aa}\dot{\mathbf{u}}_a + \mathbf{K}_{aa}\mathbf{u}_a = -(\mathbf{M}_{aa}\mathbf{R} + \mathbf{M}_{ab})\ddot{\mathbf{u}}_b - (\mathbf{C}_{aa}\mathbf{R} + \mathbf{C}_{ab})\dot{\mathbf{u}}_b$$
$$\tag{12.50}$$

Equation 12.50, written in terms of the relative displacements $\mathbf{u}_a$, is entirely equivalent to Equation 12.1, with the forcing function $\mathbf{p}(t)$ being a function of the base velocity $\dot{\mathbf{u}}_b$ and base acceleration $\ddot{\mathbf{u}}_b$. It can therefore be solved by application of the mode superposition method, using the mode shapes of a fixed-base undamped system in exactly the same manner as for a general applied force vector.

Once the displacement vector $\mathbf{u}_a$ has been determined by solving Equation 12.50, the spring forces can be obtained in the usual manner. Thus, from Equation 12.41

$$\mathbf{f}_S = \mathbf{K}_{aa}\mathbf{u}_a^t + \mathbf{K}_{ab}\mathbf{u}_b$$
$$= \mathbf{K}_{aa}\mathbf{u}_a + (\mathbf{K}_{aa}\mathbf{u}_a^s + \mathbf{K}_{ab}\mathbf{u}_b) \tag{12.51}$$
$$= \mathbf{K}_{aa}\mathbf{u}_a$$

in which $\mathbf{K}_{aa}\mathbf{u}_a^s + \mathbf{K}_{ab}\mathbf{u}_b$ has been set as equal to zero because of the relationship in Equation 12.45. The elastic forces along the support degree of freedom are obtained from

$$\mathbf{f}_{Sb} = \mathbf{K}_{ba}\mathbf{u}_a^t + \mathbf{K}_{bb}\mathbf{u}_b \tag{12.52a}$$

or

$$\mathbf{f}_{Sb} = \mathbf{K}_{ba}\mathbf{u}_a + (\mathbf{K}_{ba}\mathbf{u}_a^s + \mathbf{K}_{bb}\mathbf{u}_b) \tag{12.52b}$$

The terms within parentheses in Equation 12.52b are together equal to $\tilde{\mathbf{f}}_b^s f(t)$, as seen from Equation 12.44, and thus represent the pseudostatic elastic forces caused by the displacements of the supports. On substitution for $\mathbf{u}_a^s$ from Equation 12.45, Equation 12.52b becomes

$$\mathbf{f}_{Sb} = \mathbf{K}_{ba}\mathbf{u}_a + (\mathbf{K}_{bb} - \mathbf{K}_{ba}\mathbf{K}_{aa}^{-1}\mathbf{K}_{ab})\mathbf{u}_b \tag{12.53}$$

Equation 12.50 can be considerably simplified for certain special cases. For example, if the damping matrix is proportional to the stiffness matrix, the term $\mathbf{C}_{aa}\mathbf{R} + \mathbf{C}_{ab}$ will be zero. This can easily be verified by setting $\mathbf{C}_{aa} = \beta\mathbf{K}_{aa}$

and $\mathbf{C}_{ab} = \beta \mathbf{K}_{ab}$, where β is a constant of proportionality, and then substituting for $\mathbf{R}$ from Equation 12.46. When the damping matrix is not proportional to stiffness matrix, the contribution of the damping terms to the forcing function will not be zero. In practice, however, this contribution is relatively small and can often be neglected.

A major simplification results when the motion is identical for all the supports. In such a case, $\mathbf{u}_b$ can be expressed as

$$\mathbf{u}_b = \mathbf{1} u_g \tag{12.54}$$

where u_g is the common support motion and $\mathbf{1}$ is a vector of p elements each equal to 1, p being the number of base degrees of freedom. Equation 12.45b now gives

$$\mathbf{u}_a^s = \mathbf{K}_{aa}^{-1} \mathbf{K}_{ab} \mathbf{1} u_g$$
$$= \mathbf{r} u_g \tag{12.55}$$

where

$$\mathbf{r} = - \mathbf{K}_{aa}^{-1} \mathbf{K}_{ab} \mathbf{1} \tag{12.56}$$

It should be noted that for identical support motions, the system moves as a rigid body; vector $\mathbf{r}$ need not therefore be evaluated from Equation 12.56 but can be obtained simply by kinematic considerations. For identical support motion and a negligible damping contribution to the forcing function, Equation 12.50 becomes

$$\mathbf{M}_{aa} \ddot{\mathbf{u}}_a + \mathbf{C}_{aa} \dot{\mathbf{u}}_a + \mathbf{K}_{aa} \ddot{\mathbf{u}}_a = - \mathbf{M}_{aa} \mathbf{r} \ddot{u}_g - \mathbf{M}_{ab} \mathbf{1} \ddot{u}_g \tag{12.57}$$

Furthermore, because for a rigid-body motion, $\tilde{\mathbf{f}}_b^s$ in Equation 12.44 should be zero, Equation 12.52b for elastic support forces reduces to

$$\mathbf{f}_{Sb} = \mathbf{K}_{ba} \mathbf{u}_a \tag{12.58}$$

Finally, for a mass idealization in which the coupling terms $\mathbf{M}_{ab}$ are absent, Equation 12.57 simplifies to

$$\mathbf{M} \ddot{\mathbf{u}} + \mathbf{C} \dot{\mathbf{u}} + \mathbf{K} \mathbf{u} = - \mathbf{M} \mathbf{r} \ddot{u}_g \tag{12.59}$$

where, for simplicity, we have dropped the subscript a. Application of the mode superposition method to the solution of Equation 12.58 will provide N uncoupled equations of motion, where N is the number of superstructure degrees of freedom and it is assumed that the damping matrix possesses the orthogonality property. The nth uncoupled equation will be

$$\ddot{y}_n + 2 \xi_n \omega_n \dot{y}_n + \omega_n^2 y_n = - \frac{\boldsymbol{\phi}_n^T \mathbf{M} \mathbf{r}}{M_n} \ddot{u}_g \tag{12.60}$$

The uncoupled equations can be solved by standard methods of analysis. Response in the physical coordinates is then obtained by a superposition of the modal responses.

Example 12.3

Formulate the equations of motion for support excitation of the system shown in Figure E12.3. Obtain expressions for the motion of masses 1 and 2 when the support excitation is given by

(i) $\ddot{\mathbf{u}}_g = \begin{bmatrix} 1 \\ 2 \end{bmatrix} f(t)$

(ii) $\ddot{\mathbf{u}}_g = \begin{bmatrix} 1 \\ 1 \end{bmatrix} f(t)$

The mass, damping, and stiffness parameters are: $m_1 = 1$, $m_2 = 2$, $m_3 = \frac{1}{2}$, $m_4 = \frac{1}{2}$, $c_1 = 3.04$, $c_2 = 6.08$, $c_3 = 3.04$, $k_1 = 600$, $k_2 = 1200$, and $k_3 = 600$.

Solution

The mass, damping, and stiffness matrices are obtained quite easily and are given by

$$\mathbf{M} = \begin{bmatrix} m_1 & & & \\ & m_2 & & \\ & & m_3 & \\ & & & m_4 \end{bmatrix} \tag{a}$$

$$\mathbf{C} = \begin{bmatrix} c_1 + c_2 & -c_2 & -c_1 & 0 \\ -c_2 & c_2 + c_3 & 0 & -c_3 \\ -c_1 & 0 & c_1 & 0 \\ 0 & -c_3 & 0 & c_3 \end{bmatrix} \tag{b}$$

$$\mathbf{K} = \begin{bmatrix} k_1 + k_2 & -k_2 & -k_1 & 0 \\ -k_2 & k_2 + k_3 & 0 & -k_3 \\ -k_1 & 0 & k_1 & 0 \\ 0 & -k_3 & 0 & k_3 \end{bmatrix} \tag{c}$$

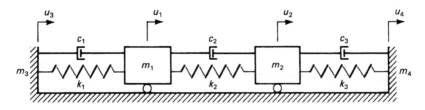

Figure E12.3. Response of a multi-degree-of-freedom system to support excitation.

On substituting the values of mass, stiffness, and damping parameters, we get

$$\mathbf{M} = \begin{bmatrix} 1 & 0 & 0 & 0 \\ 0 & 2 & 0 & 0 \\ 0 & 0 & \frac{1}{2} & 0 \\ 0 & 0 & 0 & \frac{1}{2} \end{bmatrix}$$

$$\mathbf{C} = \begin{bmatrix} 9.12 & -6.08 & -3.04 & 0.00 \\ -6.08 & 9.12 & 0.00 & -3.04 \\ -3.04 & 0.00 & 3.04 & 0.00 \\ 0.00 & -3.04 & 0.00 & 3.04 \end{bmatrix}$$

$$\mathbf{K} = \begin{bmatrix} 1800 & -1200 & -600 & 0 \\ -1200 & 1800 & 0 & -600 \\ -600 & 0 & 600 & 0 \\ 0 & -600 & 0 & 600 \end{bmatrix}$$

(i) The psuedostatic displacement is given by

$$\mathbf{u}_a^s = -\mathbf{R}\mathbf{u}_b \tag{d}$$

where

$$\mathbf{R} = -\mathbf{K}_{aa}^{-1}\mathbf{K}_{ab}$$

Substituting the submatrices

$$\mathbf{K}_{aa} = \begin{bmatrix} 1800 & -1200 \\ -1200 & 1800 \end{bmatrix}$$

$$\mathbf{K}_{ab} = \begin{bmatrix} -600 & 0 \\ 0 & -600 \end{bmatrix}$$

in Equation d, we get

$$\mathbf{R} = \frac{1}{5}\begin{bmatrix} 3 & 2 \\ 2 & 3 \end{bmatrix}$$

Because damping matrix is proportional to the stiffness matrix, we have

$$\mathbf{C}_{aa}\dot{\mathbf{u}}_a^s + \mathbf{C}_{ab}\dot{\mathbf{u}}_b = 0$$

and the equation of motion corresponding to the superstructure degrees of freedom becomes

$$\mathbf{M}_{aa}\ddot{\mathbf{u}}_a + \mathbf{C}_{aa}\dot{\mathbf{u}}_a + \mathbf{K}_{aa}\mathbf{u}_a = -\mathbf{M}_{aa}\mathbf{R}\ddot{\mathbf{u}}_b \tag{e}$$

or

$$\begin{bmatrix} 1 & 0 \\ 0 & 2 \end{bmatrix} \ddot{\mathbf{u}}_a + \begin{bmatrix} 9.12 & -6.08 \\ -6.08 & 9.12 \end{bmatrix} \dot{\mathbf{u}}_a + \begin{bmatrix} 1800 & -1200 \\ -1200 & 1800 \end{bmatrix} \mathbf{u}_a$$

$$= -\frac{1}{5} \begin{bmatrix} 7 \\ 16 \end{bmatrix} f(t)$$

To solve Equation e by the mode superposition method, we need the fixed-base undamped mode shapes of the system. These are obtained by solving the following free-vibration equations

$$\mathbf{M}_{aa} \ddot{\mathbf{u}}_a + \mathbf{K}_{aa} \mathbf{u}_a = 0 \tag{f}$$

or

$$\begin{bmatrix} 1 & 0 \\ 0 & 2 \end{bmatrix} \ddot{\mathbf{u}}_a + \begin{bmatrix} 1800 & -1200 \\ -1200 & 1800 \end{bmatrix} \mathbf{u}_a = 0$$

The eigenvalue problem of Equation f can be solved by any one of the standard methods described in Chapter 11. For a problem of small size such as this, the characteristic equation may be solved directly for the two eigenvalues. In turn, substitution of the eigenvalues in Equation f will give the corresponding eigenvectors. The following results are obtained

$$\lambda_1 = 389.5 \qquad \omega_1 = 19.74 \text{ rad/s}$$
$$\lambda_2 = 2310.5 \qquad \omega_2 = 48.07 \text{ rad/s}$$

$$\mathbf{q}_1 = \begin{bmatrix} 1 \\ 1.175 \end{bmatrix}$$

$$\mathbf{q}_2 = \begin{bmatrix} 1 \\ -0.425 \end{bmatrix}$$

The transformed matrices are obtained next

$$\tilde{\mathbf{M}} = \mathbf{Q}^T \mathbf{M} \mathbf{Q}$$

$$= \begin{bmatrix} 1 & 1.175 \\ 1 & -0.425 \end{bmatrix} \begin{bmatrix} 1 & \\ & 2 \end{bmatrix} \begin{bmatrix} 1 & 1 \\ 1.175 & -0.425 \end{bmatrix}$$

$$= \begin{bmatrix} 3.761 & 0 \\ 0 & 1.361 \end{bmatrix}$$

$$\tilde{\mathbf{C}} = \mathbf{Q}^T \mathbf{C} \mathbf{Q}$$

$$= \begin{bmatrix} 7.424 & 0 \\ 0 & 15.936 \end{bmatrix}$$

$$\tilde{\mathbf{K}} = \mathbf{Q}^T \mathbf{K} \mathbf{Q}$$

$$= \begin{bmatrix} 1465.2 & 0 \\ 0 & 3145.2 \end{bmatrix}$$

Also

$$\tilde{\mathbf{p}} = \mathbf{q}^T \mathbf{p}$$

$$= - \begin{bmatrix} 5.16 \\ 0.04 \end{bmatrix} f(t)$$

The two uncoupled equations of motion are thus

$$3.761 \ddot{y}_1 + 7.424 \dot{y}_1 + 1465.2 y_1 = -5.16 f(t)$$

$$1.362 \ddot{y}_2 + 15.936 \dot{y}_2 + 3145.2 y_2 = -0.04 f(t)$$

To obtain the damping ratios in the first mode, we note that

$$2 \xi_1 \omega_1 m_1 = c_1 \qquad\qquad\qquad (g)$$

Substituting $\omega_1 = 19.74$, $m_1 = 3.761$, and $c_1 = 7.424$, we get

$$\xi_1 = 0.05$$

In a similar manner, for the second mode

$$2 \xi_2 \omega_2 m_2 = c_2 \qquad\qquad\qquad (h)$$

Since $\omega_2 = 48.07$, $m_2 = 1.362$, and $c_2 = 15.931$, Equation h gives

$$\xi_2 = 0.1217$$

The equations of motion can now be expressed in the following alternative form

$$\ddot{y}_1 + 2 \xi_1 \omega_1 \dot{y}_1 + \omega_1^2 = \frac{k_1}{m_1} = -1.372 f(t) \qquad\qquad\qquad (i)$$

$$\ddot{y}_2 + 2 \xi_2 \omega_2 \dot{y}_2 + \omega_2^2 = \frac{k_2}{m_2} = -0.0294 f(t) \qquad\qquad\qquad (j)$$

The damped frequencies are given by

$$\omega_{D1} = \omega_1 \sqrt{1 - \xi_1^2} = 19.72$$

$$\omega_{D2} = \omega_2 \sqrt{1 - \xi_2^2} = 47.71$$

The solutions to the equations of motion can be expressed in terms of Duhamel's integrals. For the first mode

$$y_1 = -\frac{1.372}{19.72} \int_0^t f(\tau) e^{-19.74 \times 0.05(t-\tau)} \sin 19.72(t-\tau)\, d\tau$$

$$= -0.07 \int_0^t f(\tau) e^{-0.987(t-\tau)} \sin 19.72(t-\tau)\, d\tau$$

In a similar manner, for the second mode

$$y_2 = -0.000616 \int_0^t f(\tau) e^{-5.850(t-\tau)} \sin 47.71(t-\tau)\, d\tau$$

The response in the physical coordinates is given by

$$\begin{bmatrix} u_1 \\ u_2 \end{bmatrix} = \begin{bmatrix} 1 \\ 1.175 \end{bmatrix} y_1 + \begin{bmatrix} 1 \\ -0.425 \end{bmatrix} y_2$$

(ii) In this case, both supports undergo identical motion. Hence, the pseudostatic displacements of the two masses can be obtained by rigid-body kinetics. Thus

$$\mathbf{u}_a^s = \mathbf{r} u_g$$

$$= \begin{bmatrix} 1 \\ 1 \end{bmatrix} u_g \tag{k}$$

where r is the vector of pseudostatic displacements due to a unit displacement of supports. The equations of motion are given by

$$\mathbf{M}_{aa} \ddot{\mathbf{u}}_a + \mathbf{C}_{aa} \dot{\mathbf{u}}_a + \mathbf{K}_{aa} \mathbf{u}_a = -\mathbf{M}_{aa} \mathbf{r} \ddot{u}_g \tag{l}$$

or

$$\begin{bmatrix} 1 & 0 \\ 0 & 2 \end{bmatrix} \ddot{\mathbf{u}}_a + \begin{bmatrix} 9.12 & -6.08 \\ -6.08 & 9.12 \end{bmatrix} \dot{\mathbf{u}}_a + \begin{bmatrix} 1800 & -1200 \\ -1200 & 1800 \end{bmatrix} \mathbf{u}_a$$

$$= -\begin{bmatrix} 1 \\ 2 \end{bmatrix} f(t)$$

Of course, direct application of Equation e will lead to the same set of equations. Thus, the right-hand side of Equation e is obtained from

$$-\mathbf{M}_{aa}\mathbf{R}\ddot{u}_b = -\frac{1}{5}\begin{bmatrix} 1 & \\ & 2 \end{bmatrix}\begin{bmatrix} 3 & 2 \\ 2 & 3 \end{bmatrix}\begin{bmatrix} 1 \\ 1 \end{bmatrix} f(t)$$

$$= -\begin{bmatrix} 1 & \\ & 2 \end{bmatrix}\begin{bmatrix} 1 \\ 1 \end{bmatrix} f(t)$$

$$= -\begin{bmatrix} 1 \\ 2 \end{bmatrix} f(t)$$

Obviously, the use of rigid-body kinetics makes the problem much simpler. The application of mode superposition method follows lines similar to that in part (i). The frequencies, mode, shapes and damping ratios are the same. The transformed force vector is given by

$$\tilde{\mathbf{p}} = - \begin{bmatrix} 3.35 \\ 0.15 \end{bmatrix} f(t)$$

The modal displacements are obtained as

$$y_1 = -\frac{3.35}{19.72 \times 3.761} \int_0^t f(\tau) e^{-0.987(t-\tau)} \sin 19.72(t-\tau) \, d\tau$$

$$= -0.0452 \int_0^t f(\tau) e^{-0.987(t-\tau)} \sin 19.72(t-\tau) \, d\tau$$

and

$$y_2 = -0.00231 \int_0^t f(\tau) e^{-5.850(t-\tau)} \sin 47.71(t-\tau) \, d\tau$$

12.7 FORCED VIBRATION OF UNRESTRAINED SYSTEM

The mode superposition method of analysis applies without modification to the analysis of the forced-vibration response of an unrestrained system, provided that the rigid-body modes are included in the transformation. The resultant response obtained by the method automatically includes a rigid-body component of motion. A special method must, however, be used in evaluating the mode shapes. Several such methods were described in Section 11.7, and any one of them may be used as appropriate.

Example 12.4
The three-mass system shown in Figure E12.4a is suspended by a rigid wire. Describe the motion of the system in free fall after the wire suddenly snaps. Given

$$m_1 = m_2 = m_3 = m$$

$$k_1 = 2k$$

$$k_2 = k$$

Solution
When the system is at rest, mass 2 is displaced downward from the unstretched position of the spring by a distance

$$\frac{m_2 + m_3}{k_1} g = \frac{mg}{k} \tag{a}$$

where g is the acceleration due to gravity. In a similar manner, mass 3 is displaced downwards by a distance

$$\frac{m_2 + m_3}{k_1} g + \frac{m_3 g}{k_2} = \frac{2mg}{k} \tag{b}$$

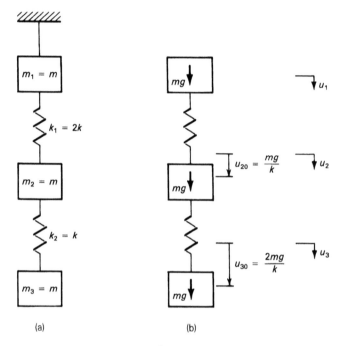

Figure E12.4. Three-mass suspended system released from rest.

We now solve for the motion of the released system shown in Figure E12.4b. The system is clearly unrestrained. The displacements calculated in Equations a and b can be treated as the initial displacements of the system. In addition, each mass of the system is subjected to a force equal to mg. The mass and stiffness matrices are given by

$$\mathbf{M} = \begin{bmatrix} m & & \\ & m & \\ & & m \end{bmatrix} \tag{c}$$

$$\mathbf{K} = \begin{bmatrix} 2k & -2k & 0 \\ -2k & 3k & -k \\ 0 & -k & k \end{bmatrix} \tag{d}$$

The system has one rigid-body mode, given by

$$\mathbf{q}_1 = \begin{bmatrix} 1 \\ 1 \\ 1 \end{bmatrix} \tag{e}$$

the corresponding eigenvalue being $\lambda_1 = 0$.

Each of the remaining two modes will satisfy the constraint equation

$$\mathbf{q}_1^T \mathbf{M} \hat{\mathbf{u}} = 0$$

or

$$\tilde{u}_1 + \tilde{u}_2 + \tilde{u}_3 = 0 \tag{f}$$

Equation f can be expressed in the form of a coordinate transformation

$$\begin{bmatrix} \tilde{u}_1 \\ \tilde{u}_2 \\ \tilde{u}_3 \end{bmatrix} = \begin{bmatrix} -1 & -1 \\ 1 & 0 \\ 0 & 1 \end{bmatrix} \begin{bmatrix} u_2 \\ u_3 \end{bmatrix} \tag{g}$$

or as

$$\tilde{\mathbf{u}} = \mathbf{Tu} \tag{h}$$

where

$$\mathbf{T} = \begin{bmatrix} -1 & -1 \\ 1 & 0 \\ 0 & 1 \end{bmatrix} \tag{i}$$

The mass and stiffness matrices in the transformed space are given by

$$\tilde{\mathbf{M}} = \mathbf{T}^T \mathbf{MT} = \begin{bmatrix} 2m & m \\ m & m \end{bmatrix} \tag{j}$$

$$\tilde{\mathbf{K}} = \mathbf{T}^T \mathbf{KT} = \begin{bmatrix} 9k & 3k \\ 3k & 3k \end{bmatrix} \tag{k}$$

The reduced eigenvalue problem becomes

$$\tilde{\mathbf{K}}\mathbf{u}_c = \lambda \tilde{\mathbf{M}}\mathbf{u}_c$$

The eigenvalues and eigenvectors in the reduced space are easily obtained and are given by

$$\lambda_2 = \tilde{\lambda}_1 = 1.268 \frac{k}{m} \quad \omega_2 = 1.128 \sqrt{\frac{k}{m}}$$

$$\lambda_3 = \tilde{\lambda}_2 = 4.732 \frac{k}{m} \quad \omega_3 = 2.175 \sqrt{\frac{k}{m}} \tag{l}$$

$$\tilde{\mathbf{q}}_1 = \begin{bmatrix} 1 \\ -3.732 \end{bmatrix}$$

$$\tilde{\mathbf{q}}_2 = \begin{bmatrix} 1 \\ -0.268 \end{bmatrix}$$

The coordinate transformation, Equation h, is used to obtain the complete mode shapes giving

$$\mathbf{q}_2 = \mathbf{T}\tilde{\mathbf{q}}_1 = \begin{bmatrix} 2.732 \\ 1 \\ -3.732 \end{bmatrix}$$

(m)

$$\mathbf{q}_3 = \mathbf{T}\tilde{\mathbf{q}}_2 = \begin{bmatrix} -0.732 \\ 1 \\ -0.268 \end{bmatrix}$$

On combining $\mathbf{q}_2$ and $\mathbf{q}_3$ with the rigid body-mode, the modal matrix $\mathbf{Q}$ is obtained as

$$\mathbf{Q} = \begin{bmatrix} 1 & 2.732 & -0.732 \\ 1 & 1 & 1 \\ 1 & -3.732 & -0.268 \end{bmatrix}$$

(m)

The equations of motion for the original system are

$$m \begin{bmatrix} 1 & & \\ & 1 & \\ & & 1 \end{bmatrix} \begin{bmatrix} \ddot{u}_1 \\ \ddot{u}_2 \\ \ddot{u}_3 \end{bmatrix} + k \begin{bmatrix} 2 & -2 & 0 \\ -2 & 3 & -1 \\ 0 & -1 & 1 \end{bmatrix} \begin{bmatrix} u_1 \\ u_2 \\ u_3 \end{bmatrix} = mg \begin{bmatrix} 1 \\ 1 \\ 1 \end{bmatrix}$$

(o)

with the initial conditions

$$u_{10} = 0$$

$$u_{20} = \frac{mg}{k}$$

(p)

$$u_{30} = \frac{2mg}{k}$$

These equations are now transformed to normal coordinates using the relationships

$$\mathbf{u} = \mathbf{Q}\mathbf{y}$$

$$\tilde{\mathbf{M}} = \mathbf{Q}^T \mathbf{M}\mathbf{Q}$$

$$\tilde{\mathbf{K}} = \mathbf{Q}^T \mathbf{K}\mathbf{Q}$$

(q)

$$\tilde{\mathbf{p}} = \mathbf{Q}^T \mathbf{p}$$

The transformed equations become

$$
m \begin{bmatrix} 3 & 0 & 0 \\ 0 & 22.39 & 0 \\ 0 & 0 & 1.608 \end{bmatrix} \begin{bmatrix} \ddot{y}_1 \\ \ddot{y}_2 \\ \ddot{y}_3 \end{bmatrix} + k \begin{bmatrix} 0 & 0 & 0 \\ 0 & 28.39 & 0 \\ 0 & 0 & 7.61 \end{bmatrix} \begin{bmatrix} y_1 \\ y_2 \\ y_3 \end{bmatrix}
$$
$$
= \begin{bmatrix} 3mg \\ 0 \\ 0 \end{bmatrix} \tag{r}
$$

The transformation has uncoupled the equations of motion and the uncoupled equations are

$$
\ddot{y}_1 = g
$$
$$
\ddot{y}_2 + 1.268 \frac{k}{m} y_2 = 0 \tag{s}
$$
$$
\ddot{y}_3 + 4.732 \frac{k}{m} y_3 = 0
$$

Before Equations s can be solved, we need the initial conditions in the normal coordinates. These are obtained by using the relationship

$$
\mathbf{Q}\mathbf{y}_0 = \mathbf{u}_0 \tag{t}
$$

Premultiplication of both sides of Equation t by $\mathbf{Q}^T\mathbf{M}$ gives

$$
\mathbf{Q}^T\mathbf{M}\mathbf{Q}\mathbf{y}_0 = \mathbf{Q}^T\mathbf{M}\mathbf{u}_0
$$
$$
\tilde{\mathbf{M}}\mathbf{y}_0 = \mathbf{Q}^T\mathbf{M}\mathbf{u}_0 \tag{u}
$$

Substitution for $\mathbf{Q}$, $\mathbf{M}$, and $\mathbf{u}_0$ gives

$$
\begin{bmatrix} 3 & 0 & 0 \\ 0 & 22.39 & 0 \\ 0 & 0 & 1.608 \end{bmatrix} \begin{bmatrix} y_{10} \\ y_{20} \\ y_{30} \end{bmatrix} = \begin{bmatrix} 3 \\ -6.464 \\ 0.464 \end{bmatrix} \frac{mg}{k}
$$

or

$$
y_{10} = \frac{mg}{k}
$$
$$
y_{20} = -0.2886 \frac{mg}{k} \tag{v}
$$
$$
y_{30} = 0.2886 \frac{mg}{k}
$$

The initial velocities in the normal coordinates are obviously zero. The solution of Equations r
with the initial conditions (Eq. v) gives

$$y_1 = \frac{gt^2}{2} + \frac{mg}{k}$$

$$y_2 = -0.2886\frac{mg}{k} \cos 1.128\sqrt{\frac{k}{m}}t \qquad\qquad (w)$$

$$y_3 = 0.2886\frac{mg}{k} \cos 2.175\sqrt{\frac{k}{m}}t$$

Finally, the displacements in the physical coordinates are given by

$$\mathbf{u} = \mathbf{Qy}$$

or

$$u_1 = \frac{gt^2}{2} + \frac{mg}{k}\left(1 - 0.7854 \cos 1.128\sqrt{\frac{k}{m}}t\right.$$

$$\left. - 0.2146 \cos 2.175\sqrt{\frac{k}{m}}t\right)$$

$$u_2 = \frac{gt^2}{2} + \frac{mg}{k}\left(2 - 0.2886 \cos 1.128\sqrt{\frac{k}{m}}t\right.$$

$$\left. + 0.2886 \cos 2.175\sqrt{\frac{k}{m}}t\right)$$

$$u_3 = \frac{gt^2}{2} + \frac{mg}{k}\left(1 + 1.077 \cos 1.128\sqrt{\frac{k}{m}}t\right. \qquad (x)$$

$$\left. - 0.077 \cos 2.175\sqrt{\frac{k}{m}}t\right)$$

SELECTED READINGS

Biggs, J.M. 1964. *Introduction to Structural Dynamics*. New York: McGraw-Hill.
Clough, R.W. & Penzien, J. 1993. *Dynamics of Structures*. New York: McGraw-Hill.
    2nd Edition.
Crandall, S.H. & McCalley, R.B. Jr. 1996. Matrix Methods of Analysis. In C.M. Harris
    (ed.) *Shock and Vibration Handbook*. New York: McGraw-Hill. 4th Edition.
Hurty, W.C. & Rubinstein, M.F. 1964. *Dynamics of Structures*. Englewood Cliffs:
    Prentice Hall.
Meirovitch, L. 1967. *Analytical Methods in Vibrations*. London: Macmillan.

PROBLEMS

12.1   An eccentric mass shaker is mounted on the third-story level of the
        frame of Problem 10.6. The shaker has a rotating weight of 800 lb

at an eccentricity of 12 in. with respect to the center of rotation. The speed of the rotor is 400 rpm. Obtain expressions for the steady state vibrations of the frame induced by the shaker.

12.2 Solve Problem 12.1 if the damping in each mode is 5% of critical.

12.3 The frame shown in Figure 8.15 is subjected to a lateral harmonic load $p_0 \sin \Omega t$ at the top floor level, where $p_0 = 45 \, \text{kN}$ and $\Omega = 5\pi/3 \, \text{rad/s}$. Derive expressions for the displacements at each floor level of the frame. Plot the top floor displacement as a function of time for the first second of response. Find the maximum displacement of the top floor during this period and the time at which the maximum occurs. Neglect damping.

12.4 For Problem 12.3, plot the modal contributions to the top floor displacements as functions of time for the first second of response. For each of the three modes, determine the maximum response and the time at which it occurs during the first second. Can the absolute values of the maximum responses be added to obtain the maximum resultant response?

12.5 For Problem 12.3, plot the modal contributions to the base shear as well as the resultant base shear as functions of time for the first second of response. In each case, determine the maximum value of the shear and the time at which it occurs.

12.6 For Problem 12.3, plot the steady state component of the top floor displacement in each of the three modes as a function of time for the first second of the response. What are the maximum values of the modal displacements and the resultant displacements? At what time do they occur?

12.7 The frame shown in Figure P8.15 is subjected to a rectangular pulse load of amplitude 50 kN and duration 0.5 s at the second floor level. Obtain expressions for the displacement at each floor level. Plot the top floor displacement as a function of time for the first second of response. Find the maximum response during this period and the time at which the maximum occurs. Neglect damping.

12.8 For Problem 12.7, plot the modal contributions to the top floor displacements as functions of time for the first second of response. For each of the three modes, determine the maximum response and the time at which it occurs during the first second.

12.9 For problem 12.7, plot the modal contributions to the interstory drift in second story for the first second of response. Also plot the resultant interstory drift. From this information find the maximum shear in the second story and the time at which it occurs.

12.10 A delicate instrument of mass m is packaged inside a rigid box of mass M (Fig. P12.10). The packaging material inside the box can be represented by a restraining spring of stiffness k and a viscous damper of coefficient c. The box is accidentally dropped from a height h. Derive expressions for the motion of the box and the instrument while in free fall.

12.11   Figure P12.11 shows a circular pipe section used in an industrial plant. The pipe is fixed to supports at a and b and has a 90° bend at c. It supports 2 heavy valves, each of mass 300 kg as indicated. The stiffness matrix of the pipe with respect to degrees of freedom 1 to 4 is given below. Find an expression for the total lateral displacement of the valve at 1 due to the following ground motions (**a**) The support at a undergoes a harmonic motion along d.o.f. 3 given by $u_g = 10 \sin 30t$, where u_g is the displacement in mm and t is the time in seconds; the support at b is fixed. (**b**) Both supports a and b have identical motion

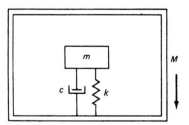

Figure P12.10.

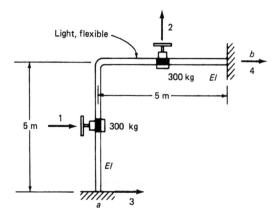

Figure P12.11.

given by $u_g = 10 \sin 30t$.

$$\mathbf{K} = \frac{EI}{L^3} \begin{bmatrix} 150.9 & -41.1 & -85.7 & -65.1 \\ -41.1 & 150.9 & 10.3 & 30.9 \\ -85.7 & 10.3 & 57.4 & 28.3 \\ -65.1 & 30.9 & 28.3 & 36.9 \end{bmatrix}$$

$E = 200,000 \text{ MPa}$

$I = 0.5 \times 10^6 \text{ mm}^4$

12.12 The flexibility and stiffness matrices of the beam of Problem 11.1 are as given below

$$\mathbf{A} = \frac{L^3}{768EI} \begin{bmatrix} 9 & 11 & 7 \\ 11 & 16 & 11 \\ 7 & 11 & 9 \end{bmatrix}$$

$$\mathbf{K} = \frac{768 \, EI}{28 \, L^3} \begin{bmatrix} 23 & -22 & 9 \\ -22 & 32 & -22 \\ 9 & -22 & 23 \end{bmatrix}$$

At $t = 0$ a constant load p_0 is suddenly applied at A and remains in contact with the beam thereafter. Using the mode superposition method, obtain the displacements, shear forces and bending moments in the beam as functions of time. Neglect damping. If $E = 200,000$ MPa; $I = 10.9 \times 10^6 \text{ mm}^4$; $m = 900 \text{ kg}$, and $L = 4 \text{ m}$, what is the bending moment under the load at $t = 0.5$ s? The load p_0 is in kN.

CHAPTER 13

Analysis of multi-degree-of-freedom systems: Approximate and numerical methods

13.1 INTRODUCTION

In characterizing the dynamic response of a system, the frequencies of vibration play a major role. The fundamental frequency of the system is often the quantity of primary interest. A great deal of attention has therefore been devoted to its determination. A particularly effective and simple method of estimating the fundamental frequency is the Rayleigh method described in Chapter 8. In this chapter, we present an extension of Rayleigh's method. Called the Rayleigh–Ritz method, it provides a means of improving the Rayleigh estimate for the fundamental frequency; in addition, it permits the determination of several higher frequencies of vibration.

While the determination of one or more frequencies may be sufficient in some cases, in others, a complete history of response over a given period of time is required. Response history can be obtained by the mode superposition method of analysis, which was described in detail in Chapters 10 and 12. The method transforms the coupled equations of motion of an N-degree-of-freedom system into N independent single-degree-of-freedom equations. The response of the system is obtained by solving the N uncoupled equations and superposing the solutions.

In the mode superposition method of analysis, if all N mode shapes of the system are used and the resulting uncoupled equations of motion are solved exactly, the exact response is obtained irrespective of the number of degrees of freedom the system may possess. In practice, even with modern powerful computers, the dynamic response analysis of a very large system with thousands of degrees of freedom is a very formidable and expensive task. In the mode superposition method, however, it is possible to obtain reasonable accuracy in the solution by including only a limited number of modes. How many modes need to be included in a particular case depends on the characteristics of the system and the applied load. In this chapter, we discuss procedures for estimating the errors introduced by mode truncation and for determining the number of modes that must be included in the analysis, so that a reasonable accuracy is obtained.

The mode superposition method described in Chapters 10 and 12 is, in fact, a special case of the more general method based on the transformation of coordinates using appropriately selected Ritz shapes. Like mode superposition, the Ritz shape methods permit a reduction in the size of the problem. The generation of suitable Ritz vectors therefore plays an important role in the numerical analysis of the response of multi-degree-of-freedom systems. In this chapter, we discuss several methods for the generation of such Ritz vectors. We also point out that many other methods of reducing the size of the problem, such as static condensation of the stiffness matrix and the mode acceleration method, belong to the category of Ritz vector methods.

The Ritz vector methods lead to a reduced set of transformed equations of motion which must be solved in order to obtain the response of the original system. In special cases, for example when undamped vibration modes shapes are used as the Ritz vectors, and the system is either undamped or has proportional damping, the transformed equations are uncoupled and can be solved by an appropriate analytical or numerical method of solving single degree-of-freedom equations. Several such methods were described in Chapters 6 through 8. When the transformed equations are coupled, they may either be uncoupled by a second transformation or solved by a simultaneous direct integration of the entire set of reduced equations. The methods of integration that may be used in the latter case are similar to those described in Chapter 8 for the solution of single degree-of-freedom systems. In this chapter, we describe the application of several such methods to the numerical integration of multiple coupled equation. It should be noted that direct numerical integration can be applied equally well to the original system of equations of motion without any transformation.

13.2   RAYLEIGH–RITZ METHOD

Rayleigh's principle and its applications have been discussed several times in previous chapters. It is instructive to summarize the different contexts in which the Rayleigh method was discussed. In Section 2.7.4, we described the application of an assumed shape function weighted by a generalized coordinate in the representation of a continuous system by a single-degree-of-freedom system. For example, by assuming that the displacement shape of a vibrating beam could be represented by $u(x,t) = z(t)\psi(x)$, where $z(t)$ was the generalized coordinate and $\psi(x)$ a shape function, we derived the following equation of motion

$$m^*\ddot{z}(t) + k^*z(t) = p^* \tag{13.1}$$

where

$$m^* = \int_0^L \bar{m}(x)\{\psi(x)\}^2 \, dx$$

$$k^* = \int_0^L EI(x)\{\psi''(x)\}^2 \, dx$$

and

$$p^* = \int_0^L \bar{p}\psi(x) \, dx$$

For free vibrations, p^* is zero and the solution of Equation 13.1 leads to the natural frequency of vibration ω. Thus

$$\omega^2 = \frac{k^*}{m^*} \tag{13.2a}$$

$$= \frac{\int_0^L EI(x)\{\psi''(x)\}^2 \, dx}{\int_0^L \bar{m}(x)\{\psi(x)\}^2 \, dx} \tag{13.2b}$$

Equation 13.2 was independently derived in Section 8.3.1 (Eq. 8.28) from energy considerations. As shown there, for free harmonic vibrations at the natural frequency, the maximum potential energy is given by

$$V_{\max} = (U)_{\max} = \frac{1}{2}z^2 \int_0^L EI(x)\{\psi''(x)\}^2 \, dx \tag{13.3}$$

while the maximum kinetic energy is given by

$$T_{\max} = \frac{1}{2}z^2\omega^2 \int_0^l \bar{m}(x)\{\psi(x)\}^2 \, dx \tag{13.4}$$

In the absence of any damping force, the energy must be conserved and hence $V_{\max}$ must be equal to $T_{\max}$. The application of this condition leads to Equation 13.2b.

The examples cited above are, in fact, applications of the Rayleigh method. The expression on the right-hand side of Equation 13.2b is called the *Rayleigh quotient*. Rayleigh's principle states that this quotient takes its minimum value when $\psi(x)$ is equal to the true vibration shape and that the minimum value of the quotient is equal to the square of the fundamental frequency. For any other value of $\psi(x)$, the Rayleigh quotient is larger than the square of the fundamental frequency.

Similar considerations apply to a multi-degree-of-freedom discrete parameter system. The application of a generalized coordinate and assumed vibration shape in the representation of such a system by an equivalent single-degree-of-freedom system was described in Section 2.8. As for a continuous system, the displacement vector in this case too is given by $u = z(t)\psi$, where ψ is now a vector. The equation of motion of the single-degree-of-freedom model is found to be

$$m^*\ddot{z} + k^*z = p^* \tag{13.5}$$

where

$$m^* = \boldsymbol{\psi}^T \mathbf{M} \boldsymbol{\psi}$$

$$k^* = \boldsymbol{\psi}^T \mathbf{K} \boldsymbol{\psi}$$

and

$$p^* = \boldsymbol{\psi}^T \mathbf{p}$$

An estimate for the fundamental frequency of the original system is obtained by solving Equation 13.5 with $p^* = 0$. This gives

$$\omega^2 = \frac{k^*}{m^*} \qquad (13.6a)$$

$$= \frac{\boldsymbol{\psi}^T \mathbf{K} \boldsymbol{\psi}}{\boldsymbol{\psi}^T \mathbf{M} \boldsymbol{\psi}} \qquad (13.6b)$$

Equation 13.6 could also be derived from energy considerations as illustrated in Example 8.3. The expression on the right-hand side of Equation 13.6b is the Rayleigh quotient and, as in the case of a continuous system, its minimum value obtained by varying $\boldsymbol{\psi}$ is equal to the square of the fundamental frequency.

In Section 10.5, we presented a formal proof of the fact that Rayleigh quotient provides an upper bound estimate of the lowest eigenvalue or the square of the fundamental frequency. We further proved that if the arbitrary shape vector $\boldsymbol{\psi}$ is selected from a subset of vectors that are orthogonal to the first $s - 1$ eigenvectors, the Rayleigh quotient provides an upper bound estimate to the sth eigenvalue. Our discussions indicated that the quality of the estimate of fundamental frequency depended on the choice of shape function $\boldsymbol{\psi}$: the closer this was to the true vibration mode shape, the better was the frequency estimate. In Section 8.4, we discussed a method of improving the frequency estimate.

An extension to the Rayleigh method suggested by Ritz and known as the *Rayleigh–Ritz method* provides an alternative method of obtaining a better estimate of the fundamental frequency. At the same time, it can be used to obtain estimates of several higher frequencies. In the Ritz extension of Rayleigh method for a discrete parameter system, displacements of the system are expressed as a superposition of several different independent shape vectors, known as *Ritz vectors*, each weighted by its own generalized coordinate. Thus

$$\begin{aligned} \mathbf{u} &= z_1(t)\boldsymbol{\psi}_1 + z_2(t)\boldsymbol{\psi}_2 + \cdots + z_M(t)\boldsymbol{\psi}_M \\ &= \boldsymbol{\Psi}\mathbf{z} \end{aligned} \qquad (13.7)$$

where $\boldsymbol{\Psi}$ is the matrix of Ritz vectors and $\mathbf{z}$ is the vector of M generalized coordinates. If we treat $\mathbf{u}$ as a possible displacement shape, the Rayleigh quotient corresponding to that shape is given by

$$\rho = \frac{\mathbf{u}^T \mathbf{K} \mathbf{u}}{\mathbf{u}^T \mathbf{M} \mathbf{u}}$$

$$= \frac{\mathbf{z}^T \mathbf{\Psi}^T \mathbf{K} \mathbf{\Psi} \mathbf{z}}{\mathbf{z}^T \mathbf{\Psi}^T \mathbf{M} \mathbf{\Psi} \mathbf{z}} \tag{13.8}$$

$$= \frac{\mathbf{z}^T \tilde{\mathbf{K}} \mathbf{z}}{\mathbf{z}^T \tilde{\mathbf{M}} \mathbf{z}}$$

where $\tilde{\mathbf{K}} = \mathbf{\Psi}^T \mathbf{K} \mathbf{\Psi}$ and $\tilde{\mathbf{M}} = \mathbf{\Psi}^T \mathbf{M} \mathbf{\Psi}$.

The value of Rayleigh quotient will change if the displacement shape is varied, which is equivalent to varying one or more of the generalized coordinates z_1, to z_M. We know that in the vicinity of an eigenvalue of the system, ρ takes a stationary value. For a discrete system, the stationary value is a minimum near all eigenvalues except the highest, where it is a maximum. The conditions of stationarity can be stated as

$$\frac{\partial \rho}{\partial z_j} = 0 \quad j = 1, 2, \ldots, M \tag{13.9}$$

For ease of reference, we denote the numerator of Equation 13.8 by v and the denominator by w. Note that both v and w are scalars. Equation 13.9 can now be expressed as

$$\frac{1}{w} \frac{\partial v}{\partial z_j} - \frac{v}{w^2} \frac{\partial w}{\partial z_j} = 0 \quad j = 1, 2, \ldots, M \tag{13.10a}$$

or as

$$\frac{\partial}{\partial z_j}(\mathbf{z}^T \tilde{\mathbf{K}} \mathbf{z}) - \frac{\mathbf{z}^T \tilde{\mathbf{K}} \mathbf{z}}{\mathbf{z}^T \tilde{\mathbf{M}} \mathbf{z}} \frac{\partial}{\partial z_j}(\mathbf{z}^T \tilde{\mathbf{M}} \mathbf{z}) = 0 \quad j = 1, 2, \ldots, M \tag{13.10b}$$

Since $\tilde{\mathbf{K}}$ and $\tilde{\mathbf{M}}$ are symmetric, the complete set of M equations represented by Equation 13.10b can be expressed as

$$\tilde{\mathbf{K}} \mathbf{z} - \frac{\mathbf{z}^T \tilde{\mathbf{K}} \mathbf{z}}{\mathbf{z}^T \tilde{\mathbf{M}} \mathbf{z}} \tilde{\mathbf{M}} \mathbf{z} = 0 \tag{13.11}$$

Now recognizing that when the condition given by Equation 13.9 is satisfied, ρ takes the value ω^2, where ω is one of the frequencies of the system, we have

$$\rho = \frac{\mathbf{z}^T \tilde{\mathbf{K}} \mathbf{z}}{\mathbf{z}^T \tilde{\mathbf{M}} \mathbf{z}} = \omega^2 \tag{13.12}$$

Substitution of Equation 13.12 in Equation 13.11 gives

$$\tilde{\mathbf{K}} \mathbf{z} = \omega^2 \tilde{\mathbf{M}} \mathbf{z} \tag{13.13}$$

Equation 13.13 will be recognized as a linearized eigenvalue problem. The frequencies ω obtained from its solution will be approximately equal to the frequencies of the original system, while the mode shapes $\mathbf{z}$ will be orthogonal to the reduced matrices $\tilde{\mathbf{K}}$ and $\tilde{\mathbf{M}}$. Denoting a normalized value of $\mathbf{z}$ by $\tilde{\phi}$, we have

$$\tilde{\phi}_i^T \tilde{\mathbf{K}} \tilde{\phi}_j = 0$$
$$\qquad\qquad i \neq j \qquad\qquad\qquad\qquad (13.14)$$
$$\tilde{\phi}_i^T \tilde{\mathbf{M}} \tilde{\phi}_j = 0$$

where $\tilde{\phi}_i$ is the ith normalized mode shapes.

Mode shapes $\tilde{\phi}$ are of size M and are not the mode shapes of the original system, which is of size N. It is, however, possible to obtain from $\tilde{\phi}_i$ the mode shapes of the original system. To do this, we apply the transformation given by Equation 13.7:

$$\phi_i = \mathbf{\Psi} \tilde{\phi}_i \quad i = 1, 2, \ldots, M \qquad\qquad\qquad (13.15a)$$

or

$$\mathbf{\Phi} = \mathbf{\Psi} \tilde{\mathbf{\Phi}} \qquad\qquad\qquad\qquad (13.15b)$$

Vectors ϕ are approximations to M mode shapes of the original system and are orthogonal with respect to both the $\mathbf{K}$ and $\mathbf{M}$ matrices. To demonstrate this, we note that

$$\phi_i^T \mathbf{K} \phi_j = \tilde{\phi}_i^T \mathbf{\Psi}^T \mathbf{K} \mathbf{\Psi} \tilde{\phi}_j$$
$$= \tilde{\phi}_i^T \tilde{\mathbf{K}} \tilde{\phi}_j \qquad i \neq j \qquad\qquad (13.16)$$
$$= 0$$

where the last step follows from Equation 13.14. In a similar manner

$$\phi_i^T \mathbf{M} \phi_j = \tilde{\phi}_i^T \tilde{\mathbf{M}} \tilde{\phi}_j$$
$$\qquad\qquad i \neq j \qquad\qquad\qquad\qquad (13.17)$$
$$= 0$$

Also

$$\frac{\phi_i^T \mathbf{K} \phi_i}{\phi_i^T \mathbf{M} \phi_i} = \frac{\tilde{\phi}_i^T \tilde{\mathbf{K}} \tilde{\phi}_i}{\tilde{\phi}_i^T \tilde{\mathbf{M}} \tilde{\phi}_i}$$
$$= \omega_i^2 \qquad\qquad\qquad\qquad (13.18)$$

Again, the last relationship holds because $\tilde{\phi}_i$ and ω_i are the eigenpairs obtained from the solution of Equation 13.13.

The frequencies and mode shapes obtained as above by application of the Rayleigh–Ritz method are only approximations to the true frequencies and mode shapes. The quality of the approximation depends on the selection of the assumed shapes $\mathbf{\Psi}$; the closer these shapes are to the true vibration mode shapes, the better are our estimates of the system frequencies and mode shapes. As proved in Section 10.5 (Eq. 10.38), the frequency estimates will always be larger than the fundamental frequency and smaller than the highest frequency. The Rayleigh–Ritz method will also provide estimates for intermediate frequencies, but what particular frequencies are estimated will depend on the selection of the Ritz shapes. In addition, the estimate of an intermediate frequency may be smaller or larger than the closest true frequency.

For a complex multi-degree-of-freedom system, the selection of vibration shapes to be used in Rayleigh–Ritz method is not simple. It is, however, possible to improve an initial estimate of the vibration shape by following a procedure very similar to that used in the Rayleigh method for obtaining the fundamental frequency of a multi-degree-of-freedom system. In the improvement to the Rayleigh method described in Section 8.4, we reasoned that the deflected shape obtained by the application of inertia forces resulting from an initial assumption of the free-vibration shape will provide a better estimate of the true free-vibration shape. Similar reasoning can be applied in application of the Rayleigh–Ritz method. Thus, if the initial assumption of the vibration shapes is $\mathbf{\Psi}^{(0)}$, the deflected shapes under the action of inertia forces resulting from $\mathbf{\Psi}^{(0)}$ will be given by an equation parallel to Equation 8.36

$$\bar{\mathbf{\Psi}}^{(1)} = \omega^2 \mathbf{K}^{-1} \mathbf{M} \mathbf{\Psi}^{(0)} \tag{13.19}$$

Since ω^2 is not known, we use vibration shapes $\mathbf{\Psi}^{(1)}$ that are proportional to $\bar{\mathbf{\Psi}}^{(1)}$ and are obtained by dropping ω^2 from Equation 13.19. Thus

$$\mathbf{\Psi}^{(1)} = \mathbf{K}^{-1} \mathbf{M} \mathbf{\Psi}^{(0)} \tag{13.20}$$

and

$$\bar{\mathbf{\Psi}}^{(1)} = \omega^2 \mathbf{\Psi}^{(1)} \tag{13.21}$$

Use of $\mathbf{\Psi}^{(1)}$ in place of $\mathbf{\Psi}$ in Equation 13.7 leads to the improved Rayleigh–Ritz equations

$$\tilde{\mathbf{K}}^{(1)} \mathbf{z} = \omega^2 \tilde{\mathbf{M}}^{(1)} \mathbf{z} \tag{13.22}$$

where

$$\begin{aligned}
\tilde{\mathbf{K}}^{(1)} &= \mathbf{\Psi}^{(1)^T} \mathbf{K} \mathbf{\Psi}^{(1)} \\
&= \mathbf{\Psi}^{(0)^T} \mathbf{M} \mathbf{K}^{-1} \mathbf{M} \mathbf{\Psi}^{(0)} \\
&= \mathbf{\Psi}^{(0)^T} \mathbf{M} \mathbf{F} \mathbf{M} \mathbf{\Psi}^{(0)}
\end{aligned} \tag{13.23}$$

and

$$\tilde{\mathbf{M}}^{(1)} = \mathbf{\Psi}^{(1)T}\mathbf{M}\mathbf{\Psi}^{(1)}$$

$$= \mathbf{\Psi}^{(0)T}\mathbf{M}\mathbf{K}^{-1}\mathbf{M}\mathbf{K}^{-1}\mathbf{M}\mathbf{\Psi}^{(0)} \qquad (13.24)$$

$$= \mathbf{\Psi}^{(0)T}\mathbf{M}\mathbf{F}\mathbf{M}\mathbf{F}\mathbf{M}\mathbf{\Psi}^{(0)}$$

in which $\mathbf{F}$ is the flexibility matrix.

The procedure outlined by Equations 13.22, 13.23, and 13.24 permits improvements in the calculated vibration shapes, so that reasonable estimates of frequencies can be obtained from crude initial assumption of the vibration shapes. In addition, it allows use of the flexibility matrix rather than the stiffness matrix in the computations, which may be an advantage in certain cases. An explicit evaluation of the flexibility matrix is not, in fact, needed; calculation of the deflections caused by applied inertia loads equal to $\mathbf{M}\mathbf{\Psi}^{(0)}$ is all that is required. As a final comment, the procedure will be recognized as the first step in the subspace iteration method described in Section 11.5.4.

Example 13.1
Obtain the stiffness and mass matrices for the five-story building frame shown in Figure E13.1a. Using the Rayleigh–Ritz method and the following displacement shapes, obtain two frequencies and the corresponding mode shapes of the building. Then by using an improved Rayleigh–Ritz

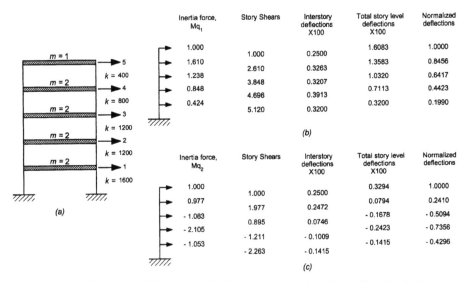

Figure E13.1. (a) Five-story building frame; (b) displacements caused by forces $\mathbf{M}\mathbf{q}_1$; (c) displacements caused by forces $\mathbf{M}\mathbf{q}_2$.

method, obtain better estimates of the frequencies and mode shapes.

$$\mathbf{\Psi}^{(0)} = \begin{bmatrix} 0.2 & -0.5 \\ 0.4 & -1.0 \\ 0.6 & -0.5 \\ 0.8 & 0.5 \\ 1.0 & 1.0 \end{bmatrix}$$

Solution

Corresponding to the degrees of freedom shown in Figure E13.1a, the mass and stiffness matrices of the frame are given by

$$\mathbf{M} = \begin{bmatrix} 2 & & & & \\ & 2 & & & \\ & & 2 & & \\ & & & 2 & \\ & & & & 1 \end{bmatrix}$$

$$\mathbf{K} = \begin{bmatrix} 2800 & -1200 & 0 & 0 & 0 \\ -1200 & 2400 & -1200 & 0 & 0 \\ 0 & -1200 & 2000 & -800 & 0 \\ 0 & 0 & -800 & 1200 & -400 \\ 0 & 0 & 0 & -400 & 400 \end{bmatrix}$$

The transformed mass and stiffness matrices are obtained as

$$\tilde{\mathbf{M}}^{(0)} = (\mathbf{\Psi}^0)^T \mathbf{M} \mathbf{\Psi}^0$$

$$= \begin{bmatrix} 3.4 & 0.2 \\ 0.2 & 4.5 \end{bmatrix} \tag{a}$$

$$\tilde{\mathbf{K}}^{(0)} = (\mathbf{\Psi}^0)^T \mathbf{K} \mathbf{\Psi}^0$$

$$= \begin{bmatrix} 208 & 40 \\ 40 & 1900 \end{bmatrix} \tag{b}$$

The reduced eigenvalue problem of Equation 13.13 becomes

$$\begin{bmatrix} 208 & 40 \\ 40 & 1900 \end{bmatrix} \mathbf{z} - \lambda \begin{bmatrix} 3.4 & 0.2 \\ 0.2 & 4.5 \end{bmatrix} \mathbf{z} = 0 \tag{c}$$

The eigenvalues can be obtained by solving the characteristic equation of the eigenproblem. Thus

$$\begin{vmatrix} 208 - 3.4\lambda & 40 - 0.2\lambda \\ 40 - 0.2\lambda & 1900 - 4.5\lambda \end{vmatrix} = 0 \tag{d}$$

or

$$\lambda^2 - 483.6\lambda + 25,790 = 0 \tag{e}$$

On solving Equation e we get

$$\lambda_1 = 61.04 \quad \omega_1 = 7.81$$
$$\lambda_2 = 422.6 \quad \omega_2 = 20.56$$

(f)

Substitution of λ_1 and λ_2, in turn, in Equation c give the corresponding eigenvectors in the reduced space

$$\mathbf{z}_1^{(1)} = \begin{bmatrix} -58.481 \\ 1 \end{bmatrix}$$

$$\mathbf{z}_2^{(1)} = \begin{bmatrix} -0.03623 \\ 1 \end{bmatrix}$$

(g)

The eigenvectors of the original system are obtained from

$$\mathbf{q}^{(1)} = \mathbf{\Psi}^{(0)} \mathbf{z}^{(1)}$$

(h)

Substitution for $\mathbf{z}$ from Equation g in Equation h gives

$$\mathbf{q}_1^{(1)} = \begin{bmatrix} -12.196 \\ -24.392 \\ -35.588 \\ -46.285 \\ -52.481 \end{bmatrix} \quad \mathbf{q}_2^{(1)} = \begin{bmatrix} -0.5072 \\ -1.0145 \\ -0.5217 \\ 0.4710 \\ 0.9638 \end{bmatrix}$$

(i)

When normalized so that the last element of the vector is 1, the eigenvectors $\mathbf{q}_1$ and $\mathbf{q}_2$ become

$$\mathbf{q}_1^{(1)} = \begin{bmatrix} 0.212 \\ 0.424 \\ 0.619 \\ 0.805 \\ 1.000 \end{bmatrix} \quad \mathbf{q}_2^{(1)} = \begin{bmatrix} -0.5263 \\ -1.0526 \\ -0.5414 \\ 0.4887 \\ 1.000 \end{bmatrix}$$

(j)

For the purpose of comparison, the exact values of the first two eigenvalues and eigenvectors are given below

$$\lambda_1 = 60.39 \quad \omega_1 = 7.77$$

$$\lambda_2 = 360.4 \quad \omega_2 = 18.98$$

$$\mathbf{q}_1 = \begin{bmatrix} 0.199 \\ 0.445 \\ 0.645 \\ 0.849 \\ 1.000 \end{bmatrix} \quad \mathbf{q}_2 = \begin{bmatrix} -0.309 \\ -0.536 \\ -0.440 \\ 0.995 \\ 1.000 \end{bmatrix}$$

(k)

The third eigenvalue is $\lambda_3 = 765.08$ or $\omega_3 = 22.66$.

We have obtained a good approximation to the first frequency. The second frequency estimate lies between the second and third true frequencies, but is closer to the second true frequency. The second frequency estimate is not as good as the first; and the estimate for the second eigenvector is, in fact, quite poor. To obtain an improved estimate of the eigenvalues and eigenvectors, we subject the frame to inertia forces caused by vibrations in each of the two mode shapes just derived. These inertia forces are proportional to $\mathbf{Mq}_1$ and $\mathbf{Mq}_2$, respectively. Next, we obtain the displacements caused by these inertia forces. The computations for obtaining the displaced shapes are straightforward and are shown in Figure E13.1b and c.

The calculations in Figure E13.1b and c are self-explanatory. Since we are interested only in a displacement shape rather than in the absolute value of the displacements, we have normalized the displacements so that the deflection at the top-story level is 1. This keeps the numbers within reasonable bounds. The resulting displaced shapes are

$$\mathbf{q}_1^{(1)} = \begin{bmatrix} 0.1990 \\ 0.4423 \\ 0.6417 \\ 0.8456 \\ 1.0000 \end{bmatrix} \qquad \mathbf{q}_2^{(1)} = \begin{bmatrix} -0.4296 \\ -0.7356 \\ -0.5094 \\ 0.2410 \\ 1.0000 \end{bmatrix}$$

Next, we use the displacement shapes just obtained as the new Ritz vectors to transform the problem to that of a reduced size. This gives

$$\tilde{\mathbf{M}}^{(1)} = \mathbf{\Psi}^{(1)^T} \mathbf{M} \mathbf{\Psi}^{(1)}$$

$$= \begin{bmatrix} 3.7241 & -0.0679 \\ -0.0679 & 3.0865 \end{bmatrix}$$

$$\tilde{\mathbf{K}}^{(1)} = \mathbf{\Psi}^{(1)^T} \mathbf{K} \mathbf{\Psi}^{(1)}$$

$$= \begin{bmatrix} 225.13 & -3.21 \\ -3.21 & 1149.97 \end{bmatrix} \tag{1}$$

in which we have used $\mathbf{\Psi}^{(1)} = [\mathbf{q}_1^{(1)} \ \mathbf{q}_2^{(1)}]$, $\mathbf{q}_1^{(1)}$ and $\mathbf{q}_2^{(1)}$ being the latest values of Ritz vectors.

The characteristic equation of the reduced eigenvalue problem becomes

$$\lambda^2 - 433.12\lambda + 22510 = 0 \tag{m}$$

Solution of Equation m leads to the following eigenvalues

$$\lambda_1 = 60.393 \quad \omega_1 = 7.771$$
$$\lambda_2 = 372.73 \quad \omega_2 = 19.306 \tag{n}$$

Substitution of λ_1 and λ_2, in turn, in the eigenvalue equation gives the following eigenvectors in the reduced space

$$\mathbf{z}_1^{(2)} = \begin{bmatrix} -1009 \\ 1 \end{bmatrix}$$

$$\mathbf{z}_2^{(2)} = \begin{bmatrix} 0.019438 \\ 1 \end{bmatrix} \tag{o}$$

The eigenvectors of the original system are obtained from

$$\mathbf{q}^{(2)} = \mathbf{\Psi}^{(1)} \mathbf{z}^{(2)} \tag{p}$$

After normalization, the new eigenvectors are given by

$$\mathbf{q}_1 = \begin{bmatrix} 0.1996 \\ 0.4435 \\ 0.6443 \\ 0.8461 \\ 1.0000 \end{bmatrix} \qquad \mathbf{q}_2 = \begin{bmatrix} -0.4174 \\ -0.7131 \\ -0.4874 \\ 0.2526 \\ 1.0000 \end{bmatrix}$$

The frequency estimates have improved for both the first and second modes.

Example 13.2

Obtain initial estimates of two of the frequencies of the system of Example 13.1 by using the following assumed vibration shapes in the Rayleigh–Ritz method.

$$\boldsymbol{\psi}_1 = \begin{bmatrix} \frac{1}{4} \\ \frac{1}{3} \\ \frac{1}{3} \\ 0 \\ -\frac{3}{4} \end{bmatrix} \qquad \boldsymbol{\psi}_2 = \begin{bmatrix} \frac{1}{3} \\ \frac{1}{3} \\ 0 \\ -\frac{1}{2} \\ \frac{1}{2} \end{bmatrix}$$

Solution

The transformed mass and stiffness matrices are given by

$$\tilde{\mathbf{M}} = \boldsymbol{\Psi}^T \mathbf{M} \boldsymbol{\Psi}$$

$$= \begin{bmatrix} 1.3190 & 0.0139 \\ 0.0139 & 1.1944 \end{bmatrix} \tag{a}$$

$$\tilde{\mathbf{K}} = \boldsymbol{\Psi}^T \mathbf{K} \boldsymbol{\Psi}$$

$$= 400 \begin{bmatrix} 1.0566 & -0.0833 \\ -0.0833 & 2.7778 \end{bmatrix} \tag{b}$$

The reduced eigenvalue problem becomes

$$400 \begin{bmatrix} 1.0566 & -0.0833 \\ -0.0833 & 2.7778 \end{bmatrix} \mathbf{z} - \lambda \begin{bmatrix} 1.1319 & 0.0139 \\ 0.0139 & 1.1944 \end{bmatrix} \mathbf{z} = 0 \tag{c}$$

On setting $\bar{\lambda} = \lambda/400$, the characteristic equation of the eigenvalue problem can be written as

$$\bar{\lambda}^2 - 2.8426\bar{\lambda} + 1.7753 = 0 \tag{d}$$

Solution of Equation d gives

$$\bar{\lambda}_1 = 0.9255 \quad \lambda_1 = 370.22 \quad \tilde{\omega}_1 = 19.24$$

$$\bar{\lambda}_2 = 1.9161 \quad \lambda_2 = 766.43 \quad \tilde{\omega}_2 = 27.68$$

The exact values of the first four frequencies are $\omega_1 = 7.77$, $\omega_2 = 18.98$, $\omega_3 = 27.66$, and $\omega_4 = 36.39$. The estimates that we have obtained, $\tilde{\omega}_1$ and $\tilde{\omega}_2$, are closest to the true frequencies ω_2 and ω_3. It is apparent from Examples 13.1 and 13.2 that the eigenvalue estimates obtained by the Rayleigh–Ritz method are greatly influenced by the selection of Ritz shapes.

Example 13.3
The two starting Ritz vectors used to obtain frequency estimates for the system of Example 13.1 happen to satisfy the following relationships

$$\psi_1 = a\phi_1 + b\phi_2$$

$$\psi_2 = c\phi_3 + d\phi_4$$

where a, b, c, and d are arbitrary constants.

(i) Prove that the first frequency estimate will lie between the first and the second true frequencies, while the second frequency estimate will lie between the third and the fourth true frequencies.
(ii) On selecting $a = 0.2$, $b = 0.9$, $c = 0.9$, and $d = 0.2$, the following trial Ritz vectors are derived

$$\psi_1 = \begin{bmatrix} 0.20 \\ 0.32 \\ 0.23 \\ -0.18 \\ -0.84 \end{bmatrix} \qquad \psi_2 = \begin{bmatrix} 0.18 \\ 0.31 \\ 0.06 \\ -0.61 \\ 0.61 \end{bmatrix}$$

in which the results have been rounded off to two decimal digits. Using these trial vectors obtain two frequency estimates.

Solution
(i) Using the orthogonality relationships the transformed mass and stiffness matrices can be shown to be

$$\tilde{M} = \begin{bmatrix} a^2 + b^2 & 0 \\ 0 & c^2 + d^2 \end{bmatrix} \tag{a}$$

$$\tilde{K} = \begin{bmatrix} a^2\omega_1^2 + b^2\omega_2^2 & 0 \\ 0 & c^2\omega_3^2 + d^2\omega_4^2 \end{bmatrix} \tag{b}$$

Because the reduced mass and stiffness matrices are both diagonal, the two frequencies of the reduced problem are obtained quite readily and are given by

$$\tilde{\omega}_1^2 = \frac{a^2\omega_1^2 + b^2\omega_2^2}{a^2 + b^2} \tag{c}$$

$$\tilde{\omega}_2^2 = \frac{c^2\omega_3^2 + b^2\omega_4^2}{c^2 + d^2} \tag{d}$$

Equation c can be expressed as

$$\tilde{\omega}_1^2 = \frac{a^2(\omega_1^2/\omega_2^2) + b^2}{a^2 + b^2}\omega_2^2 \tag{e}$$

Since ω_1 is smaller than ω_2, the numerator in Equation e is smaller than the denominator. As a result, $\tilde{\omega}_1$ is smaller than ω_2. As an alternative, Equation c can be expressed as

$$\tilde{\omega}_1^2 = \frac{a^2 + b^2(\omega_2^2/\omega_1^2)}{a^2 + b^2}\omega_1^2 \tag{f}$$

Again, because ω_2 is larger than ω_1, the numerator in Equation f is larger than the denominator and hence $\tilde{\omega}_1$ is larger than ω_1. Thus the first frequency estimate lies between the true first frequency and the true second frequency. In a similar manner, we can prove that the second frequency estimate $\tilde{\omega}_2$ will lie between the true third and fourth frequencies.

(ii) For the supplied Ritz vectors the transformed mass and stiffness matrices work out to

$$\tilde{\mathbf{M}} = \begin{bmatrix} 1.1610 & 0.0052 \\ 0.0052 & 1.3805 \end{bmatrix} \tag{g}$$

$$\tilde{\mathbf{K}} = \begin{bmatrix} 399.7 & 1.0 \\ 1.0 & 1101.6 \end{bmatrix} \tag{h}$$

which are very nearly diagonal. The solution of the reduced eigenvalue problem provides the following two estimates.

$$\tilde{\omega}_1 = 18.56 \qquad \tilde{\omega}_2 = 28.25 \tag{i}$$

The first estimate is very close to the second true frequency of 18.98, but is smaller than the true frequency. The second estimate is close to the third true frequency of 27.66, but is larger than the true frequency. It is evident that the Rayleigh–Ritz method may provide estimates for intermediate frequencies, but which true frequencies are matched depends on the selection of the starting Ritz vectors. Also, the intermediate frequency estimates may be higher or lower than the closest true frequencies. In our example the first starting vector was close to the second mode shape, hence it led to a good frequency estimate for the second frequency. However, because of contamination from the first mode shape, the estimated frequency was lower than the true frequency. It may be noted that, as proved in Section 10.5, a frequency estimate provides an upper bound to a true frequency only when the Ritz shape used to obtain the frequency estimate is orthogonal to all mode shapes lower than the one for which the frequency is being estimated. In our example the second trial shape is orthogonal to all modes lower than 3; as a result the estimate obtained for the true third frequency is an upper bound.

The Rayleigh–Ritz method is also applicable to continuous systems. Similar to Equation 13.7, the deflected shape of a continuous system can be represented by a superposition of M shape functions each weighted by a different generalized coordinate. As an example, for the lateral vibrations of a beam, the deflected shape is represented by

$$u(x, t) = z_1(t)\psi_1(x) + z_2(t)\psi_2(x) + \cdots + z_M(t)\psi_M(x)$$

$$= \boldsymbol{\psi}\mathbf{z} \tag{13.25}$$

where ψ_i, $i = 1$ to M, are Ritz shapes which are functions of the spatial coordinates x; z_i, $i = 1$ to M, are the generalized coordinates; $\boldsymbol{\psi}$ is a row vector of shapes ψ_i; and $\mathbf{z}$ is the vector of generalized coordinates. The Rayleigh quotient corresponding to the displaced shape $\mathbf{u}$ is given by

$$\rho = \frac{\int_0^L EI(x)\mathbf{z}^T[\boldsymbol{\psi}''(x)]^T\boldsymbol{\psi}''(x)\mathbf{z}\,dx}{\int_0^L \bar{m}(x)\mathbf{z}^T[\boldsymbol{\psi}(x)]^T\boldsymbol{\psi}(x)\mathbf{z}\,dx} \tag{13.26}$$

We now introduce the notations

$$\tilde{\mathbf{K}} = \int_0^L EI(x)[\boldsymbol{\psi}''(x)]^T\boldsymbol{\psi}''(x)\,dx \tag{13.27a}$$

$$\tilde{\mathbf{M}} = \int_0^L \bar{m}(x)[\boldsymbol{\psi}(x)]^T\boldsymbol{\psi}(x)\,dx \tag{13.27b}$$

so that

$$\tilde{k}_{ij} = \int_0^L EI(x)\psi_i''(x)\psi_j''(x)\,dx \tag{13.28a}$$

$$\tilde{m}_{ij} = \int_0^L \bar{m}(x)\psi_i(x)\psi_j(x)\,dx \tag{13.28b}$$

With the notations given by Equation 13.27, Equation 13.26 reduces to the form of Equation 13.8. Application of stationarity condition on the Rayleigh quotient then leads to the eigenvalue equation (Eq. 13.13).

Example 13.4
For the lateral vibrations of uniform cantilever beam shown in Figure E13.4a, determine two frequencies and mode shapes using the shape functions given below and shown in Figure E13.4b.

$$\psi_1 = 1 - \cos\frac{\pi x}{2L}$$

$$\psi_2 = 1 - \cos\frac{3\pi x}{2L}$$

Both shape functions satisfy the two geometric conditions at $x = 0$. They also satisfy the zero moment condition at $x = L$ but not the zero shear condition at the free end of the cantilever.

Solution
The eigenvalue problem is given by Equation 13.13

$$\tilde{\mathbf{K}}\mathbf{z} = \omega^2\tilde{\mathbf{M}}\mathbf{z} \tag{a}$$

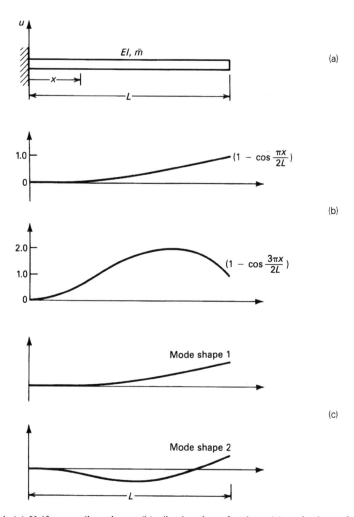

Figure E13.4. (a) Uniform cantilever beam; (b) vibration shape functions; (c) mode shapes for the lateral vibrations of uniform cantilever.

in which the elements of the stiffness and mass matrices are obtained from Equations 13.28a and 13.28b, respectively. We thus have

$$k_{11} = \frac{1}{16}\frac{\pi^4 EI}{L^4}\int_0^L \left(\cos\frac{\pi x}{2L}\right)^2 dx$$

$$= \frac{1}{32}\frac{\pi^4 EI}{L^3}$$

$$k_{12} = \frac{9}{16}\frac{\pi^4 EI}{L^4}\int_0^L \cos\frac{\pi x}{2L}\cos\frac{3\pi x}{2L}\,dx = 0$$

$$= k_{21}$$

$$k_{22} = \frac{81}{16} \frac{\pi^4 EI}{L^4} \int_0^L \left(\cos \frac{3\pi x}{2L} \right)^2 dx$$

$$= \frac{81}{32} \frac{\pi^4 EI}{L^3}$$

(b)

Also

$$m_{11} = \int_0^L \bar{m} \left(1 - \cos \frac{\pi x}{2L} \right)^2 dx$$

$$= \bar{m} \left(\frac{3L}{2} - \frac{4L}{\pi} \right)$$

$$m_{12} = \int_0^L \bar{m} \left(1 - \cos \frac{\pi x}{2L} \right) \left(1 - \cos \frac{3\pi x}{2L} \right) dx$$

$$= \bar{m} \left(L - \frac{4L}{3\pi} \right)$$

(c)

$$= m_{21}$$

$$m_{22} = \int_0^L \bar{m} \left(1 - \cos \frac{3\pi x}{2L} \right)^2 dx$$

$$= \bar{m} \left(\frac{3L}{2} + \frac{4L}{3\pi} \right)$$

Substituting Equations b and c in Equation a, we get

$$\frac{1}{32} \frac{\pi^4 EI}{L^3} \begin{bmatrix} 1 & 0 \\ 0 & 81 \end{bmatrix} \begin{bmatrix} z_1 \\ z_2 \end{bmatrix} = \omega^2 \bar{m} L \begin{bmatrix} \frac{3}{2} - \frac{4}{\pi} & 1 - \frac{4}{3\pi} \\ 1 - \frac{4}{3\pi} & \frac{3}{2} + \frac{4}{3\pi} \end{bmatrix} \begin{bmatrix} z_1 \\ z_2 \end{bmatrix}$$

(d)

On setting $\bar{\lambda} = (32 \bar{m} L^4 / \pi^4 EI) \omega^2$, we get the following characteristic equation from Equation d:

$$\begin{vmatrix} 1 - 0.2268\bar{\lambda} & -0.5756\bar{\lambda} \\ -0.5756\bar{\lambda} & 81 - 1.9244\bar{\lambda} \end{vmatrix} = 0$$

(e)

or

$$\bar{\lambda}^2 - 192.948\bar{\lambda} + 770.074 = 0$$

(f)

Solution of Equation f gives

$$\bar{\lambda}_1 = 4.0775 \quad \omega_1 = 3.523 \sqrt{\frac{EI}{\bar{m} L^4}}$$

$$\bar{\lambda}_2 = 188.87 \quad \omega_2 = 23.988 \sqrt{\frac{EI}{\bar{m} L^4}}$$

(g)

The exact values of the first two frequencies are

$$\omega_1 = 3.516 \sqrt{\frac{EI}{\bar{m} L^4}}$$

$$\omega_2 = 22.034 \sqrt{\frac{EI}{\bar{m} L^4}}$$

(h)

The approximate frequencies are fairly close to the exact values of the first two frequencies. Also, as expected, the approximate values provide upper bound estimates of the true frequencies. Substitution of $\bar{\lambda}_1$ and $\bar{\lambda}_2$, in turn, in Equation d provides the following values for the generalized coordinates

$$\begin{bmatrix} z_1 \\ z_2 \end{bmatrix} = \begin{bmatrix} 1 \\ 0.0321 \end{bmatrix}$$

$$\begin{bmatrix} z_1 \\ z_2 \end{bmatrix} = \begin{bmatrix} 1 \\ -0.3848 \end{bmatrix}$$

(i)

The mode shapes are now obtained from Equation 13.25 as

$$q_1 = \left(1 - \cos\frac{\pi x}{2L}\right) + 0.0321\left(1 - \cos\frac{3\pi x}{2L}\right)$$

$$= 1.0321 - \cos\frac{\pi x}{2L} - 0.0321\cos\frac{3\pi x}{2L}$$

(j)

$$q_2 = \left(1 - \cos\frac{\pi x}{2L}\right) - 0.3848\left(1 - \cos\frac{3\pi x}{2L}\right)$$

$$= 0.6152 - \cos\frac{\pi x}{2L} + 0.3848\cos\frac{3\pi x}{2L}$$

The two mode shapes have been plotted in Figure E13.4c.

Example 13.5
Assuming that the cantilever beam of Example 13.4 is of nonuniform section so that its moment of inertia and mass vary as follows, obtain estimates of two frequencies of the beam.

$$\bar{m}(x) = \bar{m}_0\left(1 - \frac{x}{2L}\right)$$

$$I(x) = I_0\left(1 - \frac{x}{2L}\right)$$

Solution
The elements of stiffness and mass matrices are again obtained from Equations 13.28a and 13.28b, respectively. Thus

$$k_{11} = \int_0^L \left(1 - \frac{x}{2L}\right)\frac{1}{16}\frac{\pi^4 EI_0}{L^4}\left(\cos\frac{\pi x}{2L}\right)^2 dx$$

$$= \frac{1}{16}\frac{\pi^4 EI_0}{L^3}\left(\frac{3}{8} + \frac{1}{2\pi^2}\right)$$

$$k_{12} = \int_0^L \left(1 - \frac{x}{2L}\right) \frac{9}{16} \frac{\pi^4 EI_0}{L^4} \left(\cos\frac{\pi x}{2L}\right) \left(\cos\frac{3\pi x}{2L}\right) dx$$

$$= \frac{9}{16} \frac{\pi^4 EI_0}{L^3} \left(\frac{1}{2\pi^2}\right)$$

$$= k_{21} \tag{a}$$

$$k_{22} = \int_0^L \left(1 - \frac{x}{2L}\right) \frac{81}{16} \frac{\pi^4 EI_0}{L^4} \left(\cos\frac{3\pi x}{2L}\right)^2 dx$$

$$= \frac{81}{16} \frac{\pi^4 EI_0}{L^3} \left(\frac{3}{8} + \frac{1}{18\pi^2}\right)$$

Also

$$m_{11} = \int_0^L \bar{m}_0 \left(1 - \frac{x}{2L}\right) \left(1 - \cos\frac{\pi x}{2L}\right)^2 dx$$

$$= \bar{m}_0 \left(\frac{9L}{8} - \frac{2L}{\pi} - \frac{7L}{2\pi^2}\right)$$

$$m_{12} = \int_0^L \bar{m}_0 \left(1 - \frac{x}{2L}\right) \left(1 - \cos\frac{\pi x}{2L}\right) \left(1 - \cos\frac{3\pi x}{2L}\right) dx$$

$$= \bar{m}_0 \left(\frac{3L}{4} - \frac{2L}{3\pi} - \frac{31L}{18\pi^2}\right)$$

$$= m_{21}$$

$$m_{22} = \int_0^L \bar{m}_0 \left(1 - \frac{x}{2L}\right) \left(1 - \cos\frac{3\pi x}{2L}\right)^2 dx \tag{b}$$

$$= \bar{m}_0 \left(\frac{9L}{8} + \frac{2L}{3\pi} - \frac{7L}{18\pi^2}\right)$$

The eigenvalue equation becomes

$$\frac{\pi^4 EI_0}{16L^3} \begin{bmatrix} \frac{3}{8} + \frac{1}{2\pi^2} & \frac{9}{2\pi^2} \\ \frac{9}{2\pi^2} & 81\left(\frac{3}{8} + \frac{1}{18\pi^2}\right) \end{bmatrix} \begin{bmatrix} z_1 \\ z_2 \end{bmatrix}$$

$$= \omega^2 \bar{m}_0 L \begin{bmatrix} \frac{9}{8} - \frac{2}{\pi} - \frac{7}{2\pi^2} & \frac{3}{4} - \frac{2}{3\pi} - \frac{31}{18\pi^2} \\ \frac{3}{4} - \frac{2}{3\pi} - \frac{31}{18\pi^2} & \frac{9}{8} + \frac{2}{3\pi} - \frac{7}{18\pi^2} \end{bmatrix} \begin{bmatrix} z_1 \\ z_2 \end{bmatrix}$$

or

$$\frac{\pi^4 EI_0}{16L^3} \begin{bmatrix} 0.4257 & 0.4559 \\ 0.4559 & 30.8310 \end{bmatrix} \begin{bmatrix} z_1 \\ z_2 \end{bmatrix} = \omega^2 \bar{m}_0 L \begin{bmatrix} 0.1337 & 0.3633 \\ 0.3633 & 1.2980 \end{bmatrix} \begin{bmatrix} z_1 \\ z_2 \end{bmatrix} \tag{c}$$

On setting $\bar{\lambda} = (16\bar{m}_0 L^4/\pi^4 E I_0)\omega^2$, Equation c leads to the following characteristic equation

$$\begin{vmatrix} 0.4257 - 0.1337\bar{\lambda} & 0.4559 - 0.3633\bar{\lambda} \\ 0.4559 - 0.3633\bar{\lambda} & 30.8310 - 1.2980\bar{\lambda} \end{vmatrix} = 0$$

or

$$\bar{\lambda}^2 - 104.63\bar{\lambda} + 311.176 = 0 \qquad (d)$$

Solution of Equation d gives the two frequencies

$$\bar{\lambda}_1 = 3.0638 \quad \omega_1 = 4.319\sqrt{\frac{E I_0}{\bar{m} L^4}}$$

$$\bar{\lambda}_2 = 101.57 \quad \omega_2 = 24.87\sqrt{\frac{E I_0}{\bar{m} L^4}} \qquad (e)$$

The discussion in the foregoing paragraphs and Examples 13.1 through 13.5 demonstrate that the reliability of frequency estimates obtained from the Rayleigh–Ritz method depends on a proper selection of the Ritz shapes. Unless there is some indication of the nature of the vibration shape of the system in the desired mode, the frequency estimates can be very inaccurate. In general, the Rayleigh–Ritz method is used to obtain the lowest few frequencies of the system. The selected Ritz shapes should therefore resemble the lowest modes. For complicated systems, it is extremely difficult to construct such shapes. Despite these shortcomings, the Rayleigh–Ritz method can be quite useful for estimating the lowest few frequencies whenever good Ritz shapes can be derived on the basis of experience and judgment. The method is also very useful for continuous systems, particularly for continuous systems having nonuniform properties where the exact eigenvalues are impossible to determine.

13.3   APPLICATION OF RITZ METHOD TO FORCED VIBRATION RESPONSE

The Rayleigh–Ritz method, in fact, represents a transformation of coordinates in which the transformation matrix consists of a series of linearly independent vectors called Ritz vectors or Ritz shapes. If the number of such vectors is equal to the number of degrees of freedom in the system being analyzed, the solution of the eigenproblem in the transformed space leads to the exact eigenvalues and eigenvectors of the original system. The value of the Rayleigh–Ritz method, however, lies in the fact that with appropriate choice of Ritz vectors, only a few of them may be used to represent the original system adequately. The transformation matrix is, in such a case, rectangular and the size of the transformed problem is considerably smaller than the original problem. Solution of the reduced problem then provides approximations to the true eigenvalues and eigenvectors of the original system. The position of the eigenvalues that will be approximated and the accuracy with which they are calculated depends on the choice of the Ritz vectors.

The fact that with appropriate selection of the Ritz shapes, only a few of them can be used to represent adequately the response of the original system can be utilized very effectively in the forced response analysis of multi degree-of-freedom systems. The use of Ritz method leads to a significant reduction in the size of the problem and therefore in the computations involved in the solution. However, because the quality of the results obtained is influenced strongly by the selection of Ritz shapes, methods for selecting such shapes are of considerable importance. In this section, we discuss several different approaches used in the selection of Ritz shapes, the details of the solution procedure in each case, and where applicable, the errors involved in the solution.

13.3.1 *Mode superposition method*

The mode superposition method of solution of the forced vibration response, discussed in Chapter 12, will readily be recognized as a Ritz method in which the undamped mode shapes of the system are used as the Ritz shapes. It was shown that the application of the method leads to N uncoupled single-degree-of-freedom equations given by

$$\ddot{y}_n + 2\xi_n \omega_n \dot{y}_n + \omega_n^2 y_n = \boldsymbol{\phi}_n^T \mathbf{p} \quad n = 1, 2, \ldots, N \tag{13.29}$$

in which it is assumed that mass orthonormal mode shapes have been used as the Ritz vectors and that the damping is proportional, so that the transformation diagonalizes the damping matrix.

Once the single-degree-of-freedom equations given by Equations 13.29 have been solved, the response in the physical coordinates is obtained from

$$\mathbf{u} = \sum_{n=1}^{M} \boldsymbol{\phi}_n y_n \tag{13.30}$$

in which we have superimposed the contributions from only the first M mode shapes. This implies that we need to evaluate only the first M eigenpairs and further that only the first M equations from the N equations represented by Equations 13.29 need be solved. Obviously, an important consideration is how many mode shapes should be included in the computations so as to obtain reasonable accuracy. To find an answer to this question, let us first assume that the forcing function is of the form $\mathbf{p} = \mathbf{f} \sin \Omega t$, where $\mathbf{f}$ is a time-independent amplitude vector and $\sin \Omega t$ represents the time variation. The steady-state solution to *n*th equation in Equations 13.29 is then given by (Eq. 6.27)

$$y_n = \frac{\boldsymbol{\phi}_n^T \mathbf{f}}{\omega_n^2} \frac{1}{\sqrt{(1 - \beta_n^2)^2 + (2\xi_n\beta_n)^2}} \sin(\Omega t - \theta) \tag{13.31}$$

in which $\beta_n = \Omega/\omega_n$ and $\tan\theta = 2\xi_n\beta_n/(1 - \beta_n^2)$.

It is obvious that the characteristics of the forcing function that affect the response are its amplitude vector $\mathbf{f}$ and the frequency Ω. The effect of amplitude is represented by the term

$$\gamma = \boldsymbol{\phi}_n^T \mathbf{f} \tag{13.32}$$

which is called the *participation factor*. The effect of the exciting frequency Ω is reflected in the term $A_D = 1/\sqrt{(1 - \beta_n^2)^2 + (2\xi_n\beta_n)^2}$. This term is plotted in Figure 6.4 as a function of β_n for several values of the damping fraction ξ_n. For small values of β_n, that is, when ω_n is much larger than Ω, say four times Ω or more, A_D is very nearly equal to 1. Also, as seen in Figure 6.5, the phase angle θ tends to zero as β approaches zero. Hence, the modal response y_n reduces to

$$\begin{aligned} y_n &= \frac{\boldsymbol{\phi}_n^T \mathbf{f}}{\omega_n^2} \sin \Omega t \\ &\qquad\qquad \Omega \ll \omega_n \\ &= \frac{\boldsymbol{\phi}_n^T \mathbf{p}}{k_n} \end{aligned} \tag{13.33}$$

Equation 13.33 is readily seen as the expression for static response to load $\boldsymbol{\phi}_n^T \mathbf{p}$. Since a general function $\mathbf{p}$ can be expressed as a superposition of its harmonic components, it is apparent that in the higher modes, for which ω_n is much larger than the highest frequency content in $\mathbf{p}$, the response of the system is essentially static. Furthermore, from Equation 13.33 it is evident that since ω_n increases with n, the displacement response decreases as the mode number increases, so that the contribution of the higher modes is comparatively small. However, since the modal contribution to spring force is equal to the product of the displacement response and the stiffness $k_n(= \omega_n^2)$, higher modes make a more significant contribution to the spring forces than they do to the displacements.

From the discussion in the foregoing paragraph, it is clear that while using the mode superposition method, all modes for which the participation factor $\boldsymbol{\phi}_n^T \mathbf{f}$ is significant should be included, particularly if the spring forces are to be evaluated. For modes with frequency ω_n equal to or greater than about $4\Omega_{max}$, where Ω_{max} is the highest frequency content in $\mathbf{p}$, the modal response can be taken to be equal to the static response.

The number of modes to be included in a particular case will depend on the spatial distribution and the frequency content of $\mathbf{p}$. If the forcing function is orthogonal to a particular mode ($\boldsymbol{\phi}_n^T\mathbf{f}=0$), that mode will not be excited at all. For earthquake forces, the participation factors for the higher modes are generally small and a major portion of the response is contained in only the first few modes. On the other hand, blast or shock loading may excite many more modes. To determine whether a sufficient number of modes have been included in the analysis, it is necessary to define some form of error measure. The participation factor $\boldsymbol{\phi}_n^T\mathbf{f}$ provides a measure of how significant is the contribution from a particular mode. In those cases where the forcing function is known to excite only the lower modes, the participation factor will decrease as the mode number increases. Therefore, the modal analysis can be truncated when the participation factor of the last mode included in the analysis is comparatively small, and at the same time the modal frequency is greater than, say, four times the highest-frequency content in the applied load.

In a more general case, calculation of participation factor by itself will not indicate whether the modal analysis can be truncated, because even if the participation factor of a certain mode is small, a higher mode may still have a larger participation factor. It is then useful to determine how well the forcing function is represented by the truncated series of mode shapes. This is achieved as follows.

The equations of motion to be solved are

$$\mathbf{M}\ddot{\mathbf{u}} + \mathbf{C}\dot{\mathbf{u}} + \mathbf{K}\mathbf{u} = \mathbf{f}g(t) \tag{13.34}$$

A modal coordinate transformation using M undamped mode shapes can be represented by

$$\mathbf{u}_M = \boldsymbol{\Phi}_M\mathbf{y}_M \tag{13.35}$$

in which $\boldsymbol{\Phi}_M$ is the matrix of first M mode shapes and $\mathbf{y}_M$ is a vector of M generalized coordinates. We use the subscript M on $\mathbf{u}$ to indicate that the displacement vector obtained by solving a transformed problem in which only M of the N modes are included will not be equal to the exact value denoted by $\mathbf{u}$. Equation 13.35 is, of course, identical to Equation 13.30.

Substitution of $\mathbf{u}_M$ for $\mathbf{u}$ in Equation 13.34 gives

$$\mathbf{M}\ddot{\mathbf{u}}_M + \mathbf{C}\dot{\mathbf{u}}_M + \mathbf{K}\mathbf{u}_M = \mathbf{f}_M g(t) \tag{13.36}$$

Vector $\mathbf{f}_M g(t)$ is the representation of a forcing function obtained by using a truncated series of modes. Since $\mathbf{u}_M$ is not exactly equal to $\mathbf{u}$, $\mathbf{f}_M$ is not equal to $\mathbf{f}$ and the difference between the two gives a measure of the error involved. Equation 13.36 gives

$$\mathbf{f}_M g(t) = \mathbf{M}\boldsymbol{\Phi}_M\ddot{\mathbf{y}}_M + \mathbf{C}\boldsymbol{\Phi}_M\dot{\mathbf{y}}_M + \mathbf{K}\boldsymbol{\Phi}_M\mathbf{y}_M \tag{13.37}$$

Also, with $\mathbf{M}$-orthonormal modes we have

$$\mathbf{K}\boldsymbol{\Phi}_M = \mathbf{M}\boldsymbol{\Phi}_M\boldsymbol{\Lambda}_M \tag{13.38}$$

and

$$C\Phi_M = M\Phi_M\Lambda_M \qquad (13.39)$$

where Λ_M is the diagonal matrix of the squared frequencies, ω_n^2, and Δ_M is a diagonal matrix of terms $2\xi_n\omega_n$. Substitution of Equations 13.38 and 13.39 in Equation 13.37 gives

$$f_M g(t) = M\Phi_M \ddot{y}_M + M\Phi_M \Delta_M \dot{y}_M + M\Phi_M \Lambda_M y_M \qquad (13.40)$$

The transformed set of equations given by Equations 13.29 can be expressed as

$$\ddot{y}_M + \Delta_M \dot{y}_M + \Lambda_M y_M = \Phi_M^T f g(t) \qquad (13.41)$$

On multiplying both sides of Equation 13.41 by $M\Phi_M$ and substituting the resulting equation in Equation 13.40, we get

$$f_M g(t) = M\Phi_M \Phi_M^T f g(t) \qquad (13.42)$$

Equation 13.42 is an expression for the representation of forcing function obtained by using a truncated series of mode shapes. When M approaches N, f_M will approach f. For $M < N$, an error is involved in the representation of the forcing function. If the error is denoted by $e_M g(t)$, we have

$$\begin{aligned} e_M &= f - f_M \\ &= f - M\Phi_M \Phi_M^T f \end{aligned} \qquad (13.43)$$

Using the definition of participation factor $\gamma_n = \phi_n^T f$, Equation 13.43 can be expressed as

$$e_M = f - \sum_{i=1}^{M} \gamma_i M\phi_i \qquad (13.44)$$

Using e_M, an error norm e can be defined as

$$e = \frac{f^T e_M}{f^T f} \qquad (13.45)$$

The error norm e will be zero when all the modes are included ($M=N$) and 1 when no modes are included ($M=0$). Modal analysis can thus be truncated when e becomes sufficiently small.

In the case of identical support motion, such as that due to an earthquake, the spatial distribution of the forcing function is given by

$$f = Mr \qquad (13.46)$$

where $\mathbf{r}$, the vector of pseudostatic displacements caused by simultaneous unit displacement of all supports, is obtained by kinematic consideration. The participation factors are therefore given by

$$\gamma_n = \boldsymbol{\phi}_n^T \mathbf{Mr} \tag{13.47}$$

It is customary in such a case to define the error norm as

$$
\begin{aligned}
e &= \frac{\mathbf{r}^T \mathbf{e}_M}{\mathbf{r}^T \mathbf{f}} \\
&= \frac{\mathbf{r}^T \mathbf{Mr} - \mathbf{r}^T \mathbf{M} \boldsymbol{\Phi}_M \boldsymbol{\Phi}_M^T \mathbf{Mr}}{\mathbf{r}^T \mathbf{Mr}} \\
&= \frac{\mathbf{r}^T \mathbf{Mr} - \sum_{i=1}^{M} \gamma_i^2}{\mathbf{r}^T \mathbf{Mr}}
\end{aligned}
\tag{13.48}
$$

For a diagonal mass matrix, the term $\mathbf{r}^T \mathbf{Mr}$ represents the total mass in the direction of support motion. In the special case when $\mathbf{r}$ is a unit vector

$$\mathbf{r}^T \mathbf{Mr} = \sum_{i=1}^{N} m_{ii} \tag{13.49}$$

in which m_{ii} is the ith diagonal term in the mass matrix. The term $\mathbf{r}^T \mathbf{Mr}$ is thus equal to the sum of all masses and $\sum_{i=1}^{M} \gamma_i^2$ represents the portion of system mass included in the truncated modal analysis. For $M=N$, we have $\boldsymbol{\Phi}_N \boldsymbol{\Phi}_N^T \mathbf{M} = \mathbf{I}$. Hence $\sum_{i=1}^{M} \gamma_i^2 = \mathbf{r}^T \mathbf{Mr}$ and the error norm becomes zero.

13.3.2 *Mode acceleration method*

From the discussion in the preceding section, it is evident that if the spatial distribution of the forcing function is such that the higher modes of the system are significantly excited, such modes must be included in the analysis. At the same time, if the higher-mode frequency is much larger than the highest-frequency content of the applied loading, the response in the higher mode is essentially static. Thus, if we include all modes in the analysis but consider that for mode numbers $M + 1$ to N the response is static, the total response is approximated by

$$\mathbf{u} = \sum_{j=1}^{M} \boldsymbol{\phi}_j y_j + \sum_{j=M+1}^{N} \boldsymbol{\phi}_j \frac{\boldsymbol{\phi}_j^T \mathbf{p}}{\omega_j^2} \tag{13.50}$$

in which y_j, $j=1$ to M, are obtained by solving the uncoupled Equations 13.29. The second term in Equation 13.50 has been obtained by substituting for y_j from Equation 13.33.

Equation 13.50 does not offer any particular advantage in the computations because the mode shapes, ϕ_j, and frequencies, ω_j, must still be evaluated for all values of j from 1 to N. However, if we substitute for y_j from Equations 13.29, Equation 13.50 becomes

$$\mathbf{u} = \sum_{j=1}^{M} \phi_j \left(\frac{\phi_j^T \mathbf{p}}{\omega_j^2} - \frac{\ddot{y}_j}{\omega_j^2} - \frac{2\xi_j \dot{y}_j}{\omega_j} \right) + \sum_{j=M+1}^{N} \phi_j \frac{\phi_j^T \mathbf{p}}{\omega_j^2}$$

$$= \sum_{j=1}^{N} \phi_j \frac{\phi_j^T \mathbf{p}}{\omega_j^2} - \sum_{j=1}^{M} \left(\phi_j \frac{\ddot{y}_j}{\omega_j^2} + \phi_j \frac{2\xi_j \dot{y}_j}{\omega_j} \right)$$

$$(13.51)$$

The first term on the right-hand side of Equation 13.51 can be reduced as follows

$$\sum_{j=1}^{N} \phi_j \frac{\phi_j^T \mathbf{p}}{\omega_j^2} = [\Phi][\Lambda]^{-1}[\Phi]^T \mathbf{p} \tag{13.52}$$

With mass orthonormal mode shapes, Λ is given by

$$\Lambda = \Phi^T \mathbf{K} \Phi \tag{13.53}$$

Substitution of Equation 13.53 in Equation 13.52 gives

$$\sum_{j=1}^{N} \phi_j \frac{\phi_j^T \mathbf{p}}{\omega_j^2} = [\Phi][\Phi]^{-1}[\mathbf{K}]^{-1}[\Phi^T]^{-1}[\Phi^T]\mathbf{p}$$

$$= \mathbf{K}^{-1}\mathbf{p} \tag{13.54}$$

which represents the static deflection under load $\mathbf{p}$. Equation 13.51 now becomes

$$\mathbf{u} = \mathbf{K}^{-1}\mathbf{p} - \sum_{j=1}^{M} \left(\phi_j \frac{\ddot{y}_j}{\omega_j^2} + \phi_j \frac{2\xi_j \dot{y}_j}{\omega_j} \right) \tag{13.55}$$

In using Equation 13.55 we need to obtain the modal response in only the first M modes and, therefore, only the first M frequencies and mode shapes need to be computed. Since Equation 13.55 involves the superposition of modal accelerations rather than displacements, the method is often referred to as the *mode acceleration method*. As an alternative, Equation 13.50 can be expressed as

$$\mathbf{u} = \sum_{j=1}^{N} \phi_j \frac{\phi_j^T \mathbf{p}}{\omega_j^2} + \sum_{j=1}^{M} \left(\phi_j y_j - \phi_j \frac{\phi_j^T \mathbf{p}}{\omega_j^2} \right)$$

$$= \mathbf{K}^{-1}\mathbf{p} + \sum_{j=1}^{M} \left(\phi_j y_j - \phi_j \frac{\phi_j^T \mathbf{p}}{\omega_j^2} \right) \tag{13.56}$$

Both Equations 13.55 and 13.56 include the static response solution for the higher modes. The method being discussed is therefore also referred to as the *static correction method*. It should be noted that when the forcing function is of the form $\mathbf{p} = \mathbf{f}g(t)$, term $\mathbf{K}^{-1}\mathbf{p}$ can be evaluated by calculating $\mathbf{K}^{-1}\mathbf{f}$ only once and then for each value of t for which response is required, applying the scale factor $g(t)$ to the calculated value of $\mathbf{K}^{-1}\mathbf{f}$. The vector of spring forces is obtained on multiplying Equation 13.56 by $\mathbf{K}$.

$$\mathbf{f}_S = \mathbf{K}\mathbf{u}$$

$$= \mathbf{K}\left(\mathbf{K}^{-1}\mathbf{p} + \sum_{j=1}^{M}\boldsymbol{\phi}_j y_j - \sum_{j=1}^{M}\boldsymbol{\phi}_j\frac{\boldsymbol{\phi}_j^T\mathbf{p}}{\omega_j^2}\right) \tag{13.57}$$

On noting that $\mathbf{K}\boldsymbol{\phi}_j = \omega_j^2\mathbf{M}\boldsymbol{\phi}_j$, Equation 13.57 reduces to

$$\mathbf{f}_S = \mathbf{p} + \sum_{j=1}^{M}\omega_j^2\mathbf{M}\boldsymbol{\phi}_j y_j - \sum_{j=1}^{M}\mathbf{M}\boldsymbol{\phi}_j\boldsymbol{\phi}_j^T\mathbf{p} \tag{13.58}$$

Example 13.6

The frame of Example 13.1 is subject to identical horizontal motions at each support given by $\ddot{u}_g = 1000\sin\Omega t$, where $\Omega = 22$ rad/s. Using the mode superposition method, obtain the steady-state displacement and force response. Assume that the vibrations are undamped. The frequencies and M-orthonormal mode shapes of the frame are as follows

$$[\omega] = \begin{bmatrix} 7.77 \\ 18.98 \\ 27.66 \\ 36.39 \\ 45.71 \end{bmatrix} \text{rad/s}$$

$$\Phi = \begin{bmatrix} 0.1029 & 0.2098 & 0.2688 & 0.4454 & 0.4176 \\ 0.2296 & 0.3635 & 0.2840 & 0.0552 & -0.4810 \\ 0.3333 & 0.2988 & -0.0631 & -0.4566 & 0.2955 \\ 0.4385 & -0.0675 & -0.4628 & 0.2871 & -0.0822 \\ 0.5161 & -0.6783 & 0.5077 & -0.1246 & 0.0195 \end{bmatrix}$$

Solution

If we denote the vector of story displacements relative to the base by $\mathbf{u}$, the equations of motion can be expressed as

$$\mathbf{M}\ddot{\mathbf{u}} + \mathbf{K}\mathbf{u} = -\mathbf{M}\mathbf{r}1000\sin\Omega t$$

where $\mathbf{r}$, the vector of pseudostatic displacement due to support motion, is obtained from kinematic considerations as

$$\mathbf{r}^T = [1 \quad 1 \quad 1 \quad 1 \quad 1]$$

The forcing function $\mathbf{p}(t)$ is given by

$$\mathbf{p}(t) = \mathbf{f}g(t)$$

in which

$$\mathbf{f} = \mathbf{Mr} = \begin{bmatrix} 2 \\ 2 \\ 2 \\ 2 \\ 1 \end{bmatrix}$$

$$g(t) = -1000 \sin \Omega t$$

The participation factors are now obtained from Equation 13.32

$$\gamma_j = \boldsymbol{\phi}_j^T \mathbf{f} \quad j = 1, 2, \ldots, 5$$

$$\gamma_1 = 2.7250$$

$$\gamma_2 = 0.9308$$

$$\gamma_3 = 0.5614$$

$$\gamma_4 = 0.5375$$

$$\gamma_5 = 0.3194$$

The steady-state displacement response in each mode is given by Equation 13.31 with $\xi = 0$.
First mode:

$$\beta = \frac{\Omega}{\omega} = \frac{22}{7.77} = 2.831$$

$$y_1 = \frac{-2.725}{(7.77)^2} \frac{1000}{(1 - 2.831^2)} \sin \Omega t$$

$$= 6.435 \sin \Omega t$$

Second mode:

$$\beta = \frac{22}{18.98} = 1.159$$

$$y_2 = -\frac{0.9308}{(18.98)^2} \frac{1000}{1 - (1.159)^2} \sin \Omega t$$

$$= 7.527 \sin \Omega t$$

Third mode:

$$\beta = \frac{22}{27.66} = 0.7953$$

$$y_3 = -\frac{0.5614}{(27.66)^2} \frac{1000}{1 - (0.7953)^2} \sin \Omega t$$

$$= -1.997 \sin \Omega t$$

Fourth mode:

$$\beta = \frac{22}{36.39} = 0.6046$$

$$y_4 = -\frac{0.5375}{(36.39)^2} \frac{1000}{1 - (0.6046)^2} \sin \Omega t$$

$$= -0.6397 \sin \Omega t$$

Fifth mode:

$$\beta = \frac{22}{45.71} = 0.4813$$

$$y_5 = -\frac{0.3194}{(45.71)^2} \frac{1000}{1 - (0.4813)^2} \sin \Omega t$$

$$= -0.1990 \sin \Omega t$$

The displacement response is now given by

$$\mathbf{u} = \sum_{i=1}^{5} \boldsymbol{\phi}_i y_i$$

$$= \left\{ \begin{bmatrix} 0.6622 \\ 1.4777 \\ 2.1448 \\ 2.8217 \\ 3.3211 \end{bmatrix} + \begin{bmatrix} 1.5792 \\ 2.7361 \\ 2.2491 \\ -0.5081 \\ -5.1056 \end{bmatrix} + \begin{bmatrix} -0.5368 \\ -0.5671 \\ 0.1260 \\ 0.9242 \\ -1.0139 \end{bmatrix} + \begin{bmatrix} -0.2849 \\ -0.0353 \\ 0.2921 \\ -0.1837 \\ 0.0797 \end{bmatrix} \right.$$

$$\left. + \begin{bmatrix} -0.0831 \\ 0.0957 \\ -0.0588 \\ 0.0164 \\ -0.0039 \end{bmatrix} \right\} \sin \Omega t = \begin{bmatrix} 1.3366 \\ 3.7071 \\ 4.7532 \\ 3.0705 \\ -2.7226 \end{bmatrix} \sin \Omega t$$

The force response is given by

$$\mathbf{f}_s = \mathbf{Ku} = \sum_{i=1}^{5} \omega_i^2 \mathbf{M} \boldsymbol{\phi}_i y_i$$

$$= \left\{ \begin{bmatrix} 79.96 \\ 178.40 \\ 258.96 \\ 340.70 \\ 200.58 \end{bmatrix} + \begin{bmatrix} 1137.76 \\ 1971.28 \\ 1620.42 \\ -366.06 \\ 1838.96 \end{bmatrix} + \begin{bmatrix} -821.36 \\ -867.82 \\ 192.82 \\ 1414.20 \\ -775.80 \end{bmatrix} + \begin{bmatrix} -754.60 \\ -93.52 \\ 773.58 \\ -486.42 \\ 105.54 \end{bmatrix} \right.$$

$$\left. + \begin{bmatrix} -347.26 \\ 400.00 \\ -245.86 \\ 68.36 \\ -8.11 \end{bmatrix} \right\} \sin \Omega t = \begin{bmatrix} -705.5 \\ 1588.32 \\ 2600.06 \\ 970.76 \\ -2317.75 \end{bmatrix} \sin \Omega t$$

Table E13.6. Error measure in mode truncation.

Number of modes M	$\sum_{i=1}^{M} \gamma_i^2$	$e = \dfrac{(\sum_1^N m_{ii} - \sum_1^M \gamma_i^2)}{\sum_i^N m_{ii}}$
1	7.425	0.175
2	8.291	0.079
3	8.607	0.044
4	8.896	0.012
5	9.000	0.000

The spatial distribution of the forcing function given by $\mathbf{f} = \mathbf{Mr}$ is such that the participation factor γ_i decreases with increasing mode number, γ_2 being 34.2% of γ_1 and γ_3 being 20.6% of γ_1. However, the exciting frequency lies between the second and third mode frequency. As a result, response amplification is quite high in these modes, counterbalancing the decrease due to lower participation factor. It may, however, be possible to obtain reasonable accuracy in the calculation of response with less than the full five modes. It will be of interest to calculate the error norm given by Equation 13.48 to obtain some indication of the error caused due to mode truncation. The calculations are shown in Table E13.6.

On the basis of error norm given in Table E13.6, we may choose to include only three modes in our calculation. The corresponding displacement and force responses will then be as follows

$$
\mathbf{u} = \begin{bmatrix} 1.7046 \\ 3.6467 \\ 4.5199 \\ 3.2378 \\ -2.7984 \end{bmatrix} \sin \Omega t
$$

$$
\mathbf{f}_S = \begin{bmatrix} 396.36 \\ 1281.86 \\ 2072.20 \\ 1388.84 \\ -2414.18 \end{bmatrix} \sin \Omega t
$$

The errors in the displacement response are acceptable, but those in the spring force vector are quite high. One reason is that the higher modes are needed for an adequate representation of the frequency content of the forcing function. We may be able to improve the accuracy of the spring force vector by adding static correction for the modes not included. The correction vector $\mathbf{f}_{Sc}$ can be obtained from Equation 13.58

$$
\mathbf{f}_{Sc} = \left(\mathbf{Mr} - \sum_{j=1}^{M} \mathbf{M}\boldsymbol{\phi}_j\boldsymbol{\phi}_j^T \mathbf{Mr} \right) (-1000 \sin \Omega t)
$$

$$
= \left(\mathbf{Mr} - \sum_{j=1}^{M} \gamma_j \mathbf{M}\boldsymbol{\phi}_j \right) (-1000 \sin \Omega t)
$$

Substitution for $\mathbf{M}$, $\mathbf{r}$, γ, and ϕ gives

$$\mathbf{f}_{Sc} = \begin{bmatrix} -746.8 \\ 246.9 \\ 301.9 \\ -255.5 \\ 60.0 \end{bmatrix} \sin \Omega t$$

The corrected response obtained by adding $\mathbf{f}_S$ calculated earlier and $\mathbf{f}_{Sc}$ calculated above is

$$\mathbf{f}_{Sc} = \begin{bmatrix} -350.44 \\ 1528.76 \\ 2374.10 \\ 1133.19 \\ -2354.19 \end{bmatrix} \sin \Omega t$$

The corrected response is seen to be closer to the exact value.

13.3.3 *Static condensation and Guyan's reduction*

It is of interest to note that the static condensation described in Section 3.7 can be viewed as a special case of Ritz procedure, in which the original problem of a size equal to the sum of the translational degrees of freedom in $\mathbf{u}_t$ and rotational degrees of freedom in $\mathbf{u}_\theta$ is reduced to that of the size of $\mathbf{u}_t$. The transformation is expressed as

$$\begin{bmatrix} \mathbf{u}_t \\ \mathbf{u}_\theta \end{bmatrix} = \mathbf{T}\mathbf{u}_t \tag{13.59}$$

in which the transformation matrix $\mathbf{T}$ is obtained by using the relationship given by Equation 3.106.

$$\mathbf{T} = \begin{bmatrix} \mathbf{I} \\ -\mathbf{K}_{\theta\theta}^{-1}\mathbf{K}_{\theta t} \end{bmatrix} \tag{13.60}$$

When the transformation is applied to Equation 3.103, we get

$$\tilde{\mathbf{M}}\ddot{\mathbf{u}}_t + \tilde{\mathbf{K}}\mathbf{u}_t = \tilde{\mathbf{p}}_t \tag{13.61}$$

where

$$\tilde{\mathbf{M}} = \mathbf{M}_{tt} \tag{13.62a}$$

$$\tilde{\mathbf{K}} = \mathbf{K}_{tt} - \mathbf{K}_{t\theta}\mathbf{K}_{\theta\theta}^{-1}\mathbf{K}_{\theta t} \tag{13.62b}$$

$$\tilde{\mathbf{p}}_t = \mathbf{p}_t \tag{13.62c}$$

Since the degrees of freedom eliminated from the formulation by the fore-going procedure do not have any inertia or external forces applied to them, the static condensation process does not involve any approximation. However, as an extension of this procedure, the transformation matrix given by Equation 13.60 can be used even when there are inertia or external forces acting along the eliminated θ degrees of freedom. The transformed and reduced mass matrix is in such a case given by

$$
\tilde{\mathbf{M}} = [\mathbf{I} - \mathbf{K}_{t\theta}(\mathbf{K}_{\theta\theta}^{-1})^T] \begin{bmatrix} \mathbf{M}_{tt} & \mathbf{M}_{t\theta} \\ \mathbf{M}_{\theta t} & \mathbf{M}_{\theta\theta} \end{bmatrix} \begin{bmatrix} \mathbf{I} \\ -\mathbf{K}_{\theta\theta}^{-1}\mathbf{K}_{\theta t} \end{bmatrix}
$$

$$
= \mathbf{M}_{tt} - \mathbf{M}_{t\theta}\mathbf{K}_{\theta\theta}^{-1}\mathbf{K}_{\theta t} + \mathbf{K}_{t\theta}(\mathbf{K}_{\theta\theta}^{-1})^T\mathbf{M}_{\theta\theta}\mathbf{K}_{\theta\theta}^{-1}\mathbf{K}_{\theta t} - \mathbf{K}_{t\theta}(\mathbf{K}_{\theta\theta}^{-1})^T\mathbf{M}_{\theta t}
$$

$$(13.63)$$

For symmetrical mass and stiffness matrices, this reduces to

$$
\tilde{\mathbf{M}} = \mathbf{M}_{tt} - \mathbf{M}_{t\theta}\mathbf{K}_{\theta\theta}^{-1}\mathbf{K}_{\theta t} + \mathbf{K}_{t\theta}\mathbf{K}_{\theta\theta}^{-1}\mathbf{M}_{\theta\theta}\mathbf{K}_{\theta\theta}^{-1}\mathbf{K}_{\theta t} - (\mathbf{M}_{t\theta}\mathbf{K}_{\theta\theta}^{-1}\mathbf{K}_{\theta t})^T \quad (13.64)
$$

The reduction of mass matrix as in Equation 13.64 is called *Guyan's reduction*. The reduction process now introduces approximations in the formulation, and the reliability of the results obtained depends on a judicious selection of the degrees of freedom that are to be eliminated.

Example 13.7
The uniform cantilever beam shown in Figure E13.7a is modeled by two prismatic beam elements as indicated. Using the beam element stiffness matrix given in Section 3.6.1 (Eq. 3.93), obtain the stiffness matrix of the cantilever beam corresponding to the four degrees of freedom shown in the figure. Then calculate the eigenvalues for the lateral vibrations of the cantilever by each of the following alternative procedures.

(i) Using the element mass matrix given in Equation 3.99 and a static condensation of the stiffness matrix, obtain a second-order eigenvalue problem and solve its characteristic equation.

(ii) As in part (a), except that for the element mass matrix, instead of using Equation 3.99, use the lumped mass matrix of Equation 3.101.

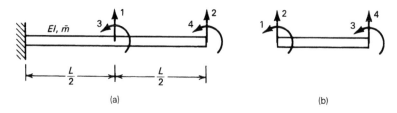

(a)   (b)

Figure E13.7. (a) Uniform cantilever beam and global coordinates; (b) beam element and local coordinates.

(iii) Obtain the mass matrix of the cantilever by using the consistent mass formulation (Eq. 3.96) for each element. Then, using Guyan's reduction, obtain a second order eigenvalue problem and solve its characteristic equation.

Solution

Figure E13.7b shows an element of the cantilever beam. Also shown are the local degrees of freedom. The stiffness matrix for each of the two elements corresponding to the local degrees of freedom shown are obtained from Equation 3.93 and are given by

$$
\mathbf{K}^{(1)} = EI
\begin{bmatrix}
\frac{8}{L} & \frac{24}{L^2} & \frac{4}{L} & -\frac{24}{L^2} \\
\frac{24}{L^2} & \frac{96}{L^3} & \frac{24}{L^2} & -\frac{96}{L^3} \\
\frac{4}{L} & \frac{24}{L^2} & \frac{8}{L} & -\frac{24}{L^2} \\
-\frac{24}{L^2} & -\frac{96}{L^3} & -\frac{24}{L^2} & \frac{96}{L^3}
\end{bmatrix}
\begin{matrix} \\ \\ 3 \\ 1 \end{matrix}
\qquad\qquad (a)
$$

$$
\qquad\qquad\qquad\qquad\quad\; 3 \qquad\quad 1
$$

$$
\mathbf{K}^{(2)} = EI
\begin{bmatrix}
\frac{8}{L} & \frac{24}{L^2} & \frac{4}{L} & -\frac{24}{L^2} \\
\frac{24}{L^2} & \frac{96}{L^3} & \frac{24}{L^2} & -\frac{96}{L^3} \\
\frac{4}{L} & \frac{24}{L^2} & \frac{8}{L} & -\frac{24}{L^2} \\
-\frac{24}{L^2} & -\frac{96}{L^3} & -\frac{24}{L^2} & \frac{96}{L^3}
\end{bmatrix}
\begin{matrix} 3 \\ 1 \\ 4 \\ 2 \end{matrix}
\qquad\qquad (b)
$$

$$
\qquad\qquad\qquad\qquad\; 3 \qquad 1 \qquad 4 \qquad 2
$$

Against the stiffness matrix of each element we have indicated the global degrees of freedom that match, in position and direction, the corresponding local degrees of freedom. For example, the local degree-of-freedom 3 of element 1 corresponds to the global degree-of-freedom 3. We have therefore placed the number 3 against each of row 3 and column 3 of $\mathbf{K}^{(1)}$. In a similar manner, local degree-of-freedom 4 of element 1 corresponds to global degree-of-freedom 1; number 1 therefore appears against row 4 and column 4 of $\mathbf{K}^{(1)}$. The 4×4 global stiffness matrix is obtained by assembling the local stiffness matrices given in Equations a and b. To carry out the assembly, each term of the element matrix is considered in turn and placed in the global matrix at the position indicated by the correspondence numbers appearing against the row and column occupied by the term in the element matrix.

Thus, the term $-24EI/L^2$ appearing in row 4, column 3 of $\mathbf{K}^{(1)}$ has correspondence numbers 1 and 3 and will be placed in row 1, column 3 of the global matrix. If a particular term in the element matrix has any of the row and the column correspondence numbers missing, that term is ignored in the assembly. The assembled global stiffness matrix is given by

$$
\qquad\qquad\qquad\qquad\quad t \qquad\qquad\quad \theta
$$

$$
\mathbf{K} = EI \quad
\begin{matrix} t \\ \\ \\ \theta \end{matrix}
\left[
\begin{array}{cc|cc}
\frac{192}{L^3} & -\frac{96}{L^3} & 0 & \frac{24}{L^2} \\
-\frac{96}{L^3} & \frac{96}{L^3} & -\frac{24}{L^2} & -\frac{24}{L^2} \\ \hline
0 & -\frac{24}{L^2} & \frac{16}{L} & \frac{4}{L} \\
\frac{24}{L^2} & -\frac{24}{L^2} & \frac{4}{L} & \frac{8}{L}
\end{array}
\right]
\qquad\qquad (c)
$$

(i) The element mass matrices obtained from Equation 3.96 are shown below along with the correspondence numbers relating the local degrees of freedom to the global degrees of freedom.

$$\mathbf{M}^{(1)} = \bar{m}L \begin{bmatrix} 0 & 0 & 0 & 0 \\ 0 & \frac{1}{6} & 0 & \frac{1}{12} \\ 0 & 0 & 0 & 0 \\ 0 & \frac{1}{12} & 0 & \frac{1}{6} \end{bmatrix} \begin{matrix} \\ 3 \\ 1 \end{matrix} \tag{d}$$

$$\begin{matrix} & & 3 & 1 \end{matrix}$$

$$\mathbf{M}^{(2)} = \bar{m}L \begin{bmatrix} 0 & 0 & 0 & 0 \\ 0 & \frac{1}{6} & 0 & \frac{1}{12} \\ 0 & 0 & 0 & 0 \\ 0 & \frac{1}{12} & 0 & \frac{1}{6} \end{bmatrix} \begin{matrix} 3 \\ 1 \\ 4 \\ 2 \end{matrix} \tag{e}$$

$$\begin{matrix} 3 & 1 & 4 & 2 \end{matrix}$$

The global mass matrix obtained by assembling the element mass matrices is

$$\mathbf{M} = \bar{m}L \begin{matrix} t \\ \theta \end{matrix} \begin{bmatrix} \frac{1}{3} & \frac{1}{12} & 0 & 0 \\ \frac{1}{12} & \frac{1}{6} & 0 & 0 \\ \hline 0 & 0 & 0 & 0 \\ 0 & 0 & 0 & 0 \end{bmatrix} \tag{f}$$

The global displacement vector **u** is partitioned as

$$\mathbf{u} = \begin{bmatrix} \mathbf{u}_t \\ \hline \mathbf{u}_\theta \end{bmatrix}$$

where

$$\mathbf{u}_t = \begin{bmatrix} u_1 \\ u_2 \end{bmatrix}$$

represents the translational degree of freedom and

$$\mathbf{u}_\theta = \begin{bmatrix} u_3 \\ u_4 \end{bmatrix}$$

represents the rotational degrees of freedom. The rotational degrees of freedom are now eliminated from the equation of motion. The condensed stiffness matrix is given by Equation 13.62b

$$\tilde{\mathbf{K}} = \mathbf{K}_{tt} - \mathbf{K}_{t\theta} \mathbf{K}_{\theta\theta}^{-1} \mathbf{K}_{\theta t} \tag{g}$$

where

$$\mathbf{K}_{tt} = EI \begin{bmatrix} \frac{192}{L^3} & -\frac{96}{L^3} \\ -\frac{96}{L^3} & \frac{96}{L^3} \end{bmatrix}$$

$$\mathbf{K}_{t\theta} = \mathbf{K}_{\theta t}^T = EI \begin{bmatrix} 0 & \frac{24}{L^2} \\ -\frac{24}{L^2} & -\frac{24}{L^2} \end{bmatrix}$$

$$\mathbf{K}_{\theta\theta} = EI \begin{bmatrix} \frac{16}{L} & \frac{4}{L} \\ \frac{4}{L} & \frac{8}{L} \end{bmatrix}$$

Inversion of $\mathbf{K}_{\theta\theta}$ gives

$$\mathbf{K}_{\theta\theta}^{-1} = \frac{L}{28EI} \begin{bmatrix} 2 & -1 \\ -1 & 4 \end{bmatrix}$$

Substitution into Equation g now leads to

$$\tilde{\mathbf{K}} = \frac{EI}{L^3} \begin{bmatrix} \frac{768}{7} & -\frac{240}{7} \\ -\frac{240}{7} & \frac{96}{7} \end{bmatrix} \tag{h}$$

The eigenvalue equation becomes

$$\frac{EI}{L^3} \begin{bmatrix} \frac{768}{7} & -\frac{240}{7} \\ -\frac{240}{7} & \frac{96}{7} \end{bmatrix} \begin{bmatrix} z_1 \\ z_2 \end{bmatrix} - \omega^2 \bar{m} L \begin{bmatrix} \frac{1}{3} & \frac{1}{12} \\ \frac{1}{12} & \frac{1}{6} \end{bmatrix} \begin{bmatrix} z_1 \\ z_2 \end{bmatrix} = \begin{bmatrix} 0 \\ 0 \end{bmatrix} \tag{i}$$

Solution of Equation i by the characteristic equation method gives the following frequencies

$$\omega_1 = 3.428 \sqrt{\frac{EI}{\bar{m}L^4}}$$

$$\omega_2 = 24.0 \sqrt{\frac{EI}{\bar{m}L^4}}$$

In comparison, the exact values are

$$\omega_1 = 3.516 \sqrt{\frac{EI}{\bar{m}L^4}}$$

$$\omega_2 = 22.034 \sqrt{\frac{EI}{\bar{m}L^4}}$$

(ii) The global mass matrix is, in this case, given by

$$\mathbf{M} = \bar{m}L \begin{bmatrix} \frac{1}{2} & 0 & 0 & 0 \\ 0 & \frac{1}{4} & 0 & 0 \\ 0 & 0 & 0 & 0 \\ 0 & 0 & 0 & 0 \end{bmatrix}$$

After static condensation, the eigenvalue problem becomes

$$\frac{EI}{L^3} \begin{bmatrix} \frac{768}{7} & -\frac{240}{7} \\ -\frac{240}{7} & \frac{96}{7} \end{bmatrix} \begin{bmatrix} z_1 \\ z_2 \end{bmatrix} - \omega^2 \bar{m} L \begin{bmatrix} \frac{1}{2} & 0 \\ 0 & \frac{1}{4} \end{bmatrix} \begin{bmatrix} z_1 \\ z_2 \end{bmatrix} = \begin{bmatrix} 0 \\ 0 \end{bmatrix} \tag{j}$$

Solution of Equation j gives the following frequencies

$$\omega_1 = 3.156\sqrt{\frac{EI}{\bar{m}L^4}}$$

$$\omega_2 = 16.258\sqrt{\frac{EI}{\bar{m}L^4}}$$

These values are less accurate than those determined in part (i).
(iii) In this case, we use consistent mass formulation for each of the two elements. The element mass matrices are shown below along with the correspondence numbers

$$\mathbf{M}^{(1)} = \frac{\bar{m}}{420}\begin{bmatrix} \frac{L^3}{2} & \frac{11L^2}{2} & -\frac{3L^3}{8} & \frac{13L^2}{4} \\ \frac{11L^2}{2} & 78L & -\frac{13L^2}{4} & 27L \\ -\frac{3L^3}{8} & -\frac{13L^2}{4} & \frac{L^3}{2} & -\frac{11L^2}{2} \\ \frac{13L^2}{4} & 27L & -\frac{11L^2}{2} & 78L \end{bmatrix} \begin{matrix} \\ \\ 3 \\ 1 \end{matrix}$$
$$\phantom{\mathbf{M}^{(1)} = \frac{\bar{m}}{420}} \quad\quad\quad\quad 3 \quad\quad 1$$

$$\mathbf{M}^{(2)} = \frac{\bar{m}}{420}\begin{bmatrix} \frac{L^3}{2} & \frac{11L^2}{2} & -\frac{3L^3}{8} & \frac{13L^2}{4} \\ \frac{11L^2}{2} & 78L & -\frac{13L^2}{4} & 27L \\ -\frac{3L^3}{8} & -\frac{13L^2}{4} & \frac{L^3}{2} & -\frac{11L^2}{2} \\ \frac{13L^2}{4} & 27L & -\frac{11L^2}{2} & 78L \end{bmatrix} \begin{matrix} 3 \\ 1 \\ 4 \\ 2 \end{matrix}$$
$$\phantom{\mathbf{M}^{(2)} = \frac{\bar{m}}{420}} \quad 3 \quad\quad 1 \quad\quad 4 \quad\quad 2$$

The global mass matrix is obtained by assembling the element matrices

$$\mathbf{M} = \frac{\bar{m}}{420} \begin{array}{c} t \\ \theta \end{array} \begin{bmatrix} 156L & 27L & 0 & -\frac{13L^2}{4} \\ 27L & 78L & \frac{13L^2}{4} & -\frac{11L^2}{2} \\ \hline 0 & \frac{13L^2}{4} & L^3 & -\frac{3L^3}{8} \\ -\frac{13L^2}{4} & -\frac{11L^2}{2} & \frac{3L^3}{8} & \frac{L^3}{2} \end{bmatrix}$$

with column headings t and θ.

The global mass matrix is partitioned along the translational and rotational degrees of freedom. Guyan reduction given by Equation 13.64 is then applied to obtain a 2×2 reduced mass matrix $\widetilde{\mathbf{M}}$

$$\widetilde{\mathbf{M}} = \frac{\bar{m}L}{420}\begin{bmatrix} 187.102 & 36.887 \\ 36.887 & 57.674 \end{bmatrix} \tag{k}$$

The eigenvalue problem is now given by

$$\frac{EI}{L^3}\begin{bmatrix} \frac{768}{7} & -\frac{240}{7} \\ -\frac{240}{7} & \frac{96}{7} \end{bmatrix}\begin{bmatrix} z_1 \\ z_2 \end{bmatrix} - \frac{\omega^2 \bar{m}L}{420}\begin{bmatrix} 187.102 & 36.887 \\ 36.887 & 57.674 \end{bmatrix}\begin{bmatrix} z_1 \\ z_2 \end{bmatrix} = \begin{bmatrix} 0 \\ 0 \end{bmatrix} \tag{l}$$

Solution of Equation 1 by the characteristic equation method gives the following frequencies

$$\omega_1 = 3.522\sqrt{\frac{EI}{\bar{m}L^4}}$$

$$\omega_2 = 22.277\sqrt{\frac{EI}{\bar{m}L^4}}$$

On comparison with the exact values, it is seen that in this case, a consistent mass formulation, even with the approximation involved in a Guyan's reduction, gives a much improved estimate of the vibration frequencies.

13.3.4 *Load-dependent Ritz vectors*

The use of undamped mode shapes of the system as Ritz vectors offers several advantages. First, the mode shape vectors are independent. Second, they automatically satisfy the geometric boundary conditions of the system. But the most useful property of the mode shapes is their orthogonality to both the mass and stiffness matrices. As a result, under a normal coordinate transformation the equations of motion become uncoupled and their solution is greatly simplified.

Notwithstanding these advantages, there are several difficulties associated with the use of eigenvectors as Ritz shapes. First, the computation of eigenvalues and eigenvectors of a large system is expensive and time consuming. Second, because the mode shapes of the system are not related to the applied load vector, a large number of shapes may be required to represent the load adequately. Several of the mode shapes may, in fact, be nearly orthogonal to the load vector and may contribute very little to the response. The effort expended in the evaluation of such mode shapes will be of no benefit, yet it is not possible, a priori, to detect whether a particular mode shape will be orthogonal to the load vector. Obviously, a desirable set of Ritz vectors will be one that can adequately represent the spatial distribution of load by the superposition of a minimum number of shapes. This criterion has motivated the development of methods for the generation of load-dependent vectors.

A method proposed by E.L. Wilson et al. begins with the assumption that the load can be represented as the product of a spatial vector and a time function, so that

$$\mathbf{p}(t) = \mathbf{f}g(t) \tag{13.65}$$

The first vector, $\bar{\mathbf{x}}_1$, in the series of Ritz vectors is then obtained by solving the equation

$$\mathbf{K}\bar{\mathbf{x}}_1 = \mathbf{f} \tag{13.66}$$

Vector $\bar{\mathbf{x}}_1$, which essentially represents the static response of the structure to the applied load, is normalized so as to be orthonormal to the mass matrix. If

$\mathbf{x}_1$ represents the normalized vector, we have

$$\mathbf{x}_1 = \frac{\bar{\mathbf{x}}_1}{(\bar{\mathbf{x}}_1{}^T \mathbf{M} \bar{\mathbf{x}}_1)^{1/2}} \tag{13.67}$$

Subsequent vectors in the series are members of the following sequence, referred to as the *Krylov sequence*

$$\mathbf{x}_1, \mathbf{K}^{-1}\mathbf{M}\mathbf{x}_1, (\mathbf{K}^{-1}\mathbf{M})^2\mathbf{x}_1, \dots, (\mathbf{K}^{-1}\mathbf{M})^j\mathbf{x}_1 \tag{13.68}$$

These vectors are obtained from the recurrence relationship

$$\mathbf{K}\bar{\mathbf{x}}_i = \mathbf{M}\mathbf{x}_{i-1} \quad i = 2, \dots, N \tag{13.69}$$

Each vector $\bar{\mathbf{x}}_i$ is then orthogonalized with respect to the vector determined previously. To achieve this, $\bar{\mathbf{x}}_i$ is expressed as

$$\bar{\mathbf{x}}_i = \tilde{\mathbf{x}}_i + \sum_{j=1}^{i-1} c_j \mathbf{x}_j \tag{13.70}$$

where $\tilde{\mathbf{x}}_i$ is the pure vector, orthogonal to all previous vectors, and terms $c_j \mathbf{x}_j$ represent components of such previous vectors present in $\bar{\mathbf{x}}_i$. To obtain c_k, both sides of Equation 13.70 are multiplied by $\mathbf{x}_k^T \mathbf{M}$. By definition, $\mathbf{x}_k^T \mathbf{M} \tilde{\mathbf{x}}_i$ must vanish. Also, all terms $\mathbf{x}_k^T \mathbf{M} \mathbf{x}_j$ vanish except $\mathbf{x}_k^T \mathbf{M} \mathbf{x}_k$, which is equal to 1. This is true because the previous vectors are supposed to have already been mass orthonormalized. Multiplication of Equation 13.70 by $\mathbf{x}_k^T \mathbf{M}$ thus gives

$$c_k = \mathbf{x}_k^T \mathbf{M} \bar{\mathbf{x}}_i \tag{13.71}$$

The orthogonalized vector $\tilde{\mathbf{x}}_i$ is now given by

$$\tilde{\mathbf{x}}_i = \bar{\mathbf{x}}_i - \sum_{j=1}^{i-1} c_j \mathbf{x}_j \tag{13.72}$$

where c_j for $j = k$ is obtained from Equation 13.71. Finally, $\tilde{\mathbf{x}}_i$ is normalized so that it is mass orthonormal

$$\mathbf{x}_i = \frac{\tilde{\mathbf{x}}_i}{(\tilde{\mathbf{x}}_i^T \mathbf{M} \tilde{\mathbf{x}}_i)^{1/2}} \tag{13.73}$$

Suppose that M Ritz vectors $\mathbf{x}_i$, $i = 1, 2, \dots, M$, have been determined as above. The following coordinate transformation is then used to reduce the set of original equations

$$\mathbf{u} = \sum_{i=1}^{M} \mathbf{x}_i y_i \tag{13.74a}$$

or

$$\mathbf{u} = \mathbf{X}\mathbf{y} \tag{13.74b}$$

in which $\mathbf{X}$ is an $N \times M$ matrix of the M Ritz vectors and $\mathbf{y}$ is a vector of M generalized coordinates. The transformed equations are given by

$$\tilde{\mathbf{M}}\ddot{\mathbf{y}} + \tilde{\mathbf{C}}\dot{\mathbf{y}} + \tilde{\mathbf{K}}\mathbf{y} = \tilde{\mathbf{p}} \tag{13.75}$$

where

$$\tilde{\mathbf{M}} = \mathbf{X}^T\mathbf{M}\mathbf{X} = \mathbf{I}$$

$$\tilde{\mathbf{C}} = \mathbf{X}^T\mathbf{C}\mathbf{X}$$

$$\tilde{\mathbf{K}} = \mathbf{X}^T\mathbf{K}\mathbf{X} \tag{13.76}$$

$$\tilde{\mathbf{p}} = \mathbf{X}^T\mathbf{p}$$

Note that the transformed mass matrix $\tilde{\mathbf{M}}$ is an identity matrix because vectors $\mathbf{x}_i$ are mass orthonormal. However, in general, both $\tilde{\mathbf{C}}$ and $\tilde{\mathbf{K}}$ are full matrices. Equation 13.75 represents a set of M coupled equations which can be solved by direct numerical integration. Since M is expected to be significantly smaller than N, the effort required in the solution of the reduced set is considerably less than that required in the solution of the original equations.

In the case of proportional damping, it is possible to uncouple the reduced set of equations by solving an eigenvalue problem of size M. This eigenvalue problem is obtained from Equation 13.75 by omitting the damping term and setting $\tilde{\mathbf{p}} = 0$. This leads to

$$\tilde{\mathbf{K}}\mathbf{z} = \tilde{\omega}^2\mathbf{z} \tag{13.77}$$

where $\mathbf{z}$ represents an eigenvector and $\tilde{\omega}$ the corresponding eigenvalue. If $\mathbf{Z}$ is the matrix of vectors $\mathbf{z}_i$, $i = 1, 2, \ldots, M$, and if the $\mathbf{z}_i$'s have been properly normalized, we will have

$$\mathbf{Z}\mathbf{Z}^T = \mathbf{Z}^T\mathbf{Z} = \mathbf{I} \tag{13.78}$$

and

$$\mathbf{Z}^T\tilde{\mathbf{K}}\mathbf{Z} = \tilde{\Lambda} \tag{13.79}$$

when $\tilde{\Lambda}$ is a matrix of the frequencies $\tilde{\omega}^2$. We now obtain a new set of Ritz vectors $\mathbf{X}^0$ from the relationship

$$\mathbf{X}^0 = \mathbf{X}\mathbf{Z} \tag{13.80}$$

It is easily deduced that vectors represented by $\mathbf{X}^0$ are orthogonal to both the mass and stiffness matrices of the original system. Thus

$$\mathbf{K}^* = (\mathbf{X}^0)^T\mathbf{K}\mathbf{X}^0 = \mathbf{Z}^T\mathbf{X}^T\mathbf{K}\mathbf{X}\mathbf{Z}$$

$$= \mathbf{Z}^T\tilde{\mathbf{K}}\mathbf{Z} \tag{13.81}$$

$$= \tilde{\mathbf{\Lambda}}$$

$$\mathbf{M}^* = (\mathbf{X}^0)^T\mathbf{M}\mathbf{X}^0 = \mathbf{Z}^T\mathbf{X}^T\mathbf{M}\mathbf{X}\mathbf{Z}$$

$$= \mathbf{Z}^T\tilde{\mathbf{M}}\mathbf{Z} \tag{13.82}$$

$$= \mathbf{I}$$

The frequencies $\tilde{\omega}$ obtained from the solution of Equation 13.77 are approximations to the true frequencies ω of the original system, while $\mathbf{X}^0$ provide estimates of M eigenvectors of the original system.

For proportional damping, matrix $\mathbf{C}^* = (\mathbf{X}^0)^T\mathbf{C}\mathbf{X}^0$ will also be diagonal. The transformation of the equations is then given by

$$\mathbf{u} = \mathbf{X}^0\mathbf{y}^0 \tag{13.83}$$

$$\mathbf{I}\ddot{\mathbf{y}}^0 + \tilde{\mathbf{\Delta}}\dot{\mathbf{y}}^0 + \tilde{\mathbf{\Lambda}}\mathbf{y}^0 = \mathbf{p}^*(t) \tag{13.84}$$

in which $\tilde{\mathbf{\Delta}}$ is a diagonal matrix of terms $2\xi_i\tilde{\omega}_i$, $\tilde{\mathbf{\Lambda}}$ is a diagonal matrix of frequencies $\tilde{\omega}^2$, and $\mathbf{p}^*$ is given by

$$\mathbf{p}* = (\mathbf{X}^0)^T\mathbf{p} \tag{13.85}$$

The algorithm for the generation of load-dependent Ritz shapes has been summarized in Table 13.1.

As in the case of a mode superposition method, the contribution of a particular Ritz vector to the response can be measured by calculating the participation factor, $\gamma_n = \mathbf{x}_n^T\mathbf{f}$, for that vector. If the participation factor is small, the vector will contribute little to the response. For certain types of loads, for example those due to earthquakes, a small value for γ_n may indicate that γ_{n+1} will be even smaller and that a sufficient number of vectors has been included. In a more general case, error norms derived from Equation 13.45 may be used in which $\mathbf{e}_M$ is given by

$$\mathbf{e}_M = \mathbf{f} - \mathbf{M}\mathbf{X}^0\mathbf{X}^{0T}\mathbf{f}$$

$$= \mathbf{f} - \mathbf{M}\mathbf{X}\mathbf{Z}\mathbf{Z}^T\mathbf{X}^T\mathbf{f} \tag{13.86}$$

$$= \mathbf{f} - \mathbf{M}\mathbf{X}\mathbf{X}^T\mathbf{f}$$

The load-dependent orthogonal vectors generated as above have several advantages over the traditional method of using undamped shapes as the Ritz

Table 13.1. Algorithm for the generation of load-dependent Ritz vectors.

1. Triangularize stiffness matrix

$$\mathbf{K} = \mathbf{L}^T \mathbf{D} \mathbf{L}$$

2. Solve for the first vector

$$\mathbf{K}\tilde{\mathbf{x}}_1 = \mathbf{f} \qquad \text{solve for } \tilde{\mathbf{x}}_1$$

$$\mathbf{x}_1 = \frac{\tilde{\mathbf{x}}_1}{(\tilde{\mathbf{x}}_1^T \mathbf{M} \tilde{\mathbf{x}}_1)^{1/2}} \qquad \text{find mass orthonormal vector } \mathbf{x}_1$$

3. Solve for additional vectors $i = 2, \ldots, M$

$$\mathbf{K}\tilde{\mathbf{x}}_i = \mathbf{M}\mathbf{x}_{i-1} \qquad \text{solve for } \tilde{\mathbf{x}}_i$$

$$c_j = \mathbf{x}_j^T \mathbf{M} \tilde{\mathbf{x}}_i \qquad \text{compute for } j = 1, 2, \ldots, i-1$$

$$\tilde{\mathbf{x}}_i = \tilde{\mathbf{x}}_i - \sum_{j=1}^{i-1} c_j \mathbf{x}_j \qquad \text{find mass orthogonalized vector } \tilde{\mathbf{x}}_i$$

$$\mathbf{x}_i = \frac{\tilde{\mathbf{x}}_1}{(\tilde{\mathbf{x}}_i^T \mathbf{M} \tilde{\mathbf{x}}_i)^{1/2}} \qquad \text{find mass orthonormalized vector } \mathbf{x}_i$$

4. Solve the reduced eigenvalue problem

$$\tilde{\mathbf{K}}\mathbf{z} = \tilde{\omega}^2 \mathbf{z}$$

where $\tilde{\mathbf{K}} = \mathbf{X}^T \mathbf{K} \mathbf{X}$ and eigenvectors $\mathbf{z}$ are normalized so that $\mathbf{Z}^T \mathbf{Z} = \mathbf{I}$

5. Obtain the final Ritz vectors

$$\mathbf{X}^0 = \mathbf{X}\mathbf{Z}$$

vectors. The computational cost of generating the load-dependent vectors is substantially smaller than that for calculating the mode shapes. Also, because the special vectors are generated from the applied load, they form a more efficient basis for representation of the load. As a result, the number of load-dependent Ritz vectors required for an adequate representation of the load may be substantially smaller than the number of mode shapes needed for the purpose.

Example 13.8
Obtain the response of the frame of Example 13.1 to a support motion given by $\ddot{u}_g = 1000 \sin 12t$ using two load-dependent Ritz vectors derived by the method of this section.

Solution
The spatial distribution of forcing function is given by

$$\mathbf{f} = \mathbf{M}\mathbf{r} = \begin{bmatrix} 2 \\ 2 \\ 2 \\ 2 \\ 1 \end{bmatrix} \qquad \qquad \text{(a)}$$

The first vector in the series is given by

$$\mathbf{K}\tilde{\mathbf{x}}_1 = \mathbf{f} \tag{b}$$

Solution of Equation b gives

$$\tilde{\mathbf{x}}_1 = \frac{1}{100} \begin{bmatrix} 0.5625 \\ 1.1458 \\ 1.5625 \\ 1.9375 \\ 2.1875 \end{bmatrix} \tag{c}$$

To make vector $\tilde{\mathbf{x}}_i$ mass orthonormal, we obtain

$$\tilde{\mathbf{x}}_1^T \mathbf{M}\tilde{\mathbf{x}}_1 = 20.434 \times \frac{1}{100^2}$$

Hence

$$\mathbf{x}_1 = \frac{\tilde{\mathbf{x}}_1}{(\tilde{\mathbf{x}}_1^T \mathbf{M}\tilde{\mathbf{x}}_1)^{1/2}}$$

$$= \begin{bmatrix} 0.1244 \\ 0.2535 \\ 0.3456 \\ 0.4286 \\ 0.4839 \end{bmatrix} \tag{d}$$

The second vector is obtained from

$$\mathbf{K}\tilde{\mathbf{x}}_2 = \mathbf{M}\mathbf{x}_1 \tag{e}$$

$$\tilde{\mathbf{x}}_2 = \frac{1}{100} \begin{bmatrix} 0.1743 \\ 0.3859 \\ 0.5553 \\ 0.7229 \\ 0.8439 \end{bmatrix} \tag{f}$$

Vector $\tilde{\mathbf{x}}_2$ is now **M**-orthogonalized with respect to vector $\mathbf{x}_1$ determined previously

$$\tilde{\mathbf{x}}_2 = \tilde{\mathbf{x}}_2 - (\mathbf{x}_1^T \mathbf{M}\tilde{\mathbf{x}}_2)\mathbf{x}_1 \tag{g}$$

Substitution for $\tilde{\mathbf{x}}_2$, $\mathbf{x}_1$, and $\mathbf{M}$ in Equation g gives

$$\tilde{\mathbf{x}}_2 = \bar{\mathbf{x}}_2 - \frac{1.6509}{100}\mathbf{x}_1$$

$$= \frac{1}{100}\begin{bmatrix} -0.0312 \\ -0.0326 \\ -0.0154 \\ 0.0153 \\ 0.0450 \end{bmatrix} \tag{h}$$

Vector $\tilde{\mathbf{x}}_2$ is now $\mathbf{M}$-orthornormalized, giving

$$\mathbf{x}_2 = \frac{\tilde{\mathbf{x}}_2}{(\tilde{\mathbf{x}}_2^T\mathbf{M}\tilde{\mathbf{x}}_2)^{1/2}}$$

$$= \begin{bmatrix} -0.3716 \\ -0.3885 \\ -0.1834 \\ 0.1826 \\ 0.5366 \end{bmatrix} \tag{i}$$

Using the Ritz vectors derived in Equations d and i, a second-order eigenvalue is formed next

$$\tilde{\mathbf{K}}\mathbf{z} = \tilde{\omega}^2\tilde{\mathbf{M}}\mathbf{z} \tag{j}$$

where

$$\tilde{\mathbf{K}} = \begin{bmatrix} \mathbf{x}_1^T \\ \mathbf{x}_2^T \end{bmatrix}[\mathbf{K}][\mathbf{x}_1 \quad \mathbf{x}_2]$$

$$= \mathbf{X}^T\mathbf{K}\mathbf{X}$$

$$= \begin{bmatrix} 61.68 & -21.794 \\ -21.794 & 429.09 \end{bmatrix}$$

and

$$\tilde{\mathbf{M}} = \mathbf{X}^T\mathbf{M}\mathbf{X}$$

$$= \mathbf{I}$$

The eigenvalue problem of Equation j leads to the following characteristic equation

$$\begin{vmatrix} 61.68 - \lambda & -21.794 \\ -21.794 & 429.09 - \lambda \end{vmatrix} = 0$$

or

$$\lambda^2 - 490.77\lambda + 22,5991.8 = 0 \tag{k}$$

Solution of Equation k gives

$$\lambda_1 = 60.39 \quad \omega_1 = 7.77$$
$$\lambda_2 = 430.38 \quad \omega_2 = 20.75$$

Substitution, in turn, of the two values of λ, in Equation j gives the following two eigenvectors

$$\tilde{\mathbf{Z}} = \begin{bmatrix} 1 & 1 \\ 0.059 & -16.92 \end{bmatrix}$$

The eigenvectors are normalized so that $\mathbf{Z}^T\mathbf{Z} = \mathbf{I}$. This gives

$$\mathbf{Z} = \begin{bmatrix} 0.9983 & 0.0590 \\ 0.0590 & -0.9983 \end{bmatrix} \tag{1}$$

The final Ritz vectors are now obtained from

$$\mathbf{X}^0 = \mathbf{XZ}$$

$$= \begin{bmatrix} 0.1244 & -0.3716 \\ 0.2535 & -0.3885 \\ 0.3456 & -0.1834 \\ 0.4286 & 0.1826 \\ 0.4839 & 0.5366 \end{bmatrix} \begin{bmatrix} 0.9983 & 0.0590 \\ 0.0590 & -0.9983 \end{bmatrix} \tag{p}$$

$$= \begin{bmatrix} 0.1023 & 0.3783 \\ 0.2301 & 0.4028 \\ 0.3342 & 0.2035 \\ 0.4386 & -0.1570 \\ 0.5147 & -0.5710 \end{bmatrix}$$

The equations of motion of the system in the physical coordinates are

$$\mathbf{M\ddot{u}} + \mathbf{Ku} = -1000\,\mathbf{Mr}\sin \Omega t \tag{n}$$

We now introduce the following transformation

$$\mathbf{u} = \mathbf{X}^0\mathbf{y}^0 \tag{o}$$

The transformation uncouples the equations of motion, giving the following two-single-degree-of-freedom equations

$$\ddot{y}_1^0 + 60.39\,y_1^0 = -1000[\mathbf{x}_1^0]^T\mathbf{Mr}\sin 12t$$
$$= -2725.3 \sin 12t$$
$$\ddot{y}_2^0 + 430.38\,y_2^0 = -1000[\mathbf{x}_2^0]^T\mathbf{Mr}\sin 12t \tag{p}$$
$$= -1148.0 \sin 12t$$

The steady-state solutions of Equations p are

$$y_1^0 = \frac{-2725.3}{60.39} \; \frac{\sin 12t}{1 - 12^2/60.39}$$

$$= 32.596 \sin 12t$$

$$y_2^0 = \frac{-1148.0}{430.38} \; \frac{\sin 12t}{1 - 12^2/430.38}$$

$$= -4.009 \sin 12t$$

Substitution for $\mathbf{y}^0$ in the transformation equation (Eq. o) gives the displacement response in the physical coordinates

$$\mathbf{u} = \begin{bmatrix} 1.8177 \\ 5.8859 \\ 10.0790 \\ 14.9274 \\ 18.8109 \end{bmatrix} \sin 12t$$

In comparison, the exact displacements obtained by mode superposition using all modes are

$$\mathbf{u} = \begin{bmatrix} 1.9355 \\ 5.7142 \\ 9.7908 \\ 14.8802 \\ 19.3315 \end{bmatrix} \sin 12t$$

It will also be of interest to compare the displacements obtained by superposing only the first two modal contributions. The resulting values are

$$\mathbf{u} = \begin{bmatrix} 2.4499 \\ 5.9170 \\ 9.5744 \\ 14.5791 \\ 19.7369 \end{bmatrix} \sin 12t$$

In this case, the Ritz vector values are as good or even better than the displacements obtained by using the first two mode shapes.

13.3.5 *Application of Lanczos vectors in the transformation of the equations of motion*

Discussion in the preceding section has indicated that a set of orthogonal vectors that forms an adequate basis for representing the applied dynamic load can also be quite effective in reducing the size of the dynamic problems. The Lanczos vectors derived in Section 11.5.5, being a set of mass orthonormal vectors, form an obvious choice for the purpose. Experience has shown that they do indeed serve as an effective set of Ritz vectors.

Like the load-dependent vectors of Wilson, Lanczos vectors are members of a Krylov sequence in which the starting vector is selected arbitrarily. Each of the subsequent vector in the sequence is mass orthogonalized with respect to the immediately preceding two vectors. However, the special procedure used in the vector generation is such that orthogonalization with respect to the preceding two vectors automatically makes the new vector orthogonal to all preceding vectors. Finally, the newly generated vector is scaled so that it is orthonormal to the mass matrix. The algorithm used in the generation of the vectors is given by Equation 11.147.

When the Lanczos vectors are applied to the transformation of a dynamic problem, the starting vector is taken as equal to the static response of the system to the applied load. Thus, assuming that the applied load is given by Equation 13.65, the first Lanczos vector is obtained from

$$\tilde{\mathbf{x}}_1 = \mathbf{K}^{-1}\mathbf{f} \tag{13.87}$$

Mass orthonormalization of $\tilde{\mathbf{x}}_1$ gives

$$\beta_1 = \left(\tilde{\mathbf{x}}_1^T \mathbf{M} \tilde{\mathbf{x}}_1\right)^{1/2} \tag{13.88a}$$

$$\mathbf{x}_1 = \frac{\tilde{\mathbf{x}}}{\beta_1} \tag{13.88b}$$

Now, let $\mathbf{X}_M$ denote the matrix of first M Lanczos vectors. Then, from Equation 11.143

$$\mathbf{K}^{-1}\mathbf{M}\mathbf{X}_M = \mathbf{X}_M\mathbf{T}_M + [\mathbf{0} \quad \beta_{M+1}\mathbf{x}_{M+1}] \tag{13.89}$$

in which $\mathbf{T}_M$ is a tridiagonal matrix given by

$$\mathbf{T}_M = \begin{bmatrix} \alpha_1 & \beta_2 & & & & \\ \beta_2 & \alpha_2 & \beta_3 & & & \\ & \cdot & \cdot & \cdot & & \\ & & \cdot & \cdot & \cdot & \\ & & & \cdot & \cdot & \cdot \\ & & & & \cdot & \beta_M \\ & & & & \beta_M & \alpha_M \end{bmatrix} \tag{13.90}$$

and the submatrix $\mathbf{0}$ is of size $N \times M - 1$.

Multiplying both sides of Equation 13.89 by $\mathbf{X}_M^T \mathbf{M}$ and using the orthonormality property of Lanczos vectors, namely $\mathbf{X}_M^T \mathbf{M} \mathbf{X}_M = \mathbf{I}_M$ and $\mathbf{X}_M^T \mathbf{M} \mathbf{x}_{M+1} = \mathbf{0}$, we get

$$\mathbf{X}_M^T \mathbf{M} \mathbf{K}^{-1} \mathbf{M} \mathbf{X}_M = \mathbf{T}_M \tag{13.91}$$

The coordinate transformation is given by

$$\mathbf{u} = \mathbf{X}_M \mathbf{y}_M \tag{13.92}$$

so that the equations of motion can be stated as

$$\mathbf{M} \mathbf{X}_M \ddot{\mathbf{y}}_M + \mathbf{C} \mathbf{X}_M \dot{\mathbf{y}}_M + \mathbf{K} \mathbf{X}_M \mathbf{y}_M = \mathbf{f} g(t) \tag{13.93}$$

On multiplying both sides of Equation 13.93 by $\mathbf{X}_M^T \mathbf{M} \mathbf{K}^{-1}$, we get

$$\mathbf{X}_M^T \mathbf{M} \mathbf{K}^{-1} \mathbf{M} \mathbf{X}_M \ddot{\mathbf{y}}_M + \mathbf{X}_M^T \mathbf{M} \mathbf{K}^{-1} \mathbf{C} \mathbf{X}_M \dot{\mathbf{y}}_M + \mathbf{X}_M^T \mathbf{M} \mathbf{X}_M \mathbf{y}_M$$
$$= \mathbf{X}_M^T \mathbf{M} \mathbf{K}^{-1} \mathbf{f} g(t) \tag{13.94}$$

On using Equations 13.87, 13.88 and 13.91 and neglecting damping, Equation 13.94 reduces to

$$\mathbf{T}_M \ddot{\mathbf{y}}_M + \mathbf{y}_M = \beta_1 \mathbf{X}_M^T \mathbf{M} \mathbf{x}_1 g(t)$$
$$= \beta_1 \mathbf{v}_1 g(t) \tag{13.95}$$

where $\mathbf{v}_1$ is the first column of an identity matrix of size M.

When damping is present, the matrix $\mathbf{X}_M^T \mathbf{M} \mathbf{K}^{-1} \mathbf{C} \mathbf{X}_M$ is not tridigonal unless $\mathbf{C}$ is of a special form. For example, in the case of Rayleigh damping we have

$$\mathbf{X}_M^T \mathbf{M} \mathbf{K}^{-1} \mathbf{C} \mathbf{X}_M = \mathbf{X}_M^T \mathbf{M} \mathbf{K}^{-1} (a_0 \mathbf{M} + a_1 \mathbf{K}) \mathbf{X}_M$$
$$= a_0 \mathbf{T}_M + a_1 \mathbf{I}_M \tag{13.96}$$

so that Equation 13.94 reduces to

$$\mathbf{T}_M \ddot{\mathbf{y}}_M + (a_0 \mathbf{T}_M + a_1 \mathbf{I}_M) \dot{\mathbf{y}}_M + \mathbf{y}_M = \beta_1 \mathbf{v}_1 g(t) \tag{13.97}$$

Equations 13.95 and 13.97 are both of tridiagonal form and of size M. They can be solved by any appropriate method of numerical integration. Alternatively, the equations can be uncoupled by a second transformation using the eigenvectors of tridiagonal matrix $\mathbf{T}_M$. In either case, the solution will lead to expressions for response in the generalized coordinates $\mathbf{y}_M$. Response in the physical coordinates is obtained by using Equation 13.92.

To determine the number of Lanczos vectors that must be included in a particular analysis to obtain reasonable accuracy, some measure is required of errors introduced in using a truncated series of Ritz vectors. As in the mode superposition method or the method using Wilson's load-dependent vectors, participation factors may be used in certain cases to indicate when the vector series can be truncated.

Recall that the jth Lanczos vector is given by

$$\beta_j \mathbf{x}_j = \mathbf{K}^{-1}\mathbf{M}\mathbf{x}_{j-1} - \alpha_{j-1}\mathbf{x}_{j-1} - \beta_{j-1}\mathbf{x}_{j-2} \qquad (13.98)$$

The participation factor of the jth vector can be obtained by multiplying both sides of Equation 13.98 by $\mathbf{f}^T$

$$\beta_j \mathbf{f}^T\mathbf{x}_j = \mathbf{f}^T\mathbf{K}^{-1}\mathbf{M}\mathbf{x}_{j-1} - \alpha_{j-1}\mathbf{f}^T\mathbf{x}_{j-1} - \beta_{j-1}\mathbf{f}^T\mathbf{x}_{j-2} \qquad (13.99)$$

From Equations 13.87 and 13.88 we get

$$\mathbf{f} = \beta_1 \mathbf{K}\mathbf{x}_1 \qquad (13.100)$$

By substituting for $\mathbf{f}$ from Equation 13.100, the first term on the right-hand side of Equation 13.99 can be reduced as follows

$$\mathbf{f}^T\mathbf{K}^{-1}\mathbf{M}\mathbf{x}_{j-1} = \beta_1 \mathbf{x}_1^T\mathbf{M}\mathbf{x}_{j-1}$$
$$= 0 \qquad (13.101)$$

Equation 13.99 now becomes

$$\beta_j\gamma_j = -\alpha_{j-1}\gamma_{j-1} - \beta_{j-1}\gamma_{j-2} \qquad (13.102)$$

The recurrence relationship given by Equation 13.102, which involves only scalar multiplications, can be used to obtain the participation factor of the jth vector in terms of the participation factors for the two preceding vectors. The vector series can be terminated when the participation factor of the latest vector is sufficiently small. The truncation criterion based on the value of participation factor will work provided that the participation factor decreases continuously with increasing vector number. In a more general case, the error norm defined by Equation 13.45 may be used, where $\mathbf{e}_M$ is obtained from

$$\mathbf{e}_M = \mathbf{f} - \mathbf{M}\mathbf{X}_M\mathbf{X}_M^T\mathbf{f} \qquad (13.103)$$

13.4   DIRECT INTEGRATION OF THE EQUATIONS OF MOTION

We have observed that the normal coordinate transformation uncouples the equations of motion, provided that the damping in proportional or is specified only at the modal level. In fact, for proportional damping, any general set of orthogonal Ritz vectors can be used to uncouple the equations of motion, provided that a reduced eigenvalue problem is solved to obtain mass and stiffness orthogonal vectors. However when the reduced eigenvalue problem is not solved, the transformed equations in the Ritz coordinates remain coupled. The equations also remain uncoupled in all cases where the damping is not proportional.

A direct integration method must be used for solution of the equations of motion whenever they are coupled. Even where uncoupling of the equations is possible, direct integration may be the more efficient method of solution provided that the damping matrix is available and the response history needs to be calculated over only a few time intervals starting with time zero.

In Chapter 8, several different methods for the integration of the equations of motion of a single-degree-of-freedom system were presented. It was pointed out that in order to ensure reasonable accuracy in the solution, the time step used in the integration must be kept sufficiently small. With some of the methods a further restriction had to be imposed on the length of the time step to ensure stability of the solution, while other methods were uncondionally stable. However, even where an integration method was only conditionally stable, the restriction on the size of the time step to ensure stability was not too severe. In the case of multi-degree-of-freedom systems, the situation is somewhat different and we need to pay greater attention to the conditions for stability.

To examine the effect the size of time step has on a direct integration of the equations of motion, it is useful to develop a relationship between the response results obtained from a mode superposition analysis and those obtained by direct integration. As stated earlier, errors are introduced in a mode superposition analysis based on two factors: 1. mode truncation, and 2. numerical integration of the transformed equations of motion using a finite-size time step. When all modes of the system are included in the analysis, the error arising from the first of the two sources vanishes and the only approximation is due to the use of a finite-size time step in integration. In such a case, if all modes are integrated with the same time step and this time step is also the one used in a direct integration of the equations of motion, the response results obtained by mode superposition and direct integration are identical. This identity allows us to draw inferences about how the step size will influence the results of a direct integration by examining the effect of step size on a modal analysis in which all modes are integrated with the same step.

To evaluate the response in any mode with reasonable accuracy, the integration time step should be a fraction of the modal period. For higher modes, whose period is very small, this imposes a very severe restriction and if the same time step is to be used in the integration of all modes, the computation cost becomes very high. On the other hand, if a comparatively larger time step is used, the lower modes may still be integrated with reasonable accuracy, but the higher mode response is inaccurate. Fortunately, this usually is not a serious concern because the contribution of the higher modes to the response is, in most cases, quite small. Furthermore, because of the idealizations involved in modeling of the system, the higher modes are seldom represented correctly, and therefore the accuracy achieved in evaluating the response in such modes is of little significance. Thus, from the point of view of accuracy, the time step size is determined not by the period of the highest mode but by the period of only the highest significant mode. On the other hand, numerical integration should remain stable for even the highest mode, because despite the fact that such a mode may contribute little to the response, instability in its integration will cause the modal response to blow up and thus render the total solution meaningless. It is apparent that for conditionally-stable integration schemes, the condition of stability would require the time step to be a certain fraction of the highest mode period. As stated earlier, for systems with very large number of degrees of freedom where the highest mode period is very small, this imposes a very severe restriction on the time step and makes the computations very expensive. Obviously, for such systems the only practical means of numerical integration is to use an unconditionally-stable integration scheme.

To illustrate the foregoing point, consider a system in which the highest significant mode has a period of 0.1 s, while the highest mode has a period of 0.001 s. When an unconditionally-stable integration scheme is used, reasonable accuracy will be obtained with a step size of, say, $\frac{1}{10}$ of 0.1 s, or 0.01 s. On the other hand, supposing that one uses the linear acceleration method to integrate the equations, the stability criterion will require that the time step be less than $0.55 \times T_N$, where T_N is the period of the highest mode, giving a time step size of 0.00055 s, which is extremely small. Obviously, the choice of a conditionally-stable integration scheme is inappropriate in this case.

It is apparent from the foregoing discussion that when an unconditionally-stable method is used in the integration, the selected time step may, in fact, be several times larger than the period of one of the higher modes. Unconditional stability of the integration method ensures that the response obtained in such a mode will remain bounded. However, it will be of interest to find out more about the nature of this response and whether it affects the accuracy of the overall response. From our discussion relating to the mode acceleration or the static correction method, it is clear that a desirable situation is one in which the calculated higher-mode response is equal to or nearly equal to the static response. In the examples that follow, we will provide illustrations of the nature of response obtained in a higher mode when an unconditionally-stable scheme is used in the integration and the step size is larger than the period of the mode being considered. As stability criterion often plays an important role in the selection of the integration scheme for a multi-degree-of-freedom system, it is more meaningful to discuss such schemes under the two categories of 1. explicit methods which are only conditionally stable, and 2. implicit methods some of which are unconditionally stable. In the following sections, we examine some of the more popular techniques in each of the two categories.

13.4.1 *Explicit integration schemes*

As discussed in Chapter 8 numerical integration is a process of marching along the time in which the response parameters: accelerations, velocities, and displacements at a given time point are evaluated from their known historic values. For a single-degree-of-freedom system, this requires three equations to determine the three unknowns. Two of these equations are usually derived from assumptions regarding the manner in which the response parameters vary during a time step. The third equation is the equation of motion written at a selected time point. When this time point represents the current time (n), the method of integration is referred to as an explicit method; on the other hand, when the equation of motion is written at the next time point in the future $(n + 1)$, the method is said to be an implicit method. A similar classification applies to the multi-degree-of-freedom systems, except that the scalar equations of the single-degree-of-freedom system now become matrix equations.

Central difference method
The explicit method that is most commonly used is the central difference method. In this method, the matrix equations used to relate the response parameters at time $n + 1$ to those at time n are similar to the equivalent scalar equations, that is, Equations 8.139 and 8.141. Thus

$$\dot{\mathbf{u}}_n = \frac{1}{2h}(\mathbf{u}_{n+1} - \mathbf{u}_{n-1}) \tag{13.104}$$

$$\ddot{\mathbf{u}}_n = \frac{1}{h^2}(\mathbf{u}_{n+1} - 2\mathbf{u}_n + \mathbf{u}_{n-1}) \tag{13.105}$$

where h is the time step used in the integration. The equation of motion is formed at time point n, giving

$$\mathbf{M}\ddot{\mathbf{u}}_n + \mathbf{C}\dot{\mathbf{u}}_n + \mathbf{K}\mathbf{u}_n = \mathbf{p}_n \tag{13.106}$$

Substitution for $\dot{\mathbf{u}}_n$ and $\ddot{\mathbf{u}}_n$ from Equations 13.104 and 13.105 into Equation 13.106 gives

$$\left(\frac{1}{h^2}\mathbf{M} + \frac{1}{2h}\mathbf{C}\right)\mathbf{u}_{n+1} = \mathbf{p}_n + \left(-\mathbf{K} + \frac{2}{h^2}\mathbf{M}\right)\mathbf{u}_n + \left(\frac{1}{2h}\mathbf{C} - \frac{1}{h^2}\mathbf{M}\right)\mathbf{u}_{n-1} \tag{13.107}$$

The displacement vector at time $n + 1$ is obtained by solving Equation 13.107. It will be noted that in the solution of Equation 13.107, factorization of $\mathbf{K}$ is not needed. In fact, if the mass matrix is diagonal and damping is negligible or the damping matrix is mass proportional, Equation 13.107 reduces to N uncoupled algebraic equations whose solution is trivial.

Before integration can begin, vector $\mathbf{u}_{-1}$ must be known. It is obtained from Equations 13.104 and 13.105 by eliminating $\mathbf{u}_{n+1}$ and substituting $n=0$. This leads to

$$\mathbf{u}_{-1} = \mathbf{u}_0 + \frac{h^2}{2}\ddot{\mathbf{u}}_0 - h\dot{\mathbf{u}}_0 \qquad (13.108)$$

The central difference method is only conditionally stable, and the condition of stability requires that the time step h be less than T/π, where T is the period of the highest mode. In some cases this may be unduly restrictive. However, in those situations where the highest mode period is not too small, so that the stability criterion can be satisfied with a reasonable step size, the central difference method may be the most appropriate because of its computational efficiency.

Example 13.9
The two-degree-of-freedom system shown in Figure E13.9 is subjected to a suddenly applied constant force of magnitude 100 along degree-of-freedom 2. Using the mode superposition method, obtain the exact response of the system at intervals of 0.1 s for the first 2 s. Next calculate the response of the same system using the central difference method and a time interval of (i) 0.1 s, and (ii) 0.4 s.

Solution
The mass matrix $\mathbf{M}$ and the stiffness matrix $\mathbf{K}$ are given by

$$\mathbf{M} = \begin{bmatrix} 2 & 0 \\ 0 & 1 \end{bmatrix} \qquad (a)$$

$$\mathbf{K} = \begin{bmatrix} 96 & -32 \\ -32 & 32 \end{bmatrix} \qquad (b)$$

The forcing function is

$$\mathbf{p} = \begin{bmatrix} 0 \\ 100 \end{bmatrix} \qquad (c)$$

The eigenvalue problem for the system is given by

$$\begin{bmatrix} 96 & -32 \\ -32 & 32 \end{bmatrix}\mathbf{q} = \omega^2 \begin{bmatrix} 2 & 0 \\ 0 & 1 \end{bmatrix}\mathbf{q} \qquad (d)$$

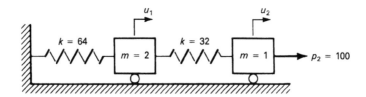

Figure E13.9. Vibrations of a two-degree-of-freedom system.

On solving Equation d by the characteristic equation method, we obtain the following frequencies and mode shapes

$$\omega_1 = 4 \quad \omega_2 = 8$$

$$\mathbf{q}_1 = \begin{bmatrix} 1 \\ 2 \end{bmatrix} \quad \mathbf{q}_2 = \begin{bmatrix} 1 \\ -1 \end{bmatrix} \tag{e}$$

Modal coordinate transformation

$$\mathbf{u} = \mathbf{Qy} \tag{f}$$

yields the following uncoupled equations of motion

$$6\ddot{y}_1 + 96y_1 = 200$$
$$3\ddot{y}_2 + 192y_2 = -100 \tag{g}$$

The exact solutions of Equations g are given by Equation 7.14 with $\xi = 0$

$$y_1 = \frac{200}{96}(1 - \cos 4t)$$
$$y_2 = -\frac{100}{192}(1 - \cos 8t) \tag{h}$$

The response in the physical coordinates is now obtained by using the transformation equation, Equation f

$$u_1 = y_1 + y_2 = 1.563 - 2.083 \cos 4t + 0.521 \cos 8t$$
$$u_2 = 2y_1 - y_2 = 4.688 - 4.167 \cos 4t - 0.521 \cos 8t \tag{i}$$

Displacement u_1 and u_2 calculated from Equation i for the first 2 s at intervals of 0.1 s are shown in Table E13.9a.
(i) Since the system starts from rest, $\mathbf{u}_0$ and $\dot{\mathbf{u}}_0$ are both null vectors. The starting accelerations are obtained from the equations of motion for $t = 0$.

$$\begin{bmatrix} 2 & 0 \\ 0 & 1 \end{bmatrix} \ddot{\mathbf{u}}_0 = \begin{bmatrix} 0 \\ 100 \end{bmatrix} \tag{j}$$

so that

$$\ddot{\mathbf{u}}_0 = \begin{bmatrix} 0 \\ 100 \end{bmatrix} \tag{k}$$

Displacement vector $\mathbf{u}_{-1}$ is calculated next using Equation 13.108

$$\mathbf{u}_{-1} = \frac{h^2}{2} \begin{bmatrix} 0 \\ 100 \end{bmatrix} \tag{l}$$

For $h = 0.1$, Equation l gives

$$\mathbf{u}_{-1} = \begin{bmatrix} 0.0 \\ 0.5 \end{bmatrix}$$

Equation 13.107 can now be used to obtain the response history of displacements, giving

$$\mathbf{M}\mathbf{u}_{n+1} = h^2\mathbf{p}_n + (2\mathbf{M} - h^2\mathbf{K})\mathbf{u}_n - \mathbf{M}\mathbf{u}_{n-1}$$

or

$$\begin{bmatrix} 2 & 0 \\ 0 & 1 \end{bmatrix}\mathbf{u}_{n+1} = \begin{bmatrix} 0 \\ 1 \end{bmatrix} + \left(\begin{bmatrix} 4 & 0 \\ 0 & 2 \end{bmatrix} - \begin{bmatrix} 0.96 & -0.32 \\ -0.32 & 0.32 \end{bmatrix} \right)\mathbf{u}_n - \begin{bmatrix} 2 & 0 \\ 0 & 1 \end{bmatrix}\mathbf{u}_{n-1}$$

(m)

Equation m leads to the following two uncoupled equations

$$u_{1(n+1)} = 1.52u_{1(n)} + 0.16u_{2(n)} - u_{1(n-1)}$$
$$u_{2(n+1)} = 1 + 0.32u_{1(n)} + 1.68u_{2(n)} - u_{2(n-1)}$$

(n)

The response calculated from Equations n for the first 2 s is shown in Table E13.9b.

For the central difference method to be stable, the time step used in the integration should satisfy the following relationship

$$h < \frac{T_2}{\pi}$$

$$h < \frac{2\pi}{8\pi} = 0.25 \text{ s}$$

(o)

The selected time step does satisfy the stability criterion. The response results are therefore bounded. However, the time step is rather coarse, being slightly over $\frac{1}{8}$ of T_2, and therefore the calculated response is expected to be somewhat imprecise. Nevertheless, the response results are reasonably close to the exact values shown in Table E13.9a.

Table E13.9a. Exact response of a two-degree-of-freedom system by mode superposition method, $h = 0.1$ s.

Time	u_1	u_2
0.1	0.006	0.487
0.2	0.096	1.800
0.3	0.423	3.562
0.4	1.103	5.329
0.5	2.089	6.762
0.6	3.144	7.714
0.7	3.929	8.209
0.8	4.160	8.330
0.9	3.748	8.107
1.0	2.848	7.487
1.1	1.780	6.390
1.2	0.867	4.836
1.3	0.294	3.027
1.4	0.052	1.350
1.5	0.002	0.247
1.6	0.000	0.043
1.7	0.018	0.798
1.8	0.160	2.288
1.9	0.584	4.096
2.0	1.367	5.792

Table E13.9b. Response of a two degree-of-freedom system
by the central difference method, $h = 0.1$ s.

Time	u_1	u_2
0.1	0.000	0.500
0.2	0.080	1.840
0.3	0.416	3.617
0.4	1.131	5.369
0.5	2.162	6.765
0.6	3.238	7.689
0.7	3.989	8.188
0.8	4.137	8.344
0.9	3.632	8.153
1.0	2.690	7.516
1.1	1.658	6.334
1.2	0.844	4.656
1.3	0.370	2.759
1.4	0.160	1.096
1.5	0.048	0.135
1.6	−0.065	0.145
1.7	−0.124	1.088
1.8	0.051	2.644
1.9	0.624	4.370
2.0	1.597	5.897

Table E13.9c. Response of a two-degree-of-freedom system
by the central difference method, $h = 0.4$ s.

Time	u_1	u_2
0.4	0.00	8.00
0.8	20.48	−8.96
1.2	−139.30	140.80
1.6	1131.00	−1127.40
2.0	−9171.10	9183.50

(ii) In this case, the selected time step fails the stability criterion for integrating the second mode. The calculated response is therefore expected to be unbounded. Details of calculations are similar to those outlined in part (i) and lead to the following expressions

$$\mathbf{u}_{-1} = \begin{bmatrix} 0 \\ 8 \end{bmatrix} \tag{p}$$

$$\begin{aligned} u_{1(n+1)} &= -5.68u_{1(n)} + 2.56u_{2(n)} - u_{1(n-1)} \\ u_{2(n+1)} &= 16 + 5.12u_{1(n)} - 3.12u_{2(n)} - u_{2(n-1)} \end{aligned} \tag{q}$$

The displacement response calculated from Equations q is tabulated in Table E13.9c for the first five time intervals.

13.4.2 *Implicit integration schemes*

In principle, these schemes are similar to the explicit schemes. However, a major difference is that in implicit schemes the equation of motion is formed at the next time point $(n+1)$ rather than at the current time point. A consequence of this is that the solution for displacement vector at $(n+1)$ requires factorization of the stiffness matrix. In the following paragraphs, we discuss several different integration schemes belonging to the implicit category.

Average acceleration method

As stated in Section 8.12.1, this method is based on the assumption that the acceleration along each coordinate of the system remains constant over the small time interval h and its value is the average of those at the beginning and end of the interval. The relationships between the response parameters at time points n and $n+1$ are similar to their scalar counterparts (Eqs. 8.122 and 8.123)

$$\ddot{\mathbf{u}}_{n+1} = -\ddot{\mathbf{u}}_n + \frac{4}{h^2}(\mathbf{u}_{n+1} - \mathbf{u}_n - h\dot{\mathbf{u}}_n) \qquad (13.109)$$

$$\dot{\mathbf{u}}_{n+1} = -\dot{\mathbf{u}}_n + \frac{2}{h}(\mathbf{u}_{n+1} - \mathbf{u}_n) \qquad (13.110)$$

A third relationship is obtained from the equation of motion written at time point $n+1$

$$\mathbf{M}\ddot{\mathbf{u}}_{n+1} + \mathbf{C}\dot{\mathbf{u}}_{n+1} + \mathbf{K}\mathbf{u}_{n+1} = \mathbf{p}_{n+1} \qquad (13.111)$$

Substitution for $\dot{\mathbf{u}}_{n+1}$ and $\ddot{\mathbf{u}}_{n+1}$ from Equations 13.110 and 13.109 in Equation 13.111 gives

$$\left(\frac{4}{h^2}\mathbf{M} + \frac{2}{h}\mathbf{C} + \mathbf{K}\right)\mathbf{u}_{n+1} = \mathbf{p}_{n+1} + \mathbf{M}\left(\frac{4}{h^2}\mathbf{u}_n + \frac{4}{h}\dot{\mathbf{u}}_n + \ddot{\mathbf{u}}_n\right)$$

$$+ \mathbf{C}\left(\frac{2}{h}\mathbf{u}_n + \dot{\mathbf{u}}_n\right) \qquad (13.112)$$

Equation 13.112 is solved for $\mathbf{u}_{n+1}$ and the resulting value is substituted in Equations 13.109 and 13.110, respectively, to obtain $\ddot{\mathbf{u}}_{n+1}$ and $\dot{\mathbf{u}}_{n+1}$. It will be noted that the solution for $\mathbf{u}_{n+1}$ requires factorization of $\mathbf{K}$. The average acceleration method is unconditionally-stable and is therefore suitable for direct integration of the equations of motion of systems with a large number of degrees of freedom.

Example 13.10

Find the response of the system of Example 13.9 using the average acceleration method and a time step h of (i) 0.1 s, and (ii) 10 s.

Solution

(i) The initial displacement and velocity vectors are zero. As in Example 13.9, the initial acceleration vector is obtained from the equation of motion at $t = 0$, giving

$$\ddot{\mathbf{u}}_0 = \begin{bmatrix} 0 \\ 100 \end{bmatrix}$$

Iterations are now carried out with Equations 13.112, 13.109, and 13.110. For $h = 0.1$, these equations reduce respectively to

$$\left\{ 400 \begin{bmatrix} 2 & 0 \\ 0 & 1 \end{bmatrix} + \begin{bmatrix} 96 & -32 \\ -32 & 32 \end{bmatrix} \right\} \mathbf{u}_{n+1}$$

$$= \begin{bmatrix} 0 \\ 100 \end{bmatrix} + \begin{bmatrix} 2 & 0 \\ 0 & 1 \end{bmatrix} (400\mathbf{u}_n + 40\dot{\mathbf{u}}_n + \ddot{\mathbf{u}}_n) \tag{a}$$

$$\ddot{\mathbf{u}}_{n+1} = -\ddot{\mathbf{u}}_n + 400(\mathbf{u}_{n+1} - \mathbf{u}_n - 0.1\dot{\mathbf{u}}_n) \tag{b}$$

$$\dot{\mathbf{u}}_{n+1} = -\dot{\mathbf{u}}_n + 20(\mathbf{u}_{n+1} - \mathbf{u}_n) \tag{c}$$

The response results obtained from Equations a, b, and c are shown in Table E13.10a for the first 2 s. Considering that the time step is rather coarse, the results are reasonably close to the exact values shown in Table E13.9a.

(ii) The initial conditions are identical to those in part (i). With $h = 10$ s, the three equations governing the iterations become

$$\left\{ 0.04 \begin{bmatrix} 2 & 0 \\ 0 & 1 \end{bmatrix} + \begin{bmatrix} 96 & -32 \\ -32 & 32 \end{bmatrix} \right\} \mathbf{u}_{n+1}$$

$$= \begin{bmatrix} 0 \\ 100 \end{bmatrix} + \begin{bmatrix} 2 & 0 \\ 0 & 1 \end{bmatrix} (0.04\mathbf{u}_n + 0.4\dot{\mathbf{u}}_n + \ddot{\mathbf{u}}_n) \tag{d}$$

$$\ddot{\mathbf{u}}_{n+1} = -\ddot{\mathbf{u}}_n + 0.04(\mathbf{u}_{n+1} - \mathbf{u}_n - 10\dot{\mathbf{u}}_n) \tag{e}$$

Table E13.10a. Response of a two-degree-of-freedom system by the average acceleration method, $h = 0.1$ s.

Time	u_1	u_2
0.1	0.017	0.464
0.2	0.121	1.728
0.3	0.437	3.458
0.4	1.062	5.241
0.5	1.970	6.733
0.6	2.978	7.746
0.7	3.799	8.260
0.8	4.156	8.343
0.9	3.912	8.067
1.0	3.130	7.446
1.1	2.058	6.450
1.2	1.012	5.082
1.3	0.249	3.454
1.4	-0.128	1.831
1.5	-0.174	0.580
1.6	-0.038	0.042
1.7	0.153	0.412
1.8	0.367	1.633
1.9	0.678	3.412
2.0	1.189	5.322

Table E13.10b. Response of a two-degree-of-freedom system by
the average acceleration method, $h = 10.0$ s.

Time	u_1	u_2
10.0	3.115	9.354
20.0	0.039	0.086
30.0	3.038	9.183
40.0	0.153	0.331
50.0	2.807	8.850
60.0	0.340	0.750
70.0	2.667	8.365
80.0	0.590	1.303
90.0	2.390	7.749
100.0	0.893	1.976

$$\dot{\mathbf{u}}_{n+1} = -\dot{\mathbf{u}}_n + 0.2(\mathbf{u}_{n+1} - \mathbf{u}_n) \tag{f}$$

The response for the first 10 time steps is shown in Table E13.10b. Even with the large time step, the response is bounded and oscillates about the static response given by

$$\begin{aligned}\mathbf{u}_s &= \mathbf{K}^{-1}\mathbf{p} \\ &= \begin{bmatrix} 1.563 \\ 4.688 \end{bmatrix}\end{aligned} \tag{g}$$

Linear acceleration method
This method is based on the assumption that the acceleration along any coordinate of the system varies linearly over the small interval of time h. Two of the three relationships used in the integration schemes are similar to their scalar counterparts (Eqs. 8.131 and 8.132)

$$\ddot{\mathbf{u}}_{n+1} = -2\ddot{\mathbf{u}}_n + \frac{6}{h^2}(\mathbf{u}_{n+1} - \mathbf{u}_n - h\dot{\mathbf{u}}_n) \tag{13.113}$$

$$\dot{\mathbf{u}}_{n+1} = -2\dot{\mathbf{u}}_n - \frac{h}{2}\ddot{\mathbf{u}}_n + \frac{3}{h}(\mathbf{u}_{n+1} - \mathbf{u}_n) \tag{13.114}$$

The third relationship is given by the equation of motion written at time point $n+1$, Equation 13.111. Simultaneous solution of Equations 13.111, 13.113, and 13.114 gives

$$\left(\frac{6}{h^2}\mathbf{M} + \frac{3}{h}\mathbf{C} + \mathbf{K}\right)\mathbf{u}_{n+1} = \mathbf{p}_{n+1} + \mathbf{M}\left(\frac{6}{h^2}\mathbf{u}_n + \frac{6}{h}\dot{\mathbf{u}}_n + 2\ddot{\mathbf{u}}_n\right)$$

$$+ \mathbf{C}\left(\frac{3}{h}\mathbf{u}_n + 2\dot{\mathbf{u}}_n + \frac{h}{2}\ddot{\mathbf{u}}_n\right) \tag{13.115}$$

Equation 13.115 is solved for $\mathbf{u}_{n+1}$ and the resulting value is substituted in Equations 13.113 and 13.114 to obtain $\ddot{\mathbf{u}}_{n+1}$ and $\dot{\mathbf{u}}_{n+1}$, respectively. The linear acceleration method is only conditionally stable and the condition of stability requires that the time step h be less than 0.55 times the highest mode period.

Example 13.11
Find the response of the system of Example 13.9 using the linear acceleration method and a time step h of (i) 0.1 s, and (ii) 0.5 s.

Solution
(i) The initial conditions are the same as in Examples 13.9 and 13.10. Equations 13.115, 13.113, and 13.114 are used in the iterations. For $h = 0.1$ s these equations reduce to the following

$$\left\{ 600 \begin{bmatrix} 2 & 0 \\ 0 & 1 \end{bmatrix} + \begin{bmatrix} 96 & -32 \\ -32 & 32 \end{bmatrix} \right\} \mathbf{u}_{n+1}$$

$$= \begin{bmatrix} 0 \\ 100 \end{bmatrix} + \begin{bmatrix} 2 & 0 \\ 0 & 1 \end{bmatrix} (600\mathbf{u}_n + 60\dot{\mathbf{u}}_n + 2\ddot{\mathbf{u}}_n) \tag{a}$$

$$\ddot{\mathbf{u}}_{n+1} = -2\ddot{\mathbf{u}}_n + 600(\mathbf{u}_{n+1} - \mathbf{u}_n - 0.1\dot{\mathbf{u}}_n) \tag{b}$$

$$\dot{\mathbf{u}}_{n+1} = -2\dot{\mathbf{u}}_n - 0.05\ddot{\mathbf{u}}_n + 30(\mathbf{u}_{n+1} - \mathbf{u}_n) \tag{c}$$

The response results obtained from Equations a, b, and c are presented in Table E13.11a for the first 2 s. Again, the results are reasonably close to the exact values shown in Table E13.9a.

Table E13.11a. Response of a two-degree-of-freedom system by the linear acceleration method, $h = 0.1$ s.

Time	u_1	u_2
0.1	0.012	0.475
0.2	0.109	1.763
0.3	0.430	3.510
0.4	1.081	5.286
0.5	2.027	6.750
0.6	3.060	7.732
0.7	3.867	8.233
0.8	4.165	8.330
0.9	3.837	8.081
1.0	2.993	7.465
1.1	1.913	6.428
1.2	0.926	4.973
1.3	0.257	3.255
1.4	0.426	1.593
1.5	0.076	0.398
1.6	0.004	0.014
1.7	0.107	0.577
1.8	0.268	1.950
1.9	0.610	3.771
2.0	1.232	5.595

Table E13.11b. Response of a two-degree-of-freedom system
obtained by the linear acceleration method,
$h = 0.5$ s.

Time	u_1	u_2
0.5	1.364	6.136
1.0	4.413	7.586
1.5	−1.213	3.913
2.0	2.949	−1.092
2.5	−1.772	13.304
3.0	11.516	−3.408
3.5	−17.178	17.259
4.0	31.867	−25.366
4.5	−51.161	63.480
5.0	99.850	−96.278
5.5	−179.780	181.040
6.0	328.640	−317.710
6.5	−587.120	596.500
7.0	1068.000	−1067.600
7.5	1934.200	1939.700
8.0	3510.200	−3497.600
8.5	6351.600	6356.100
9.0	11508.000	−11507.000
9.5	−20842.000	20857.000
10.0	37775.000	−37765.000

(ii) The stability criterion requires that h be less than 0.55 times the second mode period. The
limiting value of h is thus given by

$$h < 0.55 \frac{2\pi}{8}$$

$$= 0.432 \text{ s}$$

The selected value of h does not satisfy this requirement. The response obtained with $h = 0.5$ s
is therefore expected to be unstable. The response for the first 10 s shown in Table E13.11b. As
expected, the results are unbounded.

Wilson-θ method

The Wilson-θ method assumes that the accelerations vary linearly over an extended interval θh. The basic relationships used in the method can be obtained
by reference to their scalar counterparts (Eqs. 8.134, 8.135, and 8.136)

$$\ddot{\mathbf{u}}_{n+\theta} = -2\ddot{\mathbf{u}}_n + \frac{6}{(\theta h)^2} \{\mathbf{u}_{n+\theta} - \mathbf{u}_n - \theta h \dot{\mathbf{u}}_n\} \tag{13.116}$$

$$\dot{\mathbf{u}}_{n+\theta} = \frac{3}{\theta h}(\mathbf{u}_{n+\theta} - \mathbf{u}_n) - 2\dot{\mathbf{u}}_n - \frac{\theta h}{2}\ddot{\mathbf{u}}_n \tag{13.117}$$

$$\left(\frac{6}{(\theta h)^2}\mathbf{M} + \frac{3}{\theta h}\mathbf{C} + \mathbf{K}\right)\mathbf{u}_{n+\theta} = \mathbf{p}_{n+\theta} + \mathbf{M}\left(\frac{6}{(\theta h)^2}\mathbf{u}_n + \frac{6}{\theta h}\dot{\mathbf{u}}_n + 2\ddot{\mathbf{u}}_n\right)$$

$$+ \mathbf{C}\left(\frac{3}{\theta h}\mathbf{u}_n + 2\dot{\mathbf{u}}_n + \frac{\theta h}{2}\ddot{\mathbf{u}}_n\right) \tag{13.118}$$

As for the single-degree-of-freedom system, $\mathbf{p}_{n+\theta}$ is obtained by extrapolation from $\mathbf{p}_{n+1}$ and $\mathbf{p}_n$

$$\mathbf{p}_{n+\theta} = (1 - \theta)\mathbf{p}_n + \theta\mathbf{p}_{n+1} \tag{13.119}$$

Equation 13.118 is solved for $\mathbf{u}_{n+\theta}$. This value of $\mathbf{u}_{n+\theta}$ is substituted in Equation 13.116 to obtain $\ddot{\mathbf{u}}_{n+\theta}$. The acceleration at normal time increment h is then given by

$$\ddot{\mathbf{u}}_{n+1} = \ddot{\mathbf{u}}_n + \frac{1}{\theta}(\ddot{\mathbf{u}}_{n+\theta} - \ddot{\mathbf{u}}_n) \tag{13.120}$$

The acceleration vector $\ddot{\mathbf{u}}_{n+1}$ is now inserted in Equations 13.113 and 13.114 and those equations are solved for $\mathbf{u}_{n+1}$ and $\dot{\mathbf{u}}_{n+1}$, respectively. Using the results of Section 8.16.2, it can be stated that the Wilson-θ method will be unconditionally stable provided that θ is selected to be larger than 1.37.

Example 13.12
Find the response of the system of Example 13.9 using the Wilson-θ method and a time step of (i) 0.1 s, and (ii) 10.0 s. Select $\theta = 1.4$.

Solution
(i) Again, the initial conditions are as in Example 13.9. The displacement at time point $n + \theta$ is obtained from Equation 13.118. For $h = 0.1$ and $\theta = 1.4$, that equation reduces to

$$\left\{ 306.12 \begin{bmatrix} 2 & 0 \\ 0 & 1 \end{bmatrix} + \begin{bmatrix} 96 & -32 \\ -32 & 32 \end{bmatrix} \right\} \mathbf{u}_{n+\theta}$$

$$= \begin{bmatrix} 0 \\ 100 \end{bmatrix} + \begin{bmatrix} 2 & 0 \\ 0 & 1 \end{bmatrix} (306.12\mathbf{u}_n + 42.86\dot{\mathbf{u}}_n + 2\ddot{\mathbf{u}}_n) \tag{a}$$

Note that from Equation 13.119

$$\mathbf{p}_{n+\theta} = (1 - \theta)\begin{bmatrix} 0 \\ 100 \end{bmatrix} + \theta\begin{bmatrix} 0 \\ 100 \end{bmatrix} = \begin{bmatrix} 0 \\ 100 \end{bmatrix} \tag{b}$$

Equation a is solved for $u_{n+\theta}$. Substitution of $\mathbf{u}_{n+\theta}$ in Equation 13.116 gives $\ddot{\mathbf{u}}_{n+\theta}$

$$\ddot{\mathbf{u}}_{n+\theta} = -2\ddot{\mathbf{u}}_n + 306.12(\mathbf{u}_{n+\theta} - \mathbf{u}_n) - 42.86\dot{\mathbf{u}}_n \tag{c}$$

The accelerations at time point $n + 1$ are now obtained from Equation 13.120

$$\ddot{\mathbf{u}}_{n+1} = -2\ddot{\mathbf{u}}_n + \frac{1}{\theta}(\ddot{\mathbf{u}}_{n+\theta} - \ddot{\mathbf{u}}_n) \tag{d}$$

The displacements and velocities at t_{n+1} are calculated next from Equations 13.113 and 13.114, respectively

$$\mathbf{u}_{n+1} = 0.00167(\ddot{\mathbf{u}}_n + 2\ddot{\mathbf{u}}_n) + \mathbf{u}_n + 0.1\dot{\mathbf{u}}_n \tag{e}$$

$$\dot{\mathbf{u}}_{n+1} = -2\dot{\mathbf{u}}_n - 0.05\ddot{\mathbf{u}}_n + 30(\mathbf{u}_{n+1} - \mathbf{u}_n) \tag{f}$$

Table E13.12a. Response of a two-degree-of-freedom system obtained by the Wilson-θ method, $h = 0.1$ s.

Time	u_1	u_2
0.1	0.014	0.468
0.2	0.124	1.715
0.3	0.446	3.409
0.4	1.057	5.166
0.5	1.922	6.664
0.6	2.876	7.717
0.7	3.670	8.273
0.8	4.060	8.377
0.9	3.915	8.089
1.0	3.265	7.448
1.1	2.293	6.465
1.2	1.263	5.173
1.3	0.417	3.676
1.4	−0.102	2.183
1.5	−0.277	0.978
1.6	−0.190	0.343
1.7	0.052	0.468
1.8	0.374	1.368
1.9	0.764	2.870
2.0	1.233	4.657

Table E13.12b. Response of a two-degree-of-freedom system obtained by the Wilson-θ method, $h = 10.0$ s.

Time	u_1	u_2
10.0	1.704	1434.000
20.0	4.404	−1061.000
30.0	−4.085	857.300
40.0	9.158	−656.800
50.0	−7.004	514.500
60.0	10.330	−306.700
70.0	−6.892	304.700
80.0	9.389	−225.200
90.0	−5.478	180.700
100.0	7.760	−130.100
500.0	1.563	4.683
510.0	1.561	4.688
520.0	1.563	4.684
530.0	1.562	4.688
540.0	1.562	4.685
550.0	1.562	4.685

This completes one cycle of iteration. The displacement responses obtained as above are presented in Table E13.12a for the first 2 s. The results are reasonably close to the exact values shown in Table E13.9a.

(ii) The procedure is similar to that described in part (i). The response from 0 to 100 s and from 500 to 550 s at intervals of 10 s is shown in Table E13.12b. The response is stable and approaches the static response with the passage of time.

It is of interest to note that the displacement response for the first few time steps is very high. This is the result of using a very large time step in the integration and the presence of an initial acceleration. In fact, whenever the time step used in the integration is large in comparison to a certain mode period, the Wilson-θ method will spuriously amplify the displacement response contribution from that mode during the first few time steps. In practice, this is not likely to introduce serious errors in the total response, because the integration step will be large only in comparison to a higher mode period, and even the amplified response from such a mode will be small in comparison to the total response. However, in certain cases, this spurious amplification may be viewed as a disadvantage of the Wilson-θ method.

Houbolt's method

The expressions for $\dot{\mathbf{u}}_{n+1}$ and $\ddot{\mathbf{u}}_{n+1}$ used in this method were derived in Section 8.14.2 for a single-degree-of-freedom system. For a multi-degree-of-freedom system, the parallel equations are

$$\dot{\mathbf{u}}_{n+1} = \frac{1}{6h}(-2\mathbf{u}_{n-2} + 9\mathbf{u}_{n-1} - 18\mathbf{u}_n + 11\mathbf{u}_{n+1}) \tag{13.121}$$

$$\ddot{\mathbf{u}}_{n+1} = \frac{1}{h^2}(-\mathbf{u}_{n-2} + 4\mathbf{u}_{n-1} - 5\mathbf{u}_n + 2\mathbf{u}_{n+1}) \tag{13.122}$$

The third relationship used in Houbolt's method is the equation of motion written at time point $n + 1$, namely, Equation 13.111. Substitution for $\dot{\mathbf{u}}_{n+1}$ and $\ddot{\mathbf{u}}_{n+1}$ from Equations 13.121 and 13.122, respectively, in Equation 13.111 gives

$$\left(\frac{2}{h^2}\mathbf{M} + \frac{11}{6h}\mathbf{C} + \mathbf{K}\right)\mathbf{u}_{n+1} = \mathbf{p}_{n+1} + \mathbf{M}\left(\frac{5}{h^2}\mathbf{u}_n - \frac{4}{h^2}\mathbf{u}_{n-1} + \frac{1}{h^2}\mathbf{u}_{n-2}\right)$$

$$+ \mathbf{C}\left(\frac{3}{h}\mathbf{u}_n - \frac{3}{2h}\mathbf{u}_{n-1} + \frac{1}{3h}\mathbf{u}_{n-2}\right) \quad (13.123)$$

Before starting the integration with Houbolt's method, $\mathbf{u}_0$, $\mathbf{u}_1$, and $\mathbf{u}_2$ must be determined. Displacement $\mathbf{u}_0$ will be specified, while $\mathbf{u}_1$ and $\mathbf{u}_2$ can be calculated by one of the other methods described here, for example, the central difference method or the average acceleration method. As demonstrated in Section 8.16.4, Houbolt's method is unconditionally stable.

Example 13.13
Find the response of the system of Example 13.9 using Houbolt's method and a time step of (i) 0.1 s, and (ii) 10.0 s.

Solution
(i) In addition to the displacements at zero time, we need the displacements at $t = 0.1$ and $t = 0.2$ s to start the iterations. We select the values obtained from the average acceleration method. The displacement at the next time step is now obtained from Equation 13.123. For $h = 0.1$ s and the

Table E13.13a. Response of a two-degree-of-freedom system
obtained by Houbolt's method, $h = 0.1$ s.

Time	u_1	u_2
0.1	0.017	0.464
0.2	0.121	1.728
0.3	0.437	3.415
0.4	1.025	5.153
0.5	1.840	6.648
0.6	2.730	7.722
0.7	3.487	8.313
0.8	3.912	8.439
0.9	3.891	8.151
1.0	3.424	7.503
1.1	2.628	6.547
1.2	1.690	5.352
1.3	0.808	4.024
1.4	0.139	2.715
1.5	−0.237	1.620
1.6	−0.317	0.932
1.7	−0.149	0.800
1.8	0.197	1.265
1.9	0.657	2.278
2.0	1.182	3.665

given values of **M** and **K**, that equation reduces to

$$\left\{ 200 \begin{bmatrix} 2 & 0 \\ 0 & 1 \end{bmatrix} + \begin{bmatrix} 96 & -32 \\ -32 & 32 \end{bmatrix} \right\} \mathbf{u}_{n+1} = \begin{bmatrix} 0 \\ 100 \end{bmatrix}$$

$$+ \begin{bmatrix} 2 & 0 \\ 0 & 1 \end{bmatrix} (500\mathbf{u}_n - 400\mathbf{u}_{n-1} + 100\mathbf{u}_{n-2}) \tag{a}$$

If we are also interested in the velocity and acceleration time histories, these can be obtained from Equations 13.121 and 13.122 respectively. However if the velocities and accelerations are of no interest, the computations involved in evaluating Equations 13.121 and 13.122 can be avoided. The response results for the first 2 s are shown in Table E13.13a.

(ii) Displacement values are, in this case, needed for $t = 0$, 10, and 20 s before iterations can be begun. Displacements at $t = 0$ are given as being equal to zero; for the other two time steps, we use the values calculated from the average acceleration method. In fact, the results are quite insensitive to the starting values used.

The iteration equation (Eq. 13.123) reduces to

$$\left\{ 0.02 \begin{bmatrix} 2 & 0 \\ 0 & 1 \end{bmatrix} + \begin{bmatrix} 96 & -32 \\ -32 & 32 \end{bmatrix} \right\} \mathbf{u}_{n+1}$$

$$= \begin{bmatrix} 0 \\ 100 \end{bmatrix} + \begin{bmatrix} 2 & 0 \\ 0 & 1 \end{bmatrix} (0.05\mathbf{u}_n - 0.04\mathbf{u}_{n-1} + 0.01\mathbf{u}_{n-2}) \tag{b}$$

It is evident that Equation b is very nearly equal to

$$\mathbf{K}\mathbf{u}_{n+1} = \mathbf{p}_{n+1} \tag{c}$$

Table E13.13b. Response of a two-degree-of-freedom system obtained by Houbolt's method, $h = 10.0$ s.

Time	u_1	u_2
10.0	3.115	9.354
20.0	0.039	0.086
30.0	1.550	4.661
40.0	1.568	4.701
50.0	1.561	4.685
60.0	1.562	4.687
70.0	1.563	4.688
80.0	1.562	4.687
90.0	1.562	4.688
100.0	1.563	4.688

In fact, as h becomes larger, the iteration equation becomes closer to Equation c, and the displacement response obtained from the solution of the iteration is very nearly equal to the static response. This is seen to be the case from the results presented in Table E13.13b.

13.4.3 *Mixed methods in direct integration*

It is evident from the discussion in previous sections that explicit integration schemes are computationally more efficient, particularly when the mass matrix is diagonal and the damping is either negligible or the damping matrix is mass proportional. In such a situation, the integration scheme does not require solution of simultaneous equations. In contrast, implicit schemes invariably rely on the solution of simultaneous equations. For a large system, the computational effort involved in such a solution can be very substantial. However, despite the computational efficiency, explicit schemes cannot always be effectively utilized in the time integration of the equations of motion. This is because such schemes are only conditionally stable, and to maintain numerical stability, the time step chosen in the integration has to be smaller than a certain fraction of the highest mode period. For large systems with thousands of degrees of freedom, the highest period is usually so small that integration with a time step selected to maintain stability becomes impractical. It is apparent that for such a system, integration must be performed with an unconditionally-stable implicit integration method.

In many engineering problems related to the evaluation of dynamic response, the system is composed of two different types of components or substructures: a stiff component and a comparatively flexible component. Examples of such a system include soil-structure and fluid-structure interaction problems, where the soil or the fluid medium is very flexible compared to the structure. It is apparent that the highest mode period of the flexible medium will be significantly larger than the highest mode period of the structure. However, if an explicit scheme is to be used in the integration of the coupled system, the stability criteria and

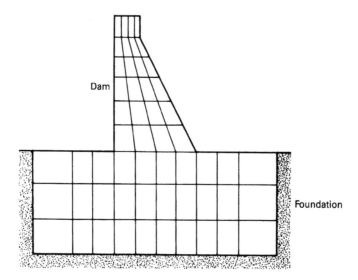

Figure 13.1. Finite element modeling of dam-foundation structure.

hence the time step to be chosen in the integration will be strongly influenced by the highest mode period of the stiff structure, making the explicit scheme impractical.

The foregoing discussion would suggest that it may be possible to improve the computational efficiency of the integration by using two different integration schemes for the two components, one for the fluid or the soil medium and another for the structure. As an example, consider the system shown in Figure 13.1, which consists of a comparatively stiff gravity dam resting on a flexible foundation. For obtaining the dynamic response of the system to an exciting force, the coupled structure is modeled as a two-dimensional system, the soil medium is assumed to be of finite extent, and both the structure and the soil are discretized by using finite elements.

Since the soil medium is flexible, its highest mode period may not be too small and it may be quite practicable to use a conditionally-stable explicit scheme for its integration. The time step chosen for such integration will presumably have to satisfy the stability criterion related to the highest mode period of the soil. The same time step can be used for the integration of the dam substructure. However, an unconditionally-stable implicit scheme must be used in the integration to maintain numerical stability.

The example cited above illustrates one possible way in which two different schemes can be combined. Several other combinations are possible. However, experience has shown that the following two combinations are the most effective:

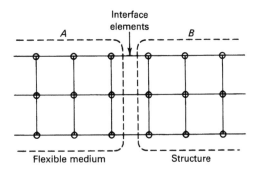

Figure 13.2. Partitioned finite element representation of medium structure system.

1. $E - I$: explicit in the flexible medium, implicit in the stiff structure, same time step used in both.
2. $E - E^m$: explicit in both the medium and the structure; time step in the flexible medium h; that in the stiff structure h/m.

In the following paragraphs, we describe briefly each of the two schemes.

Explicit–implicit integration scheme

Let the discretized structure–medium system be represented by Figure 13.2. Let nodes designated by A belong to the medium and those designated by B to the structure. At the interface, several nodes in A are connected to several nodes in B by means of interface elements. For the purpose of this discussion, it is immaterial whether these interface elements belong to the structure or the medium. The coupled equations for the system can be expressed as

$$\begin{bmatrix} \mathbf{M}_{AA} & 0 \\ 0 & \mathbf{M}_{BB} \end{bmatrix} \begin{bmatrix} \ddot{\mathbf{u}}_A \\ \ddot{\mathbf{u}}_B \end{bmatrix} + \begin{bmatrix} \mathbf{C}_{AA} & 0 \\ 0 & \mathbf{C}_{BB} \end{bmatrix} \begin{bmatrix} \dot{\mathbf{u}}_A \\ \dot{\mathbf{u}}_B \end{bmatrix}$$

$$+ \begin{bmatrix} \mathbf{K}_{AA} & \mathbf{K}_{AB} \\ \mathbf{K}_{BA} & \mathbf{K}_{BB} \end{bmatrix} \begin{bmatrix} \mathbf{u}_A \\ \mathbf{u}_B \end{bmatrix} = \begin{bmatrix} \mathbf{p}_A \\ \mathbf{p}_B \end{bmatrix} \tag{13.124}$$

We now expand Equations 13.124 and write the first one of them at time nh and the second at time $(n + 1)h$.

$$\mathbf{M}_{AA}\ddot{\mathbf{u}}_A^n + \mathbf{C}_{AA}\dot{\mathbf{u}}_A^n + \mathbf{K}_{AA}\mathbf{u}_A^n + \mathbf{K}_{AB}\mathbf{u}_B^n = \mathbf{p}_A^n \tag{13.125a}$$

$$\mathbf{M}_{BB}\ddot{\mathbf{u}}_B^{n+1} + \mathbf{C}_{BB}\dot{\mathbf{u}}_B^{n+1} + \mathbf{K}_{BA}\mathbf{u}_A^{n+1} + \mathbf{K}_{BB}\mathbf{u}_B^{n+1} = \mathbf{p}_B^{n+1} \tag{13.125b}$$

where the superscript denotes the time steps elapsed.

Let us now assume that a central difference scheme with time step h is used for the integration of partition A, while partition B is integrated by the unconditionally-stable average acceleration method but with the same time step *h.* Substitution in Equation 13.125a of central difference operators for $\dot{\mathbf{u}}_A$ and $\ddot{\mathbf{u}}_A$ given by Equations 13.104 and 13.105 respectively leads to

$$\left(\frac{1}{h^2}\mathbf{M}_{AA} + \frac{1}{2h}\mathbf{C}_{AA}\right)\mathbf{u}_A^{n+1} = \mathbf{p}_A^n + \left(\frac{2}{h^2}\mathbf{M}_{AA} - \mathbf{K}_{AA}\right)\mathbf{u}_A^n$$

$$- \mathbf{K}_{AB}\mathbf{u}_B^n + \left(-\frac{1}{h^2}\mathbf{M}_{AA} + \frac{1}{2h}\mathbf{C}_{AA}\right)\mathbf{u}_A^{n-1}$$

$$(13.126)$$

Since the right-hand side of Equation 13.126 contains only the historic values of $\mathbf{u}_A$ and $\mathbf{u}_B$, the equation can easily be solved for $\mathbf{u}_A^{n+1}$. In fact, for diagonal mass and damping matrices, Equation 13.126 are uncoupled and their solution is trivial. Substitution of the acceleration and velocity expressions of Equations 13.109 and 13.110 in Equation 13.125b gives

$$\left(\frac{4}{h^2}\mathbf{M}_{BB} + \frac{2}{h}\mathbf{C}_{BB} + \mathbf{K}_{BB}\right)\mathbf{u}_B^{n+1} = \mathbf{p}_B^{n+1} + \mathbf{M}_{BB}\left(\frac{4}{h^2}\mathbf{u}_B^n + \frac{4}{h}\dot{\mathbf{u}}_B^n + \ddot{\mathbf{u}}_B^n\right)$$

$$+ \mathbf{C}_{BB}\left(\frac{2}{h}\mathbf{u}_B^n + \dot{\mathbf{u}}_B^n\right) - \mathbf{K}_{BA}\mathbf{u}_A^{n+1}$$

$$(13.127)$$

In addition to the historic values of $\mathbf{u}_B$, $\dot{\mathbf{u}}_B$ and $\ddot{\mathbf{u}}_B$, the right-hand side of Equation 13.127 contains displacement $\mathbf{u}_A^{n+1}$. However, since the explicit partition has already been solved for time $(n + 1)h$ through Equation 13.126, $\mathbf{u}_A^{n+1}$ is available. Thus, Equation 13.127 can be solved for $\mathbf{u}_B^{n+1}$. The application of the mixed explicit–implicit scheme is, therefore, quite straightforward as long as the integration of the explicit partition precedes that of the implicit partition.

It can be shown that the stability criterion on the time step is governed by the explicit partition. Thus, in the example just cited, h should be less than T/π, where T is the highest mode period of the explicit partition. In obtaining this period, the implicit partition nodes connected to the interface elements must be taken as fully constrained.

Example 13.14

Find the response of the system of Example 13.9 using the central difference method to integrate the equation corresponding to the degree of freedom 1 and the average acceleration method to integrate the equation corresponding to the degree of freedom 2. Use a time step of 0.1 s.

Solution

Referring to Example 13.9, we have

$$M_{AA} = 2 \quad M_{BB} = 1$$

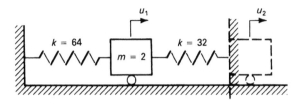

Figure E13.14. Explicit partition of a two-degree-of-freedom system.

$$K_{AA} = 96 \quad K_{BB} = 32 \quad K_{AB} = K_{BA} = -32$$

$$p_A(t) = 0 \quad p_B(t) = 100$$

The initial conditions are

$$u_A^0 \equiv u_1^0 = 0$$

$$u_B^0 \equiv u_2^0 = 0$$

$$\dot{u}_A^0 \equiv \dot{u}_1^0 = 0$$

$$\dot{u}_B^0 \equiv \dot{u}_2^0 = 0$$

$$\ddot{u}_A^0 \equiv \ddot{u}_1^0 = 0$$

$$\ddot{u}_B^0 \equiv \ddot{u}_2^0 = 100$$

To start the iterations, we will need u_A^{-1}. This is readily obtained from Equation 13.108 and is found to be equal to 0. The two equations required in the iterative solution are obtained from Equations 13.126 and 13.127

$$\frac{2}{h^2} u_1^{n+1} = \left(\frac{4}{h^2} - 96 \right) u_1^n + 32 u_2^n - \frac{2}{h^2} u_1^{n-1} \tag{a}$$

$$\left(\frac{4}{h^2} + 32 \right) u_2^{n+1} = 100 + \left(\frac{4}{h^2} u_2^n + \frac{4}{h} \dot{u}_2^n + \ddot{u}_2^n \right) + 32 u_1^{n+1} \tag{b}$$

For $h = 0.1$ s, Equations a and b reduce to

$$200 u_1^{n+1} = 304 u_1^n + 32 u_2^n - 200 u_1^{n-1} \tag{c}$$

$$432 u_2^{n+1} = 100 + 400 u_2^n + 40 \dot{u}_2^n + \ddot{u}_2^n + 32 u_1^{n+1} \tag{d}$$

For each time step, Equation c is solved first for u_1^{n+1}. Equation d is solved next for u_2^{n+1}. The response obtained as above is shown in Table E13.14 for the first 2 s.

The explicit partition used in this problem is drawn in Figure E13.14. The interface element connected to the implicit node is assumed to be fixed at that node. In other words, degree-of-freedom 2 is fully constrained. The resulting single-degree-of-freedom system of Figure E13.14 has a period $T = 2\pi \sqrt{2/96} = \pi/(2\sqrt{3})$ s. The stability criterion is therefore given by

$$h \leqslant \frac{T}{\pi} = 0.288 \text{ s}$$

Table E13.14.   Response of two-degree-of-freedom system obtained by mixed explicit–implicit integration method, $h = 0.1$ s.

Time	u_1	u_2
0.1	0.000	0.463
0.2	0.074	1.720
0.3	0.388	3.433
0.4	1.065	5.197
0.5	2.062	6.685
0.6	3.139	7.737
0.7	3.948	8.331
0.8	4.194	8.511
0.9	3.789	8.289
1.0	2.892	7.624
1.1	1.826	6.470
1.2	0.919	4.877
1.3	0.352	3.063
1.4	0.105	1.395
1.5	0.031	0.284
1.6	−0.012	0.026
1.7	−0.046	0.683
1.8	0.052	2.060
1.9	0.454	3.791
2.0	1.245	5.488

This may be compared with the limit of 0.25 s for explict integration of the entire system.

Explicit–Explicit integration scheme ($E - E^m$ method)

In this method, explicit integration is used for both components of the system. However, the time step used in the integration of the flexible component is several times larger than that used in the integration of the stiffer component. Let it be assumed that the central difference method is used for the integration of both components. The flexible component is integrated with a time step h, while the time step used in the integration of the stiff component is h/m.

The flexible partition A is integrated first. The expression used in its integration is given by Equations 13.126. For the stiff partition, the displacement at time $(n + p/m)h$ is obtained by substituting h/m for h in Equations 13.126 and writing that equation for partition B.

$$\left(\frac{m^2}{h^2} \mathbf{M}_{BB} + \frac{m}{2h} \mathbf{C}_{BB} \right) \mathbf{u}_B^{n + p/m} = \mathbf{p}_B^{n+(p-1)/m} + \left(\frac{2m^2}{h^2} \mathbf{M}_{BB} - \mathbf{K}_{BB} \right)$$

$$\mathbf{u}_B^{n+(p-1)/m} - \mathbf{K}_{BA} \mathbf{u}_A^{n+(p-1)/m}$$

$$+ \left(-\frac{m^2}{h^2} \mathbf{M}_{BB} + \frac{m}{2h} \mathbf{C}_{BB} \right) \mathbf{u}_B^{n+(p-2)/m}$$

$$(13.128)$$

In solving Equation 13.128 for $\mathbf{u}_B^{n+p/m}$, we need to know $\mathbf{u}_A^{n+(p-1)/m}$. However, vector $\mathbf{u}_A$ has been determined only at time points nh and $(n+1)h$. A linear interpolation is used to calculate $\mathbf{u}_A^{n+(p-1)/m}$ from known values of $\mathbf{u}_A^n$ and $\mathbf{u}_A^{n+1}$. Thus

$$\mathbf{u}_A^{n+(p-1)/m} = \mathbf{u}_A^n + (\mathbf{u}_A^{n+1} - \mathbf{u}_A^n)(p-1)/m \tag{13.129}$$

A theoretical value of the limit on time-step required to maintain stability with the mixed explicit–explicit $(E - E^m)$ method has not been derived. However, on the basis of studies carried out on simple systems using central difference operators, it has been found that the time step with the $E - E^m$ method can be substantially longer than that required for a straight application of central difference method to the entire system.

Example 13.15
Find the response of the system of Example 13.9 using the central difference method. In the integration of the equation corresponding to degree-of-freedom 1, use a time step h while for the equation corresponding to degree-of-freedom 2 use a time step $h/2$. Select $h = 0.1$ s.

Solution
As in Example 13.14, the iteration equation corresponding to degree-of-freedom 1 is

$$200u_1^{n+1} = 304u_1^n + 32u_2^n - 200u_1^{n-1}$$

or

$$u_1^{n+1} = 1.52u_1^n + 0.16u_2^n - u_1^{n-1} \tag{a}$$

The equation for degree-of-freedom 2 is obtained from Equation 13.128 with $m = 2$.

$$400u_2^{n+p/2} = 100 + (800 - 32)u_2^{n+(p-1)/2} + 32u_1^{n+(p-1)/2} - 400u_2^{n+(p-2)/2}$$

or

$$u_2^{n+p/2} = 0.25 + 1.92u_2^{n+(p-1)/2} + 0.08u_1^{n+(p-1)/2} - u_2^{n+(p-2)/2} \tag{b}$$

The initial conditions are $u_1^0 = 0, \dot{u}_1^0 = 0, \ddot{u}_1^0 = 0, u_2^0 = 0, \dot{u}_2^0 = 0, \ddot{u}_2^0 = 100$. To start the iteration for degree-of-freedom 1, we need u_1^{-1}. This is obtained from Equation 13.108

$$u_1^{-1} = u_1^0 - h\dot{u}_1^0 + \frac{h^2}{2}\ddot{u}_1^0$$
$$= 0 \tag{c}$$

The first step in the iteration corresponding to $n = 0$ gives

$$u_1^1 = 1.52u_1^0 + 0.16u_2^0 - u_1^{-1}$$
$$= 0 \tag{d}$$

To start the iteration for degree-of-freedom 2, we need $u_2^{-1/2}$. This is obtained from an equation similar to Equation c but with h replaced by $h/m = h/2$

$$u_2^{-1/2} = u_2^0 - \frac{h}{2}\dot{u}_2^0 + \frac{h^2}{8}\ddot{u}_2^0$$

$$= \frac{0.1^2}{8} \times 100 = 0.125$$

(e)

The first substep in the iteration, obtained by substituting $n = 0$ and $p = 1$ in Equation b, is

$$u_2^{1/2} = 0.25 + 1.92u_2^0 + 0.08u_1^0 - u_2^{-1/2}$$

$$= 0.25 - 0.125 = 0.125$$

(f)

Before we carry out the second substep of iteration with $n = 0$ and $p = 2$, we will need $u_1^{1/2}$. This is obtained by linear interpolation between u_1^0 and u_1^1

$$u_1^{1/2} = 0.5(u_1^0 + u_1^1)$$

$$= 0$$

(g)

We now write Equation b with $n = 0$ and $p = 2$

$$u_2^1 = 0.25 + 1.92u_2^{1/2} + 0.08u_1^{1/2} - u_2^0$$

$$= 0.25 + 1.92 \times 0.125 + 0.08 \times 0$$

$$= 0.49$$

(h)

The second step in the iteration for u_1 is

$$u_1^2 = 1.52u_1^1 + 0.16u_2^1 - u_1^0$$

$$= 1.52 \times 0 + 0.16 \times 0.49$$

$$= 0.078$$

(i)

Also

$$u_1^{3/2} = 0.5(u_1^1 + u_1^2)$$

$$= 0.039$$

(j)

We can now carry out two substeps of iteration for u_2.
1. $n = 1$, $p = 1$:

$$u_2^{3/2} = 0.25 + 1.92u_2^1 + 0.08u_1^1 - u_2^{1/2}$$

$$= 0.25 + 1.92 \times 0.49 + 0.08 \times 0 - 0.125 = 1.0658$$

(k)

2. $n = 1$, $p = 2$:

$$u_2^2 = 0.25 + 1.92u_2^{3/2} + 0.08u_1^{3/2} - u_2^1$$

$$= 0.25 + 1.92 \times 1.0658 + 0.08 \times 0.039 - 0.49 = 1.809$$

(l)

The response obtained by the procedure outlined above is shown in Table E13.15 for the first 2 s at intervals of 0.1 s.

Table E13.15. Response of a two-degree-of-freedom system obtained by mixed expplicit–implicit integration method, $h = 0.1$ s.

Time	u_1	u_2
0.1	0.000	0.490
0.2	0.078	1.809
0.3	0.409	3.576
0.4	1.115	5.344
0.5	2.141	6.779
0.6	3.224	7.742
0.7	3.999	8.255
0.8	4.174	8.390
0.9	3.689	8.156
1.0	2.738	7.483
1.1	1.670	6.296
1.2	0.808	4.647
1.3	0.307	2.789
1.4	0.010	1.142
1.5	0.003	0.153
1.6	−0.029	0.105
1.7	−0.056	0.997
1.8	0.104	2.545
1.9	0.621	4.323
2.0	1.532	5.934

13.5 ANALYSIS IN THE FREQUENCY DOMAIN

As for a single-degree-of-freedom system, analysis in the frequency domain may often prove to be an effective means of evaluating the response of a linear multi-degree-of-freedom system. The efficiency of frequency-domain analysis is most apparent when a numerical method of analysis must be used because the excitation is specified in the form of numerical values at regular intervals of time rather than as a mathematical function. In such a situation, frequency-domain analysis relies on the use of discrete Fourier transforms which can be evaluated very efficiently by the fast Fourier transform technique.

Application of discrete Fourier transform analysis gives the response of the system to a periodic extension of the applied load and the calculated response is not therefore, in general, equal to the true transient response. One way in which the response obtained from a discrete transform analysis can be made to be equal to the true response is by adding a grace band of zeros of appropriate duration to both the forcing function and the unit impulse function. This technique, called the fast convolution method, was described in Chapter 9 for a single-degree-of-freedom system. An alternative method of obtaining the true response is by the superposition of appropriate corrective responses on the result of a discrete Fourier analysis. Procedures for finding the corrective responses were also discussed in Chapter 9.

If the multi-degree-of-freedom system under consideration possesses classical mode shapes, the frequency domain analysis methods developed for single-degree-of-freedom systems apply without modification. When classical mode shapes do not exist, the application of frequency analysis becomes more complicated. In the following paragraphs, we discuss the analysis procedures for the two types of systems separately.

13.5.1 *Frequency analysis of systems with classical mode shapes*

The governing equations for a system with proportional damping can be transformed into a set of uncoupled single-degree-of-freedom equations by a normal coordinate transformation as described in Section 12.2. For ease of reference, the relevant steps are summarized below.

The normal coordinate transformation is given by

$$u = \sum_{n=1}^{M} \phi_n y_n \tag{13.130}$$

where only M of the N mode shapes have been used, and M may be significantly smaller than N. The nth equation in the set of transformed equation is given by

$$\ddot{y}_n + 2\xi_n \omega_n \dot{y}_n + \omega_n^2 y_n = p_n \tag{13.131}$$

along with the initial conditions

$$\begin{aligned} y_{0n} &= \phi_n^T \mathbf{M} \mathbf{u}_0 \\ \dot{y}_{0n} &= \phi_n^T \mathbf{M} \mathbf{v}_0 \end{aligned} \tag{13.132}$$

Equations 13.131 and 13.132 are the governing equations of a single-degree-of-freedom system. They can be solved in the frequency domain by any one of the methods described in Chapter 9; for example, the fast convolution method, one of the corrective response superposition methods, or the exponential window method. Once the normal coordinate response has been obtained, response in the physical coordinates can be computed by using the transformation equation (Eq. 13.130). The steps involved in the analysis are illustrated by the following simple example, which uses the superposition of a corrective response to obtain the true transient response.

Example 13.16
The two-degree-of-freedom system of Figure E13.16a is subjected to an applied force given by

$$\mathbf{p} = \begin{bmatrix} 1 \\ \frac{1}{2} \end{bmatrix} g(t)$$

where $g(t)$ is the rectangular pulse function shown in Figure E13.16b. Obtain the response of the system by using a modal analysis through frequency domain.

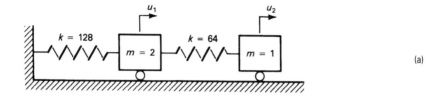

(a)

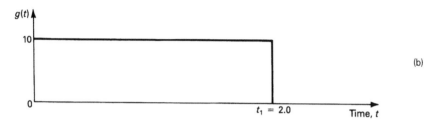

(b)

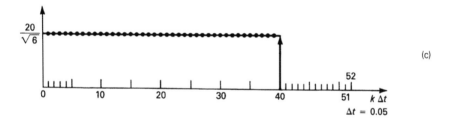

(c)

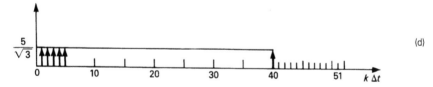

(d)

Figure E13.16. (a) Two-degree-of-freedom system; (b) forcing function; (c) modal force in mode 1; (d) modal force in mode 2.

Solution

The stiffness and mass matrices of the system are given by

$$\mathbf{K} = \begin{bmatrix} 192 & -64 \\ -64 & 64 \end{bmatrix}$$

$$\mathbf{M} = \begin{bmatrix} 2 & 0 \\ 0 & 1 \end{bmatrix}$$

The frequencies and mode shapes of the system can be obtained by any of the standard methods discussed in Chapter 11. The calculated frequencies are

$$\omega_1 = 4\sqrt{2}$$

$$\omega_2 = 8\sqrt{2}$$

The corresponding mass orthonormal mode shapes are

$$\boldsymbol{\phi}_1 = \frac{1}{\sqrt{6}} \begin{bmatrix} 1 \\ 2 \end{bmatrix}$$

$$\boldsymbol{\phi}_2 = \frac{1}{\sqrt{3}} \begin{bmatrix} 1 \\ -1 \end{bmatrix}$$

The modal forces are given by

$$p_1 = \boldsymbol{\phi}_1^T g(t) \begin{bmatrix} 1 \\ \frac{1}{2} \end{bmatrix} = \frac{2}{\sqrt{6}} g(t)$$

$$p_2 = \boldsymbol{\phi}_2^T g(t) \begin{bmatrix} 1 \\ \frac{1}{2} \end{bmatrix} = \frac{1}{2\sqrt{3}} g(t)$$

The modal equations are

$$\ddot{y}_1 + 32 y_1 = \frac{2}{\sqrt{6}} g(t)$$

$$\ddot{y}_2 + 128 y_2 = \frac{1}{2\sqrt{3}} g(t)$$

For $t < t_1$, t_1 being the duration of the rectangular pulse, the solutions of the two modal equations are given by

$$y_1 = \frac{p_0}{16\sqrt{6}} (1 - \cos 4\sqrt{2}t)$$

$$y_2 = \frac{p_0}{256\sqrt{3}} (1 - \cos 8\sqrt{2}t)$$

where $p_0 = 10$. The corresponding response in the physical coordinates is obtained from

$$\begin{bmatrix} u_1 \\ u_2 \end{bmatrix} = \boldsymbol{\phi}_1 y_1 + \boldsymbol{\phi}_2 y_2$$

$$= \frac{p_0}{768} \begin{bmatrix} 9 - 8\cos 4\sqrt{2}t - \cos 8\sqrt{2}t \\ 15 - 16\cos 4\sqrt{2}t + \cos 8\sqrt{2}t \end{bmatrix}$$

For the free vibration era, $t > t_1$, the modal response can be obtained by finding the modal displacements and velocities at $t = t_1 = 2$ s. Thus

$$y_1(t_1) = \frac{p_0}{16\sqrt{6}} (1 - \cos 8\sqrt{2}) = 0.1753$$

$$\dot{y}_1(t_1) = \frac{p_0}{4\sqrt{3}} \sin 8\sqrt{2} = -1.371$$

and

$$y_2(t_1) = \frac{p_0}{256\sqrt{3}}(1 - \cos 16\sqrt{2}) = 0.0407$$

$$\dot{y}_2(t_1) = \frac{\sqrt{2}\,p_0}{32\sqrt{3}} \sin 16\sqrt{2} = -0.1516$$

The free vibration response is given by

$$y_1 = 0.1753 \cos 4\sqrt{2}(t - 2) - \frac{1.3710}{4\sqrt{2}} \sin 4\sqrt{2}(t - 2)$$

$$y_2 = 0.0407 \cos 8\sqrt{2}(t - 2) - \frac{0.1516}{8\sqrt{2}} \sin 8\sqrt{2}(t - 2)$$

Response in the physical coordinates is obtained by superposing the modal responses

$$\begin{bmatrix} u_1 \\ u_2 \end{bmatrix} = \begin{bmatrix} \frac{1}{\sqrt{6}} \\ \frac{2}{\sqrt{6}} \end{bmatrix} y_1 + \begin{bmatrix} \frac{1}{\sqrt{3}} \\ \frac{-1}{\sqrt{3}} \end{bmatrix} y_2$$

The exact response of the system obtained as above for the entire range of t will be used for comparison with the frequency response obtained later.

The fundamentals mode period of the system is given by

$$T_1 = \frac{2\pi}{\omega_1} = 1.11 \text{ s}$$

The maximum response of the system will occur either within the duration of the rectangular pulse or in the free-vibration era after the force pulse has ceased to act. The maximum in the free-vibration era will be attained within half-a-cycle of motion at a frequency of ω_1. We therefore select a total time duration of 2.6 s to calculate the response; that time also becomes T_0, the period of the forcing function.

The augmented forcing function for the first mode is shown in Figure E13.16c. For analysis in the frequency domain, we first obtain the true periodic response to a periodic extension of the forcing function. This requires the evaluation of periodic impulse function given by

$$\bar{h}_1(t) = \frac{1}{2M_1\omega_1}\left(\sin \omega_1 t + \frac{\cos \omega_1 t \sin \omega_1 T_0}{1 - \cos \omega_1 T_0}\right)$$

where M_1 is the modal mass of the first mode. Since the mode shape have been mass orthonormalized, $M_1 = 1$.

The response in the first mode to a periodic extension of the forcing function is obtained by convolving the sampled forcing function of Figure E13.16c and $\bar{h}_1(t)$. The sampling interval is chosen as 0.05 s. The response as obtained above is not equal to the transient response and must be corrected for the initial conditions. The displacement at $t = 0$ in the periodic response is found to be 0.0213. The velocity at $t = 0$ is calculated from Equation 9.85 and works out to 1.001. The corrective responses are obtained from Equations 9.89a and 9.89b with $\Delta u(0) = -0.0213$

Table E13.16a. Modal frequency-domain analysis of two-degrees-of-freedom system.

t	Mode 1		Mode 2	
	Frequency-domain response	Exact response	Frequency-domain response	Exact response
0.0	0.0000	0.0000	0.0000	0.0000
0.1	0.0347	0.0397	0.0112	0.0089
0.2	0.1375	0.1466	0.0347	0.0306
0.3	0.2765	0.2873	0.0436	0.0403
0.4	0.4083	0.4180	0.0276	0.0267
0.5	0.4918	0.4979	0.0051	0.0055
0.6	0.5011	0.5022	0.0021	0.0011
0.7	0.4332	0.4296	0.0218	0.0185
0.8	0.3094	0.3026	0.0418	0.0373
0.9	0.1681	0.1608	0.0390	0.0367
1.0	0.0534	0.0484	0.0166	0.0166
1.1	0.0010	0.0005	0.0004	0.0005
1.2	0.0273	0.0318	0.0089	0.0069
1.3	0.1241	0.1328	0.0325	0.0285
1.4	0.2611	0.2719	0.0439	0.0404
1.5	0.3958	0.4057	0.0301	0.0290
1.6	0.4861	0.4927	0.0069	0.0074
1.7	0.5040	0.5056	0.0010	0.0004
1.8	0.4437	0.4405	0.0192	0.0160
1.9	0.3242	0.3177	0.0405	0.0364
2.0	0.1827	0.1753	0.0405	0.0380
2.1	0.0237	0.0181	0.0066	0.0052
2.2	−0.1427	−0.1448	−0.0349	−0.0363
2.3	−0.2646	−0.2625	−0.0363	−0.0361
2.4	−0.3041	−0.2985	0.0040	0.0056
2.5	−0.2488	−0.2414	0.0397	0.0408

and $\Delta \dot{u}(0) = -1.001$. Thus

$$r(t) = \cos \omega_1 t$$

$$s(t) = \frac{1}{\omega_1} \sin \omega_1 t$$

$$\eta_1(t) = \Delta u(0) r(t)$$

$$\eta_2(t) = \Delta \dot{u}(0) s(t)$$

The transient response in the first mode is obtained by superposing the corrective responses $\eta_1(t)$ and $\eta_2(t)$ on the periodic response obtained earlier. The corrected response is shown in Table E13.16a. For the purpose of comparison, the exact modal response is also shown there.

Computations for the second mode are carried out in a similar manner. The augmented forcing function for this mode is shown in Figure E13.16d. The periodic response gives an initial displacement and velocity of 0.0240 and −0.0523 respectively. The corrective responses are again obtained by using Equations 9.89a and 9.89b. Table E13.16a shows the corrected frequency domain response. For the purpose of comparison, the exact response is also shown. Response in the physical coordinates can be obtained by superposing the modal responses. The final results are shown in Table E13.16b.

Table E13.16b. Comparison of frequency-domain response and exact response in the physical coordinates

t	Frequency-domain response		Exact response	
	u_1	u_2	u_1	u_2
0.0	0.0000	0.0000	0.0000	0.0000
0.1	0.0206	0.0219	0.0237	0.0250
0.2	0.0762	0.0922	0.0812	0.0984
0.3	0.1380	0.2006	0.1429	0.2089
0.4	0.1826	0.3174	0.1861	0.3258
0.5	0.2037	0.3986	0.2057	0.4041
0.6	0.2057	0.4080	0.2067	0.4084
0.7	0.1895	0.3411	0.1892	0.3369
0.8	0.1504	0.2284	0.1487	0.2219
0.9	0.0911	0.1147	0.0881	0.1088
1.0	0.0314	0.0346	0.0287	0.0306
1.1	0.0006	0.0006	0.0003	0.0003
1.2	0.0163	0.0171	0.0191	0.0199
1.3	0.0694	0.0825	0.0743	0.0884
1.4	0.1320	0.1879	0.1369	0.1961
1.5	0.1790	0.3058	0.1826	0.3143
1.6	0.2025	0.3929	0.2046	0.3988
1.7	0.2063	0.4090	0.2074	0.4119
1.8	0.1922	0.3513	0.1921	0.3474
1.9	0.1558	0.2413	0.1542	0.2349
2.0	0.0980	0.1258	0.0951	0.1197
2.1	0.0135	0.0155	0.0104	0.0118
2.2	−0.0784	−0.0963	−0.0800	−0.0972
2.3	−0.1290	−0.1951	−0.1280	−0.1935
2.4	−0.1218	−0.2506	−0.1186	−0.2469
2.5	−0.0786	−0.2261	−0.0750	−0.2207

13.5.2 *Frequency analysis of systems without classical mode shapes*

A multi-degree-of-freedom system may not possess classical mode shapes either because the damping is nonproportional or the physical parameters of the system are frequency dependent. In either case, the system equations cannot be uncoupled by a normal coordinate transformation, and the frequency-domain analysis becomes more complex, as will be evident from discussion in the following paragraphs.

The equations of motion of a multi-degree-of-freedom system are

$$\mathbf{M\ddot{u}} + \mathbf{C\dot{u}} + \mathbf{Ku} = \mathbf{f}g(t) \tag{13.133}$$

where it is been assumed that the forcing function can be represented by the product of vector $\mathbf{f}$ representing the spatial variation and $g(t)$, a scalar function of time. As usual, frequency analysis requires the decomposition of $g(t)$ into its harmonic components through a discrete Fourier transformation. In addition, we need the complex frequency response functions corresponding to each degree of freedom in the system. Now, if the system excitation is $e^{i\Omega t}$, the response takes the form

$$\mathbf{u} = \mathbf{H}e^{i\Omega t} \tag{13.134}$$

Substitution for $\mathbf{u}$ and its derivatives from Equation 13.134 in Equation 13.133 gives

$$(-\Omega^2\mathbf{M} + i\Omega\mathbf{C} + \mathbf{K})\mathbf{H} = \mathbf{f} \tag{13.135}$$

For a given Ω, Equation 13.135 can be solved for the complex-valued vector $\mathbf{H}$. In fact, by expressing $\mathbf{H}$ in the form $\mathbf{H}_R + i\mathbf{H}_I$, where $\mathbf{H}_R$ and $\mathbf{H}_I$ are both real, Equation 13.135 can be converted into $2N$ real equations, where N is the number of degrees of freedom. It is evident that for each Ω, evaluation of $\mathbf{H}$ requires the solution of $2N$ simultaneous equations.

Element H_j of vector $\mathbf{H}$ represents the complex frequency response function corresponding to the degree-of-freedom j. It would therefore appear that the response along a physical coordinate can be obtained by taking the inverse Fourier transform of the product of H_j and G, the latter being the discrete Fourier transform of $g(t)$, so that

$$\tilde{u}_j(k\,\Delta t) = \frac{1}{2\pi} \sum_{l=0}^{L-1} H_j(l\,\Delta\Omega)G(l\,\Delta\Omega)e^{2\pi ikl/L}\Delta\Omega \quad j=1,2,\dots,N \tag{13.136}$$

where $L\,\Delta t$ represents a period of the periodic extension of forcing function, and Δt is the sampling interval.

As discussed in Chapter 9, $\tilde{u}_j$ is not the true transient response because $G(l\,\Delta\Omega)$ represents a periodic extension of the forcing function. Unless there is sufficient damping in the system and the selected period is long enough for the response to die out as the end of the period approaches, response from a previous period overlaps that in the current period. As a result, the initial conditions prescribed for the transient response are not satisfied. The response obtained from Equation 13.136 could be corrected by superposing the response produced by a series of force pulses whose magnitudes are selected so as to satisfy the $2N$ initial conditions. A number of difficulties are involved in this process. The forcing function must be augmented to locate the corrective pulses. The force pulses are usually grouped at the end of the excitation period. However, if they are located next to each other, the simultaneous equation set up to solve for the unknown magnitudes of the $2N$ force pulses may become ill-conditions. To avoid this, they may need to be located farther apart. For a system with a large number of degrees of freedom, the number of corrective pulses required to satisfy the initial conditions will be large. As a result, the length of the augmented forcing function and consequently the time required for the evaluation of discrete Fourier transforms will be increased substantially. These difficulties make the method of corrective pulses inefficient. On the other hand, the method of exponential windows does not require augmentation of the forcing function or the solution of a set of additional simultaneous equations and is very effective in obtaining the transient response. The method of corrective pulses will not, therefore, be developed further in this book.

The method of exponential window described in Section 9.16 applies without any modification to the analysis of a multi-degree-of-freedom system. The essential steps in the analysis can be summarized as follows:

1. Obtain the discrete Fourier transform of the scaled forcing function $\hat{g}(t) = e^{-at}g(t)$. As shown in Equation 9.137, transforms G and $\hat{G}$ are related so that $\hat{G}(\Omega) = G(\Omega - ia)$.

2. From the discrete version of the Fourier transform H_j obtain the discrete Fourier transform of the scaled unit impulse function $\hat{h}_j(t) = h_j(t)e^{-at}$ by noting that $\hat{H}(\Omega) = H(\Omega - ia)$.

3. Obtain the scaled response $\hat{u}_j(t)$ by taking the inverse discrete Fourier transform of the product of $\hat{G}(\Omega)$ and $\hat{H}(\Omega)$. The true response $u_j(t)$ is now given by $u_j(t) = \hat{u}_j(t)e^{at}$.

It may be noted that while the exponential window method is quite effective, substantial amount of computations are still involved in the evaluation of complex frequency response function $\mathbf{H}$ for systems that have a large number of degrees of freedom but do not possess classical modes and are excited by a forcing function having a broad frequency content. Some efficiency can be gained by calculating $\mathbf{H}$ for widely spaced frequencies in regions away from resonance where $\mathbf{H}$ does not vary rapidly with the exciting frequency. Values of $\mathbf{H}$ for intermediate frequencies required in a discrete Fourier transform may then be obtained by interpolation.

Direct numerical integration of the equations of motion may be the more efficient method for the analysis of large systems without classical damping. However, for systems with frequency-dependent parameters, direct integration of the equations is not possible and a frequency-domain analysis is the only effective procedure. Problems involving frequency-dependent parameters are not uncommon in engineering analysis. Analysis of soil-structure and fluid-structure interaction are two examples of this class of problems.

Example 13.17

Obtain the response of the system of Example 13.16 by a frequency-domain analysis but without using a normal coordinate transformation.

Solution

As in Example 13.16, we will determine the response for 3.0, so that the period of the forcing function is 3.0 s. As before, we select a sampling interval $\Delta t = 0.05$ s. The complex frequency response function is obtained by solving the following equation

$$(-\Omega^2 \mathbf{M} + \mathbf{K})\mathbf{H} = \mathbf{f}$$

(a)

Substitution for $\mathbf{M}$, $\mathbf{K}$, and $\mathbf{f}$ gives

$$\begin{bmatrix} 192 - 2\Omega^2 & -64 \\ -64 & 64 - \Omega^2 \end{bmatrix} \mathbf{H} = \begin{bmatrix} 1 \\ \frac{1}{2} \end{bmatrix} \tag{b}$$

On solving Equation b, we get

$$\mathbf{H} = \begin{bmatrix} H_1 \\ H_2 \end{bmatrix} = \frac{1}{D} \begin{bmatrix} 96 - \Omega^2 \\ 160 - \Omega^2 \end{bmatrix} \tag{c}$$

where $D = 8192 - 320\Omega^2 + 2\Omega^4$.

The complex frequency functions are evaluated for $\Omega = l\,\Delta\Omega$, $l = 0, 1, 2, \ldots,$ 0.30, with $\Delta\Omega = 2\pi/T_0 = 2.094$. Because the system is undamped, $\mathbf{H}$ is real. To generate the discrete version of frequency function, values of the function for $l > 30$ are obtained by folding $\mathbf{H}$ about a frequency of $30\Delta\Omega$. Thus

$$H_j(l\,\Delta\Omega) = H_j\{(L - l)\Delta\Omega\} \quad j = 1, 2 \quad l = 31, 32, \ldots, 59 \tag{d}$$

Table E13.17. Comparison of frequency domain displacement response obtained by the exponential window method and exact response in the physical coordinates

t	Frequency-domain response		Exact response	
	u_1	u_2	u_1	u_2
0.0	0.0008	0.0011	0.0000	0.0000
0.1	0.0247	0.0263	0.0237	0.0250
0.2	0.0813	0.0992	0.0812	0.0984
0.3	0.1423	0.2089	0.1429	0.2089
0.4	0.1853	0.3245	0.1861	0.3256
0.5	0.2051	0.4019	0.2057	0.4041
0.6	0.2061	0.4062	0.2067	0.4084
0.7	0.1887	0.3356	0.1892	0.3369
0.8	0.1482	0.2221	0.1487	0.2219
0.9	0.0883	0.1100	0.0881	0.1088
1.0	0.0296	0.0322	0.0287	0.0306
1.1	0.0016	0.0019	0.0003	0.0003
1.2	0.0202	0.0213	0.0191	0.0199
1.3	0.0745	0.0893	0.0743	0.0884
1.4	0.1364	0.1961	0.1369	0.1961
1.5	0.1818	0.3132	0.1826	0.3143
1.6	0.2039	0.3967	0.2046	0.3988
1.7	0.2068	0.4096	0.2074	0.4119
1.8	0.1916	0.3460	0.1921	0.3474
1.9	0.1537	0.2349	0.1542	0.2349
2.0	0.0946	0.1203	0.0951	0.1197
2.1	0.0105	0.0124	0.0104	0.0118
2.2	−0.0787	−0.0964	−0.0800	−0.0972
2.3	−0.1264	−0.1923	−0.1280	−0.1935
2.4	−0.1178	−0.2451	−0.1186	−0.2469
2.5	−0.0751	−0.2189	−0.0750	−0.2207
2.6	−0.0281	−0.1053	−0.0278	−0.1060
2.7	0.0140	0.0549	0.0140	0.0559
2.8	0.0598	0.1908	0.0592	0.1925
2.9	0.1100	0.2501	0.1066	0.2485

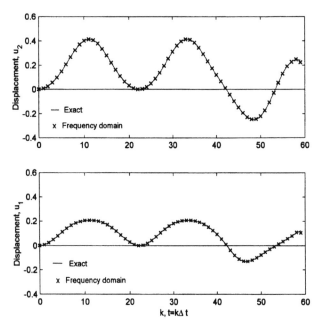

Figure E13.17. Response of a two-degree-of-freedom system to a rectangular pulse function.

where $L = 60$. The parameter a required in the exponential window method is selected as equal to 2 and the following steps are carried out to obtain the displacements in the two physical coordinates.

1. Obtain the discrete Fourier transform of the scaled function $\hat{g}(k\Delta t) = g(k\Delta t)e^{-ak\Delta t}$

$$\hat{G}(l\,\Delta\Omega) = \sum_{k=0}^{L-1} \hat{g}(k\,\Delta t)e^{-2\pi i k l/L}\,\Delta t \qquad (e)$$

2. Calculate $\hat{H}_1$ and $\hat{H}_2$ from H_1 and H_2 using the relationships

$$\hat{H}_j(l\,\Delta\Omega) = H_j(l\,\Delta\Omega - ia) \quad j = 1, 2 \qquad (f)$$

3. Obtain the scaled displacements in the physical coordinates

$$\hat{u}_1(k\,\Delta t) = \frac{1}{2\pi} \sum_{l=0}^{L-1} \hat{U}_1(l\,\Delta\Omega)e^{2\pi i k l/L}\,\Delta\Omega \qquad (g)$$

$$\hat{u}_2(k\,\Delta t) = \frac{1}{2\pi} \sum_{l=0}^{L-1} \hat{U}_2(l\,\Delta\Omega)e^{2\pi i k l/L}\,\Delta\Omega \qquad (h)$$

where

$$\hat{U}_1(l\,\Delta\Omega) = \hat{G}(l\,\Delta\Omega)\hat{H}_1(l\,\Delta\Omega)$$

$$\hat{U}_2(l\,\Delta\Omega) = \hat{G}(l\,\Delta\Omega)\hat{H}_2(l\,\Delta\Omega)$$

4. Recover the true displacements in physical coordinates from the scaled displacements by using the following relationship.

$$u_j(k\Delta t) = \hat{u}_j(k\Delta t) \qquad (i)$$

The final displacements calculated as above are shown in Table E13.17 at intervals of 0.1 s. For the purpose of comparison, exact values obtained by a mode superposition analysis are also shown. The same sets of data are presented in Figure E13.17. The two sets match very closely.

13.6 ANALYSIS OF NONLINEAR RESPONSE

As in the case of a single-degree-of-freedom system, nonlinearity may arise in a multi-degree-of-freedom system because of nonlinear geometry or nonlinear material properties, or both. The equations of motion for a system with nonlinear material properties can be expressed as

$$\mathbf{M}\ddot{\mathbf{u}}_n + \mathbf{C}\dot{\mathbf{u}}_n + \mathbf{f}_S(t_n) = \mathbf{p}_n \tag{13.145}$$

Since mass is constant, the inertia forces in Equation 13.145 are expressed as a product of the mass matrix and the vector of accelerations. Also, if the damping matrix is assumed to be constant, the forces of damping are obtained by multiplying the damping matrix and the velocity vector. However, the spring forces are not directly proportional to the displacements because the stress–strain relationship of the material changes with time. This makes the calculation of spring forces much more involved. Methods of evaluating the vector of spring forces for a nonlinear system are described in many standard textbooks on structural analysis and are not discussed here.

The analysis procedures for a nonlinear multi-degree-of-freedom system are similar to those described for a single-degree-of-freedom system. However, for a multi-degree-of-freedom system the volume of computations is greatly affected by the type of integration scheme used. In general, an explicit integration method involves significantly less computations than an implicit method. It is therefore useful to discuss the explicit and the implicit integration procedures separately.

13.6.1 *Explicit integration*

An effective procedure that belongs to the category of explicit integration techniques is the central difference method. As discussed earlier, the method employs central difference operators to express the accelerations and velocities in terms of displacements at the current and two previous time periods. When these expressions (Eqs. 13.104 and 13.105) are substituted in the equation of motion, written for the current time period (Eq. 13.145), we obtain

$$\left(\frac{1}{h^2}\mathbf{M} + \frac{1}{2h}\mathbf{C}\right)\mathbf{u}_{n+1} = \mathbf{p}_n + \frac{2}{h^2}\mathbf{M}\mathbf{u}_n + \frac{1}{2h}\mathbf{C}\mathbf{u}_{n-1} - \frac{1}{h^2}\mathbf{M}\mathbf{u}_{n-1} - \mathbf{f}_S(t_n)$$

$$\tag{13.146}$$

Equations 13.146 are different from similar expressions derived earlier (Eq. 13.107) in only the spring force term, which can no longer be expressed as $\mathbf{K}\mathbf{u}_n$.

When $\mathbf{M}$ and $\mathbf{C}$ are diagonal, Equations 13.146 are uncoupled and their solution is trivial. The central difference method is, in such cases, computationally very efficient. However, as noted earlier, stability of the method requires that the time step used in the integration is smaller than T_n/π, where T_n is the smallest period of the structure. Unlike the case of a linear system, T_n is not constant but varies with time. It is important that the selected time step remain smaller than T_n/π throughout the integration process. If this condition is violated even for a limited number of integration intervals, the accumulated errors during such intervals may render further integration seriously in error, even though the results may not blow up because the duration of instability was not long enough.

13.6.2 *Implicit integration*

In applying an implicit scheme to the integration of nonlinear equations, the latter must be expressed in an incremental form. The procedure for doing so is identical to that for a single-degree-of-freedom system. The resulting equations take the form

$$\mathbf{M}\,\Delta\ddot{\mathbf{u}} + \mathbf{C}\,\Delta\dot{\mathbf{u}} + \mathbf{f}_S(t+h) - \mathbf{f}_S(t) = \Delta\mathbf{p} \tag{13.147}$$

where, as before

$$\Delta\ddot{\mathbf{u}} = \ddot{\mathbf{u}}_{n+1} - \mathbf{u}_n$$

$$\Delta\dot{\mathbf{u}} = \dot{\mathbf{u}}_{n+1} - \mathbf{u}_n \tag{13.148}$$

$$\Delta\mathbf{p} = \mathbf{p}_{n+1} - \mathbf{p}_n$$

The remaining steps in the process depend on the details of the integration schemes used. Thus, with the average acceleration method

$$\Delta\ddot{\mathbf{u}} = \frac{4}{h^2}\left(\Delta\mathbf{u} - h\dot{\mathbf{u}}_n - \frac{h^2}{2}\ddot{\mathbf{u}}_n\right) \tag{13.149}$$

$$\Delta\dot{\mathbf{u}} = \frac{2}{h}\Delta\mathbf{u} - 2\dot{\mathbf{u}}_n \tag{13.150}$$

$$\mathbf{K}_T^*\Delta\mathbf{u} = \Delta\mathbf{p}^* \tag{13.151}$$

where

$$\mathbf{K}_T^* = \frac{4}{h^2}\mathbf{M} + \frac{2}{h}\mathbf{C} + \mathbf{K}_T$$

$$\Delta \mathbf{p}^* = \Delta \mathbf{p} + \mathbf{M}\left(\frac{4}{h}\dot{\mathbf{u}}_n + 2\ddot{\mathbf{u}}_n\right) + 2\mathbf{C}\dot{\mathbf{u}}_n \qquad (13.152)$$

and $\mathbf{K}_T$ represents the tangent stiffness matrix.

The incremental displacement vector $\Delta \mathbf{u}$ is obtained by solving Equation 13.151. Substitution of $\Delta \mathbf{u}$ in Equation 13.150 provides the value of $\Delta \dot{\mathbf{u}}$. The incremental values, $\Delta \mathbf{u}$ and $\Delta \dot{\mathbf{u}}$, are added to $\mathbf{u}_n$ and $\dot{\mathbf{u}}_n$, respectively, to obtain $\mathbf{u}_{n+1}$ and $\dot{\mathbf{u}}_{n+1}$. The acceleration at time t_{n+1} is then obtained by enforcing equilibrium between inertia, damping, and spring forces and the applied loads at time t_{n+1}. Since the average acceleration method is unconditionally stable, no restriction need be imposed on the time step to ensure stability. On the other hand, when a conditionally-stable scheme is used, the time step should meet the condition of stability with respect to the smallest period of the structure, taking into account the fact that such period may change with time.

As in the case of a single-degree-of-freedom system, use of a tangent stiffness in place of secant stiffness will introduce error in the integration process. The accumulation of such errors may cause serious difficulties. A process of iteration within a time step is therefore necessary to eliminate or minimize the errors. Again, the full Newton–Raphson method or the modified Newton–Raphson method may be used quite effectively for this purpose.

In either of the two methods mentioned above, iteration within a time step begins with

$$\mathbf{K}_T^* \Delta \mathbf{u}^{(1)} = \Delta \mathbf{p}^* \qquad (13.153)$$

Assuming that the average acceleration method is being used for the integration of equations, the steps involved in the iteration may be expressed as

$$\mathbf{K}_T^* \Delta \mathbf{u}^{(k)} = \Delta \mathbf{r}^{(k)}$$

$$\mathbf{u}_{n+1}^{(k)} = \mathbf{u}_{n+1}^{(k-1)} + \Delta \mathbf{u}^{(k)}$$

$$\Delta \mathbf{f}^{(k)} = \mathbf{f}_S^{(k)} - \mathbf{f}_S^{(k-1)} - \frac{4}{h^2}\mathbf{M}\,\Delta \mathbf{u}^{(k)} - \frac{2}{h}C\Delta \dot{\mathbf{u}}^{(k)} \qquad k=1,2,\ldots,\hat{k}$$

$$\Delta \mathbf{r}^{(k+1)} = \Delta \mathbf{r}^{(k)} - \Delta \mathbf{f}^{(k)}$$

$$(13.154)$$

where $\Delta \mathbf{r}^{(1)} = \Delta \mathbf{p}^*$

When the modified Newton–Raphson method is being used, the tangent stiffness matrix is calculated only once at the beginning of the time step, and this value is then used unchanged through all iterations within the time step. In the full Newton–Raphson method, $\mathbf{K}_T$ is updated at the beginning of each iteration. This speeds up the convergence. However, it must be noted that forming of $\mathbf{K}_T$ involves considerable computation and is likely to be very time consuming. A compromise between the full and the modified Newton–Raphson method is therefore sometimes used in which the tangent stiffness matrix is updated occasionally during the iterations but not at each new iteration.

A convergence criterion is required for the integration process. A possible selection is to require that the Euclidean norm of the residual force vector is small compared to the norm of effective applied force vector. Thus

$$\frac{\|\Delta\mathbf{r}^{(k)}\|_2}{\|\Delta\mathbf{p}^*\|_2} < \varepsilon \tag{13.155}$$

where ε is the specified tolerance.

The foregoing criterion will not be satisfactory for an elastoplastic system in which large displacements may occur for a small increase in force, so that even when the error in force is small, the displacements may still be significantly in error. An alternative is to use an error measure based on the norm of the incremental displacement

$$\frac{\|\Delta\mathbf{u}^{(k)}\|_2}{\|\Delta\mathbf{u}\|_2} < \varepsilon \tag{13.156}$$

in which $\Delta\mathbf{u}$ is the current estimate of the incremental displacements.

A difficulty is involved in using either of the two measures specified above when different elements of the force or displacement vectors are measured in different units. For example, when both rotation and displacement coordinates are included in the analysis, their units are different; also, the numerical values of the two may differ by as much as an order of magnitude. In such a case, the influence of the parameters with smaller numerical values, usually the rotations, will be completely overshadowed while taking the norm. The work done by the incremental forces serves as a useful measure of error in such a case. The corresponding error norm is defined as

$$\frac{[\Delta\mathbf{r}^{(k)}]^T \Delta\mathbf{u}^{(k)}}{[\Delta\mathbf{p}^*]^T \Delta\mathbf{u}} \tag{13.157}$$

in which $\Delta\mathbf{u}$ is again the current estimate of incremental displacements.

SELECTED READINGS

Bathe, K.J. 1996. *Finite Element Procedures*. Englewood Cliffs, New Jersey: Prentice Hall.

Bayo, E.P. & Wilson, E.L. 1983. *Numerical Techniques for the Solution of Soil-Structure Interaction Problems in the Time Domain*. Report UCB/EERC-P3/04. Berkeley, University of California: Earthquake Engineering Research Center.

Bayo, E.P. & Wilson, E.L. 1984. Use of Ritz Vectors in Wave Propagation and Foundation Response. *Earthquake Engineering and Structural Dynamics*, Vol. 12: 499–505.

Belytschko, T. & Mullen, R. 1977. Mesh Partition of Explicit–Implicit Time Integration. In K.J. Bathe, J.T. Oden & W. Wunderlich (eds) *Formulation and Computational Algorithms in Finite Element Analysis*. M.I.T. Press.

Belytschko, T., Yen, H.-J. & Mullen, R. 1979. Mixed Methods for Time Integration. *Computer Methods in Applied Mechanics and Engineering*, Vol. 17/18: 259–275.

Belytschko, T. & Mullen, R. 1978. Stability of Explicit–Implicit Mesh Partition in Time Integration. *International Journal for Numerical Methods in Engineering*, Vol. 12: 1575–1586.

Bisplinghoff, R.L., Ashley, H. & Halfman, R.L. 1955. *Aeroelasticity*. Reading, Massachusetts: Addison-Wesley.

Clough, R.W. & Wilson, E.L. 1979. Dynamic Analysis of Large Structural Systems with Local Nonlinearities. *Computer Methods in Applied Mechanics and Engineering*, Vol. 17/18: 107–129.

Cornwell, R.E., Craig, R.R. Jr. & Johnson, C.P. 1983. On the Application of Mode Acceleration Method to Structural Engineering Problems. *Earthquake Engineering and Structural Dynamics*, Vol. 11: 679–688.

Craig, R.R. Jr. 1981. *Structural Dynamics—An Introduction to Computer Methods*. New York: John Wiley.

Guyan, R.J. 1965. Reduction of Stiffness and Mass Matrices. *American Institute of Aeronautics and Astronautics Journal*, Vol. 3: 380.

Hansteen, O.E. & Bell, K. 1979. On the Accuracy of Mode Superposition Analysis in Structural Dynamics. *Earthquake Engineering and Structural Dynamics*, Vol. 7: 405–411.

Léger, P. & Wilson, E.L. 1988. Modal Summation Method for Structural Dynamic Computation. *Earthquake Engineering and Structural Dynamics*, Vol. 16: 23–27.

Léger, P., Wilson, E.L. & Clough, R.W. 1986. *The Use of Load Dependent Vectors for Dynamic and Earthquake Analysis*. Report No. EERC 86-04. Berkeley, California, University of California: Earthquake Engineering Research Center.

Maddox, N.R. 1975. On the Number of Modes Necessary for Accurate Response and Resulting Forces in Dynamic Analyses. *Journal of Applied Mechanics*, Vol. 42: 516–517. ASME.

Meek, J.W. & Veletsos, A.S. 1972. Dynamic Analysis by Extra Fast Fourier Transform. *Journal of Engineering Mechanics Division*, Vol. 98: 367–384. ASCE.

Nour-Omid, B. & Clough, R.W. 1984. Dynamic Analysis of Structures Using Lanczos Coordinates. *Earthquake Engineering and Structural Dynamics*, Vol. 12: 565–577.

Salmonte, A.J. 1982. Considerations on the Residual Contribution in Modal Analysis. *Earthquake Engineering and Structural Dynamics*, Vol. 10: 295–304.

Temple, G. & Bickley, W.G. 1956. *Rayleigh Principle and its Application to Engineering*. New York: Dover.

Wilson, E.L., Farhoomand, I. & Bathe, K.J. 1973. Nonlinear Dynamic Analysis of Complex Structures. *Earthquake Engineering and Structural Dynamics*, Vol. 1: 241–252.

Wilson, E.L., Yuan, M.W. & Dickens, J.M. 1982. Dynamic Analysis by Direct Superposition of Ritz Vectors. *Earthquake Engineering and Structural Dynamics*, Vol. 10: 813–821.

PROBLEMS

13.1 Using Rayleigh–Ritz method and the following Ritz shapes obtain estimates of 2 frequencies and the associated mode shapes of the four-story frame shown in Figure 11.14.

$$\Psi = \begin{bmatrix} 0.5 & 0.5 \\ 1.0 & 1.0 \\ 1.5 & -0.1 \\ 2.0 & -1.5 \end{bmatrix}$$

13.2 Using an improved Rayleigh–Ritz method obtain better estimates for the frequencies and mode shapes derived in Problem 13.1.

13.3 Using Rayleigh–Ritz method obtain estimates of 2 frequencies of a uniform flexible beam supported at each end by a spring of stiffness k as shown in Figure P13.3. Use the following shape functions

$$\psi_1(x) = 1$$

$$\psi_2(x) = \sin \frac{\pi x}{L}$$

Plot the mode shapes corresponding to the calculated frequencies. Note that both modes are symmetric.

13.4 Select suitable shape functions to obtain 2 unsymmetric mode shapes of the beam in Problem 13.3. Using these shape functions and Rayleigh–Ritz method calculate the two frequencies and the corresponding mode shapes.

13.5 Assume that the vibration shape of the beam in Problem 13.3 is given by $\psi(x) = 1 + d \sin(\pi x/L)$ where d is a constant. Select an appropriate value for d so as to satisfy the four boundary conditions at the two ends of the beam. Using the resulting value of $\psi(x)$ obtain the fundamental frequency and the corresponding mode shape. Compare your results with those obtained in Problem 13.3.

13.6 In Problem 12.1, suppose that the shaker is rotating at 80 rpm. Obtain the story forces under the steady state vibrations of the frame. Use the mode superposition method and include all three modes in your analysis.

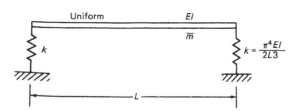

`¬13.3.`

13.7   Repeat Problem 13.6 considering only the first mode. Compute the value of error norm by using Equation 13.45. Improve the estimate of story forces obtained by you by adding static corrections for the responses in the second and third modes.

13.8   In order to reduce the size of the problem, the frame of Figure P11.14 is modeled as follows. The mass of floor 1 is distributed so that half is assigned to floor 2 and half to the fixed base. Floor 1 is then assumed to be massless. In a similar manner, half of the mass of floor 3 is assigned to floor 2 and the other half to floor 4. By a static condensation of the stiffness matrix reduce the problem to one with two degrees-of-freedom, hence obtaining the first two frequencies and mode shapes of the building. As an alternative to the procedure described above, Guyan's reduction technique is used to convert the system to one with two degrees-of-freedom. By an eigenvalue analysis of the reduced system, obtain the first two frequencies and mode shapes of the original frame. Compare your results with more exact values obtained in Problem 11.14.

13.9   Assuming that Guyan's reduction gives acceptable accuracy, obtain the response of the frame of Problem 13.8 to a lateral load of 200 kips applied suddenly to the fourth floor level. Neglect damping.

13.10   Using two load dependent Ritz vectors obtain the displacement response of the beam of Problem 12.12.

13.11   The shear frame shown in Figure P13.11 is subjected to lateral loads at each story level due to blast pressure. The variation of the loads is also shown in the figure. Obtain the displacement response of the frame for the first 1/2 s by using the central difference method. Use a time step size of 0.05 s. Neglect damping.

13.12   Obtain the displacement response history of the frame of Problem 13.11 for the first 1/2 s using the average acceleration method and a time step of 0.05 s.

13.13   Solve Problem 13.12 by the linear acceleration method. Use a time step of 0.05 s.

13.14   The frequencies of the frame of Figure P13.11 have been computed as 8.66rad/s and 17.30rad/s. Construct a damping matrix to give 5% critical damping in each mode. Obtain the response of the damped frame to the blast loads shown in Figure P13.11. Use the linear acceleration method and a time step of 0.05 s.

13.15   Obtain the response of the system of Problem 13.14 using the Wilson-θ method using a time step of 0.05 s. Select $\theta = 1.4$.

13.16   Obtain the response of the system of Problem 13.14 using the Houbolt method and a time step of 0.05 s. To start the iterations use the values of displacements calculated in Problem 13.12 for $t = 0.05$ and 0.1 s.

13.17   Obtain the response of the system in Problem 13.11 using the central difference method to integrate the equation of motion corresponding to

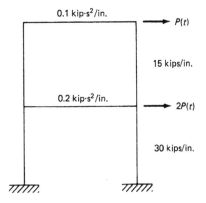

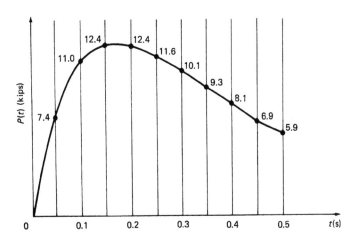

Figure P13.11.

the displacement at floor level 1 and the linear acceleration method to integrate the equation corresponding to the displacement at floor level 2.

13.18 Obtain the response of the system of Problem 13.11 for the first 0.5 s by a modal analysis through frequency domain.

13.19 Obtain the response of the system of Problem 13.11 for the first 0.5 s by an analysis in the frequency domain, but without using a normal coordinate transformation.

PART 4

CHAPTER 14

Formulation of the equations of motion: Continuous systems

14.1 INTRODUCTION

A majority of systems that we have considered so far were discrete parameter systems. As stated earlier, a discrete parameter system is an assembly of rigid-body elements having mass, and massless elements that are flexible and therefore deform under load. The elements that possess mass undergo translations and rotations as a rigid body, and a system coordinate or degree of freedom must be defined along each independent displacement component of such an element. A discrete system has a finite number of degrees of freedom, although for a complex system, the number of degrees of freedom may be quite large.

The method of modeling a system in which it is represented by an assembly of a finite number of elements that have mass and are rigid and those which have no mass but are flexible is an idealization, because all components of the system, in general, possess both mass as well as flexibility. In many systems, the displacement is, in fact, a continuous function of both space and time. Such systems have an infinite number of degrees of freedom and their motion is governed by one or more partial differential equations involving the space and time variables. The response of a continuous system is given by a solution of the governing differential equations, which must at the same time satisfy certain prescribed conditions along the boundary of the system domain.

The set of governing partial differential equations along with the prescribed boundary conditions is commonly referred to as a *boundary value problem*. In the most general case, the boundary value problem will involve three spatial coordinates and one time coordinate. The formulation of the governing equations and the boundary conditions, even for a general case, is usually quite straightforward. However, analytical solution of the problem is, in most cases, either impossible or very difficult to obtain.

Exact mathematical solutions of the boundary value problems usually exist only for very simple geometries and/or uniform material properties. As a result, most practical problems must be solved by approximate or numerical methods. In general, two different approaches are possible. In one approach, the continuous system is modeled as a multi-degree-of-freedom discrete system, by lumping the mass in a number of rigid elements which are joined together by massless but flexible elements. The behavior of the discrete model approaches that of the continuous system as the number of degrees of freedom in the former is increased. In another approach, the response of the system is expressed as a superposition of selected shape functions of the spatial coordinates, each weighted by a generalized coordinate of time. The governing equations are then expressed in terms of these generalized coordinates. Both of these approaches have been described in earlier chapters of this book.

It is apparent from the foregoing discussion that the solution of the governing partial differential equations is not a practical approach for most real-life systems. Nevertheless, solutions to such equations can be obtained in certain simple cases. These solutions provide insight into the behavior of the continuous systems. At the same time, they help in the formulation of approximate and numerical methods and in establishing the validity of such methods.

In this and the following chapters, we discuss procedures for the formulation of the boundary value problem and for its solution. We restrict our discussion to systems in which the response can be described in terms of one space coordinate and, of course, the time coordinate. The present chapter is concerned with the formulation of the governing equations and the boundary conditions for several different types of problems. While the equations of motion can be formed for any one-dimensional element whose axis follows an arbitrary curved path in space, we will deal only with straight-line elements. The free-vibration response of such elements can be obtained by solving an eigenvalue problem. Chapter 15 is devoted to the solution of the eigenvalue problem and the analysis of undamped free-vibration response. Forced vibrations of simple continuous systems are dealt with in Chapter 16. Finally, in Chapter 17, we describe a means of obtaining the response of a continuous system through wave propagation analysis.

14.2   TRANSVERSE VIBRATIONS OF A BEAM

Figure 14.1a shows a beam with flexural rigidity $EI(x)$ and mass $m(x)$ per unit length, both functions of the spatial coordinate x. For the purpose of illustration, the beam has been shown as simply supported, but other support conditions are equally admissible. The beam is undergoing transverse vibration in the plane of the paper under the action of a distributed force $p(x,t)$. The transverse displacement at any point along the beam is represented by $u(x,t)$, which is a function of both the spatial coordinate x and the time t.

A small element of the beam of length dx is shown in Figure 14.1b in its deformed position and the forces acting on the element are identified in Figure 14.1c. As indicated, these forces consist of an external force $p\,dx$

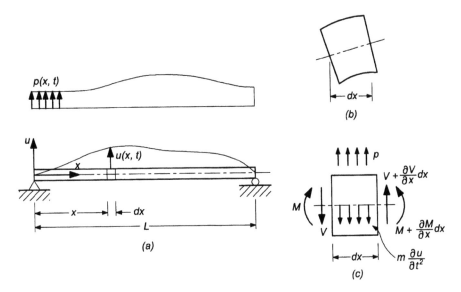

Figure 14.1. Transverse vibrations of a beam: (a) elevation of the beam; (b) small
element in its displaced position; (c) forces acting on the small element.

in the positive u direction, which is upward; the inertia force $m(\partial^2 u/\partial t^2)\,dx$ in
the downward direction; the shear force V and the moment M on the left-hand
face of the element; and shear $V + (\partial V/\partial x)\,dx$ and moment $M + (\partial M/\partial x)\,dx$
on the right-hand face. We have neglected the inertial moment caused by the
angular acceleration of the element.

The infinitesimal element is in equilibrium under the forces and moments
identified in Figure 14.1c. The shear forces, which act in a direction perpendicular to the elastic axis are slightly inclined to the vertical, but for small
displacements, their vertical components can be taken as equal to the shear
force values. For equilibrium of the element in the vertical direction, we
have

$$\frac{\partial V}{\partial x}\,dx - m\frac{\partial^2 u}{\partial t^2}\,dx + p\,dx = 0 \tag{14.1a}$$

or

$$\frac{\partial V}{\partial x} - m\frac{\partial^2 u}{\partial t^2} + p = 0 \tag{14.1b}$$

Next, equating the sum of moments about the left-hand face to zero, we get

$$\left(V + \frac{\partial V}{\partial x}\,dx\right)dx + p\,dx\frac{dx}{2} - m\frac{\partial^2 u}{\partial t^2}\,dx\frac{dx}{2} + M + \frac{\partial M}{\partial x}\,dx - M = 0 \tag{14.2}$$

On neglecting the quantities of higher order, Equation 14.2 becomes

$$V + \frac{\partial M}{\partial x} = 0 \tag{14.3}$$

If the flexural rotation is denoted by θ, then neglecting shear deformation, we have

$$\theta = \frac{\partial u}{\partial x} \tag{14.4}$$

Also, from elementary beam theory

$$
\begin{aligned}
M &= EI\frac{\partial \theta}{\partial x} \\
&= EI\frac{\partial^2 u}{\partial x^2}
\end{aligned}
\tag{14.5}
$$

Substitution of Equation 14.5 in Equation 14.3 gives

$$V = -\frac{\partial}{\partial x}\left(EI\frac{\partial^2 u}{\partial x^2}\right) \tag{14.6}$$

On differentiating Equation 14.6 and substituting in Equation 14.1b, we get

$$\frac{\partial^2}{\partial x^2}\left(EI\frac{\partial^2 u}{\partial x^2}\right) + m\frac{\partial^2 u}{\partial t^2} = p \tag{14.7}$$

Equation 14.7 is the equation governing the transverse vibration of the beam. To obtain an unique solution to this equation, we must specify four boundary conditions and two initial conditions. For a simply supported beam, the four boundary conditions are easily identified. Recognizing that both the displacements and moments should be zero at the two ends of the beam, we have

$$
\begin{aligned}
u &= 0 \quad \text{at} \quad x = 0 \\
u &= 0 \quad \text{at} \quad x = L
\end{aligned}
\tag{14.8}
$$

and

$$
\begin{aligned}
EI\frac{\partial^2 u}{\partial x^2} &= 0 \quad \text{at } x = 0 \\
EI\frac{\partial^2 u}{\partial x^2} &= 0 \quad \text{at } x = L
\end{aligned}
\tag{14.9}
$$

Other types of boundary conditions can as easily be identified, but we will postpone their discussion until we have derived the equation of motion by an alternative procedure using Hamilton's equation.

14.3   TRANSVERSE VIBRATIONS OF A BEAM: VARIATIONAL FORMULATION

As discussed in Section 4.10, the vibrations of a system are governed by Hamilton's variational formulation

$$\delta \int_{t_1}^{t_2} (T - V) \, dt = 0 \tag{14.10}$$

in which T is the kinetic energy, V is the potential energy, and δ denotes a variation.

The kinetic energy of the vibrating beam is easily shown to be

$$T = \int_0^L \frac{1}{2} m \left(\frac{\partial u}{\partial t} \right)^2 dx \tag{14.11}$$

The potential energy is given by

$$\begin{aligned}
V &= \int_0^L \frac{1}{2} \frac{M^2}{EI} \, dx - \int_0^L pu \, dx \\
&= \int_0^L \frac{1}{2} EI \left(\frac{\partial^2 u}{\partial x^2} \right)^2 dx - \int_0^L pu \, dx
\end{aligned} \tag{14.12}$$

Substitution of Equations 14.11 and 14.12 in Equation 14.10 gives

$$\delta \int_{t_1}^{t_2} \left\{ \int_0^L \frac{1}{2} m \left(\frac{\partial u}{\partial t} \right)^2 dx - \int_0^L \frac{1}{2} EI \left(\frac{\partial^2 u}{\partial x^2} \right)^2 dx + \int_0^L pu \, dx \right\} dt = 0 \tag{14.13a}$$

or

$$\begin{aligned}
\int_{t_1}^{t_2} \left\{ \int_0^L m \frac{\partial u}{\partial t} \delta \left(\frac{\partial u}{\partial t} \right) dx - \int_0^L EI \left(\frac{\partial^2 u}{\partial x^2} \right) \delta \left(\frac{\partial^2 u}{\partial x^2} \right) dx \right. \\
\left. + \int_0^L p \delta u \, dx \right\} dt = 0
\end{aligned} \tag{14.13b}$$

On changing the order of integration and recognizing that variation and differentiation are commutative, the first term in Equation 14.13b becomes

$$\int_0^L \int_{t_1}^{t_2} m \frac{\partial u}{\partial t} \frac{\partial}{\partial t} (\delta u) \, dt \, dx = \int_0^L \left(m \frac{\partial u}{\partial t} \delta u \Big|_{t_1}^{t_2} - \int_{t_1}^{t_2} m \frac{\partial^2 u}{\partial t^2} \delta u \, dt \right) dx \tag{14.14}$$

Since, in the derivation of Hamilton's equation, variation δu is taken as zero at terminal points t_1 and t_2, Equation 14.14 gives

$$\int_0^L \int_{t_1}^{t_2} m\frac{\partial u}{\partial t}\frac{\partial}{\partial t}(\delta u)\, dt\, dx = -\int_0^L \int_{t_1}^{t_2} m\frac{\partial^2 u}{\partial t^2}\,\delta u\, dt\, dx$$

$$= -\int_{t_1}^{t_2} \int_0^L m\frac{\partial^2 u}{\partial t^2}\,\delta u\, dx\, dt \tag{14.15}$$

The second integral in Equation 14.13b is reduced by repeated partial integration as follows

$$\int_0^L EI\left(\frac{\partial^2 u}{\partial x^2}\right)\delta\left(\frac{\partial^2 u}{\partial x^2}\right) dx = \int_0^L EI\frac{\partial^2 u}{\partial x^2}\frac{\partial^2}{\partial x^2}(\delta u)\, dx = EI\frac{\partial^2 u}{\partial x^2}\frac{\partial}{\partial x}(\delta u)\Big|_0^L$$

$$- \int_0^L \frac{\partial}{\partial x}\left(EI\frac{\partial^2 u}{\partial x^2}\right)\frac{\partial}{\partial x}(\delta u)\, dx$$

$$= EI\frac{\partial^2 u}{\partial x^2}\,\delta\left(\frac{\partial u}{\partial x}\right)\Big|_0^L - \frac{\partial}{\partial x}\left(EI\frac{\partial^2 u}{\partial x^2}\right)\delta u\Big|_0^L$$

$$+ \int_0^L \frac{\partial^2}{\partial x^2}\left(EI\frac{\partial^2 u}{\partial x^2}\right)\delta u\, dx \tag{14.16}$$

Substitution of Equations 14.15 and 14.16 in Equation 14.13 gives

$$\int_{t_1}^{t_2}\left[\int_0^L \left\{-m\frac{\partial^2 u}{\partial t^2} - \frac{\partial^2}{\partial x^2}\left(EI\frac{\partial^2 u}{\partial x^2}\right) + p\right\}\delta u\, dx \right.$$

$$\left. -EI\frac{\partial^2 u}{\partial x^2}\,\delta\left(\frac{\partial u}{\partial x}\right)\Big|_0^L + \frac{\partial}{\partial x}\left(EI\frac{\partial^2 u}{\partial x^2}\right)\delta u\Big|_0^L\right] dt = 0 \tag{14.17}$$

Since δu is arbitrary, the term within braces in Equation 14.17 must vanish in order that Equation 14.17 is satisfied. This gives

$$\frac{\partial^2}{\partial x^2}EI\left(\frac{\partial^2 u}{\partial x^2}\right) + m\frac{\partial^2 u}{\partial t^2} - p = 0 \tag{14.18}$$

which is identical to Equation 14.7. The remaining terms in Equation 14.17 should also vanish, so that

$$EI\frac{\partial^2 u}{\partial x^2}\,\delta\left(\frac{\partial u}{\partial x}\right)\Big|_0^L = 0 \tag{14.19a}$$

$$\frac{\partial}{\partial x}\left(EI\frac{\partial^2 u}{\partial x^2}\right)\delta u\Big|_0^L = 0 \tag{14.19b}$$

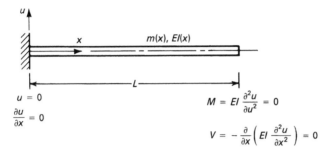

Figure 14.2. Clamped-free beam.

Equation 14.19a can be satisfied if either the slope is zero, that is, $\delta(\partial u/\partial x)$ $=0$, or the bending moment is zero, that is, $EI(\partial^2 u/\partial x^2)=0$. The condition given by Equation 14.19b can be satisfied either if the displacement is zero, so that the variation $\delta u=0$, or the shear is zero, that is

$$\frac{\partial}{\partial x}\left(EI\frac{\partial^2 u}{\partial x^2}\right)=0$$

The conditions on displacements or slopes are called *essential* or *geometric boundary conditions*, while the conditions on bending moment and shear are called *natural* or *force boundary conditions*. In any specific case of beam vibration, four boundary conditions must exist, two at each end. The set of four conditions may be made up of any combination of essential and natural conditions, as illustrated by the following examples.

1. *Both ends simply supported*
This case has already been discussed. There are two geometric conditions, corresponding to zero displacements at the supports. Equation 14.8 describes these conditions. The shear forces at the two ends given by

$$-\frac{\partial}{\partial x}\left(EI\frac{\partial^2 u}{\partial x^2}\right)$$

are nonzero and are determined once the problem solution has been obtained. The two natural conditions described by Equations 14.9 specify zero moments at the supports. The slopes $\partial u/\partial x$ at the two ends are nonzero and must be determined from the problem solution.

2. *One end clamped and other end free*
Let the end $x=0$ be rigidly fixed to a support as shown in Figure 14.2. In that case, both the displacement and the slope at that end must be zero, giving the following geometric conditions

$$
\begin{aligned}
u &= 0 \\
\frac{\partial u}{\partial x} &= 0
\end{aligned}
\qquad \text{at } x = 0
\qquad\qquad (14.20)
$$

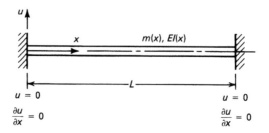

Figure 14.3. Clamped-clamped beam.

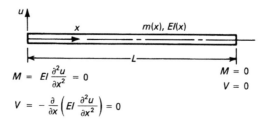

Figure 14.4. Free-free beam.

The bending moment and the shear force at the free end are zero. This leads to the following natural conditions

$$EI\frac{\partial^2 u}{\partial x^2} = 0$$

$$\frac{\partial}{\partial x}\left(EI\frac{\partial^2 u}{\partial x^2}\right) = 0 \qquad \text{at } x = L \qquad (14.21)$$

The bending moment and shear force at $x = 0$, and the displacement and slope at $x = L$ must be obtained from the problem solution.

3. *Both ends clamped*
In this case (Fig. 14.3), all four boundary conditions are geometric and are given by

$$u = 0$$

$$\frac{\partial u}{\partial x} = 0 \qquad \text{at } x = 0 \quad \text{and} \quad x = L \qquad (14.22)$$

The bending moments and shear forces at each end must be determined from the problem solution.

4. *Both ends free*

In this case, as shown in Figure 14.4, all four conditions are natural and correspond to zero moments and shear at each end

$$EI\frac{\partial^2 u}{\partial x^2}=0$$

$$\frac{\partial}{\partial x}\left(EI\frac{\partial^2 u}{\partial x^2}\right)=0$$
at $x=0$ and $x=L$ (14.23)

Other types of boundary conditions are possible, as will be seen from the following example.

Example 14.1

A uniform cantilever beam of unit mass m and flexural rigidity EI has a uniform rigid circular disk of mass M_0 and radius R rigidly attached to its end, as shown in Figure E14.1. Obtain the equations of motion and the boundary conditions for free transverse vibrations of the system.

Solution

The mass moment of inertia of the rigid circular disk for a rotation about its diameter is given by

$$I_0 = \frac{M_0 R^2}{4}$$ (a)

The kinetic energy of the system is obtained from

$$T = \int_0^L \frac{1}{2}m\left(\frac{\partial u}{\partial t}\right)^2 dx + \frac{1}{2}M_0\left(\frac{\partial u}{\partial t}\right)^2_{x=L} + \frac{1}{2}I_0\left(\frac{\partial^2 u}{\partial t\partial x}\right)^2_{x=L}$$ (b)

while the potential energy is given by

$$V = \int_0^L \frac{EI}{2}\left(\frac{\partial^2 u}{\partial x^2}\right)^2 dx$$ (c)

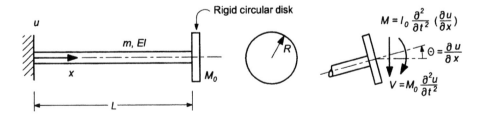

Figure E14.1. (a) Transverse vibration of a cantilever beam with rigid disk attached to its end; (b) inertia forces acting at the tip.

Substituting these values of T and V in Hamilton's expression (Eq. 14.10) and reducing the resulting integrals by the method of this section, we get

$$\int_{t_1}^{t_2} \left[\int_0^L \left\{ -m\frac{\partial^2 u}{\partial t^2} - EI\frac{\partial^4 u}{\partial x^4} \right\} \delta u \, dx \right.$$

$$-EI\frac{\partial^2 u}{\partial x^2} \, \delta\left(\frac{\partial u}{\partial x}\right)\bigg|_0^L + EI\frac{\partial^3 u}{\partial x^3} \delta u \bigg|_0^L \right] dt \qquad (d)$$

$$+\delta \int_{t_1}^{t_2} \frac{1}{2}M_0\left(\frac{\partial u}{\partial t}\right)^2_{x=L} dt + \delta \int_{t_1}^{t_2} \frac{1}{2}I_0\left(\frac{\partial^2 u}{\partial t \partial x}\right)^2_{x=L} dt = 0$$

The next-to-last term in Equation d can be reduced as follows

$$\delta \int_{t_1}^{t_2} \frac{1}{2}M_0\left(\frac{\partial u}{\partial t}\right)^2_{x=L} dt = \int_{t_1}^{t_2} M_0\left\{ \frac{\partial u}{\partial t}\frac{\partial}{\partial t}(\delta u) \right\}_{x=L} dt$$

$$= M_0\frac{\partial u}{\partial t}\delta u \bigg|_{t_1}^{t_2} - \int_{t_1}^{t_2} M_0\left\{ \frac{\partial^2 u}{\partial t^2}\delta u \right\}_{x=L} dt \qquad (e)$$

$$= -\int_{t_1}^{t_2} M_0\left\{ \frac{\partial^2 u}{\partial t^2}\delta u \right\}_{x=L} dt$$

since δu vanishes at both t_1 and t_2.

The final term in Equation d becomes

$$\delta \int_{t_1}^{t_2} \frac{1}{2}I_0\left\{ \frac{\partial}{\partial t}\left(\frac{\partial u}{\partial x}\right) \right\}^2_{x=L} dt = \int_{t_1}^{t_2} I_0\left\{ \frac{\partial}{\partial t}\left(\frac{\partial u}{\partial x}\right)\frac{\partial}{\partial t}\left(\delta\frac{\partial u}{\partial x}\right) \right\}_{x=L} dt$$

$$= I_0\left\{ \frac{\partial}{\partial t}\left(\frac{\partial u}{\partial x}\right) \delta\left(\frac{\partial u}{\partial x}\right) \right\}_{x=L}\bigg|_{t_1}^{t_2}$$

$$-I_0\int_{t_1}^{t_2} \left\{ \frac{\partial^2}{\partial t^2}\left(\frac{\partial u}{\partial x}\right) \delta\left(\frac{\partial u}{\partial x}\right) \right\}_{x=L} dt \qquad (f)$$

$$= -I_0\int_{t_1}^{t_2} \left\{ \frac{\partial^2}{\partial t^2}\left(\frac{\partial u}{\partial x}\right) \delta\left(\frac{\partial u}{\partial x}\right) \right\}_{x=L} dt$$

in which we have used the fact that $\delta(\partial u/\partial x)$ must vanish at both t_1 and t_2. Substitution of Equations e and f in Equation d gives

$$\int_{t_1}^{t_2} \left[\int_0^L \left\{ -m\frac{\partial^2 u}{\partial t^2} - EI\frac{\partial^4 u}{\partial x^4} \right\} \delta u \, dx \right.$$

$$-\left\{ EI\frac{\partial^2 u}{\partial x^2} + I_0\frac{\partial^2}{\partial t^2}\left(\frac{\partial u}{\partial x}\right) \right\} \delta\left(\frac{\partial u}{\partial x}\right)\bigg|_{x=L}$$

$$+\left\{ EI\frac{\partial^2 u}{\partial x^2} \right\} \delta\left(\frac{\partial u}{\partial x}\right)\bigg|_{x=0} + \left\{ EI\frac{\partial^3 u}{\partial x^3} - M_0\frac{\partial^2 u}{\partial t^2} \right\} \delta u \bigg|_{x=L}$$

$$-\left\{ EI\frac{\partial^3 u}{\partial x^3} \right\} \delta u \bigg|_{x=0} \right] dt = 0 \qquad (g)$$

Since δu is arbitrary, Equation g leads to the following relationships

$$m\frac{\partial^2 u}{\partial t^2} + EI\frac{\partial^4 u}{\partial x^4} = 0$$

$$EI\frac{\partial^2 u}{\partial x^2}\delta\left(\frac{\partial u}{\partial x}\right) = 0 \quad \text{at } x=0$$

(h)

or

$$\frac{\partial u}{\partial x} = 0 \quad \text{at } x=0$$

$$EI\frac{\partial^3 u}{\partial x^3}\delta u = 0 \quad \text{at } x=0$$

(i)

or

$$u = 0 \quad \text{at } x=0$$

$$\left\{-EI\frac{\partial^2 u}{\partial x^2} - I_0\frac{\partial^2}{\partial t^2}\left(\frac{\partial u}{\partial x}\right)\right\}\delta\left(\frac{\partial u}{\partial x}\right) = 0 \quad \text{at } x=L$$

(j)

or

$$EI\frac{\partial^2 u}{\partial x^2} + I_0\frac{\partial^2}{\partial t^2}\left(\frac{\partial u}{\partial x}\right) = 0 \quad \text{at } x=L$$

$$\left\{EI\frac{\partial^3 u}{\partial x^3} - M_0\frac{\partial^2 u}{\partial t^2}\right\}\delta u = 0 \quad \text{at } x=L$$

(k)

or

$$EI\frac{\partial^3 u}{\partial x^3} - M_0\frac{\partial^2 u}{\partial t^2} = 0 \quad \text{at } x=L$$

(l)

Equation h is the equation governing the motion, while Equations i through l represent the four boundary conditions. We have employed Hamilton's principle in solving the problem to demonstrate that it not only gives the equation of motion but also automatically provides the exact number of boundary conditions required for obtaining a unique solution. In the present case, however, the geometric boundary conditions are easily obtained by recognizing that the displacement and slope must be zero at the clamped end. At the free end, the bending moment and shear force must be equal, respectively, to the inertial moment and inertial force acting on the disk of mass M_0 as shown in Figure E14.1b. This directly leads to the two essential conditions. Thus, the moment M at $x=L$ is given by

$$M = -I_0\frac{\partial^2}{\partial t^2}\left(\frac{\partial u}{\partial x}\right) \quad \text{at } x=L$$

(m)

On substituting $M = EI(\partial^2 u/\partial x^2)$ in Equation m, we obtain the boundary condition given by Equation k. Also, the shear force V at $x=L$ is obtained from

$$V = -M_0\frac{\partial^2 u}{\partial t^2} \quad \text{at } x=L$$

(n)

Substitution of $V = -EI(\partial^3 u/\partial x^3)$ in Equation n leads to the boundary condition given by Equation l.

14.4   EFFECT OF DAMPING RESISTANCE ON TRANSVERSE VIBRATIONS OF A BEAM

In Sections 14.2 and 14.3, we discussed the basic case of transverse vibrations of a beam. In deriving the governing equation for the basic case, we ignored several factors that may influence the response. Such factors include damping resistances and the effect of shear deformation and rotatory inertia. In this section, we discuss the effect of damping resistance on the equations governing the response. Shear deformation and rotatory inertia are dealt with in Section 14.5.

Two different types of damping resistances can be identified. Damping can be caused by external forces opposing the vibrations, such as air resistance. Damping may also arise from internal resistance to the straining of the material. Both of these resisting forces are dissipative in nature and cause loss of energy from the system. The energy lost is converted into other forms, such as heat or sound.

The external damping can be represented by a distributed viscous damping mechanism or dashpots with a damping constant $c(x)$ per unit length, as shown in Figure 14.5a. The resulting damping force acting on the infinitesimal element is identified in Figure 14.5b. When this force is included in the vertical force balance, Equation 14.1b is modified to

$$\frac{\partial V}{\partial x} - m\frac{\partial^2 u}{\partial t^2} - c\frac{\partial u}{\partial t} + p = 0 \qquad (14.24)$$

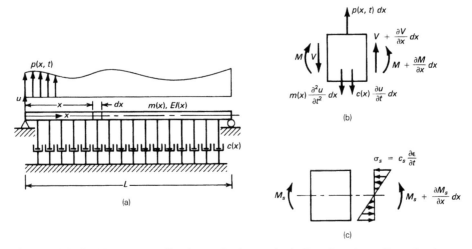

Figure 14.5. (a) Transverse vibrations of a beam, including damping effect; (b) forces acting on an element, including external damping; (c) internal damping forces and resultant moment.

Resistance to internal strain will depend on the strain rate $\partial\varepsilon/\partial t$. We define a damping coefficient c_s which converts the strain rate into a stress σ_s, so that

$$\sigma_s = c_s \frac{\partial\varepsilon}{\partial t} \tag{14.25}$$

The distribution of this stress across a section is shown in Figure 14.5c. Since according to the elementary beam theory, the strain varies linearly with the distance from the neutral axis, the strain rate and hence the damping stress are also linearly distributed. The resultant of the stress acting on the section can be expressed in terms of a moment M_s, which is derived in a manner similar to the flexural moment.

$$\begin{aligned} M_s = \int_A \sigma_s y \, dA &= \int_A c_s \frac{\partial}{\partial t}(y\kappa) y \, dA \\ &= \int_A c_s \frac{\partial}{\partial t}\left(\frac{\partial^2 u}{\partial x^2}\right) y^2 \, dA \\ &= c_s I \frac{\partial}{\partial t}\left(\frac{\partial^2 u}{\partial x^2}\right) \end{aligned} \tag{14.26}$$

in which we have used the relationship $\varepsilon = y\kappa$, where κ is the curvature given by $\kappa = \partial^2 u/\partial x^2$. In fixing the sign of stresses σ_s and the moment M_s, we must recognize that like flexural stresses, the damping stresses shown are those exerted on the beam element by the adjacent sections of the beam.

The moments caused by damping resistance are shown in Figure 14.5c. When inserted in the moment balance equations (Eqs. 14.2 and 14.3), they give

$$V + \frac{\partial M}{\partial x} + \frac{\partial M_s}{\partial x} = 0 \tag{14.27}$$

Substituting for M from Equation 14.5 and M_s from Equation 14.26 in Equation 14.27 and differentiating with respect to x, we get

$$\frac{\partial V}{\partial x} + \frac{\partial^2}{\partial x^2}\left(EI\frac{\partial^2 u}{\partial x^2}\right) + \frac{\partial^2}{\partial x^2}\left(c_s I\frac{\partial^3 u}{\partial t \partial x^2}\right) = 0 \tag{14.28}$$

Substitution of $\partial V/\partial x$ from Equation 14.28 in Equation 14.24 gives the following equation of motion

$$\frac{\partial^2}{\partial x^2}\left(EI\frac{\partial^2 u}{\partial x^2}\right) + \frac{\partial^2}{\partial x^2}\left(c_s I\frac{\partial^3 u}{\partial t \partial x^2}\right) + m\frac{\partial^2 u}{\partial t^2} + c\frac{\partial u}{\partial t} = p \tag{14.29}$$

14.5 EFFECT OF SHEAR DEFORMATION AND ROTATORY INERTIA ON THE FLEXURAL VIBRATIONS OF A BEAM

In previous sections, while formulating the equations of motion for beam vibration, we neglected the effect of both the shear deformation and the rotatory

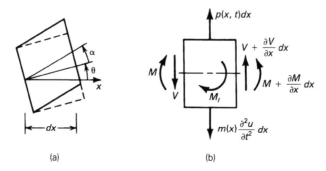

(a)   (b)

Figure 14.6. Effect of shear deformation and rotatory inertia on the transverse vibrations of a beam: (a) deformed shape of an element; (b) forces acting on an element.

inertia. For beams that are long compared to their cross-sectional dimensions, these effects are small, but for short, deep beams they may become quite significant. Figure 14.6a shows the deformed shape of an infinitesimal element of a beam for which we would like to account for the effect of both shear deformation and rotatory inertia. The slope of the elastic axis now consists of two parts: a rotation θ due to flexural deformation, and an angular deformation α due to the shear force. This can be expressed through the mathematical relationship

$$\frac{\partial u}{\partial x} = \theta + \alpha \tag{14.30}$$

Again, from elementary beam theory, the moment is related to the flexural rotation by the expression

$$M = EI\frac{\partial \theta}{\partial x} \tag{14.31}$$

while angular deformation α caused by the shear force is obtained from

$$V = k'GA\alpha \tag{14.32}$$

in which k' is a constant that depends on the shape of the cross section. This constant accounts for the fact that shear stress is not uniformly distributed across the section. Constant k' is derived for various cross-sectional shapes in standard textbooks on mechanics. As an example, for a rectangular section, $k' = \frac{5}{6}$.

The forces and moments acting on the infinitesimal element are indicated in Figure 14.6b and include an inertial moment M_I caused by the rotation of the element. The magnitude of M_I is given by

$$M_I = I_0 \ddot{\theta} = m \frac{I}{A} \ddot{\theta} \, dx$$

$$= mr^2 \ddot{\theta} \, dx \tag{14.33}$$

where $I_0 \, dx$ is the mass moment of inertia of the beam element for rotation about an axis through the centroid of the section and perpendicular to the plane of the paper, and r is the radius of gyration of the beam section. The mass moment of inertia is given by $I_0 = mI/A$. It should be noted that shear deformation does not cause a rotation of the element.

Equating the sum of vertical forces to zero, we get

$$\frac{\partial V}{\partial x} - m \frac{\partial^2 u}{\partial t^2} + p = 0 \tag{14.34}$$

which is identical to Equation 14.1b. The sum of moment about the left face of the element leads to

$$\frac{\partial M}{\partial x} + V - mr^2 \frac{\partial^2 \theta}{\partial t^2} = 0 \tag{14.35}$$

Substitution for α from Equation 14.30 in Equation 14.32 gives

$$V = k' GA \left(\frac{\partial u}{\partial x} - \theta \right) \tag{14.36}$$

On substituting in Equation 14.34 the value of $\partial V/\partial x$ obtained by differentiating Equation 14.36 with respect to x, we obtain

$$\frac{\partial}{\partial x} \left\{ k' GA \left(\frac{\partial u}{\partial x} - \theta \right) \right\} - m \frac{\partial^2 u}{\partial t^2} + p = 0 \tag{14.37}$$

Next, we substitute in Equation 14.35, V from Equation 14.36 and $\partial M/\partial x$ obtained by differentiating Equation 14.31. This gives

$$\frac{\partial}{\partial x} \left(EI \frac{\partial \theta}{\partial x} \right) + k' GA \left(\frac{\partial u}{\partial x} - \theta \right) - mr^2 \frac{\partial^2 \theta}{\partial t^2} = 0 \tag{14.38}$$

Equations 14.37 and 14.38 together govern the motion of the beam. The boundary conditions are similar to those presented in Section 14.2 and are unaffected by shear deformation and rotatory inertia.

For a uniform beam, it is possible to eliminate θ from Equations 14.37 and 14.38 to obtain a single equation of motion in terms of u and its derivatives. Thus, Equation 14.37 gives

$$\frac{\partial \theta}{\partial x} = \frac{\partial^2 u}{\partial x^2} + \frac{1}{k' GA} \left(p - m \frac{\partial^2 u}{\partial t^2} \right) \tag{14.39}$$

On differentiating Equation 14.38 with respect to x, we get

$$EI\frac{\partial^2}{\partial x^2}\left(\frac{\partial\theta}{\partial x}\right) + k'GA\left(\frac{\partial^2 u}{\partial x^2} - \frac{\partial\theta}{\partial x}\right) - mr^2\frac{\partial^2}{\partial t^2}\left(\frac{\partial\theta}{\partial x}\right) = 0 \qquad (14.40)$$

Finally, substitution for $\partial\theta/\partial x$ from Equation 14.39 in Equation 14.40 gives

$$\left\{EI\frac{\partial^4 u}{\partial x^4} + m\frac{\partial^2 u}{\partial t^2} - p\right\} - \left\{mr^2\frac{\partial^2}{\partial t^2}\left(\frac{\partial^2 u}{\partial x^2}\right)\right\}$$

$$+ \left\{\frac{EI}{k'GA}\frac{\partial^2}{\partial x^2}\left(p - m\frac{\partial^2 u}{\partial t^2}\right)\right\} - \left\{\frac{mr^2}{k'GA}\frac{\partial^2}{\partial t^2}\left(p - m\frac{\partial^2 u}{\partial t^2}\right)\right\} = 0$$

$$(14.41)$$

The term within the first set of braces will be recognized as the basic case without shear deformation or rotatory inertia. The term in the second set of braces represents the effect of rotatory inertia. This term will vanish when the mass moment of inertia $mr^2 = 0$. The term in the third set of braces arises from shear deformation; it will vanish when the beam is very rigid in shear, that is, $GA = \infty$. The term in the fourth set of braces results from a coupling between the effect of shear deformation and rotatory inertia. This term will vanish when either $mr^2 = 0$ or $GA = \infty$.

14.6   AXIAL VIBRATIONS OF A BAR

Figure 14.7a shows a bar having a cross-sectional area $A(x)$ and mass per unit length $m(x)$ clamped at one end and free at the other. It is subjected to a time-varying distributed force acting parallel to the longitudinal axis and is vibrating in the axial direction. The displacement of a point at a distance x from the clamped end of the bar is denoted by $u(x,t)$. The forces acting on an infinitesimal element of length dx are shown in Figure 14.7b. They include the axial forces acting on the two sections, the inertial force, and the applied load.

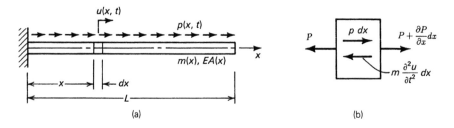

Figure 14.7. (a) Axial vibrations of a bar; (b) forces acting on an element.

For equilibrium of the element in the horizontal direction, we have

$$\frac{\partial P}{\partial x}\,dx - m\frac{\partial^2 u}{\partial t^2}\,dx + p\,dx = 0$$

or

$$\frac{\partial P}{\partial x} - m\frac{\partial^2 u}{\partial t^2} + p = 0 \tag{14.42}$$

The axial force P and the displacement u are related as follows

$$P = EA\frac{\partial u}{\partial x} \tag{14.43}$$

Substitution of Equation 14.43 in Equation 14.42 gives

$$\frac{\partial}{\partial x}\left(EA\frac{\partial u}{\partial x}\right) - m\frac{\partial^2 u}{\partial t^2} + p = 0 \tag{14.44}$$

Equation 14.44 governs the motion of the bar. The solution to Equation 14.44 must in addition satisfy the boundary conditions at the two ends of the bar. These conditions are obtained as follows. At the fixed end, the displacement of the bar must be zero

$$u = 0 \quad \text{at } x = 0 \tag{14.45}$$

Equation 14.45 represents a geometric boundary condition. In addition, there is one natural condition corresponding to zero axial force at the free end

$$EA\frac{\partial u}{\partial x} = 0 \quad \text{at } x = L \tag{14.46}$$

If a concentrated load $Q(t)$ acts at the free end, the boundary condition of Equation 14.46 will be modified to

$$EA\frac{\partial u}{\partial x} = Q \quad \text{at } x = L \tag{14.47}$$

The equations of motion and the boundary conditions can equally well be obtained from the Hamilton's expression, as illustrated in the following example.

Example 14.2
The nonuniform bar of axial rigidity $EA(x)$ and mass per unit length $m(x)$, shown in Figure E14.2, is clamped at the left end and restrained at the other end by a longitudinal spring of stiffness k. The bar is vibrating in the axial direction. Obtain the equation of motion and the boundary condition by using Hamilton's expression.

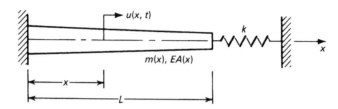

Figure E14.2. Axial vibrations of a bar restrained by a spring at one end.

Solution

The kinetic energy of the bar is given by

$$T = \int_0^L \frac{m}{2}\left(\frac{\partial u}{\partial t}\right)^2 dx \tag{a}$$

while the potential energy is obtained from

$$V = \int_0^L \frac{EA}{2}\left(\frac{\partial u}{\partial x}\right)^2 dx + \frac{k}{2}\{u(L)\}^2 \tag{b}$$

The Hamilton's expression $\delta \int_{t_1}^{t_2} (T - V)\,dt$ reduces to

$$\delta \int_{t_1}^{t_2} \left[\int_0^L \frac{m}{2}\left(\frac{\partial u}{\partial t}\right)^2 dx - \int_0^L \frac{EA}{2}\left(\frac{\partial u}{\partial x}\right)^2 dx - \frac{k}{2}\{u(L)\}^2 \right] dt = 0$$

or

$$\int_{t_1}^{t_2} \left[\int_0^L m\frac{\partial u}{\partial t}\frac{\partial}{\partial t}(\delta u)\,dx \right.$$

$$\left. - \int_0^L EA\left(\frac{\partial u}{\partial x}\right)\frac{\partial}{\partial x}(\delta u)\,dx - ku(L)\delta u(L) \right] dt = 0 \tag{c}$$

The first integral in Equation c can be further reduced as follows

$$\int_{t_1}^{t_2} \int_0^L m\frac{\partial u}{\partial t}\frac{\partial}{\partial t}(\delta u)\,dx\,dt = \int_0^L \int_{t_1}^{t_2} m\frac{\partial u}{\partial t}\frac{\partial}{\partial t}(\delta u)\,dt\,dx$$

$$= \int_0^L \left\{ m\frac{\partial u}{\partial t}\delta u\Big|_{t_1}^{t_2} - \int_{t_1}^{t_2}\left(m\frac{\partial^2 u}{\partial t^2}\right)\delta u\,dt \right\} dx$$

$$= -\int_{t_1}^{t_2}\int_0^L \left(m\frac{\partial^2 u}{\partial t^2}\right)\delta u\,dx\,dt \tag{d}$$

where the term $m(\partial u/\partial t)\delta u|_{t_1}^{t_2}$ vanishes because the variation δu is zero at both $t=t_1$ and $t=t_2$. The second integral in Equation c is reduced in a similar manner.

$$\int_{t_1}^{t_2} \left\{ \int_0^L EA\frac{\partial u}{\partial x}\frac{\partial}{\partial x}(\delta u)\,dx \right\} dt$$

$$= \int_{t_1}^{t_2} \left\{ EA\frac{\partial u}{\partial x}\delta u\Big|_0^L - \int_0^L \frac{\partial}{\partial x}\left(EA\frac{\partial u}{\partial x}\right)\delta u\,dx \right\} dt \tag{e}$$

Substitution of Equations d and e in Equation c gives

$$\int_{t_1}^{t_2} \left[\int_0^L \left\{ -\left(m\frac{\partial^2 u}{\partial t^2} \right) + \frac{\partial}{\partial x}\left(EA\frac{\partial u}{\partial x} \right) \right\} \delta u \, dx \right.$$

$$\left. - \left\{ \left(EA\frac{\partial u}{\partial x} + ku \right) \delta u \right\}_{x=L} + \left\{ EA\frac{\partial u}{\partial x}\delta u \right\}_{x=0} \right] dt = 0 \tag{f}$$

Since δu is arbitrary, the terms within each pair of braces in Equation f must vanish. This leads to the governing equation

$$\frac{\partial}{\partial x}\left(EA\frac{\partial u}{\partial x} \right) - \left(m\frac{\partial^2 u}{\partial t^2} \right) = 0 \tag{g}$$

and the two boundary conditions

$$EA\frac{\partial u}{\partial x} + ku = 0 \quad \text{at } x = L \tag{h}$$

$$u = 0 \qquad \qquad \text{at } x = 0 \tag{i}$$

14.7 TORSIONAL VIBRATIONS OF A BAR

Figure 14.8a shows a nonuniform bar of circular cross section with mass $m(x)$ per unit length and polar moment of inertia $J(x)$. The bar is clamped at its left end and is free at the right end. It is undergoing torsional vibrations about its longitudinal axis. In a general case, as the section undergoes twisting, plane sections perpendicular to the axis do not remain plane but warp. This sets up longitudinal stresses in the bar because a warping restraint is provided by the clamped end. In our case, however, since the cross section is circular, warping does not take place and plane sections remain plane as they twist.

Let $\theta(x,t)$ be the angle of twist at a distance x from the clamped end and $T(x,t)$ the torque at that point. Also, let $m_T(x,t)$ be the applied twisting moment per unit length of the bar. The moments acting on an infinitesimal element of the circular bar are as shown in Figure 14.8b. They include the torques acting on the left and right sections of the element, the applied torque $m_T \, dx$, and the

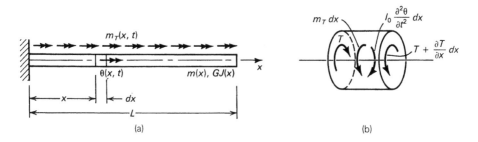

Figure 14.8. (a) Torsional vibrations of a bar; (b) moments acting on an element.

inertial moment $I_0(\partial^2\theta/\partial t^2)\,dx$, where I_0 is the mass moment of inertia per unit length about the longitudinal axis given by $I_0 = (m/A)J$. For equilibrium of the infinitesimal element

$$\frac{\partial T}{\partial x}\,dx - \frac{mJ}{A}\frac{\partial^2\theta}{\partial t^2}\,dx + m_T\,dx = 0$$

or

$$\frac{\partial T}{\partial x} - \frac{mJ}{A}\frac{\partial^2\theta}{\partial t^2} + m_T = 0 \tag{14.48}$$

The torque T is related to the angle of twist θ by the following expression

$$T = GJ\frac{\partial\theta}{\partial x} \tag{14.49}$$

Substitution of Equation 14.49 in Equation 14.48 gives

$$\frac{\partial}{\partial x}\left(GJ\frac{\partial\theta}{\partial x}\right) - \frac{mJ}{A}\frac{\partial^2\theta}{\partial t^2} + m_T = 0 \tag{14.50}$$

Equation 14.50 is the governing equation of motion. It is entirely analogous to the equation of motion for the axial vibration of a bar. The boundary conditions are obtained by noting that at the clamped end, the angle of twist θ must be zero, while at the free end, the torque must be zero

$$\theta = 0 \quad \text{at } x = 0$$

$$GJ\frac{\partial\theta}{\partial x} = 0 \quad \text{at } x = L \tag{14.51}$$

14.8   TRANSVERSE VIBRATIONS OF A STRING

A string of mass $m(x)$ per unit length is stretched in a horizontal direction, the initial tension in the string being $T(x)$. As shown in Figure 14.9a, the string is undergoing transverse vibrations under the action of distributed vertical force $p(x,t)$ per unit length. Since the cross-sectional dimensions of the string are very small compared to its length, the string has negligible flexural rigidity and the restoring forces are provided only by the initial tension in the string. We also assume that the lateral displacements are small so that the tension in the string does not undergo any appreciable change as the string vibrates. Thus, although T may vary along the length, it does not vary with time.

The forces acting on an infinitesimal section of the string are shown in Figure 14.9b. Equating the sum of vertical components of the forces acting on the element to zero, we get

$$\left(T + \frac{\partial T}{\partial x}\,dx\right)\left(\frac{\partial u}{\partial x} + \frac{\partial^2 u}{\partial x^2}\,dx\right) - T\frac{\partial u}{\partial x} + p\,dx - m\frac{\partial^2 u}{\partial t^2}\,dx = 0 \tag{14.52}$$

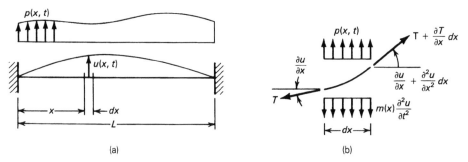

Figure 14.9. (a) Transverse vibrations of a string; (b) forces acting on a small element.

On canceling out dx and neglecting the higher-order terms, Equation 14.52 reduces to

$$\frac{\partial T}{\partial x}\frac{\partial u}{\partial x} + T\frac{\partial^2 u}{\partial x^2} - m\frac{\partial^2 u}{\partial x^2} + p = 0$$

or

$$\frac{\partial}{\partial x}\left(T\frac{\partial u}{\partial x}\right) - m\frac{\partial^2 u}{\partial x^2} + p = 0 \qquad (14.53)$$

If both ends of the string are rigidly attached to fixed supports, the transverse displacements at the two ends should be zero and we get the two geometric boundary conditions

$$u = 0 \quad \text{at } x = 0$$
$$u = 0 \quad \text{at } x = L \qquad (14.54)$$

On the other hand, if one of the two ends, say the end $x = L$, is free to slide in a transverse direction, the vertical component of the string tension at that end must be zero and we get

$$T\frac{\partial u}{\partial x} = 0 \quad \text{at } x = L \qquad (14.55)$$

Equation 14.55 represents a natural boundary condition.

14.9   TRANSVERSE VIBRATIONS OF A SHEAR BEAM

A beam in which the slope at any section is proportional to the shear acting at that section is called a *shear beam*. Referring to Figure 14.10, the relationship

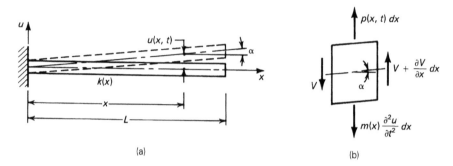

Figure 14.10. (a) Transverse vibrations of a shear beam; (b) forces acting on a small element.

between the shear force V and the slope α at a section can be expressed as

$$V = k\alpha$$

$$= k\frac{\partial u}{\partial x} \tag{14.56}$$

where k is the shear stiffness. To derive the equation of motion for a shear beam, we consider the equilibrium of an infinitesimal element of the beam. The forces acting on the element are shown in Figure 14.10b. They include the shear forces acting on the two sections, the inertial force and the applied load. For equilibrium of the beam element, we have

$$\frac{\partial V}{\partial x}dx - m\frac{\partial^2 u}{\partial t^2}dx + p\,dx = 0 \tag{14.57}$$

On substitution for V from Equation 14.56, Equation 14.57 yields

$$\frac{\partial}{\partial x}\left(k\frac{\partial u}{\partial x}\right) - m\frac{\partial^2 u}{\partial t^2} + p = 0 \tag{14.58}$$

which is the governing equation of motion. The geometric boundary condition at a clamped end is

$$u = 0 \tag{14.59}$$

while the natural boundary condition at a free end is

$$k\frac{\partial u}{\partial x} = 0 \tag{14.60}$$

Many structural systems can be idealized as a shear beam. As an example, consider the multistory building shown in Figure 14.11. If the floor beams are considered infinitely stiff in flexure and all columns in a story are of equal height, the building can be idealized as a shear beam. With stiff floors, any

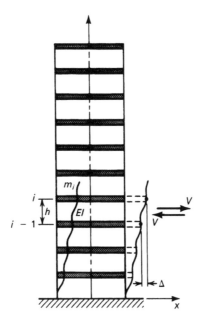

Figure 14.11. Modeling of a multistory building as a shear beam.

relative displacement between two adjacent stories is resisted by the flexural deformations of the columns. If V is the story shear and Δ is the relative story displacement, we have the relationship

$$V = \left(\sum \frac{12EI}{h^3} \right) \Delta \qquad (14.61)$$

where h is the story height, I represents the moment of inertia of a column and the summation is taken over all columns in the story. Since h is the same for all columns, we can write Equation 14.61 as

$$V = \left(\sum \frac{12EI}{h^2} \right) \frac{\Delta}{h} \qquad (14.62)$$

If we take $k = \sum(12EI/h^2)$ and $\alpha = \Delta/h$, Equations 14.56 and 14.62 become identical. The mass per unit length between floors $i - 1$ and i can be taken as $m = m_i/h$, where m_i is the mass of the ith floor. In a similar manner, the external load per unit length is given by $p = p_i/h$, where p_i is the transverse force acting at floor level i. Another example of a system that can be idealized as a shear beam is an isotropic or horizontally layered soil deposit undergoing horizontal deformations, as shown in Figure 14.12. In this case, k is the shear modulus, m the mass per unit volume, and p is the horizontal force per unit area.

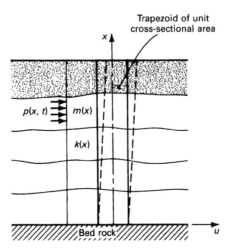

Figure 14.12. Horizontally layered soil deposit modeled as a shear beam.

It is of interest to compare the equations of motion for the axial vibrations of a bar (Eq. 14.44), the torsional vibration of a circular shape (Eq. 14.50), the transverse vibrations of a string (Eq. 14.53) and the transverse vibrations of a shear beam (Eq. 14.58). In the absence of an external force, the vibrations of all of these systems are governed by an equation of the form

$$\frac{\partial}{\partial x}\left(k\frac{\partial u}{\partial x}\right) - m\frac{\partial^2 u}{\partial t^2} = 0 \tag{14.63}$$

in which k represents a stiffness and m is a measure of the mass or the mass moment of inertia per unit length. Since the equations governing the vibrations of the aforementioned systems are similar, their response has similar characteristics and identical solution procedures apply to each case. Equation 14.63, known as a *wave equation*, is of considerable interest in the field of dynamics of structures. Its special characteristics make it possible to devise simplified procedures for its solution. These procedures are discussed briefly in Chapter 17.

14.10   TRANSVERSE VIBRATIONS OF A BEAM EXCITED BY SUPPORT MOTION

Dynamic vibrations may be set up in a continuous system due to the motion of the supports to which the system is attached. A ground-supported vertical chimney subjected to an earthquake motion of the ground can be cited as one

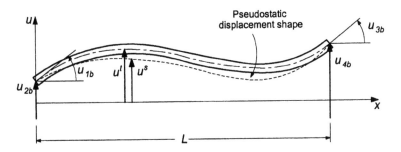

Figure 14.13. Transverse vibrations of a beam due to support motion.

example of a continuous system excited by the motion of the support. Piping system in a mechanical plant may also be subjected to vibrations due to dynamic motion of the supports.

To illustrate the effect of support motion on the response of the dynamic system, we consider the example of transverse vibrations of a beam caused by support motion. In the most general case, four different support motions are possible: two rotations and two lateral translation at each support. As indicated in Figure 14.13, we denote these motions by $u_{1b}(t)$, $u_{2b}(t)$, $u_{3b}(t)$, and $u_{4b}(t)$, respectively.

Support motion does not modify the governing equation of motion. Thus, for a beam with damping resistance included but without the effect of shear deformation and rotatory inertia, the motion is still governed by Equation 14.29, provided that u in that equation is taken as the absolute displacement with reference to a fixed frame of axis. However, it is sometimes more convenient to express the absolute or total displacement as a sum of the static displacement caused by the prescribed support motion and a displacement relative to the support displacement. Denoting the static displacement by u^s, we have

$$u^t = u^s + u \qquad\qquad (14.64)$$

where u^t is the total or absolute displacement, and u is the relative displacement. Assuming that there is no external load, the equation of motion written in terms of the total displacement becomes

$$\frac{\partial^2}{\partial x^2}\left(EI\frac{\partial^2 u^t}{\partial x^2}\right) + \frac{\partial^2}{\partial x^2}\left(c_s I\frac{\partial^3 u^t}{\partial t\partial x^2}\right) + m\frac{\partial^2 u^t}{\partial t^2} + c\frac{\partial u^t}{\partial t} = 0 \qquad\qquad (14.65)$$

The boundary conditions must be modified to account for the support motion. The modified conditions are given by

$$\frac{\partial u^t}{\partial x}(0,t) = u_{1b}(t)$$

$$u^t(0,t) = u_{2b}(t)$$

$$\frac{\partial u^t}{\partial x}(L,t) = u_{3b}(t)$$ (14.66)

$$u^t(L,t) = u_{4b}(t)$$

The static displacement caused by a support motion is obtained by assuming that the application of time-varying motion does not produce any accelerations or inertia forces, and is referred to as a pseudostatic displacement. To obtain the pseudostatic displacement caused by say $u_{1b}(t)$, we first calculate the displacement shape $\psi_1(x)$ due to the application of a unit static displacement along degree-of-freedom 1, displacements along the other three degrees of freedom being zero. The pseudostatic displacement due to $u_{1b}(t)$ is then given by

$$u^s(x,t) = \psi_1(x)u_{1b}(t)$$ (14.67)

Displacement shape $\psi_1(x)$ must also satisfy Equation 14.65 but with the velocities and accelerations equal to zero. Thus

$$\frac{\partial^2}{\partial x^2}\left(EI\frac{\partial^2\psi_1}{\partial x^2}\right) = 0$$ (14.68)

The pseudostatic displacement due to simultaneous application of u_{1b}, u_{2b}, u_{3b}, and u_{4b} is obtained by superposition

$$u^s(x,t) = \psi_1(x)u_{1b}(t) + \psi_2(x)u_{2b}(t) + \psi_3(x)u_{3b}(t) + \psi_4(x)u_{4b}(t)$$ (14.69)

A relationship similar to Equation 14.68 can now be written involving the pseudostatic displacement caused by simultaneous application of the four support motions

$$\sum_{i=1}^{4} u_{ib}\frac{\partial^2}{\partial x^2}\left(EI\frac{\partial^2\psi_i}{\partial x^2}\right) = 0$$ (14.70)

On substituting Equations 14.64 and 14.69 in Equation 14.65 and using Equation 14.70, we obtain the equation of motion in terms of the relative displacement and the specified support motions

$$\frac{\partial^2}{\partial x^2}\left(EI\frac{\partial^2 u}{\partial x^2}\right) + \frac{\partial^2}{\partial x^2}\left(c_sI\frac{\partial^3 u}{\partial t\partial x^2}\right) + m\frac{\partial^2 u}{\partial t^2} + c\frac{\partial u}{\partial t} = p_{\text{eff}}$$ (14.71)

where

$$p_{\text{eff}} = -\frac{\partial^2}{\partial x^2}\left(c_s I \frac{\partial^3 u^s}{\partial t \partial x^2}\right) - c\left(\frac{\partial u^s}{\partial t}\right) - m\frac{\partial^2 u^s}{\partial t^2} \tag{14.72}$$

The effective force components due to damping are small and are generally neglected, so that p_{eff} reduces to

$$\begin{aligned} p_{\text{eff}} &= -m\frac{\partial^2 u^s}{\partial t^2} \\ &= -m\sum_{i=1}^{4} \psi_i(x)\ddot{u}_{ib}(t) \end{aligned} \tag{14.73}$$

Equation 14.71 is the equation of motion governing the response caused by support motions. The boundary conditions are all geometric and specify that the displacements and slopes in terms of u are all zero. As an example, for the rotation at the left-hand end, we have, from Equation 14.66

$$\frac{\partial u^s}{\partial x}(0,t) + \frac{\partial u}{\partial x}(0,t) = \psi_1'(0)u_{1b}(t) \tag{14.74}$$

in which, $\psi_1'(0)=1$. Since, from Equation 14.67, $\partial u^s(0,t)/\partial x = \psi_1'(0)u_{1b}(t)$, Equation 14.74 reduces to

$$\frac{\partial u(0,t)}{\partial x} = 0 \tag{14.75}$$

For a uniform beam, the shape functions ψ_i are Hermitian polynomials given by Equation 3.79.

Example 14.3
Obtain the equation of motion for the lateral vibrations of a ground-supported vertical chimney shown in Figure E14.3 subjected to a horizontal earthquake motion $u_g(t)$. Neglect shear deformation and rotatory inertia.

Solution
The motion of the chimney is governed by the equation for the lateral vibrations of a beam. The end $x=L$ is free, while the support at $x=0$ undergoes a prescribed displacement. The pseudostatic displacement caused by the support motion is a rigid body movement in the lateral direction given by

$$u^s(x,t)=u_g(t) \tag{a}$$

Substitution in Equation 14.73 gives the effective force

$$p_{\text{eff}} = -m\ddot{u}_g(t) \tag{b}$$

The equation of motion is now obtained from Equation 14.71

$$\frac{\partial^2}{\partial x^2}\left(EI\frac{\partial^2 u}{\partial x^2}\right) + m\frac{\partial^2 u}{\partial t^2} = -m\ddot{u}_g(t) \tag{c}$$

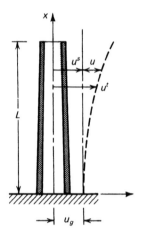

Figure E14.3. Lateral vibrations of a chimney induced by ground motion.

where u is the displacement relative to the pseudostatic displacement u^s. The boundary conditions written in terms of u are

$$u = 0 \qquad x = 0$$

$$\frac{\partial u}{\partial x} = 0 \qquad x = 0$$

$$EI\frac{\partial^2 u}{\partial x^2} = 0 \qquad x = L \tag{d}$$

$$\frac{\partial}{\partial x}\left(EI\frac{\partial^2 u}{\partial x^2}\right) = 0 \qquad x = L$$

After Equation c has been solved for u, the total motion can be obtained by adding the rigid-body displacement

$$u^t = u + u^s$$

$$= u + u_g \tag{e}$$

14.11   EFFECT OF AXIAL FORCE ON TRANSVERSE VIBRATIONS OF A BEAM

In previous sections of this chapter, we considered the transverse flexural vibrations of a beam assuming that no axial loads were present and that vibrations in the axial directions were not excited. We also considered the case of axial vibrations alone without the presence of flexural vibrations. In general, combined flexural and axial vibrations may exist. The governing equations of motion in such a case are coupled and their solution is quite complex.

The presence of an axial load will influence the transverse vibrations of a beam even if simultaneous axial vibrations did not exist, but the problem would be somewhat simplified. In fact, it is quite reasonable to ignore the presence

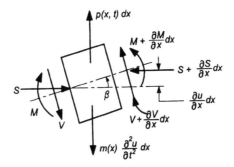

Figure 14.14. Forces acting on a small element of a beam undergoing transverse vibrations under the presence of an axial load.

of axial vibrations, provided that the axial rigidity is large compared to the flexural rigidity so that the axial deformations are comparatively small. In the following, we derive the equation of motion for the transverse vibrations of a beam in the presence of an axial load but assuming that axial deformations are negligible. For simplicity, we assume that the effect of shear deformation and rotatory inertia is also negligible.

Figure 14.14 shows an infinitesimal element of the beam under transverse vibration along with the forces acting on it. When compared to the elementary case of flexural vibrations, the only new forces are the axial force $S(x)$ on the left-hand section and the force $S + [\partial S(x)/\partial x]\,dx$ on the right-hand section, both assumed positive when they produce a compression on the section. The vertical force balance is still represented by Equation 14.1b. However, in writing the equation of moment equilibrium, we must take into account the additional anticlockwise moment contributed by the axial forces. Noting that the line of action of the axial force on the right-hand section is displaced a distance $(\partial u/\partial x)\,dx$ with respect to that on the left-hand section, the moment balance equation becomes

$$V + \frac{\partial M}{\partial x} + S(x)\frac{\partial u}{\partial x} = 0 \qquad (14.76)$$

Substitution of Equation 14.5 in Equation 14.76 gives

$$V = -\frac{\partial}{\partial x}\left(EI\frac{\partial^2 u}{\partial x^2}\right) - S(x)\frac{\partial u}{\partial x} \qquad (14.77)$$

It will be noted that the vertical force V now has two components: a beam shear

$$-\frac{\partial}{\partial x}\left(EI\frac{\partial^2 u}{\partial x^2}\right)$$

and a component arising from the presence of axial load. On differentiating Equation 14.77 with respect to x and substituting in Equation 14.1b, we get

$$\frac{\partial^2}{\partial x^2}\left(EI\frac{\partial^2 u}{\partial x^2}\right) + \frac{\partial}{\partial x}\left(S(x)\frac{\partial u}{\partial x}\right) + m\frac{\partial^2 u}{\partial t^2} = p \tag{14.78}$$

Equation 14.78 is the equation governing the transverse vibrations of a beam in the presence of an axial load. It is instructive to derive this equation by using Hamilton's expression. In doing so, we must account for the potential energy of the axial force. Referring to Figure 14.14, we note that the axial forces move toward each other a distance Δs given by

$$\Delta s = dx(1 - \cos \beta) \tag{14.79}$$

where $\beta = \partial u/\partial x$ is the slope of the elastic axis.

For small displacements, Equation 14.79 reduces to

$$\Delta s = \frac{1}{2}\left(\frac{\partial u}{\partial x}\right)^2 dx \tag{14.80}$$

The work done by the axial force in moving through the distance Δs is $\frac{1}{2}S(x)(\partial u/\partial x)^2\, dx$. Hence, the potential energy of the axial load is

$$V_s = -\int_0^L \frac{1}{2}S(x)\left(\frac{\partial u}{\partial x}\right)^2 dx \tag{14.81}$$

The variation of V_s is given by

$$\begin{aligned}
\delta V_s &= -\int_0^L S(x)\frac{\partial u}{\partial x}\frac{\partial}{\partial x}(\delta u)\, dx \\
&= -S(x)\frac{\partial u}{\partial x}\delta u\Big|_0^L + \int_0^L \frac{\partial}{\partial x}\left\{S(x)\frac{\partial u}{\partial x}\right\}\delta u\, dx
\end{aligned} \tag{14.82}$$

On adding this variation to Equation 14.17, the latter gets modified to

$$\int_{t_1}^{t_2}\left[\int_0^L\left\{-m\frac{\partial^2 u}{\partial t^2} - \frac{\partial^2}{\partial x^2}\left(EI\frac{\partial^2 u}{\partial x^2}\right) - \frac{\partial}{\partial x}\left(S(x)\frac{\partial u}{\partial x}\right) + p\right\}\delta u\, dx\right. \tag{14.83}$$
$$\left. -EI\frac{\partial^2 u}{\partial x^2}\delta\left(\frac{\partial u}{\partial x}\right)\Big|_0^L + \left\{S(x)\frac{\partial u}{\partial x} + \frac{\partial}{\partial x}\left(EI\frac{\partial^2 u}{\partial x^2}\right)\right\}\delta u\Big|_0^L\right] dt$$

Hence, the revised equation of motion becomes

$$\frac{\partial^2}{\partial x^2}\left(EI\frac{\partial^2 u}{\partial x^2}\right) + \frac{\partial}{\partial x}\left(S(x)\frac{\partial u}{\partial x}\right) + m\frac{\partial^2 u}{\partial t^2} = p \tag{14.84}$$

while the boundary conditions reduce to

$$EI\frac{\partial^2 u}{\partial x^2}\delta\left(\frac{\partial u}{\partial x}\right)\Bigg|_0^L = 0 \tag{14.85a}$$

$$\left\{\frac{\partial}{\partial x}\left(EI\frac{\partial^2 u}{\partial x^2}\right) + S(x)\frac{\partial u}{\partial x}\right\}\delta u\Bigg|_0^L = 0 \tag{14.85b}$$

Equation 14.84 is identical to Equation 14.78. The boundary condition Equation 14.85a is also the same as Equation 14.19a, but Equation 14.85b is a revised form of Equation 14.19b. The boundary condition in Equation 14.85b is satisfied provided that, on a boundary, either the displacement u is zero or the vertical force given by

$$V = -\left(\frac{\partial}{\partial x}EI\frac{\partial^2 u}{\partial x^2}\right) - S(x)\frac{\partial u}{\partial x} \tag{14.86}$$

is zero.

SELECTED READINGS

Clark, S.K. 1972. *Dynamics of Continuous Elements.* Englewood Cliffs, New Jersey: Prentice Hall.

Den Hartog, J.P. 1956. *Mechanical Vibrations.* New York: McGraw-Hill. 4th Edition.

Jacobsen, L.S. & Ayre, R.S. 1958. *Engineering Vibrations with Applications to Structures and Machinery.* New York: McGraw-Hill.

Meirovitch, L. 1967. *Analytical Methods in Vibration.* London: Macmillan.

Mindlin, R.D. & Goodman, L.E. 1950. Beam Vibrations with Time dependent Boundary Conditions. *Journal of Applied Mechanics*, Vol. 17: 377–380.

Stokey, W.F. 1996. Vibrations of Systems Having Distributed Mass and Elasticity. In C.M. Harris (ed.) *Shock and Vibration Handbook.* New York: McGraw Hill. 4th Edition.

Timoshenko, S., Young, D.H. & Weaver, W. Jr. 1990. *Vibration Problems in Engineering.* New York: John Wiley. 5th Edition.

PROBLEMS

14.1 A uniform beam of mass m per unit length and flexural rigidity EI is pinned at its left hand end and supported by a spring of stiffness k on the right hand end as shown in Figure P14.1. Using Hamilton's principle, obtain the equation of motion and the boundary conditions for free transverse vibrations of the beam.

14.2 The frame shown in Figure P14.2 is subjected to a dynamic moment $M(t)$ at joint B. Obtain the equations of motion and the boundary conditions governing the response of the frame. Assume that the members are axially rigid. (*Hint*: The slope continuity and moment equilibrium at B provide 3 out of the 12 boundary conditions.)

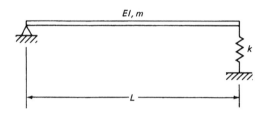

Figure P14.1.

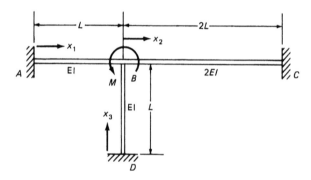

Figure P14.2.

14.3 A nonuniform bar clamped at the left end and free at the right end vibrates in an axial direction due to a motion of the clamped end parallel to the axis of the bar. If $m(x)$ is the mass per unit length, $A(x)$ the cross-sectional area, c the viscous damping coefficient per unit length, c_s the coefficient that converts the internal strain rate into a resisting stress, and u_g the support displacement, obtain the equation of motion in terms of displacement relative to the support. Also define the boundary conditions.

14.4 A uniform circular shaft of length L has a mass moment of inertia per unit length I_0 and a polar moment of inertia J (Fig. P14.4). It is clamped at its left end and is rigidly attached at the right end to a uniform flywheel of mass M_1 and radius R. A mass M_2 is suspended by an inextensible massless string wound around the flywheel. Obtain the equation of motion and the boundary conditions for torsional vibrations of the shaft. The shear modulus is G.

14.5 The concrete umbrella roof structure shown in Figure P14.5 consists of a rigid circular plate 75mm thick and 2m in radius. The supporting circular column has a radius of 150 mm and a height of 3.5 m. The mass density of concrete is 2400 kg/m^3 and the modulus of elasticity is 25,000 MPa. Obtain the equation of motion and the boundary conditions for lateral

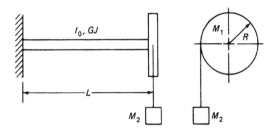

Figure P14.4.

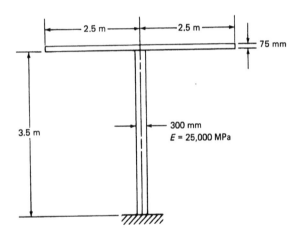

Figure P14.5.

vibrations of the structure (a) neglecting the effect of axial load due to the roof, (b) taking into account the axial load effect.

14.6 The thirty-story building with a setback at midheight shown in Figure P14.6 is idealized as a shear beam. The total dead load of each floor in the base portion is 4700 kN, in the tower portion it is 2000 kN. The floor to floor height is 3.5 m throughout. The total story stiffness in the base portion is 720, 000 kN/mm, in the tower it is 260, 000 kN/mm. Obtain the equations of motion and boundary conditions for lateral vibrations of the building.

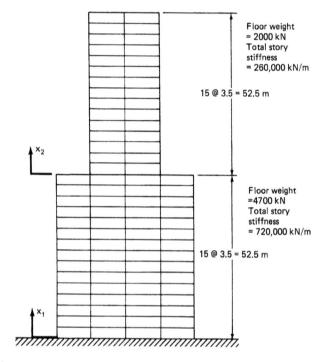

Figure P14.6.

CHAPTER 15

Continuous systems: Free vibration response

15.1 INTRODUCTION

Vibrations can be induced in a continuous system even without the presence of external forces, in the same way as in a discrete system. Free vibrations of this type are caused whenever the system is subjected to disturbances in the form of initial displacements and/or initial velocities. The equation governing the free-vibration response of a continuous system is obtained from the corresponding boundary value problem by setting the forcing function equal to zero. The resulting equation defines a problem that is referred to as an *eigenvalue problem*. Complete analogy exists between the eigenvalue problem for a continuous system and that for a discrete system. However, the eigenvalue problem for a continuous system involves a homogeneous differential equation rather than a set of algebraic equations. Solution of this differential equation leads to the determination of a set of eigenvalues and eigenfunctions which, besides being useful by themselves, can be employed in obtaining solutions to problems in forced vibration response.

In this chapter, we discuss the free-vibration response of several continuous systems. Our discussion is restricted to simple systems in which the response parameters are functions of the time coordinate and one space coordinate. Since the solution of eigenvalue problem is closely interconnected with the free-vibration response problem, and since eigenvalues and eigenfunctions play a central role in the dynamic analysis of many continuous systems, we devote considerable attention to the solution of the eigenvalue problem and to the properties of eigenvalues and eigenfunctions.

As an illustration of the nature of the problem and of the properties of the eigenvalues and eigenfunctions, the eigenvalue problem related to the free lateral vibrations of a beam is discussed first. This is followed by the presentation of a general treatment of the eigenvalue problem for simple continuous systems. The frequencies and mode shapes of a number of such systems are derived next. Finally, analysis of the free-vibration response of a continuous system is described.

15.2   EIGENVALUE PROBLEM FOR THE TRANSVERSE VIBRATIONS OF A BEAM

Neglecting shear deformation and rotatory inertia, the equation of undamped free transverse vibrations of a beam is obtained from Equation 14.7 by setting $p = 0$

$$\frac{\partial^2}{\partial x^2}\left(EI\frac{\partial^2 u}{\partial x^2}\right) + m\frac{\partial^2 u}{\partial t^2} = 0 \tag{15.1}$$

Equation 15.1 is a fourth-order linear homogeneous partial differential equation. We attempt a solution of Equation 15.1 of the form

$$u = f(x)g(t) \tag{15.2}$$

where $f(x)$ is a function of x alone and $g(t)$ is a function of t alone. Substitution of Equation 15.2 in Equation 15.1 gives

$$g(t)\frac{d^2}{dx^2}\left\{EI\frac{d^2 f(x)}{dx^2}\right\} + mf(x)\frac{d^2 g(t)}{dt^2} = 0 \tag{15.3a}$$

or

$$\frac{1}{mf(x)}\frac{d^2}{dx^2}\left\{EI\frac{d^2 f(x)}{dx^2}\right\} = -\frac{1}{g(t)}\frac{d^2 g(t)}{dt^2} \tag{15.3b}$$

The terms on the left-hand side of Equation 15.3b, including m and EI, are all functions of x alone, while the terms on the right-hand side are functions of t alone. The equality can therefore hold only provided that each of the two sides of the equation is equal to a constant, normally referred to as a *separation constant*. Let us select this constant to be equal to ω^2. The reason for selecting a positive constant will become apparent from the discussion that follows. Equation 15.3b leads to two separate equations, as follows

$$\frac{d^2 g(t)}{dt^2} + \omega^2 g(t) = 0 \tag{15.4}$$

$$\frac{d^2}{dx^2}\left\{EI\frac{d^2 f(x)}{dx^2}\right\} = \omega^2 mf(x) \tag{15.5}$$

The solution of Equation 15.4 is given by

$$g(t) = A\sin\omega t + B\cos\omega t \tag{15.6}$$

in which A and B are as yet unknown constants. They will be determined from two initial conditions: the initial displacement and velocity profiles of the beam. The selection of a positive separation constant has ensured that $g(t)$ be a harmonic function of time which implies conservation of energy in the system. A negative separation constant would, on the other hand, give a solution involving exponential growth in the displacement, and hence in the energy, which is inadmissible on physical grounds.

So far, in our solution, we have not used the four boundary conditions given by Equations 14.19a and 14.19b. When u is of the form of Equation 15.2, the boundary conditions lead to the following relationships

$$EI\frac{d^2f(x)}{dx^2}\delta\left\{\frac{df(x)}{dx}\right\}\bigg|_0^L = 0 \tag{15.7a}$$

$$\frac{d}{dx}\left\{EI\frac{d^2f(x)}{dx^2}\right\}\delta f(x)\bigg|_0^L = 0 \tag{15.7b}$$

Equation 15.7a is satisfied either by specifying that the moment $M(x)=EI\{d^2f(x)/dx^2\}$ is zero or that the slope df/dx is zero. In a similar manner, Equation 15.7b is satisfied when either the shear $V(x)=-d/dx[EI\{d^2f(x)/dx^2\}]$ is zero or the displacement $f(x)$ is zero.

Equation 15.5, along with the boundary conditions given by Equations 15.7a and b, represents an eigenvalue problem. By substitution it is easily verified that $f(x)=0$ is a possible solution of Equation 15.5. Such a solution is, however, trivial, and since it represents no motion of the system, of little interest. A nontrivial solution for Equation 15.5 is possible only for a special value of ω^2. We will find that there are an infinite number of such values separated by discrete intervals. These values are referred to as *eigenvalues* of the system and are denoted by the symbol λ. The square root of an eigenvalue is known as the *frequency* of the system. Corresponding to each eigenvalue, there exists a solution for $f(x)$, called an *eigenfunction* or a *mode shape*, which also satisfies the boundary conditions of Equations 15.7.

Let λ_i and λ_j be any two eigenvalues of the vibrating beam and let $f_i(x)$ and $f_j(x)$ be the corresponding eigenfunctions. Since the eigenfunctions must satisfy Equation 15.5, we have

$$\frac{d^2}{dx^2}\left\{EI\frac{d^2f_i(x)}{dx^2}\right\} = \lambda_i m f_i(x) \tag{15.8a}$$

$$\frac{d^2}{dx^2}\left\{EI\frac{d^2f_j(x)}{dx^2}\right\} = \lambda_j m f_j(x) \tag{15.8b}$$

We multiply both sides of Equation 15.8a by $f_j(x)$ and integrate over the length to get

$$\int_0^L f_j(x)\frac{d^2}{dx^2}\left\{EI\frac{d^2f_i(x)}{dx^2}\right\}dx = \lambda_i \int_0^L m f_j(x)f_i(x)\,dx \tag{15.9}$$

The left-hand side of Equation 15.9 can be reduced by repeated partial integration as follows

$$\int_0^L f_j(x)\frac{d^2}{dx^2}\left\{EI\frac{d^2 f_i(x)}{dx^2}\right\}dx$$

$$= f_j(x)\frac{d}{dx}\left\{EI\frac{d^2 f_i(x)}{dx^2}\right\}\Big|_0^L - \int_0^L \frac{df_j(x)}{dx}\frac{d}{dx}\left\{EI\frac{d^2 f_i(x)}{dx^2}\right\}dx$$

$$= -\frac{df_j(x)}{dx}\left\{EI\frac{d^2 f_i(x)}{dx^2}\right\}\Big|_0^L + \int_0^L \frac{d^2 f_j(x)}{dx^2}\left\{EI\frac{d^2 f_i(x)}{dx^2}\right\}dx$$

$$= \int_0^L EI\frac{d^2 f_i(x)}{dx^2}\frac{d^2 f_j(x)}{dx^2}dx \tag{15.10}$$

in which we have used the fact that $f_i(x)$ and $f_j(x)$ must satisfy the boundary conditions at the two ends of the beam so that

$$f_j(x)\frac{d}{dx}\left\{EI\frac{d^2 f_i(x)}{dx^2}\right\} \quad \text{and} \quad \frac{df_j(x)}{dx}\left\{EI\frac{d^2 f_i(x)}{dx^2}\right\}$$

are zero at both $x=0$ and $x=L$. Equation 15.9 thus reduces to

$$\int_0^L EI\frac{d^2 f_i(x)}{dx^2}\frac{d^2 f_j(x)}{dx^2}dx = \lambda_i \int_0^L m f_i(x)f_j(x)dx \tag{15.11}$$

We next multiply both sides of Equation 15.8b by $f_i(x)$ and integrate over the length, obtaining

$$\int_0^L f_i(x)\frac{d^2}{dx^2}\left\{EI\frac{d^2 f_j(x)}{dx^2}\right\}dx = \lambda_j \int_0^L m f_i(x)f_j(x)dx \tag{15.12}$$

Equation 15.12 can be reduced by partial integration in a manner identical to that used for the reduction of Equation 15.9, so that we get

$$\int_0^L EI\frac{d^2 f_i(x)}{dx^2}\frac{d^2 f_j(x)}{dx^2}dx = \lambda_j \int_0^L m f_i(x)f_j(x)dx \tag{15.13}$$

Subtraction of Equation 15.13 from Equation 15.11 gives

$$(\lambda_i - \lambda_j)\int_0^L m f_i(x)f_j(x)dx = 0 \tag{15.14}$$

or

$$\int_0^L m f_i(x)f_j(x)dx = 0 \quad \lambda_i \neq \lambda_j \tag{15.15}$$

Substitution of Equation 15.15 in Equation 15.9 or Equation 15.12 gives

$$\int_0^L f_i(x)\frac{d^2}{dx^2}\left\{EI\frac{d^2 f_j(x)}{dx^2}\right\}dx = \int_0^L f_j(x)\frac{d^2}{dx^2}\left\{EI\frac{d^2 f_i(x)}{dx^2}\right\}dx$$

$$= 0 \quad \lambda_i \neq \lambda_j \tag{15.16}$$

Equations 15.15 and 15.16 express the *orthogonality property* of the eigenfunctions, namely that eigenfunctions corresponding to two distinct eigenvalues are orthogonal to each other.

If $f_i(x)$ is a solution of Equation 15.5, it is easily verified by substitution that $\alpha f_i(x)$, where α is a constant, is also a solution of Equation 15.5. Furthermore, if $f_i(x)$ satisfies the boundary conditions given by Equation 15.7, $\alpha f_i(x)$ will also satisfy those conditions. It is thus apparent that the magnitude of the eigenfunctions is not unique and we can scale an eigenfunction in any way that we wish. The process of scaling is called *normalization*. We will find it convenient to normalize the eigenfunctions so that the normalized value denoted by $\phi_i(x)$ satisfies the relationship

$$\int_0^L m\phi_i^2(x)\,dx = 1 \tag{15.17}$$

Substitution of $\phi_i(x)$ for $f(x)$ in Equation 15.5 gives

$$\frac{d^2}{dx^2}\left\{EI\frac{d^2\phi_i(x)}{dx^2}\right\} = \omega_i^2 m\phi_i(x) \tag{15.18}$$

On multiplying both sides of Equation 15.8 by $\phi_i(x)$ and integrating over the length, we get

$$\int_0^L \phi_i(x)\frac{d^2}{dx^2}\left\{EI\frac{d^2\phi_i(x)}{dx^2}\right\}dx = \omega_i^2 \int_0^L m\phi_i^2(x)\,dx$$

$$= \omega_i^2 \tag{15.19}$$

Equation 15.19 implies that the integral on the left-hand side must always be positive; otherwise, the selection of a positive constant to represent the two sides of Equation 15.3b would not be justified. Repeated partial integration of the left-hand side of Equation 15.19 and application of boundary conditions gives

$$\int_0^L \phi_i(x)\frac{d^2}{dx^2}\left\{EI\frac{d^2\phi_i(x)}{dx^2}\right\}dx = \int_0^L EI\frac{d^2\phi_i(x)}{dx^2}\frac{d^2\phi_i(x)}{dx^2}dx$$

$$= \int_0^L EI\left\{\frac{d^2\phi_i(x)}{dx^2}\right\}^2 dx \tag{15.20}$$

The integral on the right-hand side of Equation 15.20 is obviously positive, justifying the selection of positive constant ω_i^2. This also implies that the eigenvalues of the beam are both real and positive.

15.3   GENERAL EIGENVALUE PROBLEM FOR A CONTINUOUS SYSTEM

15.3.1   *Definition of the eigenvalue problem*

The procedure that we used in the preceding section to obtain a solution of the free-vibration equation of a beam is commonly known as the method of *separation of variables*, in which it is assumed that the unknown function u can be expressed as the product of a function of spatial coordinate and a function of time. The procedure leads to two separated linear homogeneous differential equations, one of which involves the time variable t while the other involves the space variable x. The second of these equations, along with the associated boundary conditions, is called an eigenvalue problem.

In general, the eigenvalue problem for a continuous system can be expressed as

$$K\{f(x)\} = \lambda M\{f(x)\} \tag{15.21}$$

in which K and M are linear differential operators of order $2r$ and $2s$, respectively, and having the form

$$A_1 + A_2\frac{d}{dx} + A_3\frac{d^2}{dx^2} + \cdots \tag{15.22}$$

where $A_1, A_2, \ldots$ are functions of x. The order of operator M is less than the order of K; that is, $2s$ is less than $2r$. Associated with the differential equation 15.21 are r boundary conditions at each end of the length. These conditions are of the form

$$B_n\{f(x)\} = 0 \quad n = 1, 2, \ldots, r \tag{15.23}$$

where B_n are linear differential operators of the form of Equation 15.22 but of an order at most $2r - 1$.

For nonzero u, Equation 15.21 can be satisfied only for specific values of λ. These values of λ are called the eigenvalues of the system. A continuous system of finite extent has an infinite number of eigenvalues separated by discrete intervals. Associated with each eigenvalue λ_i, there is a function $u = f_i(x)$ which satisfies Equation 15.21 and its boundary conditions. This function is called an eigenfunction of the system. The eigenvalues need not all be distinct and two or more eigenvalues may be identical. It is readily seen by substitution that if $f_i(x)$ is an eigenfunction satisfying Equation 15.21, then $\alpha f_i(x)$, α being a constant, is also an eigenfunction. The magnitude of an eigenfunction is thus arbitrary and we can scale the eigenfunction in any convenient manner. The process of scaling is called normalization. We will denote a normalized eigenfunction by $\phi_i(x)$. It is usual to normalize the eigenfunction so that

$$\int_0^L \phi_i(x)M\{\phi_i(x)\}\,dx = 1 \tag{15.24}$$

Eigenfunctions normalized in such a manner are said to be *mass orthonormal*. The eigenvalues and eigenfunctions possess certain special properties which we shall discuss in the next few sections.

As an example of an eigenvalue problem, consider the lateral vibrations of a cantilever beam fixed at the left-hand end. The eigenvalue equation for the beam is given by Equation 15.5. It is easily recognized that

$$K = \frac{d^2}{dx^2}\left(EI\frac{d^2}{dx^2}\right)$$

$$= \frac{d^2(EI)}{dx^2}\frac{d^2}{dx^2} + 2\frac{d(EI)}{dx}\frac{d^3}{dx^3} + EI\frac{d^4}{dx^4} \tag{15.25}$$

so that $A_1 = 0$, $A_2 = 0$, $A_3 = d^2(EI)/dx^2$, $A_4 = 2[d(EI)/dx]$, $A_5 = EI$, and operator K is of order 4. Also

$$M = m \tag{15.26}$$

and the order of M is zero.

The boundary conditions for the example beam are

$$\left.\begin{array}{l} f(x) = 0 \\ \frac{df(x)}{dx} = 0 \end{array}\right\} \quad x = 0 \tag{15.27a}$$

$$\left.\begin{array}{l} EI\frac{d^2f(x)}{dx^2} = 0 \\ \frac{d}{dx}\{EI\frac{d^2f(x)}{dx^2}\} = 0 \end{array}\right\} \quad x = L \tag{15.27b}$$

Evidently, $B_1 = 1$, $B_2 = d/dx$, $B_3 = EI(d^2/dx^2)$, and $B_4 = (d/dx)\{EI(d^2/dx^2)\}$. The operator with the highest order is B_4, which has an order of 3.

15.3.2 *Self-adjointness of operators in the eigenvalue problem*

Let v and w be two different functions of the spatial coordinate x, each of which satisfies the boundary conditions given by Equation 15.23 but not necessarily the boundary value equation and is $2r$ times differentiable. A function of this type is known as a *comparison function*. If the operator K in the eigenvalue problem is such that

$$\int_0^L vK(w)\,dx = \int_0^L wK(v)\,dx \tag{15.28}$$

K is said to be *self-adjoint*. In a similar manner, M is self-adjoint if

$$\int_0^L vM(w)\,dx = \int_0^L wM(v)\,dx \tag{15.29}$$

The property of self-adjointness can be verified by partial integration of the integrals involved.

Eigenfunctions are comparison functions of a special kind which, in addition to satisfying the boundary conditions and being $2r$ times differentiable, also satisfy the differential equation of the boundary value problem. Obviously, for self-adjoint K and M, we have

$$\int_0^L f_i(x)K\{f_j(x)\}\,dx = \int_0^L f_j(x)K\{f_i(x)\}\,dx \tag{15.30}$$

and

$$\int_0^L f_i(x)M\{f_j(x)\}\,dx = \int_0^L f_j(x)M\{f_i(x)\}\,dx \tag{15.31}$$

For example, in the case of lateral vibrations of a beam, where $K = (d^2/dx^2)\{EI(d^2/dx^2)\}$, we had shown by partial integration that

$$\int_0^L f_i(x)\frac{d^2}{dx^2}\left\{EI\frac{d^2 f_j(x)}{dx^2}\right\}\,dx = \int_0^L f_j(x)\frac{d^2}{dx^2}\left\{EI\frac{d^2 f_i(x)}{dx^2}\right\}\,dx \tag{15.32}$$

each of them being equal to

$$\int_0^L EI\frac{d^2 f_i(x)}{dx^2}\frac{d^2 f_j(x)}{dx^2}$$

Thus, operator K in the beam vibration equation is self-adjoint. Operator M does not involve any differential, and $\int_0^L f_i(x)mf_j(x)\,dx$ can be clearly seen to be equal to $\int_0^L f_j(x)mf_i(x)\,dx$. Thus, M is also self-adjoint.

Self-adjoint operators in the eigenproblem for a continuous system are analogous to symmetric K and M matrices in the eigenproblem for a discrete system. Also, an operator M which does not involve any differentials is analogous to a diagonal mass matrix. Self-adjointness of the eigenvalue operators leads directly to the orthogonality property of the eigenfunctions, as illustrated in the next section.

15.3.3   *Orthogonality of eigenfunctions*

Let $f_i(x)$ and $f_j(x)$ represent two eigenfunctions of the eigenvalue problem of Equation 15.21. Then

$$Kf_i(x) = \lambda_i Mf_i(x) \tag{15.33}$$

and

$$Kf_j(x) = \lambda_j M f_j(x) \tag{15.34}$$

On multiplying both sides of Equation 15.33 by $f_j(x)$ and integrating from 0 to L, we have

$$\int_0^L f_j(x)K\{f_i(x)\}\,dx = \lambda_i \int_0^L f_j(x)M\{f_i(x)\}\,dx \tag{15.35}$$

Because K and M are both self-adjoint, Equation 15.35 can be restated as

$$\int_0^L f_i(x)K\{f_j(x)\}\,dx = \lambda_i \int_0^L f_i(x)M\{f_j(x)\}\,dx \tag{15.36}$$

Next, multiplying both sides of Equation 15.34 by $f_i(x)$ and integrating from 0 to L, we have

$$\int_0^L f_i(x)K\{f_j(x)\}\,dx = \lambda_j \int_0^L f_i(x)M\{f_j(x)\}\,dx \tag{15.37}$$

Subtraction of Equation 15.37 from Equation 15.36 gives

$$(\lambda_i - \lambda_j)\int_0^L f_i(x)M\{f_j(x)\}\,dx = 0 \tag{15.38}$$

or

$$\int_0^L f_i(x)M\{f_j(x)\}\,dx = 0 \quad \lambda_i \neq \lambda_j \tag{15.39}$$

Substitution of Equation 15.39 in Equation 15.36 gives

$$\int_0^L f_i(x)K\{f_j(x)\}\,dx = 0 \quad \lambda_i \neq \lambda_j \tag{15.40}$$

Equations 15.39 and 15.40 describe the orthogonality property of the eigen-functions having distinct eigenvalues. Two eigenfunctions that have identical eigenvalues are not necessarily orthogonal to each other, although they are orthogonal to all other eigenfunctions that have a different eigenvalue. Also, eigenfunctions in a set with identical eigenvalue are not unique and any linear combination of two or more members of the set is also an eigenfunction of the system. The proof of this property, which is entirely analogous to that for a discrete system, is quite straightforward. Thus, let $f_i(x)$ and $f_j(x)$ be two eigenfunctions with the same eigenvalue μ; then if a_1 and a_2 are any arbitrary constants

$$K\{a_1 f_i(x) + a_2 f_j(x)\} = a_1 K\{f_i(x)\} + a_2 K\{f_j(x)\}$$

$$= a_1 \mu M\{f_i(x)\} + a_2 \mu M\{f_j(x)\} \qquad (15.41)$$

$$= \mu M\{a_1 f_i(x) + a_2 f_j(x)\}$$

which proves that $a_1 f_i(x) + a_2 f_j(x)$ is also an eigenfunction of the system.

15.3.4 *Positive and positive definite operators*

Let v be an arbitrary comparison function; then, if the operator K is such that

$$\int_0^L vK(v)\, dx \geq 0 \qquad (15.42)$$

K is said to be *positive*. In particular, if the integral in Equation 15.42 is always greater than 0, that is

$$\int_0^L vK(v)\, dx > 0 \qquad (15.43)$$

then K is *positive definite*. If both K and M are positive definite, the eigenproblem is itself said to be positive definite, but if K is only positive, the eigenproblem is *semidefinite*. In analogy with a discrete system, a continuous system whose eigenproblem is positive definite has only positive eigenvalues. A semidefinite system has one or more zero eigenvalues and only positive other eigenvalues.

Equation 15.20 shows that a beam under lateral vibration represents either a semidefinite or a positive definite system because

$$\int_0^L \phi_i(x)K\{\phi_i(x)\}\, dx = \int_0^L EI \left\{ \frac{d^2\phi_i(x)}{dx^2} \right\}^2 dx$$

can either be zero or positive, while

$$\int_0^L \phi_i(x)M\{\phi_i(x)\}\, dx = \int_0^L m\phi_i^2(x)\, dx$$

is always greater than zero.

15.4 EXPANSION THEOREM

If $\phi_n(x)$ $n = 1, 2, \ldots, \infty$, represents the set of orthonormal modes of a system and w is any comparison function that satisfies the homogeneous boundary conditions, w can be represented by a superposition of the eigenfunctions $\phi_n(x)$

$$w = \sum_{n=1}^{\infty} y_n \phi_n(x) \tag{15.44}$$

in which y_n are constant coefficients. To determine the coefficient y_i, we apply the operator M to both sides of Equation 15.44, premultiply the two sides by $\phi_i(x)$, and integrate over the length. This gives

$$\int_0^L \phi_i(x) M(w) \, dx = \sum_{n=1}^{\infty} \int_0^L y_n \phi_i(x) M\{\phi_n(x)\} \, dx \tag{15.45}$$

On the right-hand side of Equation 15.45, all terms $\int_0^L y_i \phi_i(x) M\{\phi_n(x)\} \, dx$ for which $n \neq i$ will be zero because of the orthogonality of the eigenfunctions. Also, because the eigenfunctions have been mass orthonormalized, $\int_0^L y_i \phi_i(x) M\{\phi_i(x)\} \, dx = y_i$. Equation 15.45 thus reduces to

$$y_i = \int_0^L \phi_i(x) M(w) \, dx \tag{15.46}$$

Equation 15.44, along with Equation 15.46, defines the *expansion theorem* for a continuous system and is the counterpart of a similar theorem for discrete systems. As in the case of discrete systems, the expansion theorem forms the basis of mode superposition method of analysis for free and forced vibrations of continuous systems.

15.5 FREQUENCIES AND MODE SHAPES FOR LATERAL VIBRATIONS OF A BEAM

If shear deformation and rotatory inertia are neglected, the eigenvalue problem for the lateral vibrations of a beam is defined by Equation 15.5. For the special case of a uniform beam, Equation 15.5 takes the form

$$EI \frac{d^4 f(x)}{dx^4} = \omega^2 m f(x) \tag{15.47}$$

On setting $\omega^2 m / EI = \beta^4$, we can write Equation 15.47 in the alternative form

$$\frac{d^4 f(x)}{dx^4} - \beta^4 f(x) = 0 \tag{15.48}$$

For Equation 15.48, we try a solution given by

$$f(x) = A e^{\alpha x} \tag{15.49}$$

where A is an arbitrary constant. Substitution for $f(x)$ and its fourth-order derivative from Equation 15.49 into Equation 15.48 yields the characteristic equation

$$\alpha^4 - \beta^4 = 0 \tag{15.50a}$$

or

$$(\alpha^2 - \beta^2)(\alpha^2 + \beta^2) = 0 \tag{15.50b}$$

Equation 15.50b gives $\alpha = \pm\beta$ and $\alpha = \pm i\beta$. Hence, a general solution of Equation 15.48 is given by

$$f(x) = D_1 e^{\beta x} + D_2 e^{-\beta x} + D_3 e^{i\beta x} + D_4 e^{-i\beta x}$$

$$= C_1 \cosh \beta x + C_2 \sinh \beta x + C_3 \cos \beta x + C_4 \sin \beta x \tag{15.51}$$

where D_i and C_i are arbitrary constants to be determined from the boundary and initial conditions.

15.5.1   *Simply supported beam*

In this case, the boundary conditions are given by

$$\left.\begin{array}{l} f(x) = 0 \\ EI \frac{d^2 f(x)}{dx^2} = 0 \end{array}\right\} \quad x = 0 \quad \text{and} \quad x = L \tag{15.52}$$

Substitution for $f(x)$ and $d^2 f(x)/dx^2$ from Equation 15.51 into the boundary conditions Equation 15.52 gives the following four equations in constants C_1, C_2, C_3, and C_4

$$C_1 + C_3 = 0 \tag{15.53a}$$

$$C_1 - C_3 = 0 \tag{15.53b}$$

$$C_1 \cosh \beta L + C_2 \sinh \beta L + C_3 \cos \beta L + C_4 \sin \beta L = 0 \tag{15.53c}$$

$$C_1 \cosh \beta L + C_2 \sinh \beta L - C_3 \cos \beta L - C_4 \sin \beta L = 0 \tag{15.53d}$$

Equations 15.53a and 15.53b give $C_1 = C_3 = 0$. Substituting these in Equations 15.53c and 15.35d, we get

$$C_2 \sinh \beta L + C_4 \sin \beta L = 0 \tag{15.54a}$$

$$C_2 \sinh \beta L - C_4 \sin \beta L = 0 \tag{15.54b}$$

Equations 15.54a and 15.54b give

$$C_2 \sinh \beta L = 0 \tag{15.55}$$

and

$$C_4 \sin \beta L = 0 \tag{15.56}$$

Since $\sinh \beta L$ cannot be zero, C_2 must be zero. Equation 15.56 can be satisfied by selecting $C_4 = 0$. This will, however, lead to the trivial solution $f(x) = 0$, implying no motion. A nontrivial solution is possible only provided that

$$\sin \beta L = 0 \tag{15.57}$$

Equation 15.57, known as the *frequency equation*, determines the value of β and hence, ω. Thus

$$\beta L = n\pi \quad n = 1, 2, \ldots, \infty \tag{15.58}$$

or

$$\omega_n = \beta_n^2 \sqrt{\frac{EI}{m}} = n^2 \pi^2 \sqrt{\frac{EI}{mL^4}} \tag{15.59}$$

Corresponding to each value of β, we obtain a different solution for $f(x)$, called an eigenfunction. The eigenfunctions for the lateral vibration of a simply supported beam are thus given by

$$f_n(x) = C_4 \sin \frac{n\pi}{L} x \tag{15.60}$$

Since the magnitude of the eigenfunction is arbitrary, we can normalize the eigenfunctions so that

$$\int_0^L m f_n^2(x) \, dx = 1 \tag{15.61}$$

Equation 15.61 gives

$$C_4^2 \int_0^L m \sin^2 \frac{n\pi x}{L} dx = 1 \tag{15.62a}$$

or

$$C_4^2 = \frac{2}{mL} \tag{15.62b}$$

and the normalized eigenfunctions are given by

$$\phi_n(x) = \sqrt{\frac{2}{mL}} \sin \frac{n\pi x}{L} \quad n = 1, 2, 3, \ldots \tag{15.63}$$

In our solution of Equation 15.58, we have not included $n = 0$ because it represents a state of no motion. The first three lateral mode shapes and frequencies

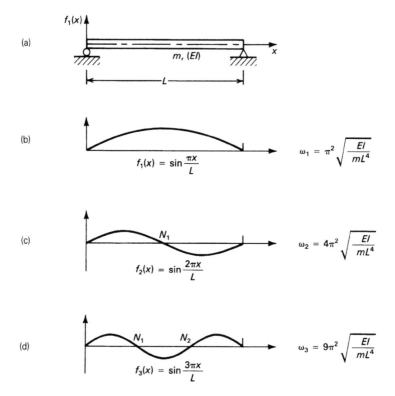

Figure 15.1. Mode shapes for the transverse vibrations of a simply supported beam.

are shown in Figure 15.1b, c, and d. A beam vibrating laterally in the first mode will at all times have the shape of a half sine wave as shown in Figure 15.1b. The amplitude of the wave will, of course, vary with time. In its second mode, the displaced shape of the vibrating beam is a full sine wave as shown in Figure 15.1c. The displaced shape of the beam crosses the x axis at point N_1. This point, called a *node*, is always stationary. The third mode shape, shown in Figure 15.1d, has two stationary points or nodes N_1 and N_2 and, in general, the nth mode has $n - 1$ nodes.

15.5.2   Uniform cantilever beam

Let the beam be clamped at its left hand end and free at the right end. The boundary conditions now become

$$\left. \begin{array}{l} f(x) = 0 \\ \frac{d f(x)}{dx} = 0 \end{array} \right\} \quad x = 0 \qquad (15.64a)$$

$$\left.\begin{array}{l} EI\frac{d^2 f(x)}{dx^2}=0 \\[2mm] EI\frac{d^3 f(x)}{dx^3}=0 \end{array}\right\} \quad x=L \qquad (15.64b)$$

Substitution for $f(x)$ and its derivatives from Equation 15.51 in Equations 15.64 leads to the following four equations in C_1 through C_4

$$C_1 + C_3 = 0 \qquad (15.65a)$$

$$C_2 + C_4 = 0 \qquad (15.65b)$$

$$C_1 \cosh \beta L + C_2 \sinh \beta L - C_3 \cos \beta L - C_4 \sin \beta L = 0 \qquad (15.65c)$$

$$C_1 \sinh \beta L + C_2 \cosh \beta L + C_3 \sin \beta L - C_4 \cos \beta L = 0 \qquad (15.65d)$$

Equation 15.65 can be expressed in matrix form as follows

$$\begin{bmatrix} 1 & 0 & 1 & 0 \\ 0 & 1 & 0 & 1 \\ \cosh \beta L & \sinh \beta L & -\cos \beta L & -\sin \beta L \\ \sinh \beta L & \cosh \beta L & \sin \beta L & -\cos \beta L \end{bmatrix} \begin{bmatrix} C_1 \\ C_2 \\ C_3 \\ C_4 \end{bmatrix} = 0 \qquad (15.66)$$

The homogeneous equations in Equation 15.66 can give nonzero values for the unknown coefficients C_i only provided that the determinant of the matrix on the left-hand side is zero. This leads to the following condition.

$$1 + \cosh \beta L \cos \beta L = 0 \qquad (15.67)$$

Equation 15.67 is the frequency equation whose solution can be obtained by a numerical method. It will lead to an infinite number of values for β and hence for the frequency ω. Corresponding to each value of β, a solution can be obtained for eigenfunction $f(x)$, by substituting for β in Equation 15.66, solving those equations for coefficients C_i and substituting the resulting values of C_i in Equation 15.51. The eigenfunctions can be shown to be

$$f_n(x) = A_n\{(\sin \beta_n L - \sinh \beta_n L)(\sin \beta_n x - \sinh \beta_n x)$$

$$+ (\cos \beta_n L + \cosh \beta_n L)(\cos \beta_n x - \cosh \beta_n x)\} \quad n = 1, 2, \ldots, \infty \qquad (15.68)$$

where A_n is an arbitrary constant. The first three mode shapes and frequencies of the cantilever beam are shown in Figure 15.2. It will be noted that as in the

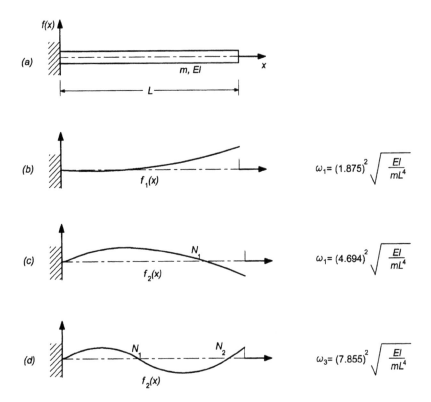

Figure 15.2. Mode shapes for the transverse vibrations of a cantilever beam.

case of a simply supported beam, the *n*th mode shape of the cantilever beam has $n - 1$ stationary points or nodes.

15.5.3 *Uniform beam clamped at both ends*

The boundary conditions for a beam clamped at both ends are given by

$$\left.\begin{array}{r}f(x)=0\\[6pt]\dfrac{df(x)}{dx}=0\end{array}\right\}\quad x=0 \tag{15.69a}$$

$$\left.\begin{array}{r}f(x)=0\\[6pt]\dfrac{df(x)}{dx}=0\end{array}\right\}\quad x=L \tag{15.69b}$$

The four equations in the unknown coefficients C_1 through C_4, obtained by substituting for $f(x)$ and $df(x)/dx$ from Equation 15.51 in Equation 15.69, can be expressed as

$$
\begin{bmatrix}
1 & 0 & 1 & 0 \\
0 & 1 & 0 & 1 \\
\cosh \beta L & \sinh \beta L & \cos \beta L & \sin \beta L \\
\sinh \beta L & \cosh \beta L & -\sin \beta L & \cos \beta L
\end{bmatrix}
\begin{bmatrix}
C_1 \\
C_2 \\
C_3 \\
C_4
\end{bmatrix}
= 0
\tag{15.70}
$$

The frequency equation is obtained by setting the determinant of the matrix on the left-hand side of Equation 15.70 equal to zero, giving

$$
1 - \cosh \beta L \cos \beta L = 0
\tag{15.71}
$$

Equation 15.71 can be solved numerically for β and will lead to an infinite number of values. Corresponding to each value of β, there exists an eigenfunction $f(x)$ which can be derived by substituting for β in Equation 15.70 and solving for the coefficients C_1 through C_4. The nth eigenfunction will be found to be

$$
f_n(x) = A_n\{(\sin \beta_n L + \sinh \beta_n L)(\sin \beta_n x - \sinh \beta_n x)
$$

$$
+ (\cos \beta_n L - \cosh \beta_n L)(\cos \beta_n x - \cosh \beta_n x)\} \quad n = 1, 2, \dots, \infty
\tag{15.72}
$$

where A_n is an arbitrary constant. It will be noted that Equation 15.71 also possesses two zero roots, $\beta_0 = 0$ and $\beta_1 = 0$. As in the case of a simply supported beam, it is possible that these roots may represent trivial solutions signifying no motion. If such is the case, we do not accept these roots. To verify whether or not the eigenfunctions corresponding to the zero roots signify any motion, we revert to the boundary value equation (Eq. 15.48). For $\beta = 0$, that equation reduces to

$$
\frac{d^4 f(x)}{dx^4} = 0
\tag{15.73}
$$

which has a solution of the form

$$
f(x) = D_1 + D_2 x + D_3 x^2 + D_4 x^3
\tag{15.74}
$$

Applying the boundary conditions of Equation 15.69 to Equation 15.74, we get

$$
\begin{aligned}
D_1 &= 0 \\
D_2 &= 0 \\
D_3 L^2 + D_4 L^3 &= 0 \\
2 D_3 L + 3 D_4 L^2 &= 0
\end{aligned}
\tag{15.75}
$$

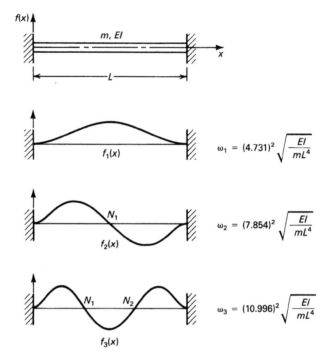

Figure 15.3. Mode shapes for the transverse vibrations of a beam clamped at both ends.

Equations 15.75 are satisfied only provided that $D_1 = D_2 = D_3 = D_4 = 0$, and therefore the zero roots of Equation 15.71 signify no motion. In Figure 15.3, we show the next three frequencies and mode shapes of the clamped beam.

15.5.4   *Uniform beam with both ends free*

The case of a beam without any support is of interest since it provides an example of a semidefinite system. The boundary conditions for this case are

$$\left.\begin{array}{l} \dfrac{d^2 f(x)}{dx^2} = 0 \\[2ex] \dfrac{d^3 f(x)}{dx^3} = 0 \end{array}\right\} \quad x = 0 \quad \text{and} \quad x = L \qquad (15.76)$$

The four equations in the unknown coefficients C_1 through C_4, obtained by substituting for $f(x)$ and $df(x)/dx$ from Equation 15.51 in Equation 15.76, can be expressed as

$$
\begin{bmatrix}
1 & 0 & -1 & 0 \\
0 & 1 & 0 & -1 \\
\cosh \beta L & \sinh \beta L & -\cos \beta L & -\sin \beta L \\
\sinh \beta L & \cosh \beta L & \sin \beta L & -\cos \beta L
\end{bmatrix}
\begin{bmatrix}
C_1 \\
C_2 \\
C_3 \\
C_4
\end{bmatrix}
= 0
\qquad (15.77)
$$

On equating the determinant of the matrix on the left-hand side of Equation 15.77 to zero, we get the following frequency equation

$$
1 - \cosh \beta L \cos \beta L = 0 \qquad (15.78)
$$

Equation 15.78 is identical to the frequency equation for the clamped beam. Thus, Equation 15.78 also possesses two zero roots, $\beta_0 = 0$ and $\beta_1 = 0$. The solution of the boundary value equation corresponding to the zero roots is given by Equation 15.74. On applying the boundary conditions of Equation 15.76, we get $D_3 = D_4 = 0$. An eigenfunction corresponding to the zero eigenvalues is

$$
f(x) = D_1 + D_2 x \qquad (15.79)
$$

in which D_1 and D_2 are arbitrary.

Equation 15.79 in fact represents two eigenfunctions, corresponding to the two zero eigenvalues. We can select the two eigenfunctions as, say D_1 and $D_2 x$, but any linear combination of these is also an eigenfunction. It will further be noted that functions D_1 and $D_2 x$ are not orthogonal. However, by an appropriate linear combination it *is* possible to construct two orthogonal eigenfunctions. Thus, let D_1 be the first eigenfunction and let $D_1 + D_2 x$ be the second function. If they have to satisfy the condition of orthogonality, we get

$$
\int_0^L m D_1 (D_1 + D_2 x)\, dx = 0 \qquad (15.80a)
$$

or

$$
D_2 = -\frac{2 D_1}{L} \qquad (15.80b)
$$

The two eigenfunctions can thus be represented by

$$
\begin{aligned}
f_0(x) &= D_1 \\
f_1(x) &= D_1 \left(1 - \frac{2x}{L} \right)
\end{aligned}
\qquad (15.81)
$$

The eigenfunctions in Equation 15.81 can, of course, be scaled in any manner, so that A_0 and $A_1 (1 - \frac{2x}{L})$ will also be eigenfunctions of the system. To mass orthonormalize the eigenfunctions, we determine A_0 and A_1 as follows

$$
\int_0^L f_0(x) m f_0(x)\, dx = 1
$$

or

$$A_0 = \frac{1}{\sqrt{mL}} \tag{15.82}$$

and in a similar manner

$$\int_0^L f_1(x) m f_1(x)\, dx = 1$$

or

$$\int_0^L m A_1^2 \left(1 - \frac{2x}{L}\right)^2 = 1$$

$$A_1 = \sqrt{\frac{3}{mL}} \tag{15.83}$$

The mass orthonormal eigenfunctions thus become

$$\phi_0(x) = \frac{1}{\sqrt{mL}} \tag{15.84a}$$

$$\phi_1(x) = \sqrt{\frac{3}{mL}} \left(1 - \frac{2x}{L}\right) \tag{15.84b}$$

Finally, the eigenfunctions given by Equation 15.81 will be recognized as rigid-body displacements: a translation in the y direction and a rotation about the center of the beam.

The nonzero roots of Equation 15.78 can be obtained numerically. Corresponding to each of these roots, there exists an eigenfunction which can be determined by solving Equation 15.77 for the coefficients C_1 through C_4, and substituting the resulting values in Equation 15.51. This gives

$$f_n(x) = A_n \{ (\sinh \beta_n L - \sin \beta_n L)(\cosh \beta_n x + \cos \beta_n x)$$

$$- (\cosh \beta_n L - \cos \beta_n L)(\sinh \beta_n x + \sin \beta_n x) \} \tag{15.85}$$

The eigenfunctions are not necessarily mass orthonormal. They are, however, orthogonal to each other and also to the two rigid-body modes given by Equation 15.81. The first four frequencies and eigenfunctions for a free–free bar are shown in Figure 15.4.

15.6   EFFECT OF SHEAR DEFORMATION AND ROTATORY INERTIA ON THE FREQUENCIES OF FLEXURAL VIBRATIONS

The governing equation for free flexural vibrations of a uniform beam taking into account the shear deformation and rotatory inertia is obtained from

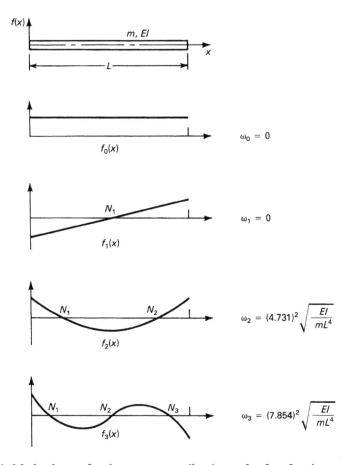

Figure 15.4. Mode shapes for the transverse vibrations of a free-free beam.

Equation 14.41 by setting $p = 0$.

$$\left\{ EI \frac{\partial^4 u}{\partial x^4} + m \frac{\partial^2 u}{\partial t^2} \right\} - \left\{ mr^2 \frac{\partial^4 u}{\partial t^2 \, \partial x^2} \right\} - \left\{ \frac{EIm}{k'GA} \frac{\partial^4 u}{\partial x^2 \, \partial t^2} \right\}$$

$$+ \left\{ \frac{m^2 r^2}{k'GA} \frac{\partial^4 u}{\partial t^4} \right\} = 0 \tag{15.86}$$

We assume that as in the case of a beam in which shear deformation and rotatory inertia are negligible, the solution of Equation 15.86 is given by

$$u = B f(x) \sin(\omega t + \theta) \tag{15.87}$$

in which B is a constant and θ is a phase angle. Substituting Equation 15.87 in Equation 15.86 and canceling out $B \sin(\omega t + \theta)$, we get

$$\left\{ EI\frac{d^4 f(x)}{dx^4} - m\omega^2 f(x) \right\} + \left\{ mr^2\omega^2\frac{d^2 f(x)}{dx^2} \right\}$$

$$+ \left\{ \frac{EIm\omega^2}{k'GA}\frac{d^2 f(x)}{dx^2} \right\} + \left\{ \frac{m^2 r^2\omega^4}{k'GA} f(x) \right\} = 0 \qquad (15.88)$$

If, as before, we set $\beta^4 = \omega^2 m/EI$, Equation 15.88 reduces to

$$\left\{ \frac{d^4 f(x)}{dx^4} - \beta^4 f(x) \right\} + \left\{ \beta^4 r^2\frac{d^2 f(x)}{dx^2} \right\} + \left\{ \frac{Er^2}{k'G}\beta^4\frac{d^2 f(x)}{dx^2} \right\}$$

$$+ \left\{ \beta^8 r^4\frac{E}{k'G} f(x) \right\} = 0 \qquad (15.89)$$

It is difficult to obtain a solution of Equation 15.89 for a general set of boundary conditions. A solution for a simply supported case is, however, straightforward. In that case, the eigenfunction is still given by

$$f_n(x) = C \sin \frac{n\pi}{L}x \qquad (15.90)$$

Substitution from Equation 15.90 into Equation 15.89 gives

$$\left\{ \left(\frac{n\pi}{L}\right)^4 - \beta^4 \right\} - \left\{ \beta^4 r^2\left(\frac{n\pi}{L}\right)^2 \right\} - \left\{ \beta^4 r^2\left(\frac{n\pi}{L}\right)^2\frac{E}{k'G} \right\}$$

$$+ \left\{ \beta^8 r^4\frac{E}{k'G} \right\} = 0 \qquad (15.91)$$

Equation 15.91 is a quadratic equation in β^4 which can be solved for β^4 once the properties of the beam have been specified. If for simplicity, we assume that the last term in Equation 15.91 is small in comparison to rotatory inertia and shear correction terms (terms in the second and third pair of braces), we obtain

$$\beta^4 = \left(\frac{n\pi}{L}\right)^4 \frac{1}{[1 + (n\pi r/L)^2\{1 + E/k'G\}]} \qquad (15.92)$$

In the elementary case without rotatory inertia and shear deformation, $\beta = n\pi/L$. The multiplier $1/[1 + (n\pi r/L)^2\{1 + E/k'G\}]$ in Equation 15.92 modifies the value of β such that both rotatory inertia and shear deformation cause a reduction in β and hence in the frequency. The effect of rotatory inertia is represented by the term $(n\pi r/L)^2$ in the denominator, while the effect of shear deformation is represented by $(n\pi r/L)^2(E/k'G)$. The shear deformation correction term is thus $E/k'G$ times as large as the rotatory inertia term. Consider, for example, a beam that is rectangular in cross section and is made of steel for which $E = 200,000$ MPa and $G = 77,000$ MPa. Since for a rectangular section, $k' = \frac{5}{6}$, we have $E/k'G = 3.12$ and the shear correction term is 3.12 times as large as the rotatory inertia correction term.

Continuing with our example of the rectangular beam, we note that for this case

$$r^2 = \frac{I}{A}$$

$$= \frac{1}{12} \frac{bd^3}{bd} \tag{15.93}$$

$$= \frac{d^2}{12}$$

Suppose that the length of example beam is 10 times the depth of its cross section. Equation 15.92 can then be used to calculate the percentage change in the frequency of the beam due to shear deformation and rotatory inertia. The change in the frequency for the first five modes is shown in Table 15.1.

To examine whether it is reasonable to neglect the last term in Equation 15.91, we rewrite that equation as follows

$$\beta^4 \left[1 + \left(\frac{n\pi r}{L} \right)^2 \left\{ 1 + \frac{E}{k'G} \right\} - \beta^4 r^4 \frac{E}{k'G} \right] = \left(\frac{n\pi}{L} \right)^4 \tag{15.94}$$

Since β^4 is nearly equal to $(n\pi/L)^4$, Equation 15.94 can be approximated as

$$\beta^4 \left[1 + \left(\frac{n\pi r}{L} \right)^2 \left\{ 1 + \frac{E}{k'G} \right\} - \left(\frac{n\pi r}{L} \right)^4 \frac{E}{k'G} \right] = \left(\frac{n\pi}{L} \right)^4 \tag{15.95}$$

When $n\pi r/L$ is very small, the last term on the left-hand side of Equation 15.95 can reasonably be neglected in comparison to the remaining terms. As an example, for the rectangular beam considered, with $n = 1$, Equation 15.95

Table 15.1. Percentage reduction in frequency ω caused by rotatory inertia and shear deformation.

Mode No.	1	2	3	4	5
Reduction caused by rotatory inertia	0.4	1.6	3.5	6.0	8.9
Reduction caused by shear deformation	1.3	4.8	9.9	15.8	22.0
Reduction caused by combined effect of shear deformation and rotatory inertia	1.6	6.2	12.5	19.5	26.4

reduces to

$$\beta^4(1 + 0.0339 - 0.0002) = \left(\frac{n\pi r}{L}\right)^4 \tag{15.96}$$

The last term on the left-hand side is obviously quite negligible in comparison to other terms.

15.7   FREQUENCIES AND MODE SHAPES FOR THE AXIAL VIBRATIONS OF A BAR

The equation of free axial vibrations of a bar is obtained by setting $p=0$ in Equation 14.44

$$\frac{\partial}{\partial x}\left(EA\frac{\partial u}{\partial x}\right) - m\frac{\partial^2 u}{\partial t^2} = 0 \tag{15.97}$$

For the special case of a uniform bar, Equation 15.97 reduces to

$$EA\frac{\partial^2 u}{\partial x^2} - m\frac{\partial^2 u}{\partial t^2} = 0 \tag{15.98}$$

As in the case of flexural vibrations of a beam, we assume that u is of the form

$$u = f(x)g(t) \tag{15.99}$$

Substituting from Equation 15.99 in Equation 15.98, we get

$$\frac{EA}{m}\frac{1}{f(x)}\left(\frac{d^2 f(x)}{dx^2}\right) = \frac{1}{g(t)}\frac{d^2 g(t)}{dt^2} \tag{15.100}$$

Since the expression on the left-hand side of Equation 15.100 is a function of x only, while that on the right-hand side is a function of t only, they can be equal only if each is a constant. We choose this separation constant as equal to $-\omega^2$. Equation 15.100 can then be expressed as

$$\frac{d^2 g(t)}{dt^2} + \omega^2 g(t) = 0 \tag{15.101}$$

$$-EA\frac{d^2 f(x)}{dx^2} = \omega^2 m f(x) \tag{15.102}$$

Equation 15.102 represents the eigenvalue equation for the axial vibrations of a bar. Comparison with Equation 15.21 shows that

$$K = -EA\frac{d^2}{dx^2} \tag{15.103a}$$

$$M = m \tag{15.103b}$$

It is easily shown that operator K is positive as well as self-adjoint. Also, $M = m$ is both positive definite and self-adjoint. Thus, if v and w are any two comparison functions

$$
\begin{aligned}
\int_0^L vK(w)\,dx &= -\int_0^L EAv\frac{d^2w}{dx^2}\,dx \\
&= -EAv\frac{dw}{dx}\Big|_0^L + \int_0^L EA\frac{dv}{dx}\frac{dw}{dx}\,dx \\
&= \int_0^L EA\frac{dv}{dx}\frac{dw}{dx}\,dx
\end{aligned}
\tag{15.104}
$$

where we have used the boundary conditions of the problem, which signify that either v, w or dv/dx, dw/dx are zero at either end of the bar.

In a similar manner, we can show that

$$
\int_0^L wK(v)\,dx = \int_0^L EA\frac{dv}{dx}\frac{dw}{dx}\,dx
\tag{15.105}
$$

From Equations 15.104 and 15.105, we have

$$
\int_0^L vK(w)\,dx = \int_0^L wK(v)\,dx
\tag{15.106}
$$

proving the adjointness of the operator K. By setting $v = w$ in Equation 15.104, we get

$$
\int_0^L vK(v)\,dx = \int_0^L EA\left(\frac{dv}{dx}\right)^2 dx
\tag{15.107}
$$

Since the right-hand side of Equation 15.107 can be either zero or positive, K is a positive operator. It follows from the self-adjointness of the operators K and M that the eigenfunctions of the system are orthogonal. The boundary value problem of Equation 15.102 can be expressed as

$$
\frac{d^2 f(x)}{dx^2} + \beta^2 f(x) = 0
\tag{15.108}
$$

where $\beta^2 = \omega^2 m/EA$. Equation 15.108 has a solution of the form

$$
f(x) = C_1 \cos \beta x + C_2 \sin \beta x
\tag{15.109}
$$

where C_1 and C_2 are arbitrary constants to be determined from boundary and initial conditions.

15.7.1 *Axial vibrations of a clamped–free bar*

The boundary conditions for this case are given by

$$f(x) = 0 \qquad x = 0$$
$$\frac{df(x)}{dx} = 0 \quad x = L \tag{15.110}$$

Substitution for $f(x)$ and $df(x)/dx$ from Equation 15.109 in Equation 15.110 gives the following expressions

$$C_1 = 0 \tag{15.111a}$$

$$C_2 \cos \beta L = 0 \tag{15.111b}$$

Equation 15.111b provides the following solution for β

$$\beta_n = \frac{(2n - 1)\pi}{2L} \quad n = 1, 2, 3, \ldots, \infty \tag{15.112}$$

so that the frequencies are given by

$$\omega_n = \frac{(2n - 1)\pi}{2} \sqrt{\frac{EA}{mL^2}} \quad n = 1, 2, 3, \ldots, \infty \tag{15.113}$$

The corresponding eigenfunctions are

$$f_n(x) = A_n \sin \beta_n x \quad n = 1, 2, 3, \ldots, \infty \tag{15.114}$$

where A_n is an arbitrary constant. Eigenfunction $f_n(x)$ has not been normalized. To obtain a mass-orthonormal eigenfunction, we use the relationship

$$\int_0^L m f_n^2(x)\, dx = 1 \tag{15.115a}$$

or

$$A_n^2 m \int_0^L \sin^2 \beta_n x\, dx = 1$$
$$A_n = \sqrt{\frac{2}{mL}} \tag{15.115b}$$

Substituting for A_n in Equation 15.114, we get

$$\phi_n(x) = \sqrt{\frac{2}{mL}} \sin \beta_n x \quad n = 1, 2, \ldots, \infty \tag{15.116}$$

The first three frequencies and eigenfunctions for the axial vibrations of a clamped–free uniform bar are shown in Figure 15.5.

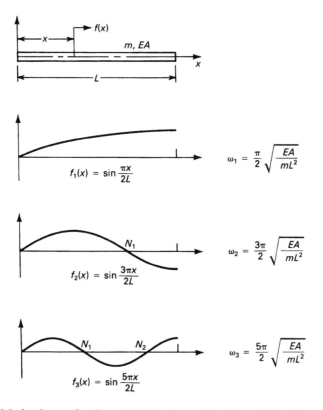

Figure 15.5. Mode shapes for the axial vibrations of a clamped-free bar.

15.7.2 *Axial vibrations of a free–free bar*

A free–free bar under axial vibrations provides another illustration of a semidefinite system. The boundary conditions in this case are given by

$$\frac{df(x)}{dx} = 0 \quad x = 0 \text{ and } x = L \tag{15.117}$$

Substituting for $df(x)/dx$ from Equation 15.109, we get

$$C_2 = 0$$
$$C_1 \sin \beta L = 0 \tag{15.118}$$

Equation 15.118 yields the following values for β

$$\beta = \frac{n\pi}{L} \quad n = 0, 1, 2, \ldots, \infty \tag{15.119}$$

With $\beta = 0$, Equation 15.108 reduces to

$$\frac{d^2 f(x)}{dx^2} = 0 \qquad\qquad (15.120)$$

Equation 15.120 has a solution of the form

$$f(x) = D_1 + D_2 x \qquad\qquad (15.121)$$

Application of the boundary conditions (Eq. 15.117) gives $D_2 = 0$. Hence, the eigenfunction corresponding to the zero frequency is given by

$$f(x) = D_1 \qquad\qquad (15.122)$$

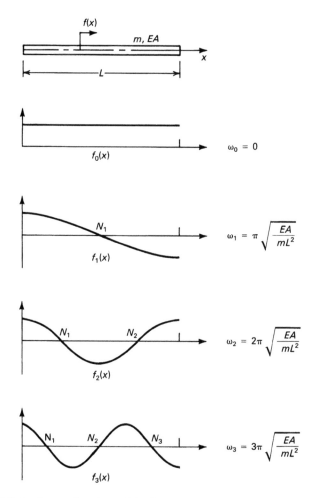

Figure 15.6. Mode shapes for the axial vibrations of a free-free bar.

which represents a rigid-body translation. The remaining eigenfunctions are obtained from Equation 15.109 with $C_2 = 0$. The normalized values of the eigenfunctions can be shown to be

$$\phi_n(x) = \sqrt{\frac{2}{mL}} \cos \beta_n x \qquad (15.123)$$

The first four mode shapes and frequencies for the axial vibrations of a free–free bar are shown in Figure 15.6.

Example 15.1
The uniform bar shown in Figure E15.1a has axial rigidity EA and mass per unit length m. It is clamped at the left-hand end and restrained at the other by a longitudinal spring of stiffness k. The bar is vibrating in the axial direction. Given $k = EA/L$, obtain the frequencies and mode shapes of the system and prove the orthogonality of modes.

Solution
The equation of motion for the bar derived from the results of Example 14.2 is given by

$$EA\frac{\partial^2 u}{\partial x^2} - m\frac{\partial^2 u}{\partial t^2} = 0 \qquad (a)$$

along with the boundary conditions

$$u = 0 \quad \text{at} \quad x = 0$$

$$EA\frac{\partial u}{\partial x} = -ku \quad \text{at} \quad x = L \qquad (b)$$

The method of separation of variables leads to the following boundary value problem (Eq. 15.102)

$$-EA\frac{d^2 f(x)}{dx^2} = \omega^2 m f(x) \qquad (c)$$

along with the boundary conditions

$$f(x) = 0 \quad \text{at} \quad x = 0$$

$$EA\frac{d f(x)}{dx} = -kf(x) \quad \text{at} \quad x = L \qquad (d)$$

By setting $\beta^2 = \omega^2 m/EA$, Equation c can be expressed as

$$\frac{d^2 f(x)}{dx^2} + \beta^2 f(x) = 0 \qquad (e)$$

Equation e has a solution of the form

$$f(x) = C_1 \cos \beta x + C_2 \sin \beta x \qquad (f)$$

Applying the boundary conditions to Equation f, we get

$$C_1 = 0$$

$$EAC_2\beta \cos \beta L = -kf(L) \qquad (g)$$

$$= -kC_2 \sin \beta L$$

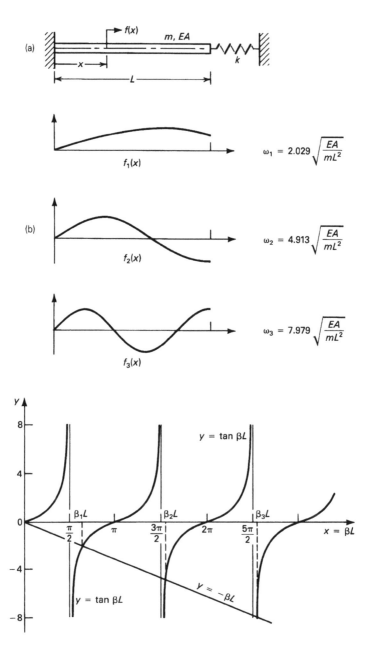

Figure E15.1. (a) Axial vibrations of a bar clamped at one end and restrained at the other; (b) first three mode shapes; (c) Roots of the frequency equation.

or

$$\tan \beta L = -\frac{EA\beta}{k} = -\beta L \tag{h}$$

Equation h can be solved numerically. Ignoring the root $\beta = 0$, which will give $f(x) = 0$, signifying no motion, the first three roots are

$$\beta_1 L = 2.0288$$

$$\beta_2 L = 4.9132$$

$$\beta_3 L = 7.9787$$

Figure E15.1c shows two curves, one for $y = -\beta L$ and the other for $y = \tan \beta L$. The intersections of the two give the roots $\beta_1 L, \beta_2 L, \dots, \beta_n L$. It is evident that as n increases, $\beta_n L$ approaches $(2n - 1)\pi/2$. The mode shapes are given by

$$f_n(x) = A_2 \sin \beta_n x \tag{i}$$

The first three mode shapes and frequencies are shown in Figure E15.1b. From Equation c we have

$$-EA\frac{d^2 f_i(x)}{dx^2} = \omega^2 m f_i(x) \tag{j}$$

$$-EA\frac{d^2 f_j(x)}{dx^2} = \omega^2 m f_j(x) \tag{k}$$

On multiplying both sides of Equation j by $f_j(x)$ and integrating from 0 to L, we get

$$-\int_0^L EA f_j(x)\frac{d^2 f_i(x)}{dx^2} dx = \int_0^L \omega_i^2 m f_i(x) f_j(x) dx \tag{l}$$

The left-hand side of Equation l can be reduced by partial integration as follows

$$\begin{aligned} -\int_0^L EA f_j(x)\frac{d^2 f_i(x)}{dx^2} dx &= -EA f_j(x)\frac{d f_i(x)}{dx}\Big|_0^L \\ &\quad + \int_0^L EA \frac{d f_j(x)}{dx}\frac{d f_i(x)}{dx} dx \\ &= k f_i(L) f_j(L) + \int_0^L EA \frac{d f_i(x)}{dx}\frac{d f_j(x)}{dx} dx \end{aligned} \tag{m}$$

in which we have used the boundary conditions given by Equation d. Substituting Equation m in Equation l, we get

$$\int_0^L \omega_i^2 m f_i(x) f_j(x) dx = k f_i(L) f_j(L) + \int_0^L EA\frac{d f_i(x)}{dx}\frac{d f_j(x)}{dx} dx \tag{n}$$

Starting with Equation k and proceeding in a similar manner, we will get

$$\int_0^L \omega_j^2 m f_i(x) f_j(x) dx = k f_i(L) f_j(L) + \int_0^L EA\frac{d f_i(x)}{dx}\frac{d f_j(x)}{dx} dx \tag{o}$$

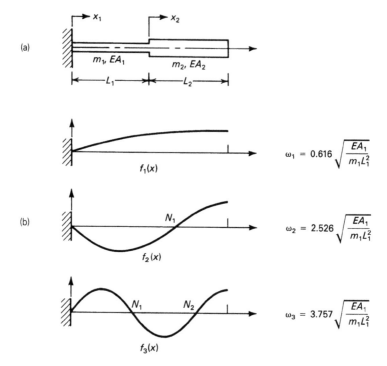

Figure E15.2. (a) Nonuniform clamped-free bar under axial vibrations; (b) first three mode shapes and frequencies.

Subtracting Equation o from Equation n, we get

$$(\omega_i^2 - \omega_j^2) \int_0^L m f_i(x) f_j(x)\, dx = 0 \tag{p}$$

or

$$\int_0^L m f_i(x) f_j(x)\, dx = 0 \quad \omega_i^2 \neq \omega_j^2 \tag{q}$$

Equation q expresses the orthogonality condition for the eigenfunctions. By substituting in Equation l, we get

$$\int_0^L EA f_j(x) \frac{d^2 f_i(x)}{dx^2}\, dx = 0 \quad \omega_i^2 \neq \omega_j^2 \tag{r}$$

Example 15.2
Obtain the frequency equation for the axial vibrations of the system shown in Figure E15.2a.

(i) For $A_2 = 2A_1$, $m_2 = 2m_1$, and $L_2 = L_1$, calculate the first three mode shapes and frequencies.
(ii) Assuming that L_2 approaches zero, so that $m_2 L_2$ tends to a mass M, show that the frequency equation reduces to the form

$$\tan \beta_1 L_1 = \frac{m_1 L_1}{M} \frac{1}{\beta_1 L_1}$$

Solution

The boundary value problem for section AB with spatial coordinate x_1 measured from A is given by

$$-EA_1 \frac{d^2 f_1(x_1)}{dx_1^2} = \omega^2 m_1 f_1(x_1) \tag{a}$$

Equation a has a solution of the form

$$f_1(x_1) = C_1 \cos \beta_1 x_1 + C_2 \sin \beta_1 x_1 \tag{b}$$

where $\beta_1^2 = \omega^2 m_1 / EA_1$ and C_1, C_2 are arbitrary constants.

For section BC, the boundary value problem written in terms of the spatial coordinate x_2 measured from B is

$$-EA_2 \frac{d^2 f_2(x_2)}{dx_2^2} = \omega^2 m_2 f_2(x_2) \tag{c}$$

The solution of Equation c is given by

$$f_2(x_2) = C_3 \cos \beta_2 x_2 + C_4 \sin \beta_2 x_2 \tag{d}$$

The boundary condition at A is

$$f_1(0) = 0 \tag{e}$$

The continuity of displacement and axial force at B gives

$$f_1(L_1) = f_2(0) \tag{f}$$

$$EA_1 \frac{df_1}{dx_1}(L_1) = EA_2 \frac{df_2}{dx_2}(0) \tag{g}$$

Finally, since the free end should not have any stress

$$EA_2 \frac{df_2}{dx_2}(L_2) = 0 \tag{h}$$

Substituting for $f_1(x_1)$ and $f_2(x_2)$ from Equations b and d in the boundary condition Equations e through h, we get

$$
\begin{aligned}
& C_1 = 0 \\
& C_2 \sin \beta_1 L_1 - C_3 = 0 \\
& C_2 A_1 \beta_1 \cos \beta_1 L_1 - C_4 A_2 \beta_2 = 0 \\
& -C_3 \sin \beta_2 L_2 + C_4 \cos \beta_2 L_2 = 0
\end{aligned}
\tag{i}
$$

The last three equations can be expressed in the matrix form

$$
\begin{bmatrix}
\sin \beta_1 L_1 & -1 & 0 \\
A_1 \beta_1 \cos \beta_1 L_1 & 0 & -A_2 \beta_2 \\
0 & -\sin \beta_2 L_2 & \cos \beta_2 L_2
\end{bmatrix}
\begin{bmatrix}
C_2 \\
C_3 \\
C_4
\end{bmatrix} = 0
\tag{j}
$$

For a nontrivial solution, the determinant of the matrix on the left-hand side of Equation j should be zero. This condition gives the required frequency equation

$$\tan \beta_1 L_1 \tan \beta_2 L_2 = \frac{A_1 \beta_1}{A_2 \beta_2} \tag{k}$$

(i) In this case, $\beta_1 = \omega \sqrt{m_1/A_1 E}$ and $\beta_2 = \omega \sqrt{m_2/A_2 E} = \omega \sqrt{2m_1/2A_1 E} = \beta_1$. Also, $L_1 = L_2$, so that $\beta_1 L_1 = \beta_2 L_2$ and the frequency equation reduces to

$$\tan^2 \beta_1 L_1 = \frac{1}{2} \tag{l}$$

Equation l can be solved numerically. The first three solutions are

$$\beta_1^{(1)} L_1 = 0.6155 \quad \omega_1 = 0.6155 \sqrt{\frac{EA_1}{m_1 L_1^2}}$$

$$\beta_1^{(2)} L_1 = 2.5261 \quad \omega_2 = 2.5261 \sqrt{\frac{EA_1}{m_1 L_1^2}} \tag{m}$$

$$\beta_1^{(3)} L_1 = 3.7571 \quad \omega_3 = 3.7571 \sqrt{\frac{EA_1}{m_1 L_1^2}}$$

The corresponding mode shapes are obtained by substituting for $\beta_1 L_1$ in Equation j and solving the latter for two of the coefficients C_2, C_3, and C_4 in terms of the third one.

Mode shape 1:

$$\beta_1 L_1 = \beta_2 L_2 = 0.6155$$

$$C_2 = 1.7320 C_3$$

$$C_4 = 0.7071 C_3$$

$$f_1(x_1) = C_3 \left[1.7320 \sin 0.6155 \frac{x_1}{L_1} \right]$$

$$f_2(x_2) = C_3 \left[\cos 0.6155 \frac{x_2}{L_2} + 0.7071 \sin 0.6155 \frac{x_2}{L_2} \right]$$

Mode shape 2:

$$\beta_1 L_1 = \beta_2 L_2 = 2.5261$$

$$C_2 = 1.7320 C_3$$

$$C_4 = -0.7071 C_3$$

$$f_1(x_1) = C_3 \left[1.7320 \sin 2.5261 \frac{x_1}{L_1} \right]$$

$$f_2(x_2) = C_3 \left[\cos 2.5261 \frac{x_2}{L_2} - 0.7071 \sin 2.5261 \frac{x_2}{L_2} \right]$$

Mode shape 3:

$$\beta_1 L_1 = \beta_2 L_2 = 3.7571$$

$$C_2 = -1.7320 C_3$$

$$C_4 = 0.7071 C_3$$

$$f_1(x_1) = C_3 \left[-1.7320 \sin 3.7571 \frac{x_1}{L_1} \right]$$

$$f_2(x_2) = C_3 \left[\cos 3.7571 \frac{x_2}{L_2} + 0.7071 \sin 3.7571 \frac{x_2}{L_2} \right]$$

The first three mode shapes and frequencies are shown in Figure E15.2b.
(ii) We have

$$\beta_2 L_2 = \omega \sqrt{\frac{m_2}{A_2 E}} L_2$$

$$= \omega \sqrt{\frac{M}{A_2 E}} \sqrt{L_2}$$

(p)

Thus as L_2 tends to zero, $\beta_2 L_2$ tends to zero and hence $\tan \beta_2 L_2$ tends to $\beta_2 L_2$. The frequency equation (Eq. k) reduces to

$$\tan \beta_1 L_1 = \frac{A_1 \beta_1}{A_2 \beta_2^2 L_2}$$

$$= \frac{A_1 \beta_1}{A_2} \frac{A_2 E}{\omega^2 m_2 L_2}$$

(q)

$$= \frac{m_1 L_1}{M} \frac{1}{\beta_1 L_1}$$

15.8 FREQUENCIES AND MODE SHAPES FOR THE TRANSVERSE VIBRATION OF A STRING

The equation of motion for the transverse vibrations of a string with initial tension T is entirely analogous to that for the axial vibrations of a bar. The tension T in the former replaces the axial rigidity EA in the latter. The boundary value problem for the transverse vibrations of the string can thus be derived by a procedure similar to that described for the axial vibrations of a bar.

For constant tension T, the boundary value problem for the transverse vibration of a string is given by

$$\frac{d^2 f(x)}{dx^2} + \beta^2 f(x) = 0$$

(15.124)

where $\beta^2 = m\omega^2 / T$.

As before, Equation 15.124 has a solution of the form

$$f(x) = C_1 \cos \beta x + C_2 \sin \beta x \tag{15.125}$$

where C_1 and C_2 are arbitrary constants to be determined from boundary and initial conditions.

15.8.1   Vibrations of a string tied at both ends

The boundary conditions for this case are given by

$$f(x) = 0 \quad x = 0 \text{ and } x = L \tag{15.126}$$

Substitution for $f(x)$ from Equation 15.125 gives

$$C_1 = 0 \tag{15.127a}$$

$$C_2 \sin \beta L = 0 \tag{15.127b}$$

Equation 15.127b yields the following solutions for β

$$\beta_n = \frac{n\pi}{L} \quad n = 1, 2, \dots, \infty \tag{15.128}$$

The corresponding frequency is given by

$$\omega_n = n\pi \sqrt{\frac{T}{mL^2}} \tag{15.129}$$

It is easily verified that the solution corresponding to $n = 0$ leads to no motion. The lowest or the fundamental frequency $\omega_1 = \pi \sqrt{T/mL^2}$. The higher frequencies are seen to be integer multiples of the fundamental frequency and are referred to as the higher *harmonics*. When the frequencies hold this special relationship, the vibrations produce a very pleasant sound. Hence, the vibrating string is very widely used in the construction of musical instruments. The eigenfunctions of the vibrating string are given by

$$f_n(x) = A_n \sin \frac{n\pi}{L} x \tag{15.130}$$

Mass orthonormal eigenfunctions can be derived from $f_n(x)$ by appropriate scaling and are given by

$$\phi_n(x) = \sqrt{\frac{2}{mL}} \sin \frac{n\pi x}{L} \quad n = 1, 2, 3, \dots, \infty \tag{15.131}$$

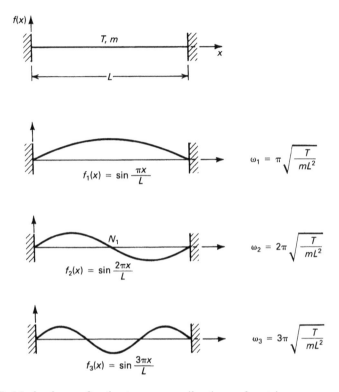

Figure 15.7. Mode shapes for the transverse vibrations of a string.

The first three frequencies and eigenfunctions of the vibrating string are shown in Figure 15.7.

15.9 BOUNDARY CONDITIONS CONTAINING THE EIGENVALUE

In our discussion so far, we have assumed that the homogeneous boundary conditions are of the form of Equation 15.23. The definition of self-adjointness (Eqs. 15.28 and 15.29) and the orthogonality relationships (Eqs. 15.39 and 15.40) were based on the foregoing assumption regarding the boundary condition. At times, the homogeneous boundary condition may involve the eigenvalue of the system. In such a case, the definition of self-adjointness as well as of orthogonality must be modified. In the following paragraphs, we illustrate the type of modification involved by means of an example.

Consider the vibrations of a cantilever beam with a rigid disk of mass M_0, and mass moment of inertia for rotation about a diameter I_0, rigidly attached to its free end as shown in Figure 15.8. This system was considered in Example 14.1.

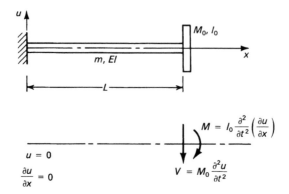

Figure 15.8. Transverse vibrations of a cantilever beam with a rigid disk attached to the free end.

The motion of the system is governed by the following equation

$$EI\frac{\partial^4 u}{\partial x^4} + m\frac{\partial^2 u}{\partial t^2} = 0 \qquad (15.132)$$

along with the boundary conditions

$$\left.\begin{array}{c} u = 0 \\[2mm] \dfrac{\partial u}{\partial x} = 0 \end{array}\right\} \quad x = 0 \qquad (15.133a)$$

and

$$\left.\begin{array}{c} EI\dfrac{\partial^2 u}{\partial x^2} = -I_0\dfrac{\partial^2}{\partial t^2}\left(\dfrac{\partial u}{\partial x}\right) \\[4mm] -EI\dfrac{\partial^3 u}{\partial x^3} = -M_0\dfrac{\partial^2 u}{\partial t^2} \end{array}\right\} \quad x = L \qquad (15.133b)$$

By using the method of separation of variables, we obtain, as before, the following separated equations

$$\frac{d^2 g(t)}{dt^2} + \omega^2 g(t) = 0 \qquad (15.134)$$

$$EI\frac{d^4 f(x)}{dx^4} = \lambda m f(x) \qquad (15.135)$$

where $\lambda = \omega^2$ is the separation constant. Equation 15.134 has a solution of the form

$$g(t) = A \sin(\omega t + \theta) \tag{15.136}$$

where A and θ are arbitrary constants to be determined from the initial conditions.

The boundary conditions at $x = 0$ can be expressed as

$$\left. \begin{aligned} f(x) &= 0 \\ \frac{df(x)}{dx} &= 0 \end{aligned} \right\} \quad x = 0 \tag{15.137}$$

At the free end, $x = L$, the boundary condition corresponding to the moment becomes

$$EIg(t)\frac{d^2 f(x)}{dx^2} = -I_0 \frac{df(x)}{dx}\frac{d^2 g(t)}{dt^2}$$

$$= I_0 \omega^2 g(t)\frac{df(x)}{dx} \quad x = L \tag{15.138}$$

in which we have used Equation 15.134 to substitute for $d^2 g(t)/dt^2$. On canceling out $g(t)$ from Equation 15.138, we get

$$EI\frac{d^2 f(x)}{dx^2} = I_0 \omega^2 \frac{df(x)}{dx} \quad x = L \tag{15.139}$$

The boundary condition corresponding to shear gives

$$-EIg(t)\frac{d^3 f(x)}{dx^3} = -M_0 f(x)\frac{d^2 g(t)}{dt^2}$$

$$= M_0 \omega^2 f(x)g(t) \quad x = L \tag{15.140}$$

By canceling out $g(t)$ from Equation 15.140, we get

$$-EI\frac{d^3 f(x)}{dx^3} = M_0 \omega^2 f(x) \quad x = L \tag{15.141}$$

Equation 15.135, along with Equations 15.137, 15.139, and 15.141, defines the boundary value problem for this case.

Now let $f_i(x)$ and $f_j(x)$ be two different eigenfunctions of the system. Then, we have

$$\int_0^L EI f_i(x) \frac{d^4 f_j(x)}{dx^4} dx = f_i(x) EI \frac{d^3 f_j(x)}{dx^3}\bigg|_0^L - \int_0^L EI \frac{d f_i(x)}{dx} \frac{d^3 f_j(x)}{dx^3} dx$$

$$= f_i(x) EI \frac{d^3 f_j(x)}{dx^3}\bigg|_0^L - \frac{d f_i(x)}{dx} EI \frac{d^2 f_j(x)}{dx^2}\bigg|_0^L$$

$$+ \int_0^L EI \frac{d^2 f_i(x)}{dx^2} \frac{d^2 f_j(x)}{dx} dx \qquad (15.142)$$

On substituting the boundary conditions from Equations 15.137, 15.139, and 15.141 into Equation 15.142, we get

$$\int_0^L EI f_i(x) \frac{d^4 f_j(x)}{dx^4} dx + \lambda_j M_0 f_i(x) f_j(x)\bigg|_{x=L} + \lambda_j I_0 \frac{d f_i(x)}{dx} \frac{d f_j(x)}{dx}\bigg|_{x=L}$$

$$= \int_0^L EI \frac{d f_i^2(x)}{dx} \frac{d f_j^2(x)}{dx} dx \qquad (15.143)$$

In a similar manner, we can show that

$$\int_0^L EI f_j(x) \frac{d^4 f_i(x)}{dx^4} dx + \lambda_i M_0 f_j(x) f_i(x)\bigg|_{x=L} + \lambda_i I_0 \frac{d f_j(x)}{dx} \frac{d f_i(x)}{dx}\bigg|_{x=L}$$

$$= \int_0^L EI \frac{d f_j^2(x)}{dx} \frac{d f_i^2(x)}{dx} dx \qquad (15.144)$$

From Equations 15.143 and 15.144, we get

$$\int_0^L EI f_i(x) \frac{d^4 f_j(x)}{dx^4} dx + \lambda_j M_0 f_i(x) f_j(x)\bigg|_{x=L} + \lambda_j I_0 \frac{d f_i(x)}{dx} \frac{d f_j(x)}{dx}\bigg|_{x=L}$$

$$= \int_0^L EI f_j(x) \frac{d^4 f_i(x)}{dx^4} dx + \lambda_i M_0 f_i(x) f_j(x)\bigg|_{x=L}$$

$$+ \lambda_i I_0 \frac{d f_i(x)}{dx} \frac{d f_j(x)}{dx}\bigg|_{x=L} \qquad (15.145)$$

Equation 15.145 can be viewed as a modified definition of the self-adjointness of operator K. Compare this with Equation 15.32. From Equation 15.135, we have

$$EI\frac{d^4 f_i(x)}{dx^4} = \lambda_i m f_i(x)$$

$$EI\frac{d^4 f_j(x)}{dx^4} = \lambda_j m f_j(x)$$

(15.146)

On substituting Equation 15.146 in Equation 15.145, we get

$$(\lambda_j - \lambda_i)\left\{\int_0^L m f_j(x) f_i(x) dx + M_0 f_i(x) f_j(x)\Big|_{x=L}\right.$$

$$\left. + I_0 \frac{d f_j(x)}{dx}\frac{d f_i(x)}{dx}\Big|_{x=L}\right\} = 0$$

(15.147)

or

$$\left\{\int_0^L m f_j(x) f_i(x)\, dx + M_0 f_i(x) f_j(x)\Big|_{x=L}\right.$$

$$\left. + I_0 \frac{d f_j(x)}{dx}\frac{d f_i(x)}{dx}\Big|_{x=L}\right\} = 0 \quad \lambda_i \neq \lambda_j$$

(15.148)

Equation 15.148 is the modified orthogonality relationship. It can be interpreted to mean that the work done by the inertia forces in mode i on the displacements in mode j is zero. The first term in Equation 15.148 is proportional to the work done on the distributed inertia forces, while the last two terms are proportional to the work done by the inertia force and inertia moment due to the rigid disk on their corresponding displacements.

Example 15.3
The uniform bar shown in Figure E15.3a has axial rigidity EA and mass per unit length m. The bar is clamped at one end and free at the other. A rigid disk of mass M is rigidly attached to the free end. If $M = mL$, find the first three frequencies and mode shapes for the axial vibrations of the rod. Express the orthogonality relationship between mode shapes.

Solution
Referring to Example 15.1, the boundary value equation for axial vibrations of the uniform bar is given by

$$-EA\frac{d^2 f(x)}{dx^2} = \omega^2 m f(x)$$

(a)

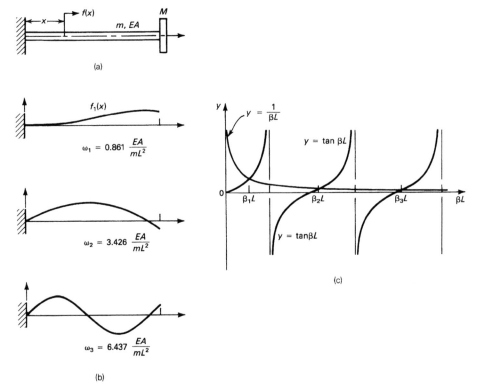

Figure E15.3. (a) Axial vibrations of a bar with rigid disk attached to the free end; (b) first three modes; (c) roots of the frequency equation.

The boundary condition at x is

$$f(x) = 0 \quad x = 0 \tag{b}$$

At $x = L$, the axial force in the bar should be equal to the inertia force on the disk, so that

$$EA\frac{\partial u}{\partial x} = -M\frac{\partial^2 u}{\partial t^2} \quad x = L \tag{c}$$

or

$$EAg(t)\frac{df(x)}{dx} = -Mf(x)\frac{d^2g(t)}{dt^2}$$

$$= M\omega^2 g(t)f(x) \quad x = L \tag{d}$$

Canceling $g(t)$ from Equation d, we get

$$EA\frac{df(x)}{dx} = M\omega^2 f(x) \quad x = L \tag{e}$$

The solution of Equation a is of the form

$$f(x) = C_1 \cos \beta x + C_2 \sin \beta x \tag{f}$$

where $\beta^2 = \omega^2 m / EA$.

Applying the boundary conditions at the fixed end, we get $C_1 = 0$, hence

$$f(x) = C_2 \sin \beta x \tag{g}$$

Applying the boundary condition of Equation e, we obtain

$$EAC_2 \beta \cos \beta L = M\omega^2 C_2 \sin \beta L \tag{h}$$

or

$$
\begin{aligned}
\tan \beta L &= \frac{EA}{m\omega^2} \frac{\beta}{L} \frac{mL}{M} \\
&= \frac{mL}{M} \frac{1}{\beta L} \\
&= \frac{1}{\beta L}
\end{aligned}
\tag{i}
$$

Equation i can be solved for βL numerically. The first three values are

$$\beta_1 L = 0.8605$$

$$\beta_2 L = 3.4256 \tag{j}$$

$$\beta_3 L = 6.4373$$

The first three frequencies and the corresponding mode shapes are shown in Figure E15.3b.

Figure E15.3c shows the curves $y = 1/\beta L$ and $y = \tan \beta L$. The intersections of these curves provide the roots of the frequency equation. It is evident that as n increases, $\beta_n L$ tends to $(n-1)\pi$. For large βL, therefore, $\sin \beta L$ approaches zero and the higher mode shapes have a node very close to the end of the bar. The rigid disk thus remains stationary in the higher modes.

15.10   FREE-VIBRATION RESPONSE OF A CONTINUOUS SYSTEM

The general partial differential equation for undamped free vibrations of a continuous system can be expressed as

$$K(u) + \frac{\partial^2}{\partial t^2} M(u) = 0 \tag{15.149}$$

where K and M are homogeneous linear differential operators of the spatial coordinate defined earlier. Using the method of separation of variables, we express $u(x,t)$ as the product of a function $f(x)$ of the spatial coordinate and function $y(t)$ of time.

$$u(x,t) = f(x)y(t) \tag{15.150}$$

Substitution of Equation 15.150 in Equation 15.149 leads to the following two separated equations

$$\ddot{y}(t) + \omega^2 y(t) = 0 \tag{15.151a}$$

$$K\{f(x)\} = \omega^2 M\{f(x)\} \tag{15.151b}$$

where ω^2, known as an eigenvalue, is the separation constant.

Equation 15.151b, along with the associated boundary conditions, is known as a boundary value problem. In the preceding sections, we solved this problem for several different cases. In each case, we found that Equation 15.151 leads to an infinite number of solutions for the frequency ω_n. Associated with each frequency, there is an eigenfunction $f_n(x)$ determined to within an arbitrary constant. Since the magnitude of $f_n(x)$ is arbitrary, we can scale it so that it is mass orthonormal. We denote the mass orthonormalized eigenfunction by $\phi_n(x)$. Referring to Equation 15.150, we can now express the general solution for $u(x,t)$ as

$$u(x,t) = \sum_{n=1}^{\infty} \phi_n(x)y_n(t) \tag{15.152}$$

where $y_n(t)$ is determined by the solution of an equation similar to Equation 15.151a

$$\ddot{y}_n(t) + \omega_n^2 y_n(t) = 0 \tag{15.153}$$

Equation 15.153 has the following solution

$$y_n(t) = y_n(0)\cos \omega_n t + \frac{\dot{y}_n(0)}{\omega_n} \sin \omega_n t \tag{15.154}$$

where $y_n(0)$ and $\dot{y}_n(0)$ are, respectively, the initial modal displacement and initial modal velocity, which must be determined from the prescribed values of the initial system displacement $u(x,0)$ and initial system velocity $\dot{u}(x,0)$. Now, using Equation 15.152, we obtain

$$u(x,0) = \sum_{n=1}^{\infty} \phi_n(x)y_n(0) \tag{15.155}$$

Following the procedure used to derive the expansion theorem, we apply the operator $\phi_i(x)M$ to both sides of Equation 15.155 and integrate from 0 to L, obtaining

$$\int_0^L \phi_i(x)M\{u(x,0)\}\,dx = \sum_{n=1}^{\infty} y_n(0) \int_0^L \phi_i(x)M\{\phi_n(x)\}\,dx \qquad (15.156)$$

Because of the orthogonality property of the eigenfunctions, all terms on the right-hand side except the one corresponding to $n=i$ vanish, and since eigenfunctions $\phi_n(x)$ are mass-orthonormal, $\int_0^L \phi_i(x)M\{\phi_i(x)\}\,dx = 1$. Equation 15.156 thus reduces to

$$y_i(0) = \int_0^L \phi_i(x)M\{u(x,0)\}\,dx \qquad (15.157)$$

In a similar manner

$$\dot{y}_i(0) = \int_0^L \phi_i(x)M\{\dot{u}(x,0)\}\,dx \qquad (15.158)$$

Equations 15.152, 15.154, 15.157, and 15.158 together provide the free-vibration response of the system.

It will be noted that the procedure is entirely analogous to the mode super-position method used in the solution of free-vibration problem for a discrete system. As in the case of a discrete system, the procedure relies on the determination of the frequencies ω_n and the eigenfunctions $\phi_n(x)$. Unlike a discrete system, however, the determination of the mode shapes and frequencies of a continuous system is not simple. In general, solutions can be obtained only for systems with simple geometry and boundary conditions and uniform properties.

15.11 UNDAMPED FREE TRANSVERSE VIBRATIONS OF A BEAM

The procedure used for the solution of the free-vibration problem of the flexural vibrations of a beam follows the steps outlined in the preceding section. For negligible damping and no shear deformation or rotatory inertia effects, we have

$$K = \frac{\partial^2}{\partial x^2}\left(EI\frac{\partial^2}{\partial x^2}\right)$$

$$M = m \qquad (15.159)$$

The mode shapes and frequencies are obtained by solving Equation 15.5. The solution of this equation is, however, difficult or impossible for a nonuniform beam section and general boundary conditions. For the simple case of a uniform, simply supported beam, the frequencies are given by Equation 15.59, while the normalized mode shapes are given by Equation 15.63. The free-vibration solutions are thus obtained from

$$u(x,t) = \sum_{n=1}^{\infty} \sqrt{\frac{2}{mL}} y_n(t) \sin \frac{n\pi x}{L} \tag{15.160}$$

$$y_n(t) = y_n(0) \cos n^2 \pi^2 \sqrt{\frac{EI}{mL^4}} t + \frac{\dot{y}_n(0)}{n^2 \pi^2} \sqrt{\frac{mL^4}{EI}} \sin n^2 \pi^2 \sqrt{\frac{EI}{mL^4}} t \tag{15.161}$$

$$y_n(0) = \int_0^L \sqrt{\frac{2m}{L}} u(x,0) \sin \frac{n\pi x}{L} dx \tag{15.162}$$

$$\dot{y}_n(0) = \int_0^L \sqrt{\frac{2m}{L}} \dot{u}(x,0) \sin \frac{n\pi x}{L} dx \tag{15.163}$$

Example 15.4
A load P is suspended from the midspan of a simply supported uniform beam as shown in Figure E15.4. The beam has a length L, mass per unit length m, and flexural rigidity EI. If the wire supporting the load suddenly snaps, describe the ensuing vibrations of the beam.

Solution
The initial displacement due to central load P is given by

$$u(x,0) = \frac{P}{48EI}(4x^3 - 3L^2 x) \quad 0 \le x \le L/2 \tag{a}$$

The displacement is symmetrical about the midspan.
   The initial modal displacement is obtained from Equation 15.162. For n even, the mode shape $\sqrt{2m/L}(\sin n\pi x/L)$ is antisymmetric and the integral in Equation 15.162 vanishes. For n odd, the

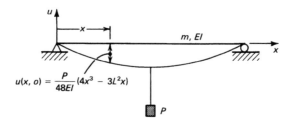

Figure E15.4. Vibrations of a simply supported beam from initial displacements.

value of the integral is given by

$$y_n(0) = \frac{P}{24EI}\sqrt{\frac{2m}{L}}\int_0^{L/2}(4x^3 - 3L^2x)\sin\frac{n\pi x}{L}dx \quad n = 1, 3, 5, \ldots, \infty$$

$$= -\sqrt{\frac{2m}{L}}\frac{PL^4}{EI}\frac{1}{(n\pi)^4} \quad n = 1, 5, 9, \ldots, \infty \tag{b}$$

$$= +\sqrt{\frac{2m}{L}}\frac{PL^4}{EI}\frac{1}{(n\pi)^4} \quad n = 3, 7, 11, \ldots, \infty$$

The displacement response is now obtained by substitution in Equations 15.161 and 15.160

$$u(x,t) = \frac{2PL^3}{\pi^4 EI}\left[-\sin\frac{\pi x}{L}\cos\pi^2\sqrt{\frac{EI}{mL^4}}t\right.$$

$$+\frac{1}{81}\sin\frac{3\pi x}{L}\cos 9\pi^2\sqrt{\frac{EI}{mL^4}}t$$

$$\left.-\frac{1}{625}\sin\frac{5\pi x}{L}\cos 25\pi^2\sqrt{\frac{EI}{mL^4}}t + \cdots\right] \tag{c}$$

Compared to the first term in Equation c, the second term has an amplitude of only 1.23%, while the third term has an amplitude of only 0.16%. As a check, we may obtain the displacement at the center of the beam at $t = 0$

$$u(L/2) = \frac{2PL^3}{\pi^4 EI}\left(-1 - \frac{1}{81} - \frac{1}{625} - \cdots\right) \tag{d}$$

The series converges very rapidly to $-PL^3/48EI$. Taking the first term alone, the midspan deflection works out to $-PL^3/48.705EI$, which is 98.6% of the exact deflection, $-PL^3/48EI$.

If desired, the internal moments or shear can be obtained from Equation c. Thus, the moment M is given by

$$M = EI\frac{\partial^2 u}{\partial x^2} = \frac{2PL}{\pi^2}\left[\sin\frac{\pi x}{L}\cos\pi^2\sqrt{\frac{EI}{mL^4}}t\right.$$

$$-\frac{1}{9}\sin\frac{3\pi x}{L}\cos 9\pi^2\sqrt{\frac{EI}{mL^4}}t$$

$$\left.+\frac{1}{25}\sin\frac{5\pi x}{L}\cos 25\pi^2\sqrt{\frac{EI}{mL^4}}t - \cdots\right] \tag{e}$$

This series converges less rapidly when compared to the series in Equation c, and more terms must therefore be included to obtain equivalent accuracy. The midspan moment at $t = 0$ is obtained from

$$M(L/2, 0) = \frac{2PL}{\pi^2}\left(1 + \frac{1}{9} + \frac{1}{25} + \cdots\right)$$

Taking the first 12 terms in the series, we get $M = PL/4.069$ which is 98.3% of the exact moment $PL/4$.

15.12   DAMPED FREE TRANSVERSE VIBRATIONS OF A BEAM

Neglecting shear deformation and rotatory inertia, but including the effect of damping, the equation of free transverse vibrations of a beam are obtained from Equation 14.29 with $p = 0$.

$$\frac{\partial^2}{\partial x^2}\left(EI\frac{\partial^2 u}{\partial x^2}\right) + \frac{\partial}{\partial t}\frac{\partial^2}{\partial x^2}\left(c_s I\frac{\partial^2 u}{\partial x^2}\right) + c\frac{\partial u}{\partial t} + m\frac{\partial^2 u}{\partial t^2} = 0 \tag{15.164}$$

In this case, the method of separation of variables will fail to uncouple the spatial and time variation. Classical modes and frequencies therefore do not exist. To obtain a solution of Equation 15.164, we express the displacement $u(x,t)$ as a superposition of the mode shapes of the associated undamped system. The transformation is given by Equation 15.152

$$u(x,t) = \sum_{n=1}^{\infty} \phi_n(x)y_n(t) \tag{15.152}$$

Substitution of Equation 15.152 in Equation 15.164 gives

$$\sum_{n=1}^{\infty} y_n(t)\frac{d^2}{dx^2}\left\{EI\frac{d^2\phi_n(x)}{dx^2}\right\} + \sum_{n=1}^{\infty} \dot{y}_n(t)\frac{d^2}{dx^2}\left\{c_s I\frac{d^2\phi_n(x)}{dx^2}\right\}$$

$$+ \sum_{n=1}^{\infty} c\dot{y}_n(t)\phi_n(x) + \sum_{n=1}^{\infty} m\ddot{y}(t)\phi_n(x) = 0 \tag{15.165}$$

On multiplying Equation 15.165 by $\phi_i(x)$, integrating from 0 to L, and using the orthogonality of mode shapes, we get

$$\omega_i^2 y_i(t) + \sum_{n=1}^{\infty} \dot{y}_n(t)\int_0^L \phi_i(x)\frac{d^2}{dx^2}\left\{c_s I\frac{d^2\phi_n(x)}{dx^2}\right\}dx$$

$$+ \sum_{n=1}^{\infty} \dot{y}_n(t)\int_0^L \phi_i(x)c\phi_n(x)\,dx + \ddot{y}(t) = 0 \quad i = 1,2,\ldots,\infty \tag{15.166}$$

Obviously, the equations are coupled because of the presence of damping terms. However, if we consider the special case in which c_s is proportional to E so that $c_s = \beta E$ and c is proportional to m so that $c = \alpha m$, the damping terms also satisfy the orthogonality relationship and Equation 15.166 reduces to

$$\ddot{y}_i(t) + (\alpha + \beta\omega_i^2)\dot{y}_i(t) + \omega_i^2 y_i(t) = 0 \quad i = 1,2,3,\ldots,\infty \tag{15.167}$$

Equations 15.167 are now uncoupled and are identical to the uncoupled single degree-of-freedom equations of a discrete system. In analogy with a discrete system, we can express the damping term as $2\xi_i\omega_i\dot{y}_i(t)$ where ξ_i is defined as the damping ratio in the ith mode. Equations 15.167 now reduces to

$$\ddot{y}_i(t) + 2\xi_i\omega_i\dot{y}_i(t) + \omega_i^2 y_i(t) = 0 \quad i = 1, 2, 3, \ldots, \infty \qquad (15.168)$$

Constants α and β can be selected to give desired levels of damping in any two of the modes. Thus, if ξ_j and ξ_k are damping ratios for the jth and kth mode, respectively

$$2\xi_j\omega_j = \alpha + \beta\omega_j^2$$
$$2\xi_k\omega_k = \alpha + \beta\omega_k^2 \qquad (15.169)$$

Equation 15.169 can be solved for α and β giving

$$\begin{bmatrix} \alpha \\ \beta \end{bmatrix} = \frac{2\omega_j\omega_k}{\omega_k^2 - \omega_j^2} \begin{bmatrix} \omega_k & -\omega_j \\ -\frac{1}{\omega_k} & \frac{1}{\omega_j} \end{bmatrix} \begin{bmatrix} \xi_j \\ \xi_k \end{bmatrix} \qquad (15.170)$$

The modal responses can be obtained from Equation 15.168 by, say, the Duhamel integral solution once the initial conditions in the physical coordinates have been translated to the initial modal displacements and velocities using Equations 15.157 and 15.158, respectively. The response in the physical coordinates is then obtained by using the transformation equation (Eq. 15.152).

SELECTED READINGS

Clark, S.K. 1972. *Dynamics of Continuous Elements*. Englewood Cliffs, New Jersey: Prentice Hall.
Den Hartog, J.P. 1956. *Mechanical Vibrations*. New York: McGraw-Hill. 4th Edition.
Jacobsen, L.S. & Ayre, R.S. 1958. *Engineering Vibrations with Applications to Structures and machinery*. New York: McGraw-Hill.
Meirovitch, L. 1967. *Analytical Methods in Vibration*. London: Macmillan.
Stokey, W.F. 1996. Vibrations of Systems Having Distributed Mass and Elasticity. In C.M. Harris (ed.) *Shock and Vibration Handbook*. New York: McGraw-Hill. 4th Edition.
Timoshenko, S., Young, D.H. & Weaver, W. Jr. 1990. *Vibration Problems in Engineering*. New York: John Wiley. 5th Edition.

PROBLEMS

15.1 For the beam of Problem 14.1, the spring stiffness $k = \pi^4 EI/(2L^3)$. Obtain the frequency equation. By a graphical method or by trial and error, determine the fundamental frequency. Plot the corresponding mode shape.

15.2 Obtain the frequency equation for torsional vibrations of the circular shaft of Problem 14.4. If $(M_1/2 + M_2)R^2 = I_0 L$, calculate the first two frequencies. Plot the corresponding mode shapes.

15.3 Find the first three frequencies and mode shapes of the setback building of Problem 14.6.

15.4 Derive the frequency equation for the frame shown in Figure P15.4. Determine the fundamental frequency and the corresponding mode shape.

15.5 The shaft of Problem 15.2 is given an initial twist by pulling down the mass M_2 a distance Δy, and then released. Obtain an expression for the ensuing vibrations of the shaft.

15.6 A simply supported beam of span L has mass per unit length m and flexural rigidity EI. The right-hand end of the beam is turned anticlockwise through an angle θ and then released. Obtain an expression for the subsequent vibrations of the beam.

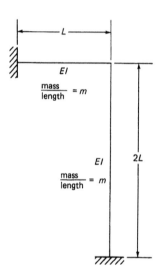

Figure P15.4.

CHAPTER 16

Continuous systems: Forced-vibration response

16.1 INTRODUCTION

The vibrations of a continuous system under the action of time-varying forces are governed by a non-homogeneous partial differential equation. The solution to the governing equation must, in addition, satisfy certain prescribed conditions on the boundary of the domain. Procedures for the formulation of governing equations and boundary conditions were presented in Chapter 14. Free-vibration response was discussed in Chapter 15. It was indicated there that the equation governing undamped free vibrations reduces to an eigenvalue problem, the solution of which provides the eigenvalues or frequencies and eigenfunctions or mode shapes of the system. In this chapter, we focus on techniques for the analysis of forced-vibration response. As in previous chapters, we restrict our discussion to systems in one-dimensional space in which the response is a function of one spatial coordinate and the time variable.

The methods of solution can be broadly divided into two categories: 1. integral transform methods, and 2. the mode superposition method. Among the transform methods are the Laplace transform and the Fourier transform. Both of these transforms remove the time dependence of the response and convert the problem to that of a single-degree-of-freedom differential equation in the spatial coordinate. The transformation is also applied to the associated boundary conditions so that the transformed conditions are independent of time. A solution to the transformed differential equation is obtained that at the same time satisfies the boundary conditions. An inverse transformation is then applied to the solution to obtain the response of the original system. The success of the transform method, in fact, depends on being able to obtain the inverse transform. This is not always easy, and in many cases, the inverse transform may be impossible to evaluate except possibly through numerical methods.

As in the case of a discrete system, the mode superposition method of solving the equation of motion for a continuous system relies on a transformation to normal coordinates in which the response of the system is expressed as a superposition of the eigenfunctions or mode shapes of the free vibrations of an associated undamped system. In theory, there are an infinite number of eigenfunctions, and the superposition must include all of them. In practice, sufficient accuracy may be obtained by using only the first few eigenfunctions. If the system being analyzed is undamped, the transformation converts the partial differential equation into a series of uncoupled single-degree-of-freedom equations which can be solved by standard procedures. When the system is damped, uncoupling will not be achieved unless the damping is of a special nature, so that the damping operator satisfies the orthogonality relationship.

In this chapter, we concentrate on the mode superposition method of analysis, also known as modal analysis or analysis through normal coordinate transformation. It should be noted that although modal analysis is straightforward, once the mode shapes and frequencies have been determined, the evaluation of the latter is nontrivial, and indeed, except for simple systems with uniform properties and simple boundary conditions, it is quite difficult and often impossible to determine the exact mode shapes and frequencies. In such cases, one must rely on approximate methods of analysis which use appropriately selected Ritz shapes rather than the mode shapes of the system to carry out a transformation. The Ritz method of analysis of continuous systems was discussed in detail earlier.

16.2 NORMAL COORDINATE TRANSFORMATION: GENERAL CASE OF AN UNDAMPED SYSTEM

The general partial differential equation governing the vibrations of a one-dimensional continuous system with negligible damping is given by

$$K\{u(x,t)\} + M\left\{\frac{\partial^2}{\partial t^2}u(x,t)\right\} = p(x,t) \tag{16.1}$$

in which K and M are homogeneous linear differential operators containing partial derivatives with respect to the spatial coordinate x up to order $2r$ and $2s$, respectively, and $p(x,t)$ is a forcing function which is a function of both x and t and may include one or more time varying concentrated forces. Associated with the differential equation of motion, there are r boundary conditions at each end of the one-dimensional domain. These boundary conditions are of the form

$$B_i\{u(x,t)\} = 0 \quad i = 1, 2, \ldots, r \tag{16.2}$$

in which B_i is a linear differential operator containing partial differentials with respect to the spatial coordinate up to an order of at most $2r - 1$. The boundary conditions given by Equation 16.2 are also linear and homogeneous, and for simplicity we assume that they do not contain the eigenvalue of the system.

The undamped free-vibration equation for the system is obtained from Equation 16.1 by setting $p = 0$. The resulting equation can then be solved by the method of separation of variables. This leads to the eigenvalue problem given by

$$K\{f(x)\} = \lambda M\{f(x)\} \tag{16.3}$$

in which $\lambda = \omega^2$ is the separation constant called the eigenvalue. Equation 16.3, along with its associated separated boundary conditions, gives an infinite number of solutions for λ. Associated with each solution for the eigenvalue, λ, there is a value of $f(x)$ called an eigenfunction. The eigenfunctions are determined only to within a constant; that is, they may be scaled arbitrarily and will satisfy Equation 16.3. An eigenfunctions normalized in this manner satisfy the following orthogonality relationships

$$\int_0^L \phi_j(x) K\{\phi_i(x)\}\, dx = \delta_{ij}\lambda_i \tag{16.4}$$

$$\int_0^L \phi_j(x) M\{\phi_i(x)\}\, dx = \delta_{ij} \tag{16.5}$$

where δ_{ij} is the Kronecker delta, which is equal to 1 for $i = j$ and is zero otherwise.

In a normal coordinate transformation, we express the response $u(x,t)$ as a superposition of the mode shapes $\phi(x)$, each multiplied by a weighting function of time. This weighting function is called the modal coordinate or the normal coordinate. Thus

$$u(x,t) = \sum_{i=1}^{\infty} \phi_i(x) y_i(t) \tag{16.6}$$

Substitution of Equation 16.6 in Equation 16.1 gives

$$\sum_i^{\infty} K\{\phi_i(x)\} y_i(t) + \sum_i^{\infty} M\{\phi_i(x)\} \ddot{y}_i(t) = p(x,t) \tag{16.7}$$

On multiplying both sides of Equation 16.7 by $\phi_j(x)$ and integrating from 0 to L, we get

$$\sum_{i=1}^{\infty} \int_0^L \phi_j(x) K\{\phi_i(x)\}\, dx\, y_i(t) + \sum_{i=1}^{\infty} \int_0^L \phi_j(x) M\{\phi_i(x)\}\, dx\, \ddot{y}_i(t)$$

$$= \int_0^L \phi_j(x) p(x,t)\, dx \tag{16.8}$$

Because of the orthogonality relationship given by Equations 16.4 and 16.5, such terms in the summations in Equation 16.8 for which $i = j$ drop out.

Equation 16.8 thus reduces to

$$\ddot{y}_j(t) + \omega_j^2 y_j(t) = p_j(t) \tag{16.9}$$

where p_j, called the jth modal force, is given by

$$p_j = \int_0^L \phi_j(x) p(x,t)\, dx \tag{16.10}$$

Equation 16.9 is an uncoupled single-degree-of-freedom differential·equation in the normal coordinate $y_j(t)$. There are an infinite number of such equations, one corresponding to each value of $j = 1, 2, 3, \ldots, \infty$. Once these equations are solved, the system response is obtained by using the transformation of Equation 16.6.

The solution of Equation 16.10 can be obtained by Duhamel's integral

$$y_j(t) = y_j(0) \cos \omega_j t + \frac{\dot{y}_j(0)}{\omega_j} \sin \omega_j(t)$$

$$+ \frac{1}{\omega_j} \int_0^t p_j(\tau) \sin \omega_j(t - \tau)\, d\tau \tag{16.11}$$

in which $y_j(0)$ and $\dot{y}_j(0)$ are the initial modal displacement and velocity, respectively obtained from the given initial system displacement and velocity by using the relationships in Equations 15.157 and 15.158 and repeated here for ease of reference

$$y_j(0) = \int_0^L \phi_j(x) M\{u(x,0)\}\, dx \tag{16.12}$$

$$\dot{y}_j(0) = \int_0^L \phi_j(x) M\{\dot{u}(x,0)\}\, dx \tag{16.13}$$

A forcing function of particular interest is a concentrated load $P_r(t)$ applied at a distance x_r from the origin. Mathematically, such a force can be expressed as

$$P_r(t)\delta(x - x_r) \tag{16.14}$$

where δ represents a special function called a delta function. The delta function is zero everywhere except at $x = x_r$. In the vicinity of x_r its magnitude is such that

$$\int_0^L \delta(x - x_r)\, dx = \int_{x_r - \varepsilon}^{x_r + \varepsilon} \delta(x - x_r)\, dx$$

$$= 1 \tag{16.15}$$

where ε is a small quantity. On multiplying Equation 16.14 by $\phi_j(x)$ and integrating from 0 to L, we get

$$p_j(t) = P_r(t) \int_0^L \phi_j(x)\delta(x - x_r)\,dx$$

$$= P_r(t)\phi_j(x_r)$$

(16.16)

The modal superposition analysis procedure described in the foregoing paragraphs is perfectly general and applies to the forced undamped vibration analysis of all one-dimensional continuous systems. In fact, its extension to systems with more than one spatial dimension is quite straightforward. As seen in Chapter 15, the eigenvalue problems for each system may be unique in some respect. However, once the frequencies and mode shapes have been determined, the remaining steps in the modal analysis are identical. In the following sections, we provide further demonstration of this by applying the modal analysis procedure to the solution of several different continuous systems.

16.3 FORCED LATERAL VIBRATION OF A BEAM

Consider the lateral vibrations of a uniform simply supported beam. The frequencies and normalized mode shapes of the beam are given by

$$\omega_n = n^2\pi^2\sqrt{\frac{EI}{mL^4}} \quad n = 1, 2, 3, \ldots, \infty$$

(16.17)

$$\phi_n(x) = \sqrt{\frac{2}{mL}}\sin\frac{n\pi x}{L} \quad n = 1, 2, 3, \ldots, \infty$$

(16.18)

Using an eigenfunction expansion, the equation of motion can be transformed into a set of infinite number of single-degree-of-freedom differential equations of motion in the normal coordinates. The jth equation in the series is given by Equation 16.9. As an example, suppose that the beam is subjected to a step function load P applied at quarter point as shown in Figure 16.1. The effective modal force p_n is obtained from Equation 16.10

$$p_n = \sqrt{\frac{2}{mL}}\mu(t)\int_0^L P\delta\left(x - \frac{L}{4}\right)\sin\frac{n\pi x}{L}\,dx$$

$$= P\sqrt{\frac{2}{mL}}\sin\frac{n\pi}{4}\mu(t)$$

(16.19)

in which $\mu(t)$ represents a unit step function applied at $t = 0$.

The nth modal equation thus reduces to

$$\ddot{y}_n(t) + \omega^2 y_n(t) = P\sqrt{\frac{2}{mL}}\mu(t)\sin\frac{n\pi}{4}$$

(16.20)

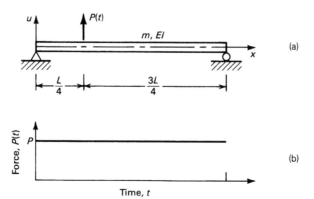

Figure 16.1. Lateral vibration of a simply supported beam: (a) elevation of beam; (b) forcing function.

For zero initial conditions, Equation 16.20 has a solution given by

$$y_n(t) = P\sqrt{\frac{2}{mL}}\frac{1}{\omega_n^2}\sin\frac{n\pi}{4}(1 - \cos\omega_n t)$$

(16.21)

The total response is obtained by superposing the modal responses

$$u(x,t) = \sum_{n=1}^{\infty}\sqrt{\frac{2}{mL}}\,y_n(t)\sin\frac{n\pi x}{L}$$

$$= \sum_{n=1}^{\infty}\frac{2PL^3}{\pi^4 EI}\frac{1}{n^4}\sin\frac{n\pi}{4}\sin\frac{n\pi x}{L}(1 - \cos\omega_n t)$$

$$= \frac{2PL^3}{\pi^4 EI}\left\{(1 - \cos\omega_1 t)\frac{1}{\sqrt{2}}\sin\frac{\pi x}{L} + \frac{1 - \cos\omega_2 t}{2^4}\sin\frac{2\pi x}{L}\right.$$

$$\left. + \frac{1 - \cos\omega_3 t}{3^4}\frac{1}{\sqrt{2}}\sin\frac{3\pi x}{L} + \cdots\right\}$$

(16.22)

As an example, the static displacement under the load can be obtained by solving Equation 16.20 for modal coordinate y_n with $\ddot{y}_n$ assuming to be zero and superposing the modal responses. The resulting value is

$$u\left(\frac{L}{4}\right) = \sum_{n=1}^{\infty}\frac{2PL^3}{\pi^4 EI}\frac{1}{n^4}\sin\frac{n\pi}{4}\sin\frac{n\pi x}{L} \qquad x = \frac{L}{4}$$

$$= \frac{2PL^3}{\pi^4 EI}\left\{\frac{1}{2} + \frac{1}{2^4} + \frac{1}{3^4}\cdot\frac{1}{2} + \cdots\right\}$$

(16.23)

Taking the first two terms in this series, we get

$$u\left(\frac{L}{4}\right) = 0.01155\frac{PL^3}{EI} \tag{16.24}$$

which is 98.6% of the exact value of $0.01172(PL^3/EI)$.

The bending moment along the length of the beam resulting from the application of dynamic load is given by

$$M(x,t) = EI\frac{\partial^2 u(x,t)}{\partial x^2}$$

$$= EI\sum_{n=1}^{\infty} -\sqrt{\frac{2}{mL}}\left(\frac{n\pi}{L}\right)^2 y_n(t)\sin\frac{n\pi x}{L} \tag{16.25a}$$

Substitution of $y_n(t)$ from Equation 16.21 yields

$$M(x,t) = -\frac{2PL}{\pi^2}\sum_{n=1}^{\infty}\frac{1}{n^2}\sin\frac{n\pi}{4}\sin\frac{n\pi x}{L}(1-\cos\omega_n t)$$

$$= -\frac{2PL}{\pi^2}\left\{\frac{1}{\sqrt{2}}(1-\cos\omega_1 t)\sin\frac{\pi x}{L}\right.$$

$$+\frac{1}{2^2}(1-\cos\omega_2 t)\sin\frac{2\pi x}{L} \tag{16.25b}$$

$$\left.+\frac{1}{3^2}\frac{1}{\sqrt{2}}(1-\cos\omega_3 t)\sin\frac{3\pi x}{L}+\cdots\right\}$$

The static load moment at $x = L/4$ can be obtained by solving Equation 16.20 for y_n with acceleration $\ddot{y}_n = 0$ and substituting the resulting value of y_n in Equation 16.25a.

$$M\left(\frac{L}{4}\right)$$

$$= -\frac{2PL}{\pi^2}\left\{\frac{1}{2}+\frac{1}{2^2}+\frac{1}{2}\frac{1}{3^2}+0+\frac{1}{2}\frac{1}{5^2}+\frac{1}{6^2}+\frac{1}{2}\frac{1}{7^2}+0+\cdots\right\} \tag{16.26}$$

The result obtained by taking the first eight terms of the series in Equation 16.26 is $-0.175PL$, which is 93.33% of the exact value of $-0.1875PL$. Obviously, the series in Equation 16.26 converges a bit more slowly than the series in Equation 16.23. The rapid convergence of the series of Equation 16.23 is due to the term n^4 in the denominator. This is replaced by n^2 in the series for moment. The shear force is given by $-EI(\partial^3 u/\partial x^3)$. The series representing the shear force will contain terms that decrease only inversely as n and will therefore converge much more slowly.

16.4   TRANSVERSE VIBRATIONS OF A BEAM UNDER TRAVELING LOAD

The transverse vibrations induced by a load traveling along a beam are of considerable engineering importance. A practical situation in which such vibrations may be induced is that of a hoisting crane, in which the hoist tolley travels along the span of the crane and the crane itself along with the trolley travels on the supporting girders. Another example is provided by wheel loads traveling across a bridge deck. The traveling wheels contribute both a concentrated mass which varies in position, and a moving load, which varies in position and may also vary in magnitude with time. In the following discussion, we assume that the mass contributed by the moving load is small compared to the system mass and does not appreciably alter the vibration response of the system.

Consider a load $g(t)$ traveling at a constant velocity v across a uniform, simply supported beam. The moving load can be represented mathematically as

$$p(x,t) = \begin{cases} \delta(x - vt)g(t) & 0 \le t \le L/v \\ 0 & t > L/v \end{cases} \tag{16.27}$$

where $\delta(x - vt)$ is a delta function centered at $x = vt$ and $g(t)$ represents the time variation. If $g(t)$ is a step function load given by $g(t) = P\mu(t)$, the effective modal force is obtained from

$$p_n(t) = \begin{cases} \displaystyle\int_0^L P\delta(x - vt)\mu(t)\phi_n(x)\,dx & 0 \le t \le L/v \\ 0 & t > L/v \end{cases} \tag{16.28}$$

On substituting $\phi_n(x) = \sqrt{2/mL}\,\sin(n\pi x/L)$ in Equation 16.28, we get

$$p_n(t) = \sqrt{\frac{2}{mL}} \int_0^L P\delta(x - vt)\mu(t)\sin\frac{n\pi x}{L}\,dx \quad 0 \le t \le L/v$$
$$= \sqrt{\frac{2}{mL}}P\sin\frac{n\pi vt}{L}\mu(t) \tag{16.29}$$
$$p_n(t) = 0 \qquad\qquad\qquad\qquad\qquad t > L/v$$

The modal response obtained by Duhamel's integral solution is

$$y_n(t) = \frac{P}{\omega_n}\int_0^t \sqrt{\frac{2}{mL}}\sin\frac{n\pi v\tau}{L}\sin\omega_n(t - \tau)d\tau$$
$$= \sqrt{\frac{2}{mL}}\frac{P}{(n\pi v/L)^2 - \omega_n^2}\left\{\frac{n\pi v}{L\omega_n}\sin\omega_n t - \sin\frac{n\pi vt}{L}\right\} \tag{16.30}$$
$$0 \le t \le L/v$$

Substitution of the modal response in the transformation Equation 16.6 gives

$$u(x,t) = \frac{2P}{mL} \sum_{n=1}^{\infty} \frac{1}{(n\pi v/L)^2 - \omega_n^2} \left\{ \frac{n\pi v}{L\omega_n} \sin \omega_n t - \sin \frac{n\pi vt}{L} \right\} \sin \frac{n\pi x}{L}$$

$$n = 1, 2, 3, \ldots, \infty \quad 0 \le t \le L/v \tag{16.31}$$

Equations 16.30 and 16.31 are valid provided that $n\pi v/L \ne \omega_n$. For $n\pi v/L = \omega_n$, evaluation of the integral in Equation 16.30 gives

$$y_n(t) = \frac{P}{\omega_n} \sqrt{\frac{2}{mL}} \int_0^t \sin \omega_n \tau \sin \omega_n(t - \tau) \, d\tau$$

$$\tag{16.32}$$

$$= \frac{P}{2\omega_n} \sqrt{\frac{2}{mL}} \left\{ \frac{1}{\omega_n} \sin \omega_n t - t \cos \omega_n t \right\} \quad 0 \le t \le L/v$$

and in the series given by Equation 16.31, the term corresponding to $n\pi v/L = \omega_n$ becomes

$$u_n(x,t) = \frac{P}{\omega_n^2 mL} \{ \sin \omega_n t - \omega_n t \cos \omega_n t \} \sin \frac{n\pi x}{L} \tag{16.33a}$$

$$= \frac{PL^3}{\pi^4 EI} \frac{1}{n^4} \left\{ \sin \frac{n\pi v}{L} t - \frac{n\pi v}{L} t \cos \frac{n\pi v}{L} t \right\} \sin \frac{n\pi x}{L} \tag{16.33b}$$

It is evident from Equation 16.33b that $u_n(x,t)$ grows indefinitely with t.

Example 16.1
A bridge with a single span of 80 ft has a deck of uniform cross section with mass $m = 200$ lb · s^2/ft^2 and flexural rigidity $EI = 1.328 \times 10^{10}$ lb · ft^2. A single wheel load P travels across the bridge at a uniform velocity of 50 mph, as shown in Figure E16.1a.
  (i) Obtain the influence line for deflection at midspan of the bridge considering only the first mode of vibration.
 (ii) Obtain the impact factor defined as the ratio of maximum midspan deflection from part (a) above and the maximum static deflection.
(iii) How much would the next-higher mode contribute to the maximum midspan deflection?

Solution
(i) From the data supplied, we get

$$\omega_n = n^2 \pi^2 \sqrt{\frac{EI}{mL^4}}$$

$$= 12.566 n^2 \text{ rad/s} \tag{a}$$

$$v = 73.33 \text{ ft/s}$$

$$\frac{n\pi v}{L} = 2.88n$$

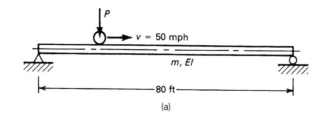

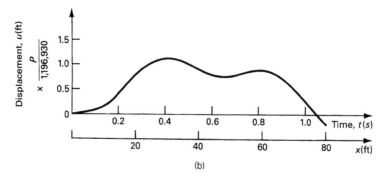

(b)

Figure E16.1. (a) Single wheel load traveling across a bridge deck; (b) lateral displacement at mid span.

The dynamic displacement caused by the traveling wheel load is obtained from Equation 16.31. Substitution for ω_n and $n\pi v/L$ from Equation a in Equation 16.31 gives

$$u(x,t) = \frac{P}{8000} \sum_{n=1}^{\infty} \frac{1}{(2.88n)^2 - (12.566n^2)^2}$$

$$\times \left\{ \frac{0.2292}{n} \sin 12.566n^2 t - \sin 2.88nt \right\} \sin \frac{n\pi x}{L} \qquad \text{(b)}$$

For $n = 1$ and $x = L/2$, Equation b reduces to

$$u\left(\frac{L}{2}, t\right) = \frac{P}{1,196,930} \left\{ -0.2292 \sin 12.566t + \sin 2.88t \right\} \qquad \text{(c)}$$

The wheel load will take $L/v = 1.091$ s to travel across the deck. Equation c is therefore valid for $0 \le t \le 1.091$ s. The values of midspan deflection calculated from Equation c at intervals of 0.1 s are shown in Table E16.1.

(ii) To obtain the magnitude of displacement, the figures in column 4 of Table E16.1 must be multiplied by $P/1,196,930$. The resultant displacement at midspan has been plotted in Figure E16.1b. The maximum midspan displacement occurs at $t = 0.4$ s when the wheel load is at a distance of $x = vt = 29.33$ ft from the left-hand support. The magnitude of maximum displacement is $u_{max}(L/2) = 1.1315(P/1,196,930)$. The maximum static midspan deflection is given by

$$\Delta_{max} = \frac{PL^3}{48EI} = \frac{P}{1,245,000} \qquad \text{(d)}$$

Table E16.1. Mid-span deflection in the first mode due to traveling load.

Time	$0.2292 \sin 12.5662t$	$\sin 2.88t$	$-0.2292 \sin 12.5662t$ $+ \sin 2.88t$
0.1	0.2180	0.2840	0.0661
0.2	0.1347	0.5446	0.4099
0.3	−0.1347	0.7604	0.8951
0.4	−0.2180	0.9135	1.1315
0.5	0.0000	0.9914	0.9914
0.6	0.2179	0.9877	0.7697
0.7	0.1347	0.9026	0.7679
0.8	−0.1347	0.7431	0.8778
0.9	−0.2180	0.5225	0.7405
1.0	0.0000	0.2588	0.2588
1.091	0.2085	0.0000	−0.2085

The ratio of maximum dynamic deflection to the maximum static deflection is

$$\frac{u_{max}}{\Delta_{max}} = 1.177 \tag{e}$$

The static load should thus be increased by 17.7% to account for dynamic effect.
(iii) The midspan deflection in the second mode is zero because the shape function $\sin(n\pi x/L)$ is zero for $n=2$ and $x=L/2$. The contribution due to the third mode is obtained from Equation b by substituting $n=3$ and $x=L/2$.

$$u\left(\frac{L}{2},t\right) = \frac{-P}{101.7 \times 10^6}(0.0764 \sin 113.1t - \sin 8.64t) \tag{f}$$

For $t=0.4$ s, Equation f gives

$$u\left(\frac{L}{2},0.4\right) = \frac{-P}{101.7 \times 10^6} \times 0.3817 \tag{g}$$

The additional impact factor at $t=0.4$ s becomes

$$\frac{u(L/2,0.4)}{\Delta_{max}} = -0.0047 \tag{h}$$

This may be added to the value obtained in Equation e to give a total impact factor of 1.172. The contribution from the third mode is comparatively very small.

16.5   FORCED AXIAL VIBRATIONS OF A UNIFORM BAR

The axial vibration of a bar under the action of a force $p(x,t)$ are governed by

$$-\frac{\partial}{\partial x}\left(EA\frac{\partial u}{\partial x}\right) + m\frac{\partial^2 u}{\partial t^2} = p(x,t) \tag{16.34}$$

Again, normal coordinate transformation results in an infinite series of single-degree-of-freedom equations of the form of Equation 16.9. Consider the vibrations of a uniform bar clamped at $x = 0$ and free at $x = L$ subjected to a tip load $F(t)$. The frequencies and mode shapes of the bar are given by

$$\omega_n = \frac{(2n-1)\pi}{2}\sqrt{\frac{EA}{mL^2}} \tag{16.35}$$

$$\phi_n(x) = \sqrt{\frac{2}{mL}}\sin\frac{(2n-1)\pi x}{2L} \tag{16.36}$$

The applied force can be represented by

$$p(n,t) = F(t)\delta(x-L) \tag{16.37}$$

The effective modal force is obtained from Equation 16.10

$$\begin{aligned}
p_n &= \int_0^L \phi_n(x)F(t)\delta(x-L)\,dx \\
&= F(t)\phi_n(L) \\
&= F(t)\sqrt{\frac{2}{mL}}\sin\frac{(2n-1)\pi}{2}
\end{aligned} \tag{16.38}$$

The modal response for zero initial conditions is given by

$$y_n(t) = \frac{1}{\omega_n}\int_0^t F(\tau)\sqrt{\frac{2}{mL}}\sin\frac{(2n-1)\pi}{2}\sin\omega_n(t-\tau)\,d\tau \tag{16.39}$$

Suppose that the applied load is a step function load $F\mu(t)$. On substituting this value in Equation 16.39 and carrying out the integration, we get

$$\begin{aligned}
y_n(t) &= \sqrt{\frac{2}{mL}}\frac{F}{\omega_n^2}\sin\frac{(2n-1)\pi}{2}(1-\cos\omega_n t) \\
&= (-1)^{n-1}\sqrt{\frac{2}{mL}}\frac{F}{\omega_n^2}(1-\cos\omega_n t)
\end{aligned} \tag{16.40}$$

The displacement in the physical coordinates is obtained by substituting Equation 16.40 in Equation 16.6

$$\begin{aligned}
u(x,t) &= \frac{2F}{mL}\sum_{n=1}^{\infty}(-1)^{n-1}\frac{1}{\omega_n^2}(1-\cos\omega_n t)\sin\frac{(2n-1)\pi x}{2L} \\
&= \frac{8FL}{\pi^2 EA}\sum_{n=1}^{\infty}\frac{(-1)^{n-1}}{(2n-1)^2}(1-\cos\omega_n t)\sin\frac{(2n-1)\pi x}{2L}
\end{aligned} \tag{16.41}$$

The axial force $P = EA\partial u/\partial x$ is obtained by differentiating Equation 16.41 with respect to x and multiplying by EA

$$P(x,t) = \frac{4F}{\pi}\sum_{n=1}^{\infty}(-1)^{n-1}\frac{1-\cos\omega_n t}{2n-1}\cos\frac{(2n-1)\pi x}{2L} \tag{16.42}$$

The displacement and axial force for specified values of x and t can be obtained by summing the series in Equations 16.41 and 16.42, respectively. The series converge quite slowly, particularly the one related to axial force P, and a large number of terms must be included to obtain reasonable accuracy.

As an example, the displacement at the free end is obtained by substituting $x = L$ in Equation 16.41

$$u(L,t) = \frac{8FL}{\pi^2 AE} \sum_{n=1}^{\infty} \frac{1}{(2n-1)^2}(1 - \cos \omega_n t) \tag{16.43}$$

Equation 16.43 indicates that the free end vibrates about an average displacement position given by

$$
\begin{aligned}
u_{avg} &= \frac{8FL}{\pi^2 EA} \sum_{n=1}^{\infty} \frac{1}{(2n-1)^2} \\
&= \frac{8FL}{\pi^2 EA} \left\{ 1 + \frac{1}{9} + \frac{1}{25} + \frac{1}{49} + \frac{1}{81} + \cdots \right\}
\end{aligned}
\tag{16.44}
$$

The series within the braces converges to $\pi^2/8$, so that the average displacement is FL/AE, which is equal to the static displacement under an axial load F. In Equations 16.41 and 16.42, the time parameter $\omega_n t$ can be expressed as

$$
\begin{aligned}
\omega_n t &= \frac{(2n-1)\pi}{2} \sqrt{\frac{EA}{m}} \frac{t}{L} \\
&= \frac{(2n-1)\pi}{2} \frac{ct}{L}
\end{aligned}
\tag{16.45}
$$

where $c = \sqrt{EA/m}$. Parameter c, which has the units of velocity, plays an important role in characterizing the response of the system. It is known as the *velocity of wave propagation* and its physical meaning will become apparent when we discuss the wave propagation analysis technique in Chapter 17.

Example 16.2
The uniform shear beam shown in Figure E16.2a has a mass m per unit length and shear stiffness k. It is subjected to a rectangular ground acceleration pulse as shown in Figure E16.2b. Obtain expressions for the maximum displacement and shear in each mode of vibration. By taking the absolute sum of the modal values, obtain an upper bound estimate for the base shear.

Solution
The equation of motion for a shear beam subjected to a support excitation $u_g(t)$ is given by

$$-k\frac{\partial^2 u}{\partial x^2} + m\frac{\partial^2 u}{\partial t^2} = -m\ddot{u}_g(t) \tag{a}$$

where u is the displacement relative to the support. The uncoupled modal equations are

$$\ddot{y}_n(t) + \omega_n^2 y_n(t) = p_n \tag{b}$$

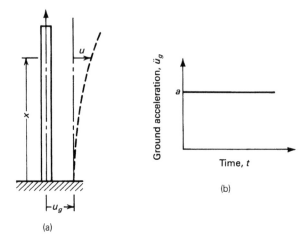

(a)

(b)

Figure E16.2. (a) Transverse vibrations of a shear beam under ground motion; (b) ground acceleration pulse.

where

$$p_n = \int_0^L -m\ddot{u}_g(t)\phi_n(x)\,dx \qquad\qquad (c)$$

and

$$\phi_n(x) = \sqrt{\frac{2}{mL}}\sin\frac{(2n-1)\pi x}{2L} \qquad\qquad (d)$$

$$\omega_n = \frac{(2n-1)\pi}{2}\sqrt{\frac{k}{mL^2}} \qquad\qquad (e)$$

On substituting for $\ddot{u}_g$ and $\phi_n(x)$ in Equation c and ignoring the negative sign, we get

$$p_n = ma\sqrt{\frac{2}{mL}}\int_0^L \sin\frac{(2n-1)\pi x}{2L}\,dx$$

$$= ma\sqrt{\frac{2}{mL}}\frac{2L}{(2n-1)\pi} \qquad\qquad (f)$$

Equation b has the solution

$$y_n = \frac{p_n}{\omega_n^2}(1-\cos\omega_n t) \qquad\qquad (g)$$

Transformation to physical coordinates gives

$$u(x,t) = \sum_{n=1}^{\infty} u_n \qquad\qquad (h)$$

where

$$u_n = y_n \phi_n(x)$$

$$= \frac{4a}{(2n-1)\pi\omega_n^2} \sin \frac{(2n-1)\pi x}{2L} (1 - \cos \omega_n t) \tag{i}$$

The modal shears are given by

$$V_n = k \frac{\partial u_n}{\partial x}$$

$$= \frac{2ak}{L\omega_n^2} \cos \frac{(2n-1)\pi x}{2L} (1 - \cos \omega_n t) \tag{j}$$

Substitution for ω_n gives

$$V_n = \frac{8amL}{\pi^2(2n-1)^2} \cos \frac{(2n-1)\pi x}{2L} (1 - \cos \omega_n t) \tag{k}$$

The maximum modal displacement is obtained from Equation i with $\cos \omega_n t = -1$, so that

$$(u_n)_{\text{max}} = \frac{8a}{(2n-1)\pi\omega_n^2} \sin \frac{(2n-1)\pi x}{2L} \tag{l}$$

On substituting for ω_n and setting $x = L$ in Equation l to obtain the maximum displacement at the top, we get

$$\{u_n(L)\}_{\text{max}} = \pm \frac{32amL^2}{\pi^3 k} \frac{1}{(2n-1)^3} \tag{m}$$

The maximum value of V_n is obtained from Equation j with $\cos \omega_n t = -1$:

$$(V_n)_{\text{max}} = \frac{16amL}{\pi^2(2n-1)^2} \cos \frac{(2n-1)\pi x}{2L} \tag{n}$$

The corresponding base shear is

$$(V_{0n})_{\text{max}} = \frac{16amL}{\pi^2(2n-1)^2} \tag{o}$$

The absolute sum of the maximum base shears in all modes is given by

$$(V_0)_{\text{max}} = \frac{16amL}{\pi^2} \sum_{n=1}^{\infty} \frac{1}{(2n-1)^2} \tag{p}$$

The series in Equation p converges to $\pi^2/8$. Hence

$$(V_0)_{\text{max}} = 2amL \tag{q}$$

Example 16.3
The 30-story building shown in Figure E16.3a has a uniform mass and story stiffness throughout the height. The building mass can be assumed to be lumped at the floor levels and is 16.4 metric

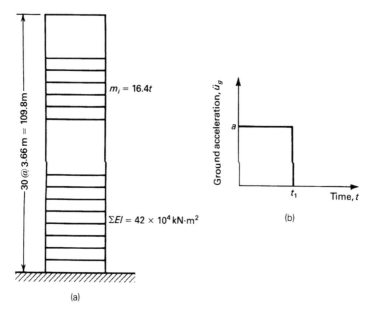

(a)

(b)

Figure E16.3. (a) Thirty-story building subjected to ground motion; (b) ground acceleration pulse.

ton per floor. The floor to floor height is 3.66 m and the sum of the flexural rigidities of column in a story is 42×10^4 kN·m^2. The building, which can be idealized as a shear beam, is subjected to a specified ground motion.

(i) If the ground motion consists of an acceleration pulse of duration $t_1 = 0.60$ s, as shown in Figure E16.3b, obtain the maximum base shear and the maximum top-story deflection in the first mode of vibration. Also determine the time at which these maximum values are attained. Then obtain the base shear and top displacements at this instant of time in the second and third modes of the building. By superposing the modal responses for the three modes, obtain an estimate of the total response. An upper bound estimate of the total response in the first three modes can be obtained by taking the sum of the absolute maximum values in each of the three modes. Compare the upper bound estimate of the base shear with the estimate obtained earlier.

(ii) Solve the example in part (a) with $t_1 = 1.5$ s.

Solution

The governing equation and the modal equations are given by Equations a and b of Example 16.2 respectively. The effective modal force p_n is obtained from Equation f of that example. When the ground acceleration a is in the form of a rectangular pulse, the solution to the modal equations are similar to those derived in Section 7.7.1

$$y_n(t) = \frac{p_n}{\omega_n^2}(1 - \cos \omega_n t) \quad t \leq t_1 \tag{a}$$

$$y_n(t) = \frac{p_n}{\omega_n^2}\{\cos \omega_n(t - t_1) - \cos \omega_n t\} \quad t > t_1 \tag{b}$$

The response spectrum for a rectangular pulse function was also derived in Section 7.7.1. For $\omega_n t_1 / 2\pi \geq 1/2$, the maximum displacement response occurs at $t_p = \pi/\omega_n$ and its value is given by

$$(y_n)_{\max} = \frac{2 p_n}{\omega_n^2} \qquad \text{(c)}$$

In this case, the maximum response is the same as that for a rectangular pulse of infinite duration. When $\omega_n t_1 / 2\pi < 1/2$, the maximum response occurs at $t_p = \pi/2\omega_n + t_1/2$ and its value is given by

$$(y_n)_{\max} = \frac{2 p_n}{\omega_n^2} \sin \frac{\omega_n t_1}{2} \qquad \text{(d)}$$

For the given building, we have

$$k = \frac{\sum 12EI}{h^2} = \frac{12 \times 42 \times 10^4}{(3.66)^2} = 376,240 \text{ kN}$$

$$m = \frac{m_i}{h} = \frac{14.6}{3.66} = 3.989 \text{ metric tons/m}$$

$$\omega_n = \frac{(2n-1)\pi}{2} \sqrt{\frac{k}{mL^2}}$$

$$= \frac{(2n-1)\pi}{2} \sqrt{\frac{376,240}{3.989 \times 109.8^2}}$$

$$= (2n-1)4.393 \text{ rad/s}$$

Hence $\omega_1 = 4.393$, $\omega_2 = 13.18$, and $\omega_3 = 21.97$.

(i) For the first mode

$$\frac{\omega_1 t_1}{2\pi} = \frac{4.393 \times 0.6}{2\pi}$$

$$= 0.419 < 0.5$$

Hence the maximum occurs at

$$t_p = \frac{\pi}{2\omega_1} + \frac{t_1}{2}$$

$$= 0.657 \text{ s}$$

and the value of the maximum modal response is

$$(y_1)_{\max} = \frac{4mLa}{\pi\omega_1^2} \sqrt{\frac{2}{mL}} \sin \frac{\omega_1 t_1}{2} \qquad \text{(e)}$$

The corresponding response in the physical coordinate is

$$(u_1)_{\max} = \frac{8a}{\pi\omega_1^2} \sin \frac{\omega_1 t_1}{2} \sin \frac{\pi x}{2L} \qquad \text{(f)}$$

The top displacement is obtained by substituting $x = L$ in Equation f

$$\{u_1(L)\}_{\max} = \frac{32amL^2}{\pi^3 k} \sin \frac{\omega_1 t_1}{2} \qquad \text{(g)}$$

Equation g may be compared with Equation m of Example 16.2. Substitution of appropriate values in Equation g gives

$$\{u_1(L)\}_{\text{max}} = 0.1319a \sin \frac{\omega_1 t_1}{2} = 0.1277a \text{ m}$$

For the second mode, the response at $t = 0.657$ s is obtained from Equation b

$$y_2(0.657) = \frac{2Lma}{3\pi} \sqrt{\frac{2}{mL}} \frac{1}{\omega_2^2} \{\cos(0.657 - 0.6)\omega_2 - \cos 0.657\omega_2\}$$

The response in the physical coordinate is given by

$$u_2(x, 0.657) = \phi_2(x)y_2(0.657)$$

$$= \frac{4a}{3\pi\omega_2^2}(\cos 0.057\omega_2 - \cos 0.657\omega_2) \sin \frac{3\pi x}{2L}$$

$$= 0.00355a \sin \frac{3\pi x}{2L}$$

The top displacement becomes

$$u_2(L, 0.657) = -0.00355a \text{ m}$$

Also, for the second mode

$$\frac{\omega_2 t_1}{2\pi} = 1.258 > 0.5 \text{ s}$$

Hence the maximum response in the second mode occurs at $t_p = \pi/\omega_2 = 0.238$ s and its value is given by

$$(u_2)_{\text{max}} = \frac{8a}{3\pi\omega_2^2} \sin \frac{3\pi x}{2L}$$

The top displacement is

$$\{u_2(L)\}_{\text{max}} = \frac{32amL^2}{27\pi^3 k}$$

$$= 0.00489a \text{ m}$$

For the third mode, the response at $t = 0.657$ s is obtained from

$$y_3(0.657) = \frac{2Lma}{5\pi} \sqrt{\frac{2}{mL}} \frac{1}{\omega_3^2} \{\cos(0.657 - 0.6)\omega_3 - \cos 0.657\omega_3\}$$

$$u_3(x, 0.657) = \phi_3(x)y_3(0.657)$$

$$= \frac{4a}{5\pi\omega_3^2}(\cos 0.057\omega_3 - \cos 0.657\omega_3) \sin \frac{5\pi x}{2L}$$

$$= 0.00032a \sin \frac{5\pi x}{2L}$$

The top displacement is given by

$$u_3(L, 0.657) = 0.00032a \text{ m}$$

The maximum response in the third mode occurs at $t_p = \pi/\omega_3 = 0.143$ s and its value is

$$(u_3)_{\text{max}} = \frac{8a}{5\pi\omega_3^2} \sin \frac{5\pi x}{2L}$$

The top displacement is

$$\{u_3(L)\}_{\text{max}} = \frac{32amL^2}{125\pi^3 k}$$

$$= 0.00106a \text{ m}$$

The top-story deflection at $t = 0.657$ s considering the first three modes is given by

$$u(L, 0.657) = (0.1277 - 0.00355 + 0.00032)a$$

$$= 0.1245a \text{ m}$$

The estimate obtained by taking the absolute sum of the modal maxima is

$$\{u(L)\}_{\text{max}} = (0.1277 + 0.00489 + 0.00106)a$$

$$= 0.1338a \text{ m}$$

The base shear is obtained from

$$V_0 = \left(k \frac{\partial u}{\partial x} \right)_{x=0}$$

The base shear at $t = 0.657$ s is therefore given by

$$V(0, 0.657) = \left(0.1277 \times \frac{\pi k}{2L} - 0.00353 \times \frac{3\pi k}{2L} \right.$$

$$\left. + 0.00032 \times \frac{5\pi k}{2L} \right) a$$

$$= 639.2a \text{ kN}$$

while the estimate obtained by taking the absolute sum of the modal shears is

$$V_{\text{max}} = (0.1277 + 3 \times 0.00489 + 5 \times 0.00106) \frac{\pi k}{2L} a$$

$$= 795.4a \text{ kN}$$

(ii) In this case, $\omega_1 t_1/2\pi = 1.048$ is greater than 0.5. Hence, the maximum response in the first mode occurs at $t_p = \omega_1/\pi = 1.398$ s and is given by

$$\{u_1(L)\}_{\text{max}} = \frac{32amL^2}{\pi^3 k}$$

$$= 0.1319a \text{ m}$$

The modal response in the second mode at $t = 1.398$ s is obtained from Equation a

$$y_2(t) = \frac{p_2}{\omega_2^2}(1 - \cos \omega_2 t)$$

$$= \frac{2Lma}{3\pi}\sqrt{\frac{2}{mL}}\frac{1}{\omega_2^2}(1 - \cos 1.398\omega_2)$$

The response in the physical coordinate is

$$u_2(x, 1.398) = \frac{4a}{3\pi\omega_2^2}(1 - \cos 1.398\omega_2)\sin\frac{3\pi x}{2L}$$

$$= 0.000216 \sin\frac{3\pi x}{2L}$$

The top displacement becomes

$$u_2(L, 1.398) = -0.000216a \text{ m}$$

The modal response in the third mode at $t = 1.398$ s is given by

$$y_3(t) = \frac{2Lma}{5\pi}\sqrt{\frac{2}{mL}}\frac{1}{\omega_3^2}(1 - \cos 1.398\omega_3)$$

The response in the physical coordinate is

$$u_3(x, 1.398) = \frac{4a}{5\pi\omega_3^2}(1 - \cos 1.398\omega_3)$$

$$= 0.000125a \text{ m}$$

The top-story displacement at $t = 1.398$ s considering the first three modes is given by

$$u(L, 1.398) = (0.1319 - 0.000216 + 0.000125)a$$

$$= 0.1318a \text{ m}$$

The estimate obtained by taking the absolute sum of the modal maxima is

$$\{u(L)\}_{max} = (0.1319 + 0.00489 + 0.00106)a$$

$$= 0.1379 \text{ m}$$

The base shear at $t = 1.398$ s is given by

$$V(0, 1.398) = (0.1319 - 3 \times 0.000126 + 5 \times 0.000125)\frac{\pi k}{2L}a$$

$$= 711.3a \text{ kN}$$

The estimate obtained by taking the absolute sum of the modal shear is

$$V(0)_{max} = (0.1319 + 3 \times 0.00489 + 5 \times 0.00106)\frac{\pi k}{2L}a$$

$$= (709.9 + 78.96 + 28.53)a$$

$$= 817.4a \text{ kN}$$

When all modes of the system are included, the base shear is obtained from Equation q of Example 16.2

$$V(0)_{max} = 2amL$$

$$= 876.0a \text{ kN}$$

Thus, the first three modes together account for 93.3% of the total base shear, while the first mode alone accounts for 81.0%.

16.6 NORMAL COORDINATE TRANSFORMATION, DAMPED CASE

When damping is included, the partial differential equation governing the vibrations of a continuous system takes the form

$$K\{u(x,t)\} + C\left\{\frac{\partial}{\partial t}u(x,t)\right\} + M\left\{\frac{\partial^2}{\partial t^2}u(x,t)\right\} = p(x,t) \qquad (16.46)$$

Introduction of normal coordinate transformation given by Equation 16.6 in Equation 16.46 gives

$$\sum_{i=1}^{\infty} K\{\phi_i(x)\}y_i(t) + \sum_{i=1}^{\infty} C\{\phi_i(x)\}\dot{y}_i(t)$$

$$+ \sum_{i=1}^{\infty} M\{\phi_i(x)\}\ddot{y}_i(t) = p(x,t) \qquad (16.47)$$

Multiplying both sides of Equation 16.47 by $\phi_j(x)$, and integrating from 0 to L, and using the orthonormality property (Eqs. 16.4 and 16.5), we get

$$\ddot{y}_j(t) + \sum_{i=1}^{\infty} \int_0^L \phi_j(x)C\{\phi_i(x)\}\,dx + \omega_j^2 y_j(t) = p_j(t)$$

$$j = 1, 2, 3, \ldots, \infty \qquad (16.48)$$

Because, in general, the damping operator C does not possess the orthogonality property, Equation 16.48 are coupled. However, if we assume that operator C is of the form

$$C = a_0 M + a_1 K \qquad (16.49)$$

Equations where a_0 and a_1 are constants, the integral in Equation 16.48 involving the damping term will reduce to the following

$$\sum_{i=1}^{\infty} \int_0^L \phi_j(x)C\{\phi_i(x)\}\,dx = \sum_{i=1}^{\infty} \int_0^L a_0\phi_j(x)M\{\phi_i(x)\}$$

$$+ a_1\phi_j(x)K\{\phi_i(x)\}\,dx$$

$$= a_0 + a_1\omega_j^2 \qquad (16.50)$$

If we define the damping fraction in the jth mode as ξ_j, we have

$$2\xi_j\omega_j = a_0 + a_1\omega_j^2 \tag{16.51}$$

and Equation 16.40 is reduced to

$$\ddot{y}_j(t) + 2\xi_j\omega_j\dot{y}_j(t) + \omega_j^2 y_j(t) = p_j(t) \quad j = 1, 2, 3, \ldots, \infty \tag{16.52}$$

The condition for damping orthogonality given by Equation 16.49 was derived in the case of free-vibration response of a beam in Chapter 15. As demonstrated there, parameters a_0 and a_1 can be selected to provide specified amounts of damping in any two modes. Once a_0 and a_1 have been so determined, the damping in any other mode can be derived from Equation 16.51 by substituting the appropriate value of ω_j. The foregoing treatment of the damped vibrations of a continuous system is quite general and applies to any one-dimensional system. It is not necessary for the system to be uniform and its properties may vary along the length, as long as the mode shapes and frequencies of the nonuniform system are used in Equations 16.47 through 16.52.

As an example of a damped system, consider a beam undergoing lateral vibrations. In this case, the operators K, M, and C are given by

$$K = \frac{\partial^2}{\partial x^2}\left\{EI(x)\frac{\partial^2}{\partial x^2}\right\}$$

$$M = m(x) \tag{16.53}$$

$$C = \frac{\partial^2}{\partial x^2}\left\{c_s(x)I(x)\frac{\partial^2}{\partial x^2}\right\} + c(x)$$

If C is to satisfy the orthogonality property, we must have

$$c(x) = a_0 m(x) \tag{16.54}$$

$$c_s(x) = a_1 E$$

For the axial vibration of a bar

$$K = -\frac{\partial}{\partial x}\left\{EA(x)\frac{\partial}{\partial x}\right\}$$

$$M = m(x) \tag{16.55}$$

$$C = -\frac{\partial}{\partial x}\left\{c_s A(x)\frac{\partial}{\partial x}\right\} + c(x)$$

and the conditions for damping orthogonality are again given by Equation 16.54.

Example 16.4

A precast concrete T-beam used to support fixed seating in a stadium is shown in Figure E16.4. The cross-sectional properties of the beams are indicated in the figure. The beam is simply supported. The superimposed load on account of fixed seats and spectators can be taken as 1.829 kN/m. Clapping and stamping by spectators during a sporting event impose a harmonic dynamic load on the beam. The load amplitude is 0.305kN/m and the frequency is 3Hz. Damping can be taken as uniformly distributed along the length of the beam, and measurements have shown that the damping ratio in the first mode is 3% of critical. Determine the first three natural frequencies of the system and the maximum dynamic deflection at the midspan considering the first three modes of vibration.

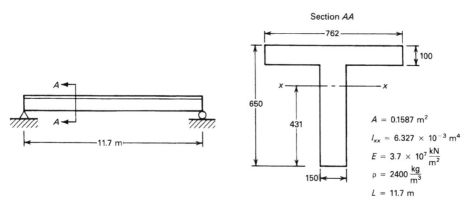

Figure E16.4. Precast concrete T-beam to support stadium seating.

Solution

The problem is similar to Example 6.1, where it was solved by assuming a vibration shape. The natural frequencies and mode shapes of the loaded beam are given by

$$\omega_n = n^2 \pi^2 \sqrt{\frac{EI}{mL^4}} \tag{a}$$

$$\phi_n(x) = \sqrt{\frac{2}{mL}} \sin \frac{n\pi x}{L}$$

Since the mode shapes have been mass-orthonormalized, the generalized mass $m_n^* = 1$ and the generalized stiffness $k_n^* = \omega_n^2$. The uniformly distributed harmonic load is given by

$$p = p_0 \sin \Omega t \tag{b}$$

where $p_0 = 0.305$ kN/m and $\Omega = 6\pi$ rad/s. The effective modal force is obtained from

$$p_n = \int_0^L p\phi_n(x)\, dx$$

$$= \sqrt{\frac{2}{mL}} \frac{2L}{n\pi} p_0 \sin \Omega t \quad n = \text{odd}$$

$$= 0 \quad n = \text{even} \tag{c}$$

The uncoupled modal equations are given by

$$\ddot{y}_n(t) + 2\xi_n\omega_n\dot{y}_n(t) + \omega_n^2 y_n(t) = \sqrt{\frac{2}{mL}}\frac{2L}{n\pi}p_0\sin\Omega t \tag{d}$$

Equation d has the solution

$$y_n = \sqrt{\frac{2}{mL}}\frac{2L}{n\pi}p_0\frac{1}{\omega_n^2 D_n}\sin(\Omega t - \theta_n) \tag{e}$$

where

$$D_n = \frac{1}{\sqrt{(1 - \beta_n^2)^2 + (2\xi_n\beta_n)^2}}$$

$$\theta_n = \tan^{-1}\frac{2\xi_n\beta_n}{1 - \beta_n^2} \tag{f}$$

$$\beta_n = \frac{\Omega}{\omega_n}$$

The displacement is given by

$$u(x,t) = \sum_{n=1}^{\infty} y_n\phi_n(x)$$

$$= \sum_{n=1}^{\infty} \frac{4p_0 L^4}{\pi^5 EI}\frac{D_n}{n^5}\sin(\Omega t - \theta_n)\sin\frac{n\pi x}{L}$$

From the data supplied

$$E = 3.7 \times 10^7 \text{ N/m}^2$$

$$I = 6.32 \times 10^{-3} \text{ m}^4$$

$$L = 11.7 \text{ m}$$

Mass per unit length:
Due to mass of the beam $0.1587 \times 2.4 = 0.3809 \times 10^3$ kg/m
Due to seats and spectators $\frac{1.829}{9.81} = 0.1864 \times 10^3$ kg/m
Total $= 0.5673 \times 10^3$ kg/m
First mode response:

$$\omega_1 = \pi^2\sqrt{\frac{EI}{mL^4}}$$

$$= 46.315 \text{ rad/s}$$

$$= 7.374 \text{ Hz}$$

$$\xi_1 = 0.03$$

$$\beta_1 = \frac{3}{7.374} = 0.4068$$

$$D_1 = \{(1 - \beta_1^2)^2 + (2\xi\beta_1)^2\}^{-1/2}$$

$$= 1.1978$$

$$\{u(L/2)\}_{\max} = \frac{4 p_0 L^4}{\pi^5 EI} D_1$$

$$= \frac{4 \times 0.305 \times (11.7)^4 \times 1.1978}{\pi^5 \times 3.7 \times 10^7 \times 6.32 \times 10^{-3}}$$

$$= 3.284 \times 10^{-4} \text{ m}$$

The maximum response in the first mode occurs when

$$\sin(\Omega t - \theta_1) = 1$$

so that

$$\Omega t - \theta_1 = \frac{\pi}{2}$$

Also

$$\tan \theta_1 = \frac{2\xi_1 \beta_1}{1 - \beta_1^2}$$

or

$$\theta_1 = 0.02925 \text{ rad}$$

Hence, the time at maximum is given by

$$t_m = \frac{1}{\Omega} \left(\frac{\pi}{2} + \theta_1 \right)$$

$$= 0.0849 \text{ s}$$

Second mode response:

$$\omega_2 = 4\pi^2 \sqrt{\frac{EI}{mL^4}}$$

$$= 185.3 \text{ rad/s}$$

$$= 29.5 \text{ Hz}$$

Since the effective force in the second mode is zero, that mode does not contribute anything to the response.

Third mode response:

$$\omega_3 = 9\pi^2 \sqrt{\frac{EI}{mL^4}}$$

$$= 416.8 \text{ rad/s}$$

$$= 66.37 \text{ Hz}$$

The damping is mass proportional; hence, from Equation 16.43

$$a_0 = 2\xi_1 \omega_1$$

$$a_0 = 2\xi_3 \omega_3$$

or

$$\xi_3 = \xi_1 \frac{\omega_1}{\omega_3}$$

$$= \frac{1}{9} \times 0.03 = 0.0033$$

$$\beta_3 = \frac{3}{66.37} = 0.0452$$

$$D_3 = \{(1 - \beta_3^2)^2 + (2\xi_3 \beta_3)^2\}^{-1/2}$$

$$= 1.002$$

The absolute value of the maximum midspan deflection in the third mode is given by

$$\{u(L/2)\}_{max} = \frac{4 p_0 L^4}{3^5 \pi^5 EI} D_3$$

$$= 1.317 \times 10^{-6} \text{ m}$$

which is quite small compared to the deflection in the first mode. An upper bound estimate of the displacement in the first three modes can be obtained by adding the absolute values in the first and third modes. This gives

$$\Delta_{max} = 3.824 \times 10^{-4} + 1.317 \times 10^{-6}$$

$$= 3.837 \times 10^{-4} \text{ m}$$

A more precise estimate of the deflection is likely to be obtained by calculating the deflection in the third mode at $t_m = 0.0849$ s, at which time the deflection in the first mode is a maximum, and adding the calculated third-mode value to that obtained in the first mode. The midspan deflection in the third mode at $t = t_m = 0.0849$ s is given by

$$u_3(L/2, t_m) = \frac{4 p_0 L^4}{3^5 \pi^5 EI} D_3 \sin(\Omega t_m - \theta_3) \sin \frac{3\pi}{2}$$

where

$$\tan \theta_3 = \frac{2\xi_3 \beta_3}{1 - \beta_3^2}$$

$$= \frac{2 \times 0.0033 \times 0.0452}{1 - 0.0452^2}$$

$$= 2.989 \times 10^{-4}$$

$$\theta_3 = 2.989 \times 10^{-4} \text{ rad}$$

On substituting this value of θ_3 and $t_m = 0.0849$, we get

$$u_3(L/2, t_m) = -1.316 \times 10^{-6}\,\text{m}$$

Hence the total deflection in the first three modes works out to

$$\Delta_{\max} = 3.824 \times 10^{-4} - 1.316 \times 10^{-6}$$

$$= 3.811 \times 10^{-4}\,\text{m}$$

SELECTED READINGS

Clark, S.K. 1972. *Dynamics of Continuous Elements.* Englewood Cliffs, New Jersey: Prentice Hall.

Den Hartog, J.P. 1956. *Mechanical Vibrations.* New York: McGraw-Hill. 4th Edition.

Jacobsen, L.S. & Ayre, R.S. 1958. *Engineering Vibrations with Applications to Structures and Machinery.* New York: McGraw-Hill.

Meirovitch, L. 1967. *Analytical Methods in Vibration.* London: Macmillan.

Stokey, W.F. 1996. Vibrations of Systems Having Distributed Mass and Elasticity. In C.M. Harris (ed.) *Shock and Vibration Handbook.* New York: McGraw-Hill. 4th Edition.

Timoshenko, S., Young, D.H. & Weaver, W. Jr. 1990. *Vibration Problems in Engineering.* New York: John Wiley. 5th Edition.

PROBLEMS

16.1 A simply supported beam of span L, mass per unit length m, and flexural rigidity EI vibrates under the following support motions. Obtain expressions for the displacement response of the beam. (a) Both the left and the right hand ends undergo identical lateral motions given by $u_g = A_0 \sin \Omega t$. (b) The left hand end of the beam translates in a lateral direction according to the equation $u_g = A_0 \sin \Omega t$.

16.2 The dynamic deflection of a simply supported bridge deck due to a moving load is given by Equation 16.31. Show that when only the first term in the series is taken into account, that equation can be expressed as

$$u(x,t) = \frac{2PL^3}{\pi^4 EI} \left(\frac{1}{1-\alpha^2} \sin 2\pi\alpha \frac{t}{T} - \frac{\alpha}{1-\alpha^2} \sin 2\pi \frac{t}{T} \right) \sin \frac{\pi x}{L}$$

where $T = $ fundamental period of the bridge and $\alpha = $ speed parameter $= vT/2L$. The 'crawl' deflection $u_s(x,t)$ is defined as the deflection produced at a point on the bridge as the moving load travels across the bridge at a very slow speed so that $\alpha \approx 0$. The deflection impact factor DI is defined as

$$DI = \frac{u(x,t) - u_s(x,t)}{(u_s)_{\max}}$$

Derive expressions for $u_s(x,t)$ and DI.

16.3   An indication of the deflections and stresses produced in the bridge deck due to a traveling vehicle can be obtained by considering that the vehicle imposes a force of the form

$$W\left(1 + A\cos 2\pi \frac{t}{T_v}\right)$$

where W is the gross weight of the vehicle and T_v is the vehicle heave period. The deflection due to a constant traveling force is given by Equation 16.31. Show that when only the first term in the series is considered, the deflection due to the alternating force $WA\cos(2\pi t/T_v)$ is given by

$$u_a(x,t) = \frac{2WL^3 A}{\pi^4 EI}\left(D_1 \sin 2\pi\alpha\frac{t}{T} \cos 2\pi\frac{t}{T_v}\right.$$

$$\left. + D_2 \cos 2\pi\alpha\frac{t}{T} \sin 2\pi\frac{t}{T_v} - \alpha D_3 \sin 2\pi\frac{t}{T}\right)\sin\frac{\pi x}{L}$$

where T is the fundamental period of the bridge $\alpha = vT/2L$ and

$$\phi = \frac{T}{T_v}$$

$$\Delta = \{1 - (\phi + \alpha)^2\}\{1 - (\phi - \alpha)^2\}$$

$$D_1 = \{1 - (\phi^2 + \alpha^2)\}/\Delta$$

$$D_2 = 2\alpha\phi/\Delta$$

$$D_3 = \{1 + (\phi^2 - \alpha^2)\}/\Delta$$

16.4   The bar of Example 15.2 is vibrating in an axial direction under the action of a constant axial force of magnitude P_0 applied suddenly at a distance L from the fixed end. Obtain an expression for the displacement response using the first three modes of vibration.

16.5   The vibrations of the bar in Problem 16.4 are resisted by damping forces that are viscous in nature. Assuming that the distributed damping coefficient is proportional to the mass per unit length and that damping in the first mode is 5% of critical, derive the damping ratios for the second and the third modes. Then, using the first three modes obtain an expression for the damped axial vibrations of the bar.

16.6   The building shown in Figure P14.6 is excited into lateral vibration by an eccentric motor mounted at the top floor level. The exciting force can be represented by $p = p_0 \cos \Omega t$ where $p_0 = 400$ kN and Ω is twice the fundamental frequency of the building. Obtain the steady state response of the frame and the maximum base shear using the first three mode shapes. Neglect damping.

CHAPTER 17

Wave propagation analysis

17.1 INTRODUCTION

Theoretically, mode superposition is an effective method of obtaining the free
or forced vibration response of a continuous system of finite extent. In practice,
several difficulties may arise in the application of the method. First, it may not
be possible to determine the mode shapes and frequencies of the continuous
system being analyzed. This is particularly so for systems of complex geometry, or/and nonuniform properties. Second, the mode superposition response
is obtained in the form of an infinite series, and although the series usually
converges quite rapidly when the displacement response is being calculated,
convergence may be quite slow for other parameters of interest, such as for
example, the bending moment and shear force in a beam undergoing transverse
vibrations, and axial force in a bar undergoing axial vibrations. In such a case,
many more terms of the series must be included in the computations to obtain
reasonable accuracy.

The mode superposition method is no longer applicable if the given system
is of semi-infinite or infinite extent. In contrast to a system of finite extent,
where the frequencies are spaced at discrete intervals, a system of infinite extent has a continuous band of frequencies and the term mode shape looses its
meaning. Obviously, modal analysis is not possible in such a case. However,
an alternative method known as wave propagation analysis may prove to be
quite effective in obtaining the response of the system.

The method of wave propagation may also be applied to the response analysis of certain special finite extent systems. In fact, it may in some cases,
prove to be simpler than a mode superposition analysis. In this chapter, we
describe the application of wave propagation method to the analysis of simple
one-dimensional systems in which the response is a function of one spatial coordinate and the coordinate of time is. Several special terms have become part
of the accepted vocabulary related to analysis by wave propagation. Such terms
include: traveling and standing waves, harmonic waves, wave velocity, wave
length and frequency, reflection and refraction of waves, and wave dispersion.
These terms are defined as they appear in the discussion. It is shown that for the

systems considered, the response can be represented by a superposition of one or more waves whose shapes satisfy the equation governing the free vibration response, known as the wave equation. Methods of finding the wave shapes for given initial disturbances are described and the phenomenon of reflection and refraction of such waves when they arrive at a boundary of the system is discussed.

17.2   THE PHENOMENON OF WAVE PROPAGATION

The term *wave* is commonly understood to mean a disturbance that travels through a medium with or without change of form, usually at a constant velocity. Several examples of wave propagation can be found in our common experience. Ripples on the surface of water, sound waves traveling through air, the vibrations of a stretched string, are all instances of wave propagation.

For a more precise definition of a wave, consider a uniform bar of infinite length lying along the x axis. Let the bar be subjected to a disturbance which is a function of both the spatial coordinate x and the time t. More specifically, let this disturbance be an axial displacement given by

$$u = f_1(x - ct) \tag{17.1}$$

in which c is a constant with units of velocity so that the product ct has the units of space. The shape of the disturbance at $t = 0$ is given by $u = f_1(x)$ and is shown in Figure 17.1a. For the purpose of illustration, we have assumed that the origin of the space coordinate is at the beginning of the disturbance.

After time t, the displacement shape will be given by Equation 17.1. By defining a new coordinate, $\tilde{x} = x - ct$, it is easily recognized that $u = f(\tilde{x})$ is of the same shape as $f(x)$ except that the spatial distance is now measured from a new origin which is shifted by a distance ct to the right of the previous origin. The shape of the disturbance at time t is indicated by dashed lines in Figure 17.1a which shows that the disturbance has traveled a distance ct in the positive direction of x axis without change of shape. In other words, function $f_1(x - ct)$ represents a wave traveling at a constant velocity c in the positive direction of x. A wave of this type is known as a *traveling wave*.

As another example of wave propagation, let the displacement of the bar be given by

$$u = f_2(x + ct) \tag{17.2}$$

The displaced shape of the bar at $t = 0$ is shown in Figure 17.1b. After elapse of time t, the displaced shape will appear as shown by dashed lines in Figure 17.1b. It is seen that the disturbance has traveled to the left a distance ct. Function $f_2(x + ct)$ thus represents a wave traveling at a constant velocity c in the negative direction of x.

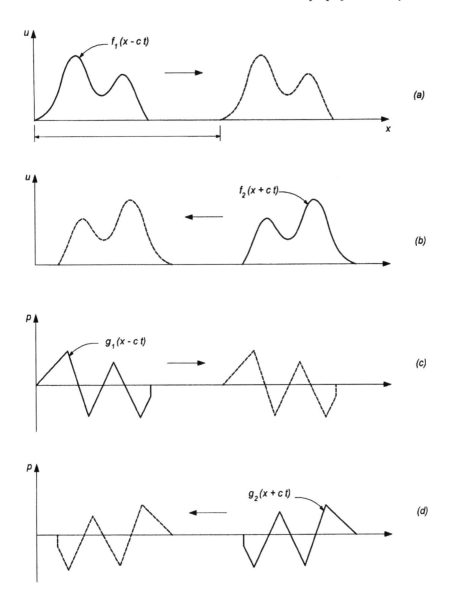

Figure 17.1. Wave propagation in a uniform bar vibrating in an axial direction: (a) forward-moving displacement wave; (b) backward-moving displacement wave; (c) forward-moving pressure wave; (d) backward-moving pressure wave.

The waves represented by Equations 17.1 and 17.2 are displacement waves, propagating, respectively, in the positive and negative directions of the x axis. The displacement caused by these waves gives rise to an axial pressure in the bar. For example, the distribution of the axial pressure caused by the forward moving displacement wave is given by

$$
\begin{aligned}
p &= E\frac{du}{dx} \\
&= E f_1'(x - ct) \\
&= g_1(x - ct)
\end{aligned}
\tag{17.3}
$$

where $g(x-ct)$ is another function of $x-ct$. Equation 17.3 represents a forward moving pressure wave as shown in Figure 17.1c. In a similar manner, the displacement defined by Equation 17.2 gives rise to a backward moving pressure wave

$$
p = g_2(x + ct)
\tag{17.4}
$$

where $g_2(x+ct) = E f_2'(x+ct)$. The wave represented by Equation 17.4 is shown in Figure 17.1d.

17.3   HARMONIC WAVES

A traveling wave of special interest is a *harmonic wave* given by

$$
u(x, t) = \sin(kx - bt)
\tag{17.5}
$$

A harmonic wave, for example, can be produced in a long string by moving one of its ends up and down in a special manner. The resulting disturbances in the string will be as shown in Figure 17.2a. The first curve in that figure shows the disturbed shape of the string at say $t = 0$. It is obtained by substituting $t = 0$ in Equation 17.5, so that

$$
u(x, t) = \sin kx
\tag{17.6}
$$

It is easily verified that the wave shape repeats itself after a distance $\lambda = 2\pi/k$. This distance is known as the *wave length*.

The remaining curves in Figure 17.2a show the displaced shape of the string at several different intervals of time separated by $\pi/4b$. They are obtained by substituting for t its different values in Equation 17.2. The displaced shape or the wave is seen to move progressively to the right. By tracing the movement of a point with a specified amplitude of displacement as the wave travels, it can be concluded that it takes a time $2\pi/b$ for the wave to travel a distance λ. The *wave velocity* c is thus given by

$$
\begin{aligned}
c &= \frac{\lambda b}{2\pi} \\
&= \frac{b}{k}
\end{aligned}
\tag{17.7}
$$

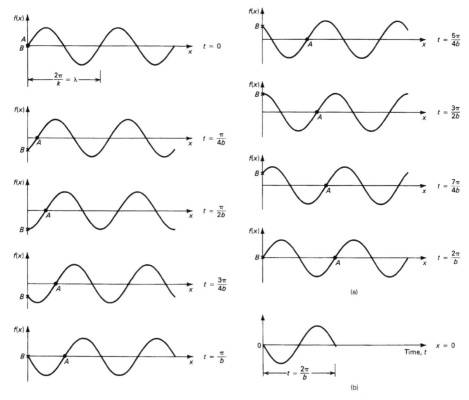

Figure 17.2. Harmonic traveling wave in a stretched string: (a) displaced shape of the string at different instances of time; (b) motion of end $x = 0$.

It should be noted that a point such as A whose movement is being traced refers to different locations along the string and that the curves in Figure 17.2a are, in fact, snapshots of the string at different instances of time. On the other hand, if we observe the motion of a specific point on the string, say B, it will describe a simple harmonic motion as shown in Figure 17.2b. The period of this motion is given by $T = 2\pi/b$ so that circular frequency is $\omega = b$ and the frequency in cycles per second is $f = \omega/2\pi$. From Equation 17.7, we obtain the following important relationship

$$c = \frac{\lambda\omega}{2\pi} = \lambda f \qquad (17.8)$$

signifying that the wave velocity is the product of wave length and frequency. Also

$$k = \frac{\omega}{c} \qquad (17.9)$$

where parameter k is referred to as the *wave number*.

The equation for a harmonic wave can thus be expressed in several alternative ways

$$u(x, t) = \sin(kx - \omega t) \tag{17.10a}$$

$$= \sin k(x - ct) \tag{17.10b}$$

$$= \sin \frac{2\pi}{\lambda}(x - ct) \tag{17.10c}$$

In many instances, we can represent the motion of a system by a superposition of two or more waves. In particular, the superposition of harmonic waves leads to some interesting results. As an example, suppose that the motion is given by the sum of two identical harmonic waves moving in opposite directions, so that

$$u(x, t) = \sin(kx - \omega t) + \sin(kx + \omega t)$$
$$= 2 \sin kx \cos \omega t \tag{17.11}$$

Equation 17.11 shows that the superposition has changed the character of the displacement function, so that, instead of being a function of $(x \pm ct)$, it is now the product of two separate functions one of which depends on time alone and the other on the space coordinate alone. The displaced shape given by Equation 17.11 is plotted in Figure 17.3 as a function of x, for different values of t. The displacement wave does not now move in space but remains stationary; only the magnitude of displacement varies with time. A wave of this nature is called a *standing wave*. It is interesting to note that points such as $N1, N2$ in Figure 17.3 do not move at all. Such points are called *nodes*. A point midway between two nodes attains the maximum displacement and is called an *antinode*.

To provide a physical meaning to a standing wave, consider again the case of vibrating string of length L fixed at its two ends. If the vibrations of a string were to be represented by a standing wave, we must have a node located at each $x = 0$ and $x = L$ because these two points on the string remain stationary. This is only possible provided k in Equation 17.11 takes one of the values $k = n\pi/L$, $n = 1, 2, 3 \cdots \infty$. Corresponding to each value of k, we have a standing wave solution given by

$$u_n = 2 \sin \frac{n\pi x}{L} \cos \omega_n t \tag{17.12}$$

In a general case, we can represent the displacement of a string by a superposition of a number of solutions of the type given by Equation 17.12. Shape $\sin n\pi x/L$ will be recognized as a mode shape of the vibrating string and the

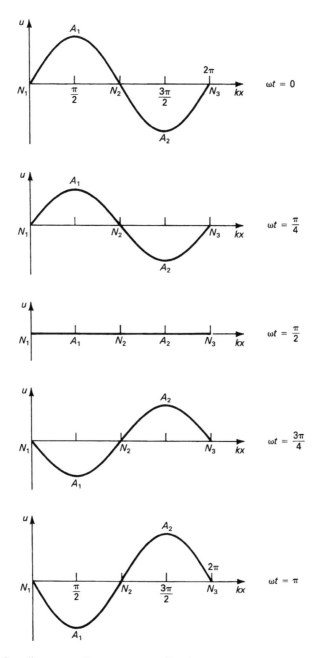

Figure 17.3. Standing wave for transverse vibrations of a string.

solution expressed by a superposition of standing-wave patterns is equivalent to a mode superposition analysis.

17.4   ONE-DIMENSIONAL WAVE EQUATION AND ITS SOLUTION

The wave propagation method is most effective in the analysis of free vibration response of those systems whose motion is governed by a wave equation. Examples of one-dimensional systems that fall in this category are: a stretched string undergoing transverse vibrations, a bar in axial or torsional vibrations, and a shear beam in lateral vibrations. Although the wave analysis is most appropriate for systems of infinite extent, it can also be used in the analysis of finite-extent systems. In this section, we discuss the essential steps in wave propagation analysis by considering the vibrations of a shear beam. The procedure, however, applies without modification to all systems whose motion is governed by a one-dimensional wave equation.

The equation governing the free vibrations of a uniform shear beam is given by

$$k\frac{\partial^2 u}{\partial x^2} = m\frac{\partial^2 u}{\partial t^2} \tag{17.13}$$

where k is the shear stiffness and m is mass per unit length. Equation 17.13 can be expressed in the alternative form

$$\frac{\partial^2 u}{\partial x^2} = \frac{1}{c^2}\frac{\partial^2 u}{\partial t^2} \tag{17.14}$$

where $c = \sqrt{k/m}$. Parameter c which has the units of velocity is known as the wave velocity and Equation 17.14 is referred to as a *one-dimensional wave equation*. It is readily seen that the vibrations of all of the systems referred to in this section are governed by Equation 17.14, provided c is interpreted appropriately in each case. Thus, for axial vibrations of a uniform bar

$$c = \sqrt{\frac{EA}{m}} \tag{17.15}$$

where A is the cross-sectional area and m is the mass per unit length. For torsional vibrations of a uniform bar with circular cross-section

$$c = \sqrt{\frac{GJ}{I_0}} \tag{17.16}$$

where G is the shear modulus, J is the polar moment of inertia and I_0 is the mass moment of inertia per unit length about a longitudinal axis. For transverse vibrations of a stretched string

$$c = \sqrt{\frac{T}{m}} \tag{17.17}$$

where T is the uniform tension in the string and m is again the mass per unit length. In each of the cases cited above, c is a constant that depends only on the physical properties of the system and has the units of velocity.

For Equation 17.14, we attempt a solution of the form

$$u = f_1(x - ct) \tag{17.18}$$

If $\xi \equiv x - ct$, it is easily seen that

$$\frac{\partial u}{\partial x} = \frac{df_1}{d\xi}; \qquad \frac{\partial^2 u}{\partial x^2} = \frac{d^2 f_1}{d\xi^2} \tag{17.19a}$$

$$\frac{\partial u}{\partial t} = -c\frac{df_1}{d\xi}; \qquad \frac{\partial^2 u}{\partial t^2} = c^2\frac{d^2 f_1}{d\xi^2} \tag{17.19b}$$

Substitution of Equation 17.19 in Equation 17.14 shows that Equation 17.18 is indeed a solution of Equation 17.14. In a similar manner, we can show that the following is also a valid solution

$$u = f_2(x + ct) \tag{17.20}$$

A general solution of Equation 17.14 is, in fact, given by

$$u = f_1(x - ct) + f_2(x + ct) \tag{17.21}$$

which is a superposition of two different waves, a forward moving wave and a backward moving wave, each of which travels without change of shape and at a constant velocity c.

So far, we have placed no restrictions on the type of functions f_1 and f_2. These two functions are, in fact, uniquely determined by the two initial conditions: initial displacement profile $u_0(x)$ and the initial velocity profile $v_0(x)$. Substituting the initial displacement condition in Equation 17.21, we have

$$u_0(x) = f_1(x) + f_2(x) \tag{17.22}$$

On differentiating Equation 17.21 with respect to t, we get

$$\frac{\partial u}{\partial t} = -cf_1'(x - ct) + cf_2'(x + ct) \tag{17.23}$$

where f' represents differential with respect to the argument of the function. Substitution of initial velocity condition in Equation 17.23 gives

$$v_0(x) = -cf_1'(x) + cf_2'(x) \tag{17.24}$$

Integration of Equation 17.24 with respect to x gives

$$\frac{1}{c} \int_{x_0}^{x} v_0(\eta)d\eta = -f_1(x) + f_2(x) \tag{17.25}$$

where, for the sake of clarity, we have replaced x within the integral by the integration variable η, and integration limit x_0 accounts for the constant of integration. On subtracting Equation 17.25 from Equation 17.22, we get

$$f_1(x) = \frac{1}{2}u_0(x) - \frac{1}{2c} \int_{x_0}^{x} v_0(\eta)d\eta \tag{17.26}$$

Hence

$$\begin{aligned} f_1(x - ct) &= \frac{1}{2}u_0(x - ct) - \frac{1}{2c} \int_{x_0}^{x-ct} v_0(\eta)d\eta \\ &= \frac{1}{2}u_0(x - ct) + \frac{1}{2c} \int_{(x-ct)}^{x_0} v_0(\eta)d\eta \end{aligned} \tag{17.27}$$

Addition of Equations 17.25 and 17.22 gives

$$f_2(x) = \frac{1}{2}u_0(x) + \frac{1}{2c} \int_{x_0}^{x} v_0(\eta)d\eta \tag{17.28}$$

so that

$$f_2(x + ct) = \frac{1}{2}u_0(x + ct) + \frac{1}{2c} \int_{x_0}^{x+ct} v_0(\eta)d\eta \tag{17.29}$$

Finally, on adding Equations 17.27 and 17.29, we get

$$\begin{aligned} u(x, t) &= f_1(x - ct) + f_2(x + ct) \\ &= \frac{1}{2}u_0(x - ct) + \frac{1}{2}u_0(x + ct) + \int_{x-ct}^{x+ct} v_0(\eta)d\eta \end{aligned} \tag{17.30}$$

Equation 17.29 fully determines the motion of an infinite shear beam due to prescribed initial displacement and velocity profiles.

Example 17.1

The stretched string of infinite length shown in Figure E17.1 is subjected to an initial displacement given by

$$u_0 = 2(1-x) \quad 0 \leq x \leq 1$$
$$u_0 = 2+x \quad -2 \leq x \leq 0 \tag{a}$$

The string is released from this position and allowed to vibrate. If the velocity of wave propagation in the string is $c = 1/2$, draw the displaced shapes at $t = 1, 2, 3$ and 4.

Solution

The initial displacement of the string is shown in Figure E17.1a. It can be represented by a superposition of two wave shapes $f_1 = f_2 = \frac{1}{2}u_0$. Waves f_1 and f_2 are also indicated in Figure E17.1a.

At the end of $1s$, wave f_1 has moved to the right without change in shape a distance $ct = 1/2$; while f_2 has moved to the left the same distance. Shapes $f_1(x-1/2)$ and $f_2(x+1/2)$ along with $u = f_1 + f_2$ are shown in Figure E17.1b.

At $2s$, wave f_1 has moved a distance $ct = 1$ to the right while f_2 has moved the same distance to the left. The resultant displacement of the string is shown in Figure E17.1c. Figure E17.1d

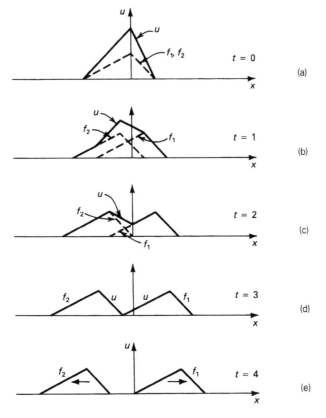

Figure E17.1. Traveling displacement waves in an infinite string.

and e shows, respectively, the displaced shapes at $t = 3$ and $t = 4$. It is evident that for $t \geq 3$, the two waves do not overlap.

Expressions for wave shapes can also be derived from Equation a by noting that

$$f_1(x - ct) = \frac{1}{2} u_0(x - ct) \tag{b}$$

$$f_2(x + ct) = \frac{1}{2} u_0(x + ct) \tag{c}$$

Substitution of $x - ct$ for x in Equation a will provide $u_0(x - ct)$. When the latter is substituted in Equation b, we get

$$f_1(x - ct) = \begin{cases} 1 - x + \frac{t}{2} & \frac{t}{2} \leq x \leq 1 + \frac{t}{2} \\ 1 + \frac{x}{2} - \frac{t}{4} & \frac{t}{2} - 2 \leq x \leq \frac{t}{2} \end{cases} \tag{d}$$

In a similar manner, by obtaining $u_0(x + ct)$ from Equation a and substituting it in Equation c, we get

$$f_2(x + ct) = \begin{cases} 1 - x - \frac{t}{2} & -\frac{t}{2} \leq x \leq 1 - \frac{t}{2} \\ 1 + \frac{x}{2} + \frac{t}{4} & -2 - \frac{t}{2} \leq x \leq -\frac{t}{2} \end{cases} \tag{e}$$

The displaced shape at any given time t can be obtained by adding f_1 and f_2 calculated from Equations d and e, respectively.

Example 17.2

A string of infinite length stretched along the x axis is imparted an initial transverse velocity as shown by the profile in Figure E17.2a. Assuming that the velocity of wave propagation in the string is $c = 1/2$, obtain the displaced shape of the string at $t = 1, 2, 3, 4$ and 5.

Solution

The displaced shape of the string at any time t is obtained from Equation 17.30 with $u_0 = 0$.

$$u(x, t) = \int_{x - ct}^{x + ct} v_0(\eta) d\eta \tag{17.1}$$

The integral in Equation a represents the area under the initial velocity curve between ordinates at $x - ct$ and $x + ct$. Since the initial velocity profile v_0 has a value of zero except between $-2 < x < 2$, the following situations arise

1. $x + ct < -2$; For this case, $u(x, t) = 0$.
2. $x - ct > 2$; For this case too, $u(x, t) = 0$.
3. For all other cases, the lower limit is changed to -2 if it is less than -2, and the upper limit is changed to $+2$ if it is greater than 2.

The displaced shape is obtained by evaluating the integral in Equation a with the limits adjusted as above. The results are shown in Figure E17.2b for the five different values of t. The maximum amplitude of displacement is reached at $t = 4$ when the string displacement at $x = 0$ is $4A$. After this, the displacement spreads on both sides of $x = 0$ and as time passes, a larger and larger portion of the string attain a displacement of $4A$.

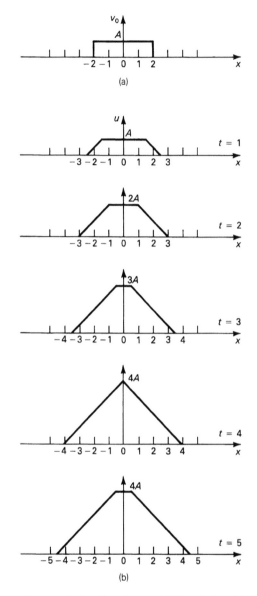

Figure E17.2. Response of a stretched string due to initial velocity: (a) initial velocity profile; (b) displacement response for different values of t.

17.5 PROPAGATION OF WAVES IN SYSTEMS OF FINITE EXTENT

In our discussion so far, we have assumed that the system was of infinite extent. The traveling wave in such a system moves away towards infinity as the time passes. In a system of finite extent, however, the wave will eventually arrive

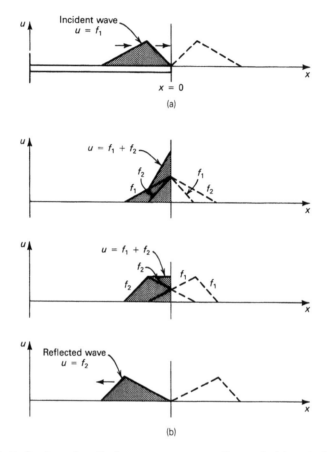

Figure 17.4. Reflection of a displacement wave at a free end: (a) arrival of forward wave at free end; (b) resultant displacement due to superposition of incident and reflected waves.

at a boundary of the system. The question we must address is: what happens to the wave when it arrives at such a boundary? Obviously, the answer will depend on the nature of the boundary.

As an example of wave propagation in a system of finite extent, consider the longitudinal vibrations of a uniform rod which is free at its right hand end. The vibrations of the rod are still governed by Equation 17.14. As shown earlier, the general solution of Equation 17.14 is obtained by the superposition of a forward traveling wave and a backward traveling wave. Let us for the present limit our consideration to the forward traveling wave. At a certain instant of time, let the forward wave given by $u = f_1(x - ct)$ arrive at the free end as shown in Figure 17.4a. The arrival of the forward wave gives rise to a *reflected wave* traveling backwards from the free end. Let this wave be given by $u = f_2(x + ct)$. For all subsequent times after the arrival of the forward wave, the displacement response will be given by a superposition of waves f_1 and f_2. For simplicity and without loss of generality, let us now assume the free end is located at $x = 0$. Since the axial force at the free end should be zero, we have

$$E\frac{\partial f_1}{\partial x} + E\frac{\partial f_2}{\partial x} = 0 \quad x = 0 \tag{17.31}$$

Since

$$\frac{\partial}{\partial x}f_1(x - ct) = f_1'$$
$$\frac{\partial}{\partial x}f_2(x + ct) = f_2' \tag{17.32}$$

Equation 17.31 leads to

$$f_1'(-ct) + f_2'(ct) = 0 \tag{17.33}$$

If we set $ct = \xi$ so that $dt = \frac{1}{c}d\xi$, then integration of Equation 17.33 with respect to time gives

$$-\frac{1}{c}\int\frac{df_1(-\xi)}{d\xi}d\xi + \frac{1}{c}\int\frac{df_2(\xi)}{d\xi}d\xi = 0 \tag{17.34a}$$

or

$$-\frac{1}{c}f_1(-\xi) + \frac{1}{c}f_2(\xi) = 0 \tag{17.34b}$$

To obtain f_2 we replace ξ in Equation 17.34 by $x + ct$, so that

$$f_2(x + ct) = f_1(-x - ct) \tag{17.35}$$

Equation 17.35 implies that the reflected wave is a mirror image of f_1 about $x = 0$, the free end, and the displacement at the free end, being a superposition of f_1 and f_2 is double that due to the incident wave. Figure 17.4b indicates the positions of the incident and reflected waves at several different intervals of time after the arrival of the incident wave at the free end. It would be of interest to study the propagation of the force waves corresponding to the displacement waves described above. The incident force wave is given by

$$p = E\frac{\partial f_1}{\partial x}$$
$$= g_1(x - ct) \tag{17.36}$$

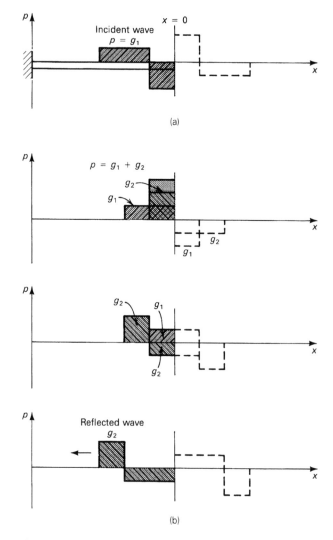

Figure 17.5. Reflection of a pressure wave at a free end: (a) arrival of forward wave at a free end; (b) resultant pressure due to superposition of incident and reflected waves.

and is shown in Figure 17.5a. The reflected force wave is

$$P = E\frac{\partial f_2}{\partial x}$$

$$= -Ef_1'(-x - ct)$$

$$= -g_1(-x - ct)$$

(17.37)

The reflected force wave is also a mirror image of the incident pressure wave but its magnitude is negative of the latter. Therefore, at the free end, the two waves will cancel each other and the net force would be zero. Figure 17.5b shows the position of the incident and reflected force waves at several different intervals of time after the arrival of the incident wave at the free end.

Let us now consider the events after the arrival of a displacement wave $f_1(x - ct)$ at a fixed end. The reflected wave should, in this case, be such that the net displacement at the fixed end is zero. Considering again that the fixed end is located at $x = 0$ we have

$$f_1(-ct) + f_2(ct) = 0 \qquad (17.38)$$

or

$$f_2(\xi) = -f_1(-\xi) \qquad (17.39)$$

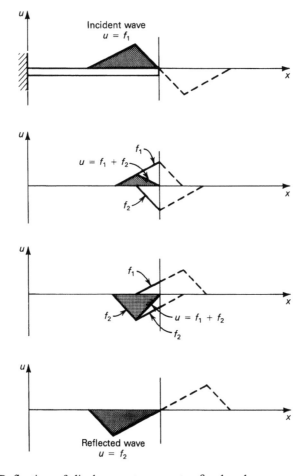

Figure 17.6. Reflection of displacement wave at a fixed end.

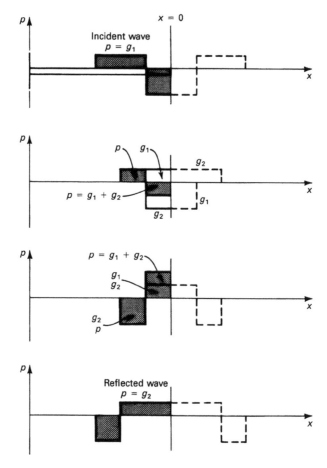

Figure 17.7. Reflection of a pressure wave at a fixed end.

Replacing ξ by $x + ct$, we obtain from Equation 17.39

$$f_2(x + ct) = -f_1(-x - ct) \tag{17.40}$$

Equation 17.40 implies that in this case also, the reflected displacement wave is a mirror image of the incident wave about $x = 0$, the fixed end, but the amplitude of f_2 is negative of the amplitude of f_1. As a result, the two waves cancel each other at the fixed end and the net displacement is zero. It is readily seen that the incident and reflected force waves at the fixed end will be the mirror images of each other and of the same sign. The force at the free end would therefore be double that due to the incident wave alone. The shapes of the incident and reflected displacement and pressure waves are shown in Figures 17.6 and 17.7, respectively.

Example 17.3

A multistory building idealized as a shear beam is subjected to a ground displacement pulse as shown in Figure E17.3a. The properties of the building are also shown in the figure. Obtain the displacement and shear profiles for $t = 1, 3, 5, 6, 8$ and 9s. The ground displacement is given by

$$u_g(t) = \begin{cases} A_0 t & 0 \leq t \leq 1 \\ A_0(\frac{3}{2} - \frac{1}{2}t) & 1 \leq t \leq 3 \end{cases} \tag{a}$$

Solution

The displacement pulse gives rise to a forward moving wave $u = f_1(x - ct)$, where c is the wave velocity. At $x = 0$, $u = u_g$, hence

$$f_1(-ct) = \begin{cases} -\frac{A_0}{c}(-ct) & 0 < -\frac{1}{c}(-ct) < 1 \\ A_0\{\frac{3}{2} + \frac{1}{2c}(-ct)\} & 1 \leq -\frac{1}{c}(-ct) \leq 3 \end{cases} \tag{b}$$

On replacing $-ct$ by $x - ct$ in Equations b, we get

$$f_1(x - ct) = \begin{cases} -\frac{A_0}{c}(x - ct) & 0 < -\frac{1}{c}(x - ct) < 1 \\ A_0\{\frac{3}{2} + \frac{1}{2c}(x - ct)\} & 1 \leq -\frac{1}{c}(x - ct) \leq 3 \end{cases} \tag{c}$$

The equivalent stiffness of the building is obtained from

$$k = \frac{12 \sum EI}{h^2} = \frac{12 \times 24.0 \times 10^4}{12^2} \tag{d}$$

$$= 2 \times 10^4 \text{ kips}$$

and the mass per unit length is given by

$$m = \frac{m_i}{h} = \frac{24}{12}$$

$$= 2 \text{ kip.s}^2/\text{ft}^2$$

where h is the story height, m_i is the mass of any one floor and $\sum EI$ is the sum of the flexural rigidities of columns in a story. The wave propagation velocity is obtained from

$$c = \sqrt{\frac{k}{m}}$$

$$= \left(\frac{2 \times 10^4}{2}\right)^{\frac{1}{2}}$$

$$= 100 \text{ ft/s}$$

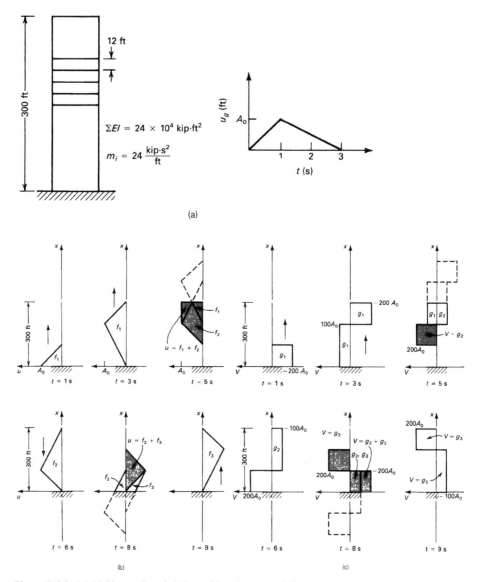

Figure E17.3. (a) Multistory shear building subjected to ground displacement; (b) Displacement profile of the shear building; (c) shear profile of multistory shear building.

Substitution for wave velocity in Equation c gives

$$f_1(x - ct) = \begin{cases} A_0(t - \frac{x}{100}) & 100t - 100 \le x \le 100t \\ A_0(\frac{3}{2} + \frac{x}{200} - \frac{t}{2}) & 100t - 300 \le x \le 100t - 100 \end{cases} \tag{e}$$

The displaced shape obtained from Equation e for $t = 1$ and 3 s are shown in Figure E17.3b. At $t = 3$ s, the displacement wave traveling at 100 ft/s has just arrived at the free end. At 5 s, the displaced shape is given by superposition of the incident and reflected waves. As shown in Figure E17.3b, the reflected wave is a mirror image of the incident wave and is of the same sign as the incident wave but is moving in a reverse direction. The displacement at the free end is thus double that due to the incident wave. At $t = 6$ s, the displacement is due entirely to the reflected wave which has just reached the fixed end. For the subsequent time intervals, we treat this wave as a wave that is incident on the fixed end. Reflection at the fixed end gives rise to another wave that is a mirror image of the incident wave, is of opposite sign, and is moving in the positive direction of x. At $t = 8$ s, the displacement is a superposition of the wave incident on the fixed end and the resulting reflected wave. At $t = 9$ s, the displacement throughout the length of the building is negative. The distribution of shear in the building is given by

$$V = k\frac{\partial u}{\partial x} \qquad (f)$$

Using Equation e, we get

$$V = -200A_0 \qquad 100t - 100 \le x \le 100t$$

$$V = 100A_0 \qquad 100t - 300 \le x \le 100t - 100$$

The shear profile for the first 9 s is shown in Figure E17.3c. At $t = 3$ s, the shear wave arrives at the free end. During subsequent intervals of time, the incident wave gets reflected at the free end, such that the reflected wave is a mirror image of the incident wave but is of opposite sign. Superposition of the two waves gives the total shear at $t = 5$ s. At $t = 6$ s, the shear is caused entirely due to the reflected wave. For the subsequent intervals of time, this wave will become a wave incident on the fixed end. Reflection at the fixed end gives rise to a wave that is a mirror image of the incident wave and is of the same sign. For $t = 8$ s, the total shear is a sum of the shears due to the incident and the reflected waves.

Example 17.4
A uniform bar that is fixed at its base and free at the top is subjected to a suddenly applied axial force P as shown in Figure E17.4a. If the cross-sectional area of the bar is A and the velocity of propagation of a longitudinal wave c, obtain the distribution of pressure and displacement at four different times given by $0 < ct < L$; $L < ct < 2L$; $2L < ct < 3L$; $3L < ct < 4L$.

Solution
The pressure distribution in the four phases is shown in Figure E17.4a. For $0 \le ct \le L$, the pressure is caused by a wave propagating towards the base. As this wave reaches the base, it gives rise to the first reflected wave which has the same sign and amplitude as the incident wave but is moving towards the tip. For $L \le ct \le 2L$, the total pressure is the sum of two components: one due to the original wave and the second due to the first reflected wave.

At $ct = 2L$, the first reflected wave arrives at the free end and is, in turn, reflected there. This second reflected wave has the same amplitude as the first reflected wave but is of opposite sign and is moving towards the base. For $2L \le ct \le 3L$, the total pressure is the sum of the original wave, the first reflected wave, and the second reflected wave. The second reflected wave, in turn, reaches the base at $ct = 3L$ and is reflected from there with the same sign and amplitude. For $3L \le ct \le 4L$, the total response is the sum of the original wave, the first reflected wave, the second reflected wave and the third reflected wave. As the third reflected wave travels towards the free end, the bar is progressively relieved of the axial pressure and at $ct = 4L$, the entire bar has zero pressure. The complete cycle then repeats itself. The displacement profile is obtained by

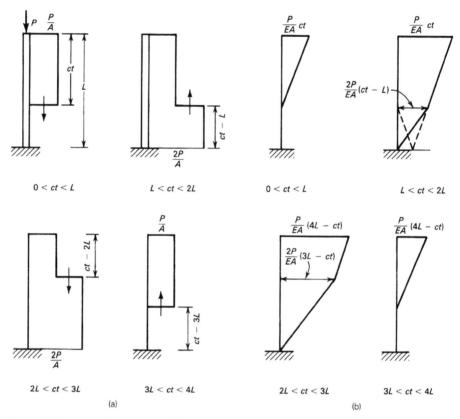

Figure E17.4. (a) Axial pressure profile in a uniform bar subjected to suddenly applied axial force; (b) displacement profile in a uniform bar subjected to suddenly applied axial force.

integrating the pressure curve from 0 to x and dividing the integral by E. The resulting profiles in the four phases are shown in Figure E17.4b. This problem was solved earlier in Section 16.5 by the mode superposition method. It is evident that in the present case, wave propagation analysis is more straightforward.

17.6   REFLECTION AND REFRACTION OF WAVES AT A DISCONTINUITY IN THE SYSTEM PROPERTIES

When a wave arrives at a point of discontinuity in the system properties, it gives rise to a reflected wave propagating in a direction opposite to the incident wave and a *refracted wave* that propagates in the same direction as the incident wave. The reflected and refracted waves can be determined by considering compatability and equilibrium at the point of discontinuity.

As an illustration of wave reflection and refraction, let us again consider the case of a shear beam with a discontinuity at say $x = 0$ as shown in Figure 17.8.

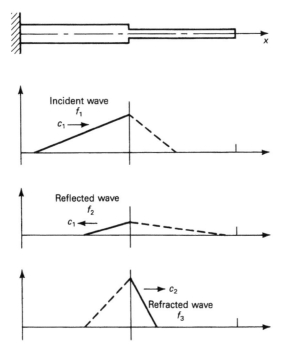

Figure 17.8. Reflection and refraction of a wave at a discontinuity in system property.

In the section of the beam left of the point of discontinuity, the shear stiffness $k = k_1$ and the mass per unit length $m = m_1$. To the right of the discontinuity, $k = k_2$ and $m = m_2$. The incident displacement wave arriving at the discontinuity is given by $u = f_1(x - c_1 t)$, the refracted wave is $u = f_3(x - c_2 t)$ and the reflected wave is $u = f_2(x + c_1 t)$, where c_1 and c_2 are the wave velocities for the sections on the left and the right, respectively. Compatability of displacements at $x = 0$ requires that

$$f_1(-c_1 t) + f_2(c_1 t) = f_3(-c_2 t) \qquad (17.41)$$

Further, for equilibrium we must have

$$k_1 \frac{\partial f_1}{\partial x} + k_1 \frac{\partial f_2}{\partial x} = k_2 \frac{\partial f_3}{\partial x}; \quad x = 0 \qquad (17.42)$$

or

$$k_1 f_1'(-c_1 t) + k_1 f_2'(c_1 t) = k_2 f_3'(-c_2 t) \qquad (17.43)$$

Integration of both sides of Equation 17.43 with respect to t gives

$$-\frac{k_1}{c_1} f_1(-c_1 t) + \frac{k_1}{c_1} f_2(c_1 t) = -\frac{k_2}{c_2} f_3(-c_2 t) \qquad (17.44a)$$

or

$$-f_1(-c_1 t) + f_2(c_1 t) = -\frac{k_2 c_1}{k_1 c_2} f_3(-c_2 t) \qquad (17.44b)$$

We now define a parameter α given by

$$\alpha = \frac{k_2 c_1}{k_1 c_2}$$
$$= \sqrt{\frac{k_2 m_2}{k_1 m_1}} \qquad (17.45)$$

On solving Equations 17.41 and 17.44b for f_3 and using the definition of α, we obtain

$$f_3(-c_2 t) = \frac{2}{1+\alpha} f_1 \left\{ \frac{c_1}{c_2} (-c_2 t) \right\} \qquad (17.46)$$

Replacing $-c_2 t$ by $x - c_2 t$ in Equation 17.46, we get

$$f_3(x - c_2 t) = \frac{2}{1+\alpha} f_1 \left\{ \frac{c_1}{c_2} (x - c_2 t) \right\} \qquad (17.47)$$

In a similar manner, on solving Equations 17.41 and 17.44b for f_2, we obtain

$$f_2(c_1 t) = \frac{1-\alpha}{1+\alpha} f_1(-c_1 t) \qquad (17.48)$$

and hence

$$f_2(x + c_1 t) = \frac{1-\alpha}{1+\alpha} f_1 \{ -(x + c_1 t) \} \qquad (17.49)$$

Parameter α is a measure of the amount of discontinuity and controls the amplitude and sign of the refracted and reflected waves. When the two sections of the beam have identical properties, $\alpha = 1$, no reflected wave exists, and the refracted wave is identical to the incident. For $\alpha > 1$, which for example will be the case when the right hand section is stiffer than the left hand section, the reflected wave is opposite in sign to the incident wave. In the limit, if the right hand section is comparatively very stiff, the left hand section can be considered fixed at the point under consideration and α tends to ∞. Equation 17.49 then becomes identical to Equation 17.40, so that the reflected wave has the same amplitude as the incident but is opposite in sign. On the other hand, when α tends to 0 simulating a free end, Equation 17.49 reduces to Equation 17.35 and the reflected wave is identical in amplitude and sign to the incident wave. It is evident that reflections at a fixed and a free end can be viewed as limiting cases of reflections at a discontinuity in the system property.

Example 17.5

A shear wave propagates vertically through a layered medium consisting of a stiffer layer overlying a softer layer as shown in Figure E17.5a. For the purpose of wave propagation analysis, both layers are idealized as shear beams, the stiffness of the upper layer being k_2 and that of the lower layer k_1. It is given that $k_2 = 9k_1$ and that the mass densities of the two layers are identical. The profile of the shear wave propagating through the softer medium is shown in Figure E17.5b. Assuming that the velocity of wave propagation in the softer medium is c, obtain the displacement profiles of the layered medium at times $ct = 1, 2,$ and 3 after the wave arrives at the interface of the two layers.

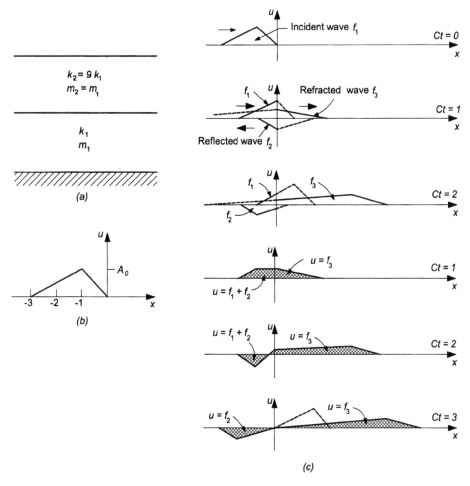

Figure E17.5. (a) Layered soil medium; (b) profile of shear wave; (c) wave refraction and reflection in a layered medium.

Solution

The velocity of wave propagation in the stiffer medium is given by

$$c_2 = \sqrt{\frac{k_2}{m_2}} = 3\sqrt{\frac{k_1}{m_1}} = 3c$$

Also

$$\alpha = \frac{c_1 k_2}{c_2 k_1} = \frac{9 k_1 c}{3 k_1 c} = 3$$

The equation of the wave profile is given by

$$f_1(\xi) = \begin{cases} -A_0 \xi & -1 \le \xi \le 0 \\ A_0(\frac{3}{2} + \frac{\xi}{2}) & -3 \le \xi \le -1 \end{cases} \qquad \text{(a)}$$

The wave shape at any time t is obtained from Equation a on replacing ξ by $x - ct$

$$f_1(x - ct) = \begin{cases} -A_0(x - ct) & ct - 1 \le x \le ct \\ A_0\{\frac{3}{2} + \frac{1}{2}(x - ct)\} & ct - 3 \le x \le ct - 1 \end{cases} \qquad \text{(b)}$$

The shape of the reflected wave is obtained from Equation 17.49. Substituting $\alpha = 3$, we get

$$f_2(x + ct) = -\frac{1}{2} f_1\{-(x + ct)\} \qquad \text{(c)}$$

Using the definition of $f_1(\xi)$ in Equation a, we obtain

$$f_2(x + ct) = \begin{cases} A_0(-\frac{x}{2} - \frac{ct}{2}) & -ct \le x \le 1 - ct \\ A_0(-\frac{3}{4} + \frac{x}{4} + \frac{ct}{4}) & 1 - ct \le x \le 3 - ct \end{cases} \qquad \text{(d)}$$

The refracted wave is given by Equation 17.47

$$f_3(x - c_2 t) = \frac{1}{2} f_1 \left\{ \frac{1}{3}(x - 3ct) \right\} \qquad \text{(e)}$$

Again, using the definition of $f_1(\xi)$ from Equation a, we get

$$f_3(x - c_2 t) = \begin{cases} A_0(-\frac{x}{6} + \frac{ct}{2}) & 3ct - 3 \le x \le 3ct \\ A_0(\frac{3}{4} + \frac{x}{12} - \frac{ct}{4}) & 3ct - 9 < x < 3ct - 3 \end{cases} \qquad \text{(f)}$$

The incident, reflected and refracted waves obtained from Equations b, d and f, respectively, are shown in Figure E17.5c for $ct = 0, 1, 2$ and 3. Also shown are the resultant displacement shapes obtained by superposing the three waves.

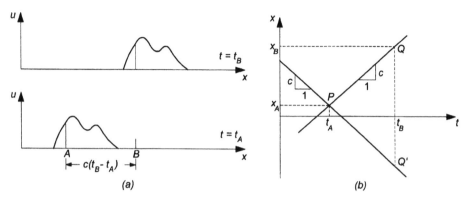

Figure 17.9. Characteristics of the wave equation: (a) forward-traveling wave; (b) characteristics of forward and backward waves.

17.7 CHARACTERISTICS OF THE WAVE EQUATION

We have seen that the response of a system whose vibrations are governed by the wave equation (Eq. 17.14) can be represented by the superposition of two waves that travel along the length at a constant velocity and without change of shape. Consider, for example, a single wave traveling in the positive direction of x axis at a velocity c. Figure 17.9a shows the position of the wave at two different instances of time. As the traveling wave passes through point B, that point experiences the same disturbance that a point A located to the left of B experienced some time earlier. The time lag is, in fact, equal to $(x_B - x_A)/c$ where x_B and x_A are the coordinates of points B and A, respectively. This characteristics of the motion is more easily represented by showing the wave travel in the x-t plane as in Figure 17.9b.

By drawing a line through A, sloping to the right, the slope of line being c, we can easily find the time at which point B will experience the motion that A experienced at time t_A. Conversely, we can locate the point along x at which a certain section of the wave will arrive at time t_B if the same wave section passed through A at time t_A. A line such as PQ is called a *characteristic* of the wave equation. It is readily seen that the equation of line PQ is

$$x - x_A = c(t - t_A) \tag{17.50}$$

The characteristic of a wave traveling in the negative direction of x is a straight line on the x-t plane sloping to the left and having a slope c. As an example, line $Q'P$ in Figure 17.9b is a characteristic representing a backward traveling wave that passed through A at time t_A. The equation of $Q'P$ is

$$x - x_A = -c(t - t_A) \tag{17.51}$$

As long as the system properties are uniform, a characteristic is a straight line with constant slope. At a boundary representing a discontinuity in the system properties, the characteristic splits into two lines, one representing the reflected wave and the other a refracted wave. The characteristic of the reflected wave has a slope that is of the same magnitude as the characteristic of the incident wave but is of opposite sign. The slope of the characteristic representing the refracted wave is equal to the velocity of wave propagation in the section beyond the discontinuity. The application of characteristics in solving some simple problems is illustrated in the examples given below.

Example 17.6

For the uniform bar of Example 17.4, draw the characteristic for the stress wave due to a suddenly applied load P and obtain the stress history for the fixed base.

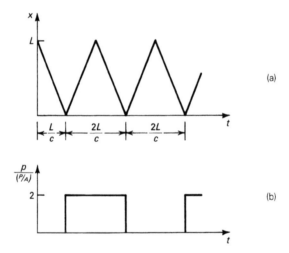

Figure E17.6. Axial vibrations of a bar under suddenly applied tip force: (a) characteristics of the stress wave; (b) stress history at the base.

Solution

The characteristics of the stress wave consists of a series of straight line with slope c as shown in Figure E17.6a. When the stress wave arrives at the fixed base, it is reflected with the same amplitude as the incident wave and travels upwards at velocity c. The slope of the characteristic changes sign as shown in the figure. On arriving at the free end, the wave is reflected again, this time with a change in the sign of the amplitude. The slope of the characteristic also changes sign. The ratio of the stress induced at the base to the static stress at the same location is shown in Figure E17.6b.

Example 17.7

A rectangular velocity pulse v propagates as a transverse shear wave through the layered medium of Example 17.5. The bedrock is infinitely stiff so that the lower layer can be assumed to be fixed at $x = 0$. Draw the characteristic of the pulse. Also, obtain the observed velocity at free field as a function of time.

Wave mark	Amplitude	Wave mark	Amplitude
LR1	V	UR1	1/2 V
LL1	-1/2 V	UR2	1/4 V
LL2	3/4 V	UR3	1/8 V
LL3	3/8 V	UR4	1/16V + 1/4 V = 5/16 V
LL4	3/16 V - 1/4 V = - 1/16V	UR5	5/32 V - 3/8 V = - 7/32 V
LR2	1/2 V	UL1	1/2 V
LR3	- 3/4 V	UL2	1/4 V
LR4	-3/8 V	UL3	1/8 V
		UL4	5/16 V
		UL5	-7/32 V

(a)

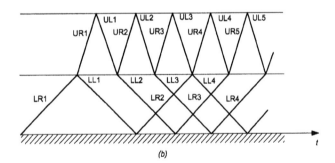

(b)

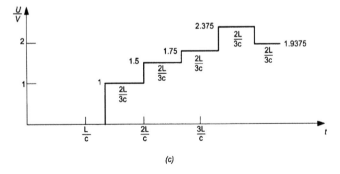

(c)

Figure E17.7. (a) Amplitudes of reflected and refracted waves in a layered medium; (b) characteristics of the velocity wave; (c) measured velocity at the free field as a function of time.

Solution
A wave propagating upwards from the bedrock with velocity c undergoes reflection and refraction at the interface of the two layers. Considering that the amplitude of incident wave is v_1, the reflected wave has an amplitude of

$$v_1 \frac{1-\alpha}{1+\alpha} = -(1/2)v_1$$

and travels downward with velocity c. The refracted wave has an amplitude of

$$v_2 = \frac{2}{1+\alpha} v_1 = \frac{1}{2} v_1$$

and travels upward at a velocity $3c$. Arriving at the free field, this wave is reflected back with an amplitude v_2. When this latter wave arrives back at the interface, it is refracted into the lower layer with an amplitude $v_3 = 2v_2/(1+1/\alpha) = \frac{3}{2}v_2$ and is also reflected into the upper layer with an amplitude $v_2(1 - 1/\alpha)/(1 + 1/\alpha) = (\frac{1}{2})v_2$. The wave of amplitude v_3 traveling through the lower layer and arriving at the bedrock is reflected with an amplitude $-v_3$. Reflections and refractions of this type continue indefinitely. The wave amplitudes are shown in Figure E17.7a, and the characteristics are drawn in Figure E17.7b. The observed velocity at the free field is obtained from Figure E17.7a and b and is shown in Figure E17.7c.

17.8   WAVE DISPERSION

The propagation of waves without change in shape is a useful property of certain types of media. Not all media, however, possess this property and the propagating waves, in general, undergo a change in shape. The phenomenon of change in the wave shape is called *wave dispersion* and the medium in which propagating waves undergo such a change is called a *dispersive medium*. As an example of wave dispersion, consider the free flexural vibrations of a uniform beam. The equation of free vibration is, in this case, given by

$$EI \frac{\partial^4 u}{\partial x^4} + m \frac{\partial^2 u}{\partial t^2} = 0 \tag{17.52}$$

or

$$\frac{\partial^4 u}{\partial x^4} + \frac{1}{a^2} \frac{\partial^2 u}{\partial t^2} = 0 \tag{17.53}$$

where $a^2 = \frac{EI}{m}$

On comparing Equation 17.53 with Equation 17.14, we notice two important differences. First, in Equation 17.53 the differential of u with respect to the space coordinate is of fourth order rather than of order 2 as in Equation 17.14. Further, parameter a does not have the units of velocity. It is also readily seen that Equation 17.53 does not have a solution of the type of Equation 17.21. In other words, flexural vibration response cannot be represented as a superposition of waves traveling with a constant velocity and without change of shape.

Let us now consider a solution of Equation 17.53 in the form of a harmonic wave of length λ given by Equation 17.10c

$$u(x,t) = \sin \frac{2\pi}{\lambda}(x - ct) \tag{17.10c}$$

Substitution of u and its partial derivatives with respect to x and t from Equation 17.10c in Equation 17.53 shows that Equation 17.10c will be a valid solution provided

$$c = \frac{2\pi a}{\lambda} \tag{17.54}$$

Equation 17.54 implies that in flexural vibration of a beam, a harmonic wave will travel without change of shape, but with a velocity that is dependent on the wave length unlike in a system governed by the one-dimensional wave equation for which the wave velocity is a constant. A harmonic wave is, in fact, the only type of wave that can travel in a dispersive medium without a change in shape. A wave of any general shape can be represented by a superposition of a series of harmonic waves of different wave lengths. Since the wave lengths of these component waves are different, they will travel with different velocities along the length of the beam and therefore, the shape of the original wave represented by a superposition of its harmonic components will change as the wave propagates along the length. Flexural vibrations of a beam are thus dispersive in nature.

The velocity with which a harmonic wave travels in a dispersive medium, is referred to as a *phase velocity*, to distinguish it from another kind of velocity called a *group velocity* which is more representative of the propagation of a group of harmonic waves. To illustrate the concept of group velocity, consider two harmonic waves of wavelengths λ_1 and λ_2 propagating through a dispersive medium. At time $t = 0$, the two wave shapes are given by

$$u_1 = \sin\left(\frac{2\pi}{\lambda_1}\right) x \tag{17.55}$$
$$= \sin k_1 x$$

and

$$u_2 = \sin\left(\frac{2\pi}{\lambda_2}\right) x \tag{17.56}$$
$$= \sin k_2 x$$

At the end of time t, the first wave would have moved to the right a distance c_1t, so that the equation describing its shape will become

$$u_1 = \sin\frac{2\pi}{\lambda_1}(x - c_1t)$$
$$= \sin(k_1x - \omega_1t)$$

(17.57)

in which c_1, k_1 and ω_1 are all functions of λ_1. Similarly

$$u_2 = \sin(k_2x - \omega_2t)$$

(17.58)

The total displacement at time t is given by

$$u = u_1 + u_2$$
$$= \sin(k_1x - \omega_1t) + \sin(k_2x - \omega_2t)$$

(17.59)

Using the trigonometric identity

$$\sin A + \sin B = 2\sin\frac{A+B}{2}\cos\frac{A-B}{2}$$

Equation 17.59 can be reduced to

$$u = 2\sin\left(\frac{k_1+k_2}{2}x - \frac{\omega_1+\omega_2}{2}t\right)\cos\left(\frac{k_1-k_2}{2}x - \frac{\omega_1-\omega_2}{2}t\right)$$

(17.60)

Let us now assume that the wave lengths of the two waves in the group are only slightly different, so that $\lambda_2 = \lambda_1 + \Delta\lambda$; $k_2 = k_1 + \Delta k$ and $\omega_2 = \omega_1 + \Delta\omega$, where $\Delta\lambda$, Δk and $\Delta\omega$ are small. We then have

$$\frac{k_1+k_2}{2} = \frac{2k_1+\Delta k}{2}$$
$$\simeq k_1$$
$$\frac{\omega_1+\omega_2}{2} = \frac{2\omega_1+\Delta\omega}{2}$$
$$\simeq \omega_1$$

(17.61)

and Equation 17.60 reduces to

$$u = 2\sin(k_1x - \omega_1t)\cos\left(\frac{\Delta k}{2}x - \frac{\Delta\omega}{2}t\right)$$

(17.62)

The corresponding expression for $t = 0$ becomes

$$u = 2\sin(k_1x)\cos\left(\frac{\Delta k}{2}x\right)$$

(17.63)

Equation 17.63 shows that the displacement at time $t = 0$ is the product of two harmonic waves, one with the wave length $2\pi/k_1$ and the other with a much larger wave length $4\pi/\Delta k$. These two waves and their product are shown in Figure 17.10a. The wave of length $2\pi/k_1$ is referred to as the *carrier wave* and that of length $4\pi/\Delta k$ as the *modulating wave*. The resultant displacement has the shape of the carrier wave with its amplitude modified by the modulating wave. The modulating wave envelopes the resultant response as shown in Figure 17.10a. After time t, the response is still given by the product of the carrier wave and the modulating wave, except that the carrier wave has moved to the right by a distance $c_1 t$, the wave velocity being equal to the phase velocity, and the modulating wave has moved a distance $\frac{\Delta\omega}{\Delta k} t$. The velocity of propagation of the modulating wave is called the group velocity and may be denoted by c_g. We thus have

$$c_g = \frac{\Delta\omega}{\Delta k}$$

(17.64)

$$= \frac{d\omega}{dk} \quad \text{as } \Delta k \text{ tends to zero.}$$

The position of carrier and modulating waves as well as the resultant response at time t are shown in Figure 17.10b. The group velocity may be smaller than, equal to, or larger than the phase velocity. If the two velocities are equal, no dispersion takes place and all waves propagate at a constant speed without change of shape.

Consider the example of a beam in flexure. For this case, the phase velocity is given by

$$c = \frac{2\pi a}{\lambda}$$

(17.65)

so that

$$\omega = \frac{2\pi}{\lambda} c = \left(\frac{2\pi}{\lambda}\right)^2 a$$

(17.66)

and since $k = \frac{2\pi}{\lambda}$, we have $\omega = ak^2$. The group velocity is now obtained as follows

$$c_g = \frac{d\omega}{dk}$$

(17.67)

$$= 2ak$$

From Equation 17.65, the phase velocity $c = ak$. Hence, the group velocity is twice the phase velocity.

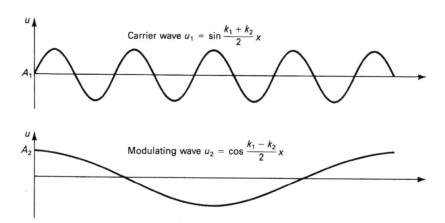

Carrier wave $u_1 = \sin \dfrac{k_1 + k_2}{2} x$

Modulating wave $u_2 = \cos \dfrac{k_1 - k_2}{2} x$

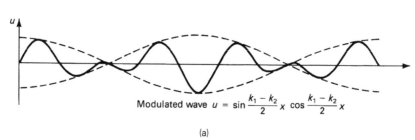

Modulated wave $u = \sin \dfrac{k_1 - k_2}{2} x \cos \dfrac{k_1 - k_2}{2} x$

(a)

Carrier wave $u_1 = \sin \left(\dfrac{k_1 + k_2}{2} x - \dfrac{\omega_1 + \omega_2}{2} t_1 \right)$

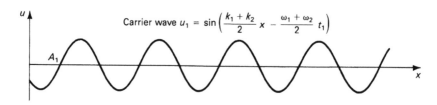

Modulating wave $u_2 = \cos \left(\dfrac{k_1 - k_2}{2} x - \dfrac{\omega_1 - \omega_2}{2} t_1 \right)$

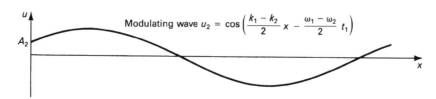

Modulated wave $u = u_1 \times u_2$

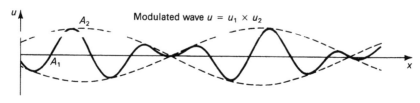

(b)

Equation 17.66 for a beam in flexural vibrations shows that high frequency waves will have a short wave length. Equations 17.65 and 17.67 thus imply that for very high frequency waves, both the phase velocity and the group velocity will tend to infinity. Physically, this conclusion appears irrational. The anomaly arises because of the fact that for short wave lengths approaching the depth of the beam cross section, rotatory inertia and shear deformation effects cannot be neglected and the simple beam theory is no longer valid. Consider the free transverse vibrations of a uniform beam, including the effect of rotatory inertia and shear deformation. The equation governing the motion is

$$ EI\frac{\partial^4 u}{\partial x^4} + m\frac{\partial^2 u}{\partial t^2} - mr^2\frac{\partial^4 u}{\partial x^2 \partial t^2} - \frac{EIm}{k'AG}\frac{\partial^4 u}{\partial t^2 \partial x^2} + \frac{m^2 r^2}{k'AG}\frac{\partial^4 u}{\partial t^4} = 0 \qquad (17.68) $$

For a harmonic wave given by Equation 17.10, substitution of u and its derivatives in Equation 17.68 gives

$$ \frac{mr^2}{k'AG}\left(\frac{2\pi}{\lambda}\right)^2 c^4 - \left\{\frac{EI}{k'AG}\left(\frac{2\pi}{\lambda}\right)^2 + \left(\frac{2\pi}{\lambda}\right)^2 r^2 + 1\right\} c^2 $$

$$ + a^2\left(\frac{2\pi}{\lambda}\right)^2 = 0 \qquad (17.69a) $$

or

$$ \frac{E}{k'G}\left(\frac{c^4 r^4}{a^4}\right) - \left\{\frac{E}{k'G} + 1 + \left(\frac{\lambda}{2\pi r}\right)^2\right\}\left(\frac{c^2 r^2}{a^2}\right) + 1 = 0 \qquad (17.69b) $$

Equation 17.69 can be solved for c^2 and will yield two values. It is evident from Equation 17.69 that as λ approaches zero, wave velocity c remains finite. As an example, for $E/(k'G) = 3$ and $\lambda = 0$, the two values of c are given by

$$ c = \sqrt{\frac{EI}{mr^2}} $$

and

$$ c = \sqrt{\frac{EI}{3mr^2}} \qquad (17.70) $$

Figure 17.10. Propagation of a group of harmonic waves in a dispersive medium: (a) carrier and modulating waves at $t = 0$; (b) carrier and modulating waves at $t = t_1$.

From the foregoing discussion it is evident that, wave dispersion makes an analysis through wave propagation quite complex and the practical utility of the method is therefore somewhat limited.

SELECTED READINGS

Achenbach, J.D. 1973. *Wave Propagation in Elastic Solids*. Amsterdam: North-Holland.

Clough, R.W. & Penzien, J. 1967. *Dynamics of Structures*. New York: McGraw-Hill. 2nd Edition.

Meriovitch, L. 1971. *Analytical Methods in Vibrations*. London: Macmillan.

Newmark, N.M. & Rosenblueth, E. 1971. *Fundamentals of Earthquake Engineering*. Englewood Cliffs, New Jersey: Prentice Hall.

Stephens, R.W.B. & Bate, A.E. 1950. *Wave Motion and Sound*. London, U.K.: Edward Arnold.

PROBLEMS

17.1   A string of mass 0.5 kg/m is stretched across a length of 60 m with a tension of 0.4 kN. The left-hand end of the string is imparted a harmonic motion in the lateral direction given by

$$u(0,t) = A_0 \sin 2\pi t$$

Determine (a) the velocity of wave propagation in the string; (b) the wave length; (c) the time at which the first disturbance arrives at the right end. Plot the displaced shapes of the string at 2 s and 3 s.

17.2   A concrete pile 600 mm in diameter and 30 m long is being driven into the foundation soil. The pile driving hammer imparts a force impulse which can be assumed to vary as a half sine wave of amplitude 2500 kN and period 0.012 s. To monitor the stresses arising in the pile during the driving process, strain measuring devices are attached at 5 m from the pile tip and at the midheight of the pile. Determine the maximum tensile and compressive stresses recorded by the gauges (a) when the pile is being driven through a very soft soil which offers no resistance to the penetration of the pile, (b) when the pile tip encounters rigid bedrock. For concrete $E = 20,000$ MPa, $\rho = 2300$ kg/m^3

17.3   A long metal wire stretched with an initial tension is imparted the following disturbances

$$u_0(x) = 0$$

$$v_0(x) = \begin{cases} \cos x & |x| \leq \pi/2 \\ 0 & \text{elsewhere} \end{cases}$$

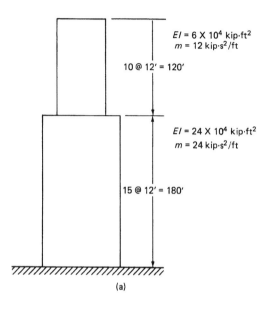

$EI = 6 \times 10^4$ kip·ft²
$m = 12$ kip·s²/ft

10 @ 12' = 120'

$EI = 24 \times 10^4$ kip·ft²
$m = 24$ kip·s²/ft

15 @ 12' = 180'

(a)

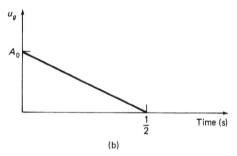

u_g

A_0

$\frac{1}{2}$

Time (s)

(b)

Figure P17.5.

If c is the velocity of wave propagation in the string, obtain the displaced shape of the string when $ct = (2/3)\pi$. What is the displacement at $x = -\pi/3$?

17.4 The wire of Problem 17.3 is given the following disturbances

$$u_0(x) = \begin{cases} -(x+2) & -2 \leq x \leq -1 \\ x & -1 \leq x \leq 1 \\ (x-2) & 1 \leq x \leq 2 \\ 0 & \text{elsewhere} \end{cases}$$

$$v_0(x) = 0$$

Plot the displaced shape of the string at $ct = 0.75$. What is the displacement at $x = -0.25$?

17.5 The multistory building with a setback as shown in Figure P17.5a is idealized as a shear beam. The properties of the building are indicated in the figure. The building is subjected to the lateral ground displacement pulse shown in Figure P17.5b. Obtain the displacement profile along the height of the building at 1-s intervals for the first 6 s. What is the displacement at the level of setback at 2 s?

17.6 For the building of Problem 17.5, obtain the shear force profile at 1-s intervals for the first 7 s.

ANSWERS TO SELECTED PROBLEMS

CHAPTER 2

2.1 Springs of stiffness 60 and 90 N/mm in series in association with spring of stiffness 40 N/mm in parallel.

2.3 $\left(M + \frac{3m}{2}\right)\ddot{u} + 4ku = 0$, where u is the vertical displacement of the rigid block.

2.5 $M\left(L_1^2 + \frac{3}{2}R^2 + 2L_1R\right)\ddot{\theta} + cd^2\dot{\theta} = kL_2^2\theta$, where θ is the angle of rotation of rigid link.

2.7 $m^* = 0.03016\,\bar{m}L, \quad k^* = 7.2EI/L^3$

2.9 $m^* = 0.2357\,\bar{m}L, \quad k^* = 3EI/L^3, \quad p^* = -0.375\,\bar{m}L\ddot{u}_g, \quad k_G^* = 0.375\,\bar{m}g$

2.10 $\dfrac{\bar{m}L}{2}\ddot{z} + \dfrac{\pi^4 EI}{2L^3}z = F\sin\frac{\pi vt}{L}$

2.11 $m^* = 19 \text{ kip s}^2/\text{in.}, \quad k^* = 1250 \text{ kips/in.}, \quad p^* = -1170\sin(6\pi t) \text{ kips}$

CHAPTER 3

3.1 $\mathbf{M\ddot{u} + C\dot{u} + Ku} = 0$

$$\mathbf{M} = \frac{W}{g}\begin{bmatrix} 1 & & & \\ & \frac{r^2}{a^2} & & \\ & & \frac{w}{W} & \\ & & & \frac{w}{W} \end{bmatrix} \qquad \mathbf{C} = c_s\begin{bmatrix} 2 & 0 & -1 & -1 \\ 0 & 2 & 1 & -1 \\ -1 & 1 & 1 & 0 \\ -1 & -1 & 0 & 1 \end{bmatrix}$$

$$\mathbf{K} = k_S\begin{bmatrix} 2 & 0 & -1 & -1 \\ 0 & 2 & 1 & -1 \\ -1 & 1 & 1+\frac{k_T}{k_S} & 0 \\ -1 & -1 & 0 & 1+\frac{k_T}{k_S} \end{bmatrix} \qquad \mathbf{u}^T = [q_1 \quad aq_2 \quad q_3 \quad q_4]$$

where q_1 is the vertical displacement of the body, q_2 the anticlockwise rotation of the body about the centre of mass, q_3 the vertical motion of the rear axle and q_4 the vertical motion of front axle.

3.3 $\mathbf{M\ddot{q}} + \mathbf{Kq} = 0$

$$\mathbf{M} = m \begin{bmatrix} 1 & & \\ & 1 & \\ & & \frac{a^2+b^2}{12} \end{bmatrix} \qquad \mathbf{K} = \begin{bmatrix} 2k_1 & 0 & 0 \\ 0 & k_2 & \frac{k_2a}{2} \\ 0 & \frac{k_2a}{2} & \frac{2k_1b^2+k_2a^2}{4} \end{bmatrix}$$

3.5 $\mathbf{M\ddot{q}} + \mathbf{Kq} = 0$

$$\mathbf{K} = \begin{bmatrix} 440.7 & -29.3 & 14.7 \\ -29.3 & 59.6 & -29.3 \\ 14.7 & -29.3 & 440.7 \end{bmatrix} \frac{\text{kips}}{\text{in.}}$$

$$\mathbf{M} = \begin{bmatrix} 0.0356 & 0 & 0 \\ 0 & 0.0906 & 0 \\ 0 & 0 & 0.0356 \end{bmatrix} \frac{\text{kip. s}^2}{\text{in}}$$

$\mathbf{q}^T = [q_1 \quad q_2 \quad q_3]$ in which q_1, q_2 and q_3 are respectively the vertical displacements of the left-end, center and right-end of the crane.

3.7 $\ddot{\theta}_1 + \frac{k}{m_1}(\frac{a}{l})^2(\theta_1 - \theta_2) + \frac{g}{l}\theta_1 = 0$

$\ddot{\theta}_2 + \frac{k}{m_2}(\frac{a}{l})^2(\theta_2 - \theta_1) + \frac{g}{l}\theta_2 = 0$

3.9

$$\bar{m}L \begin{bmatrix} 0.2268 & 0.5755 \\ 0.5755 & 1.9244 \end{bmatrix} \begin{bmatrix} \ddot{z}_1 \\ \ddot{z}_2 \end{bmatrix} + \left\{ \frac{\pi^4 EI}{32L^3} \begin{bmatrix} 1 & 0 \\ 0 & 81 \end{bmatrix} \right.$$

$$\left. - \bar{m}g \begin{bmatrix} 0.3669 & 0 \\ 0 & 5.3017 \end{bmatrix} \right\} \begin{bmatrix} z_1 \\ z_2 \end{bmatrix} = -\bar{m}L\ddot{u}_g \begin{bmatrix} 0.3634 \\ 1.2122 \end{bmatrix}$$

3.10 $\mathbf{M} = \dfrac{m}{53,760} \begin{bmatrix} 13L^3 & -9L^3 & 152L^2 & 0 & 0 & 0 \\ -9L^3 & 75L^3 & 472L^2 & -48L^3 & 416L^2 & 0 \\ 152L^2 & 472L^2 & 16,128L & -416L^2 & 3456L & 0 \\ 0 & -48L^3 & -416L^2 & 75L^3 & -472L^2 & -9L^3 \\ 0 & 416L^2 & 3456L & -472L^2 & 16,128L & -152L^2 \\ 0 & 0 & 0 & -9L^3 & -152L^2 & 13L^3 \end{bmatrix}$

$$\mathbf{K} = EI \begin{bmatrix} \frac{40}{L} & \frac{16}{L} & \frac{-224}{L^2} & 0 & 0 & 0 \\ \frac{16}{L} & \frac{32}{L} & \frac{-136}{L^2} & \frac{4}{L} & \frac{-24}{L^2} & 0 \\ \frac{-224}{L^2} & \frac{-136}{L^2} & \frac{1632}{L^3} & \frac{24}{L^2} & \frac{-96}{L^3} & 0 \\ 0 & \frac{4}{L} & \frac{24}{L^2} & \frac{32}{L} & \frac{136}{L^2} & \frac{16}{L} \\ 0 & \frac{-24}{L^2} & \frac{-96}{L^3} & \frac{136}{L^2} & \frac{1632}{L^2} & \frac{224}{L^2} \\ 0 & 0 & 0 & \frac{16}{L} & \frac{224}{L^2} & \frac{40}{L} \end{bmatrix}$$

3.13
$$\begin{bmatrix} m_v & 0 \\ 0 & \frac{\bar{m}L}{2} + m_t \sin^2 \frac{\pi vt}{L} \end{bmatrix} \begin{bmatrix} \ddot{u}_v \\ \ddot{z}_b \end{bmatrix}$$

$$+ \begin{bmatrix} c & -c \sin \frac{\pi vt}{L} \\ -c \sin \frac{\pi vt}{L} & c \sin^2 \frac{\pi vt}{L} + 2m_t \frac{\pi v}{L} \sin \frac{\pi vt}{L} \cos \frac{\pi vt}{L} \end{bmatrix} \begin{bmatrix} \dot{u}_v \\ \dot{z}_b \end{bmatrix}$$

$$+ \begin{bmatrix} k & -k \sin \frac{\pi vt}{L} - c \frac{\pi v}{L} \cos \frac{\pi vt}{L} \\ -k \sin \frac{\pi vt}{L} & k \sin^2 \frac{\pi vt}{L} + c \frac{\pi v}{L} \sin \frac{\pi vt}{L} \cos \frac{\pi vt}{L} - m_t (\frac{\pi v}{L})^2 \sin^2 \frac{\pi vt}{L} + \frac{\pi^4 EI}{2L^3} \end{bmatrix}$$

$$\times \begin{bmatrix} u_v \\ z_b \end{bmatrix} = \begin{bmatrix} -m_v g \\ -m_t g \sin \frac{\pi vt}{L} \end{bmatrix}$$

where u_v is the total vertical displacement of mass m_v, and $z_b \sin(\frac{\pi x}{L})$ is the displacement of the deck.

3.15 $\mathbf{M}\ddot{\mathbf{q}} + \mathbf{K}\mathbf{q} = 0$

$$\mathbf{M} = \begin{bmatrix} 49.89 & -40.82 \\ -40.82 & 365.56 \end{bmatrix} \text{ kg.m}^2 \quad \mathbf{K} = 10^4 \begin{bmatrix} 26.26 & 15.76 \\ 15.76 & 82.47 \end{bmatrix} \text{ N.m}$$

$\mathbf{q}^T = [q_1 \quad q_2]$ where q_1 is the rotation at the hinge and q_2 is the rotation at the knee.

CHAPTER 4

4.1 $\left\{\frac{r_1^2 + 3r_2^2}{2r_2^2}(R - r_2)\right\} \ddot{\theta} + g\theta = 0$

where θ is the angle that the radius from the center of the hollow cylinder to the center of cylindrical surface makes with the vertical.

4.3
$$\begin{bmatrix} \frac{\bar{m}L}{2} & 0 \\ 0 & M \end{bmatrix} \begin{bmatrix} \ddot{u}_1 \\ \ddot{u}_2 \end{bmatrix} + \begin{bmatrix} \frac{\pi^4 EI}{2L^3} + k & -k \\ -k & k \end{bmatrix} \begin{bmatrix} u_1 \\ u_2 \end{bmatrix} = \begin{bmatrix} 0 \\ 0 \end{bmatrix}$$

4.5   For small vibrations

$$\ddot{\theta}_1 + \frac{l_2}{2l_1}\ddot{\theta}_2 + \frac{g}{l_1}\theta_1 = 0$$

$$\frac{1}{2}\ddot{\theta}_1 + \frac{1}{3}\frac{l_2}{l_1}\ddot{\theta}_2 + \frac{1}{2}\frac{g}{l_1}\theta_2 = 0$$

4.7   For small vibrations

$$(M+m)\ddot{u} + ml\ddot{\theta} + 2ku = 0$$

$$m\ddot{u} + ml\ddot{\theta} + mg\theta = 0$$

4.9 $$\ddot{\theta} + \frac{\cos\theta}{l}\dot{\theta}\dot{u}_g + \frac{g}{l}\sin\theta\left(1 + \frac{\ddot{u}_g}{g}\right) = 0$$

CHAPTER 5

5.1   11.93 kips/ft; 1.49 kip.s/ft

5.3   0.439 m including deflection due to self weight.

5.5   (**a**) 50.00 mm, (**b**) 43.10 mm

5.7   Mass of vehicle $= 980.4$ kg; stiffness of spring $= 395,130$ N/m; damping constant $= 13,545$ N.s/m

5.9   0.0822 m

5.11 $$H = \frac{1}{2\omega^2}\left[g + \sqrt{g^2 - \frac{4J_0\omega^4}{M}}\right]$$

5.13   35.9 mm, 30.0 mm

5.15   (**a**) 2.79, (**b**) 3.52

CHAPTER 6

6.1   (**a**) 6.42 rad/s, (**b**) 3.47 mm

6.3   1467 N.s/m, 0.805 mm

6.5   (**a**) 0.0428 mm, (**b**) 0.1457, (**c**) 52.46 N

6.7 $\omega = 48.11$ rad/s, $c = 90.94$ N.s/m. Calibration factor of 1157.2 will convert reading in mm to acceleration in mm/s^2. Error at 5 Hz: 7.4%.

6.9 Calibration factor $= 1,600$. Measured amplitudes: 4.094, 3.099.

6.12 3.468 mm

6.14 $\xi = 0.105$

6.16 0.116 in.

CHAPTER 7

7.3 $t_1 < \dfrac{T}{6}, u_{max} = \left(\dfrac{4P_0}{k\omega t_1}\right) \sin \omega t_1 \sin \dfrac{\omega t_1}{2}$

7.5 $u_{max} = 0.9267(v_0/\omega)$ for $\xi = 0.05$

$u_{max} = 0.8626(v_0/\omega)$ for $\xi = 0.10$

7.7 $u = -\dfrac{v_0}{\omega} \sin \omega t \quad 0 < t \le t_1$

$u = -\dfrac{v_0}{\omega} \sin \omega t + \dfrac{v_0}{\omega^2 t_1}[1 - \cos \omega(t - t_1)] \quad t_1 \le t < 2t_1$

$u = -\dfrac{v_0}{\omega} \sin \omega t + \dfrac{v_0}{\omega^2 t_1}[\cos \omega(t - 2t_1) - \cos \omega(t - t_1)] \quad 2t_1 \le t$

where u is the displacement relative to the ground.

7.10 $u = \dfrac{P_0}{k}[\cos \omega(t - t_1) - \cos \omega t + \cos \omega(t - 2t_1 - t_0) - \cos \omega(t - t_1 - t_0)]$

The response beyond $t = 2t_1 + t_0$ is zero when $t_0 = (\pi/\omega) - t_1$.

7.12 Maximum relative displacement while the car is on the hump $= -45$ mm. Maximum spring force 1800 N. Maximum absolute displacement is in the free-vibration era $= 43.7$ mm. Maximum absolute acceleration occurs when the car is on the hump $= 1285.6$ mm/s^2.

CHAPTER 8

8.1 $\omega = \sqrt{\dfrac{4k}{M + \frac{3m}{2}}}$

8.3 $\omega = \sqrt{\dfrac{2k}{m}\left(\dfrac{a}{l}\right)^2 - \dfrac{g}{l}}$

8.5 With an assumed vibration shape $\psi(x) = \sin \pi x/L$; hoist at 1.2 m, $\omega = 77.62$ rad/s; maximum frequency when hoist at support $\omega = 149.3$ rad/s; minimum frequency when hoist at mid-span $\omega = 69.62$ rad/s.

8.7 $\omega = 1.61\sqrt{\dfrac{EI}{\bar{m}L^4}}$

8.9 The applied load is $50\,(e^{-4t} - e^{-8t})$. The exact solution for zero initial conditions is: $u = -0.754\cos\omega t - 0.574\sin\omega t + 3.205e^{-4t} - 2.451e^{-8t}$. This solution can be used to compare the results obtained.

8.15 Maximum base shear at $0.5\,\mathrm{s} = 103.8\,\mathrm{kN}$.

CHAPTER 9

9.1 $u(t) = \dfrac{p_0}{2k} - \dfrac{4p_0}{\pi^2 k}\left[\dfrac{1}{1 - \beta_1^2}\cos\Omega t + \dfrac{1}{9}\dfrac{1}{(1 - \beta_3^2)}\cos 3\Omega t\right.$

$$\left. + \dfrac{1}{25}\dfrac{1}{(1 - \beta_5^2)}\cos 5\Omega t + \cdots\right]$$

where $\beta_n = \dfrac{n\Omega}{\omega}$

9.2 $u(t) = \dfrac{2p_0}{k\pi}\sum_{n=1}^{\infty}\left[\dfrac{(-1)^{n-1}}{n(1 + \beta_n^4 - 1.96\beta_n^2)}\{(1 - \beta_n^2)\sin n\Omega t - 0.2\beta_n\cos n\Omega t\}\right]$

where $\beta_n = \dfrac{n\Omega}{\omega} = 1.5n$

9.6 Continuous convolution

$$g(t) * h(t) = \begin{cases} t - 1 + e^{-t} & 0 < t \leq 1.5 \\ -2e^{(1.5-t)} + e^{-t} + 4 - t & 1.5 < t \leq 3 \\ -2e^{(1.5-t)} + e^{(3-t)} + (4 - t)e^{-3} & 3 < t \leq 4.5 \\ e^{(3-t)} + (t - 7)e^{-3} & 4.5 < t \leq 6 \\ 0 & 6 < t \end{cases}$$

9.8 Closed form solution

$$u = 0.0494(\sin 6.283t - 0.4353\sin 14.43t) \qquad 0 < t < \tfrac{1}{2}$$

$$u = -0.0173\cos 14.43(t - \tfrac{1}{2}) - 0.0343\sin 14.43(t - \tfrac{1}{2}) \quad t > \tfrac{1}{2}$$

CHAPTER 10

10.1 $\omega_1 = 24.72\,\mathrm{rad/s}$ $\omega_2 = 74.8\,\mathrm{rad/s}$

$$\mathbf{q}_1 = \begin{bmatrix} 1.00 \\ 0.92 \end{bmatrix} \quad \mathbf{q}_2 = \begin{bmatrix} 1.00 \\ -1.09 \end{bmatrix}$$

$u_1 = 0.542 \cos 24.72t + 0.458 \cos 74.8t$

$u_2 = 0.5 \cos 24.72t - 0.5 \cos 74.8t$

10.3 (a) $u_1 = -0.9965 \cos 13.79t - 0.0036 \cos 31.51t$

$u_2 = 0.01502 \cos 13.79t - 0.01502 \cos 31.51t$

(b) $u_1 = -0.1375 \cos 13.79t + 0.0263 \cos 31.51t$

$u_2 = 0.0021 \cos 13.79t + 0.1089 \cos 31.51t$

(c) $u_1 = -0.859 \cos 13.79t - 0.030 \cos 31.51t$

$u_2 = 0.013 \cos 13.79t - 0.1242 \cos 31.51t$

where u_1 is the vertical displacement at G in inches and u_2 the rotation about G.

10.5 $u_1 = -23.72e^{-1.343t} \sin 18.273t + 11.38e^{-5.955t} \sin 38.13t$ mm

$u_2 = -51.23e^{-1.343t} \sin 18.273t - 8.23e^{-5.955t} \sin 38.13t$ mm

10.7 $u_1 = e^{-1.06t}[0.1056 \cos 10.55t + 0.0106 \sin 10.55t]$

$\quad + e^{-3.37t}[0.2451 \cos 33.53t + 0.0245 \sin 33.53t]$

$\quad + e^{-5.92t}[-0.3510 \cos 58.90t - 0.0353 \sin 58.90t]$ in.

$u_2 = e^{-1.06t}[0.2409 \cos 10.55t + 0.0242 \sin 10.55t]$

$\quad + e^{-3.37t}[0.1882 \cos 33.53t + 0.0188 \sin 33.53t]$

$\quad + e^{-5.92t}[0.3201 \cos 58.90t + 0.0322 \sin 58.90t]$ in.

$u_3 = e^{-1.06t}[0.3339 \cos 10.55t + 0.0336 \sin 10.55t]$

$\quad + e^{-3.37t}[-0.2137 \cos 33.53t - 0.0214 \sin 33.53t]$

$\quad + e^{-5.92t}[-0.1204 \cos 58.90t - 0.0121 \sin 58.90t]$ in.

10.9 $\mathbf{C} = \begin{bmatrix} 0.3131 & -0.1347 & 0.0234 \\ -0.1347 & 0.2627 & -0.0898 \\ 0.0234 & -0.0898 & 0.1370 \end{bmatrix}$

$\xi_2 = 0.0375$

10.10 $$\mathbf{C} = \begin{bmatrix} 0.2327 & -0.1965 & 0.0937 \\ -0.1965 & 0.2150 & -0.0359 \\ 0.0937 & -0.0359 & 0.0758 \end{bmatrix}$$

10.11

$$\mathbf{C} = \begin{bmatrix} 0.3670 & 0.6523 & 0.5826 & 0.0331 & -0.4576 \\ 0.6523 & 1.1669 & 1.0627 & 0.1274 & -0.7517 \\ 0.5823 & 1.0627 & 1.0235 & 0.3017 & -0.5022 \\ 0.0331 & 0.1274 & 0.3017 & 0.6320 & 0.5251 \\ -0.4576 & -0.7517 & -0.5022 & 0.5251 & 1.0807 \end{bmatrix}$$

CHAPTER 11

11.1 $$\mathbf{F} = \frac{L^3}{768EL} \begin{bmatrix} 9 & 11 & 7 \\ 11 & 16 & 11 \\ 7 & 11 & 9 \end{bmatrix} \qquad \mathbf{M} = m \begin{bmatrix} 1 & 0 & 0 \\ 0 & 1 & 0 \\ 0 & 0 & 1 \end{bmatrix}$$

$$\omega_1 = 4.933 \sqrt{\frac{EI}{mL^3}} \qquad \omega_2 = 19.596 \sqrt{\frac{EI}{mL^3}} \qquad \omega_3 = 41.606 \sqrt{\frac{EI}{mL^3}}$$

11.2 $$\mathbf{K} = \frac{768EI}{28L^3} \begin{bmatrix} 23 & -22 & 9 \\ -22 & 32 & -22 \\ 9 & -22 & 23 \end{bmatrix}$$

11.3 Frequencies as in Problems 11.1

$$\mathbf{Q} = \begin{bmatrix} 1 & 1 & 1 \\ \sqrt{2} & 0 & -\sqrt{2} \\ 1 & -1 & 1 \end{bmatrix}$$

11.5 33.8 rad/s.

11.7 $\omega_1 = 10.02$ rad/s $\omega_2 = 33.81$ rad/s $\omega_3 = 59.25$ rad/s.

11.9 $\omega_1 = 0$ rad/s $\omega_2 = 9.83$ rad/s $\omega_3 = 17.02$ rad/s

$$\mathbf{q}_1 = \begin{bmatrix} 1 \\ 1 \\ 1 \end{bmatrix} \qquad \mathbf{q}_2 = \begin{bmatrix} -1 \\ 0 \\ 1 \end{bmatrix} \qquad \mathbf{q}_3 = \begin{bmatrix} -\frac{1}{2} \\ 1 \\ -\frac{1}{2} \end{bmatrix}$$

11.13 The first two exact frequencies

$$\omega_1 = 7.77 \text{ rad/s} \qquad \omega_2 = 18.98 \text{ rad/s}$$

The corresponding mode shapes

$$\phi_1 = \begin{bmatrix} 0.1028 \\ 0.2296 \\ 0.3333 \\ 0.4385 \\ 0.5164 \end{bmatrix} \qquad \phi_2 = \begin{bmatrix} -0.2098 \\ -0.3635 \\ -0.2989 \\ 0.0672 \\ 0.6783 \end{bmatrix}$$

CHAPTER 12

12.1 $u_1 = 0.3374 \sin 41.88t$ $u_2 = 0.05656 \sin 41.88t$ $u_3 = -0.5022 \sin 41.88t$

12.3 Displacements in mm at each of the three floors are as given below:

$u_1(t) = 10.2 \sin(5\pi t/3) - 7.963 \sin 7.995t + 0.6947 \sin 18.42t$
$\qquad - 0.0891 \sin 26.82t$

$u_2(t) = 19.2 \sin(5\pi t/3) - 13.89 \sin 7.995t + 0.4478 \sin 18.42t$
$\qquad + 0.0792 \sin 26.82t$

$u_3(t) = 33.1 \sin(5\pi t/3) - 18.66 \sin 7.995t - 1.260 \sin 18.42t$
$\qquad - 0.0419 \sin 26.82t$

Maximum values of top-floor displacement: 30.75 mm at 0.47 s; -48.83 mm at 0.95 s.

12.5 Maximum values of base shear are as given below:

Mode 1: 107.3 kN at 0.47 s; -170.3 kN at 0.95 s

Mode 2: -27.02 kN at 0.27 s; 23.57 kN at 0.8 s

Mode 3: 4.25 kN at 0.39 s; -4.39 kN at 0.98 s

Total: 100.5 kN at 0.47 s; -159.8 kN at 0.95 s

12.8 Maximum top-floor displacements are as follow:

Mode 1: 26.9 mm at 0.39 s

Mode 2: -3.2 mm at 0.17 s

Mode 3: 0.87 mm at 0.12 s

Total: 25.96 mm at 0.39 s

12.9 Maximum interstory drift in the second story: 11.82 mm at 0.36 s. Maximum shear in 2nd story: 103.6 kN.

12.10 $u_1 = \dfrac{gt^2}{2} + \dfrac{m^2}{M+m}\dfrac{g}{k} - e^{-\xi\omega t}\left(\dfrac{m^2}{M+m}\dfrac{g}{k}\right)\left(\cos\omega_d t + \dfrac{\omega\xi}{\omega_d}\sin\omega_d t\right)$

$u_2 = \dfrac{gt^2}{2} + \dfrac{m^2}{M+m}\dfrac{g}{k} + e^{-\xi\omega t}\left(\dfrac{Mm}{M+m}\dfrac{g}{k}\right)\left(\cos\omega_d t + \dfrac{\omega\xi}{\omega_d}\sin\omega_d t\right)$

where u_1 is the vertical displacement of the box, and u_2 the vertical displacement of package measured from the unstretched initial position of the spring.

12.12 $u_1 = \dfrac{p_0 L^3}{EI}(0.01172 - 0.01027 \cos \omega_1 t - 0.0013 \cos \omega_2 t$

$$- 0.00014 \cos \omega_3 t)\,\text{m}$$

$$u_2 = \dfrac{p_0 L^3}{EI}(0.01433 - 0.01453 \cos \omega_1 t + 0.00020 \cos \omega_3 t)\,\text{m}$$

$$u_3 = \dfrac{p_0 L^3}{EI}(0.00911 - 0.01027 \cos \omega_1 t + 0.00130 \cos \omega_2 t$$

$$- 0.00014 \cos \omega_3 t)\,\text{m}$$

where $\omega_1 = 4.933\sqrt{\dfrac{EI}{mL^3}}$, $\omega_2 = 19.596\sqrt{\dfrac{EI}{mL^3}}$, $\omega_3 = 41.606\sqrt{\dfrac{EI}{mL^3}}$

Bending moment $M = 1.375\,p_0$ kN m.

CHAPTER 13

13.1 $\omega_1 = 9.56$ rad/s, $\omega_2 = 27.55$ rad/s.

$$\phi_1 = \begin{bmatrix} 0.1777 \\ 0.3555 \\ 0.4498 \\ 0.5286 \end{bmatrix} \quad \phi_2 = \begin{bmatrix} -0.2285 \\ -0.4569 \\ 0.0470 \\ 0.6882 \end{bmatrix}$$

13.3 $\omega_1 = 1.025\sqrt{\dfrac{k}{\bar{m}L}}$, $\omega_2 = 4.48\sqrt{\dfrac{k}{\bar{m}L}}$

$$f_1(x) = 1 + 1.414 \sin \frac{\pi x}{L} \quad f_2(x) = 1 - 1.414 \sin \frac{\pi x}{L}$$

13.5 $d = \dfrac{\pi}{2}$, $\omega = 1.027\sqrt{\dfrac{k}{\bar{m}L}}$

13.7 In first mode $\mathbf{f}_S = \begin{bmatrix} 0.907 \\ 2.067 \\ 2.865 \end{bmatrix} \sin \Omega t \quad \dfrac{\mathbf{f}^T \mathbf{e}_M}{\mathbf{f}^T \mathbf{f}} = 0.383$

$$\text{After static correction } \mathbf{f}_S = \begin{bmatrix} 0.570 \\ 1.300 \\ 3.544 \end{bmatrix} \sin \Omega t$$

13.9 $u_1 = 0.1667 - 0.1679 \cos 9.71t + 0.0012 \cos 27.55t$

$u_2 = 0.3333 - 0.3357 \cos 9.71t + 0.0024 \cos 27.55t$

$u_3 = 0.4000 - 0.3999 \cos 9.71t - 0.0001 \cos 27.55t$

$u_4 = 0.5000 - 0.4962 \cos 9.71t - 0.0038 \cos 27.55t$

CHAPTER 14

14.1 $EI \dfrac{\partial^4 u}{\partial x^4} + m \dfrac{\partial^2 u}{\partial t^2} = 0$

$$\left. \begin{array}{l} u = 0 \\ EI \frac{\partial^2 u}{\partial x^2} = 0 \end{array} \right\} x = 0$$

$$\left. \begin{array}{l} EI \frac{\partial^2 u}{\partial x^2} = 0 \\ EI \frac{\partial^3 u}{\partial x^3} = ku \end{array} \right\} x = L$$

14.3 $\dfrac{\partial}{\partial x} \left(EA \dfrac{\partial u}{\partial x} \right) + c_s \dfrac{\partial^2}{\partial t \partial x} \left(A \dfrac{\partial u}{\partial x} \right) - m \dfrac{\partial^2 u}{\partial t^2} - c \dfrac{\partial u}{\partial t} = m \dfrac{\partial^2 u_g}{\partial t^2}$

$u = 0$ at $x = 0$ $\quad \dfrac{\partial u}{\partial x} = 0$ at $x = L$.

14.5 Ignoring the axial load effect

$$EI \dfrac{\partial^4 u}{\partial x^4} + m \dfrac{\partial^2 u}{\partial t^2} = 0$$

$$x = 0 \left\{ \begin{array}{l} \frac{\partial u}{\partial x} = 0 \\ u = 0 \end{array} \right.$$

$$x = L \left\{ \begin{array}{l} E \frac{\partial^2 u}{\partial x^2} + I_0 \frac{\partial^2}{\partial t^2} \left(\frac{\partial u}{\partial x} \right) = 0 \\ EI \frac{\partial^3 u}{\partial x^3} - M_0 \frac{\partial^2 u}{\partial t^2} = 0 \end{array} \right.$$

where M_0 is the mass of the umbrella roof and I_0 is the mass moment of inertia about a diameter.

Considering axial load effect

$$EI\frac{\partial^4 u}{\partial x^4} + S\frac{\partial^2 u}{\partial x^2} + m\frac{\partial^2 u}{\partial t^2} = 0$$

$$x = 0 \begin{cases} \frac{\partial u}{\partial x} = 0 \\ u = 0 \end{cases}$$

$$x = L \begin{cases} S\frac{\partial u}{\partial x} + EI\frac{\partial^3 u}{\partial x^3} - M_0\frac{\partial^2 u}{\partial t^2} = 0 \\ EI\frac{\partial^2 u}{\partial x^2} + I_0\frac{\partial^2}{\partial t^2}(\frac{\partial u}{\partial x}) = 0 \end{cases}$$

$$S = 22,190 \text{ N}, \qquad EI = 9.94 \times 10^6 \text{ N.m}^2, \qquad I_0 = 2262 \text{ kg.m}^2$$

$$M_0 = 2262 \text{ kg}, \qquad m = 169.6 \text{ kg/m}$$

CHAPTER 15

15.1 $\omega = \beta^2\sqrt{\dfrac{EI}{m}}, \quad \beta_1 L = 2.83$

$$f_1(x) = \sinh \beta_1 x + \frac{\sinh \beta_1 L}{\sin \beta_1 L}\sin \beta_1 x$$

15.3 Mode 1

$$\beta_1 L = 0.9692, \quad \omega_1 = 2.505 \text{ rad/s}$$

$$f_1^{(1)} = \sin 0.9692\frac{x_1}{L}$$

$$f_2^{(1)} = 0.8244 \cos 1.0521\frac{x_2}{L} + 1.4338 \sin 1.0521\frac{x_2}{L}$$

Mode 2

$$\beta_2 L = 2.0464, \quad \omega_2 = 5.289$$

$$f_1^{(2)} = \sin 2.0464\frac{x_1}{L}$$

$$f_2^{(2)} = 0.8890 \cos 2.2214\frac{x_2}{L} - 1.1680 \sin 2.2214\frac{x_2}{L}$$

15.5 $\theta(x, t) = \displaystyle\sum_{n=1}^{\infty} \frac{2\Delta y}{R}\frac{1}{\beta_n^2 L^2}\frac{\sin \beta_n L - \beta_n L \cos \beta_n L}{1 - (\sin 2\beta_n L/2\beta_n L)} \cos \omega_n t \sin \beta_n x$

$$\beta_1 L = 0.8605, \qquad \beta_2 L = 3.4256, \qquad \beta_3 L = 6.4373$$

$$\beta_n L \approx (n-1)\pi \text{ for } n \text{ large}, \quad \beta_n = \omega_n\sqrt{\frac{I_0}{GJ}}$$

CHAPTER 16

16.1 (a) $u(x,t) = \dfrac{4A_0}{\pi}\left(\sum \dfrac{\beta_n^2}{1-\beta_n^2}\dfrac{1}{n}\sin\dfrac{n\pi x}{L}\right)\sin\Omega t \quad n = 1,3,5,\ldots$

(b) $u(x,t) = \dfrac{2A_0}{\pi}\left(\sum \dfrac{\beta_n^2}{1-\beta_n^2}\dfrac{1}{n}\sin\dfrac{n\pi x}{L}\right)\sin\Omega t \quad n = 1,2,3,4,\ldots$

$\omega_n = n^2\pi^2\sqrt{EI/mL^4}$, $\beta_n = \Omega/\omega_n$, and u is the lateral displacement relative to the pseudo-static displacement imposed by support motion.

16.5 $\xi_1 = 0.05, \quad \xi_2 = 0.0122, \quad \xi_3 = 0.082$

$$\omega_{d1} = 0.6147\sqrt{\dfrac{EA_1}{m_1 L^2}}$$

$$\omega_{d2} = 2.5620\sqrt{\dfrac{EA_1}{m_1 L^2}}$$

$$\omega_{d3} = 3.7570\sqrt{\dfrac{EA_1}{m_1 L^2}}$$

Modal responses

$$y_1 = 1.0776\dfrac{P_0 L}{EA_1}\left[1 - e^{-\xi_1\omega_1 t}\left(\cos\omega_{d1}t + \dfrac{\xi_1\omega_1}{\omega_{d1}}\sin\omega_{d1}t\right)\right]$$

$$y_2 = -0.06397\dfrac{P_0 L}{EA_1}\left[1 - e^{-\xi_2\omega_2 t}\left(\cos\omega_{d2}t + \dfrac{\xi_2\omega_2}{\omega_{d2}}\sin\omega_{d2}t\right)\right]$$

$$y_3 = -0.02892\dfrac{P_0 L}{EA_1}\left[1 - e^{-\xi_3\omega_3 t}\left(\cos\omega_{d3}t + \dfrac{\xi_3\omega_3}{\omega_{d3}}\sin\omega_{d3}t\right)\right]$$

$$u(x,t) = \sum_{i=1}^{3} f^{(i)}(x)y_i(t)$$

where $f^{(i)}(x)$ are as in Example 15.2.

CHAPTER 17

17.1 $c = 28.28$ m/s, $\lambda = 28.28$ m, $t = 2.12$ s

17.3 Displacement at $-\pi/3$: 0.933

17.5 Displacement at the level of setback at 2 s: $0.8866\,A_0$ ft.

Index

Milton Keynes UK
Ingram Content Group UK Ltd.
UKHW051925141024
449569UK00027B/1356